Island, Ocean and Deep-Sea Biology

Developments in Hydrobiology 152

Series editor
H. J. Dumont

Island, Ocean and Deep-Sea Biology

Proceedings of the 34th European Marine Biology Symposium, held in Ponta Delgada (Azores), Portugal, 13–17 September 1999

Edited by

M.B. Jones,[1] J.M.N. Azevedo,[2] A.I. Neto,[2]
A.C. Costa[2] & A.M. Frias Martins[2]

[1] *Plymouth Environmental Research Centre (Department of Biological Sciences), University of Plymouth, Plymouth PL4 8AA, U.K.*

[2] *Secção de Biologia Marinha, Departamento de Biologia, Universidade dos Açores, Rua da Mãe de Deus 58, 9501-801 Ponta Delgada, Portugal*

Reprinted from Hydrobiologia, volume 440 (2000)

34 EMBS

Kluwer Academic Publishers
Dordrecht / Boston / London

Library of Congress Cataloging-in-Publication Data

A C.I.P. Catalogue record for this book is available from the Library of Congress.

ISBN 0-7923-6846-0

Published by Kluwer Academic Publishers,
P.O. Box 17, 3300 AA Dordrecht, The Netherlands

Sold and distributed in North, Central and South America
by Kluwer Academic Publishers,
101 Philip Drive, Norwell, MA 02061, U.S.A.

In all other countries, sold and distributed
by Kluwer Academic Publishers Group,
P.O. Box 322, 3300 AH Dordrecht, The Netherlands

Printed on acid-free paper

Printed in the Netherlands

TABLE OF CONTENTS

Preface ix

List of participants xi–xii

Ecology and Evolution on Island Shores

INVITED KEYNOTE ADDRESS
Thoughts on the ecology and evolution of the intertidal biota of the Azores and other Atlantic islands
S.J. Hawkins, H.B.S.M. Corte-Real, F.G. Pannacciulli, L.C. Weber, J.D.D. Bishop 3–17

A provisional classification of algal characterized rocky shore biotopes in the Azores
I. Tittley, A.I. Neto 19–25

Variability at different spatial scales of a coralligenous assemblage at Giannutri Island (Tuscan Archipelago, northwest Mediterranean)
F. Ferdeghini, S. Acunto, S. Cocito, F. Cinelli 27–36

Diversity, recruitment and competition on island shores at south-polar localities compared with lower latitudes: encrusting community examples
D.K.A. Barnes 37–44

Are kelp holdfasts islands on the ocean floor? – indication for temporarily closed aggregations of peracarid crustaceans
M. Thiel, J.A. Vásquez 45–54

A sponge diversity centre within a marine 'island'
J.J. Bell, D.K.A. Barnes 55–64

An offshore buoy as a small artificial island and a fish aggregating device (FAD) in the Mediterranean
G. Relini, M. Relini, L. Relini Orsi, M. Montanari 65–80

Periodic movement, recruitment and size-related distribution of the sea cucumber *Holothuria scabra* in Solomon Islands
A. Mercier, S.C. Battaglene, J.-F. Hamel 81–100

A "minimum stress inflexion" in relation to environmental and biotic influences on the dynamics of subtidal encrusting communities?
B.C. Maughan, D.K.A. Barnes 101–109

Shell occupancy of the intertidal hermit crab *Clibanarius erythropus* (Decapoda, Diogenidae) on São Miguel (Azores)
A.Z. Botelho, A.C. Costa 111–117

Direct and indirect effects of *Littorina littorea* (L.) on barnacles growing on mussel beds in the Wadden Sea
C. Buschbaum 119–128

Age and growth, reproduction and diet of a sublittoral population of the rock goby *Gobius paganellus* (Teleostei, Gobiidae)
J.M.N. Azevedo, A.M.V. Simas 129–135

Luminous capabilities and life-history traits of the polychromatic ophiuroid *Amphipholis squamata* in two isolated populations from Oliveri-Tindari lagoon system (Sicily)
S. Dupont, J. Mallefet 137–144

Physical control of plankton population abundance and dynamics in intertidal rock pools
M.P. Johnson 145–152

New records of brown algae (Phaeophyta) from the Azores
M.I. Parente, R.L. Fletcher, A.I. Neto 153–157

The Open Ocean

INVITED KEYNOTE ADDRESS
Quantification of diel vertical migration by micronektonic taxa in the northeast Atlantic
M.V. Angel, P.R. Pugh 161–179

Colonization of floating seaweed by pelagic and subtidal benthic animals in south-western Iceland
A. Ingólfsson 181–189

Semper's (zoanthid) larvae: pelagic life, parentage and other problems
J.S. Ryland, S. de Putron, R.S. Scheltema, P.J. Chimonides, D.G. Zhadan 191–198

The Deep Ocean

INVITED KEYNOTE ADDRESS
A review of the distribution of hydrothermal vent communities along the northern Mid-Atlantic Ridge: dispersal vs. environmental controls
D. Desbruyères, A. Almeida, M. Biscoito, T. Comtet, A. Khripounoff, N. Le Bris, P.-M. Sarradin, M. Segonzac 201–216

The effect of water depth on bacterial numbers, thymidine incorporation rates and C:N ratios in northeast Atlantic surficial sediments
J.L. Dixon, C.M. Turley 217–225

Distribution of live benthic Foraminifera at three oceanographically dissimilar sites in the northeast Atlantic: preliminary results
J.A. Hughes, A.J. Gooday, J.W. Murray 227–238

Foraminiferan (Protozoa) epizoites on deep-water arcturids (Crustacea, Isopoda)
J. Svavarsson, S.H. Ólafsdóttir 239–247

Resting stages in a submarine canyon: a component of shallow–deep-sea coupling?
L. Della Tommasa, G. Belmonte, A. Palanques, P. Puig, F. Boero 249–260

Patterns in deep-sea macrobenthos at the continental margin: standing crop, diversity and faunal change on the continental slope off Scotland
J.D. Gage, P.A. Lamont, K. Kroeger, G.L.J. Patterson, J.L. Gonzalez Vecino 261–271

Lipid profiles of *Nematocarcinus gracilis*, a deep-sea shrimp from below the Arabian Sea oxygen minimum zone
C.E. Allen, P. Tyler, M.S. Varney 273–279

Community structure of deep-sea demersal fish in the Northern Aegean Sea (north-eastern Mediterranean)
M. Labropoulou, C. Papaconstantinou 281–296

Open Session

Influence of light and nitrogen on the phlorotannin content of the brown seaweeds *Ascophyllum nodosum* and *Fucus vesiculosus*
H. Pavia, G. Toth 299–305

Nutrification impacts on coral reefs from northern Bahia, Brazil
O.S. Costa Jr., Z.M.A.N. Leão, M. Nimmo, M.J. Attrill 307–315

Stable isotope evidence of benthic microalgae-based growth and secondary production in the suspension feeder *Cerastoderma edule* (Mollusca, Bivalvia) in the Marennes-Oléron Bay
P.-G. Sauriau, C.-K. Kang 317–329

Annual growth of the cockle *Clinocardium ciliatum* in the Norwegian Arctic (Svalbard area)
M.E. Tallqvist, J.H. Sundet 331–338

Distribution and abundance of co-occurring chthamalid barnacles *Chthamalus montagui* and *Chthamalus stellatus* (Crustacea, Cirripedia) on the southwest coast of Portugal
E.B. Sousa, T. Cruz, J.J. Castro 339–345

Predation by juvenile *Platichthys flessus* (L.) on shelled prey species in a bare sand and a drift algae habitat
K. Aarnio, J. Mattila 347–355

Feast or famine for intertidal grazing molluscs: a mis-match between seasonal variations in grazing intensity and the abundance of microbial resources
R.C. Thompson, M.F. Roberts, T.A. Norton, S.J. Hawkins 357–367

Bioturbation as a mechanism for setting and maintaining levels of diversity in subtidal macrobenthic communities
S. Widdicombe, M.C. Austen, M.A. Kendall, R.M. Warwick, M.B. Jones 369–377

Detecting anthropogenic disturbance in an environment with multiple gradients of physical disturbance, Manukau Harbour, New Zealand
J.I. Ellis, D.C. Schneider, S.F. Thrush 379–391

Preface

This volume presents some of the papers given at the 34th European Marine Biology Symposium held in Ponta Delgada (Azores) between 13th and 17th September 1999. As this was the last EMBS of the Millennium, our aim was to attract and encourage young scientists to present their work, get them involved in the EMBS philosophy, and thus ensure the continued success of these meetings into the next Millennium. The themes of the meeting, directed by the position of the host institution in the Atlantic Ocean, were: (1) Ecology and Evolution on Island Shores, (2) The Open Ocean and, (3) The Deep Ocean; there was also a general "Open Session". In addition to the 78 oral presentations, 127 posters were presented, mostly by students, providing much opportunity for fruitful exchange of ideas and opportunity to forge new friendships and research partnerships.

The symposium was co-organized by the University of the Azores and Sociedade "Afonso Chaves", and was hosted by the Biology Department, University of the Azores, and attracted ca. 260 scientists from 28 countries. The Organizing and Scientific Committees comprised members of the University of the Azores in both Ponta Delgada (A.M. Frias Martins, J.M.N. Azevedo, A.I. Neto, A.C. Costa and F. Wallenstein Macedo) and Faial (R.S. Santos), together with Museu Municipal do Funchal, Madeira (Manuel Biscoito) and the University of Plymouth, UK (Malcolm B. Jones). Financial and logistic support for the meeting was provided by Governo Regional dos Açores, Universidade dos Açores, Secretaria Regional da Economia – Delegação Regional de Turismo, FLAD – Fundação Luso Americana para o Desenvolvimento, Câmara Municipal da Lagoa, Biblioteca Municipal de Ponta Delgada, SATA Internacional, Caixa Geral de Depósitos, Consulmar Açores, Banco Pinto & Sotto Mayor, Banco Comercial dos Açores, Prolacto – Lacticínios de São Miguel, Fábrica Mello Abreu and Companhia de Seguros Açoreana. Financial support for the publication was received from FCT – Fundação para a Ciência e Tecnologia. We thank all our sponsors for, without their help, the meeting would not have been possible

We emphasize that each paper presented in this volume received rigorous review from at least two internationally renowned scientists. The Editors thank these reviewers for their time and scientific expertise, given freely; their efforts have enhanced the quality of the papers in this volume. We thank also the authors for responding so positively, and within tight time schedules, to our requests for revision.

Finally, we hope that this Proceedings Volume brings back many happy memories of the 34 EMBS held on the Azores. In addition, we hope the volume maintains the high standards set by its predecessors and stimulates further research on the topics covered.

MALCOLM B. JONES
JOSÉ MANUEL N. AZEVEDO
ANA ISABEL NETO
ANA CRISTINA COSTA
ANTÓNIO FRIAS MARTINS

Guest Editors

Hydrobiologia **440**: xi–xii, 2000.
M.B. Jones, J.M.N. Azevedo, A.I. Nedo, A.C. Costa & A.M. Frias Martins (eds), Island, Ocean and Deep-Sea Biology.

List of participants

Australia
Greenwood, Jack
Greenwood, Joan

Austria
Ott, Joerg
Ott, Renate

Belgium
Dupont, Samuel
Nunes, Cristina de A. P.

Canada
Hamel, Jean-François
Mercier, Annie

Chile
Thiel, Martin

Croatia
Radosevic, Maja

Estonia
Kotta, Jonne
Kukk, Henn
Lips, Urmas
Martin, Georg
Murumets, Kai
Orav, Helen

Finland
Aarnio, Katri
Bonsdorff, Erik
Boström, Christoffer
Engström, Jonna
Kilpi, Eva Sandberg
Kilpi, Mikael
Olin, Ninny
Tallqvist, Minna
Viherluoto, Maiju
Viitasalo, Markku
Westerbom, Mats

France
Bris, Nadine Le
Colaço, Ana
Desbruyeres, Daniel
Gillet, Patrick
Pradillon, Florence
Sauriau, Pierre-Guy
Sorbe, Jean Claude

Germany Anger, Klaus
Borowski, Christian
Buschbaum, Christian
Gatti, Susanne
Giere, Olav
Grebel-Koehler, Doerthe
Guenther, Carmen-Pia
Liebezeit, Gerd
Menn, Iris
Schewe, Ingo
Soltwedel, Thomas
Stecher, Jens
Ziebis, Wiebke

Greece
Labropoulou, Mary

Iceland
Ingolfsson, Agnar
Svavarsson, Jorundur
Wendel, Linda

Ireland
Barnes, David
Bell, James John
Costello, Mark J.
Crook, Anne
Delany, Jane
Maughan, Ben
O'Ceidigh, Eileen
O'Ceidigh, Mairin
O'Ceidigh, Padraig
Shaw, Claire

Italy
Arculeo, Marco
Chessa, Lorenzo
Cocito, Silvia
Corriero, Giuseppe
Della Tommasa, Luigi
Ferdeghini, Francesca
Liaci, Lidia Scalera
Longo, Caterina
Mercurio, Maria
Pais, Antonio
Pansini, Maurizio
Relini, Giulio
Relini, Lidia Orsi
Ugolini, Alberto
Ungaro, Nicola

Japan
Dazai, Akihiro
Dazai, Kyoko

Lebanon
Lakkis, Sami

Netherlands
Janssen, Gerard

New Zealand
Cryer, Martin
Ellis, Joanne
Norkko, Alf
Norkko, Joanna

Norway
Christiansen, Bengt
Christiansen, Marit Ellen
Fosshagen, Harald
Frantzen, Sylvia
Knabenschuh, Tanja
Kongsrud, Jon
Sekkingstad, Torgeir
Sundet, Jan H.
Suneetha, Kristogu Baduge
Vassenden, Gisle

Poland
Jurewicz, Ewa Styczynska

Portugal
Abreu, Maria Helena T.
Albergaria, André
Alexandre, Bruno
Almeida, Maria João
Álvaro, Nuno Vaz
Alves, Filipe
Amaral, André Santos
Amaro, Teresa
Amen, Rui
Ávila, Sérgio
Azevedo, José M.N.
Baptista, Rita Gomes
Barreiros, João
Bento, Rita
Biscoito, Manuel
Boaventura, Diana
Borges, Isabel Mateus
Borges, Luísa
Botelho, Andrea Zita
Brito, Carlos
Brum, João Medeiros
Campos, Carlos
Cardoso, Joana
Carvalho, Susana I.A.
Castro, João
Cepeda, Helena Isabel
Cerqueira, Pedro
Constância, João Paulo
Costa, Ana Cristina
Costa, Manuel João T.M.
Costa, Maria Helena
Couto, Rubén Pacheco
Cruz, Ana Paula
Cruz, Teresa
Cunha, Marina Ribeiro da
Cúrdia, João Luís Silva
Dinis, Maria Filomena M.
Ferraz, Rogério

Frada, Miguel José
Fraga, Rafael
Garcia, Vasco
Gonçalves, Maura
Gouveia, Isabel
Henrique, Marco
Lacerda, Susana
Leonardo, Margarida Vieira
Luís, Paulo Santos
Macedo, Francisco W.
Magalhães, Sara
Martins, António M. Frias
Martins, Gustavo Meneses
Matias, Miguel Graça
Medeiros, Alexandra
Mello, Paulo da Rocha
Monteiro, Sandra
Neto, Ana Isabel
Neuparth, Maria Teresa
Norberto, Rita
Olim, Sonia
Oliveira, Gabriela
Parente, Manuela
Patarra, Ana Rita
Pereira, Ana Martins
Pereira, Joana Micael
Pinheiro, Pedro Alvim
Porteiro, Filipe
Queiroga, Henrique
Quintela, Adriano
Rabaça, João
Ramos, Manuela Fangueiro
Ramos, Pedro Silveira
Ravara, Ascensão A.
Regala, João
Rocha, Eduardo
Sampaio, Leandro
Santos, João Oliveira
Santos, Marco D.
Santos, Sónia
Santos, Ricardo Serrão
Silva, Teresa
Sousa, Fernando
Sousa, Rui
Sousa, Sónia
Sprung, Martin
Tavares, João
Terra, Marlene
Toste, Marisa
Wirtz, Peter
Xavier, Joana

Russia
Kisliuk, Daniil

South Africa
Pakhomov, Evgeny

Spain
Barreiro, Angélica F.
Carbonell, Aina
Cividanes, Sira Pereira
Dominguez, Sandra Beloso
Filgueira, Mónica Incera
Gálvez, Laura Prieto
Huz, Charo de la
Magro, Ana Bartual
Medina, Lidia
Moreira, Juan
Patiño, José G. Rodríguez
Quintas, Patricia
Quinteiro, Javier
Troncoso, Jesús S.
Troncoso, Nicolás
Vidal, Rodrigo
Zabala, Laura

Sweden
Pavia, Henrik
Skold, Mattias
Toth, Gunilla

Ukraine
Shulman, Georgy
Tokarev, Yuriy

United Kingdom
Angel, Martin V.
Baker, Maria
Baker, Trevor
Billett, David
Copley, Cathy Allen
Costa, Ozeas
Dando, Margaret
Dando, Paul
Dixon, David
Dixon, Joanna
Dixon, Linda R.J.
Fenton, Joanne
Fenton, Simon
Fletcher, Robert
Gage, John
Garnacho, Eva
Gibson, Robin N.
Hawkins, Steve J.
Hughes, James Alan
Johnson, Mark
Jones, Emma
Jones, Malcolm
Kelmo, Francisco
Llodra, Eva Ramirez
Lopez Acuna, Lus
Mills, Suzanne
Morritt, David
Narayanaswamy, Bhavani
Navarro, Nuria
Putron, Samantha de
Roberts, J. Murray
Rogers, Alex David
Rogers, Candida
Ryland, Christine
Ryland, John S.
Stockley, Bruce
Thompson, Richard
Tittley, Ian
Tyler, Paul
Widdicombe, Steve
Wigham, Benjamin
Williams, Mark
Williams, Robert

United States of America
Riewald, Brian
Scheltema, Amélie
Scheltema, Rudolf
Young, Craig
Young, Robyn
Lima, Arthur Edward

Ecology and Evolution on Island Shores

Invited Keynote Address:

Thoughts on the ecology and evolution of the intertidal biota of the Azores and other Atlantic islands

S.J. Hawkins[1], H.B.S.M. Corte-Real, F.G. Pannacciulli, L.C. Weber & J.D.D. Bishop

[1]*The Marine Biological Association of the U.K.*
The Laboratory
Citadel Hill
Plymouth PL1 2PB
United Kingdom

Hydrobiologia **440:** 3–17, 2000.
M.B. Jones, J.M.N. Azevedo, A.I. Neto, A.C. Costa & A.M. Frias Martins (eds), Island, Ocean and Deep-Sea Biology.

Thoughts on the ecology and evolution of the intertidal biota of the Azores and other Atlantic islands *

S.J. Hawkins[1,2,3], H.B.S.M. Corte-Real[2], F.G. Pannacciulli[2,3], L.C. Weber[2] & J.D.D. Bishop[2,3,4]
[1]*Biodiversity and Ecology Division, School of Biological Sciences, University of Southampton, Southampton SO17 1BJ, U.K.*
[2]*Port Erin Marine Laboratory, University of Liverpool, Port Erin, Isle of Man IM9 6JA, U.K.*
[3]*The Marine Biological Association of the U.K., The Laboratory, Citadel Hill, Plymouth PL1 2PB, U.K.*
[4]*Department of Biological Sciences, University of Plymouth, Plymouth PL4 8AA, U.K.*

Key words: islands, Macaronesia, speciation, ecology, *Patella*, *Chthamalus*, intertidal

Abstract

The special features of the intertidal ecosystems of remote islands are reviewed briefly before focusing on the littoral zone of Macaronesia (Azores, Madeira and Canaries). Distribution patterns are briefly compared with other European shores. Species missing from the Azores compared with Madeira, the Canaries and continental European shores are listed. The degree of genetic differentiation within and amongst selected important species of gastropods (*Patella* spp.) and barnacles (*Chthamalus* spp.) in Macaronesia is reviewed. The role of isolated islands in allopatric speciation of these organisms is discussed in relation to dispersal. Possible threats to the populations and communities of rocky shores in Macaronesia are considered in the context of the endemic nature of certain species. The importance of studies of basic ecology and genetics to inform resource management and conservation is highlighted. Future research directions are indicated, emphasizing the usefulness of Macaronesia as a model system to explore speciation in intertidal organisms.

Introduction

Islands, as well as being of intrinsic interest, have featured strongly in the development of evolutionary and ecological ideas from the earliest formal studies (e.g. Darwin, 1859). Most attention in this respect has been given to terrestrial organisms, especially with regard to theories of colonization (e.g. MacArthur & Wilson, 1967) and speciation (Mayr, 1963, 1970). Mainland terrestrial landscapes are often highly fragmented and particular areas of habitat are often likened to islands. Freshwater systems, especially lakes and ponds, can also be considered as inverted islands (Turner, 1999). By contrast, the sea is generally viewed as being more homogeneous, being interconnected via ocean currents and broad-scale dispersal mechanisms of the biota (e.g. Valentine & Jablonski, 1983; Palumbi, 1992, 1994; May, 1994). There are, however, many exceptions where isolated marine ecosystems (e.g. sea lochs, lagoons) or habitat patches (e.g. hydrothermal vents, rock pools, macroalgae) can be considered as islands. These are highlighted elsewhere in this volume.

In the littoral zone of true islands, organisms of the intertidal and shallow subtidal zone (= littoral) are isolated from other areas of suitable habitat by intervening deep water. This is particularly the case for rocky shore organisms as most hard substrata give way at some depth to sedimentary habitats, and even on continental coastlines, rocky areas are usually interspersed with sandy beaches. Many littoral organisms have broadly dispersing propagules or larval stages, teleplanic in some instances (e.g. Scheltema, 1971, 1986), which may mean that shores are less isolated than the terrestrial habitats above the high-tide mark.

* During writing of this paper, an inconsistency between Figure 7 in Santos et al. (1995) and the text describing it became apparent. The text in our paper is a more accurate account of this figure.

In the rest of this paper, we review briefly the special features of island shore communities in general, before turning attention to Macaronesia (Azores, Madeira, Canaries), focusing particularly on the Azores. This sets the scene to consider the relationships between populations of common and ecologically important rocky shore species (barnacles, limpets) occurring in Macaronesia, and conspecific or closely related forms on continental shores. Finally, conservation is discussed because there is such intensive exploitation of Macaronesian shores. Much of this paper summarises work done by collaborators and ourselves over the last 15 years or so; as such, it is meant to be illustrative rather than exhaustive.

Island shores

Islands close to the mainland, or occurring in chains or archipelagos emanating outwards from a continent, generally show very similar shore communities to those on nearby continental coasts. However, surprising absences or reductions in abundance can occur on small islands quite close to the mainland. Crisp & Southward (1958) commented on absences and reductions in abundance in the Isle of Scilly (U.K.) and Ushant (France) of common species with larval dispersal (*Patella depressa* – absent in Scilly, rare in Ushant; *Littorina littorea* rare in both; *Mytilus* spp. uncommon in both). *Patella depressa* also failed to reach Ireland at the end of the last glaciation and is absent from its whole coastline (Southward & Crisp, 1954). More recently, Hawkins & Hiscock (1983) showed similar reductions in abundance of several species of gastropods from the island of Lundy just 18 km offshore from the Devon (U.K.) coastline. The isolated islet of Rockall (U.K.) has long been the subject of attention (e.g. Charcot, 1921; Crisp, 1956) and the absence of all intertidal species with planktonic larvae (limpets, barnacles, mussels) has been established (Moore, 1977). Paradoxically, animals with direct development can reach and do establish populations on isolated islands such as Rockall, including members of the *Littorina saxatilis* complex (Johanneson, 1988). Populations belonging to this complex have also been found in the Azores on São Miguel (C. Brito, pers. comm.), and in Flores and Faial (Reid, 1996). Clearly, rafting can be an effective dispersal method for animals with direct development (Highsmith, 1985; Johanneson, 1988). Once founded, via transport of one gravid individual or at least two different-sexed individuals, a population can establish and, more importantly, persist. Interestingly, the population of *L. saxatilis* in São Miguel (in 1999) appears to be at a precarious stage of colonization; for example, the population present at Vila Franco do Campo in the early 1990s could not be found in 1999 (Carlos Brito, pers. comm. at EMBS 34).

Although, in many cases, species can be absent or have reduced abundance on islands due to poor recruitment, in other cases isolated islands with little human impact can become important refuges. *Patella ferruginea* was once spread throughout the Mediterranean, as evidenced by sub-fossil material in raised beaches and in caves (Fischer-Piette, 1935; Christiaens, 1973). Now it is extremely rare on much of the European mainland of the Mediterranean, especially in Italy. It can be found on some islands such as parts of Sardinia and Corsica, and also in the Tuscan archipelago, especially amongst islands which are used for prisons (e.g. Gorgone) or have been designated as nature reserves (e.g. Capraia). Provided the islands are close enough to each other, larval exchange can occur and populations can resist extinction and even be reservoirs for recolonization of the mainland (SJH found one individual of *P. ferruginea* at Livorno in 1994).

The absence or reduced abundance of certain species will inevitably influence community composition and possibly function. This will be especially the case for key space occupiers, such as barnacles or mussels, important grazers (limpets, littorinids, trochids, echinoderms) and predators (dogwhelks, crabs).

The colonization of islands by organisms can be considered as an invasive process. In recent years, this has been accelerated by anthropogenic vectors such as shipping and aquaculture as well as deliberate introductions (e.g. the bivalve *Ruditapes decussatus* in the Lagao Santo Cristo, São Jorge may be one such introduction: Morton et al., 1998). Shipping is probably the most common vector (for the Azores, see Monniot & Monniot, 1983; Cornelius, 1992), whether the immigrants travel as fouling organisms or in ballast water. The invasion of Hawaii by an alien barnacle from the Atlantic has recently been documented by Southward et al. (1998); *Chthamalus proteus* has invaded an unoccupied niche high on sheltered shores in the Hawaiian archipelago, which are considered to have an impoverished fauna resulting from their isolation and age (Kay & Palumbi, 1987 for review).

Shores of Macaronesia – especially the Azores

The name Macaronesia has been applied to the archipelagoes of Cape Verde, Canaries, Madeira and Azores. In this paper, the Cape Verde islands are given very little attention as we focus primarily on the Canaries, Madeira and particularly the Azores with which we have greatest familiarity.

The various archipelagoes and their islands all have differing degrees of isolation from the continent, with Flores and Corvo (Azores) on the far side of the mid-Atlantic ridge being the most remote. This isolation from the continent, and hence difficulties for colonization, results from both distance and current patterns. Surface currents generally move from the Azores to Madeira and hence the continent (summarised in Santos et al., 1995; Morton et al., 1998), although under some conditions, they can reverse, particularly in the autumn and winter (reviewed in Santos et al., 1995). Deep-water counter-currents do flow towards the Azores from the Mediterranean, but many larval forms live in surface waters. Additionally, deep-water 'meddies' of Mediterranean water move out from the Straits of Gibraltar and track off across the Atlantic passing Madeira, the Canaries and the Azores (Pingree et al., 1999). These would be potential occasional vectors of long-lived deep-water larvae.

In interpreting colonization of these remote islands, we must also consider that it may have occurred under very different circumstances, with different sea level (lower during ice ages), and current patterns, and with the Mediterranean either open or closed. Present day sea-mounts may have been potential 'stepping stones' for colonization by intertidal organisms when sea levels were lower.

An account of the coastal ecology of the Azores has been published recently (Morton et al., 1998). This summarises much of the published information as well as including detailed qualitative descriptions of many shores, plus comparisons with Bermuda, Madeira, Europe and America. Similar general accounts can be found of the shores of the Canaries (e.g. Brito in Aranega et al., 1984; Sanchez & Batet, 1990), coupled with more detailed publications (e.g. Lawson & Norton, 1971). Less information exists for Madeira (but see Wirtz, 1996; Morton et al., 1998). Our account is based on the above sources plus our own and collaborators' published work (e.g. Hawkins et al., 1990a,b; Corte-Real, 1992; Corte-Real et al., 1992, 1996a,b; Santos et al. 1995; Pannacciulli et al., 1997) and much unpublished experience (largely SJH).

Floristic and faunistic affinities of the Azores

Formal biogeographic comparisons based on species lists of particular groups are limited because of taxonomic uncertainty. For fish and algae, there has been extensive sampling and their taxonomy is reasonably stable, enabling comparison. The account below is largely based on Santos et al. (1995) and Morton et al. (1998).

For fish, Santos et al. (1995) reviewed the literature which led to a recent checklist of Azorean fishes (Santos et al., 1997; see also Lloris et al., 1991; Azevedo & Heemstra, 1995; Arruda, 1997). For similar work in the Canaries see Brito et al. (1996). A biogeographic analysis of 20 families of nearshore and littoral fish showed that the Azores clusters out with Madeira and the Canaries, but not with the Cape Verde Islands (Fig. 7 in Santos et al., 1995; see footnote at front of paper). There are also strong affinities with the Mediterranean (see Briggs, 1970, 1974), but also both cold and warm temperate European and north african Atlantic coastlines.

Turning to the algae, various checklists have been compiled and phytogeographic analyses have been undertaken (Feldman, 1946; Fralick & Hehre, 1990; Fredericq et al., 1992; Neto, 1992, 1994; reviewed in Morton et al., 1998). Prud'homme Van Reine (1988) showed that the Azorean algal species largely also occur in the other Macaronesian islands, coasts of Africa, Europe and the western Mediterranean. There is also a substantial North American element, which also reached the shores of the eastern Atlantic. Tittley et al. (1990) also stressed affinities with warm temperate North America. The Canaries show stronger affinities with the tropical western Atlantic (45% of marine flora: Van Den Hoek, 1987). Diversity of algae in the Azores is much lower than in the rest of Macaronesia and there are few endemic species (at most 10) in difficult taxonomic groups which casts doubt on whether they are true endemics (Tittley & Neto, 1995).

In other groups, strong Mediterranean affinities have been suggested for the Azorean biota, including sponges (Boury-Esnault & Lopes, 1985; Weerdt, 1989; Moss, 1992) and Crustacea (Wirtz & Martins, 1993; Lopes et al., 1993). Hydroids, in contrast, have strong affinities with the Caribbean and American coasts, although a Mediterranean component is also evident (Rees & White, 1966; Cornelius, 1992).

Clearly, it is difficult to generalize, with differences between different groups reflecting their life-history characteristics, especially powers of dispersal. The Azores have many affinities with the rest of Macaronesia and there are sufficient similarities for the various archipelagoes to be viewed as a useful biogeographical unit, although it is difficult to say how the Cape Verde islands fit, in part due to lack of information. The Azores have more in common with Europe than America, as do the Canaries and Madeira, but the presence of circum-tropical species means that the latter show more similarities with the Caribbean. The Azores can be considered to be at a crossroads in the middle of the Atlantic (Santos et al., 1995). Their young age (Azevedo et al., 1991) and the recent effects of sea-level and climatic changes associated with ice ages probably account for the lower level of endemism than the rest of Macaronesia (Briggs, 1966). This is certainly the case for fish where there are at most one or two endemics [*Scorpaena azorica* Patzner et al. (1992) has recently been found in the Mediterranean and is part of a difficult group; Azevedo (pers. comm.) has also recently found an endemic species of labrid]. In contrast, Lopez et al. (1993) state that there are 48 species of gammaridean and caprellid amphipods known only from the Azores out of the 122 recorded.

Ecological patterns

Turning to a more ecological approach, Table 1 summarizes the occurrence of several important space occupiers, grazers and predators in the Azores, Madeira and Canaries to enable comparisons with the European mainland. The species highlighted in Table 1 have been chosen because of their importance in the structure and function of rocky shore communities and because of our own taxonomic and population genetic work on some of them.

When comparing the Azores with other parts of Macaronesia and mainland shores, it is clear that certain guilds are impoverished or missing. The lack of large intertidal trochids [*Osilinus (Monodonta)* and *Gibbula* spp.] in the intertidal of the Azores is a good example, although one small species of *Gibbula*, as yet unidentified, has been found recently on the north coast of Sao Miguel (Avila, pers. comm.). Space-occupying animals are reduced to only *Chthamalus stellatus* on the mid to upper shore of the Azores and Madeira, in contrast to the mainland and Canaries where mussels, stalked barnacles (*Pollicipes pollicipes*), and additional barnacle species (*C. montagui, Balanus perforatus* on mainland) are all present (see Crisp et al., 1981). Mussels (*Mytilus* spp.) are virtually absent in the Azores and Madeira although *Perna perna* is present in the Canaries. The presence of *P. perna* reflects the proximity of the Canaries to Africa. Occasionally, individual *Mytilus* spp. raft in or are transported on ships and yachts, having been found in the harbour of Ponta Delgada (São Miguel) and in the marina at Horta. A small population of *Mytilus* has become established in the harbour at São Miguel but is yet to spread (Avila, pers. comm). There also remains the possibility that food supply is limiting in the Azores for filter feeders such as mussels. However, it would be interesting to compare growth rates of these mussels with mainland conspecifics. The Azores only has two species of patellid limpets, whereas there are three (or more?) in the Canaries and Madeira, four in the Mediterranean proper and four on the continental coasts of Africa and southern Europe.

In contrast to the molluscs, the major guild of grazing sea urchins is well represented in the Azores (*Paracentrotus lividus, Sphaerechinus granularis, Arbacia lixula*) and the rest of Macaronesia with additional pan-tropical species such as *Diadema antillarum* being present in the Canaries and Madeira.

Patterns of zonation

The shores of Macaronesia broadly conform to the Stephensonian (Stephenson & Stephenson, 1972) three zone system (see Chapman, 1954; Hawkins et al., 1990a; Morton et al., 1998 for Azores; Lawson & Norton, 1971, for Canaries). A red algal turf-dominated region (Chapman, 1955) predominates either side of low water; above this is a barnacle-dominated region with limpets giving way at the top of the shore to a zone with littorinids, lichens and Cyanobacteria. Below low tide, the red algal-dominated area gives way to a sea urchin-dominated (*Arbacia lixula, Sphaerechinus granularis*) area with calcareous algae. In sheltered conditions, this can extend up to the low-water mark, restricting the width of the red algae. In exposed conditions, the sea urchin zone is usually deeper except where large numbers of *Paracentrotus lividus* form burrows, being particularly numerous in soft volcanic tuff. In more exposed conditions, there can be large numbers of large barnacles (*Megabalanus azoricus*) just below the low-tide mark. In areas of moderate wave action, dense canopies of *Cystoseira* sp. can be found. The expansion of the littorinid zone in exposed conditions is clearly shown

Table 1. Occurrence of some common shore organisms in Macaronesia and adjacent areas

Species	Location					Comments	Refs
	Portugal (continental)	Madeira	Canaries/ Salvages	Azores	Mediterranean		
Melaraphe neritoides	✓	✓	✓	✓	✓	Very widespread species	1,2
Littorina striata	x	✓	✓	✓	x	Macaronesian endemic	1,2,9
Littorina saxatilis	✓	?	?	✓	x	V. small form found in Faial, São Miguel	1
Balanus perforatus	✓	x	x	x	✓	Also absent Ireland	
Chthamalus stellatus	✓	✓	✓	✓	✓	Very widespread species	1,7,8
Chthamalus montagui	✓	x?	✓?	x	✓	Not found by SJH in Madeira	1,7,8
Megabalanus azoricus	x	✓	✓	✓	x	Status uncertain in rest of Macaronesia	1,2,8
Tesseropora atlantica	x	✓	✓	✓	x	At Bermuda, probably not same species	8,10,11
Pollicipes pollicipes	✓	x?	✓	x	x	Shown in Morton et al. (1998) but not present in Madeira	1
Gibbula spp.	✓	✓ *G.candei*	✓ *G.candei*	x	✓	Validation of endemic status needed	2
Osilinus spp (= *Monodonta*)	✓ *O. lineatus+ colubrina*	✓ *O. trappei+ atratus*	✓ *O. trappei+ atratus*	x	✓ *O.turbinatus*	Validation of endemic status needed	2
Patella piperata	x	✓	✓	x	x	Some uncertainty but probably two good,	3,5
Patella rustica	✓	x	x	x	✓	but recently diverged, species	3,5
Patella aspera/ ulyssiponensis	✓	✓	✓	✓	✓	*P.aspera* probably macaronesian species. *P.ulyssiponensis* probably continental species	5
Patella candei complex	x	*P.c.crenata*	*P.c.crenata* *P.c.candei*	*P.c.gomesii* (endemic)	x	Status of *P.c.candei* and relation to other *P.candei* species needs confirmation	4,5
Siphonaria spp.	✓	x	✓	x	x	Does not penetrate far into Med.	
Mytilus spp.	✓	x	*Perna perna* present	Drift specimens? Rare	✓	Small population establishing in Porto Delgado, São Miguel	1
Stramonita (Thais) haemostoma	✓	✓	✓	✓	✓	Also in Florida and Caribbean. Badly affected by TBT in Azorian harbours	6

References: 1. Morton et al. (1998); 2. Sanchez & Batet (1990); 3. Corte-Real et al. (1996a); 4. Corte-Real et al. (1996b); 5. Christiaens (1973); 6. Spence et al. (1989); 7. Pannaccuilli et al. (1997a); 8. Southward (1998); 9. Backeljau et al. (1995); 10. Newman & Ross (1977); 11. Young (1998).

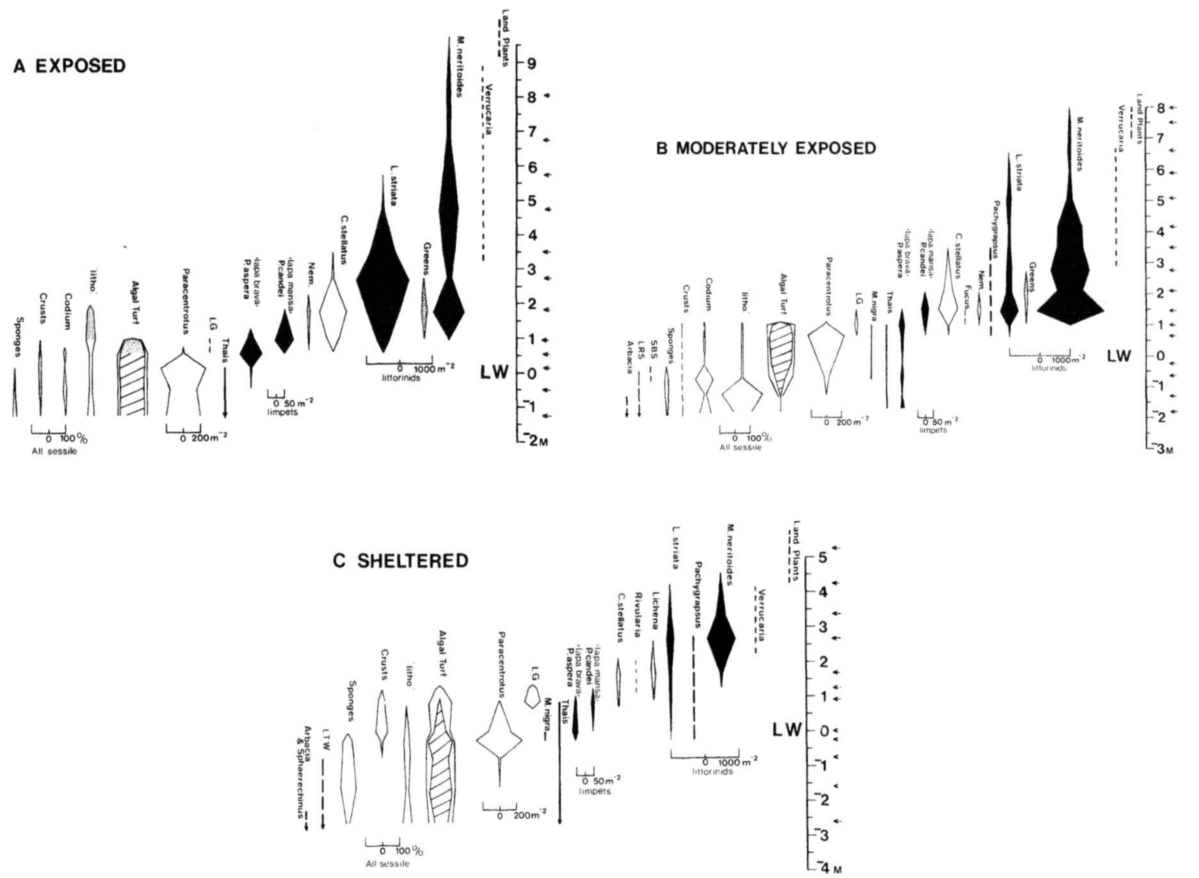

Figure 1. Zonation of rocky shore organisms at Caloura, on the south coast of São Miguel (from Hawkins et al., 1990a). Major organisms were quantified by up to five 0.5×0.5 m quadrats at various levels/depths. Animals such as littorinids were sub-sampled using smaller quadrats (0.1×0.1m). Abundance of algae and sessile animals is expressed as % cover, mobile animals as densities. Three locations in a small cove (Exposed A, Moderately exposed B, Sheltered C) were sampled (shown as plate 3–5 page 53 of Morton et al., 1998).

in Figure 1A. Barnacles become more abundant and have a wider distribution in more exposed conditions. Where limpets have not been totally removed by over-exploitation, *Patella candei gomesii* occurs on the mid and high shore and also subtidally, especially on boulders. *Patella aspera* can occur in both the intertidal, amongst clearings in the red algal turf, and below the low-water mark.

Broadly similar patterns are found in Madeira (see Morton et al., 1998) and in the Canaries (Lawson & Norton, 1971; Sanchez & Batet, 1990). In both places, there are also numerous *Patella piperata* high on the shore, plus *Osilinus* spp. (*O. atratus* and *O. trappei*, taxonomy under review), except where the latter have been heavily exploited. Dense populations of *Diadema antillarum* have led to extensive grazed barren zones subtidally in the Canaries and to a lesser extent in Madeira.

Relationships between selected Macaronesian rocky shore organisms and continental congeners and conspecifics

Barnacles are major space occupiers on moderately exposed and exposed shores throughout the north-east Atlantic (Hawkins et al., 1992) and are particularly prevalent on Macaronesian shores, where mussels are generally lacking (except in the Canaries). Patellid limpets are the dominant grazers throughout the region, and can be considered key-stone species (Hawkins & Hartnoll, 1983; Raffaelli & Hawkins, 1996). Limpets are also heavily exploited throughout Macaronesia. Their ecological and economic (limpets only) importance prompted studies of their taxonomy and population genetics, especially of interchange amongst islands within an archipelago, between archipelagoes, and with the continental coastline. Definition of stocks is particularly important for management of heavily exploited species such as limpets. The relationship between Macaronesian and continental populations of conspecific and closely related species is also a good system for understanding allopatric speciation.

Chthamalus *and other barnacles*

The work of some of us on barnacles (Pannacciulli et al., 1997) was prompted primarily by uncertainties surrounding the relationship between nominally conspecific Atlantic and Mediterranean populations of *Chthamalus stellatus* and *C. montagui.* Previous allozyme analysis had shown that in *C. montagui*, distinct Mediterranean and Atlantic forms existed (Dando & Southward, 1981). We extended this work to *C. stellatus* by resolving additional loci and including populations sampled in Madeira and the Azores. *Chthamalus montagui* does not reach Madeira and the Azores, although it has been reported from the Canaries (see Fig. 2). *Chthamalus montagui* has a shorter larval life than *C. stellatus* (Burrows, 1988) which probably explains its absence from the Azores. *Chthamalus montagui* shows (Fig. 3) much greater apparent differentiation between Atlantic and Mediterranean populations than *C. stellatus.* Atlantic-Mediterranean differentiation is, however, discernible in *C. stellatus*, but interestingly the Atlantic populations, including the Azores and Madeira, are strikingly uniform. This must reflect the considerable dispersive power of *C. stellatus* and/or the relatively recent colonization of the islands.

Other puzzles still remain to be solved. What exactly is the status of *Megabalanus azoricus* in the Azores compared to the rest of Macaronesia (Southward, 1998)? Of greater interest is the degree of genetic isolation between islands in the Azorean archipelago (and possibly also in Madeira) of *Tesseropera atlantica*, a species which broods its nauplii and only releases larvae at the cyprid stage (Southward, 1998). Its relationship with supposed conspecifics in Bermuda also warrants further investigation (Southward, 1998).

Patellid limpets

Work on limpet taxonomy and population genetics was prompted by concerns about stocks of limpets ('*lapas*') during the mid 1980s (Martins et al., 1987; Santos et al., 1990). When SJH first started work in the Azores, the taxonomic status of limpets was confused, with a variety of names being applied, especially by ecologists (Hawkins et al., 1990b). Of the two major reviews written on limpets based on traditional taxonomic methods, Powell (1973) was unfamiliar at first hand with north-east Atlantic material, whilst Christiaens (1973) was, it has turned out, broadly correct about the Macaronesian species. The lack of certainty about the taxonomy and phylogeny of north-east Atlantic patellids, and their relationships with other patellids world-wide, has prompted a series of studies. These have used morphology supported by scanning electron microscopy, and morphometrics analysed using multivariate methods, coupled with allozyme-based biochemical genetics (e.g. Corte-Real et al., 1992, 1996a,b; Hodgson et al., 1996), and more modern molecular genetics (Koufopanou et al., 1999). The phylogenetic work has also used a cladistic approach (Ridgeway et al., 1998; Koufopanou et al., 1999).

The high-shore limpet of the Canaries and Madeira, *Patella piperata,* is very similar morphologically to *Patella rustica.* Christiaens (1973) was certain of its separate specific status. Corte-Real et al. (1996a) showed that one of Christiaens' (1973) distinguishing radular features was not so clear when viewed with the SEM. Their allozyme work showed clear separation of the Canaries and Madeira forms from the mainland except for a population from Majorca which grouped with the Macaronesian populations. Subsequent molecular work has pointed to specific status (Koufopanou et al., 1999) and it looks likely that *P. piperata* is a recently diverged Macaronesian form. The status of the Balearic populations needs checking and this work is planned.

The *Patella candei* complex has also been the subject of considerable uncertainty. Christiaens (1973) recognised several subspecies in the Azores (*P. candei gomesii*), Madeira and Canaries (*P. candei crenata, P. candei candei*), some of which were possible allopatric sibling species. Co-occurring sub-species are more difficult to justify (i.e. *P. candei crenata, P. candei candei*). The very large *P. candei candei* was formerly very common throughout Madeira and the Canaries, and is now confined to Fuertaventura and the Selvagem Islands. Corte-Real et al. (1996b) provide convincing evidence that *P. candei gomesii* in the Azores ('*lapa mansa*' – the tame limpet as it is known locally) is clearly not *Patella caerulea* as named in the past (Morton, 1967, 1990). Furthermore, when compared to *P. candei crenata* from Madeira and the Canaries there was as much genetic differentiation

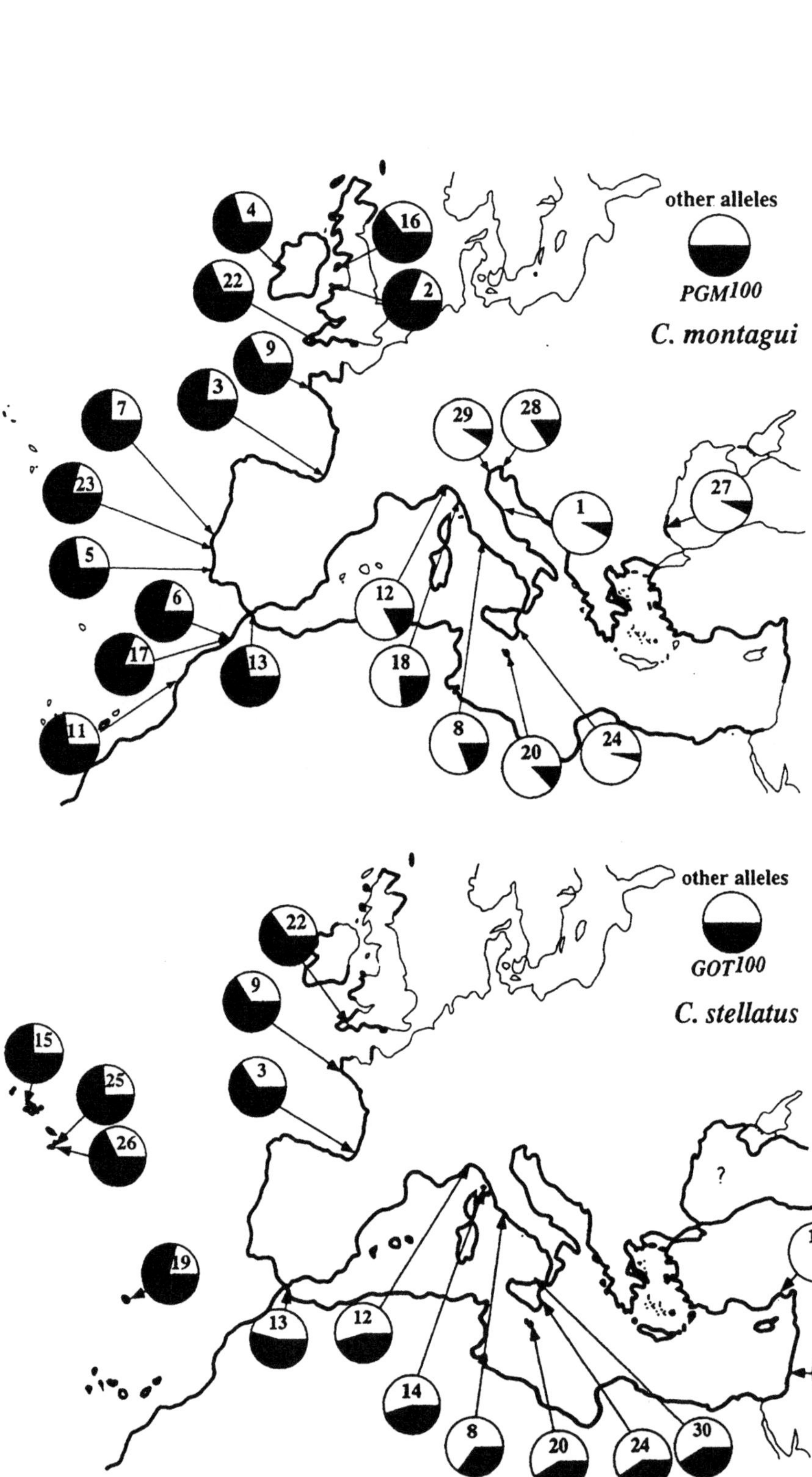

Figure 2. Geographical distribution of *Chthamalus* spp. With variation in allele frequencies at *PGM (C. montagui)* and *GOT (C. stellatus)* loci. (from Pannacciulli et al., 1997).

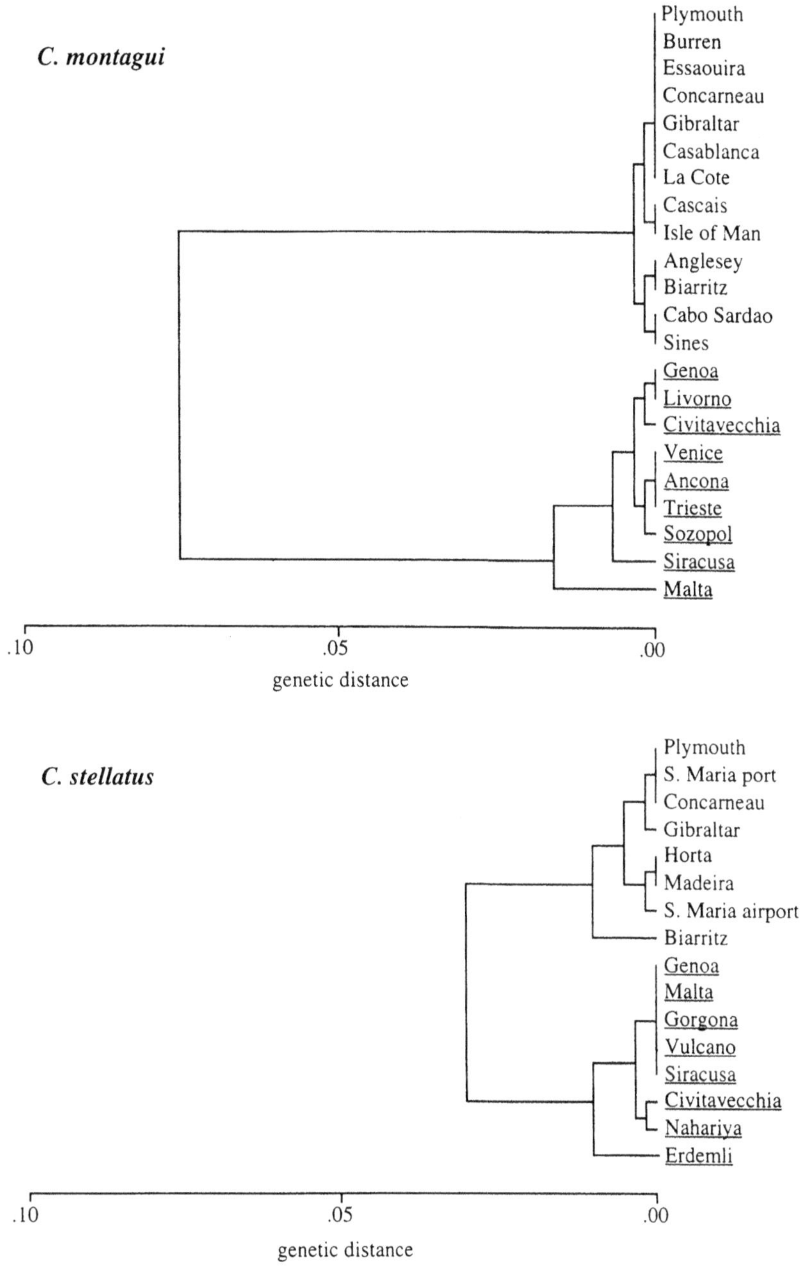

Figure 3. UPGMA dendrogram (based on Nei's unbiased genetic distance) showing relationships amongst Atlantic and Mediterranean populations of the two species of *Chthamalus*.

as between *Patella depressa* Pennant (also called *P. intermedia*) and *P. caerulea*, which most authorities would accept as good and clearly separable species (Fig. 4). Thus, *P. candei gomesii* is probably a separate species and an Azorean endemic, and awaits formal description (but see Titselaar, 1998 who designated it as separate species largely on the basis of work by Corte-Real et al., 1996b). The relationship of the, now very rare, *P. candei candei* to the rest of the complex still remains to be resolved using genetic approaches. It is most likely to be a separate species as widely considered in the Canaries (e.g. Brito, 1984).

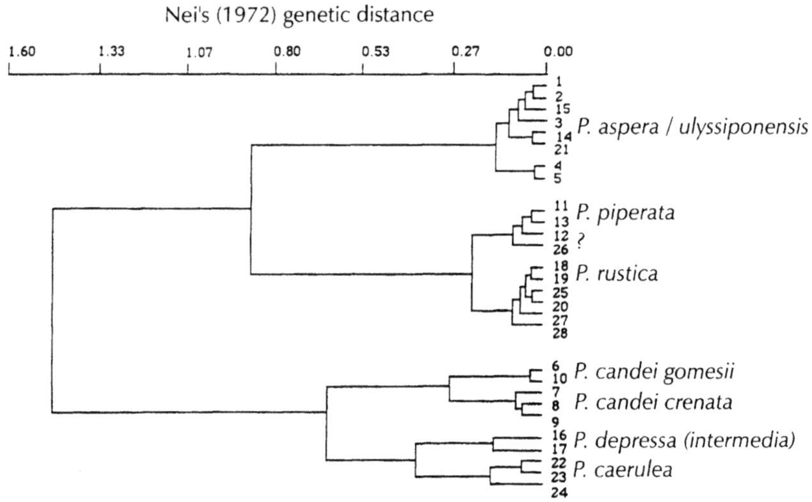

Figure 4. UPGMA dendrogram (based on Nei's unbiased genetic distance) produced from the allele frequencies of ten loci from 28 populations of several species of *Patella* (from Corte-Real, 1992). Numbers refer to populations of various species at sampled localities: *P. aspera*? at 1. (intertidal), 2. (subtidal) Faial, Azores; 3. El Golfo, Lanzarote, Canaries; 4. Porta da Cruz and 5. Porta Moniz both Madeira. *P. candei gomesii* 6. Faial, 10. Mosterios, São Miguel both Azores. *P. candei crenata* 7. Porta Moniz , Madeira 8. (1988) 9. (1991) El Golfo, Lanzarote, Canaries. *P. piperata* 11. Porta Moniz, 12. Porta da Cruz both Madeira. 13. El Golfo, Lanzarote, Canaries. *P. ulyssiponensis* 14. Estoril, Portugal, 15. Tangiers, Morocco, 21. Estepona, Spain. *P. depressa (intermedia)* 16. Estoril, Portugal; 22. Estepona, Spain. 17. Tangiers, Morocco. *P. rustica* 18. Estoril, Portugal; 19. Tangiers, 20. Cap Spartel, 27. Jbel Musa, 28. Tetuan, all Morocco. 26. *P. rustica/P. piperata*? Mal pasa, Majorca *P. caerulea* 23. Estepona 24. Porto Petro, Majorca.

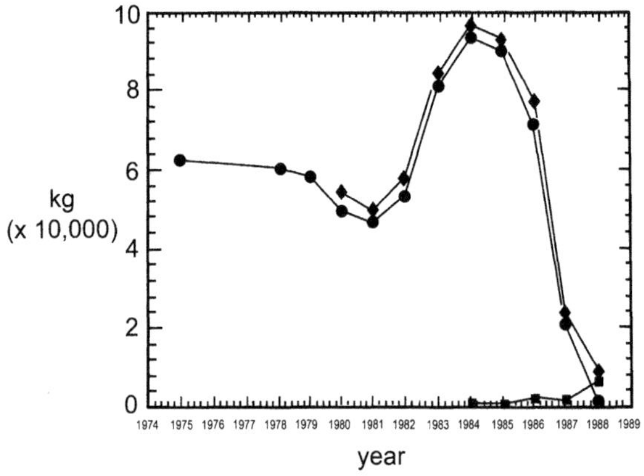

Figure 5. Collapse of the Sao Miguel limpet fishery (from Santos et al., 1990). Symbols: diamonds all islands; circles São Miguel, squares other islands.

Patella ulyssiponensis (or *P. aspera* as it was once more widely known) occurs throughout Macaronesia. It is very similar morphologically in Macaranosia and on the continental coasts. In the Azores, there are both intertidal and subtidal populations, which although showing some morphological differences are genetically the same (Corte-Real et al., 1992). Corte-Real (1992) did not show any clearly defined cluster in the islands compared to the continent on the basis of the ten loci examined. More recently (Weber & Hawkins, unpublished), a study employing 21 allozyme loci showed clear differences, with one diagnostic locus. Probably the name *P. aspera* should be used for the Macaronesian populations which represent a separate endemic species, whilst *P. ulyssiponensis* should be used for the continental populations. This usage will be provisionally adopted in this paper.

Conservation of living littoral resources

The shores of Macaronesia have been heavily exploited for food since the first colonization of the islands by people: the Guanche, and subsequently the Spanish in the Canaries; and the Portuguese in the Azores and Madeira. In consequence, many stocks are currently overfished (reviewed in Santos et al., 1995; Morton et al., 1998) including various species such as lobsters (*Scyllarus arctus* and the larger *Scyllarides latus, 'cavaco'*), barnacles (*Megabalanus azoricus, 'cracas'*), trochids (*Osilinus* sp., *'caramujos'*) and limpets (*Patella* sp., *'lapas'*). Seaweeds (*Pterocladia*) have also been collected for agar (Fralick & Andrade, 1981; Santos & Duarte, 1991). For most species, management measures have been put in place, although in many cases, enforcement has lagged behind legislation. The fishery for limpets in the Azores is a good example which is worth focusing upon.

In the early and mid 1980s, exploitation and commercial gathering of limpets gathered pace (Martins et al., 1987; Santos et al., 1990, 1995). Large numbers were exported to Azorian ex-patriots on the east coast of the U.S.A. The north coast of São Miguel was the site of the most intensive organised commercial activity, but subsistence and informal commercial activity occurred throughout the central group of islands and even in the less populated western group. In 1983, a '*doencia*' (illness) amongst the limpets throughout the central group was reported by the local people. Whether this was a real disease or more likely just a scapegoat for over-collection was never established. Whatever the cause, limpets became much rarer, prompting various investigations funded by the regional government (Santos et al., 1990; Menezes, 1991). In 1988, the fishery in São Miguel collapsed (Fig. 5). Ultimately, a ban on limpet collection was imposed for 1 year in 1989 throughout the archipelago of the Azores. Although probably only reducing fishing effort by a half, this measure probably allowed the stocks to avoid catastrophic over-exploitation. Since then, progressively tighter fisheries regulations have been introduced, but limpet populations are still very much reduced in many of the islands.

One likely reason for their vulnerability is that they are probably protandrous hermaphrodites, especially *P. aspera*. In heavily exploited populations, the average size decreases and there will inevitably be fewer of the larger-sized females, which in turn could lead to increased likelihood of recruitment failure due to lowered reproductive output. Genetically, it would also reduce the effective population size considerably. External fertilizers always face problems when population densities are low as this reduces the opportunities for sperm and eggs to meet (Jamieson,1993; Levitan & Sewell, 1998). Clearly, protandry makes matters worse in heavily cropped populations.

Similar heavy exploitation occurs in the Madeira archipelago, and there are much higher populations in the Desertas than on Madeira and in Porto Santo (R. S. Santos & SJH, unpublished observations). Madeiran limpets are now on sale in São Miguel (SJH, pers obs. at EMBS 34) which can only increase pressure on the source populations.

Recent work in Madeira (Weber et al., 1998) has indicated small-scale genetic differentiation in *P. aspera* and to a lesser extent in *P. candei crenata*. Translating these results to the Azores it would seem that larval exchange between the various parts of the archipelago (Western, Central and Eastern island groups) is likely to be limited, although this remains to be tested. The short length of patellid larval life has led to speciation between the Macaronesian archipelagoes (Corte-Real et al., 1996b). As endemics, *P. candei gomesii* and *P. candei crenata* are probably vulnerable. There are no populations outside their archipelagoes to act as reservoirs for recruitment during recovery. Fortunately, they seem to grow faster than *P. aspera* and reach maturity earlier. Probably the best way forward for these animals is a network of no-take zones coupled with the more traditional fisheries measures now being put in place (closed seasons, size limits, licensing of fishermen and endeavours to get better management information). In both the Azores and Madeira (the Desertas), various reserves are now in place or being designated (Santos, 1992; Santos et al., 1995). The Portuguese-administered Salvage Islands, between Madeira and the Canaries, are a particularly valuable reserve, especially as an important refuge for the endangered *P. candei candei* (as well as birds and seals).

Overview

Ecologically, Macaronesian shores resemble those of the adjacent continents of Europe and Africa. There are various absences which become more pronounced in the geographically younger and more remote Azores. In functional terms, the most important absence is probably of large mussel beds in the Azores and Madeira. Reduced diversity amongst space-occupying barnacles and grazing molluscs may also have implications for community function.

The Mediterranean/Atlantic interface, which includes Macaronesia, has clearly generated considerable differentiation of populations leading to allopatric speciation, including the generation of some endemic species. Greater differentiation has occurred in limpets with their shorter length of larval life (1–2 weeks) than in barnacles (3–6 weeks). The Azores (and possibly Madeira) is also a refuge for at least one tethyan relict species (*Tesseropora atlantica*) which may also have endemic status. During glaciation, it is likely that many temperate species retreated to northwest Africa and the Canaries, from which refuge they may have subsequently expanded outwards and northwards, including into the Mediterranean Sea.

The isolated nature of island populations, including likely low recruitment due to larval loss, makes many species vulnerable to over-exploitation – especially given the Latin taste for seafood. The risk of recruitment failure is increased further by protandric life histories in patellid limpets. The risk of complete extinction is possible given the endemic nature of *P. candei gomesii* in the Azores, but given its early sexual maturity and less marked protandry compared to *Patella aspera,* it is less at risk. *Patella aspera* is potentially vulnerable, but at least has a broad range throughout Macaronesia. Genetic work has important implications for conservation and management. It certainly validates *post hoc* the decision of the Azorean regional government to ban limpet collection in 1989. Human predation is a significant factor in the Azorian littoral zone.

Much work remains to be done on the shores of Macaronesia. Communities have been described, but field experiments are needed to understand the processes structuring them — especially the interplay of physical and biological interactions. Larval dispersal and subsequent recruitment also needs to be studied: such work would inform population genetics, but also be of value to studies of community interactions and the dynamics of fished populations. Macaronesia is a superb natural laboratory to understand speciation in intertidal organisms. Basic studies of the ecology and population genetics of exploited shore organisms are vital for rational management of stocks. Sustainable use of resources will be helped greatly by public education programmes so that intertidal organisms are appreciated for their own sake, as well as for their distinctive place in a regional culinary tradition.

Acknowledgements

This paper would not have been written without the opportunity to come to the 34th EMBS, and we thank the organising committee of the symposium. Work in the Azores, Madeira and Canaries has been funded by a variety of sources including NERC, JNICT, The British Council (especial thanks to Manuela D'Oliveira), The Royal Society, EU, Rotary Foundation and most importantly the regional governments of the Azores and Madeira. None of this work by SJH and colleagues would have been possible without the initial invitation to visit the Azores from Helen Martins at the Department of Oceanography and Fisheries of the University of the Azores and much subsequent collaboration with her and Ricardo Santos and other colleagues in Horta. Norberto Serpa deserves special thanks for the samples of limpets and much amusement. SJH has also been fortunate to have been invited to visit São Miguel and work with the Department of Biology – thank you Antonio Frias Martins, Ana Neto and Ze Avezedo. In the Canaries, Alberto Brito of the University of la Laguna, Tenerife provided a base for field work and advice. Various people have helped in Madeira, especially Thomas Dillinger, Manuel Biscoito and Peter Wirtz. At Port Erin, this work benefited from the inputs of Richard Nash and John Thorpe.

References

Arruda, L. M., 1997. Checklist of the marine fishes of the Azores. Arquivos do Museu Bocage, Nova Série 3: 13–164.

Azevedo, J. M. M., M. R. Portugal Ferreira & J. A. Martins, 1991. The emergent volcanism of Flores Island, Azores, Portugal. Arquipélago, Life and Earth Sciences 9: 37–46.

Azevedo, J. M. N. & P. C. Heemstra, 1995. New records of fishes from the Azores. Arquipélago, Life and Marine Sciences 13A: 1–10.

Backeljau, T., K. Breugelmans, C. Brito, L. de Bruyn, H. de Wolf & J.-M. Timmermans, 1995. Macrogeographic genetic homogeniety in *Littorina striata* from the Azores (Mollusca: Prosobranchia). Açoreana, suppl: 159–171.

Boury-Esnault, N. & M. T. Lopes, 1985. Les démosponges littorales de l'archipel des Açores. Ann. Inst. Oceanogr, Paris, n.s. 61: 149–225.

Briggs, J. C., 1966. Oceanic islands, endemism and marine paleotemperatures. Syst. Zool. 15: 153–163.

Briggs, J. C., 1970. A faunal history of the north Atlantic Ocean. Syst. Zool. 19: 19–34.

Briggs, J. C., 1974. Marine Zoogeography. McGraw-Hill, New York, 475 pp.

Brito, A., 1984. In Aránega J. J. B. & L. Santana (eds), Fauna (Marina y Terrestre) Del Archipielago Canario. Edirca, Islas Canarias: 27–41.

Brito, A., I. J. Lozano, J. M. Falcon, F. M. Rodriguez & J. Mena, 1996. Biogeographic analysis of the ichthyofauna of the Canary Islands. In Llinas O., J. A. Gonzalez & M. J. Rueda (eds), Oceanography and Marine Resources in the Eastern Central Atlantic.

Burrows, M. T., 1988. The comparative biology of *Chthamalus stellatus* (Poli) and *Chthamalus montagui* Southward. Ph.D Thesis. Faculty of Science, University of Manchester, Manchester.

Christiaens, J., 1973. Revision de genre *Patella* (Mollusca: Gastropoda). Bull. Mus. natn. Hist., Paris 182: 1305–1392.

Chapman, G., 1954. Aspects of the fauna and flora of the Azores I. Introduction and notes on the littoral conditions. Ann. Mag. nat. Hist., Series 12, 7: 673–677.

Chapman, G., 1955. Aspects of the fauna and flora of the Azores VI. The density of animal life in the coralline algae zone. Ann. Mag. nat. Hist. Series 12, 8: 801–805.

Charcot, J. B., 1921. Sur une expédition du *Pourquoi-Pas?* à Rockall. Compte rendu hebdomadaire des séances de l'Académie de sciences 173: 290–292.

Cornelius, P. F. S., 1992. The Azores hydroid fauna and its origin, with discussion of rafting and medusa suppression. Arquipélago, Life and Earth Sciences 10: 75–99.

Côrte-Real, H. B. S. M., 1992. Taxonomy and population genetics of exploited species of *Patella* in the Azores, Madeira and Canaries. Ph.D. Thesis, University of Liverpool.

Côrte-Real, H. B. S. M., S. J. Hawkins & J. P. Thorpe, 1992. Genetic confirmation that intertidal and subtidal morphs of *Patella ulyssiponensis aspera* Röding (Mollusca, Gastropoda, Patellidae) are conspecific. Arquipélago, Life and Earth Sciences 10: 55–66.

Côrte-Real, H. B. S. M., S. J. Hawkins & J. P. Thorpe, 1996a. An interpretation of the taxonomic relationships between the limpets *Patella rustica* and *P. piperata*. J. mar. biol. Ass. U.K. 76: 717–732.

Côrte-Real, H. B. S. M., S. J. Hawkins & J. P. Thorpe, 1996b. Population differentiation and taxonomic status of the exploited limpet *Patella candei* in the Macaronesian islands (Azores, Madeira, Canarias). Mar. Biol. 125: 141–152.

Crisp, D. J., 1956. The intertidal zoology of Rockall. In: Fisher, J. (eds), Rockall G. Bles, London: 177–170.

Crisp, D. J. & A. J. Southward, 1958. The distribution of intertidal organisms along the coasts of the English Channel. J. mar. biol. Ass. U.K. 37: 1037–1048.

Crisp, D. J., A. J. Southward & E. C. Southward, 1981. On the distribution of the intertidal barnacles *Chthamalus stellatus, Chthamalus montagui* and *Euraphia depressa*. J. mar. biol. Ass. U.K. 61: 359–380.

Dando, P. R. & Southward, 1981. Existence of Atlantic and Mediterranean forms of *Chthamalus montagui* (Crustacea, Cirrepedia) in the Western Mediterranean. Mar Biol. Lett 2: 239–248.

Darwin, C., 1859. On the Origin of Species by Means of Natural Selection, or the Preservation of Favoured Races in the Struggle for Life. Murray. London.

de Weerdt, W. H. 1989. Phylogeny and vicariance biogeography of the north Atlantic Chalinidae (Haplosclerida, Demospongiae). Beaufortia 39: 55–88

Fischer-Piette, E., 1935. Systématique et biogéographie – les patelles d'Europe et d'Afrique du nord. J. Conchyl. 79: 5–66.

Feldmann, J., 1946. La Flore marine des iles atlantides. Mémoires de la Société de Biogéographie 28: 395–435.

Fralick, R. A. & F. Andrade, 1981. The growth, reproduction, harvesting and management of *Pterocladia pinnata* (Rhodophyceae) in the Azores, Portugal. In Levring, T. (eds), Proceedings of the Tenth International Seaweed Symposium, Goteborg (Sweden), 1980. Walter de Gruyter: 289–295.

Fralick, R. A. & E. J. Hehre, 1990. Observations on the marine algal flora of the Azores II: an annotated checklist of the Chlorophycota from the Azores. Arquipélago, Life and Earth Sciences 8: 11–17.

Fredericq, S., E. Serrão & J. Norris, 1992. New records of marine red algae from the Azores. Archipelago, Life and Earth Sciences 10: 1–4.

Hawkins, S. J., L. P. Burnay, A. I. Neto, R. Tristào da Cunha & A. M. F. Martins, 1990a. A description of the zonation patterns of molluscs and other biota on the south coast of São Miguel, Azores. In Martins, A. M. F. (eds), Proceedings of the First International Workshop of Malacology and Marine Biology, Vila Franca do Campo, São Miguel, Azores, 1988. Açoreana, Suplemento (2), Outubro de 1990: 21–38.

Hawkins, S. J., H. B. S. M. Côrte-Real, H. Rost Martins, R. S. Santos & A. M. F. Martins, 1990b. A note on the identity of *Patella* in the Azores. In Martins, A. M. F. (ed.), Proceedings of the First International Workshop of Malacology and Marine Biology, Vila Franca do Campo, São Miguel, Azores, 1988. Açoreana, Suplemento (2), Outubro de 1990: 167–173.

Hawkins, S. J. & R. G. Hartnoll, 1983. Grazing of intertidal algae by marine invertebrates. Oceanogr. Mar. Biol. Ann. Rev. 21: 195–282.

Hawkins, S. J. & K. Hiscock, 1983. Anomalies in the abundance of common eulittoral gastropods with planktonic larvae on Lundy Island, Bristol Channel. J. moll. Stud. 49: 86–88.

Hawkins, S. J., R. G. Hartnoll, J. M. Kain, & T. A. Norton, 1992. Plant–animal interactions on hard substrata in the north-east Atlantic. In John, D. M., S. J. Hawkins & J. H. Price (eds), Plant-Animal Interactions in the Marine Benthos. Clarendon Press, Oxford. Systematics Association Special Volume No. 46. 1–32.

Highsmith, R. C., 1985. Floating and larval rafting as potential dispersal mechanisms in brooding invertebrates. Mar. Ecol. Prog. Ser. 25: 169–179.

Hodgson, A. N., S. Ridgway, G. M. Branch & S. J. Hawkins, 1996. Spermatozoan morphology of 19 species of prosobranch limpets (Patellogastropoda) with a discussion of patellid relationships. Phil. Trans. r. Soc. B., Lond. 351: 339–347.

Jamieson, G. S., 1993 Marine invertebrate conservation: evaluation of fisheries over-exploitation concerns. Am. Zool. 33: 551–567.

Johannesson, K., 1988. The paradox of Rockall: Why is a brooding gastropod (*Littorina saxatilus*) more widespread than one having a planktonic larval dispersal stage (*L. littorea*)? Mar. Biol. 99: 507–513.

Kay, E. A. & S. R. Palumbi, 1987. Endemism and evolution in Hawaiian marine invertebrates. Trends Ecol. Evol. 2: 183–187.

Koufapanou, V., D. G. Reid, S. A. Ridgway & R. H. Thomas, 1999. A molecular phylogeny of the patellid limpets (Gastropoda: Patellidae) and its implications for the origins of their antitropical distribution. Mol. Phylogen. Evol. 11: 138–156.

Lawson, G. W. & T. A. Norton, 1971. Some observations on littoral and sublittoral zonation at Teneriffe (Canary Islands). Bot. Mar. 14: 116–120.

Levitan, D. R. & M. A. Sewell, 1998. Fertilization success in free-spawning marine invertebrates: review of the evidence and fisheries implications. Canadian Special Papers in Fisheries and Aquatic Science 125: 159–164.

Lloris, D., J. Rucabado & H. Figueroa, 1991. Biogeography of the Macaronesian ichthyofauna (The Azores, Madeira, The Canary Islands, Cape Verde and the African enclave). Bol. Mus. Munic. Funchal 43: 191–241.

Lopes, M. F. R., J. C. Marques & D. Bellan-Santini, 1993. The benthic amphipod fauna of the Azores (Portugal): an up-to-date annotated list of species, and some biogeographic considerations. Crustaceana 65: 204–217.

MacArthur, R. H. & E. O. Wilson, 1967. The Theory of Island Biogeography. Princeton University Press, Princeton, N.J.

Martins, H. R., R. S. Santos & S. J. Hawkins, 1987. Exploitation of limpets (*Patella* spp.) in the Azores, with a preliminary analysis of the stocks. ICES, C.M: K53: 1–18.

May, R. M., 1994. Biological diversity: differences between land and sea. Phil. Trans. r. Soc., Lond. B. 343: 105–111.

Mayr, E., 1963. Animal Species and Evolution. Harvard University Press, Cambridge, Mass.

Mayr, E., 1970. Populations, Species and Evolution. Harvard University Press, Cambridge, Mass.

Menezes, G. M. M., 1991. Aspectos da Biologia e ecologia de *Patella ullyssiponensis aspera* e *P. candei candei* (Mollusca, Gastropoda, Patellidae) dos Açores. Potencial da sua exploração. Relatório de Estágio da Licenciatura de Recursos Faunísticos e Ambientes. Faculdade de Ciencias de Lisboa.

Monniot, C. & F. Monniot, 1983. Navigation ou courants? La colonisation des Açores et des Bermudes par les ascidies (tuniciers benthiques). C.R Soc. Biogéogr. 59: 53–58.

Moore, P. G., 1977. Additions to the littoral fauna of Rockall, with a description of *Areolaimus penelope* sp. nov. (Nematoda: Axonolaimidae). J. mar. biol. Ass. U.K. 57: 191–200.

Morton, B., 1967. Malacological Report. In Final Report, Chelsea College Azores Expedition, 1965. London: Chelsea College, University of London: 30–39.

Morton, B., 1990. The intertidal ecology of Ilhéu de Vila Franca – a drowned volcanic crater in the Azores. In Marins, A. M. F. (ed.), Proceedings of the First International Workshop of Malacology and Marine Biology, Vila Franca do Campo, São Miguel, Azores, 1988. Açoreana, Suplemento [2], Outubro de 1990: 3–20.

Morton, B., 1995. Coastal conservation in the Azores. Mar. Poll. Bull. 30: 430–431.

Morton, B., J. C. Britton & A. M. De Frias Martins, 1998. Coastal Ecology of the Açores. Sociedade Afonso Chaves, Ponta Delgada.

Moss, D., 1992. A summary of the Porifera collected during 'Expedition Azores 1989". Arquipélago, Life and Earth Sciences 10: 45–53.

Nei, M., 1978. Estimation of average heterozygosity and genetic distance from small number of individuals. Genetics 89: 583–590.

Neto, A. I., 1992. Contribution to the taxonomy and ecology of the Azorean benthic marine algae. Biol. J. linn. Soc. 46: 163–176.

Neto, A. I., 1994. Checklist of the benthic marine macroalgae of the Azores. Arquipélago, Life and Earth Sciences 12A: 15–34.

Newman, W. A. & A. Ross, 1977. A living *Tesseropora* (Cirrepedia, Balanomorpha) from Bermuda and the Azores, first records from the Atlantic since the Oligocene. Transactions of the San Diego society of Natural History 18: 207–218.

Palumbi, S. R., 1992. Marine speciation on a small planet. Trends Ecol. Evol. 7: 114–118.

Palumbi, S. R., 1994. Genetic divergence, reproductive isolation, and marine speciation. Ann. Rev. Ecol. Syst. 25: 547–572.

Pannacciulli, F. G., J. D. D. Bishop & S. J. Hawkins, 1997. Genetic structure of populations of two species of *Chthamalus* (Crustacea: Cirripedia) in the north-east Atlantic and Mediterranean. Mar. Biol. 128: 73–82.

Patzner, R. A., R. S. Santos, P. Ré & R. D. M. Nash, 1992. Littoral fishes of the Azores: an annotated checklist of fishes observed during the 'Expedition Azores 1989'. Arquipélago, Life and Earth Sciences 10: 101–111.

Pingree, R. D., C. Garcio-Soto & B. Sinha, 1999. Position and structure of the subtropical/Azores front region from combined Lagrangian and remote sensing (IR/altimeter/SeaWiFS) measurements. J. mar. biol. Ass. U.K. 79: 769–792.

Powell, A., 1973. The patellid limpets of the world. Indo-Pacific Mollusca 3: 75–99.

Prud'homme Van Reine, W. F., 1988. Phytogeography of seaweeds of the Azores. Biogeography of Marine Benthic Algae 42: 165–185.

Prud'homme Van Reine, W. F., 1990. Biogeography of Macaronesian seaweeds. 8th contribution to Fauna and Flora of the Cape Verde Islands. Results of the 4th symposium. Lobin, W. (ed.), pp. 55–73.

Raffaelli, D. & S. J. Hawkins 1996. Intertidal Ecology. Kluwer Academic Publishers, Dordrecht, The Netherlands: 356 pp.

Rees, W. J. & E. White, 1966. New records and fauna list of hydroids from the Azores. Ann. Mag. Nat. Hist. Ser 13, 9: 271–284.

Reid, D. G., 1996. Systematics and Evolution of *Littorina*. The Ray Society, Andover: 463 pp.

Ridgway, S. A., D. G. Reid, J. D. Taylor, G. M. Branch & A. N. Hodgson, 1998. A cladistic phylogeny of the family Patellidae (Mollusca: Gastropoda). Phil. Trans. Roy. Soc., Lond. B. 353: 1645–1671.

Sanchez, M. J. P. & E. M. Batet, 1990. Invertebrados Marinos de Canarias. Cabidilo Insular de Gran Canaria.

Santos, R. & P. Duarte, 1991. Marine plant harvest in Portugal. J. appl. Phycol. 3: 11–18.

Santos, R. S., 1992. Protecção e conservação do meio marinho nos Açores. In Saldanha, L., P. Ré & A. F. Martins (eds), Centenaire de la Dernière Campagne Océanographique du Prince Albert de Monaco aux Açores à bord de L'Hirondelle. Açoreana, Suppl: 107–122.

Santos, R. S., F. M. Porteiro & J. P. Barreiros, 1997. Marine fishes of the Azores: an annotated checklist and bibliography. Arquipélago, Life and Marine Sciences, Supplement 1: 244 pp.

Santos, R. S., R. D. M. Nash & S. J. Hawkins, 1994. Fish assemblages on intertidal shores on the island of Faial, Azores. Arquipélago, Life and Marine Sciences 12A: 87–100.

Santos, R. S., S. Hawkins, L. R. Monteiro, M. Alves & E. J. Isidro, 1995. Marine research, resources and conservation in the Azores. Aquat. conserv. Mar. Freshwat. Ecos. 5: 311–354.

Santos, R. S., H. R. Martins & S. J. Hawkins, 1990. Relatório de estudos sobre o estado das populações de lapas do Arcipélago dos Açores e da Ilha da Madeira. Relatório da X Semana das Pescasdos Açores, Universidade dos Açores, Horta, Açores 137–161.

Scheltema, R. S., 1971. Larval dispersal as a means of genetic exchange between geographically separated populations of shoal-water benthic marine gastropods. Biol. Bull. 140: 284–322.

Scheltema, R. S., 1986. On dispersal of benthic invertebrates: An eclectic overview and summary of problems. Bull. mar. Sci. 39: 290–322.

Southward, A. J., 1998. New observations on barnacles (Crustacea: Cirripedia) of the Azores Region. Arquipélago 16A: 11–27.

Southward, A. J., R. S. Burton, S. L. Coles, P. R. Dando, R. DeFelice, J. Hoover, P. E. Parnell, T. Yamaguchi & W. A. Newman 1998. Invasion of Hawiian shores by an Atlantic barnacle. Mar. Ecol. Prog. Ser. 165: 119–126.

Southward, A. J. & D. J. Crisp, 1954. The distribution of certain intertidal animals around the Irish coast. Proc. R. Irish Acad. 57B: 1–29.

Spence, S. K., S. J. Hawkins & R. S. Santos, 1990. The mollusc *Thais haemastoma* – an exhibitor of 'imposex' and potential biological indicator of tributylin pollution. Mar. Ecol. (PSZN) 11: 147–156.

Stephenson, T. A. & A. Stephenson, 1972. Life between tide marks on rocky shores. W.H. Freeman & Company, San Francisco.

Titselaar, F. L. M., 1998. A review of the recent European patellids (Mollusca: Gastropoda). The patellids of the Azores, Madeira, the de Selvagens and the Canary Islands. Vita Marina 45: 21–62.

Tittley, I., P. Moss & A. I. Neto, 1990. Azores Expedition 1989: taxonomy and ecology of the marine littoral, algal studies. In Expedição Açores/89. Ecologia e Taxonomia do Litoral Marinho dos Açores – Relatório Preliminar – Parte II. Departamento de Oceanographia e Pescas, Universidade dos Açores, Horta, Açores.

Tittley, I. & A. I. Neto, 1995. Marine algal flora of the Azores and its biogeographical affinities. Bol. Mus. Mun. Funchal. Suppl: 747–766.

Tsuchiya, M. & C. Retiere, 1992. Zonation of intertidal organisms and community structure of small animals associated with patches of the mussel *Mytilus edulis* L. along the rocky coast of Dinard, Brittany, France. Bull. Coll. Sci. Univ. Ryukus Ryudai Rigakubu Kiyo. 54: 47–81.

Turner, G. F., 1999. Explosive speciation of African cichlid fishes. In Magurran, A. E. & R. M. May (eds), Evolution of Biological Diversity. Oxford University Press, Oxford: 113–129.

Valentine, J. W. & D. Jablonski, 1983. Speciation in the shallow sea: General patterns and biogeographic controls. Palaeontology 14: 201–226.

Van Den Hoek, C., 1984. World-wide latitudinal and longitudinal seaweed distribution patterns and their possible causes, as illustrated by the distribution of Rhodophytan genera. Helgölander wiss. Meersunters. 38: 227–257.

Van Den Hoek, C., 1987. The possible significance of long-range dispersal for the biogeography of seaweeds. Helgölander wiss. Meersunters. 41: 261–272.

Weber L. I., J. P. Thorpe, R. S. Santos & S. J. Hawkins, 1998. Identification of stocks of the exploited limpets *Patella aspera* and *P. candei* at Madeira Archipelago by allozyme electrophoresis. J. Shellfish Res. 17: 945–953.

Wirtz, P., 1996. Underwater Guide Madeira-Canary Islands-Azores: Invertebrates. S. Naglschmid Verlag, Stuttgart.

Wirtz, P. & H. R. Martins, 1993. Notes on some rare and little known marine invertebrates from the Azores with a discussion of the zoogeography of the region. Arquipélago, Life and Marine Sciences 11A: 55–63.

Young, P. S., 1998. Cirripedia (Crustacea) from the 'Campagne Biocores' in the Azores region, including a generic revision of the Verrucidae. Zoosystema 20: 32–91.

Hydrobiologia **440:** 19–25, 2000.
M.B. Jones, J.M.N. Azevedo, A.I. Neto, A.C. Costa & A.M. Frias Martins (eds), Island, Ocean and Deep-Sea Biology.

A provisional classification of algal-characterised rocky shore biotopes in the Azores

Ian Tittley[1] & Ana I. Neto[2]
[1]*Department of Botany, The Natural History Museum, London SW7 5BD, U.K.*
Fax: +44 207 5529. E-mail: I.Tittley@ nhm.ac.uk
[2]*Universidade dos Açores, 9500 Ponta Delgada, Açores, Portugal*
Fax: +351 296 653455. E-mail: aneto@alf.uac.pt

Key words: algae, Azores, biotope, community structure, rocky shores

Abstract

Recent studies of the rocky shores of the Azores archipelago have provided information on community structure allowing provisional identification of plant-characterised biotopes (habitats and their associated communities). Although the Azores share some littoral and sublittoral biotopes with the Atlantic coast of mainland Europe, shores in the archipelago mostly lack the functionally important 'leathery macrophyte' communities of fucoids and laminarians widespread in the North Atlantic. Intertidal biotopes are mainly turfs typical of warm-temperate and tropical regions, and characterised by articulated Corallinaceae or by non-coralline algae such as *Cladophora* spp., *Gelidium* spp., *Pterocladiella capillacea*, *Stypocaulon scoparia*, and *Valonia utricularis*. Subtidal algal biotopes are characterised by *Dictyota* spp., *Halopteris filicina*, *Sphaerococcus coronopifolius* and, most commonly, *Zonaria tournefortii*.

Introduction

Until the 1990s, ecological investigations that described marine algal community structure in the Azores were few (Schmidt, 1931; Pryor, 1967; Castro & Viegas, 1983; Marques, 1984; Castro & Viegas, 1987). The most thorough study was that of Schmidt (1931) who defined the algal associations (Table 1). In the past decade, detailed surveys have been undertaken throughout the archipelago and, as a result, there is now a more comprehensive body of information on community structure (Neto, 1992; Neto & Tittley, 1996; Morton et al., 1998; Tittley et al., 1998; Neto et al., 2000). Tittley et al. (1998) provisionally identified some algal-characterised biotopes of the island of Flores following Hiscock (1995). The present paper provides, for the first time, a provisional classification of algal-characterised biotopes drawn from field studies on the islands of Flores, Faial and São Miguel.

A biotope is defined as 'a seashore or sea-bed habitat (i.e. the environment's physical and chemical characteristics, and the particular conditions of wave exposure, salinity, tidal streams and other factors which contribute to the overall nature of the location) and its associated community of species, operating together' (Connor et al., 1997a, b). In this paper, Azorean littoral and sublittoral biotopes characterised by algae are defined provisionally following the methods of the Marine Nature Conservation Review (MNCR) (Connor et al., 1997a, b). The MNCR classification aims to be comprehensive, and includes biotopes for artificial, polluted or barren areas and is intended to complement other marine classifications (e.g. in France, Dauvin, 1995). This paper focuses on rocky shores (as opposed to sedimentary shores), the primary division of the biotope classification. Generally, rocky shores provide the stable substratum required for algal growth.

Materials and methods

Schmidt (1931) applied the phytosociological methodology of Braun-Blanquet used also by other workers to classify marine vegetation on continental European coasts [e.g. Den Hartog (1959) for the Netherlands]. There are similarities between the phytosociological

Table 1. Azores algal associations of Schmidt (1931) and biotope equivalents as defined in this paper

Association	Biotope
Open Littoral	
Bangia-Cyanophyta	Cyanobacteria-*Bangia*
Lichina pygmaea	*Lichina pygmaea*
Polysiphonia	?
Nemalion-Grateloupia-Bryopsis-Cladophora	Soft algal turf complex
Enteromorpha	*Enteromorpha*
Fucus spiralis	*Fucus spiralis*
Gelidium microdon	*Gelidium microdon*
Laurencia obtusa	Soft algal turf complex
Haliptylon virgatum	Coralline algal turf complex
Caulacanthus-Chondracanthus-Ceramiaceae	Soft algal turf complex
Gelidium-Codium adhaerens	*Codium adhaerens*
Corallina elongata	Coralline algal turf
Cystoseira	*Cystoseira*
Osmundea pinnatifida	Soft algal turf complex
Shaded littoral	
Gymnogongrus-Lomentaria-Rhodymenia	Shade turf
Catenella	Shade turf-*Catenella*
Griffithsia phyllamphora	?
Haliptylon squamatum	?
Crustose algae	Shade crustose spp.
Sublittoral	
Corallina officinalis	*Corallina* turf complex
Asparagopsis	*Asparagopsis* turf
Sargassum-Cladostephus-Nitophyllum-Plocamium	?
Crustose algae	Crustose algae

and biotope methods in defining marine algal communities. Schmidt's (1931) sublittoral observations were limited as SCUBA diving had not been developed at that time and sublittoral areas were sampled by dredging, an inexact method of surveying the vegetation of such habitats. Although many of the transect and quadrat studies recorded in recent surveys have provided data that support Schmidt's (1931) observations, the MNCR biotope classification was taken as a starting point for this paper.

The MNCR biotope classification is hierarchical and based on a matrix of environmental features including substratum type, wave exposure conditions, and shore level (Connor et al., 1997a, b). Within each of the major divisions of substratum, 'biotope complexes' (a broad feature often comprising several biotopes) were identified for each of the main shore zones (littoral, infralittoral, circalittoral) and four points on a scale of wave exposure conditions from very exposed to sheltered. Other biotope complexes (rock pools, overhangs, caves) were classified separately. For each biotope, a name is given together with a coded classification representing substratum type, wave exposure conditions and the characterising species. A description is provided also for each biotope, the characterising and other associated species are listed. For this provisional assessment of Azorean biotopes, we provide only a name.

The present study used both descriptive and quantitative data (obtained from quadrats along transect lines) to define provisionally algal-characterised biotopes. Field studies were undertaken at littoral and sublittoral levels (to 30 m depth) in a range of wave exposure conditions from sheltered to exposed.

Results

As the focus of this paper is on algal biotopes, some animal-characterised biotopes have been omitted from the classification. Table 1 presents the 24 algal associations defined by Schmidt (1931) with, where possible, a modern biotope or biotope complex equivalent. Of the associations identified by Schmidt (1931), two were restricted to supralittoral levels, 12 occurred in the littoral in full light conditions, a further five occupied shaded littoral situations and five were recorded from sublittoral areas.

Tables 2a, b present the biotope complexes in the Azores using the matrix structure of the MNCR biotope classification (cf. Table 5.2 in Connor et al., 1997a, b). At both littoral and sublittoral levels, these complexes in the Azores are similar irrespective of wave action. An important caveat is the definition of the scale of wave action in the Azores; there are probably no truly sheltered habitats in the archipelago comparable to those of sea lochs and inlets in Britain.

Tables 2a, b also present actual biotopes within the complexes (cf. Table 5.5 'Littoral rock habitat matrix' in Connor et al. 1997a, b). On wave exposed shores at supralittoral levels, the principal biotopes are characterised by various yellow lichens; at supralittoral fringe levels, black lichens (*Verrucaria* spp.) and Cyanobacteria crusts characterise a set of biotopes. Littorinids [*Melarhaphe neritoides* Linnaeus (1758) and *Littorina* spp.] often form a distinct animal biotope at upper midlittoral levels overlapping at lower levels with patches of the marine lichen *Lichina pygmaea* (Lightfoot) C. Agardh which may also form a distinct biotope. A feature of most Azorean shores (from very wave exposed to more sheltered situations) is an eulittoral 'barnacle-limpet' biotope complex. This may be interrupted by green algal (*Blidingia* spp., *Enteromorpha* spp.) biotopes according to season, wave action, presence of pools and draining water. The crustose coralline alga *Tenarea tortuosa* (Esper) Me Lemoine also forms a biotope in the 'barnacle-limpet' complex. On less wave exposed, more sheltered shores, the biotopes are broadly similar. In spring, the Cyanobacteria biotope may contain noticeable amounts of the filamentous red alga *Bangia atropurpurea* (Roth) C.Agardh and *Porphyra umbilicalis* (Linnaeus) Kützing. The only fucoid in the Azores, *Fucus spiralis* Linnaeus, occurs patchily as small plants and locally defines a biotope.

On both wave exposed and more sheltered shores, lower littoral levels commonly support algal 'turfs'. A turf is a convenience term to describe compact, sometimes densely intertwined mats of algae (Neto & Tittley, 1996). We have identified two broad types of turf biotope complex, those characterised by soft (mainly non-calcified) algae [e.g. *Gelidium* spp., *Pterocladiella capillacea* (S.G.Gmelin) Santelices & Hommersand, *Stypocaulon scoparia* (Linnaeus) Kützing, *Cladophora* spp., *Valonia utricularis* (Roth) C. Agardh], and those by calcified, articulated Corallinaceae (e.g. *Corallina* spp., *Jania* spp., *Haliptylon* spp.). Coralline turfs are a distinct feature of many rocky shores in the Azores, comprise several biotopes and extend to deep sublittoral levels. The soft algal turf is often species rich and comprises a complex of biotopes. The prostrate *Codium adhaerens* C. Agardh characterises a distinct biotope at low-water level on many shores. Crustose coralline algae are the dominant life forms and form a biotope on strongly wave washed rocky shores in the surge zone at shallow sublittoral levels. At sublittoral fringe levels, prolific growths of *Asparagopsis armata* Harvey often define a distinct summer biotope.

Occasionally, large brown algae may form biotopes at lower littoral or sublittoral levels. Morton et al. (1998) reported a carpet of *Cystoseira tamariscifolia* (Hudson) Papenfuss on the island of Santa Maria.

Intertidal areas are often modified to form rock pools, overhangs and caves and distinct algal communities are associated with these. Shaded habitat biotopes may have *Lomentaria articulata* (Hudson) Lyngbye or *Catenella caespitosa* (Withering) L. Irvine as the dominant species, or the crustose red alga *Hildenbrandia* sp. Pool biotopes are often characterised by the large brown algae *Cystoseira* spp. and *Sargassum* spp.

At subtidal levels, foliose algae are more prominent; the green alga *Ulva rigida* C. Agardh, and the brown alga *Padina pavonica* (Linnaeus) Thivy, are locally, seasonally, predominant at São Vicente on the island of São Miguel. Another brown alga, *Zonaria tournefortii* (Lamouroux) Montagne, forms extensive stands throughout the archipelago often to considerable depths. The red algae *Sphaerococcus coronopifolius* Stackhouse and *Plocamium cartilagineum* (Linnaeus) P.S. Dixon characterise biotopes in deep water. On the island of Faial, the *Corallina* turf extended to 30 m depth. The Formigas islets are unique in the Azores in the occurrence there of the leathery macrophyte kelp, *Laminaria ochroleuca* De La Pylaie; this single recorded example of a kelp-characterised

Table 2a. Habitat matrix in the Azores. Biotope complexes and (indented) species characterising probable biotopes

	Very exposed	Exposed
Supra-littoral	Yellow & grey lichens	Yellow & grey lichens
Upper Littoral Fringe	Cyanobacteria-*Verrucaria*	Cyanobacteria-*Verrucaria*
Lower Littoral Fringe	Littorinids	Littorinids
	Lichina pygmaea	*Lichina pygmaea*
Upper Eulittoral	Barnacles-limpets	Barnacles-limpets
	Enteromorpha spp.	*Enteromorpha* spp.
Mid Eulittoral	Barnacles-limpets	*Fucus spiralis*
		Gelidium microdon
	Tenarea tortuosa	*Tenarea tortuosa*
Lower Eulittoral	Soft algal turf	Soft algal turf
	Gelidium latifolium	*Gelidium latifolium*
	Cladophora prolifera	*Osmundea pinnatifida*
	Chondracanthus acicularis	*Rhodymenia pseudopalmata*
		Codium adhaerens
Rockpool	Rockpool	Rockpool
	Corallina spp.	*Corallina* spp.
	Sargassum spp.	*Sargassum* spp.
	Cystoseira spp.	*Cystoseira* spp.
Sublittoral Fringe	Coralline turf	Coralline turf
	Crustose Corallinaceae	Crustose Corallinaceae
		Asparagopsis armata
Sublittoral		Soft algal turf
		Stypocaulon scoparium
	Foliose algae	Foliose algae
	Ulva rigida	*Ulva rigida*
	Zonaria tournefortii	*Zonaria tournefortii*
	Dictyota spp.	*Dictyota* spp.
Deep Sublittoral	Red algae	Red algae
	Sphaerococcus coronopifolius	*Sphaerococcus coronopifolius*
	Plocamium cartilagineum	*Plocamium cartilagineum*
	Kelp stand	
	Laminaria ochroleuca	

Table 2b. Habitat matrix in the Azores (biotope complexes and (indented) keystone species characterising probable biotopes)

	Moderately exposed	Less exposed
Supra-littoral	Yellow & grey lichens	Yellow & grey lichens
Upper Littoral Fringe	Cyanobacteria-*Verrucaria*	Cyanobacteria-*Verrucaria* *Bangia*-Cyanophyta
Lower Littoral Fringe	Littorinids *Lichina pygmaea*	Littorinids *Lichina pygmaea*
Upper Eulittoral	Barnacles-limpets *Enteromorpha* spp.	Barnacles-limpets *Enteromorpha* spp.
Mid Eulittoral	*Fucus spiralis* *Gelidium microdon*	*Fucus spiralis* *Gelidium microdon*
Lower Eulittoral	Soft algal turf *Gelidium latifolium* *Rhodymenia pseudoplamata* *Pterocladiella capillacea*	Soft algal turf *Valonia utricularis* *Chondracanthus acicularis* *Rhodymenia pseudopalmata* *Pterocladiella capillacea* *Stypocaulon scoparium* *Codium fragile*
Rockpool	Rockpool *Corallina* spp. *Sargassum* spp. *Cystoseira* spp.	Rockpool *Codium fragile* *Corallina* spp. *Sargassum* spp. *Cystoseira* spp.
Sublittoral fringe	Coralline turf	Coralline turf *Asparagopsis armata*
Sublittoral	Soft algal turf *Stypocaulon scoparium* Foliose algae *Ulva rigida* *Zonaria tournefortii* *Dictyota* spp.	Soft algal turf *Stypocaulon scoparium* Foliose algae *Ulva rigida* *Zonaria tournefortii* *Dictyota* spp.
Deep sublittoral	Red algae *Sphaerococcus coronopifolius* *Plocamium cartilagineum*	Red algae *Sphaerococcus cornonopifolius* *Plocamium cartilagineum*

biotope occurs only in deep waters (30–40 m and below).

Discussion

A major distinction between the algal-characterised biotopes of the Azores and those of the northern North Atlantic in both Europe and America is the near absence of the leathery macrophyte communities of fucoids and laminarians. Where algae predominate in the Azores, they are often turf forming. Algal turf biotopes also occur in the northern North Atlantic Ocean. For example, an *Osmundea pinnatifida* (Hudson) Stackhouse and *Gelidium pusillum* (Stackhouse) Le Jolis turf occurs on moderately exposed mid-littoral rock in Britain where the fucoid canopy is reduced or absent (Connor et al., 1997a, b); both species are turf-forming constituents in the Azores but such turfs in Britain are less common and less species rich. The non-coralline algal turf biotope is typical of other Macronesian islands and also tropical West Africa where Lawson & John (1987) described a red algal turf as a principal feature of intertidal shores. A biotope defined by '*Corallina officinalis* on very exposed eulittoral rock' (Connor et al., 1997a, b) occurs widely in the northern North Atlantic and also in the Azores, usually as pure stands of *C. officinalis* Linnaeus in its typical form. A second *Corallina* turf biotope also occurs in the Azores and comprises a compact and tightly entangled species-rich mat in which *Corallina* is less recognisable and mixed with other articulated Corallinaceae, this biotope closely resembles *Corallina* turfs described from the Canary Islands (Lawson & Norton, 1971), West Africa (Lawson & John, 1977), Mexico (Stewart, 1989) and Brazil (Oliveira & Mayral, 1976). The predominance at sublittoral levels of the foliose *Zonaria tournefortii* biotope contrasts with the predominance of leathery macrophyte kelp biotope complexes of the North Atlantic but is a feature of Madeira (Levring, 1974) and the Mediterranean (Garcia & Carrascosa, 1987, Drago et al., 1997).

Although sea urchins graze areas clear of non-calcified algae to reveal a biotope characterised by crustose Corallinaceae, such cleared areas are restricted in extent in the Azores and contrast with localities on Madeira where this 'urchin barren' biotope is widespread and extends into deep waters (pers. obs). In the Azores, the littoral limpets have been removed by human predation (and thus the biotope destroyed) allowing algal assemblages to develop in areas that would otherwise lack such growth (Martins et al., 1987).

Conclusions

The improving body of knowledge of algal community structure and its dynamics has encouraged the application of the MNCR biotope method to marine community classification in the Azores. Although the biotope method was developed in Britain as a tool to aid management and conservation of marine habitats, its general principles can be applied to other geographical areas. In the present study, the method has helped identify differences in algal community structure between Britain and the Azores. It will also provide a useful monitoring tool to assess change in marine communities in the Azores whether natural or anthropogenic.

References

Castro, M. L. & M. C. Viegas, 1983. Estudo dos povamentos de algas fotófilas da ilha de S. Miguel (Açores). 1 – Resultados preliminaires sobre facies de *Corallina elongata* Ellis & Solander. Arquipélago 4: 7–30.

Castro, M. L. & M. C. Viegas, 1987. Contribuição para o estudo da zona intertidal (substrato rochoso) da ilha de São Miguel-Açores. Fácies de *Corallina elongata* Ellis & Solander. Resultados preliminares. Cuad. Marisq. 11: 59–69.

Connor, D., D. P. Brazier, T. O. Hill & K. O. Northen 1997a. Marine Nature Conservation Review: marine biotope classification for Britain and Ireland. Volume 1. Littoral biotopes. Version 97.06. JNCC Report, No. 229: 362 pp.

Connor, D., M. J. Dalkin, T. O. Hill, R. H. F. Holt & W. G. Sanderson, 1997b. Marine Nature Conservation Review: marine biotope classification for Britain and Ireland. Volume 2. Sublittoral biotopes. Version 97.06. JNCC Report, No. 230: 448 pp.

Dauvin, J.-C., 1995. The development of the marine ZNIEFF (Zones Naturelles d'Intérêt Ecologique, Faunistique et Floristique) in France. In Hiscock, K. (ed.), Classification of Benthic Marine Biotopes of the North East Atlantic. Proceedings of a Biomar-Life workshop held in Cambridge 16–18 November 1994. Joint Nature Conservation Committee, Peterborough: 47–53.

Den Hartog, C., 1959. The epilithic marine algal communities of the Netherlands coast. Wentia 1: 1–241.

Drago, D., A. M. Mannino & S. Sortino, 1997. La vegetazione sommersa dei mari siciliani Lineanmenti iconografici. L'EPOS, Palermo: 117 pp.

Garcia Carrascosa, A. M., 1987. El bentos de los alrededores de las Islas Columbretes. Elementos para a sua cartografia bionómica. In Alonso M. L. A., J. L. Carretero & A. M. G. Carrascosa (eds), Islas Columbretes. Contribución al Estudio de su Medio Natural. Generalitat Valenciana, Valencia: 477–507.

Hiscock, K., 1995. Classification of benthic marine biotopes of the north east Atlantic. Proceedings of a Biomar-Life workshop held in Cambridge 16–18 November 1994. Joint Nature Conservation Committee, Peterborough: 105 pp.

Lawson, G. W. & D. M. John, 1977. The marine flora of the Cape Blanc peninsula, its distribution and affinities. Bot. J. linn. Soc. 75: 99–118.

Lawson, G. W. & D. M. John, 1987. The marine algae and coastal environment of tropical West Africa (Second Edition). Beih. Nova Hedwigia 83: i–vi + 1–415.

Lawson, G. W. & T. A. Norton, 1971. Some observations on littoral and sublittoral zonation at Tenerife, Canary Isles. Bot. Mar. 14: 116–120.

Levring, T., 1974. The marine algae of the archipelago of Madeira. Bol. Mus. Mun. Funchal 28: 1–111.

Marques, V. M., 1984. Effects of *Arbacia lixula* (L.) (Echinoidea) on the algal communities of São Miguel and Graciosa Islands (Azores Archipelago, Portugal). Arq. Mus Boc., 2ª sér. B, 13: 103–108.

Martins, H., R. S. Santos & S. J. Hawkins, 1987. Exploitation of limpets (*Patella* spp.) in the Azores with a preliminary analysis of the stocks. Report to ICES Selfish Committee (C. M. 1987/S3): 14 pp.

Morton, B., J. C. Britton & A. M. De Frias Martins, 1998. Coastal ecology of the Azores. Sociedade Afonso Chaves, Ponta Delgada. viii+ 249 pp.

Neto, A. I., 1992. Contribution to the taxonomy and ecology of the Azorean benthic marine algae. Biol. J. linn. Soc. 46: 163–176.

Neto, A. I. & I. Tittley, 1996. Structure and zonation of algal turf communities on the Azores: a numerical approach. Bol. Mus. Mun. Funchal Suppl. No. 4: 487–504.

Neto, A. I., I.Tittley, A. Levi & W. F. Farnham, 2000. Structure and zonation of algal communities in the bay of São Vicente (São Miguel, Azores). Arquipélago Suppl. 2 (Part A): 63–69.

Oliveira Filho, E. C. De & E. M. Mayral, 1976. Seasonal distribution of intertidal organisms at Ubatuba, São Paulo (Brazil). Rev. Bras. Biol. 36: 305–316.

Pryor, J., 1967. Intertidal marine algae of São Jorge. In Chelsea College Azores Expedition July – October 1965. Final Report, London: 17–30.

Schmidt, O. C., 1931. Die marine Vegetation der Azoren in ihren Grundzügen dargestellt. Bibl. Bot. 24: 1–116.

Stewart, J. G., 1989. Establishment, persistence and dominance of *Corallina* (Rhodophyta) in algal turf. J. Phycol. 25: 436–446.

Tittley, I., A. I. Neto & W. F. Farnham, 1998. Marine algae of the island of Flores, Azores: ecology and floristics. Bol. Mus. Mun. Funchal Suppl No. 5: 463–479.

Hydrobiologia **440:** 27–36, 2000.
M.B. Jones, J.M.N. Azevedo, A.I. Neto, A.C. Costa & A.M. Frias Martins (eds), Island, Ocean and Deep-Sea Biology.

Variability at different spatial scales of a coralligenous assemblage at Giannutri Island (Tuscan Archipelago, northwest Mediterranean)

Francesca Ferdeghini[1,2], Stefano Acunto[2], Silvia Cocito[1] & Francesco Cinelli[2]
[1]*Marine Environment Research Centre ENEA S. Teresa, P.O. Box 316, I-19100 La Spezia, Italy*
[2]*Dipartimento di Scienze dell'Uomo e dell'Ambiente, Università di Pisa, Via Volta 6, I-56126 Pisa, Italy*

Key words: coralligenous assemblages, algae, sessile invertebrates, spatial scale, variability, Tyrrhenian Sea

Abstract

This study was carried out on the rocky cliffs of Giannutri Island (Tyrrhenian Sea, Italy) to test the hypothesis that coralligenous assemblages are consistent within the bathymetric range considered (25–35 m depth) over three different spatial scales (1000 m, 100 m and 10 m). A multi-factorial sampling design was used to assess patterns of vertical distribution in the studied area. Data on the percent cover of algae and invertebrates were collected at three depths (25, 30 and 35 m) using a photographic method, and percentage cover was obtained using a visual method. Analysis of the results using ANOVA indicated that the distribution and abundance of algae and some invertebrates of the sublittoral assemblages are clearly heterogeneous. Algae, sponges and bryozoans showed significant variability in distribution and abundance at different depths, but this variability was not consistent amongst transects. Ascidians did not show any variability, while the abundance of anthozoans differed significantly amongst transects. We concluded that heterogeneity in the distribution and abundance of the taxa analysed is related to the smallest spatial scale investigated (10's of m). Some of the possible causes of the observed variability are discussed.

Introduction

Natural species assemblages are intrinsically variable. Quantitative description of this variability is a fundamental requirement before explanatory models for the patterns of observed distribution can be proposed and tested (Underwood, 1991). Variability in the distribution of marine organisms can occur at different spatial scales, ranging from biogeographic to local (Barry & Dayton, 1991; Levin, 1992; Underwood & Petraitis, 1993). Documenting the spatial scales at which variability in the abundance of organisms has been found can help to focus attention on the relative importance of different ecological processes that can determine the patterns of distribution observed (Underwood & Chapman, 1996). As the scale of observation may influence recognition of patterns and processes of an assemblage, multi-scale investigations are needed to understand how scale affects the structure of assemblages (Hewitt et al., 1998; Blanchard & Bourget, 1999). Variability in the distribution and abundance of organisms has been studied at different spatial scales in intertidal assemblages (Underwood & Petraitis, *op. cit.*; Archambault & Bourget, 1996; Schoch & Dethier, 1996; Underwood & Chapman, *op. cit.*; Menconi et al., 1999) and in sublittoral habitats (Foster, 1990; Kennelly & Underwood, 1992; Benedetti-Cecchi et al., 1998), but no previous study has examined spatial patterns on sublittoral rocky assemblages across different scales in the Mediterranean.

Mediterranean sublittoral rocky walls below the limit of seagrass meadows (*Posidonia oceanica*) and photophilic algal assemblages are characterised by the so-called 'coralligenous formations'. Coralligenous formations refer to the secondary hard bottoms formed by concretioning organisms and the biota inhabiting them (Sarà, 1969). A fundamental feature of coralligenous assemblages is the occurrence of concretions due to the growth of organisms with calcareous skeletons, especially encrusting coralline algae and *Peyssonnelia* spp. (Pérès & Picard, 1964; Sarà, *op. cit.*; Ros et al., 1985). Because of the building activity due to calcareous algae, and the important vertical stratification of the epibiota associated with

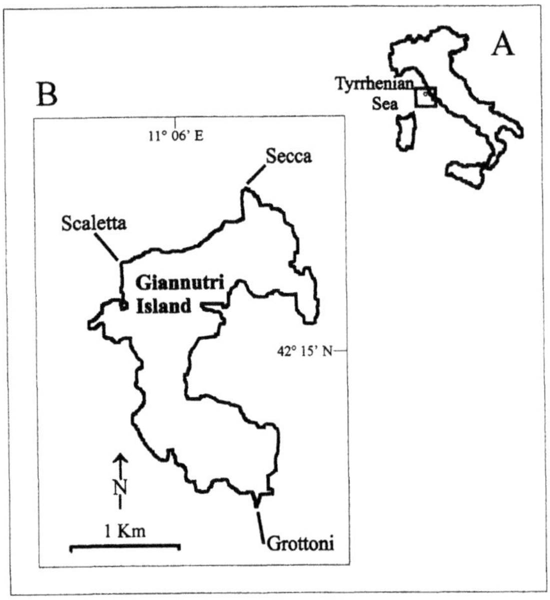

Figure 1. Study area: (A) Italy; (B) Giannutri Island and locations.

them, coralligenous assemblages display high spatial complexity. This spatial complexity contributes to the diversity and richness of these assemblages, which have been compared to those of tropical coral reefs (Ros et al., *op. cit.*).

Few quantitative descriptions of the distribution and abundance patterns of algae and sessile invertebrates within coralligenous assemblages have been carried out (Abbiati et al., 1996; Garrabou et al., 1998). Present knowledge of coralligenous assemblages is based mainly on classic studies of zonation with depth of Mediterranean benthic communities (Péres & Picard, *op. cit.*), and on the distribution and abundance of the most representative species (Laborel, 1961; Laubier, 1966; Sarà, *op. cit.*; True, 1970; Augier, 1982; Hong, 1982). More recently, Ros et al. (*op. cit.*) gave a description of the coralligenous assemblages of Medes Islands (Spain) and identified several *facies*, based on qualitative criteria, thus proposing a scheme of vertical distribution of organisms.

Giannutri (Tyrrhenian Sea, Italy) is a small island (3.6 km^2) with little anthropogenic impact and, since 1989, it has been included in the Italian National Park of the Tuscan Archipelago. For these reasons, it can be considered a natural laboratory, enabling spatial patterns in natural assemblages to be assessed (Whittaker, 1998). All around the island, below 25 m depth, apparently homogeneous coralligenous assemblage occur on vertical walls. This study, carried out on the rocky cliffs of Giannutri Island, tested the hypothesis that the coralligenous assemblage structure is consistent within the bathymetric range considered (25–35 m depth) at different spatial scales. Distribution and abundance of algae and sessile invertebrates on vertical walls were examined using a non-destructive method. A multifactorial sampling design was used to assess: (1) whether the distribution and abundance of the dominant species differed among depths; and (2) whether the vertical patterns were homogeneous at different spatial scales. To test the hypothesis that the structure of the assemblages is constant along the bathymetric range, data on the percentage cover of algae and invertebrates were collected at three different depths (25, 30 and 35 m) on the rocky cliffs of Giannutri Island (Fig. 1). To test whether patterns of vertical distribution were consistent in space, we examined different locations (one to a few km's apart), different sites within locations (100's of m apart) and replicate transects within sites (over a distance of 10's of m).

Materials and methods

Study site and organisms

The study was carried out on the rocky cliffs of Giannutri Island (42° 15′ N, 11° 06′ E) in June 1998 (Fig. 1). The island is exposed to waves (highest frequency ranging from 0.1 to 1.25 m height) coming from the south and, to a lesser extent, from the northeast (data from the Istituto Idrografico della Marina – Genoa). Preliminary observations found the substratum to be characterised generally by rocky vertical cliffs at depths between 35 and 25 m. Above 25 m, the slope decreased and assemblages were dominated by photophilic species. We analysed the distribution and abundance of the major space occupiers of the coralligenous assemblages on vertical walls between 25 and 35 m, including heavily calcified red algae (constituting the encrusting layer) and all the taxa represented in the erect layer. Of the sessile invertebrates, five taxa were analysed: sponges, anthozoans, bryozoans, polychaetes and ascidians.

Sampling design and analysis of data

Patterns of vertical distribution were analysed at different spatial scales along the Giannutri coast, using a multifactorial design. The largest spatial scale of some

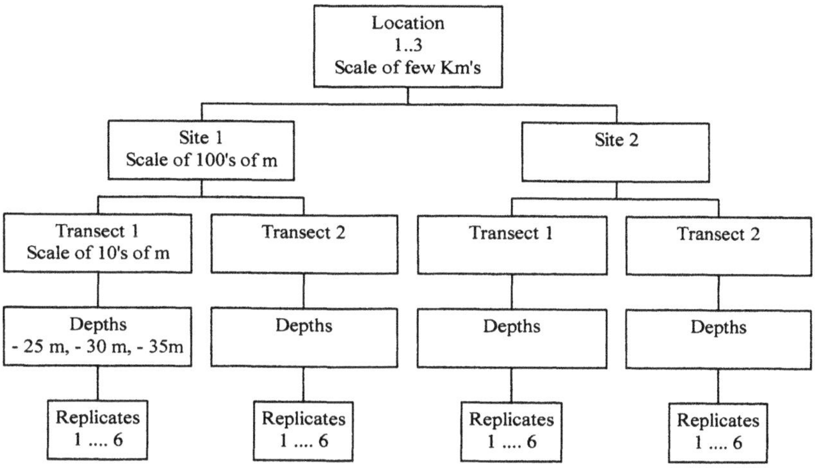

Figure 2. Scheme of the sampling design.

kilometres comprised locations, selected a few kilometres apart: Secca, Scaletta, Grottoni (Fig. 1). Each location comprised a stretch of coast about 1 km wide, representing a random sample from a set of all the possible locations supporting a coralligenous assemblage. Within each of the three locations, two replicate sites were chosen randomly. Sites were spaced several hundreds of m apart and each represented a stretch of coastline about 100 m long. Within each site, two replicated transects (each comprising a suitable section of substratum about 10 m wide) were positioned at a distance of 10's of m, and sampled at three depths (35 m, 30 m and 25 m) within the range of distribution of coralligenous assemblages (Pérès & Picard, *op. cit*; Ros et al., *op. cit.*) and checked in previous surveys. Six replicated plots of 600 cm^2 were randomly sampled at each depth (Fig. 2).

Sampling was carried out non-destructively by taking photographs of the 600 cm^2 plots using a Nikonos V camera with 35 mm lens fitted with an additional lens and with a framer. Estimates of cover of each taxon were obtained by the visual method described by Dethier et al. (1993) and used in a sublittoral Mediterranean habitat by Benedetti-Cecchi et al. (1996). Slides were projected onto a grid of 24 quadrats; each quadrat was given a cover value ranging from 0 to 4. Cover values for each species were summed and expressed as percentages (Abbiati et al., *op. cit.*).

Spatial distributional patterns of algae and invertebrates were analysed using a 4-factor mixed model ANOVA, with depth as fixed factor, and location, site, and transect as random factors. Transect (2 levels) was nested into site (2 levels), which was nested into location (3 levels), while depth (3 levels) was orthogonal to the other factors. Homogeneity of variances was assessed using Cochran's C test and transformations were applied to the data when necessary (Underwood, 1997). When appropriate, SNK tests were used for multiple comparisons of the means (at $\alpha = 0.05$). Anderson's test (Winer et al., 1991) was used to test for the significance of differences in the frequency with which a factor was given a particular rank by SNK test in a series of multiple comparisons.

Results

Amongst the calcified algae that formed the prostrate layer, encrusting coralline algae, *Peyssonnelia* spp. and the green alga *Palmophyllum crassum* (Naccari) Rabenhorst were most abundant. The green algae *Flabellia petiolata* (Turra) Nizamuddin and *Halimeda tuna* (Ellis & Solander) Lamouroux were the main constituents of the erect layer. Turf-forming algae such as *Cladophora prolifera* (Roth) Kützing, *Sphacelaria* spp., *Ceramium* spp., *Polysiphonia* spp. and *Womersleyella setacea* (Hollenberg) R.E. Norris were also abundant. As component turf species were not identifiable easily in the photographs, turf was analysed as a unit.

Amongst sessile invertebrates, sponges included encrusting species such as *Spirastrella cunctatrix* Schmidt, *Pleraplysilla spinifera* (Schulze), *Oscarella lobularis* (Schmidt) and *Phorbas tenacior* Topsent, and massive forms such as *Spongia officinalis* Linnè, *Aplysina* sp., *Chondrosia reniformis* Nardo, *Acanthella acuta* Schmidt and *Agelas oroides* (Schmidt).

Anthozoans were represented by the gorgonians *Eunicella cavolinii* (Koch), *Eunicella singularis* (Esper), *Paramuricea clavata* (Risso) (forming the erect stratum of the assemblage), and the solitary and carpeting madreporian *Leptopsammia pruvoti* Lacaze – Duthiers, *Caryophyllia* sp., *Balanophyllia* sp. (forming the lower stratum). Bryozoans comprised erect species such as *Myriapora truncata* (Pallas), *Margaretta cereoides* (Ellis & Solander) and *Cellaria salicornioides* Lamouroux, massive species such as *Pentapora fascialis* (Pallas), *Reteporella grimaldii* (Jullien, in Jullien & Calvet), *Smittina cervicornis* (Pallas), and encrusting species such as *Schizomavella* spp. Bryozoan species were living either on primary substratum or as epibionts on gorgonians. Polychaetes were represented only by a few species: serpulids such as *Serpula vermicularis* Linnaeus and *Salmacina dysteri* (Huxley), and sabellids including *Protula tubularia* (Montagu), *Sabella pavonina* Savigny and *Myxicola infundibulum* (Renier). *Clavelina* sp. and *Halocynthia papillosa* (Linnaeus) mainly represented the ascidians.

ANOVA results indicated that the distribution and abundance of algae, and some invertebrates, were clearly heterogeneous (Tables 1 and 2). The algal turf always occupied a high proportion of space as did other algal components of the assemblage. The significant interaction between depth × transect suggests that patterns of distribution observed along the bathymetric gradient were not consistent at small spatial scales. Moreover, algal turf showed a heterogeneous distribution at the scale of hundreds of m as supported by the significant differences among sites (Table 1; Figs 3–5).

The abundance of *H. tuna* and *F. petiolata*, forming the erect algal layer of the assemblage, showed a high significant depth × transect interaction. The same occurred for the major space occupiers forming the prostrate layer (i.e. *Peyssonnelia* spp. and encrusting coralline algae) (Table 1). SNK tests within the depth × transect interaction never showed a clear bathymetric pattern of distribution. For example, at Secca, the green alga *H. tuna* was significantly more abundant at 25 m along transect 1 at site 2, but was more abundant at 30 m along transect 2 of the same site; no significant differences were found between depths along the two transects at site 1 (Fig. 3). At Scaletta and Grottoni, significant differences between depths were found only once at each location (respectively at transect 2 of site 2 and transect 1 of site 1) where *H. tuna* was more abundant at 35 m depth (Figs 4 and 5).

Coralline algae showed a pattern of vertical distribution amongst transects (SNK test: 35 m > 30 m > 25 m) (Figs 3 and 4). Consistency of the pattern was tested with Anderson's test within all depth × transect interactions. The model of distribution proved not to be consistent, probably due to the absence of vertical patterns at Grottoni (Fig. 5). For all other algal taxa which did not display patterns of abundance after the SNK test, Anderson's test for rank order of abundance was not applied. Further variability in the distribution of encrusting coralline algae was observed in some locations, at the largest spatial scale considered (Table 1).

Amongst sessile invertebrates, significant depth × transect interactions were observed for sponges and bryozoans (Table 2). The results showed that there were differences in the percentage cover of these two taxa amongst depths, and differences varied from one transect to the other. No consistent patterns of vertical distribution arose from the SNK tests. For example, at Secca, sponges were significantly more abundant at 30 m along transect 1 at site 1, while they were more abundant at 35 m at transect 2 of site 2; no differences between depths were observed for other transects (Fig. 3). Variability in pattern of distribution was observed at Scaletta (Fig. 4) and Grottoni (Fig. 5).

Anthozoans displayed significant differences in percentage cover between transects (Table 2; Figs 3–5). Gorgonians were the principal anthozoan group present. Separate analysis of gorgonian distribution showed significant differences in percentage cover between transects, indicating that patterns observed for anthozoans were due to the distribution of gorgonians.

Ascidians showed no significant differences in abundance at any of the spatial scales examined (Table 2). Analysis of the distribution of the principal space occupier belonging to this taxon (*H. papillosa*), produced a different pattern. In fact, *H. papillosa* exhibited significant differences in percentage cover between locations (i.e. the largest spatial scale analysed).

ANOVA results on percentage cover of polychaetes are not presented since variances were very heterogeneous and also transformations failed to reduce such heterodasticity.

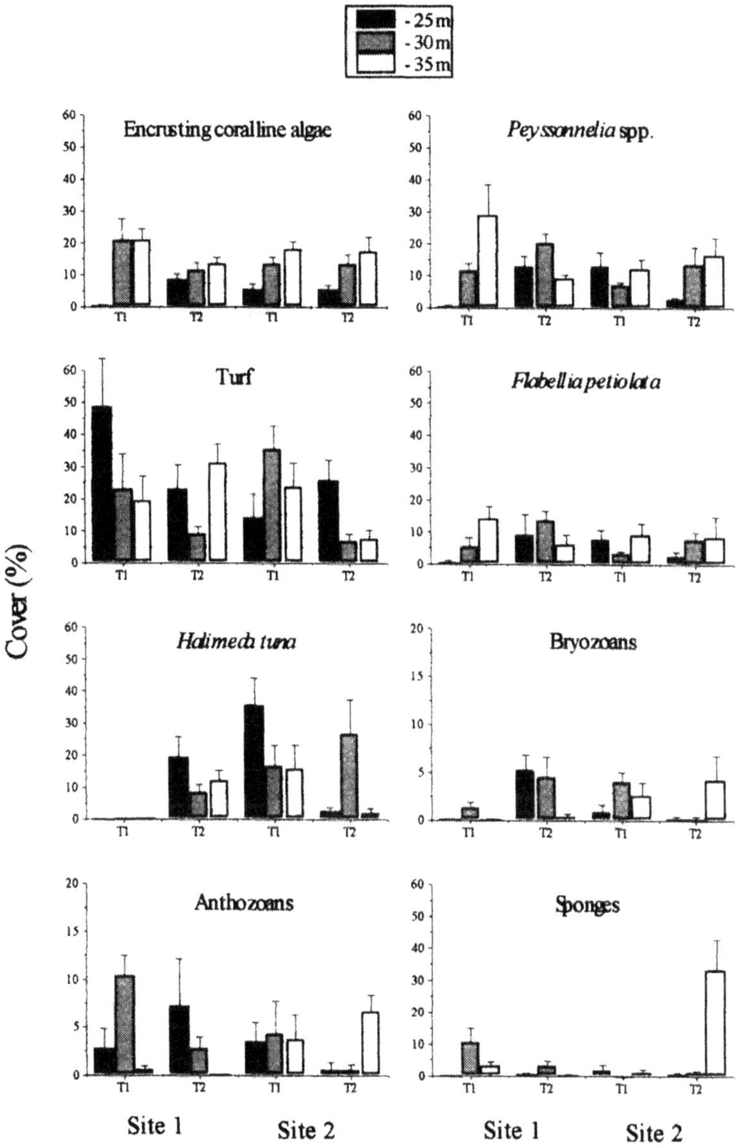

Figure 3. Mean percentage cover of the more abundant algae and sessile invertebrates at each of the two sites at Secca. Each bar represents the mean of six replicated plots sampled at the same depth. Error bars indicate ± SE (n = 6).

Discussion

Distribution and abundance of algae, and almost all the invertebrates of the sublittoral assemblages within the depth range studied at Giannutri Island, were highly variable. The analysis of vertical patterns of distribution indicated that variability in abundance, for most taxa, occurred across the depth range, even if a general pattern of distribution along the bathymetric gradient was not evident. Variability in the vertical patterns of distribution and abundance was not consistent at the smallest spatial scales considered (transects) for all the algal taxa and for some of the invertebrates. No consistency in the vertical structure of an assemblage, and consequently significant differences between areas at the same depth, have been found in other sublittoral (Foster, *op. cit.*; Kennelly & Underwood, *op. cit.*) and intertidal habitats (Benedetti-Cecchi & Cinelli, 1997; Menconi et al., *op. cit.*). Various spatially-different ecological processes may cause the variability observed.

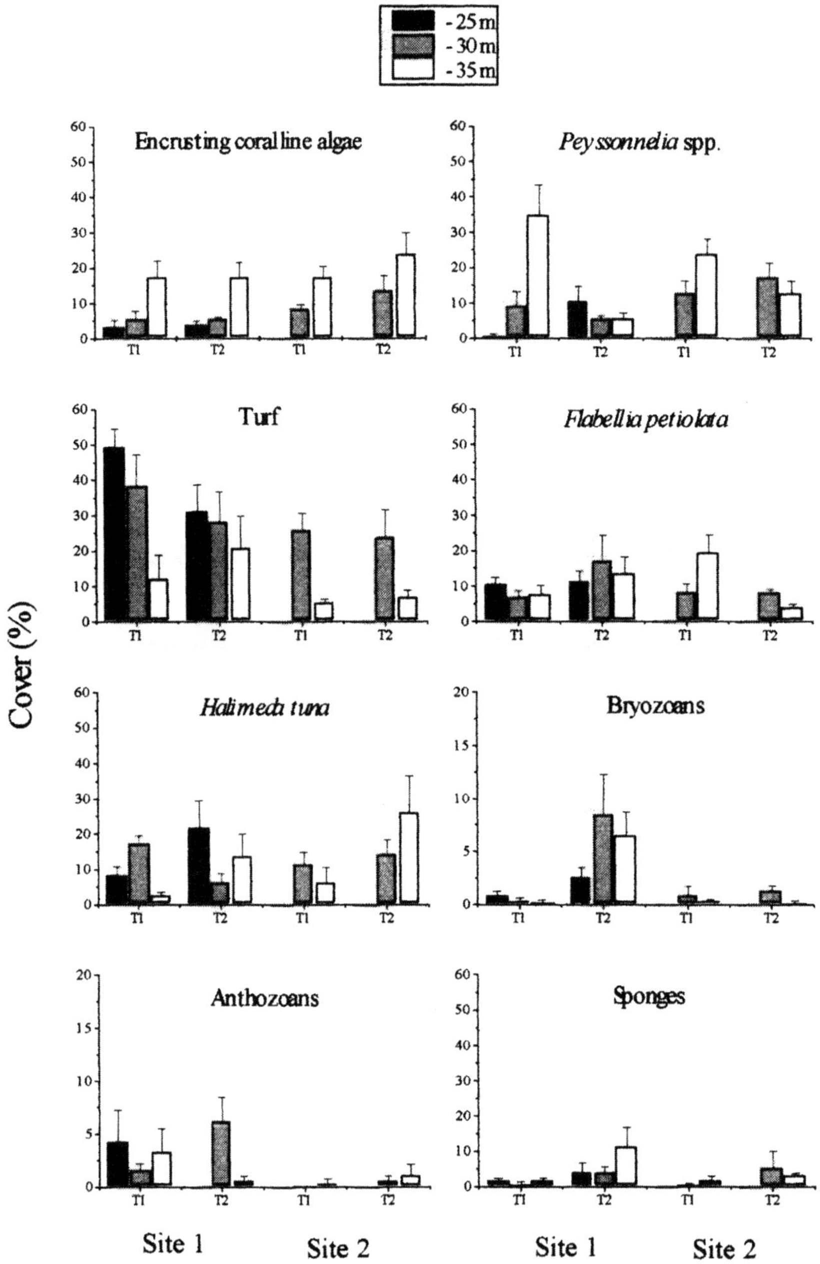

Figure 4. Mean percentage cover of the more abundant algae and sessile invertebrates at each of the two sites at Scaletta. Each bar represents the mean of six replicated plots sampled at the same depth. Error bars indicate ± SE (*n*= 6).

Light is considered the major factor influencing the occurrence and distribution of coralligenous assemblages in the Mediterranean Sea (Pérès & Picard, *op. cit.*; Ros et al., *op. cit.*), however, our results suggest that different abiotic factors (such as hydrodynamics and heterogeneity of substratum), biotic factors (such as competition and predation), or the interaction of physical and biological factors, may be possible causes of the variability observed.

Our data indicate that algal species were more abundant than sessile invertebrates. This result is comparable to those of Abbiati et al. (*op. cit.*), who sampled on a rocky cliff in the Ligurian Sea (Italy). They found that algal species were the major space oc-

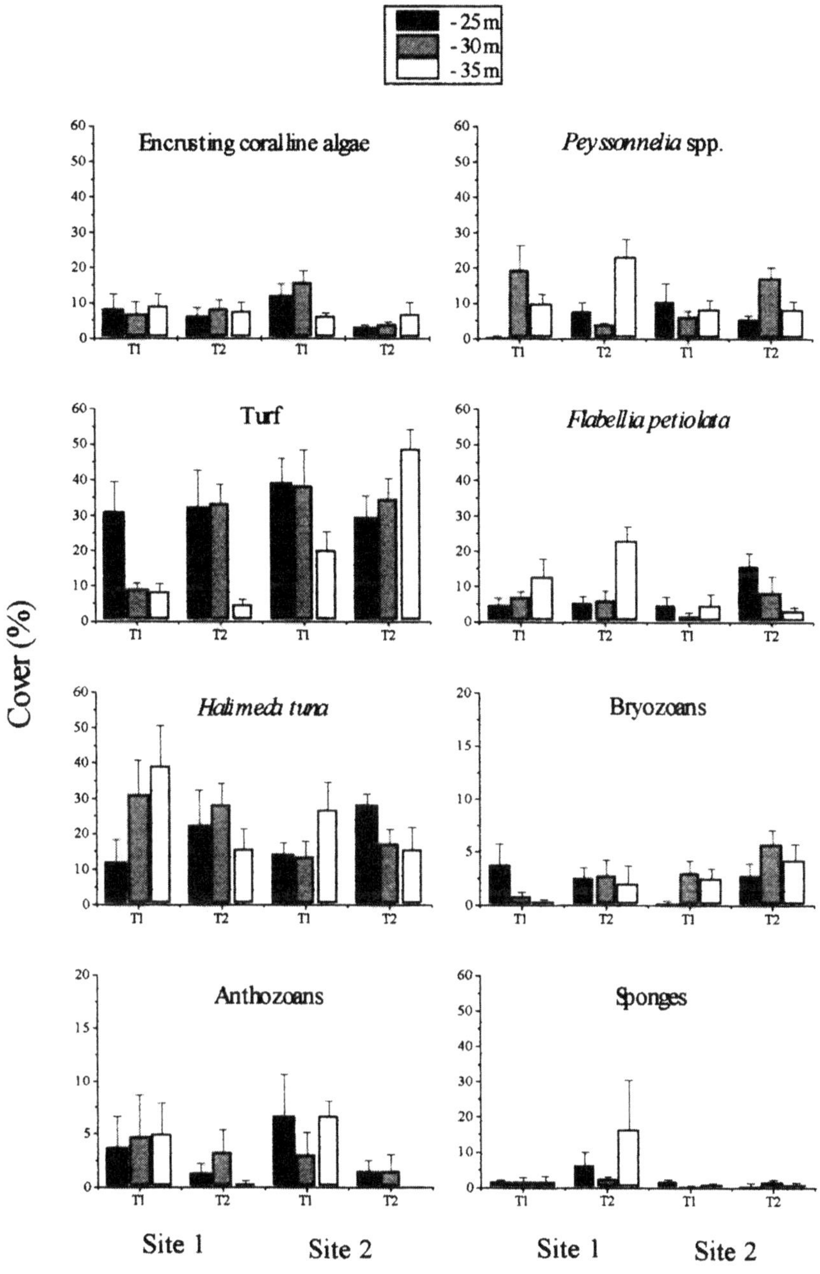

Figure 5. Mean percentage cover of the more abundant algae and sessile invertebrates at each of the two sites at Grottoni. Each bar represents the mean of six replicated plots sampled at the same depth. Error bars indicate ± SE (n= 6).

cupants on vertical walls, whereas the fauna was more abundant in crevices where light was reduced (Abbiati et al., *op.cit.*). The differences reported by Abbiati et al. (*op. cit.*), however, were not so sharp suggesting that the distribution of species in the coralligenous assemblages could not be explained completely by differences in light gradients. Glasby (1999) found evidence that not only shading but also other factors (or combinations of factors) affected the distribution of sublittoral epibiotic assemblages.

Amongst the algal taxa analysed at Giannutri Island, turf-forming algae were the most abundant organisms. Airoldi (1998) documented the dominance by algal turf in rocky sublittoral habitats in the Lig-

Table 1. Analyses of variance of mean percentage algal cover in relation to Location, Site, Transect and Depth. L = Location; Si = Site; Tr = Transect; De = Depth. ns = not significant, * = $p < 0.05$, ** = $p < 0.01$, *** = $p < 0.001$. Significant values are in bold

Source of variation	df	*Flabellia petiolata*		*Halimeda tuna*		*Encrusting coralline algae*		*Peyssonnelia* spp.		Turf	
		MS	F	MS	F	MS	F	MS	F	MS	F
Location=L	2	16.641	0.05	2937.403	3.81	299.8257	**92.95****	125.7047	5.63	1025.3002	0.41
Si (L)	3	322.8621	1.81	770.6441	1.17	3.2257	0.02	22.3359	0.32	2489.3352	**12.38***
Tr (Si(L))	6	178.4695	**2.74***	660.5122	**3.25****	136.2734	**2.69***	69.2636	1.13	201.0084	1.40
Depth=De	2	522.097	4.26	83.1557	0.30	2215.9351	3.60	3009.8523	**14.36***	898.1386	1.29
De×L	4	122.6789	0.30	272.7326	0.48	615.6393	3.58	209.5512	0.78	696.0673	0.62
De×Si(L)	6	407.7818	2.36	569.6577	0.93	172.1468	1.83	268.4275	0.56	1116.7353	2.53
De×Tr(Si(L))	12	172.6524	**2.65****	615.7173	**3.03*****	93.8332	**1.85***	476.9161	**7.76*****	441.9773	**3.09*****
RES	180	65.0391		203.5345		50.6011		61.469		143.079	
TOT	215										
Cochran's C test		$C = 0.099$; P = ns		$C = 0.109$; P = ns		$C = 0.099$; P = ns		$C = 0.111$; P = ns		$C = 0.107$; P = ns	
Transformation		Arcsine		none		Arcsine		Arcsine		Arcsine	

urian Sea (Italy). Algal turfs are the dominant flora of many tropical and temperate intertidal and sublittoral shores, and, during the last few years, have spread as a result of their ability to occupy substrata over a large range of environmental conditions and to prevent colonisation of the substratum by other species (Airoldi, *op. cit.*). Variability sometimes occurred at the larger spatial scale considered. For example, encrusting coralline algae showed significant differences in abundances between locations. This result could be attributed to differences in recruitment and mortality at the scale of a few kilometres.

Of the sessile invertebrates, sponges were the most abundant taxon. Because of their ability to display different morphologies, sponges were distributed in different layers, and were well represented in the lower and upper coralligenous layer (Sarà, 1999). Water current, and thus food availability, is an important factor regulating the distribution and abundance of these suspension feeders (Gili & Ballesteros, 1991; Gili & Coma, 1998).

Significant differences in the abundance of anthozoans were found between transects, but not between depths. Gili & Ros (1985) underlined the aggregation tendency of anthozoans, especially gorgonians, which may contribute to patchiness in distribution. At Giannutri Island, lack of significant differences amongst depths could be interpreted as a result of aggregation phenomena.

Differences of topography, at small scale, may lead to considerable spatial variability in a very localised area (Underwood & Petraitis, *op. cit.*). Variability at small spatial scale (10's of m) in coralligenous assemblages at Giannutri might, therefore, be due to the effect of substratum complexity. Significant differences in the distribution of organisms at several spatial scales in relation to heterogeneity of the substratum have already been found in sublittoral habitats (Bourget et al., 1994; Archambault & Bourget, *op. cit.*; Schoch & Dethier, *op. cit.*). Small-scale topographical complexity could significantly influence predation, larval settlement and community characteristics, such as abundance of organisms (Guichard & Bourget, 1998; Blanchard & Bourget, *op. cit.*).

Heterogeneity of the substratum is characteristic of coralligenous assemblages where differential rates of growth of calcareous algae result in the formation of small canals, crevices and irregularities (Ros et al., *op. cit.*). Calcareous algae have been defined as 'ecosystem engineers' (Lawton & Jones, 1995); by changing the environment via their own physical structures, they supply microhabitats to other elements of the community (Sarà, *op. cit.*). Variation at the smallest spatial scale may be explained by the ability of calcareous algae to create suitable microhabitats scattered across the assemblage (Underwood & Chapman, *op. cit.*).

Experiments are needed to explain the observed patterns of distribution of sessile organisms in coralligenous assemblages of Giannutri Island. Whether the variability documented in the present study is distinctive of the island of Giannutri, or is a general feature of coralligenous assemblages, requires additional studies at other localities in the Mediterranean.

Table 2. Analyses of variance of mean percentage cover of sessile invertebrates in relation to Location, Site, Transect and Depth. L = Location; Si = Site; Tr = Transect; De = Depth. ns = not significant, * = $p < 0.05$, ** = $p < 0.01$, *** = $p < 0.001$. Significant values are in bold

Source of variation	df	Sponges		Bryozoans		Anthozoans		Ascidians	
		MS	F	MS	F	MS	F	MS	F
Location=L	2	0.1704	0.06	2.5163	0.80	254.148	1.97	35.071	0.06
Si (L)	3	2.7324	0.92	3.1631	0.82	128.741	0.97	4.496	0.95
Tr (Si(L))	6	2.9836	4.22	3.8526	8.79	133.12	**2.51***	4.7521	0.41
Depth=De	2	5.6384	4.11	2.1997	16.60	43.46	1.04	50.984	5.46
De×L	4	1.3711	0.42	0.1325	0.06	41.83	0.24	9.343	0.40
De×Si(L)	6	3.2895	1.90	2.1238	2.44	176.53	2.00	23.305	2.44
De×Tr(Si(L))	12	1.7295	**2.44****	0.8696	**1.98***	88.14	1.66	9.551	0.82
RES	180	0.7075		0.4382		52.96		11.664	
TOT	215								
Cochran's C test		$C = 0.0973$; P = ns		$C = 0.1051$;P = ns		$C = 0.0981$; P = ns		$C = 0.1041$;P = ns	
Transformation		Ln (x+1)		Ln (x+1)		Arcsine		Arcsine	

Acknowledgements

We are grateful to S. Sgorbini (La Spezia), N. Biagini (Pisa) and A. Proietti-Zolla (Viterbo) for field work, to L. Piazzi (Pisa) for identifying algae, and to D. K. A. Barnes (Cork) for reviewing the English text. Special thanks to L. Benedetti-Cecchi (Pisa) for helping with sampling design, statistical analyses and suggestions on the manuscript. This work forms part of PhD theses by F. F. and S. A.

References

Abbiati, M., M. Virgilio & J. Querci, 1996. Spatial and temporal variability of species distribution on a sublittoral rocky cliff in the Ligurian Sea. S. It. E. Proceedings 17: 337–340.

Airoldi, L., 1998. Roles of disturbance, sediment stress, and substratum retention on spatial dominance in algal turf. Ecology 79 (8): 2759–2770.

Archambault, P. & E. Bourget, 1996. Scales of coastal heterogeneity and benthic intertidal species richness, diversity and abundance. Mar. Ecol. Prog. Ser. 136: 111–121.

Augier, H., 1982. Inventaire et classification des biocenoses marines benthiques de la Mediterranée. Collection Sauvegarde de la Nature, Conseil de l'Europe, Strasbourg, 25: 1–59.

Barry, J. P. & P. K. Dayton, 1991. Physical heterogeneity and the organization of marine communities. In Kolasa J. & S. T. A. Pickett (eds), Ecological Heterogeneity. Springer Verlag, Berlin: 269–320.

Benedetti-Cecchi, L. & F. Cinelli, 1997. Spatial distribution of algae and invertebrates in the rocky intertidal zone in the Strait of Magellan: are patterns general? Polar Biology 18: 337–343.

Benedetti-Cecchi, L., L. Airoldi, M. Abbiati & F. Cinelli, 1996. Estimating the abundance of benthic invertebrates: a comparison of procedures and variability between observers. Mar. Ecol. Prog. Ser. 138: 93–101.

Benedetti-Cecchi, L., L. Airoldi, M. Abbiati & F. Cinelli, 1998. Spatial variability in the distribution of sponges and cnidarians in a sublittoral marine cave with sulphur-water springs. J. mar. biol. Ass. U.K. 78: 43–58.

Blanchard, D. & E. Bourget, 1999. Scales of coastal heterogeneity: influence on intertidal community structure. Mar. Ecol. Prog. Ser. 179: 163–173.

Bourget, E., J. DeGuise & G. Daigle, 1994. Scales of substratum heterogeneity, structural complexity, and the early establishment of a marine epibenthic community. J. exp. mar. Biol. Ecol. 181: 31–51.

Dethier, M. N., E. S. Graham, S. Cohen & L. M. Tear, 1993. Visual versus random-point percent cover estimations: 'objective' is not always better. Mar. Ecol. Prog. Ser. 110: 9–18.

Foster, M. S., 1990. Organisation of macroalgae assemblages in the Northwest Pacific: the assumption of homogeneity and the illusion of generality. Hydrobiologia 192: 21–33

Garrabou, J., J. Riera & M Zabala, 1998. Landscape pattern indices applied to Mediterranean subtidal rocky benthic communities. Landsc. Ecol. 13 (4): 225–247

Gili, J. M. & E. Ballesteros, 1991. Structure of cnidarian populations in Mediterranean sublittoral benthic communities as a result of adaptation to different environmental conditions. In Ros J. D. & N. Prat (eds), Homage to Ramòn Margalef or Why There is Such Pleasure in Study Nature. Oecol. Acquat. 10: 243–254.

Gili, J. M. & R. Coma, 1998. Benthic suspension feeders: their paramount role in littoral marine food webs. TREE 13: 316–321.

Gili, J. M. & J. D. Ros, 1985 Estudio cuantitativo de tres poblaciones circalitorales de cnidarios bentónicos. Inv. Pesq. 49 (3): 323–352.

Glasby, T. M., 1999. Effects of shading on subtidal epibiotic assemblages. J. exp. mar. Biol. Ecol. 234: 275–290.

Guichard, F. & E. Bourget, 1998. Topographic heterogeneity, hydrodynamics and benthic community structure: a scale-dependent cascade. Mar. Ecol. Prog. Ser. 171: 59–70.

Hewitt, J. E., S. F. Thrush, V. J. Cummings & S. J. Turner, 1998. The effect of changing sampling scales on our ability to detect effects of large-scale processes on communities J. exp. mar. Biol. Ecol. 227(2): 251–264.

Hong, J. S., 1982. Contribution à l'étude des peuplements d'un fond de concrétionnement coralligéne dans la région marseilleise en

Méditerranée nord-occidentale. Reprinted Bull. KORDI 4 (1-2): 27–51.

Kennelly, S. J. & A. J. Underwood, 1992. Fluctuations in the distributions and abundances of species in sublittoral kelp forests in New South Wales. Aust. J. Ecol. 17: 367–382.

Laborel, J., 1961. Le concrétionnement algal 'coralligène' et son importance géomorphologique en Méditerranée. Rec. Trav. St. Mar. Endoume 23 (37): 37–60.

Laubier, L., 1966. Le Coralligène des Albères, monographie biocénotique. Ann. Institut océanogr., Paris 43 (2): 137–201.

Lawton, J. H. & C. G. Jones, 1995. Linking species and ecosystems: organisms as ecosystem engineers. In Jones, C. J. & J. H. Lawton (eds), Linking Species & Ecosystems. Chapman & Hall, New York: 141–150.

Levin, S. A., 1992. The problem of pattern and scale in ecology. Ecology 73 (6): 1943–1967.

Menconi, M., L. Benedetti-Cecchi & F. Cinelli, 1999. Spatial and temporal variability in the distribution of algae and invertebrates on rocky shores in the Northwest Mediterranean. J. exp. mar. Biol. Ecol. 233: 1–25.

Pérès, J. M. & J. Picard, 1964. Nouveau manuel de bionomie bentique de la Mer Mediterranée. Rec. Trav. Stat .mar. Endoume 31 (=47): 123 pp.

Ros, J. D., J. Romero, E. Ballestreros & J. M. Gili, 1985. Diving in blue water. The benthos. In Margalef, R. (ed.), Western Mediterranean. Pergamon Press, Oxford: 233–295.

Sarà, M., 1969. Research on coralligenous formations: problems and perspectives. Reprinted from the Pubbl. Staz. Zool. Napoli 37 suppl.: 124–134.

Sarà, M., 1999. The role of sponges in the biodiversity, structure and dynamics of the Mediterranean coralligenous formations. Biol. Mar. Medit. 6 (1): 144–150.

Schoch, G. C. & M. N. Dethier, 1996. Scaling up: the statistical linkage between organismal abundance and geomorphology on rocky intertidal shorelines. J. exp. mar. Biol. Ecol. 201: 37–72.

True, M. A., 1970. Étude quantitative de quatre peuplements sciaphiles sur substrat rocheaux dans la région marseilleise. Bull. Inst. Océanogr. Monaco 69: 48 pp.

Underwood, A. J., 1991. The logic of ecological experiments: a case history from studies of the distribution of macro-algae on rocky intertidal shores. J. mar. biol. Ass. U.K. 71: 841–866.

Underwood, A. J., 1997. Experiments in ecology: their logical design and interpretation using analysis of variance. Cambridge University Press, Cambridge: 504 pp.

Underwood, A. J. & M. G. Chapman, 1996. Spatial patterns of distribution of intertidal invertebrates. Oecologia 107: 212–224.

Underwood, A. J. & P. S. Petraitis, 1993. Structure of intertidal assemblages in different locations: how can local processes be compared? In Rickleffs, R. E. & D. Schluter (eds), Species Diversity in Ecological Communities: Historical and Geographical Perspectives. University of Chicago Press, Chicago: 39–51.

Whittaker, R. J., 1998. Island Biogeography. Oxford University Press, New York: 1–6.

Winer, B. J., D. R. Brown & K. M. Michels, 1991. Statistical Principles in Experimental Design. 3rd edn. McGraw-Hill, New York: 1057 pp.

Hydrobiologia **440:** 37–44, 2000.
M.B. Jones, J.M.N. Azevedo, A.I. Neto, A.C. Costa & A.M. Frias Martins (eds), Island, Ocean and Deep-Sea Biology.

Diversity, recruitment and competition on island shores at south-polar localities compared with lower latitudes: encrusting community examples

David K. A. Barnes
Department of Zoology and Animal Ecology, University College Cork, Cork, Ireland

Key words: diversity, recruitment, spatial competition, intertidal, south-polar

Abstract

Comparisons of temperate and tropical shores have yielded considerable debate as to whether the former really are less benign, diverse and structured by different ecological processes. Studies of comparable boulder communities have shown high within region variability. Equivalent polar assemblages, from island shores compared here, show much reduced within region variability and considerably reduced numbers of phyla and species encrusting boulders. The rate of colonisation (compared from settlement panel studies) was an order of magnitude higher in warmer water, but did vary with isolation (near vs offshore islands). Comparison of the most ubiquitous taxon, the bryozoans, between polar and non polar sites shows a decrease in the proportion of inter-specific competition, indeterminate competitor (species) pairs and incidence of tied outcomes in competition. These three parameters all increased with depth at the localities studied, whilst no obvious influence of isolation was found.

Introduction

In the last few decades, the investigation of latitudinal trends into south-polar waters has changed from being largely theoretical to practical. Hypothesised lower growth rates, greater longevity, reduced basal metabolic rates and lowered tempo of reproduction, with respect to warm water (non-polar) ecological equivalents, have all been found to be generally true (Pearse et al., 1991). Certain exceptions have been found. Growth rates of a couple of south-polar sponges and ascidians are as fast or faster than typical low-latitude representatives of the same taxa (Dayton et al., 1974; Rauschert, 1991). Thorson's rule (Mileikovsky, 1971), concerning the polar predominance of a suite of reproductive and larval characters, has proved true for only few taxa (Stanwell-Smith et al., 1999 and refs therein). The duration of phytoplankton abundance is greater than previously believed, thus explaining the findings of near year round activity of polar suspension feeders (Barnes & Clarke, 1995). Antarctic fauna have long been considered very stenothermal and tests on some organisms support this (e.g. Peck, 1989) but others experience and tolerate regular fluctuations of both temperature and salinity (Barnes et al., 1996). There has been considerable debate as to the reason for a general latitudinal cline in species diversity, most recently whether it is explained by area (Rosenweig, 1995; Rhode, 1997). What appears certain is that the pattern in the sea differs in the northern and southern hemispheres as many Antarctic localities are highly species rich (Clarke, 1992; Brey et al., 1994). Of the sessile encrusting fauna, some can be relatively species rich in south-polar waters (e.g. polychaetes and bryozoans) whilst others are much more speciose in warmer waters (e.g. cnidarians, barnacles and bivalves). A preliminary study by Barnes & Arnold (1999) of encrusting boulder fauna on three polar island shores, suggested that south-polar assemblage diversity, growth, mortality and tempo of reproduction may increase with latitude – all the converse of typical patterns.

Boulder communities have proved an important testing ground for many ecological ideas, particularly those centred on the influences of disturbance. Boulders are characterised by being discrete (unlike sediment samples), portable (unlike solid rock surfaces) and with a relative lack of variability. Of course, as with any comparison of different sites, no two boulder fields are identical, but problems of variance in spatial heterogeneity can be minimised by selection of substrata which are similar in size, wear (roundness) and smooothness. Of key significance to the ecologist is that (within a region) disturbance is the single most important factor determining faunal assemblage (Sousa, 1979; McGuiness, 1987). Typically, disturbances create space by lethal or sublethal destruction of colonists, thus high frequencies reduce diversity. At intermediate levels, however, disturbance may prevent superior competitors from monopolising resources (Connell, 1978; Huston, 1979).

Comparisons of temperate and tropical shores have yielded considerable debate as to whether the former really are less benign, diverse and structured by different ecological processes (Pianka, 1966; Moore, 1972; Menge & Lubchenco, 1981; McGuiness, 1990). The findings of high levels of variability within regions, and differing results dependent on spatial scale examined, suggest that, overall, we do not have the data to conclude that there are significant differences. The wide geographical range of boulder community and artificial simulation studies have, however, rarely included polar regions until recently (Barnes et al., 1996; Barnes & Clarke, 1998; Barnes & Arnold, 1999). The present study, using boulder substrata, conveys advantages in comparing a similar habitat, over similar areas, using similar sampling design and over a number of sites.

There have been few studies comparing fouling/encrusting taxa across latitudinal or isolation gradients with similar sampling methodology and more than one site at each latitude. Schoener et al. (1978) and Barnes (1996) found that colonisation of artificial panels was much faster at low latitudes, at least in the early stages of community development. In addition, species diversity was much higher at warm water sites, although exceptions have been found to both these trends – probably due to local conditions (Schoener & Schoener, 1981). Long & Rucker (1970) and Hughes & Jackson (1992) compared near and offshore sites, and Holmes et al. (1997) compared two sites at each of two latitudes. The present study sought to examine how phyletic richness, species richness and the bryozoan (the most abundant taxon) species richness at Antarctic and Subantarctic localities compared with other island study sites around the world, with a consideration of isolation and depth. In addition, this study investigated whether rates of colonisation vary significantly: (1) between broad latitudinal regions, and (2) with isolation by comparing the data from four polar, nine temperate and eight tropical studies. Finally, this study aimed at answering a number of questions concerning competition: are there relationships between latitude, depth or isolation and the incidence of inter- vs intraspecific competition of the outcomes of interspecific competition? If there are relationships, are they simple and do they have an obvious biological interpretation? Quinn (1982) considered that the differences he observed between competiton in his temperate study and those in tropical studies were due to differing taxa and more specifically that a major taxon, the bryozoans, added intransitivity to assemblage interactions. Here, interactions were studied and extracted from the literature involving a single taxon, the bryozoans, common to all studies.

Materials and methods

Study and literature sites

Intertidal and shallow subtidal zone encrusting species in boulder communities were compared between sites on islands. Particular reference is made to bryozoans as the one taxon present (and abundant) at all locations and studies. Boulder faunas examined during this study included southern hemisphere localities of polar, subpolar, temperate and tropical latitudes (Table 1). Various components of boulder fauna competition were investigated and calculated by drawing up or interpreting existing contact matrices. A contact matrix is a table in which various types of pair-wise interactions between different competitor identities can be displayed. Such a matrix (Table 2) can represent a subset or all competitors and is typically organised such that for each competitor pairing the number of meetings, wins, losses and ties are apparent. A matrix of meetings or interactions can thus be established for the suite of competitors at each site. From each matrix (site), the proportions of inter- and intraspecific encounters, the proportions of tied outcomes, indeterminate species pairs and competitive loops in inter-specific encounters were calculated. Standoffs, redirected growth on meeting or overgrowth by both competitors, were recorded as tied outcomes (Russ, 1982; McKinney, 1992). Overall indeterminate species pairs refers to encounters by competitors such that neither competitor won all encounters, so even if each encounter produced a decided outcome one species won some encounters, the second species, others. At three locations, the above data were also recorded for a variety of depths; 0 and 6 m (South Georgia), 0, 6, 12 and 25 m (Adelaide Is., Antarctica) and 0, 6,

Table 1. Islands investigated or previously studied in the literature from which the data for this study were taken. The latitude, longitude, ocean and near or offshore status are shown together with the reference material

	Study islands	Latitude and longitude	Near/Offshore & Ocean	Reference
Antarctic	Adelaide	(68°S, 67°W)	Near, Southern	(Barnes & Arnold, 1999)
	Gellindez	(66°S, 63°W)	Near, Southern	●This study
	Dobrowlski	(64°S, 62°W)	Near, Southern	●This study
	Signy	(61°S, 45°W)	Off, Southern	+Barnes & Rothery (1996)
	Coronation	(60°S, 45°W)	Off, Southern	●This study
Subantarctic	South Georgia	(54°S, 37°W)	Off, South Atlantic	*Barnes & Arnold
	Bird	(54°S, 38°W)	Off, South Atlantic	●This study
	Falkland	(52°S, 60°W)	Off, South Atlantic	●This study
	Tierra del Fuego	(53°S, 67°W)	Near, South Atlantic	●This study
Southern temperate	South	(40°S, 173°E)	Off, South Pacific	@Gordon (1980)
	Inhaca	(32°S, 33°E)	Near, Indian	●This study
	Long Reef	(33°S, 153°E)	Near, South Pacific	#McGuinness & Underwood (1986)
	Lord Howe	(31°S, 159°E)	Off, South Pacific	○Holmes et al. (1997)
	Solitary	(30°S, 153°E)	Near, South Pacific	○Holmes et al. (1997)
Tropical	Heron	(23°S, 151°E)	Off, Pacific	◇Ryland (1974)
	Green	(17°S, 146°E)	Off, Pacific	○Holmes et al. (1997)
	Quirimba	(12°S, 42°E)	Near, Indian	●This study
	Sencar	(12°S 42°E)	Near, Indian	●This study
	Bali	(8°S, 115°E)	Off, Indian	℧Winston & Heimberg (1986)
	Galapagos	(0°S, 90°W)	Off Pacific	♈Hastings (1930)

12, 25, 32, 42 and 50 m (Signy Is., Antarctica). As well as species richness, diversity of assemblages was compared by counting the number of different species pairs meeting in inter-specific competition for each site.

Results and discussion

Species and interaction richness

The number of encrusting/sessile phyla present (occupying >1% of space) on boulders decreased towards the south-polar sites whilst there was little, if any, difference between tropical and temperate sites (Figure 1a). Typically, in non-polar water, 3–6 phyla were present on boulders while only two (polychaetes and bryozoans) were present at the high latitude sites. The total number of species at each site is shown in Figure 1b and the total number of bryozoans in Figure 1c. Both plots show the number of species increased with decreasing latitude (i.e. towards the tropics from south-polar shores). The highest points were in the mid-latitude temperate or subArctic localities. The most striking difference with latitude was the comparative lack of variability (in numbers of phyla/species per site) in south-polar sites. There was considerable variation between sites at comparable latitudes – indicating the importance of local conditions (Schoener & Schoenern, 1981; McGuiness, 1990). At higher latitudes, ice scour (scraping of substratum by floating ice) severely reduces shallow-water colonisation and species richness (Barnes, 1999). Below the level of ice scour, Antarctic benthic diversity may reach generally high values (White, 1984; Arntz et al., 1997). Whilst there are various agents of disturbance (e.g. storms, floods and bioturbators) at non-polar localities causing a wide range of impact levels (McGuiness, 1990), very few rank alongside the magnitude of that caused by ice (Gutt & Starmans, in press).

There was no obvious correlation between the level of site isolation and species richness. Mainland, near- and off-shore island sites showed similar patterns (Fig. 1a, b, c), although this is partly due to a lack of studies of both island categories at certain latitudes. Whilst some workers (e.g. Long &

Table 2. Contact matrix of five competititor identities (at Dobrowlski Island, Antarctica). The values in each cell are: number of tied outcomes between competitors A and B (top-left), top-right (wins by B [= losses by A]) and bottom-left (wins by A [= losses by B]). Value in bottom-right of each cell = total number of observed interactions for that competitor-pair. The proportion of interspecific competition for this matrix is (9+3+11+2+5+6+3+6+1)/(9+3+11+2+5+6+3+6+1+5+6+46) = 44.7%. The proportion of indeterminate species pairs is 2/7 = 28.6% (*I. nutrix* vs *M. brevissima* and *F. rugula* vs *C. bougainvillei*. The proportion of tied outcomes is (1+1)/(9+3+11+2+5+6+3+6+1) = 4.3%. Three competitor pairs do not meet (*M. brevissima* vs *M. brevissima*, *Spirorbis* sp. vs *Spirorbis* sp. and *Spirorbis* sp. vs *C. bougainvillei*

	M. brevissima	*I. nutrix*	*C. bougainvillei*	*F. rugula*	*Spirorbis* sp.
Micropora brevissima		1 1 / 7 9	0 0 / 3 3	0 0 / 11 11	0 0 / 2 2
Inversiula nutrix		4 1 / 1 5	0 0 / 5 5	0 0 / 6 6	0 0 / 3 3
Celleporella bougainvillei			4 2 / 2 6	1 2 / 3 6	
Fenestrulina rugula				46 0 / 0 46	0 0 / 1 1
Spirorbis sp.					

Rucker, 1970; Holmes et al., 1997) have found reduced shallow marine benthic diversity on offshore islands compared with comparable latitude near shore islands, others (e.g. Dinesen, 1983; Hughes & Jackson, 1992) found the converse. Variations in many factors, such as proximity of breeding adults (larval supply), current conditions, water quality and small-scale parameters may mask or even reverse patterns between near and offshore sites. There was also no apparent relationship between island size and number of taxa found (in samples) as might be predicted by standard island biogeography models (MacArthur & Wilson, 1967).

The number of different competitor (species) pairs meeting in interspecific competition followed similar patterns to those of species richness. The two highest values (of number of different competitor pairings) were the same sites/studies exhibiting the peaks in species richness (plot not shown). The top values were all from the Subantarctic or temperate regions, with a decline to both tropical and polar regions. A large number of different species interaction pairs at a site not only demonstrates the presence of many species, but a degree of evenness such that so many different competitors meet. This could be further used as quantifying evenness in competition by modifying diversity indices to evaluate the spread of interactions between competitors (compared to more usually measuring the spread of different species present).

Colonisation and recruitment

Experiments involving artificial panels (of similar size and material) have been conducted at localities spanning most latitudes. Although some studies have shown initial rates (<6 months) of colonisation and species diversity were higher at tropical than temperate sites, the majority have found little difference after 10 months (Schoener et al., 1978). Recent panel studies in Antarctic waters have shown that both colonisation rates and species richness may be nearly two orders of magnitude lower than in warm water experiments (Dayton, 1989; Barnes, 1996; Stanwell-Smith & Barnes, 1997). Variation in panel colonisation in excess of an order of magnitude is dependant on oceanic conditions (Winston & Jackson, 1984), light conditions (Nandakumar, 1995), depth (Barnes, 1996), isolation (Holmes et al., 1997) and sedimentation rates (Maughan & Barnes, unpublished data). Annual or even super-annual variability in recruitment onto artificial panels may be substantial both in non-polar (Kendall et al., 1985) and polar environments (Dayton, 1989).

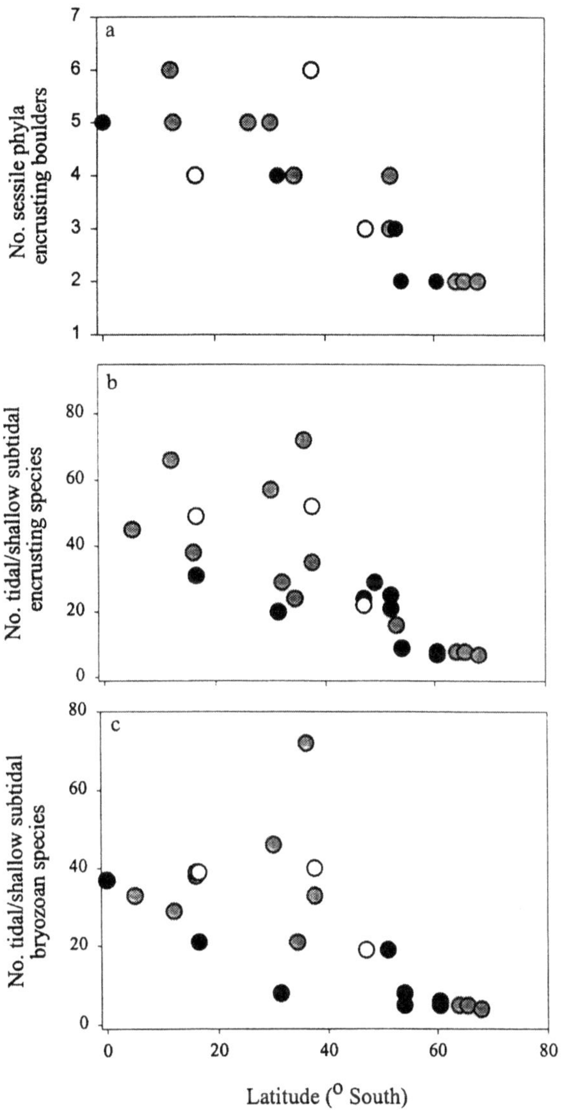

Figure 1. Number of sessile phyla (a), subtidal encrusting species (b), and subtidal bryozoan species (c) with latitude south. The symbols are mainland (○), near-shore islands (◐) and off-shore islands (●). Data are from the present study and taken from the literature (see Table 1).

Monthly rates of panel colonisation (Table 3), calculated from the literature, show lower rates at polar than warmer latitudes, despite the many complicating factors, but no significant difference was found between temperate and tropical levels. Both temperate and tropical levels of recruitment halved from coastal to isolated sites, whilst, curiously, polar levels doubled. The small number of polar studies carried out, and the potential confounding factors of annual variability and physical environmental variables, means such a finding must be treated with caution.

Table 3. Monthly colonisation, expressed as % cover, of shallow-water panels in coastal or offshore studies. Data are calculated for mean cover after initial 6 months deployment, from the following literature; Sutherland & Karlson (1977), Schoener & Schoener (1981), Jackson & Winston (1982), Dayton (1989), Rauschert (1991), Barnes (1996) and references therein, Holmes et al. (1997), Stanwell-Smith & Barnes (1997), Hextall (unpublished data). Values are presented as means ± Standard error with sample size

Region	Coastal/near shore	Offshore
Antarctic	0.05 (0.05) $n = 2$	0.17 (0.05) $n = 2$
Temperate	10.9 (1.7) $n = 6$	4.6 (1.9) $n = 3$
Tropical	9.9 (2.1) $n = 5$	3.3 (1.6) $n = 3$

Competition

The data suggest that the itensity of interspecific interactions decreases towards the pole in the southern hemisphere sites (Fig. 2A) and thus the level of intraspecific competition must proportionally increase. This result is consistent with, and must be partly caused by, the reduction in numbers of encrusting species with latitude (Fig. 1b). The proportion of interspecific interactions increased with depth from the intertidal zone to 50 m, although virtually all of this increase occurred over the depth range 0–12 m (Fig. 3A). There is a suggestion of a increase in the proportion of determinate species pairings (in which one competitor wins all encounters against another competitor species). Typical proportions of determinate species pairings in warm waters were 70% compared to around 30% in polar waters (Fig. 2B). The proportion of tied outcomes also decreased with latitude (Fig. 3C). As any tied outcomes between a pair of species competing makes such a pairing indeterminate, any trend in tied outcomes must influence any pattern of determinate species pairings with latitude. The trend of increased determinate species pairs (Fig. 3B) and tied outcomes (Fig. 3C) with depth followed a similar pattern to that of the increased prevalence of interspecific competition (Fig. 3A). There appears to be little or no influence of isolation, but this may need more studies at comparable latitudes such as that of Holmes et al. (1997).

Jackson (1979) suggested that increasing the number of taxa should increase the number of methods

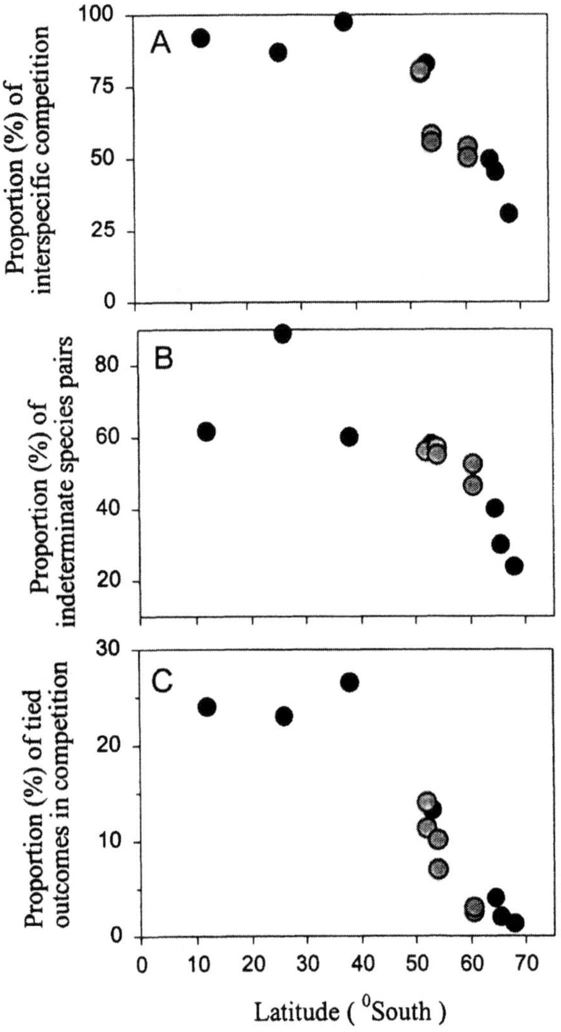

Figure 2. Proportion of inter-specific competition (c.f. intra-specific competition), proportion of indeterminate species pairs and proportion of tied outcomes (in inter-specific competition) with latitude South. The symbols are near-shore islands (●) and off-shore islands (◍). Data are calculated as shown in Table 2, and values are from the present study and taken from the literature (see Table 1).

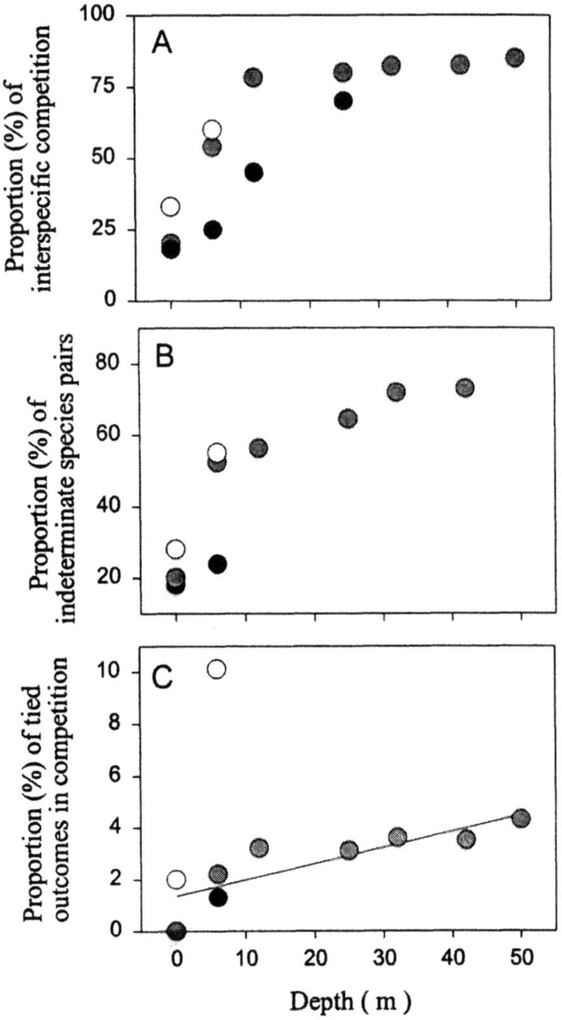

Figure 3. Proportion of inter-specific competition (c.f. intra-specific competition), proportion of indeterminate species pairs and proportion of tied outcomes (in inter-specific competition) with depth (m). The symbols are South Georgia (○), Signy Island (◍) and Adelaide Island (●). Data calculated from contact matrices (e.g. Table 2) from the present study and Barnes & Rothery (1996).

of interaction and, therefore, increase the probability of varied outcomes (so indeterminate species pairs) and competitive loops. However, virtually all studies have found interphyletic interactions to be predictably ranked in which ascidians ≥ sponges > ctenostomatid bryozoans > cheilostomatid bryozoans > cyclostomatid bryozoans > polychaetes, barnacles and hydroids (Russ, 1982; Quinn, 1982). Although only few taxa were involved in competitive interactions on polar boulders, the data analysed here suggest that, as with warmer water assemblages, a ranking of sponges > cheilostomatid bryozoans > cyclostomatid bryozoans > polychaetes.

Conclusions

The polar data collected, and temperate/tropical data analysed here, suggest a number of trends in encrusting communities, more specifically the bryozoan component, with latitude and depth but not obviously isolation. Species richness and colonisation rates follow approximately parabolic relationships in southern hemisphere assemblages, though possibly with highest points at high temperate/subpolar latitudes in the former. Also, in southern hemisphere assemblages, intraspecific competition becomes more prevalent with increasing latitude and less with depth. There are suggestions of distinct changes in the ecological structure of competition with latitude in the southern hemisphere, that is the outcomes of interactions between competitor pairs alters (the proportion of tied interactions and indeterminate species pairs both increase).

Potentially, the three factors have similar paths of influence as, typically, the number of species increases towards the tropics and with depth (within the context of the studies here), and decreases with isolation. Patterns of species richness have, in the case of bryozoans, been shown to be questionable along both latitudinal (Clarke, 1992) and isolation gradients (Hughes & Jackson, 1992). Another potential common link for a *modus operandi* of latitude, isolation and depth on the workings of encrusting communities is via disturbance. There was no evidence of the high within region variability, characteristic of the temperate and tropical regions (McGuiness, 1990), in the south-polar sites studied. This is probably due to the high variation in exposure of warm-water shores whereas even the most sheltered south-polar shores are highly disturbed by ice scour. Water-born disturbance, probably in terms of wave action (trans-global fetch), but definitely in terms of ice scour, increases with latitude and isolation but decreases with depth (Barnes, 1999). So, increased levels and/or frequency of disturbance may be important for generation and maintenance of species diversity at high latitudes and very shallow depths. An obvious question raised by this study is whether patterns in the southern and northern hemispheres are different? Whilst diversity patterns are believed to be different (Clarke, 1992; Rosenweig, 1995), colonisation and recruitment rates may be very similar (Barnes, 1999). Many studies of competition in boulder assemblages have been carried out in the northern hemisphere, but there are few data from the Arctic. Arctic patterns might be expected to be different from the Antarctic findings for a number of geographical, bathymetric and historical reasons. Land surrounded by ocean would clearly be subject to very different influences than ocean surrounded by land (such as freshwater runoff). The proportion of continental shelf is much greater in Antarctica, most of the Arctic is deep and Antarctica is much older, so patterns of invasion and speciation have had longer to proceed.

References

Arntz, W. E., J. Gutt & M. Klages, 1997. Antarctic marine biodiversity: an overview. In, Battaglia, B., J. Valencia & D. W. H. Walton (eds), Antarctic Communities. Cambridge University Press: 3–14.

Barnes, D. K. A., 1996. Low levels of colonisation in Antarctica: the role of bryozoans in early community development. Bryozoans in space and time. Proceedings of the International Bryozoology Association 10th Conference. N.I.W.A., Wellington, New Zealand: 19–28.

Barnes, D. K. A., 1999. The influence of ice on polar benthos. J. mar. biol. Ass. U.K. 79: 401–408.

Barnes, D. K. A. & R. J. Arnold, 1999. Possible latitudinal clines in Antarctic intertidal and subtidal zone communities encrusting ephemeral hard substrata. J. Biogeog. 26: 207–213.

Barnes, D. K. A. & A. Clarke, 1995. Seasonality of feeding ecology in polar suspension feeders. Polar Biol. 15: 335–340.

Barnes, D. K. A. & A. Clarke, 1998. The ecology of an assemblage dominant: the encrusting bryozoan *Fenestrulina rugula*. Invert. Biol. 117: 331–340.

Barnes, D. K. A. & P. Rothery, 1996. Competition in encrusting Antarctic bryozoan assemblages: outcomes influences and implications. J. exp. mar. Biol. Ecol. 196: 267–284.

Barnes, D. K. A., P. Rothery & A. Clarke, 1996. Colonisation and development in encrusting communities from the Antarctic intertidal and sublittoral. J. exp. mar. Biol. Ecol. 196: 251–265.

Brey, T., M. Klages, C. Dahm, M. Gorny, J. Gutt, S. Hain, M. Stiller, W. E. Arntz, J.-W. Wägele & A. Zimmerman, 1994. Antarctic benthic diversity. Nature, London 368: 297 pp.

Clarke, A. 1992. Is there a latitudinal diversity cline in the sea? Trends Ecol. Evol. 7: 286–287.

Connell, J. H., 1978. Diversity in tropical rain forests and coral reefs. Science 199: 1302–1310.

Dayton, P. K., 1989. Interdecadal variation in an Antarctic sponge and it's predators from oceanographic climate shifts Science, 245: 1484–1486.

Dayton, P. K., G. A. Robilliard, R. T. Paine & L. B. Dayton, 1974. Biological accommodation in the benthic community at McMurdo Sound, Antarctica. Ecol. Monogr., 44: 105–128.

Dinesen, Z. D., 1983. Patterns in the distribution of soft corals across the Great Barrier Reef. Coral Reefs 1: 229–236.

Gordon, D. P., 1980. Bryozoa of the Cape Rodney to Okakari Point marine reserve: an identification manual. New Zealand Oceanographic Institute, Wellington. Unpublished report: 137 pp.

Gutt, J. & A. Starmans, 2000. Quantification of iceberg impact and benthic recolonisation patterns in the weddell sea (Antarctica). Polar Biol in press.

Hastings, A. B., 1930. Cheilostomatous Bryozoa from the vicinity of the Panama Canal collected by Dr C. Crossland on the cruise of the S.Y. St George. Proc. Zool. Soc. Lond. 1929: 1–17.

Holmes, N. J., V. J. Harriott & S. A. Banks, 1997. Latitudinal variation in patterns of colonisation of cryptic calcareous marine organisms. Mar. Ecol. Prog. Ser. 155: 103–113.

Hughes, D. J. & J. B. C. Jackson, 1992. Distribution and abundance of cheilostome bryozoans on the Carribean reefs of central Panama. Bull. mar. Sci. 51: 443–465.

Huston, M., 1979. A general hypothesis of species diversity. Am. Nat. 113: 81–101.

Jackson, J. B. C., 1979. Morphological strategies of sessile animals. In Larwood, G. & B. R. Rosen (eds), Biology and Systematics of Colonial Organisms. Academic press, London, p. 499–555.

Jackson, J. B. C. & J. E. L. Winston, 1982. Ecology of cryptic coral reef communities. I. Distribution and abundance of major groups of encrusting organisms. J. exp. mar. Biol. Ecol. 57: 135–147.

Kendall, M. A., R. S. Bowman, P. Williamson & J. R. Lewis, 1985. Annual variation in the recruitment of *Semibalanus balanoides* on the north Yorkshire coast. J. mar. biol. Ass. U.K. 65: 1009–1030.

Long, E. R. & J. B. Rucker, 1970. Offshore marine cheilostome bryozoans from Fort Lauderdale, Florida. Mar. Biol. 6: 18–25.

MacArthur, R.H. & E. O. Wilson, 1967. The Theory of Island Biogeography. Princeton University Press, Princeton, New Jersey.

McGuiness, K. A., 1987. Disturbance and organisms on boulders II. Causes of patterns in diversity and abundance. Oecologia 71: 420–430.

McGuiness, K. A., 1990. Physical variability, diversity gradients and the ecology of temperate and tropical reefs. Aust. J. Ecol. 15: 465–476.

McGuiness, K. A. & A. J. Underwood, 1986. Habitat structure and the nature of communities on intertidal boulders. J. exp. mar. Biol. Ecol. 104: 97–123.

McKinney, F. K., 1992. Competitive interactions between related clades: evolutionary implications of overgrowth interactions between encrusting cyclostome and cheilostome bryozoans. Mar. Biol. 114: 645–652.

Menge, B. A. & J. Lubchenco., 1981. Community organisation in temperate and tropical intertidal habitats: prey refuges in relation to consumer pressure gradients. Ecol. Monogr. 46: 355–393.

Mileikovsky, S. A., 1971. Types of larval development in marine bottom invertebrates, their distribution and ecological significance: a re-evaluation. Mar. Biol. 10: 193–213.

Moore, H.B., 1972. Aspects of stress in the tropical marine environment. Adv. Mar. Biol. 10: 217–269.

Nandakumar, K., 1995. Competition amongst sessile organisms in Tomioka Bay, South Japan: importance of light conditions on the panel surface. Mar. Biol. 121: 713–719.

Pearse, J. S., J. B. McClintock & I. Bosch, 1991. Reproduction in Antarctic benthic marine invertebrates: tempos, modes and timing. Am. Zool., 31: 65–80.

Peck, L. S., 1989. Temperature and basal metabolism in two Antarctic marine herbivores. J. exp. mar. Biol. Ecol. 127: 1–12.

Pianka, E. R., 1966. Latitudinal gradients in species diversity: A review of concepts. Am. Nat. 100: 33–46.

Quinn, J. F., 1982. Competitive hierarchies in marine benthic communities. Oecologia 54: 129–135.

Rauschert, M., 1991. Ergebnisse der faunistischen Arbeiten im Benthal von King George Island (Südshetlandinseln, Antarktis). Ber. Polarforsch. 76: 1–75.

Rhode, K., 1997. The larger area of the tropics does not explain latitudinal gradients in species diversity. Oikos 79: 169–172.

Rosenweig, M. L., 1995. Species Diversity in Space and Time. Cambridge University Press, Cambridge.

Russ, G. R., 1982. Overgrowth in a marine epifaunal community: competitive hierarchies and competitive networks. Oecologia 53: 12–19.

Ryland, J. S., 1974. Bryozoa in the Great Barrier Reef province. Proc 2nd Int Symp Coral Reefs 1: 341–348.

Schoener, A., E. R. Long & J. R. DePalma, 1978. Geographic variation in artificial island colonisation curves. Ecology 59: 367–382.

Schoener, A. & T. W. Schoener, 1981. The dynamics of the species-area relation in marine fouling systems: 1. Biological correlates of changes in the species-area slope. Am. nat. 118: 339–360.

Sousa, W., 1979. Experimental investigations of disturbance and ecological succession in a rocky intertidal algal community. Ecol. Monogr. 49: 227–254.

Stanwell-Smith, D. & D. K. A. Barnes, 1997. Benthic community development in Antarctica: recruitment and growth on settlement panels at Signy Island. J. exp. mar. Biol. Ecol. 212: 61–79.

Stanwell-Smith, D., L. S. Peck, A. Clarke, A. W. A. Murray & C. D. Todd, 1999. The distribution, abundance and seasonality of pelagic marine invertebrate larvae in the maritime Antarctic. Phil Trans Roy Soc 354.

Sutherland, J.P. & R.H. Karlson, 1977. Development and stability of the fouling community at Beaufort, North Carolina. Ecol. Monogr. 47: 425–446.

White, M. G., 1984. Marine benthos. Anarct Ecol, Laws, R.M. (ed.) 2: 421–461.

Winston, J. E. & B. G. Heimberg, 1986. Bryozoans from Bali, Lombok and Komodo. Am. Mus. Novit. 2847: 1–49.

Winston, J. E. & J. B. C. Jackson, 1984. Ecology of cryptic coral reef communities. IV. Community development and life histories of encrusting Cheilostome Bryozoa. J. exp. mar. Biol. ecol. 76: 1–21.

Hydrobiologia **440:** 45–54, 2000.
M.B. Jones, J.M.N. Azevedo, A.I. Neto, A.C. Costa & A.M. Frias Martins (eds), Island, Ocean and Deep-Sea Biology.

Are kelp holdfasts islands on the ocean floor? – indication for temporarily closed aggregations of peracarid crustaceans

Martin Thiel & Julio A. Vásquez
Departamento de Biologia Marina, Universidad Católica del Norte, Campus Guayacan, Coquimbo, Chile
Fax: 56 51 209 812. E-mail: thiel@nevados.cecun.ucn.cl

Key words: algal holdfasts, Peracarida, reproduction, dispersion, benthos

Abstract

During the colonisation process of islands, newly immigrating species often arrive as single individuals. Islands that have received single colonisers may subsequently harbour large populations of a species, while other islands may completely lack this species. Exchange between islands is limited, thereby strongly affecting evolutionary processes. While this concept is widely used in the context of oceanic islands or habitat patches on the mainland, it is rarely used to explain and examine the distribution patterns of marine invertebrates. Benthic marine organisms inhabiting patches with island-like features may also be restricted in their movements between patches. Once established in a patch, it may be more favourable to remain there rather than moving to another patch. Juveniles of species with direct development may recruit to the island patch of their parents. Herein, we examined the peracarid fauna in patches that have island-like features, i.e. kelp holdfasts. The number of peracarid species within an individual holdfast increased with its size. Similarly, the number of individuals per holdfast increased with holdfast size. However, several peracarid species showed a strongly aggregated distribution pattern, being highly abundant in some holdfasts and almost completely absent in others. Our results suggest that these aggregations of conspecifics may be a consequence of the peracarid reproductive biology: fully developed juveniles emerge from the female's marsupium and recruit to the immediate vicinity of their mother, showing little or no tendency to emigrate towards other patches. At present, while it is not known how long peracarid aggregations within kelp holdfasts persist, our data suggest that some juveniles may remain with the natal holdfast and possibly reproduce therein. It is concluded that, during certain time periods, reproduction rates of peracarids in a holdfast may exceed their migration rates between holdfasts.

Introduction

Many marine microhabitats may have island-like characteristics and organisms inhabiting such microhabitats may, for certain time periods, be spatially isolated from conspecifics inhabiting other nearby microhabitats. Specifically, organisms may immigrate to or emigrate from these islands and, once established, may use them to conduct all their activities including feeding and reproduction. Founder effects may occur, particularly in well isolated microhabitats (=islands). Few individuals may reproduce and build an island population that has only limited exchange with neighbouring islands. These founder effects are well known from terrestrial species with individuals that reproduce and build local populations within a particular island or patch. In the marine environment, founder effects have received comparatively little attention, possibly because of the large proportion of species with planktonic larval stages. Reproductive isolation is considered less likely in these marine organisms than in terrestrial organisms since larvae (= gene-carrier) are easily transferred and exchanged between microhabitats. However, many marine invertebrates such as polychaetes, molluscs, echinoderms and crustaceans have direct development. For example, snapping shrimp release their offspring within their sponge colony and juveniles of several species remain therein, forming large colonies of closely related individuals (Duffy, 1996; Duffy & Macdonald, 1999). Another group of crustaceans without any free-living larval stages is the peracarids. Peracarid species are common inhabitants of many marine microhabitats such as sponges, pieces of wood, corals and macroalgae. Development in peracarids is direct and it is increasingly recognised that juveniles may recruit to the immediate vicinity of their parents (Flach, 1992; Thiel, 1999). Single ovigerous females immigrating into a patch may potentially be able to serve as founders for large aggregations of closely related individuals within one microhabitat island. Reproductive females, when present in a patch, may be able to sustain a local group, while in patches without reproductive females, a species may disappear quickly unless it is repeatedly replenished by newly immigrating individuals. When reproduction rates in a patch are not sufficiently high to counterbalance losses due to death and migration, local populations within a patch may become extinguished (see e.g. Gunnill, 1982, 1983).

Several marine microhabitats have island-like features, such as sponges, ascidians, corals, mangrove trees and macroalgae. Herein, we examined whether distinct peracarid aggregations occur in holdfasts of three common kelp species along the Chilean coast.

Materials and methods

Macroalgae, *Lessonia trabeculata*, *L. nigrescens* and *Macrocystis integrifolia*, were sampled during the austral fall (April 1999) in the shallow subtidal zone around Punta de Choros, Chile (29.25 S, 71.55 W) (Fig. 1). The algal density in the kelp beds was determined by transect and quadrat counts. For *L. trabeculata* and *M. integrifolia*, all plants in 2 m-wide and 10 m-long transect strips were counted (10 strips counted for each species). For *L. nigrescens,* all plants in 17 randomly chosen quadrats (1 m^2) were counted.

Thirty-one individual plants of *L. trabeculata*, 36 of *L. nigrescens* and 27 plants of *M. integrifolia* were collected at the respective sampling sites (Fig. 1). Plants of *L. trabeculata* and *M. integrifolia* were collected 7–10 m below MLW (mean low water), and plants of *M. nigrescens* in the low intertidal zone. Whole plants were separated from the substratum and stored immediately in mesh bags (1 mm mesh diameter). On the shore, whole plants were placed in large trays and the blades separated from the holdfasts. For preservation and subsequent transport to the laboratory, each holdfast, together with all associated macroinvertebrates, was placed in a large sealable plastic bag and formalin was added until a 10% solution was achieved. In the laboratory, holdfasts were carefully retrieved from the bag and washed over large trays with freshwater. Holdfasts were dissected and surveyed carefully for macrofauna organisms. The contents remaining in the trays and bags were washed over a 0.5 mm sieve, which retained all remaining macrofauna organisms. All macrofauna was sorted from the residue and identified to the lowest taxonomic level. Herein, we distinguish two main groups: the non-peracarid macroinvertebrates and the peracarid macroinvertebrates (amphipods, isopods and tanaids). The total volume of the kelp holdfasts was determined by placing all holdfast pieces in a beaker and measuring the volume of water displaced. Previous studies had revealed that holdfast volume is best suited to describe species abundance and diversity in kelp holdfasts (Vásquez & Santelices, 1984). The relationship between holdfast volume and the species composition was examined by a regression analysis searching for the curve that best fitted the data (utilizing the program SYSTAT8.0).

Results

The average density of kelp plants was 3.7 ($\pm$ 0.61 SE) plants m^{-2} (*Lessonia nigrescens*), 3.5 ($\pm$ 0.30 SE) (*Macrocystis integrifolia*) and 2.1 ($\pm$ 0.25 SE) plants m^{-2} (*Lessonia trabeculata*). The number of invertebrate species per kelp holdfast was strongly correlated with the volume of the holdfast, both for other macroinvertebrates and for peracarids (Fig. 2). The number of other macroinvertebrate species increased rapidly in small and medium-sized holdfasts but appeared to approach an equilibrium in larger holdfasts. The number of peracarid species increased over the whole size range of holdfasts of *L. trabeculata* and *M. integrifolia* (Fig. 2). Some holdfasts harboured comparatively large numbers of peracarids, but holdfasts of similar sizes contained only few individuals.

Many peracarid species occurred exclusively within the holdfasts of one algal species (Table 1). The most common species was the boring amphipod *Perampithoe femorata,* which occurred in about 70% of all samples. The suspension feeding amphipod *Aora typica* occurred in about 40% of all samples. The only other species that were found in holdfasts of all three algal species were the amphipods *Ericthonius brasiliensis*, *Gammaropsis typica*, *Ventojassa frequens* and *Maera incerta*. Many isopod and tanaid species were found only in holdfasts of one algal species.

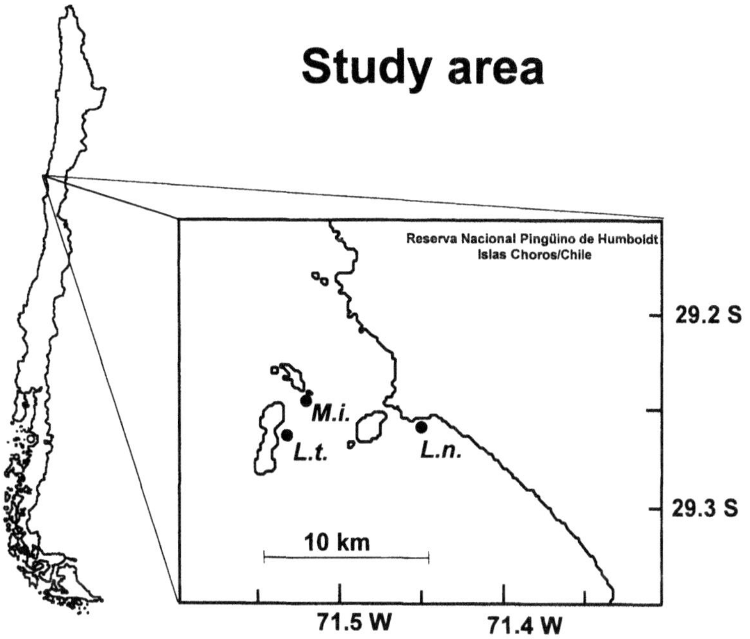

Figure 1. Study area at the northern-central coast of Chile showing sites of collection for individual plants of the macroalgae *Lessonia trabeculata* (*L.t.*), *L. nigrescens* (*L.n.*), and *Macrocystis integrifolia* (*M.i.*) during the austral fall (April 1999).

For most of the common invertebrate species, the number of individuals was significantly correlated with the volume of the holdfasts (Fig. 3). In particular, species with pelagic larvae (*Brachidontes granulata*, *Semimytilus algosus*, Nereidae sp., Nemertea sp., *Parantheopsis cruentata*) or with highly mobile adults (*Pachycheles grossimanus*, *Gaudichaudia gaudichaudi*) occurred in densities related to the volume of the holdfasts. This trend was less clear for the most abundant peracarid species. These appeared to occur in similar numbers in holdfasts of all sizes (see e.g. *Perampithoe femorata* and *Aora typica* in Figure 4). However, there were always a few holdfasts in which unusually high numbers of individuals occurred (Fig. 4). For example, most holdfasts contained only a few individuals of the tube-building amphipod *Ericthonius brasiliensis,* but there were a few holdfasts with > 20 individuals per holdfast (Fig. 4). Similarly, for the amphipods *Leucothoe* sp. A and *Ventojassa frequens*, most holdfasts contained between 0 and 5 individuals, but a few harboured > 20 individuals per holdfast. In most cases, the typical clutch size of these species ranged between 10 and 30 individuals (see shaded areas in Figure 4). This suggests that, in holdfasts with high numbers of individuals, females had released offspring that had established subsequently in those holdfasts.

Discussion

The number of macroinvertebrate species and individuals per holdfast was correlated with holdfast volume. An increase of individuals per holdfast with increasing holdfast volume was found for single macroinvertebrate species that recruit to holdfasts as larvae and that remain mobile throughout their adult life. A different distribution pattern was found for many common peracarid species. On several occasions, dense peracarid aggregations were found in the kelp holdfasts that did not correspond to the holdfast size. Some holdfasts harboured large numbers of conspecifics, while others contained only single individuals. These aggregations could indicate that juvenile peracarids recruit to their natal holdfast.

Macroinvertebrates in kelp holdfasts

Kelp holdfasts represent an important microhabitat for a variety of macroinvertebrates (Vásquez & Santelices, 1984; Moore, 1986; Smith & Simpson, 1992). Inhabitants of kelp holdfasts receive shelter from predation (Cancino & Santelices, 1984) and access to food and mates (Dayton, 1985; Tegner et al., 1995). Most species are not associated exclusively with holdfasts, and occur also in surrounding habitats under cobble stones, on rock surfaces, or on other secondary sub-

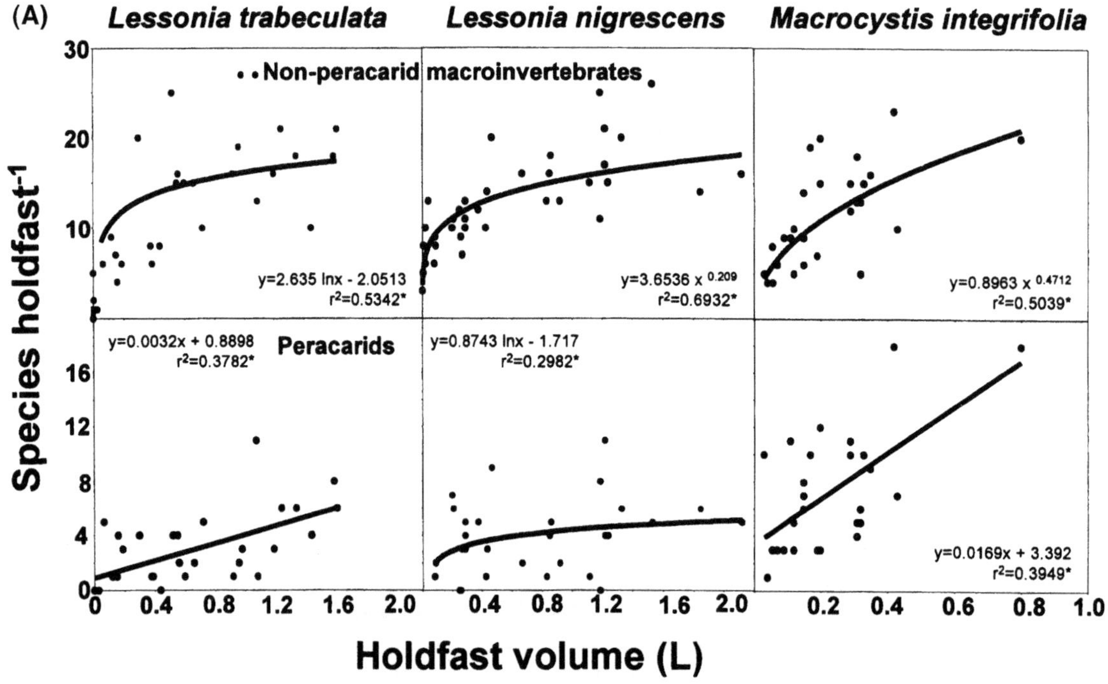

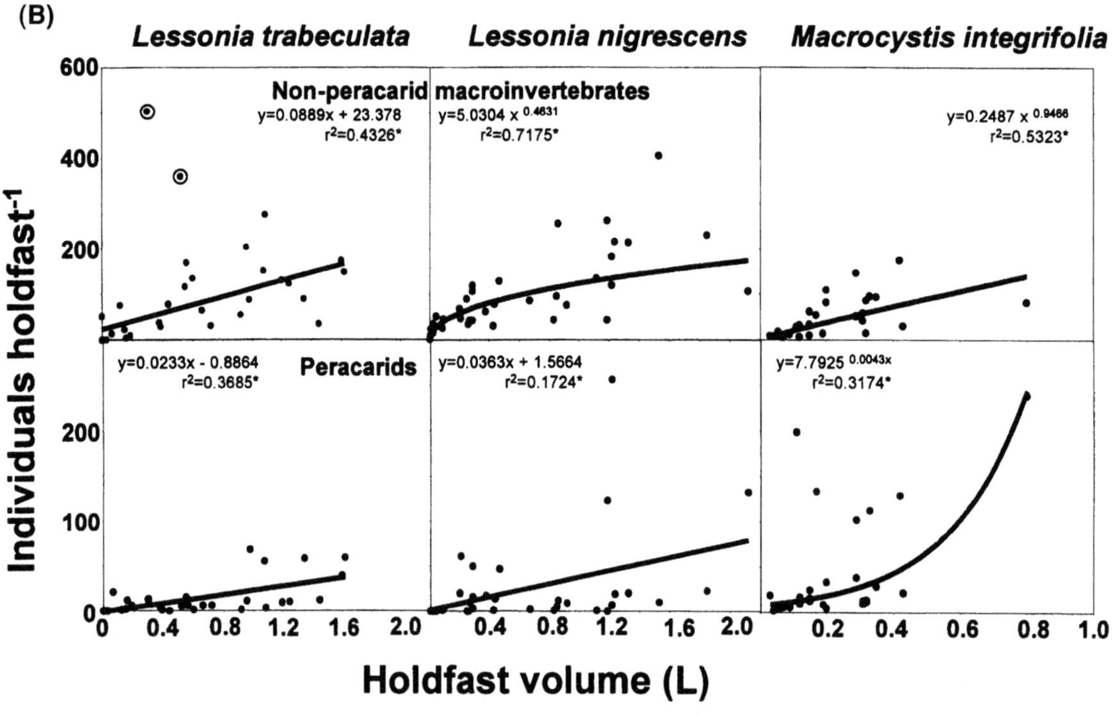

Figure 2. Relationship between holdfast volume and (A) number of species and (B) number of individuals for non-peracarid and peracarid macroinvertebrates found in *Lessonia trabeculata*, *L. nigrescens* and *Macrocystis integrifolia*; $n = 31$ holdfasts of *L. trabeculata*, 36 holdfasts of *L. nigrescens* and 27 of *M. integrifolia* were analysed; regression equations are based on units of ml.

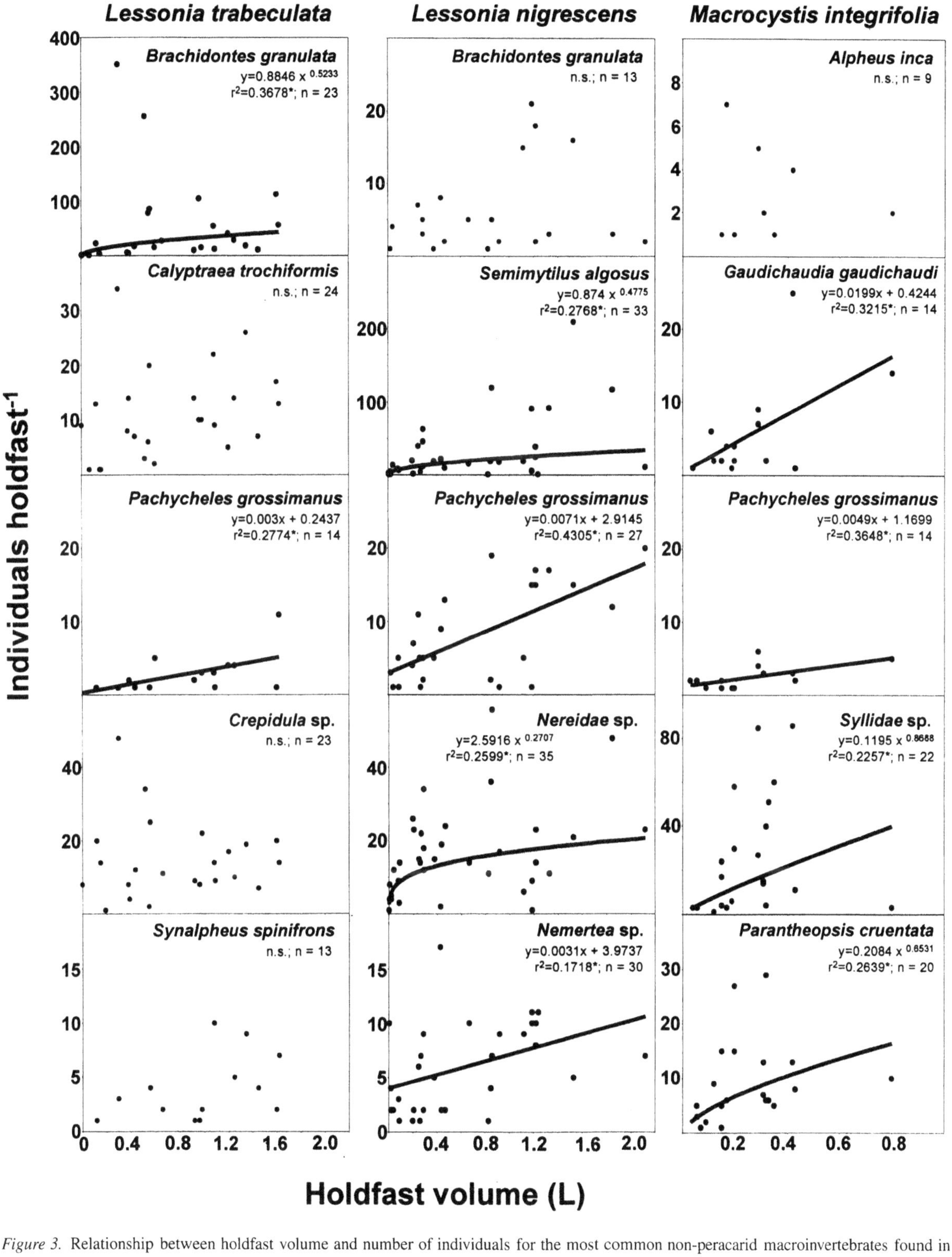

Figure 3. Relationship between holdfast volume and number of individuals for the most common non-peracarid macroinvertebrates found in *Lessonia trabeculata*, *L. nigrescens*, and *Macrocystis integrifolia*; *n* = number of holdfasts in which the respective macroinvertebrate species was found – only these holdfasts were used for the regression analysis; regression equations are based on units of ml.

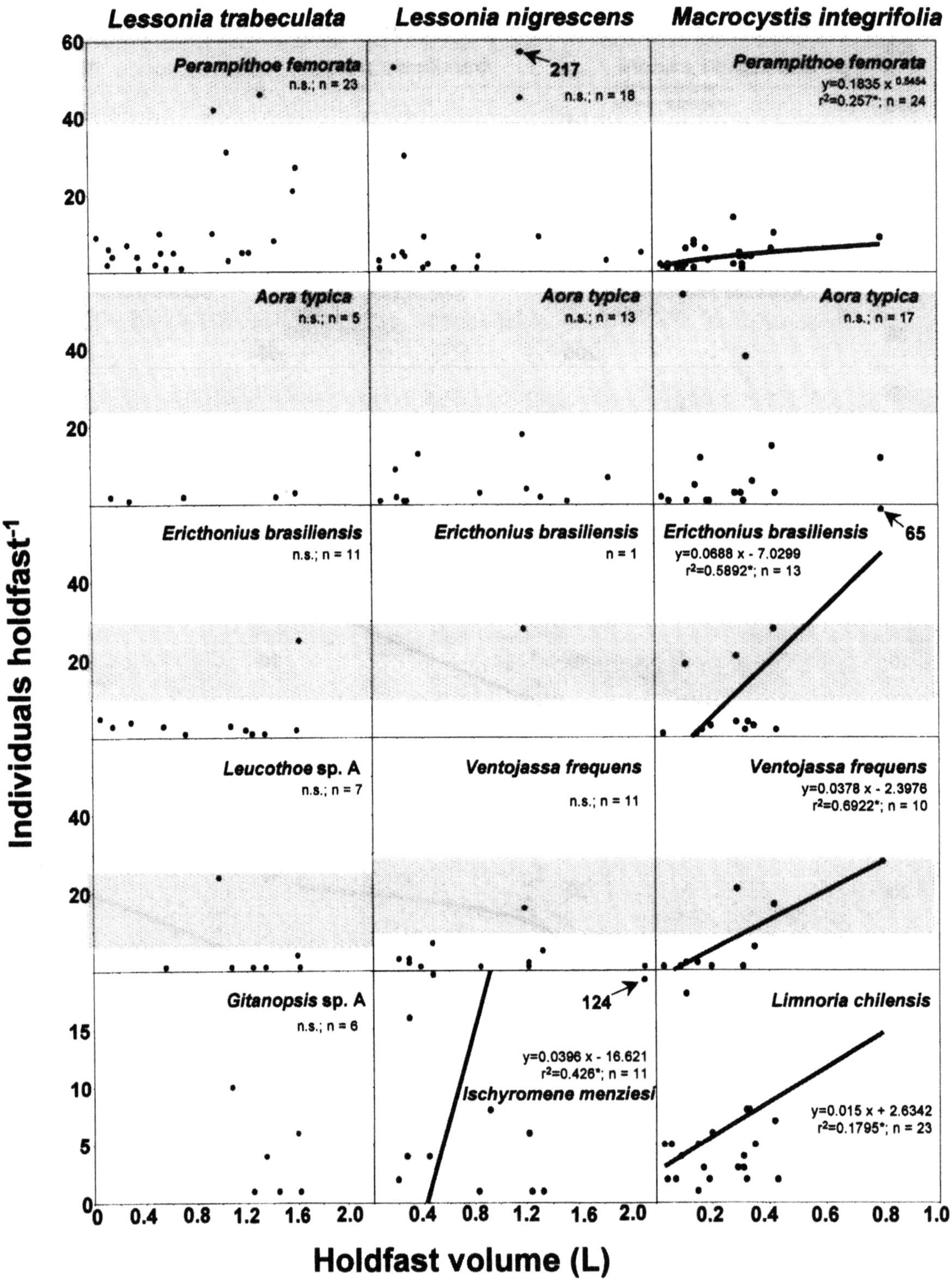

Figure 4. Relationship between holdfast volume and number of individuals for the most common peracarid macroinvertebrates found in *Lessonia trabeculata*, *L. nigrescens*, and *Macrocystis integrifolia*; n = number of holdfasts in which the respective peracarid species was found - only these holdfasts were used for the regression analysis; shaded areas indicate typical clutch sizes for the species for which information was available (own data and from Sainte-Marie, 1991); regression equations are based on units of ml.

Table 1. Percentage of samples in which the respective peracaid species occurred.

	Functional group	Taxon	*Lessonia trabeculata*	*Lessonia nigrescens*	*Macrocystis integrifolia*
			$n = 31$	$n = 36$	$n = 27$
Aora typica	s	a	16.1	36.1	63.0
Aora sp. A	s	a			11.1
Ericthonius brasiliensis	s	a	35.5	2.8	48.1
Ischyrocerus longimanus	s	a	12.9		22.2
Gammaropsis typica	s	a	6.5	2.8	25.9
Gammaropsis typica, var. Y	s	a			3.7
Gammaropsis sp. A	s	a		8.3	3.7
Ventojassa frequens	s	a	25.8	30.6	37.0
Ampelisca sp. A	s	a			11.1
Corophium sp. A	s	a	3.2		
Caprella equilibra	s	a	12.9		7.4
Caprella sp. A	s	a	3.2		
Caprella penantis	s	a	3.2		3.7
Tanais (marmoratus)	s	t		19.4	
Tanaid sp. B	s	t		8.3	
Tanaid sp. C	s	t	3.2		22.2
Hyale hirtipalma	g	a		19.4	3.7
Hyale (grandicornis)	g	a		2.8	
Ianiropsis (chilensis)	g	i	9.7	25.0	
Joeropsis bidens	g	i		47.2	14.8
Santia dimorphis	g	i		5.6	
Amphoroidea typa	g	i	3.2	5.6	
Ischyromene menziesi	g	i		30.6	
Cymodocella foveolata	g	i		13.9	
Perampithoe femorata	b	a	74.2	50.0	88.9
Bircenna fulva	b	a			29.6
Limnoria chilensis	b	i			85.2
Elasmopus chilensis	?	a	9.7		55.6
Elasmopus sp. A	?	a	3.2		11.1
Stenothoe sp. A	?	a	12.9		
Lysianassa sp. A	?	a			7.4
Leucothoe sp. A	?	a	22.6		
Paramoera sp. A	?	a			59.3
Maera incerta	?	a	9.7	8.3	44.4
Melita inaequistylis	?	a			37.0
Gitanopsis sp. A	?	a	19.4		
unidentified amphipod	?	a			22.2

s – suspension-feeder; g – grazer; b – borer; a – amphipod; i – isopod; t – tanaid.

strata such as sponges and ascidians (e.g. Cancino & Santelices, 1984). In the present study, we found that both the number of species and the number of individuals increased with holdfast volume. While the number of species of other macroinvertebrates appeared to reach an equilibrium in the largest holdfasts, the number of individuals continued to increase even in the largest holdfasts (see also Vásquez & Santelices, 1984). For sponges, Westinga & Hoetjes (1981) found similar results, with the number of invertebrate taxa increasing logarithmically with sponge volume (reaching an equilibrium) and the total number of individuals being directly proportional to sponge volume (not reaching an equilibrium). Based on these results, the authors concluded that "the intra-sponge-fauna indeed is an ecological community, in which, however, interrelationships are not yet clear." A similar conclusion can be drawn for the fauna inhabiting kelp holdfast as is suggested by the distinct relationship between microhabitat characteristics (here: holdfast volume) and the number of species and individuals inhabiting individual holdfasts. However, little is known about the general biology of many of the macroinvertebrates inhabiting the holdfasts, let alone the biotic interactions among them.

Peracarid aggregations in kelp holdfasts

The peracarid crustaceans inhabiting kelp holdfasts comprise a variety of functional groups such as grazing, boring and suspension feeding species, and possibly predators and scavengers. All peracarid species share a common reproductive feature – the female brood pouch from which fully developed juveniles emerge. After becoming independent from their parents, these juveniles are capable of establishing a home (e.g. a tube or a burrow) and feeding in a way similar to the adults. Several studies have indicated that juvenile amphipods may recruit in the immediate vicinity of their parents (Flach, 1992; Thiel et al., 1997). Similarly, it has been observed that juveniles of kelp-boring amphipods and isopods excavate their burrows as offshoots of the parental burrow (Menzies, 1957; Jones, 1971; Conlan & Bousfield, 1982; Conlan & Chess, 1992). Aggregations of conspecifics may thus be a consequence of the reproductive biology of peracarids, yet it is not known how long these aggregations persist. In situations where juvenile individuals find themselves released into a suitable microhabitat, they may remain therein without any intention of leaving. Many peracarids inhabiting kelp holdfasts are boring and suspension feeding species that have a relatively sedentary life style associated with tubes or burrows. Duffy & Hay (1994) found that a tube-dwelling amphipod was much less mobile than a free-living species. Several studies of soft-bottom dwelling amphipods indicate that emigration rates are low in disturbance-free environments but substantially increase when disturbers are present (Flach & De Bruin, 1994). While most of the species found in the present study are potentially good swimmers, they may not leave their natal kelp holdfast if not disturbed by conspecifics or benthic predators. Furthermore, fish predation on peracarids is high in the shallow coastal waters of the central Chilean coast (Muñoz & Ojeda, 1997, 1998) and it is, therefore, unlikely that peracarids voluntarily leave the shelter of a kelp holdfast.

While the above considerations suggest that many peracarids remain in the natal holdfast if conditions are favourable, other studies on algal-dwelling peracarids indicate that they are highly mobile. Studies on epifaunal amphipods inhabiting the algal canopy indicated a high turnover of individuals within time periods of days (Edgar, 1992; Taylor, 1998), thus indicating that individuals may easily move from one plant to another. Epifaunal peracarids that forage on algal fronds may be highly mobile, moving easily between algae that provide food and those that offer shelter (e.g. Buschmann, 1990). In contrast to these epifaunal amphipods, species that inhabit holdfasts may find both food and shelter in these microhabitats, thus having little need to change their location.

Do the aggregations represent closely related individuals? The fact that peracarid crustaceans may recruit in the immediate vicinity of their parents suggests that on a small scale (here: within individual holdfasts), some individuals are closely related to each other. Our data suggest that, for certain time periods (weeks to months), holdfasts may harbour aggregations of closely related individuals. A consequence of this 'neighbourhood recruitment' would be high genetic variation over relatively small distances (several m, e.g. between holdfasts). Unfortunately, we are not aware of any studies on marine invertebrates addressing the question of genetic variation over these small distances (but for marine plants see e.g. Reusch et al., 1999). There are, however, several molecular studies on amphipod populations within bay systems that indicate high genetic variation between individuals separated by distances of only a few km (Stanhope, 1993; Wilson et al., 1997). The main cause for this high genetic variation could be the fact that amphipods have direct development and that juveniles establish home in the vicinity of their parents. Studies on other crustaceans and a gastropod with direct development also revealed high genetic variation over similar distances (Knight et al., 1987; Johannesson et al., 1993; Duffy, 1996). Poulin & Féral (1995) found dense aggregations of the brood-protecting echinoid *Abatus cordatus* that has direct development - juveniles immediately recruit to the patch they were born in. The authors hypothesize that patches may remain stable for several years, resulting in individuals within a patch being more closely related to each other than those between patches (Poulin & Féral, 1994, 1998). While these studies have not addressed the question of genetic variation over even smaller distances (several m), they indicate that, in marine invertebrates with direct development, high genetic variation may occur over surprisingly small distances. In contrast, comparatively low genetic variation was found on similar spatial scales for marine molluscs with indirect development and a pelagic larval phase (De Wolf et al., 1998; Skalamera et al., 1999). Since most amphipod and other peracarid species are relatively mobile, aggregations of closely related individuals may not be stable for very long time periods. However, during certain time periods, migration rates may be strongly exceeded by reproduction rates within individual holdfasts. We suggest that aggregations of peracarids in holdfasts may occasionally persist sufficiently long for siblings to attain sexual maturity. As a consequence, the likelihood that closely related individuals mate in these holdfasts (and other biotic microhabitats) would be high. It is not known whether abnormalities found in typical algal dwelling amphipods (e.g. Vader, 1968; Moore, 1973) are a consequence of inbreeding in island-like algal patches. Molecular and experimental studies are required to confirm our suspicion that temporarily closed aggregations of peracarids occur in kelp holdfasts and other biotic microhabitats such as sponges, ascidians or wood.

Acknowledgements

We thank A. Carvacho, E. Gonzalez, R. Heard, B. Kensley and D. Lancellotti for their help with the identification of species. A very special thanks to Alonso Vega and his crew for processing and analysing the samples – without their unconditional support, this study would not have been possible. As always, thanks also to S. Sampson for improving the English in the final version of the manuscript. Funding for this study was provided by FONDECYT 3980002 and FONDAP O&BM.

References

Buschmann, A. H., 1990. Intertidal macroalgae as refuge and food for Amphipoda in Central Chile. Aquat. Bot. 36: 237–245.

Cancino, J. & B. Santelices, 1984. Importancia ecologica de los discos adhesivos de *Lessonia nigrescens* Bory (Phaeophyta) en Chile central. Rev. Chil. Hist. Nat. 57: 23–33.

Conlan, K. E. & E. L. Bousfield, 1982. The amphipod superfamily Corophioidea in the Northeastern Pacific region. Family Ampithoidae: Systematics and Distributional Ecology. Publs. biol. Oceanogr. Nation. Mus. Canada 10: 41–75.

Conlan, K. E. & J. R. Chess, 1992. Phylogeny and ecology of a kelp-boring amphipod, *Peramphithoe stypotrupetes*, new species (Corophioidea: Amphithoidae). J. Crust. Biol. 12: 410–422.

Dayton, P. K., 1985. Ecology of kelp communities. Annu. Rev. Ecol. Syst. 16: 215–245.

De Wolf, H., T. Backelhau & R. Verhagen, 1998. Spatio-temporal genetic structure and gene flow between two distinct shell morphs of the planktonic developing periwinkle *Littorina striata* (Mollusca: Prosobranchia). Mar. Ecol. Prog. Ser. 163: 155–163.

Duffy, J. E., 1996. Eusociality in a coral-reef shrimp. Nature 381: 512–514.

Duffy, J. E. & M. E. Hay, 1994. Herbivore resistance to seaweed chemical defense: the roles of mobility and predation risk. Ecology 75: 1304–1319.

Duffy, J. E. & K. S. Macdonald, 1999. Colony structure of the social snapping shrimp *Synalpheus filidigitus* in Belize. J. Crust. Biol. 19: 283–292.

Edgar, G. J., 1992. Patterns of colonization of mobile epifauna in a Western Australian seagrass bed. J. exp. mar. Biol. Ecol. 157: 225–246.

Flach, E. C., 1992. The influence of four macrozoobenthic species on the abundance of the amphipod *Corophium volutator* on tidal flats of the Wadden Sea. Neth. J. Sea Res. 29: 379–394.

Flach, E. C. & W. De Bruin, 1994. Does the activity of cockles, *Cerastoderma edule* (L.) and lugworms, *Arenicola marina* L., make *Corophium volutator* Pallas more vulnerable to epibenthic predators: a case of interaction modification? J. exp. mar. Biol. Ecol. 182: 265–285.

Gunnill, F. C., 1982. Macroalgae as habitat patch islands for *Scutellidium lamellipes* (Copepoda: Harpacticoida) and *Amphithoe tea* (Amphipoda: Gammaridae). Mar. Biol. 69: 103–116.

Gunnill, F. C., 1983. Seasonal variation in the invertebrate faunas of *Pelvetia fastigiata* (Fucaceae): effects of plant size and distribution. Mar. Biol. 73: 115–130.

Johannesson, K, B. Johannesson & E. Rolán-Alvarez, 1993. Morphological differentiation and genetic cohesiveness over a microenvironmental gradient in the marine snail *Littorina saxatilis*. Evolution 47: 1770–1787.

Jones, L. G., 1971. Studies on selected small herbivorous invertebrates inhabiting *Macrocystis* canopies and holdfasts in southern California kelp beds. Beiheft Nova Hedwigia 32: 343–367.

Knight, A. J., R. N. Hughes & R. D. Ward, 1987. A striking example of the founder effect in the mollusc *Littorina saxatilis*. Biol. J. linn. Soc. 32: 417–426.

Menzies, R. J., 1957. The marine borer family Limnoridae (Crustacea, Isopoda). Part I: Northern and Central America: systematics, distribution and ecology. Bull. Mar. Sci. Gulf. and Carib. 7: 101–200.

Moore, P. G., 1973. The larger Crustacea associated with holdfasts of kelp (*Laminaria hyperborea*) in North-East Britain. Cah. biol. Mar. 16: 493–518.

Moore, P. G., 1986. Levels of heterogeneity and the amphipod fauna of kelp holdfasts. In Moore, P. G. & R. Seed (eds), The Ecology of Rocky Coasts. Columbia University Press, New York: 274–289.

Muñoz, A. A. & F. P. Ojeda, 1997. Feeding guild structure of a rocky intertidal fish assemblage in central Chile. Env. Biol. Fish. 49: 471–479.

Muñoz, A. A. & F. P. Ojeda, 1998. Guild structure of carnivorous intertidal fishes of the Chilean coast: implications of ontogenetic dietary shifts. Oecologia 114: 563–573.

Poulin, E. & J.-P. Féral, 1994. The fiction and facts of Antarctic brood protecting: population genetics and evolution of schizasterid echinoids. In David, B. A. Guille, J.-P. Féral & M. Roux (eds), Echinoderms Through Time, Proc. 8th Intn. Echinoderm Conf., Dijon, France. Balkema, Rotterdam: 837–843.

Poulin, E. & J.-P. Féral, 1995. Pattern of spatial distribution of a brood-protecting schizasterid echinoid, *Abatus cordatus*, endemic to the Kerguelen Islands. Mar. Ecol. Prog. Ser. 118: 179–186.

Poulin, E. & J.-P. Féral, 1998. Genetic structure of the brooding sea urchin *Abatus cordatus*, an endemic of the Subantarctic Kerguelen Island, and the origin of the diversity of Antarctic echinoids. In Mooi, R. & M. Telford (eds), Echinoderms, Proc. 9th Intn. Echinoderm Conf., San Francisco, U.S.A. Balkema, Rotterdam: 793–795.

Reusch, T. B. H., W. T. Stam & J. L. Olsen, 1999. Size and estimated age of genets in eelgrass, *Zostera marina*, assessed with microsatellite markers. Mar. Biol. 133: 519–525.

Sainte-Marie, B., 1991. A review of the reproductive bionomics of aquatic gammaridean amphipods: variation of life history traits with latitude, depth, salinity and superfamily. Hydrobiologia 223: 189–227.

Skalamera, J.-P., F. Renaud, M. Raymond & T. de Meeus, 1999. No evidence for genetic differentiation of the mussel *Mytilus galloprovincialis* between lagoons and the seaside. Mar. Ecol. Prog. Ser. 178: 251–258.

Smith, S. D. A. & R. D. Simpson,1992. Monitoring the shallow sublittoral using the fauna of kelp (*Ecklonia radiata*) holdfasts. Mar. Poll. Bull. 24: 46–52.

Stanhope, M. J., 1993. Molecular phylogeographic evidence for multiple shifts in habitat preference in the diversification of an amphipod species. Molec. Ecol. 2: 99–112.

Taylor, R. B., 1998. Short-term dynamics of a seaweed epifaunal assemblage. J. exp. mar. Biol. Ecol. 227: 67–82.

Tegner, M. J, P. K. Dayton, P. B. Edwards & K. L. Riser, 1995. Sea urchin cavitation of giant kelp (*Macrocystis pyrifera* C. Agardh) holdfasts and its effects on kelp mortality across a large California forest. J. exp. mar. Biol. Ecol. 191: 83–99.

Thiel, M., 1999. Parental care behaviour in crustaceans – a comparative overview. Crustacean Issues 12: 211–226.

Thiel, M., S. Sampson & L. Watling, 1997. Extended parental care in two endobenthic amphipods. J. nat. Hist. 31: 713–725.

Vader, W., 1968. Notes on norwegian marine amphipoda. 4. Bifurcation of the gnathopod dactylus in a specimen of *Parajassa pelagica* (Ischyroceridae). Sarsia 33: 109–112.

Vásquez, J. A. & B. Santelices, 1984. Comunidades de macroinvertebrados en discos adhesivos de *Lessonia nigrescens* Bory (Phaeophyta) en Chile central. Rev. Chil. Hist. nat. 57: 111–122.

Westinga, E. & P. C. Hoetjes, 1981. The intrasponge fauna of *Spheciospongia vesparia* (Porifera, Demospongiae) at Curacao and Bonaire. Mar. Biol. 62: 139–150.

Wilson, A. B., J. S. Boates & M. Snyder, 1997. Genetic isolation of populations of the gammaridean amphipod, *Corophium volutator*, in the Bay of Fundy, Canada. Mol. Ecol. 6: 917–923.

Hydrobiologia **440:** 55–64, 2000.
M.B. Jones, J.M.N. Azevedo, A.I. Neto, A.C. Costa & A.M. Frias Martins (eds), Island, Ocean and Deep-Sea Biology.

A sponge diversity centre within a marine 'island'

James J. Bell & David K.A. Barnes
Department of Zoology and Animal Ecology, University College Cork, Cork, Ireland
E-mail: sponge@indigo.ie

Key words: sponge diversity, Lough Hyne, sublittoral, temperate

Abstract

The exposed and gulf-stream warmed south-west coast of Ireland has a Lusitanean fauna composed of elements of the colder waters to the north and east, and others from the warmer Mediterranean Sea. Lough Hyne, a small marine body, is unusual on this coast in being very sheltered, but also in being characterised by many different niches within a small space (1 km^2). Sponges are particularly abundant, morphologically varied and more than 100 species have been described. Species diversity was measured at 6 m intervals on vertical and inclined profiles (to a maximum of 30 m) at six sites, spanning a range of flow rate and sedimentation regimes. Diversity, richness, evenness and density varied significantly with both flow regime and depth, but was much lower on the surrounding Atlantic coast. Four different sponge communities were differentiated on the basis of sponge species assemblages which correlated with different environmental conditions. At sites of turbulent and fast flow conditions, sponge diversity and richness were lowest, with the highest values being found at the sites of moderate and high sedimentation. Significant differences were observed in all four ecological variables with respect to substratum angle with the exception of the site experiencing the most turbulent flow conditions. Lough Hyne was found to possess the second highest sponge species diversity (H=3.626) and richness (77 species) of all available figures from temperate, polar and tropical areas (of similar sized sampling area). The uniqueness, diversity and species composition of the sponge community at this location suggests Lough Hyne is, biologically, a marine island within the island of Ireland.

Introduction

Islands may exist in two forms: true islands which are geographic land masses (e.g. Ireland) or conceptual 'islands' such as FADs (Fish Aggregating Devices), hydrothermal vent fields and even boulders. Lough Hyne is one such conceptual 'marine island' due to its limited exchange with the adjacent Atlantic Ocean, with currents inside the Lough only being detected during inflow of water. The conditions within Lough Hyne are extremely different from the exposed coastline outside the Lough and, as a result, so are the sublittoral cliff communities (Picton, 1990).

The intertidal and sublittoral sponge communities at Lough Hyne are known to be extremely rich with over 100 species being reported (Van Soest & Weinberg, 1980; Van Soest et al., 1981; Picton, 1990). Several sponge species found within the shallow waters of Lough Hyne have only been found elsewhere in deep-water environments (e.g. *Eurypon* sp.). This is thought to be due both to the sheltered nature of the site and the large number of niches within such a small area (1 km^2). Quantification of sponge species diversity, comparable to other marine ecosystems (see Diaz et al., 1990; Schmahl, 1990), has not been obtained previously at Lough Hyne, even though species richness is thought to be extremely high (Lilly et al., 1953; Van Soest & Weinberg, 1980; Van Soest et al., 1981; Picton, 1990). Many sponge species, including those from the genera *Polymastia* and *Eurypon*, have not yet been described with up to 8 new species from the genus *Eurypon* and 5 from *Polymastia* awaiting description (Bernard Picton, pers com.).

Specific sponge species diversity within tropical areas is generally thought to be high (Alcolado, 1990) providing Shannon Wiener diversity indices ranging from ≈ 0.5 (Diaz et al., 1990) in Venezuela to 3.39 in Southern Florida (Schmahl, 1990). Published sponge species diversity values from Polar regions have not been reported, although sponge species richness appears to be high (Sarà et al., 1992), area effects, however, need to be considered. Sponges are known to form a conspicuous part of the sessile fauna in many sublittoral temperate communities such as Lundy Island (Hiscock et al., 1983) and Strangford Lough (Erwin, 1976). However, these sites are much larger than Lough Hyne. Quantitative estimates of sponge species diversity within sublittoral temperate regions have not been produced. The reason for so few studies providing quantitative information of this nature may be due to the problems of field recognition of sponge species (Ackers & Moss, 1987; Bell & Barnes, unpublished data) and the time required for such surveys.

Water flow is important to many suspension feeding organisms including sponges (Vogel, 1974, 1977, 1978). It is thought that inhalant water flow through sponges is enhanced by increased ambient flow (Vogel, 1977). Wave action and current speed is also believed to be important as has been shown by transplant experiments, resulting in slower growth of sponges in weak currents (Wilkinson & Vacelet, 1979), and increased stiffness and lower profiles in wave exposed areas (Palumbi, 1984). Sponges have also been found to show morphological variation in relation to flow and depth at Lough Hyne (Bell & Barnes, 2000) with more delicately branching sponges being found where disturbance is moderate to low.

This study details sponge species diversity, richness, evenness and density from an area of high sponge diversity in a temperate region. This information is compared with data from both the literature and those collected by the present authors for sponge communities found within tropical, polar and other temperate ecosystems. This study also investigates changes in the aforementioned population statistics within Lough Hyne as a function of site, flow regime and depth.

Study site

Lough Hyne (Fig. 1) is a small ($\approx$1 km^2) semi-enclosed sea lough on the south-west coast of Ireland (51° 29′ N, 9° 18′ W) attached to the adjacent Atlantic coast by a narrow ($\approx$25 m wide) tidal rapids. Water flow into the lough is essentially unidirectional due to a sill in the rapids, meaning the incoming tide must reach the level of the sill before inflow of water can begin. This results in inflow lasting only four hours with currents reaching >35 cms^{-1} and outflow lasting eight hours when currents are slight (>5 cms^{-1}). As the in-flowing water moves across the lough, current speed decreases, resulting in a sedimentation gradient from east to west (Bassindale et al., 1957). Five sites within Lough Hyne were sampled: Whirlpool Cliff, Glannafeen, Goleen, West Cliff and Labhra Cliff and one, Bullock Island, on the adjacent Atlantic coast (Fig. 1). Bullock Island is subject to severe wave action especially during strong winds and winter storms, therefore, encrusting organisms are dominant at this site. Whirlpool Cliff is subjected to strong but predictable unidirectional flow conditions (during inflow), hence the environment is favourable to many organisms. Glannafeen lies between the extremes of the sedimentation gradient and, therefore, experiences moderate current flow (<15 cms^{-1}) which leads to considerable sedimentation upon cliff surfaces. Cliff surfaces at Goleen, Labhra and West Cliff are highly sedimented as only slight water currents (<5 cms^{-1}) move across these cliff surfaces. The sublittoral communities, including sponges found upon the cliff sites within Lough Hyne, are described in detail by Picton (1990).

Materials and methods

Ten 0.25 m^2 quadrats were sampled at 6 m depth intervals on both vertical ($\approx$ 90°) and inclined ($\approx$ 45°) surfaces at each of the sites. The number of each species of sponge in each quadrat was recorded. Photographs and samples were taken of sponges that could not be identified *in situ* for laboratory analysis. Patterns of species diversity were described with the Shannon and Weiner (Krebs, 1989) information function $H' = -\Sigma p_i \ln p_i$. Paired *t*-tests were used to examine differences between rock surface types.

Bray–Curtis similarity analysis, using hierarchical agglomerative group average clustering, was performed upon data using the unweighted pair group method using arithmetic averages (UPGMA) using the PRIMER program (Plymouth Marine Laboratory). Data were log (x+1) transformed to reduce the importance of extreme values. As sites were sampled to different depths. All data on each surface type from each depth sampled were combined and then standardised by dividing by total abundance at each site, to account for different population sizes. Ordination of samples was performed using Correspondence Analysis upon untransformed data, but standardisation, as used for Bray Curtis Analysis, was performed upon species composition data.

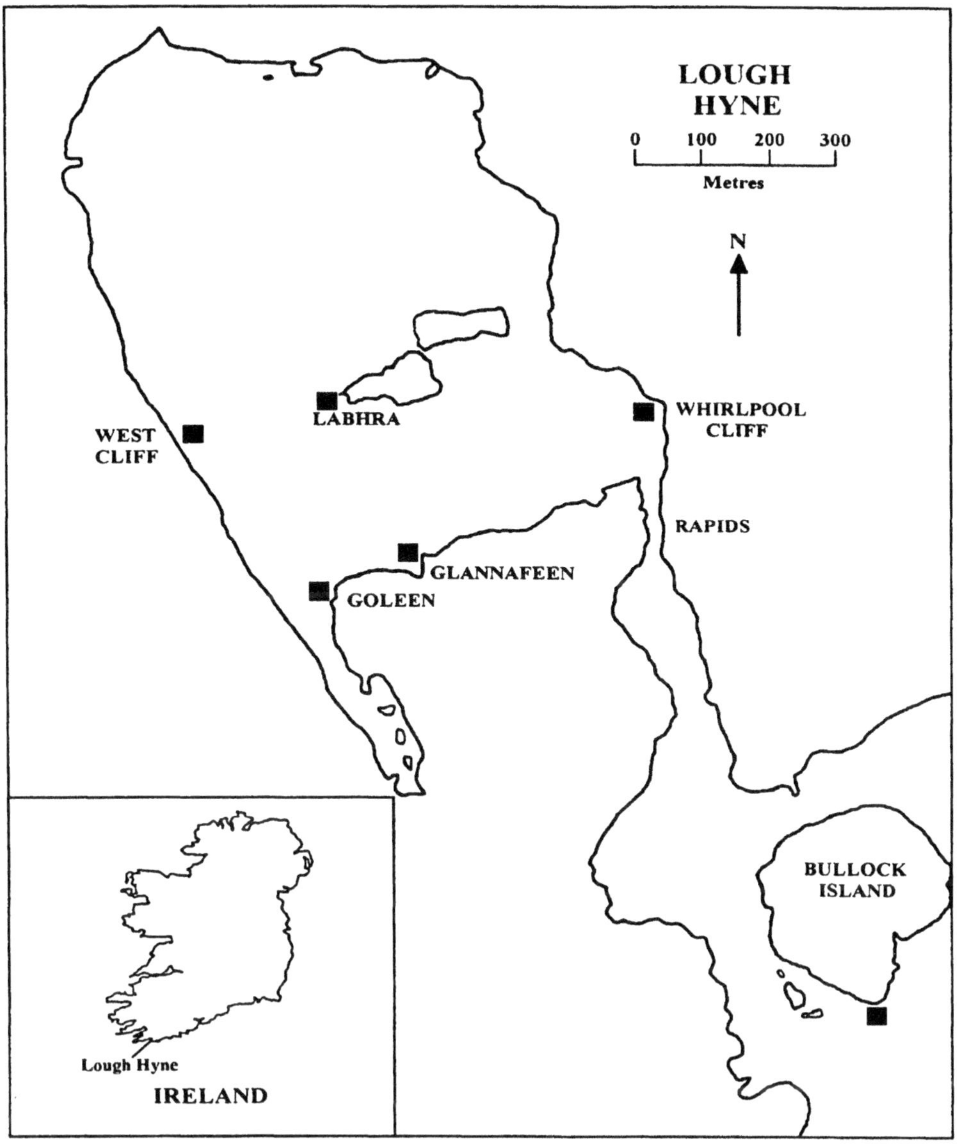

Figure 1. The location of Lough Hyne Marine Nature Reserve in Ireland (Long 51° 29′ N, Lat 9° 18′ W), with details of study sites where sponge species diversity and richness was sampled.

Results

Seventy seven species of sponge were identified at Lough Hyne and along the adjacent Atlantic coast (Table 1), of which only one species (*Plakortis simplex*) has not been identified previously from Lough Hyne. Twenty-eight of the 77 species were found on the Atlantic coast (Bullock Island) of which only three species were exclusive to this site (*Iophonopsis nigricans*, *Leuconia nivea* and *Haliclona* sp 5). Species diversity, richness, evenness and sponge density varied with both depth and flow regime at Lough Hyne MNR (Figs 2a–d).

At Bullock Island (Fig. 2a), sponge species diversity and evenness were moderate compared to the other five sites sampled. However, sponge richness and density were the lowest of the six sites sampled. All four variables remained relatively constant with depth. Sponge species diversity, evenness, richness and density showed no significant variation between surface type ($p>0.05$ for each t-value). Sponge density at this site was low, with only 20 sponges m^{-2}

Table 1. Sponges found at Lough Hyne Marine Nature Reserve and upon the adjacent Atlantic coast. (* species found within Lough Hyne and on the surrounding Atlantic coast. ** species found exclusively on the adjacent Atlantic coast)

*Antho involvens**	*Iophon ingalli**
Aplysilla rosea	*Iophonopsis nigricans***
*Aplysilla sulfurea**	*Laxosuberites incrustans*
Axinella damicornis	*Leucosolenia complicata***
Axinella dissimilis	*Leuconia nivea***
Biemna variantia	*Mycale contarenii*
Clathrina coriacea	*Mycale macilenta*
*Cliona celata**	*Mycale rotalis*
*Dysidea fragilis**	*Myxilla rosacea**
Dysidea pallescens	*Myxilla incrustans*
*Esperiopsis fucorum**	*Myxilla fimbriata*
Eurypon sp 1	*Pachymatisma johnstonia**
Eurypon sp 2	*Paratimea constellata**
Eurypon sp 3	*Phakellia* sp
Eurypon sp 4	*Phorbas fictitius**
Eurypon sp 5	*Plakortis simplex*
Eurypon sp 6	*Polymastia* sp 1
Eurypon sp 7	*Polymastia* sp 2
Eurypon sp 8	*Polymastia* sp 3*
Halichondria bowerbanki	*Polymastia* sp 4*
*Halichondria panicea**	*Polymastia* sp 5
Haliclona cinerea	*Polymastia* sp 6
*Haliclona fistulosa**	*Polymastia* sp 7
Haliclona simulans	*Polymastia* sp 8
Haliclona sp 1*	*Pseudosuberites sulphureus**
Haliclona sp 2*	*Rhaphidostyla kitchingi*
Haliclona sp 3*	*Raspailia hispida*
Haliclona sp 4	*Raspailia ramosa*
Haliclona sp 5**	*Scypha cilatum**
Haliclona urceolus	*Stelligera rigida**
Haliclona viscosa	*Stelligera stuposa**
Halicnemia patera	*Suberites carnosus*
Halisarca sp	*Suberites* sp 1
*Hemimycale columella**	*Suberites* sp 2*
*Hymedesmia paupertus**	*Suberites* sp 3*
*Hymeniacidon perleve**	*Suberites* sp 4*
Hymeraphia stellifera	*Tethya aurantium**
Iophon hyndmani	

being found at each depth. Whirlpool Cliff (Fig. 2a) showed variation in sponge diversity, evenness and richness with depth. All three variables remained relatively constant with depth until approximately 12 m, after which they increased. Sponge species diversity and richness (Fig. 2a) was found to be significantly higher on vertical surfaces ($p>0.05$ for each t-value), however, sponge density was significantly higher upon inclined surfaces. Density of sponges at Whirlpool Cliff remained constant with depth, with a mean of 50 sponges m^{-2} being found upon vertical surfaces and 100 sponges m^{-2} on inclined surfaces. Species diversity was low at this site, but species richness was moderate.

At Glannafeen and Goleen (Fig. 2b), the relationships between species diversity, richness, evenness and density with depth were very similar at both sites: higher values occurred on vertical than on inclined surfaces and all increased with depth reaching a maximum at 18 m. However, only sponge species diversity, richness and density are significantly higher ($p<0.05$ for each t-value) on vertical than inclined surfaces. Evenness increased from 0 to 6 m and then remained relatively constant. Sponge ecology at West Cliff (Fig. 2c) showed similar trends to those found at Goleen and Glannafeen, except that this cliff extended to 24 m, rather than 18 m at Goleen and Glannafeen. At West Cliff, sponge species diversity, richness and density increased to 18 m then became constant, but evenness declined slightly at 12 m. Species richness at West Cliff was the highest ($\approx$40 species) of all the sites sampled (at a depth of 24 m). Of the four variables, only sponge density was significantly ($p<0.05$ for each t-value) higher on vertical surfaces. At Labhra Cliff, which extended to 30 m, the same increases with depth occurred with sponge species diversity, evenness, richness and density initially. However, a sharp decrease in all measured variables occurred between 24–30 m (e.g. from $\approx H=3$ to $H=1.5$). Higher variables also occurred on vertical surfaces [although only sponge density was significantly ($p>0.05$, $t=3.55$) higher], which was less pronounced with depth. The highest sponge diversity ($H=\approx3$), richness ($\approx$ 30 species) and density ($\approx$150 sponges m^{-2}) at this site was observed at a depth of 24 m.

Sponge species diversity at Bullock Island was moderate but richness was the lowest of the six sites sampled (both of which remained constant with depth). The pattern at Whirlpool Cliff was the converse with moderate species richness and low diversity. At Whirlpool Cliff, variation in diversity and richness with depth was evident. There was little difference in species diversity and richness between the remaining four sites (Goleen, Glannafeen, West Cliff and Labhra), with the exception of West Cliff which had a higher species richness than the other three sites.

Diversity and richness values at 12 m depth at the six sites (Table 2) showed the highest diversity occurring at Glannafeen where sedimentation and water

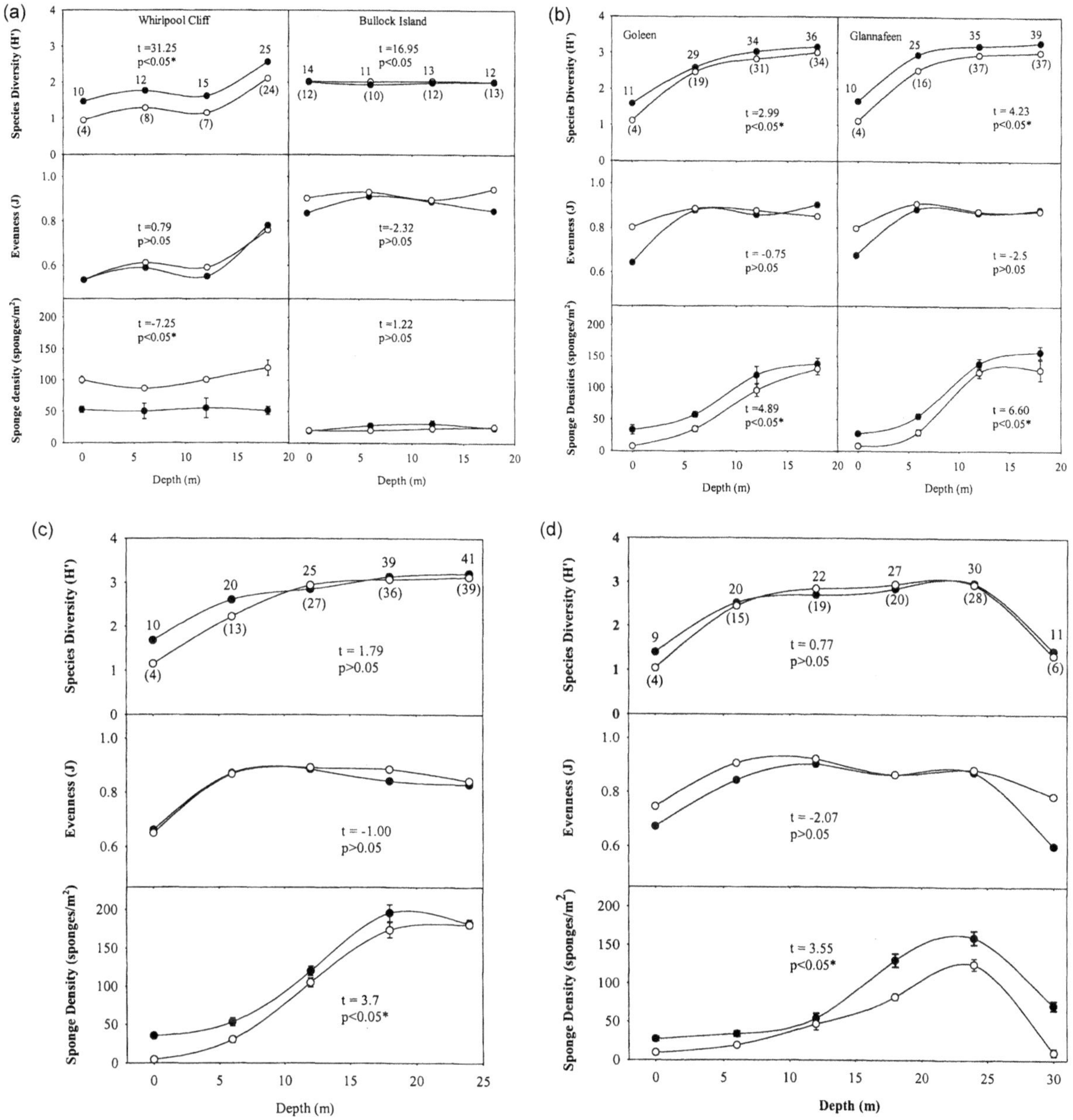

Figure 2. (a–d) Sponge species diversity (Shannon–Weiner), evenness, richness and density with depth at Bullock Island & Whirlpool Cliff (a), Goleen & Glannafeen (b), West Cliff (c) and Labhra Cliff (d). Filled circles represent data from vertical surfaces and open circles from inclined surfaces. Significant differences ($p<0.05$) between vertical and inclined surfaces are indicated by an asterisk (*). Numbers on species diversity graphs indicate species richness at each depth, with bracketed numbers representing communities from inclined surfaces and non-bracket numbers from vertical surfaces.

movement is moderate (authors, unpublished data). The lowest diversity occurred at Whirlpool Cliff where flow was strong and unidirectional, but sedimentation was extremely slight (authors, unpublished data). Moderate diversity and richness was found at Bullock Island where turbulent conditions prevail and sedimentation was not observed. An overall sponge diversity of 3.626 and richness of 77 species (at a depth of 12 m) was calculated within Lough Hyne to allow comparison with other global sites where these variables have been measured.

Table 2. Sponge species diversity (H') and richness at six sites sampled on both vertical (*) and inclined (†) surfaces within Lough Hyne at 12 m

Site	Diversity		Richness		Current flow	Sedimentation	Disturbance
Bullock Island	2.021*	2.002†	10*	11†	Turbulent	Low	Very High
Whirlpool Cliff	1.8010*	1.192†	10*	7†	Fast Unidirectional	Low	High
Glannafeen	3.211*	2.912†	34*	31†	Moderate Unidirectional	Moderate	Moderate
Goleen	3.056*	2.823†	34*	31†	Slight	High	Low
West Cliff	2.789*	2.874†	27*	29†	Slight	High	Low
Labhra Cliff	2.798*	2.895†	21*	18†	Slight	High	Low

Four different sponge communities were identified using Bray–Curtis similarity analysis (Fig. 3a). (1) Sponges on vertical and inclined surfaces at Bullock Island, (2) sponges on vertical and inclined surfaces at Whirlpool Cliff, (3) sponges on vertical surfaces and (4) inclined surfaces at Glannafeen, Goleen, West Cliff and Labhra Cliff. Clustering of inclined and vertical surfaces at Whirlpool Cliff and Bullock Island showed a lower similarity (≈60%) than did the sites and surfaces within other sponge communities identified (>70%). Correspondence Analysis (Fig. 3b) showed a similar pattern to that of Bray Curtis Similarity Analysis, with the same four communities being identified. Species compositions and their corresponding abundances were significantly dependent upon site (Chi-square 13192, $p<0.05$, d.f. 825). The first three Eigen Values explain 36%, 29% and 7% of the total inertia respectively, and therefore the initial triplot (Fig. 3b) explains 72% of the total inertia.

Discussion

Sponges are known to be important contributors to most subtidal marine benthic communities spanning tropical reefs (Alvarez et al., 1990; Diaz et al., 1990; Schmahl, 1990), Polar regions (Sarà et al., 1992), the deep sea (Barthel et al., 1991; Barthel & Tendal, 1993) and hard bottom temperate communities (Lilly et al., 1953; Erwin, 1976; Hiscock et al., 1983). The reasons for such widespread occurrence of sponges is due to a number of factors including: morphological plasticity (Palumbi, 1984, 1986), efficient filter feeding (Pile et al., 1996), symbiotic associations (Vacelet & Donadey, 1977; Burlando et al., 1988), toxicity to other marine organisms (Porter & Targett, 1988; Huysecom et al., 1990) and fast regenerative capabilities (Ayling, 1983).

The density of sponges found in the present study (up to 200 sponges m^{-2}) has shown sponges to be important contributors to the sublittoral cliffs at Lough Hyne. However, variation between sites experiencing different flow regimes and sedimentation was evident. At the site experiencing the most turbulent conditions (Bullock Island), diversity was not the lowest of the sites sampled (Table 2) as might have been expected, with the lowest density occurring at Whirlpool Cliff. The most likely reason for this apparent discrepancy is that fast flowing conditions at Whirlpool Cliff, with an almost total absence of wave action, means conditions for growth are favourable for many different types of organisms (including anthozoans and bryozoans). Overall community diversity is high at Whirlpool Cliff (Maughan & Barnes, unpublished data) and so is the total coverage of organisms (Hammer, 1996). The moderate level of disturbance probably maintains the high community diversity at this site. Sponges are generally effective competitors in benthic communities, however, they are often overgrown by ascidians, soft corals and some species of algae (Russ, 1982; Quinn, 1982; Nandakumar, 1995), all of which are

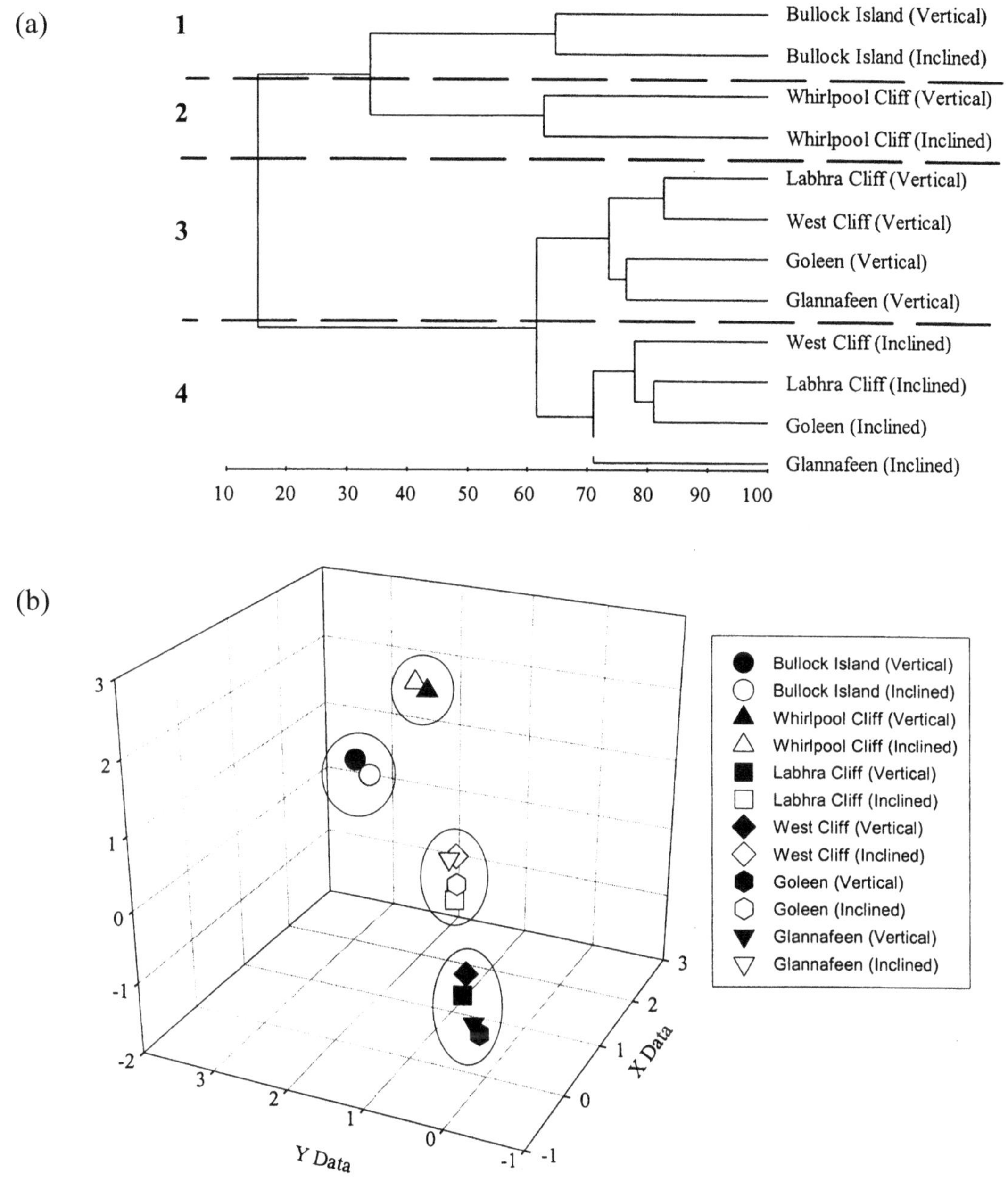

Figure 3. Bray–Curtis Similarity (a) and Correspondence Analysis (b) for sublittoral sponge populations on vertical and inclined surfaces at six sites of Lough Hyne.

very common at Whirlpool Cliff (authors' pers. obs.). We suggest space acquisition by superior competitors and lack of available space for colonisation causes sponge diversity to be low compared with other sites sampled. Although sponges do not appear most diverse in conditions which seem most favourable, this becomes less of a paradox when the entire sublittoral community is considered. At Whirlpool Cliff, sponge species diversity, richness and evenness were found to increase between 12 and 18 m. Current flow rate decreases between these depths (authors' pers. obs.), resulting in a reduction in disturbance, enabling more delicate species (including branching species), with smaller basal attachment areas (less space required) to become more abundant (Bell & Barnes, 2000). At Bullock Island, sponge species diversity, richness, evenness and density are all relatively constant with depth. The reason for this is that disturbance is high at all depths sampled and, despite some reduction in disturbance with depth, it is still high at all depths. Wave action is known to affect sponge morphological types (Palumbi, 1984, 1986), where many of the more delicately branching species are destroyed by drag and tissue fatigue in high energy environments (Hiscock, 1983). Therefore, limits of morphological adaptation for any particular sponge species may reduce species richness at sites of high wave exposure (e.g. Bullock Island), with encrusting and robust species being more suited to this environment. Evenness remained constant with depth at Bullock Island, showing no one species dominated at any of the depths sampled. At Whirlpool Cliff, evenness remains low (<0.6) until 12 m, suggesting that only a few species dominate (richness values of 10 species on vertical and <5 on inclined surfaces). Below 12 m, evenness then increases showing no single species to dominate.

At West Cliff, Labhra Cliff and Goleen, where currents are slight and sedimentation is high, sponge species diversity, richness, evenness and density all increased with depth. The low sponge community composition in shallow water is most likely due to fast growing calcareous algae, which smother sponges and other sessile organisms. As depth increases, so does sedimentation (Bell & Barnes, 2000) and hence light penetration is reduced. Sponges then become a more conspicuous part of the sessile fauna. Sponges become more branched with depth to prevent sediment accumulation upon their surfaces as is seen with other sessile organisms (Chappell, 1980). Water currents at the Glannafeen site are moderate (<15 cm s^{-1}) as is sedimentation, and sponge communities are very similar to the highly sedimented sites of West Cliff and Labhra Cliff. The Bray Curtis and Correspondence analyses performed in the present study indicated that only three different sponge communities existed within Lough Hyne and one outside on the adjacent Atlantic coast. The sudden decrease (between 24 and 30 m at Labhra Cliff) which occurred in all four variables may be due to a seasonal thermocline that develops within the deep basin on the western side of the Lough (Kitching et al., 1976; Kitching, 1987). The thermocline develops at approximately 24 m during summer months, below which an anoxic layer forms which kills off most of the organisms below this depth. Many subtidal sponges are slow growing (Ayling, 1983; Fowler & Laffoley, 1993) and so cannot recolonise cliff surfaces between the annual formation of the thermocline. Slight currents at this site may also be responsible for slow growth rates (Wilkinson & Vacelet, 1979). Some sponges are present below the thermocline, namely those from the genus *Eurypon*, which appear to be able to survive seasonal immersion in this anoxic layer. Sponges which may survive heat stress, by means of a smaller resistant stage (Gaino et al., 1996), may employ similar strategic solutions to survive anoxic stress. At Labhra Cliff, West Cliff, Goleen and Glannafeen, evenness is low at the surface (≈0.6), suggesting that several fast growing sponges dominate, which can compete with the fast growing calcareous algae. Evenness increases quickly between 0 and 6 m, meaning no single species dominates as depth increases.

Table 3. Global values for sponge species diversity (H') and richness. All values shown are taken from a depth of ≈ 12 m over similar sized sampling areas on hard substrata

Location	Species diversity	Species Richness	Source
Mozambique (Quirimba)	4.07	73	Bell & Barnes (unpub.)
Ireland (Lough Hyne)	3.63	77	Bell & Barnes (unpub.)
Southern Florida	3.23	43	Schmahl (1990)
Cuba (Havana)	2.86	80	Alcollado (1990)
Panama (San Blas)	2.45	17	Bell & Barnes (unpub.)
Britain (Sussex)	2.22	12	Bell & Barnes (unpub.)
Antarctica (Signy)	2.29	19	Bell & Barnes (unpub.)
Ireland (Cork)	2.12	13	Bell & Barnes (unpub.)
Cape Verde (Sal)	1.94	9	Bell & Barnes (unpub.)
Britain (Cornwall)	1.40	6	Bell & Barnes (unpub.)
Venezuela	≈1.00	15	Diaz et al. (1990)

At Goleen, Glannafeen and Whirlpool Cliff, species diversity was higher on vertical than inclined surfaces. At Goleen and Glannafeen, this probably relates to the differences in sedimentation falling upon the different surface types, with larger amounts settling upon inclined surfaces. Increased sediment is thought to reduce water pumping rates in sponges which is likely to be detrimental to the animal (Gerrodette & Flechsig, 1979). Other effects of sediment upon sponges are currently unknown. Sponge morphology is know to be related to the degree of sedimentation and flow regime, but whether these variables affect other aspects of sponge ecology still remains undiscovered. At Whirlpool Cliff, we suggest the higher diversity on vertical surfaces results from decreased algal populations (not quantified) on vertical surfaces, relating to reduction in light intensity. No significant differences were found between species diversity upon vertical and inclined surfaces at Labhra Cliff or West Cliff. The reason for this is unclear, but may relate to differences in the type of sediment falling upon cliff surfaces at the different sites. Labhra Cliff and West Cliff are further along the sedimentation gradient than Goleen or Glannafeen, meaning larger particles will fall out of suspension at the latter two sites. The smaller more cohesive particles may be falling out of suspension at West Cliff and Labhra Cliff, and settle on vertical, as well as inclined surfaces, making sedimentation rates on both surface types high.

In shallow sublittoral sponge communities in Cuba (Acolado, 1990), Southern Florida (Schmahl, 1990), and Venezuela (Diaz et al., 1990), sponge diversity also increased with depth, with maxima occurring between 15 and 30 m. However, such values in tropical reefs, where siltation is not excessive, differ from Lough Hyne in which maximum diversity occurs in highly sedimented areas (Table 2). In tropical reef areas, structural complexity appears to be more important to sponge diversity (Diaz et al., 1990). When sponge species diversity and richness at Lough Hyne is compared with other global locations (Table 3), it has the second highest species diversity and richness of all the sites from which data are available. Care should be taken when comparing the figures for sponge species diversity and richness from Lough Hyne with other future studies, as the size effect of study sites should not be overlooked. The high diversity and richness of sponges at Lough Hyne, compared to tropical (e.g. Mozambique), Antarctic (e.g. Signy) and temperate regions, such as its adjacent Atlantic coastline, makes Lough Hyne a true 'Marine Island' with respect to sponges.

Acknowledgements

Work at Lough Hyne was carried out under permission of Declan'O Donnell from the Irish Wildlife Service (Permit No. R6/99). Enterprise Ireland provided financial assistance. Fieldwork was made possible by the help of Claire Shaw to whom we express gratitude. Thanks also go to Amy Greenwood and Bob McNamara for technical support and assistance during fieldwork. Sponge identification was aided by Bernard Picton (Ulster Museum, Belfast) to whom we also owe our appreciation.

References

Ackers, R. G. & D. Moss, 1987. Current problems with field recognition and species differentiation in shallow water British sponges. In Earll, R. & S. Gubbay (eds), Marine Recordings. Marine Conservation Society, U.K. 2: 22–27.

Alcolado, P. M., 1990. General features of Cuban sponge communities. In: Rutzler, K. (ed.), New Perspectives in Sponge Biology. Smithsonian Institute Press: 351–357.

Alvarez, B., M. C. Diaz & R. A. Laughlin, 1990. The sponge fauna on a fringing coral reef in Venezuela, I: composition, distribution and abundance. In Rützler, K. (ed.), New Perspectives in Sponge Biology. Smithsonian Institute Press: 358–366.

Ayling, A., 1983. Growth and regeneration rates in thinly encrusting demospongia from temperate waters. Biol. Bull. 165: 343–252.

Barthel, D & O. S. Tendal, 1993. The association of the abyssal Norwegian-Greenland Sea. Species composition, substrate relationships and distribution. Sarsia 78: 83–96.

Barthel, D., J. Gutt & O. S. Tendal, 1991. New information on the biology of Antarctic deep water sponges derived from underwater photography. Mar. Ecol. Prog. Ser. 69: 303–307.

Bassindale, R., E. Davenport, F. J. Ebling, J. A. Kitching, M. A. Sleigh & J.F. Sloane, 1957. The ecology of Lough Hyne rapids with special reference to water currents. VI. Effects of the rapids on the hydrography of the south basin. Ecology 45: 879–900.

Bell J. J. & D. K. A. Barnes, 2000. Morphological variation in sublittoral sponge populations in relation to flow regime at Lough Hyne MNR. J. Mar. Biol. Ass. U.K. 80, 707–718.

Burlando, B., M. A. Sabatini & E. Gaino, 1988. Association between calcareous Clathrina cerebrum (Haeckel) and bacteria: electron microscope study. J. Exp. Mar. Biol. Ecol. 116: 35–42.

Chappell, J., 1980. Coral morphology, diversity and reef growth. Nature 286: 249–252.

Diaz, M. C., B. Alvarez & R. A. Laughlin, 1990. The sponge fauna on a fringing coral reef in Venezuela, II: community structure. In Rützler, K. (ed.), New Perspectives in Sponge Biology. Smithsonian Institute Press: 367–375.

Erwin, D. G., 1976. A diving survey of Strangford Lough: the benthic communities and their relation to substrate – a preliminary account. In Keegan, B. F., Ceidigh, P. O. & P. J. S. Boaden (eds), Biology of Benthic Organisms: 215–224.

Fowler, S. & D. Laffoley, 1993. Stability in Mediterranean-Atlantic sessile epifaunal communities at the northern limits of there range. J. Exp. Mar. Biol. Ecol. 172: 109–127.

Gaino, E., G. Bavestrello, C. Cerrano & M. Sarà, 1996. Survival of the calcareous sponge *Clathrina cerebrum* (Haeckel, 1872) on a vertical cliff during the summer crisis. Ital. J. Zoo. 63: 41–46

Gerrodette, T. & A. O. Flechsig, 1979. Sediment-induced reduction in the pumping rate of the tropical sponge *Verongia lacunosa*. Mar. Biol. 55: 103–110

Hammer, J., 1996. Competition and grazing as processes structuring a sessile, subtidal community. Unpublished MSc. Thesis. University of Wales, Bangor.

Hiscock, K., 1983. Water movement. In Earll, R & D. G. Erwin (eds), Sublittoral Ecology. The Ecology of the Shallow Sublittoral Bethos. Clarendon Press, Oxford: 58–96.

Hiscock, K., S. Stone & J. D. George, 1983. The marine fauna of Lundy. Porifera (sponges): a preliminary study. Rep. Lundy Field Soc. 34: 16–35.

Huysecom, J., G. Van de Vyer, J.C. Braekman & D. Dalzone, 1990. Chemical defence in sponges from north Brittany. In Rutzler, K. (ed.), New Perspectives in Sponge Biology. Smithsonian Institute Press: 115–118.

Kitching, J. A, F. J. Ebling, J. C. Gamble, R. Hoare, Q. R. McLeod & T. A. Norton, 1976. The Ecology of Lough Ine XIX. Seasonal changes in the Western Trough. J. Anim. Ecol. 45: 731–758.

Kitching, J. A., 1987. Ecological studies at Lough Hyne. Adv. Ecol. Res. 17: 115–185.

Krebs, C. J., 1989. Ecological methodology. Harper Collins. London: 328–369.

Lilly, S. J., J.F. Sloane, R. Bassindale, F. J. Ebling & J. A. Kitching, 1953. The ecology of Lough Ine rapids with special reference to water currents. IV The sedentary fauna of sublittoral boulders. J. Anim. Ecol. 22(1): 87–122.

Lilly, S. J., F. Sloane, R. Bassindale, F. J. Ebling & J. A. Kitching, 1953. The ecology of the Lough Hyne rapids with special reference to water currents. IV. The sedentary fauna of sublittoral boulders. J. Anim. Ecol. 22: 87–122.

Nandakumar, K., 1995. Competitive interactions among sessile organisms in Tomioka Bay, south Japan: importance of light conditions on the panel surface. Mar. Biol. 121: 713–719.

Palumbi, S. R., 1984. Tactics of acclimation: morphological changes of sponges in an unpredictable environment. Science 225: 1478–1480.

Palumbi, S. R., 1986. How body plans limit acclimation: responses of a demosponge to wave force. Ecology 67 (1): 208–214.

Porter, J. W. & N. M. Targett, 1988. Allelochemical interactions between sponges and corals. Biol. Bull. 175: 230–239.

Picton, B. E., 1990. The sessile fauna of sublittoral cliffs. In Myers, A. A., C. Little, M. J. Costello & J. C. Partridge (eds), Theecology of Lough Hyne: Proceeding of a Conference 4–5th Sep 1990. Royal Irish Academy: 139–142.

Pile, A. J., M. R. Patterson & J.D. Witman, 1996. *In situ* grazing on plankton <10 μm by the boreal sponge *Mycale lingua*. Mar. Ecol. Prog. Ser. 141: 95–102.

Quinn, J.F., 1982. Competitive hierarchies in marine benthic communities. Oecologia 54: 129–135.

Russ, G. R., 1982. Overgrowth in a marine epifaunal community: competitive hierarchies and competitive networks. Oecologia 53: 12–19.

Sarà, M., A. Balduzzi, M. Barbieri, G. Bavestrello & B. Burlando, 1992. Biogeographic traits and checklist of Antarctica demosponges. Polar Biol. 12: 559–585.

Schmahl, G. P., 1990. Community structure and ecology of sponges associated with four southern Florida coral reefs. In Rutzler, K. (ed.), New Perspectives in Sponge Biology. Smithsonian Institute Press: 376–383.

Vacelet, J. & C. Donadey, 1977. Electron microscope study of the association between some sponges and bacteria. J. Exp. Mar. Biol. Ecol. 116: 35–42.

Van Soest, R. M. W. & S. Weinberg, 1980. A note on the sponges and octocorals from Sherkin Island and Lough Ine, Co. Cork. Irish Nat J. 20 (1): 1–15.

Van Soest, R. M. W., Guiterman, J.D. & M.S., Sayer, 1981. Sponges from Roaringwater Bay and Lough Ine J. Sherkin Island 12: 35–49.

Vogel, S., 1974. Current induced flow through the sponge, Halichondria. Biol. Bull. 47: 443–456.

Vogel, S., 1977. Current-induced flow through living sponges in nature. Proc. Natl. Acad. Sci. U.S.A. 74(5): 2069–2071.

Vogel, S., 1978. Organisms that capture currents. Sci. Amer. 239(2): 128–139.

Wilkinson, C. R. & J. Vacelet, 1979. Transplantation of marine sponges to different conditions of light and current. J. Exp. Mar. Biol. Ecol. 37: 91–104.

Hydrobiologia **440:** 65–80, 2000.
M.B. Jones, J.M.N. Azevedo, A.I. Neto, A.C. Costa & A.M. Frias Martins (eds), Island, Ocean and Deep-Sea Biology.

An offshore buoy as a small artificial island and a fish-aggregating device (FAD) in the Mediterranean

Giulio Relini[1], Marco Relini[1] & Manuela Montanari[2]
[1]*DIP.TE.RIS., University of Genoa, Via Balbi 5, 16126 Genova, Italy*
[2]*Istituto per la Corrosione Marina dei Metalli ,CNR, Via de Marini 6, 16149 Genova, Italy*
E-mail: biolmar@unige.it

Key words: macrofouling, settlement, offshore, buoy, fishes, FAD, Ligurian Sea

Abstract

The ODAS Italia 1 oceanographic buoy is moored in the Ligurian Sea, 37 nm from Genoa, along the Genoa-Cape Corse transect (43° 48.90′ N–09° 06.80′ E), over a 1270 m deep sea bottom. The underwater portion of the buoy is 37 m long and 0.60 m in diameter, acting as a small island for colonization of fouling organisms and as a fish-aggregating device (FAD). The role of the buoy in attracting and maintaining fish assemblages was investigated by visual censuses in different seasons at depths of 0–40 m. Fish from seven families, comprising 12 species, of which three are benthic, were recorded with maximum abundance in summer. Fouling was studied from samples collected on the buoy and on immersed panels. The fouling community of the buoy consisted of 34 algae and 100 animal species, including three fish. The settlement processes of the fouling community on the panels, in particular on those exposed for over 70 months at 12 m and 33 m depth, are described based on counts of settled organisms, the covering index of each taxa and biomass assessments. On the panels, 63 species were identified. The fouling biomass, on the panel submerged for 70 months, assessed as wet weight, reached 2.8 kg/m^2 at 12 m depth and 4.8 kg/m^2 at 33 m depth.

Observations of benthic organisms settled directly on the buoy were made between 1988 and 1989 and when the buoy was retrieved and brought back to shore on April 15, 1991 after 52 months at sea. At this time, the fouling community along the full 37 m length of the buoy was sampled, and 91 taxa, including 83 species, were identified. Several of the species present on the buoy are shallow, coastal species, some with a very short larval period. Possible ways of colonization by such species are discussed. Despite seasonal changes, the pelagic fish community was more stable over the period of 11 years of study than the benthic community settled on the buoy (that is still developing).

Introduction

Artificial substrata immersed for long periods in offshore waters offer many opportunities to investigate both benthic and nectonic organisms. The substratum, in our case a floating one, represents an artificial island and is colonized by coastal species transported by sea currents or through other dispersal strategies. The artificial island is used by pelagic necton in different ways: as shelter (similar to the function generally offered by floatsam); as a reference point during large displacements and as a moving pelagic association (when the current is strong, the moored substratum behaves like a large swimming pelagic organism). These effects of the artificial structure on fish behaviour are grouped together under the name of FAD (fish-aggregating device). Most of the Mediterranean papers on offshore fouling generally deal with platforms in neritic waters (Relini, 1993; Relini & Relini, 1994). Fouling of floating or suspended objects and structures such as oceanographic equipment (Tsikhon-Lukanina et al., 1977; Aliani et al., 1998; Aliani & Meloni, 1999) has been described only after short exposure periods. Papers on macrofouling of buoys are also relatively rare (Redfield & Hutchins, 1952; Thompson, 1977 – quoted by Haderlie, 1984; Yan et al., 1994).

More data are available on fishery FADs such as ‘kannizzati’, ‘canizzi’ or ‘capcers’: these are floats

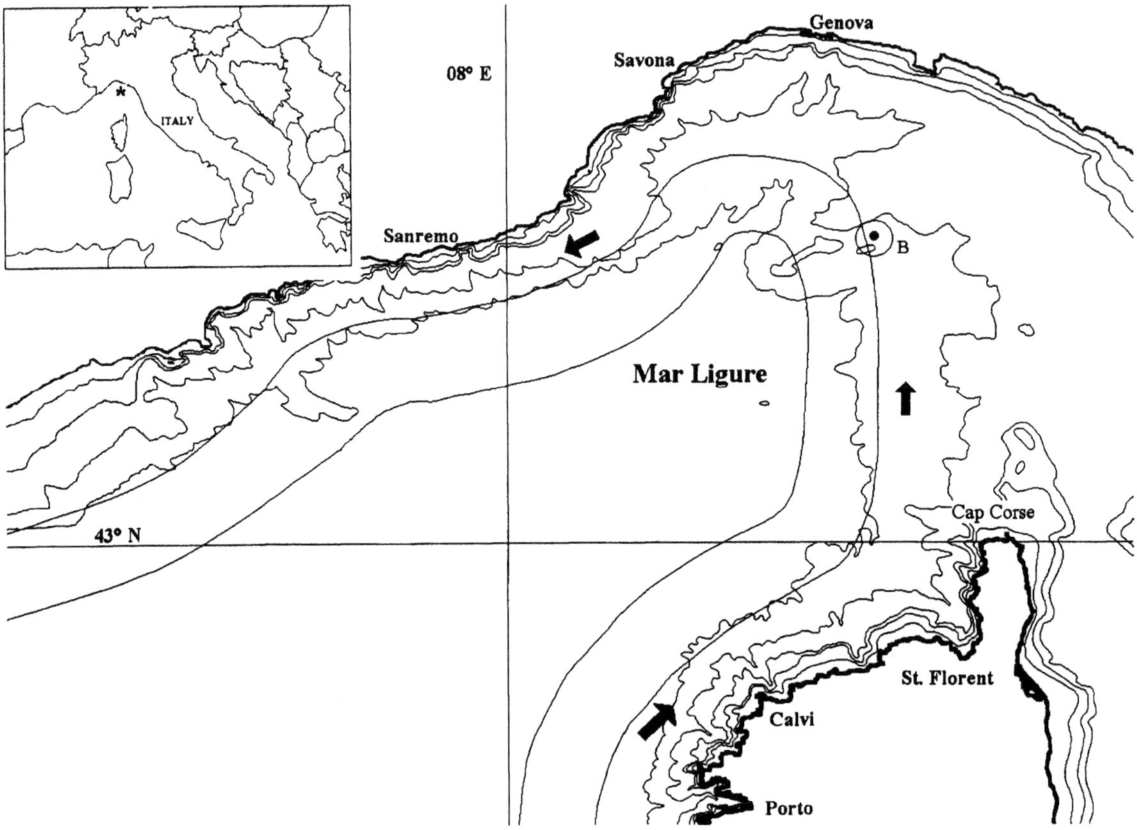

Figure 1. The site of mooring of the buoy ODAS Italia 1 in the Ligurian sea at 43°48.90′ N, 09°06.80′ E, 37 nm from Genoa, on a 1270 m deep sea bottom. The arrows show the main direction of the Corso-Ligurian-Provençal current.

consisting of large assemblies of cork slabs, palm leaves or cork and lentisk branches, also moored in deep water in the Mediterranean (Galea, 1961; Morales-Nin & Massuti, 1992; Potoschi & Sturiale, 1996, Bono et al., 1998; Renones et al., 1998), or buoys and other floating structures for tuna fisheries outside the Mediterranean (Matsumoto et al., 1981; Anon, 1983; Druce & Kingsford, 1995).

The presence in the Ligurian Sea (north-west Mediterranean Sea) of an oceanographic buoy moored 37 nm (nautical miles) from Genoa (Fig. 1) since 1987, has offered the opportunity to study the fouling of this structure and its FAD behaviour. In 1991, a group consisting of the National Research Council and the University in Genoa undertook a multidisciplinary investigation of the buoy.

This paper reports on the development of the community settled on the buoy (fouling), and on the fish influenced by the FAD effect generated by the buoy, and discusses the role of this floating structure as a 'miniature island'. We also provide a critical review of data collected from 1987 to 1998 to explain the colonization patterns of fouling and FAD effects of offshore artificial structures moored far from the coast in deep water and subjected to a long period of observation.

Materials and methods

The study site

The study area has hydrological peculiarities such as a frontal region and upwelling areas (Jacques, 1994). The environment around the buoy is influenced by the Corso-Ligurian-Provençal current, which is 30 km wide and 150 m deep, and has an average annual flow of 1.4×10^6 m^{-3} sec^{-1} (Bethoux et al., 1982). Due to its large front, primary production is higher in offshore than in neritic inshore waters (Innamorati et al., 1982; Magazzù et al., 1989). A concise description of the environment around the buoy is given by Siccardi & Montanari (1997).

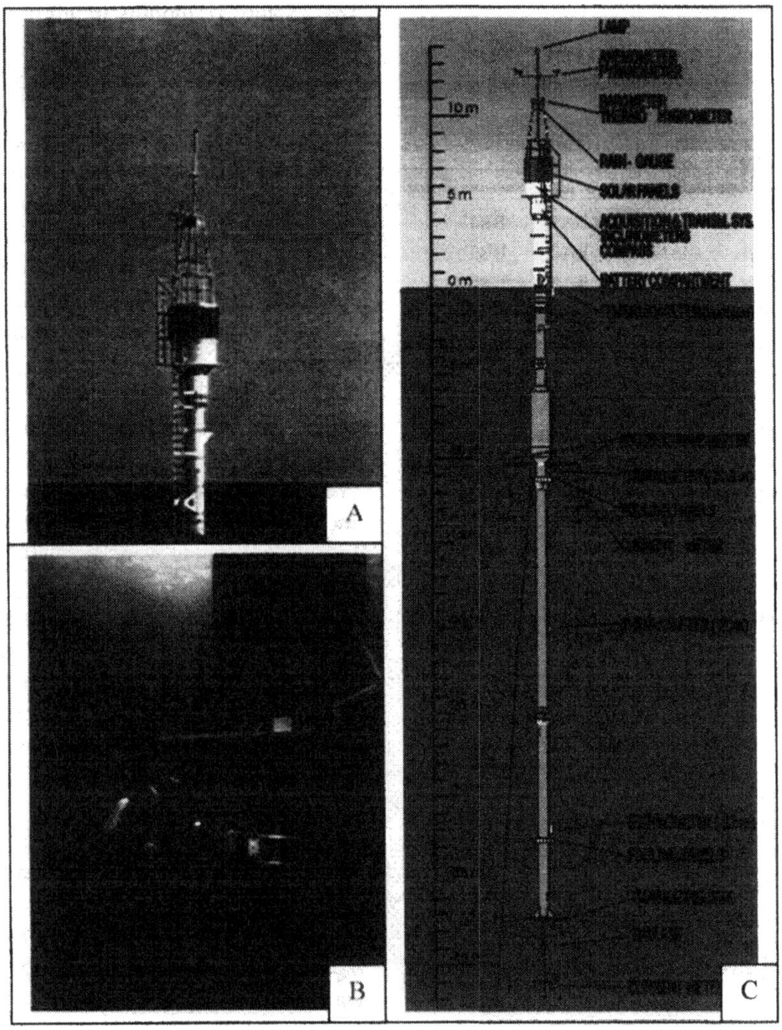

Figure 2. The emerging portion of the buoy ODAS Italia 1 (A). One of the frames fixed to the buoy and holding panels for fouling settlement studies (B). An overall schematic view of the buoy (C).

The buoy system

The oceanographic buoy ODAS Italia 1 (Oceanographic Data Acquisition System) is operated by IAN-CNR (Institute for Ship Automation of the Italian National Research Council) and consists of a 52 m-long tubular steel structure (Fig. 2); 5 m above water (Fig. 2A) and 37 m submerged (Fig. 2B). The pole body is divided into four segments joined by flanges. The three upper segments (Ø 60 cm) are water-proofed, and the middle one has a thicker portion (Ø 120 cm and 200 cm in height) for greater stability and buoyancy. The fourth segment is filled with water, with a stabilizing disk (Ø 200 cm) designed to moderate the buoy's vertical motion. A 500 kg ballast hangs from below this disk (Fig. 2C). The buoy is anchored to the sea bed (1270 m deep) by means of an elastic 1900 m-long mooring cable which allows the buoy to move around the mooring for about half a nautical mile (nm) and keeps it afloat even under strong side thrusts. The buoy was first moored on 9 January 1987, 38 nm offshore (43° 51.07′ N, 09° 06.00′ E) to a 1125 m depth bottom, very close to its current position (43° 48.90′ N, 09° 06.80′ E) on a 1270 m deep sea bed, 37 nm from Genoa. It was brought back to shore for maintenance on 15 April 1991, after 52 months at sea, and put to sea again on 28 June 1992, after repairs and restyling.

Table 1. Panels used for fouling studies during the two periods

Months of immersion	Immersion	Recovery	Depth 12 m	Depth 33 m
First period (from 02/05/89 to 15/04/91)				
1A – Set of panels with different immersion time				
3	02/05/89	04/08/89	BA1	BB1
	02/05/89	04/08/89	BA2	BB2
4	04/08/89	09/12/89	BA6	BB6
6	04/08/89	05/02/90	BA5	BB5
7	02/05/89	09/12/89	BA3	BB3
Second period (from 28/06/92, still in progress)				
1B - Set of panels with increasing immersion time				
3	28/06/92	03/10/92	BA1	BB1
9	28/06/92	29/03/93	BA3	BB3
14	28/06/92	21/08/93	BA2	BB2
18	28/06/92	19/12/93	BA5	BB5
24	28/06/92	02/06/94	BA6	BB6
40	28/06/92	02/10/95	BA10	BB7
51	28/06/92	18/09/96	Lost	BB10
57	28/06/92	03/03/97	BA1	BB2
70	28/06/92	24/04/98	BA2	BB8
1C - Set of panels with intermediate immersion time				
3	19/12/93	12/03/94	BA18	BB18
	12/03/94	02/06/94	BA20	BB20
4	21/08/93	19/12/93	BA14	BB14
	02/06/94	11/10/94	BA22	BB22
5	28/03/93	21/08/93	BA12	BB12
	28/03/93	21/08/93	BA13	BB13
6	03/10/92	28/03/93	BA11	BB11
7	21/08/93	12/03/94	BA15	BB15
10	21/08/93	02/06/94	BA16	Lost
14	28/06/92	21/08/93	BA4	BA4
	12/03/94	02/06/95	BA21	BB21
37	21/08/93	18/09/96	BA17	BB16

Macrofouling

Samples and colour photographs of fouling were collected by scuba divers; panels (20 × 30 × 0.4 cm) were fixed in frames secured to the body of the buoy at two levels for increasing periods of time (from some months to several years) and were retrieved. Further samples and direct observations were made on land when the buoy was recovered for maintainance. Sixteen dives were performed from 9 January 1987, when the buoy was positioned and the first period of observation started. Scraped samples were collected after 18, 23 and 31 months of immersion. Evaluation of macrofouling of the buoy was made on 15 April 1991, after 52 months of immersion, when the buoy, after the retrieval of the mooring system, was towed to Genoa harbour at a speed of 3 knots and then transferred to a dockyard wharf. The fouling along the full 37 m length of the immersed portion of the buoy was sampled every 2 m by scraping a 50 × 50 cm^2 surface. Additional samples were taken from both sides of the stabilizing disk. Each sample was photographed, weighed, examined visually and fixed in formalin (5% in sea water).

Three sets of panels (Table 1) are considered in this paper: one from 2 May 1989 to 15 April 1991 (first period) and two from 28 June 1992 (second period). The first series of panels was immersed at about 15 m and 30 m depth from 2 May 1989 to 15 April 1991, but only a few panels were available for fouling studies (Table 1A).

The second period of observation started on 28 June 1992, when another set of panels was placed in round panel-holding racks secured to the body of the buoy at two depths (−12 and −33 m). Investigations of the panels are still in progress. Substrata were recovered after 3, 9, 14, 18, 24, 40, 51, 57 and 70 months of immersion (Table 1B), allowing us to describe the fouling settlement and development over a relatively long time, something which, unfortunately, is unusual in fouling or benthic colonization studies when artificial substrata are employed. Also, the fouling of panels exposed for intermediate periods of time was considered (Table 1C).

Settlement on panels was examined both qualitatively (species identification) and quantitatively. The Covering Index (C.I.) was calculated according to Boudouresque's definition (1971). Covering (C) means the approximate percentage of the substratum surface covered by the projection of the total number of individuals of a species. This index has been applied for each species or other taxonomic level. The C.I. is expressed using Braun-Blanquet's (Boudouresque 1971) scale of abundance-dominance coefficient (+ = negligible presence; 1 = <5% surface covered; 2 = 5–25%; 3 = 25–50%; 4 = 50–75%; 5 = >75%). Starting from data on the C.I., the average covering index (A.C.I.) was calculated for each taxon. For example, the value 2 referred to a coverage of between 5% and 25% results in an the A.C.I. of 15%. According to Boudouresque's definition (1971), average covering means the average conventional value indicated below, corresponding to the class centre of each interval

Table 2. Fouling taken from scraped standard areas (20 × 20 cm) on the buoy

Date	Depth	Wet weight		Dominant
		g/4 dm^2	g/dm^2	organisms
	6	210	52.5	M-GA-A
22/07/88	12	430	107.5	M-As-A
	25	360	90.9	M-As-A
	3	65	16.2	A-O-D
	6	565	141.2	M-A
09/12/88	12	440	110.0	M-D
	30	845	211.2	M-O-As
	36	1000	250.0	M-P-O
	6	Spoiled	–	BYSSUS
	12	720	180.0	M-S-B-O
06/08/89	30	440	110.0	M-S-D-B
	39	680	170.0	M-O-P-S

GA = Green Algae. M = Mussels. S = Serpulids. O = Other Bivalves. P = Other Polychaetes. B = Bryozoans. A = Amphipods. As = Ascidians.

in the abundance-dominance coefficient (Covering Index). They are as follows: Absence = 0; + = 0.1%; 1 = 2.5%; 2 = 15%; 3 = 37.5%; 4 = 62.5%; 5 = 87.5%.

Density (no/dm^2) and sizes of the main taxa were also determined (not all these data are given in this paper). Biomass was expressed in non-decalcified wet weights, but dry and ash weights were also measured.

In October 1995, a special sampling programme to analyze the algae was conducted at seven different depths, between 3 and 30 m.

When counting the number of taxa at species level, the genus of non-identified species was considered as a species when no other congeneric species could be identified. For example, in Table 3, *Piliolaria* sp. is counted as a species, while *Hydroides* sp. is not, because there are other species of the same genus.

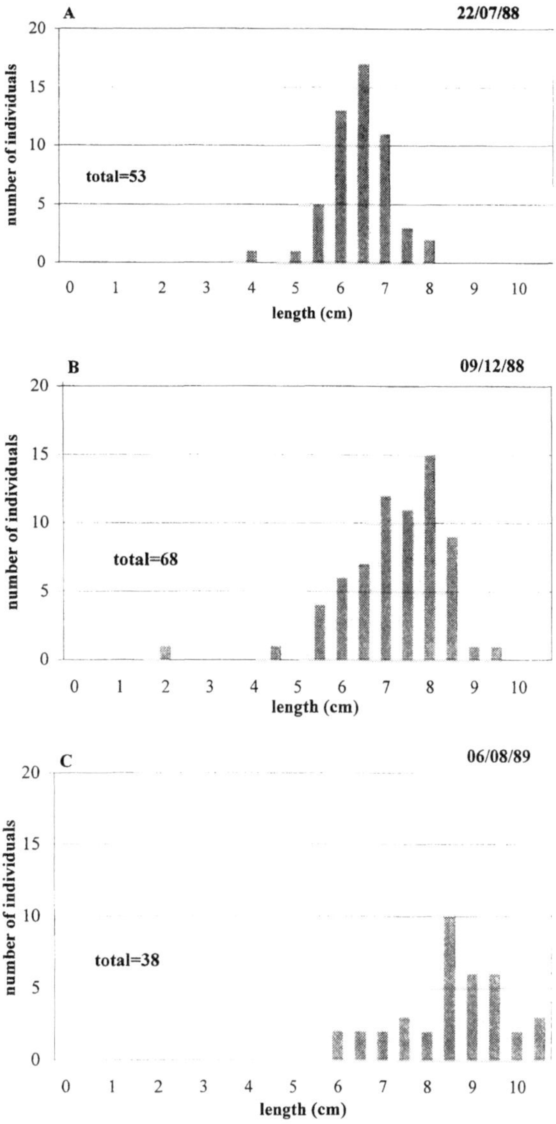

Figure 3. Length/frequency distribution of mussels collected on the buoy on different dates.

Fishes and FAD study

Since 1987, diving records of the types of fish present on the buoy, or around it, have been taken using a 35 mm camera and, more recently, video cameras. During the July 1988–April 1991 period, a detailed study of the fish assemblages was carried out; 20 underwater inspections were performed by two or three divers down to 40 m depth. The team generally operated late in the morning for about 40 min. Fish close to the buoy and around it were identified and counted by visual census. Sampling, using spear and lines, was limited to six cases only and when the taxonomic identity of some uncommon fish like Centrolophidae had to be verified.

Results

Macrofouling study

Data from samples scraped at different levels of the buoy between 1988 and 1989 are summarized in Table 2; for a detailed description, see Relini et al. (1990). The fouling community was characterized

Table 3. List of taxa and number of species of animals (excluding fish) found directly on the buoy or on panels (within each taxon, species are listed in alphabetical order)

	Buoy				Panels			
	Pole body		Disk		0–12 months		13–70 months	
	1988–89	1991	Upper	Lower	12–15 m	30–33 m	12 m	33 m
PROTOZOA FORAMINIFERA								
Miniacina miniacea (Pallas)		▼	▼	▼			▼	▼
Protozoa not ident.		▽	▽	▽	▽	▽	▽	▽
PROTOZOA FOLLICULINIDA								
Folliculina sp.		▼	▼	▼	▼	▼	▼	▼
PORIFERA								
Sycon sp.		▼	▼	▼	▼		▼	▼
HYDROIDA								
° *Bougainvillia ramosa* (Van Beneden)	▼	▼			▼	▼		▼
° * *Clytia hemisphaerica* (L.)				▼	▼		▼	
Clytia sp.				▽				
Eudendrium capillare Alder		▼	▼			▼	▼	▼
Garveia franciscana (Torrey)	▼							
Hydranthea magarica (Hincks)		▼	▼				▼	
Lafoeina tenuis M. Sars		▼	▼		▼	▼	▼	▼
Obelia bidentata Clarke		▼		▼	▼	▼	▼	
° *Obelia dichotoma* (L.)	▼	▼	▼	▼	▼	▼		▼
° *Tubularia crocea* Agassiz		▼	▼	▼			▼	▽
Hydroids not ident.						▽		
ALCYONARIA								
Alcyonium palmatum Pallas	▼			▼				
ZOANTHARIA								
Condylactis aurantiaca (Delle Chiaje)			▼					
Corynactis viridis Allman		▼	▼	▼			▼	▼
ENDOPROCTA								
Endoprocta not ident.			▽	▽				
POLYCHAETA SERPULOIDEA								
° * *Hydroides elegans* (Haswell)					▼		▼	
* *Hydroides norvegicus* Gunnerus					▼			
Hydroides stoichadon Zibrowius		▼						
* *Hydroides pseudouncinatus* Zibrowius					▼			
Hydroides sp.		▽	▽	▽	▽	▽	▽	▽
° *Pomatoceros lamarckii* (Quatrefages)		▽			▽		▽	▽
° *Pomatoceros triqueter* (L.)	▼	▼		▼	▼	▼	▼	
Semivermillia cribrata (O.G.Costa)	▼				▼	▼		
Serpula concharum Langerhans		▼	▼	▼	▼	▼	▼	▼
Serpula vermicularis L.		▼			▼			
Spirobranchus polytrema (Philippi)	▼	▼	▼	▼	▼	▼	▼	▼

Continued on p. 71

Table 3. Continued

	Buoy				Panels			
	Pole body		Disk		0–12 months		13–70 months	
	1988–89	1991	Upper	Lower	12–15 m	30–33 m	12 m	33 m
Serpulids not ident.					▽			
° * *Pileolaria* sp.							▼	▼
* *Simplaria pseudomilitaris* (Thiriot-Quièvreux)					▼		▼	▼
* *Spirorbis* sp.					▼		▼	▼
NON SERPULOIDEA								
Adyte pellucida (Ehlers)		▼					▼	▼
Chaetopterus variopedatus (Renier)			▼	▼				▼
* *Dorvillea rubrovittata* (Grube)								▼
Eunice harassii Audouin & Milne-Edwards		▼						▼
Eunice schizobranchia (Claparéde)		▼						▼
* *Glycera* sp.								▼
Harmothoe sp.		▼						
Lumbrineris sp.		▼						
Lysidice ninetta Audouin & Milne-Edwards		▼						
Marphysa sanguinea (Montagu)		▼						
Nereididae		▽				▽	▽	▽
Nereis rava (Ehlers)		▼						
Polynoidae		▽						
Polyophthalmus pictus (Dujardin)		▼						
Sabella spallanzanii (Gmelin)		▼		▼				▼
Scoletoma funchalensis (Kinberg)		▼						
Syllis gracilis Grube		▼					▼	
* *Syllis* sp.								▽
PYCNOGONIDA								
Achelia echinata Hodge		▼		▼	▼		▼	
* *Endeis spinosa* Montagu				▼	▼	▼	▼	
Pycnogonida not ident.			▽	▽	▽	▽		▽
CIRRIPEDIA								
Lepas anatifera L.		▼	▼					
Lepas hilli (Leach)		▼						
Lepas pectinata Spengler		▼						
DECAPODA								
Athanas nitescens (Leach)		▼						
Eriphia verrucosa (Forsskal)			▼					
Pilumnus hirtellus (L.)	▼	▼		▼			▼	▼
Pilumnus villosissimus (Rafinesque)		▼						
* *Thoralus cranchii* (Leach)								▼
TANAIDACEA								
Tanaidacea not ident.		▽						
AMPHIPODA								
Caprella andreae Mayer	▼	▼	▼		▼	▼	▼	▼
Jassa marmorata Holmes	▼	▼	▼	▼	▼		▼	▼
GASTROPODA								
Bittium reticulatum (Da Costa)		▼	▼				▼	▼
Doto sp.		▼		▼		▼		
Epitonium pulchellum (Ant. Bivona)			▼					

Continued on p. 72

Table 3. Continued

	Buoy				Panels			
	Pole body		Disk		0–12 months		13–70 months	
	1988–89	1991	Upper	Lower	12–15 m	30–33 m	12 m	33 m
Eubrachus sp.	▼							
Lamellaria perspicua L.	▼							
Patella ulyssiponensis Gmelin	▼							
Patella caerulea L.	▼							
BIVALVIA								
Aequipecten opercularis (L.)		▼		▼			▼	▼
° *Anomia ephippium* L.	▼	▼	▼	▼	▼	▼	▼	▼
Arca noae L.		▼						▼
Arca tetragona Poli		▼	▼	▼		▼	▼	
Chama gryphoides L.		▼	▼					
Chlamys multistriata (Poli)		▼	▼	▼				▼
Chlamys varia (L.)		▼					▼	▼
Chlamys sp.	▽					▽	▽	
Corbula gibba (Olivi)		▼	▼					
Hiatella arctica (L.)		▼	▼	▼	▼		▼	▼
Hiatella rugosa (L.)	▼	▼	▼		▼		▼	▼
∗ *Hiatella* sp.								
Lima hians (Gmelin)		▼						▼
Lima lima (L.)	▼							
Modiolarca subpicta (Cantraine)		▼	▼	▼	▼		▼	▼
Modiolus barbatus (L.)		▼	▼	▼			▼	▼
o *Mytilus galloprovincialis* Lamarck	▼	▼	▼	▼	▼		▼	▼
Neopycnodonte cochlear (Poli)				▼				▼
o *Ostrea edulis* L.			▼				▼	
Palliolum incomparabile (Risso)		▼						▼
Parvicardium exiguum (Gmelin)		▼						
Pecten jacobaeus (L.)		▼						
Plagiocardium papillosum (Poli)			▼					
Pododesmus patelliformis (L.)		▼	▼	▼				▼
BRYOZOA								
Aetea truncata (Landsborough)		▼	▼	▼	▼	▼	▼	▼
Scruparia ambigua (D'Orbigny)		▼	▼	▼	▼	▼	▼	▼
Scruparia chelata (L.)		▼						
ECHINODERMATA								
Arabacia lixula (L.)	▼	▼					▼	
Echinus acutus Lamarck	▼							
Paracentrotus lividus (Lamarack)	▼							
Psammechinus microtuberculatus (Blainville)		▼						
Sphaerechinus granularis (Lamarck)		▼						
ASCIDIACEA								
o ∗ *Botryllus schlosseri* (Pallas)					▼		▼	
o ∗ *Ciona intestinalis* (L.)								▼
o *Diplosoma listerianum* (Milne-Edwards)	▼	▼	▼	▼	▼	▼	▼	
o ∗ *Styela canopus* (Savigny)						▼		
Number of species	27	63	34	28	28	20	39	54

oFouling species commonly found in italian harbours (from Relini, 1980).
∗Species found only on panels.
▽Taxon not counted as species.

by mussels (*Mytilus galloprovincialis*), which ranged from 3 to 42 m (on rope) depth mainly down to 10 m, with a high growth rate (Relini, 1990). Modal lengths (shell length) of mussels were 6.5 cm in July 1988 (Fig. 3A), and 7–8 cm in December 1988 (Fig. 3B). Samples collected in August 1989 (after 31 months' exposure) (Fig. 3C) showed the presence of two cohorts: 6–7 cm-long mussels, more recently settled, and the oldest ones with lengths of up to 10.5 cm. The 1987 cohort appeared stronger than the following one. On the basis of an algorithm proposed for length-based fish stock assessment (Sparre, 1987), the following Von Bertalanffy growth curve was established (Fig. 4):

$$L_t = 12.5[1 - e^{-0.048(t+4.1)}]. \quad (1)$$

By the beginning of August 1989, almost all the mussels down to 10 m depth had unacountably disappeared, terminating our investigations. Interestingly, in 1989 no new mussel settlement took place, neither directly on the buoy, nor on the series of panels immersed for fouling studies. This indicates that either no larvae arrived or were produced by mussels settled on the buoys, or larvae were not successful after metamorphosis.

Considering the full study period (1987–1998), more than 100 animal taxa (Table 3), of which 97 were identified at species level (fish not included) and 34 algae (Table 4), comprised the fouling community. A total of 83 animal species was collected: 32 only on the buoy and three species only on the disk.

Molluscs were the dominant taxon with 30 species (23 bivalves and seven gastropods), followed by polychaetes (28 species), and hydroids (nine species). Algae (13 green algae, 13 red algae and eight brown algae) listed in Table 4 were collected during sampling on 2 October 1995. No other species was found on panels or in other samples.

On panels immersed from 3 to 70 months, at 12–15 m and 30–33 m depths, 63 species were identified. The 14 species indicated with an asterisk in Table 3 were found only on panels and not directly on the buoy. Except for the 3 and 9-month periods, the highest number of species was always found at −33 m (Fig. 5). At −12 and −15 m, 44 species were found, seven of which were restricted to this depth range, while at −30 and −33 m, 55 species were found, 18 of which exclusive to this depth. Only three species were present on substrata immersed for 12 months or less, while the number of species increased on panels with more than 13-month exposure and 30 species were found only on panels with a long immersion period.

Table 4. List of algae collected along the body buoy, at 3–30 m, mainly during October 1995

Chlorophyta
Chaetomorpha aerea (Good. ex Dill) Kuetz.
Cladophora albida (Nees) Kuetz.
Cladophora cfr. *albida* (Nees) Kuetz.
Cladophora dalmatica Kuetz.
Cladophora liniformis Kuetz.
Cladophora sericea (Huds.) Kuetz.
Cladophora socialis Kuetz.
Cladophora cfr.*socialis* Kuetz.
Cladophora sp.
Enteromorpha compressa (L.) Grev.
Enteromorpha crinita (Roth) J. Ag.
Enteromorpha intestinalis (L.) Link
Enteromorpha prolifera (Muller) J. Agardh ssp. *prolifera*
Enteromorpha torta (Mertens) Reinbold
Enteromorpha sp.
Rhizoclonium sp.
Ulothrix flacca (Dill.) Thuret
Phaeophyta
Acinetospora crinita (Charm. ex Harv.) Sauv.
Dictyopteris membranacea (Stackh.) Batters
Ectocarpus confervoides (Roth) Kjell. var. *confervoides*
Ectocarpus siliculosus (Dyllwyn) Lyngbye
Felmania globifera (Kuetz.) Hamel
Feldmania sp.
Myriactula rivulariae (Suhr) Feldm.
Sphacelaria cirrosa (Roth) C. Ag.
Sphacelaria tribuloides Kuetz.
Sphacelaria sp.
Rhodophyta
Ceramium fastigiatum (Roth) Harv.
Agloathamnion bipinnatum (Cr. & Cr.) Feld.-Mazoyer
Bangia atropurpurea (Roth) C. Ag.
Ceramium circinatu (Kuetz.) J. Ag.
Ceramium codii (Rich.) Feld.-Mazoyer
Ceramium diaphanum (Lightf.) Roth
Dasya rigidula (Kuetz.) Ardiss.
Lophosiphonia subadunca (Kuetz.) Falkenb.
Polysiphonia atlantica (Kapraun et J.N. Norris)
Polysiphonia denudata (Dill.) Grev. ex Harv. in Hook.
Polysiphonia elongata (Huds.) Spreng.
Polysiphonia elongella Harv.
Polysiphonia furcellata (C. Ag.) Harv.
Polysiphonia sp.

There was a clear, direct correlation between total biomass values and immersion period. This correlation, however, was sometimes lost because some

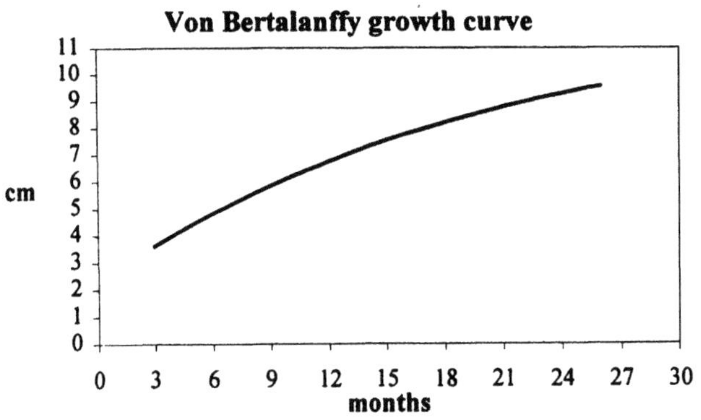

Figure 4. The Von Bertalanffy growth curve for mussels settled on the buoy (length of the valve in cm).

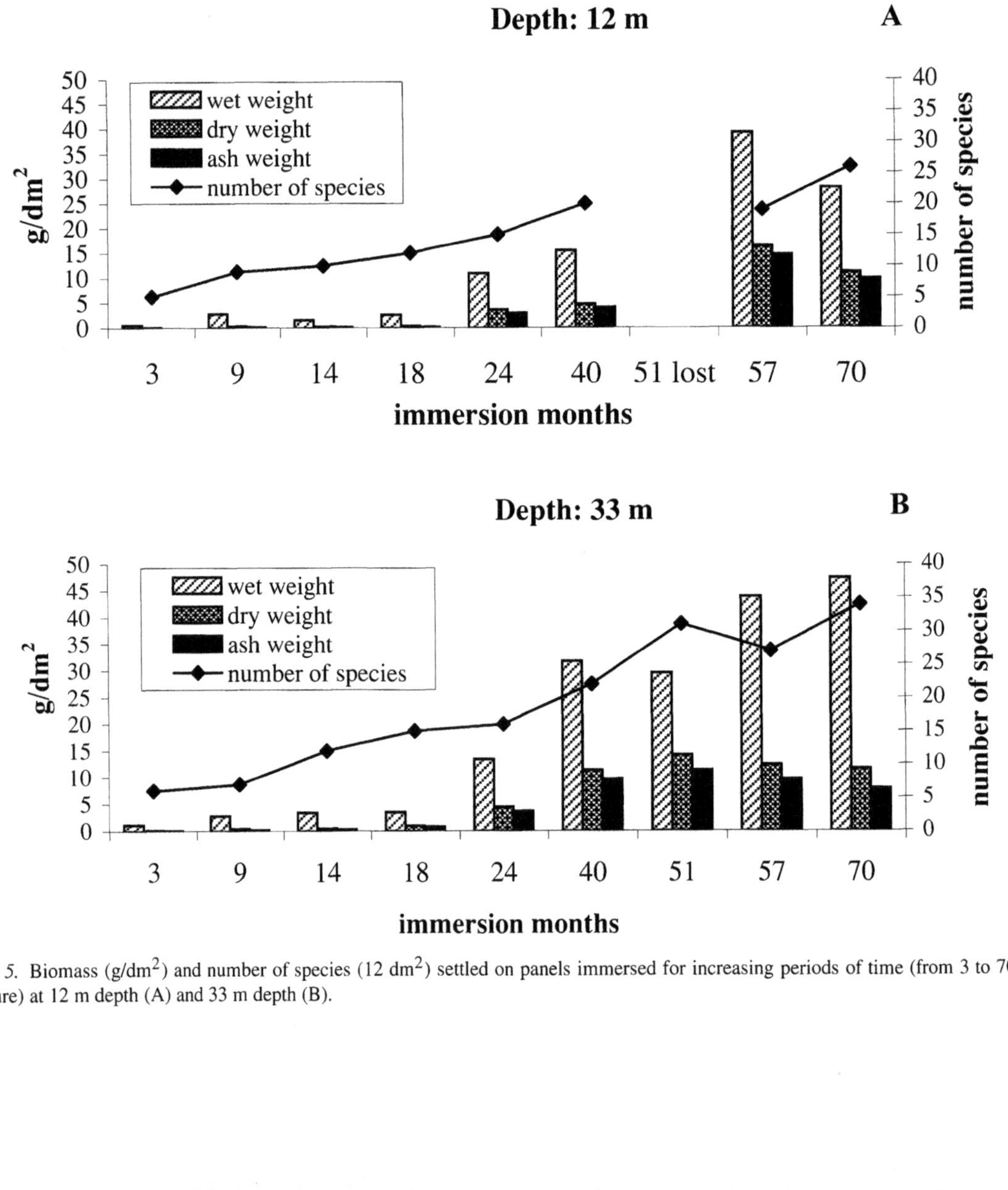

Figure 5. Biomass (g/dm^2) and number of species (12 dm^2) settled on panels immersed for increasing periods of time (from 3 to 70 months exposure) at 12 m depth (A) and 33 m depth (B).

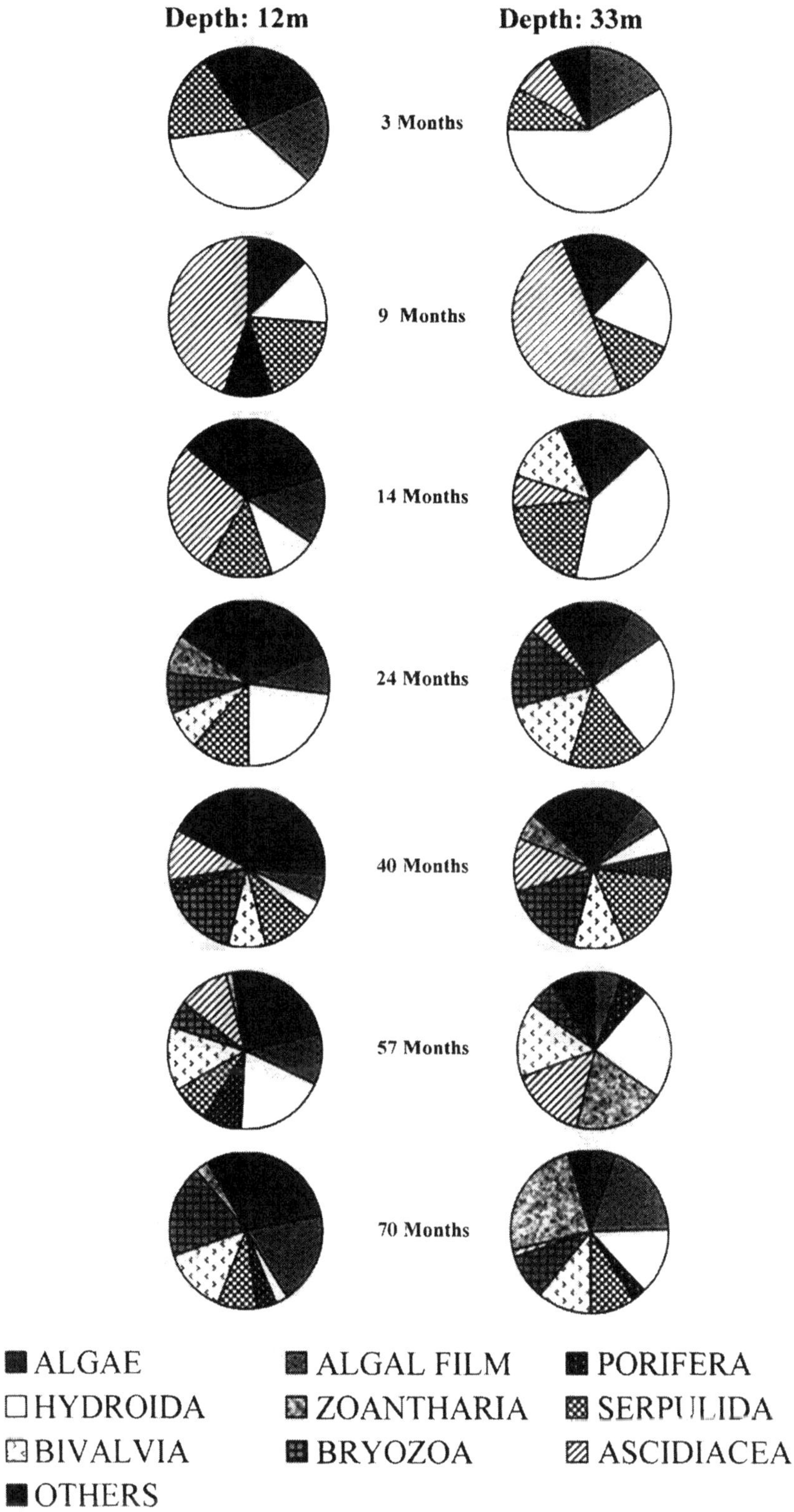

Figure 6. Development of fouling communities on panels immersed for increasing periods of time (from 3 to 70 months exposure) at 12 m depth and 33 m depth.

settled organisms had been dislodged, as was most likely the case with the 70-month panel closer to the sea surface. The histogram shows that biomass too has always at its greatest on the panels at −33 m, with values of 47, 39 g/dm^2 (Fig. 5B). The development of communities on panels exposed at two different levels for 70 months is shown in Fig. 6, representing the surface settled by the main taxonomic groups evaluated according to the Average Covering Index.

During the first 9 months of immersion, the settlement of substrata exposed at two different depths was quite similar and comprised algal film, algae, hydroids, serpulids and compound ascidians, the latter organisms being dominant on the 9-month panels. From 14 to 40 months of exposure, the number of taxa increased at both depths. With the settlement of bivalves, bryozoans, sponges and zoantharians, the importance of hydroids and ascidians decreased. In 57- and 70-month exposure substrata, the community became more complex due to the presence of further species. The main difference between the communities settled at the two depths was in the amount of algae (22%) at 12 m depth and in the prevailing presence of sea anemones (24%) at 30 m depth.

The lists of species found at different depths on the buoy are pooled and given in Table 3, while the number of species found in each sample is recorded in Figure 7. Seventy eight taxa, of which 71 at species level, were identified and evaluated in the samples collected on the retrieved buoy; 30 species were found only on the body of the buoy, and eight only on the stabilizing disk. On the latter structure, a total of 41 species was found, 34 on the upper and 28 on the lower side. On the cylinder of the buoy, the highest number of species was found at −19 m (27 species) and the lowest at −1 and −3 m (10 species) (Fig. 7).

Some species were found at all depths, while others occurred only at one specific site on the stabilizing disk. The following species were found only on the latter structure during the 1991 sampling: the polychaete *Chaetopterus variopedatus* and endoprocta (which were present on both sides of the disk), the polychaete *Sabella spallanzani*, the bivalve *Ostrea edulis*, the gastropods *Plagiocardium papillorum* and *Epitonium pulchellum* (found only on the upper side), the soft coral *Alcyonium palmatum*, the sea anemone *Condylactis aurantiaca*, and the bivalve *Neopycnodonte cochlear* (found only on the lower side).

The stabilizing disk was the part of the buoy with the highest settlement, both in terms of number of species and biomass (Fig. 7). Interestingly, there were no marked differences between the two sides (upper and lower) of the stabilizing disk, unlike what normally happens to horizontal surfaces. On the upper side, the wet weight reached 40.87 g/dm^2 while on the lower side it was 42.44 g/dm^2. Conversely, dry weight was higher on the upper side, due to the significant presence of bivalves. Along the body of the buoy, from the surface down to −11 m, there was a decreasing trend in wet weight of fouling with a high water content, up to 80% lower in terms of dry weight. From −11 to −20 m, this value increased up to a maximum absolute value of 20.81 g/dm^2, followed by a slight decrease down to −27 m. Between −27 and −34 m, wet weight values were variable.

Fishes and FAD

The buoy attracted different species of fish that can be divided into two categories: benthic fish living in the fouling community (such as like blennies and wrasse), and pelagic fish attracted by drifting items to which they like to stay close. As mentioned already, the ODAS buoy consists of a floating vertical cylinder, which has all the structural features of this kind of device and, therefore, acts as a FAD. Moreover, since it has remained in the same place for a long period of time, it has gathered an interesting group of nectonic organisms around it. Targeted observations made it possible to describe the composition of the community, seasonal changes and the duration of fish-floating substratum association.

In total, 12 species belonging to seven families (Table 5) were recorded on or near the buoy. The Blennidae and Labridae belong to the benthic community formed on the substratum (fouling). *Parablennius incognitus* and *P. rouxi* were identified and censused several times, hence they are assumed to have been living on the buoy for a long period, though the two blennies were observed for the first time after the buoy had been immersed for more than 2 years. The presence of other blennoid species, which are difficult to recognize in the midst of well-developed macrofouling during diving inspections, cannot be ruled out. The wrasse was seen only recently and only in one census. For this reason, they have been set apart. Hence, the FAD effect of the buoy was considered to be confined to nine species only (Table 6). Both *Trachurus* species were pooled together, because species identification in the sea is difficult.

Some species of pelagic fish were present for a short period (Table 6), such as *Trachurus* sp. in sum-

Table 5. Fish species found on and around the buoy

Family	Scientific name	Common name
Balistidae	*Balistes carolinensis* (Gmelin)	Grey trigger fish
Blennidae	*Parablennius incognitus* (Bath)	Blenny
	Parablennius rouxi (Cocco)	Blenny
Carangidae	*Naucrates ductor* (L.)	Pilot fish
	Seriola dumerili (Risso)	Greater amberjack
	Trachurus trachurus (L.)	Atlantic horse mackerel
	Trachurus mediterraneus (Steindachner)	Mediterranean horse mackerel
Centrolophidae	*Centrolophus niger* (Gmelin)	Blackfish
	Schedophilus ovalis (Cuvier)	Imperial blackfish
Labridae	*Symphodus tinca* (L.)	Peacock wrasse
Scombridae	*Thunnus thynnus* (L.)	Northern bluefin tuna
Serranidae	*Polyprion americanus* (Bloch & Scheneider)	Wreckfish

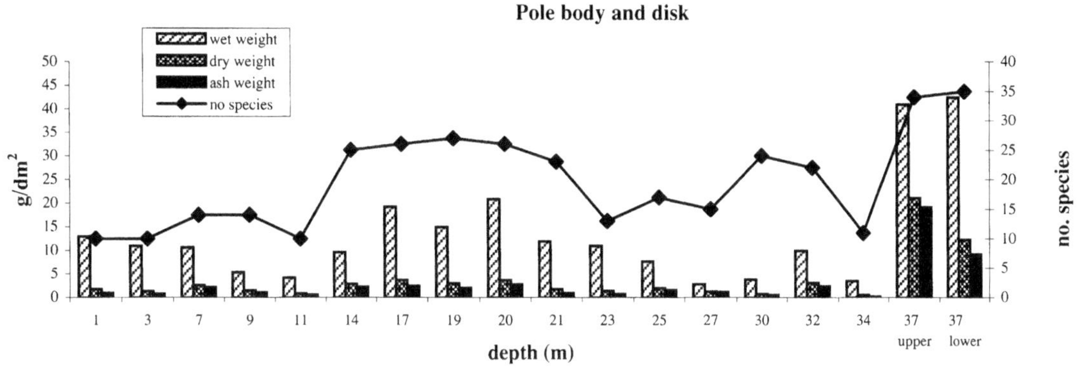

Figure 7. Biomass (g/dm^2) and number of species (25 dm^2) collected, at various depths, along the pole body of the buoy and on both sides of the stabilizing disk at 37 m depth.

Table 6. Seasonal pattern of association and general rank

Species	Average number of individuals censused in different seasons				
	Spring	Summer	Autumn	Winter	Pooled
Naucrates ductor	48.6	32.3	116	100	61.3
Thunnus thynnus	==	50	16	==	20
Trachurus sp.	0.1	28.6	==	==	10.05
Centrolophus niger	8.3	4.7	1	0.6	4.8
Schedophilus ovalis	3.8	6.6	4	2.3	4.6
Seriola dumerilii	0.3	1.4	0.3	0.6	1.10
Polyprion americanus	0.4	1.4	0.3	0.6	0.75
Balistes carolinensis	==	0.3	==	0.6	0.15

mer and part of spring, or a little longer, such as *Thunnus thynnus*. Others stay for longer, but only *Naucrates ductor* was recorded each month, as reported by Relini et al. (1994). Although *N. ductor* was seen sometimes at deeper levels (30 m) near the body of the buoy, it preferred to stay within the first 6 m depth around the buoy, where it actively caught plankton organisms. When the pilot fish was frightened, or when the sea current was particularly strong, it formed compact shoals near the surface, finding shelter in the main body of the buoy. Another fish that remained very close to the buoy for protection against predators was the horse mackerel *(Trachurus* spp.), which was present mainly in August with a compact school of juveniles of about 6 cm total lenght (TL). The wreck fish *Polyprion americanus* was found more frequently in the lower portion of the buoy, particularly on the stabilizing disk. It was represented by a few specimens, of 30–40 cm TL all year round, but mainly in August. At a radius of 5–6 m from the buoy, frequently mixing with *N. ductor*, two Centrolophidae and *Seriola dumerilii* were also found. The greater amberjack *Seriola dumerilii* was present all year and the smallest 10–12 cm TL specimens, with yellow colouring, were detected in August, very close to the sea surface. From autumn onwards, they acquired a silvery colour and moved in a larger vertical range. The smallest specimens of *Schedophilus ovalis* (25 cm TL) were seen in spring, the largest in late autumn. *Centrolophus niger* was found seasonally, beginning in spring. The smallest specimens, 25–30 cm TL, had iridescent silvery colours with white spots; some were blackish. Bluefin tuna, *Thunnus thynnus*, was seen only during July-October, its presence being perhaps most notable in the area around the buoy. The sizes were mainly around 50–60 cm fork length, which corresponds to an about 1-year-old fish. However, both smaller and bigger individuals were present. Sometimes they were seen to catch small *Trachurus*. *Balistes carolinensis* was detected sporadically with small individuals, about 12 cm TL in winter and 25 cm TL in summer.

Discussion

Within the buoy fouling, including panels, a total of more than 100 animal taxa (protozoa and invertebrates), of which 96 at species level, three fish types and 34 species of algae, was identified over 11 years of study. Macrofouling included many species typical of shallow, coastal waters, such as the sea urchins *Sphaerechinus granularis*, *Arbacia lixula*, *Psammechinus microtubercolatus*, the crabs *Eriphia verrucosa* and *Pilumnus hirtellus*, and the gastropods *Patella caerulea* and *Bittium reticulatum*. Some live on soft bottoms, like some Pectinidae, while others are common harbour fouling organisms in the Mediterranean Sea (Relini, 1980), such as, for example, *Clytia hemisphaerica*, *Obelia dichotoma*, *Hydroides elegans*, *Botryllus schlosseri* and *Ciona intestinalis*. Interestingly, cirripedes were represented by goose barnacles of the genus *Lepas*. Acorn barnacles were never found throughout the long observation period, although *Balanus* species are very common in the Mediterranean basin and sometimes adults have been found settled on stranded floating objects. The same is true of bryozoans (Cheilostomata, Ascosphora) of the genera *Cryptosula*, *Schizoporella* and *Watersipora*, which are very common fouling organisms.

It is well known that the fouling community species composition, and development, depends on the arrival of larvae, on recruitment and settlement success, and also on the transport mechanisms to the site of investigations (Butman, 1987; Pechenik, 1985, 1990) in the form of larvae, adults, or resting stages, such as cysts and propagules. The dispersal of marine organisms through their larval stages has been studied by many scientists. Some of them (e.g. Thorson, 1950; Mileikovsky, 1968, 1971; Scheltema, 1971, 1992) believe that larval duration is one of the main factors responsible for the distribution of organisms: the longer the period, the wider the distribution of organisms. Pelagic life span increases from lecithotrophic to teleplanic larvae. Free spawning with planktotrophic development seems to be the most widespread strategy in marine invertebrates with a known life cycle (Giangrande et al., 1994).

There are other mechanisms that have an impact on dispersal, such as human activities like shipping (ballast waters, fouling), drifting (Lane et al., 1985) (post-larvae and adults transported by sea currents) and rafting (De Vantier, 1992), namely the transport of adult individuals on floating objects, pieces of plastic, wood and mainly algae (Highsmith, 1985). Therefore, currents can play a very important role in the dispersal and distribution of marine organisms at different stages of their life cycle. Marine currents, in our case, have a dual function: to supply organisms to the buoy that is exposed to the prevalent current coming from Corsica, while a branch of strong current parallel to the Ligurian coast exercises a barrier effect, preventing organisms from the Ligurian coast from reaching this study station.

The buoy acted as a FAD for nine fish species, four of which are the fish normally caught by fishermen off Malta, Sicily and Majorca, where the 'shade effect' of 'Kannizzati', 'Canizzi', 'Capcer' is employed. The dolphin fish *Coryphaena hippurus*, which is one of the main target species of the abovementioned fisheries, has never been observed around the ODAS buoy. The presence of large stromateoids, *C. niger* and *S. ovalis*, was probably the most original observation with regard to the FAD effect of the buoy, because these fishes have a juvenile phase associated with Cnidaria rather than with flotsam. Just like the wreckfish *P. americanus*, these centrolophids have an adult phase living close to the sea bed in deep water. In particular, *S. ovalis* seems to be associated with sea mounts: the only catches that we know of in the Ligurian Sea are those made with lines called 'filaccioni' in offshore waters close to deep rocks, west of the buoy.

Furthermore, offshore artificial structures like buoys seem to be interesting, not only as a way of promoting fishing activities, but also monitoring and possibly even farming poorly-known species. However, some basic problems are still unresolved: we need to know more about fish behaviour in relation to FADs, and we need to adopt a precautionary approach in the light of increasing fishing activities of this kind.

In conclusion, we observed that the ODAS Italia 1 buoy, which we have been studying for 11 years, acts as a 'miniature island' and is an important aggregation point for a peculiar community which ranges from pelagic fish species, which react to the FAD effect of the structure, to many fouling benthic organisms. The pelagic community, despite seasonal variations, was more stable during the decade-long observations than the benthic assemblages that had settled on the buoy, the latter community being in constant development. Some species disappeared, new ones arrived but, in general, there was an increase in the number of species and in biomass with time. The case of the mussel *Mytilus galloprovincialis* is intriguing because it seems that mussels are able to settle with new individuals on the buoy only in certain years. It also seems that larvae produced by adult mussels on the buoy drift away by current, so they probably do not contribute to colonization of this artificial structure. Therefore, the buoy is an extremely interesting site, not only for studies on the fouling of offshore floating structures, but also as a way of investigating basic ecological issues, such as colonization and the dispersal of marine organisms.

Acknowledgements

We are grateful to Professor Guido Bressan and Dr Annalisa Falace of the University of Trieste for their help in identifying algae, and to Dr Giovanni Torchia for carrying out the fish censuses. We also wish to thank Dr Antonella Garofalo and Miss Chiara Barisione for their help in the collection and processing of data.

References

Aliani, S., C. De Asmundis, R. Meloni, M. Borghini & G. P. Gasparini, 1998. Transport of benthic species in the Sicily Channel: preliminary observation. In Piccazzo, M. (ed.), Proceedings of the 12th Associazione Italiana di Oceanografia e Limnologia Symposium (Isola di Vulcano, 18–21 Settembre 1996) – Genova A.I.O.L. 1998 II: 173–182.

Aliani, S. & R. Meloni, 1999. Dispersal strategies of benthic species and water current variability in the Corsica Channel (Western Mediterranean).Sci. Mar. 63 (2): 137–145.

Anon, 1983. Fish attraction buoys deployed off California. Mar. Fish. Rev. 45 (10–12): 60.

Bethoux, J. P., L. Prieur & F. Nyffeler, 1982. The water circulation in the NW Mediterranean sea, its relations with wind and atmospheric pressure. In Nihoul, J. C. (ed.), Hydrodynamics of Semi-enclosed Seas. Elservier, Oxford, U.K.: 129–142.

Bono, G., L. Cannizzaro, S. Gancitano & P. Rizzo, 1998. La pesca sui cannizzati: aspetti quali-quantitativi. Biol. Mar. Medit. 5 (1): 661–664.

Boudouresque, C. F., 1971. Méthodes d'etude qualitative et quantitative du benthos (en particulier du phytobenthos). Tethys 3: 79–104.

Butman, C. A., 1987. Larval settlement of soft-sediment invertebrates: the spatial scales of pattern explained by active habitat selection and the emerging role of hydrodynamical processes. Oceanography and Marine Biology: an Annual Rewiew 25: 113–165.

De Vantier, L. M., 1992. Rafting of tropical marine organisms on buoyant coralla. Mar. Ecol. Prog. Ser. 86: 301–302.

Druce, B. E. & M. J. Kingsford, 1995. An experimental investigation on the fishes associated with drifting objects in coastal waters of temperate Australia. Bull. mar. Sci. 57 (2): 378–392.

Galea, J. A., 1961. The 'Kannizzati' Fishery. Proc. Gen. Fish. Counc. Med. 6: 85–91.

Giangrande, A, S. Geraci & G. Belmonte, 1994. Life-cycle and life-history diversity in marine invertebrates and the implications in community dynamics. Oceanography and Marine Biology: an Annual Review 32: 305–333.

Haderlie, E. C., 1984. A brief overview of the effects of macrofouling. In Costlow, J. D. & R. C. Tipper (eds), Marine Biodeterioration: An Interdisciplinary Study. U.S. Naval Institute Press, Annapolis Maryland: 163–166.

Highsmith, R. C., 1985. Floating and algal rafting as potential dispersal mechanisms in brooding invertebrates. Mar. Ecol. Prog. Ser. 25: 169–179.

Innamorati, M., G. Buracchi, M. Mannucci & P. Senesi, 1982. Chiazze termiche e fitoplanctoniche nel Mar Ligure. Campagna Ligure VI - Giugno 1981. Boll. Mus. Ist. biol. Univ. Genova 50 (suppl.): 228–235.

Jacques, G., 1994. Nouvelles vues sur le système pélagique de Mer Ligure. Biol. Mar. Medit. 1 (1): 65–82.

Lane, D. J. W., A. R. Beaumont & J. R. Hunter, 1985. Byssus drifting and the drifting threads of the young post-larval mussel *Mytilus edulis*. Mar. Biol. 84: 301–308.

Magazzù, G., V. Bruni, F. Decembrini & S. Pannella, 1989. La produzione primaria del picoplancton fotosintetico dei nostri mari. Oebalia 15 (1 N.S.): 463–478.

Matsumoto, W. M., T. K. Kazama & D. C. Aasted, 1981. Anchored Fish Aggregating Devices in Hawaiian Waters. Mar. Fish. Rev. 43 (9): 1–13.

Mileikovsky, S. A., 1968. Some common features in the drift of pelagic larvae and juvenile stages of bottom invertebrates with marine currents in temperate regions. Sarsia 34: 209–216.

Mileikovsky, S. A., 1971. Types of larval development in marine bottom invertebrates, their distribution and ecological significance, a re-evaluation. Mar. Biol. 10: 193–213.

Morales-Nin B. & E. Massuti, 1992. Dolphin-Fish *Coryphaena hippurus* (Linnaeus, 1758) biology and fishery in Majorca Island (N-W Mediterranean). Rapp. Comm. Int. Mer Médit. 33: 302 pp.

Pechenik, J. A., 1985. Delayed metamorphosis of marine molluscan larvae: current status and directions for future research. American Malacological Bulletin Spec. Edn No. 1: 85–91.

Pechenik, J. A., 1990. Delayed metamorphosis by larvae of benthic marine invertebrates: does it occur? Is there a price to pay? Ophelia 32: 63–94.

Potoschi A. & P. Sturiale, 1996. Distribuzione dei 'cannizzi' nel Tirreno Meridionale e nello Ionio per la cattura della lampuga *Coriphaena hippurus* L. 1758. Biol. Mar. Medit. 3 (1): 384–386.

Redfield, A. C. & L. W. Hutchins, 1952. The effects of fouling (navigation buoys). In Marine fouling and its prevention. U.S. Naval Institute Press, Annapolis Maryland: 3–20.

Relini, G., 1980. Insediamento di organismi marini di substrato duro in ambienti portuali mediterranei. Mem. Biol. Marina e Oceanogr. 10 (suppl.): 61–70.

Relini, G., 1993. Mediterranean Macrofouling. Oebalia 19 (suppl.): 103–154.

Relini, G. & M. Relini, 1994. Macrofouling on offshore structures in the Mediterranean Sea. In Garg K. L., N. Garg & K.G. Mukerji (eds), Recent Advances in Biodeterioration and Biodegradation. Naya Prokash, Calcutta, India 2: 307–326.

Relini, G., M. Relini & M. Montanari, 1990. Offshore buoy fouling in the Ligurian Sea. Rapp. Comm. int. Mer Médit. 32 (1): 26 pp.

Relini, M., 1990. Growth of *Mytilus galloprovincialis* in offshore waters of the Ligurian Sea. Rapp. Comm. int. Mer Médit. 32 (1): 25 pp.

Relini M., L. Relini Orsi & G. Relini, 1994. An offshore buoy as a FAD in the Mediterranean. Bull. mar. Sci. 55 (2–3): 1099–1105.

Renones, O., E. Massuti, S. Deudero & B. Morales-Nin, 1998. Biological characterization of pilotfish (*Naucrates ductor*) from the FADs (fish aggregation devices) fishery of the island of Mallorca (western Mediterranean). Bull. mar. Sci. 63 (1): 249–256.

Scheltema, R. S., 1971. Larval dispersal as a mean of genetic exchange between geographically separated populations of shallow-water benthic marine gastropods. Biol. Bull. 140 (2): 284–322.

Scheltema, R. S., 1992. Passive dispersal of planktonic larvae and the biogeography of tropical sublittoral invertebrate species. In Colombo G., I. Ferrari, V. U. Ceccherelli & R. Rossi (eds), Marine Eutrophication and Population Dynamics. Olsen & Olsen, Fredensborg: 195–202.

Siccardi, A. & M. Montanari, 1997. Open Sea Laboratory ODAS Italia 1. Special publication by IAN-CNR: 1–40, I–VIII appendix.

Sparre P., 1987. Computer programs for fish stock assesment. FAO Fish. tech. Paper, 101 Suppl.(2): 218 pp.

Thorson, G., 1950. Reproduction and larval ecology of marine bottom invertebrates. Biol. Rev. 25: 1–45.

Tsikhon-Lukanina, Ye. A., I. N. Soldatova, I. A. Kuznetsova & I. I. Il'in, 1977. Macrofouling community in the strait of Tunisia (Sicily). Oceanology 16: 519–522.

Yan Wenxia, Yu Dong, Huajie Wang, Yan Yan, Tao Yan & Guanhe Liang, 1994. Methods of marine fouling investigation in offshore areas. Trop. Oceanol. Redai Haiyang 13 (4): 81–86.

Hydrobiologia **440:** 81–100, 2000.
M.B. Jones, J.M.N. Azevedo, A.I. Neto, A.C. Costa & A.M. Frias Martins (eds), Island, Ocean and Deep-Sea Biology.

Periodic movement, recruitment and size-related distribution of the sea cucumber *Holothuria scabra* in Solomon Islands

Annie Mercier[1,2,3], Stephen C. Battaglene[2] & Jean-François Hamel[1]
[1]*Society for the Exploration and Valuing of the Environment (SEVE), 655 rue de la Rivière, Katevale (Québec), Canada J0B 1W0*
Tel/Fax: +819 843-3466. E-mail: seve@sympatico.ca
[2]*International Center for Living Aquatic Resources Management (ICLARM), Coastal Aquaculture Centre, P.O. Box 438, Honiara, Solomon Islands*
[3]*Institut des Sciences de la Mer de Rimouski (ISMER), Université du Québec, 310 allée des Ursulines, Rimouski (Québec), Canada G5L 3A1*

Key words: Holothuria scabra, sea cucumber, holothurian, distribution, movement, population structure

Abstract

Field studies of the sea cucumber *Holothuria scabra* conducted in Kogu Veke, Solomon Islands, showed monthly recruitment of newly-settled juveniles on seagrass and indicated that size distribution was a function of substratum type and depth. Adults >250 mm body length were found mainly on sand, with <5% organic matter (OM), at depths of >1–3 m. Individuals >10–250 mm were found mostly in 30–120 cm of water, on mud and muddy sand with OM content between 5 and 10%. Specimens >40–150 mm were also found in the intertidal zone, sometimes burrowed on exposed sandflats at low tide. *Holothuria scabra* avoided substrata of fine silt or shell and coral pebbles, and sediment with an organic content ≥30%. Juveniles ≤100 mm burrowed at sunrise and surfaced at sunset, whereas individuals >100 mm burrowed and surfaced a few hours earlier. *Holothuria scabra* tended to burrow when salinity decreased, whereas increased water temperatures reduced normal burrowing behaviour. Spatial distribution, observed during tank experiments, suggested that adult *H. scabra* aggregated prior to spawning and in response to the lunar cycle. The formation of pairs, trios or larger groups increased during the new moon and was most common just before the full moon. Newly-settled juveniles up to ca. 9 mm were found on seagrass leaves. Typically, maximum densities and smallest recruits were observed a couple of weeks after the full moon, lower densities and slightly larger recruits were found a few days later. Juveniles with a mean length around 65 mm released on sand moved less and grew faster than juveniles released in seagrass beds or on substrata of shells and crushed coral.

Introduction

Holothuria scabra Jaeger is a deposit feeding holothurian that occurs mainly in sandy environments throughout the tropical Indo-Pacific. In recent years, *H. scabra* populations have been depleted severely by over harvesting (Preston, 1990; Holland, 1994; Lokani et al., 1996) and stock enhancement programmes have been initiated in a few countries (Battaglene & Bell, 1999). To date, biological studies on *H. scabra* have focused mainly on the population structure of adults, the reproductive cycle, spawning, juvenile growth (Krishnan, 1968; Ong Che & Gomez, 1985; Shelley, 1985; Baskar, 1994; James et al., 1994; Conand & Tuwo, 1996; Long & Skewes, 1997; Battaglene et al., 1999; Tuwo, 1999) and daily activity cycles (Mercier et al., 1999). Some reports discuss small specimens found in shallow waters or mention that the young and the adults may have different habitat preferences, but observations of juveniles are rare and their recruitment has not been investigated (Shelley, 1985; James & James, 1994; James et al., 1994; Lokani et al., 1996).

One reason why so little information exists on young holothurians in general is that they are rarely found in the wild (Seeto, 1994; Hamel & Mercier, 1996a; Long & Skewes, 1997). Small recruits of the temperate species *Parastichopus californicus* (Cameron & Fankboner, 1989) and *Cucumaria frondosa* (Hamel & Mercier, 1996a) have been studied, and Wiedemeyer (1994) examined the movements and substratum preferences of young *Actinopyga echinites* collected on a Japanese reef after a storm. However, most of the existing knowledge on tropical holothurians is focused on adults, and is often gathered within large surveys of species abundance and distribution in coastal habitats (Lawrence, 1979; Massin & Doumen, 1986; Kerr et al., 1993; Conand, 1994; Long & Skewes, 1997). Substratum characteristics, light intensity, energy level, food availability, salinity fluctuations and presence of predators or conspecific adults are cited as the most important variables influencing distribution of holothurians (Webb et al., 1977; Conand & Chardy, 1985; Bulteel et al., 1992; Conand, 1994; Hamel & Mercier, 1996a; Rodgers & Bingham, 1996; Zhou & Shirley, 1996). The distribution patterns of several tropical species are also explained by hydrodynamic factors (Sloan & Von Bodugen, 1980; Massin & Doumen, 1986; Kerr et al., 1993; Uthicke, 1994).

Holothuria scabra colonize shallow coastal habitats and are exposed to a wide range of physico-chemical factors. Juvenile *H. scabra* respond to low salinities by burrowing into the sediment and stop feeding until conditions return to normal (Mercier et al., 1999). Previous authors have suggested that *H. scabra* prefer brackish waters and areas of high organic matter (James & James, 1994; James et al., 1994), although these claims are mostly speculative and not supported by salinity or substratum richness data.

Our studies have been designed to gather information on aspects of the biology of tropical holothurians that have received little attention, and to assist in the protection and restoration of *H. scabra* stocks through future stock enhancement. The aim of the current study was to describe the distribution and population structure of newly-settled juveniles, young and adults of *H. scabra* with respect to the characteristics of their habitat, and to investigate the aggregative and burrowing behaviours that might influence the population structure and spatial distribution. Recruitment over a yearly period, and movement and growth of juveniles on different substratum types were also monitored to determine their influence on the spatial distribution of *H. scabra* in the Solomon Islands.

Materials and methods

Distribution

The first field study monitored distribution and population structure of juvenile and adult *H. scabra* with respect to the physico-chemical characteristics of the habitat. Observations were conducted between September and November 1997, and between February and April 1998 in Kogu Veke, Vonavona lagoon, along the western coast of Kohinggo Island (8° 10′ S, 157° 11′ E), Solomon Islands (Fig. 1). The site was chosen because it was one of the *H. scabra* populations in Solomon Islands that had not been overfished. The depth was established using a ruler for the very shallow areas and an electronic depth meter for deeper sections at different times over the tidal cycle. Light intensity was measured with a LI-COR light meter model Li-189, and temperature, pH, turbidity and salinity were monitored using a U-10 Horiba Water Quality Meter. Coarseness of the substratum was assessed by successive sieving and observation under a binocular microscope of ca. 80 samples using the classification outlined in Mercier et al. (1999). Current velocity and direction were established using methylene-blue stain released on the surface of the water and followed for ca. 3 m and replicated in 30–50 locations depending on the tide level. All data were collected in wind conditions <5 knots. The organic matter (OM) in representative sediment samples (ca. 80 samples of 500 g) was determined after drying in an oven at 55 °C for 48 h and burning in a muffle furnace at 500 °C for 48 h.

Given the relatively low densities of sea cucumbers observed, individuals larger than ca. 100 mm body length were located by surveying the entire bay. In October 1997 and February 1998, three different observers gathered the data by walking on sand flats at low tide, and using snorkeling or diving equipment where appropriate. Sea cucumbers were located visually and 3-m long plastic rods were used to delimit successive sample areas perpendicular to the coastline. Upon finding a sea cucumber, its length was recorded as the distance between the mouth and anus in a contracted posture, as used by Seeto (1994) and Hamel & Mercier (1996a). The exact location of each individual was noted on a detailed map of the study site. To establish the average density, the number of individuals was divided by the total surface area for the whole bay and for each substratum type (i.e. fine silt, mud, muddy sand, sand, coral pebbles, seagrass bed). Burrowed an-

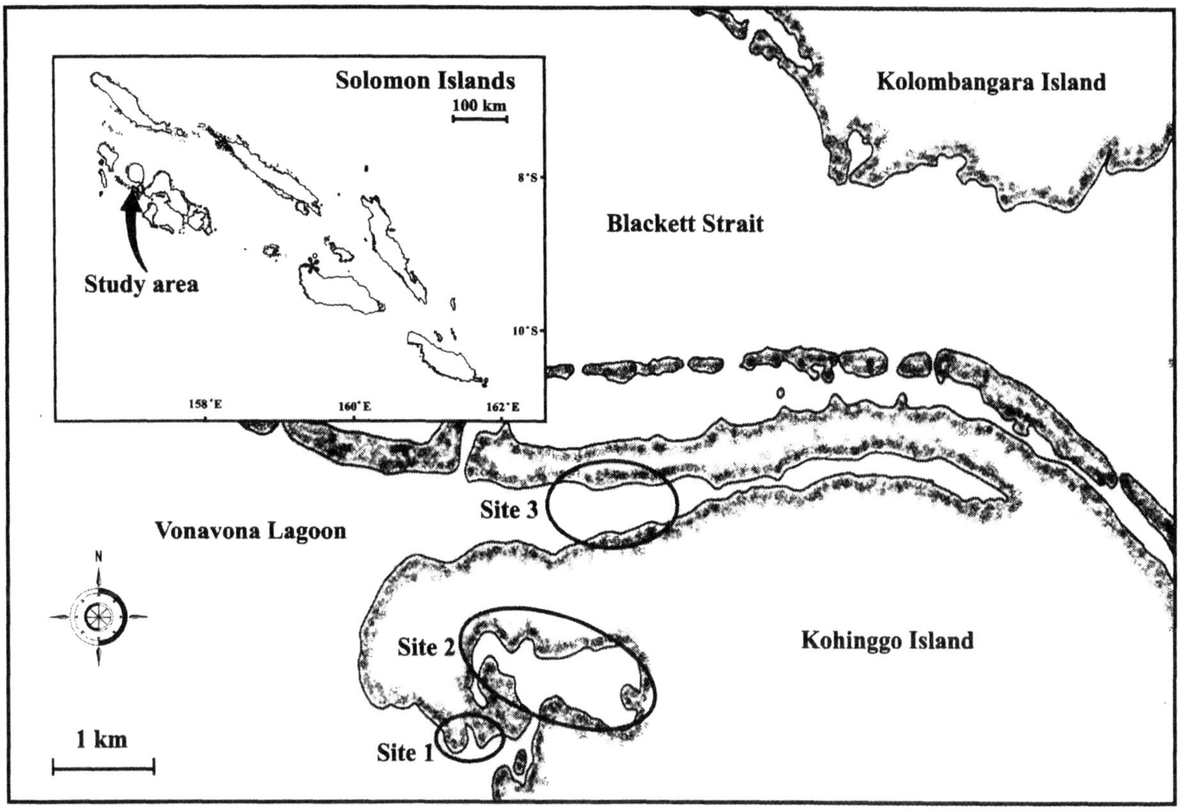

Figure 1. Map of Solomon Islands showing the location of the three study sites in the Vonavona lagoon: (1) Kogu Veke, (2) Kogu Halingi and, (3) Malmaragiri Inlet. The asterisk on the map identifies the site on Guadalcanal where the study of juvenile sea cucumber movements took place.

imals were detected by the faeces and/or the anus that protruded above the sediment. Individuals larger than 150 mm were sexed on several occasions using a syringe to collect gametes from the gonadal tubules, and then returned to their initial location. Because smaller specimens <100 mm were more difficult to locate visually, stratified random sampling was used: 325 one square meter quadrats were examined randomly with an equal effort applied to the different substratum types (i.e. 3% of the surface area) in October 1997 and February 1998. For each quadrat, the first 10 cm of the sediment was sieved with an air lift pump powered by compressed air. Juveniles burrow within the first 8–10 cm of substratum (Mercier et al., 1999) and the collector bag or sieve retrieved all material larger than ca. 5 mm. All *H. scabra* found were measured and their location described as above.

For comparative purposes, the population structure of *H. scabra* was also monitored in October 1997 and February 1998 at two nearby sites along Kohinggo Island: Malmaragiri Inlet and Kogu Halingi (Fig. 1). These sites had been fished heavily for sea cucumbers during the months that preceded collection. Characterisation and sampling of these two sites was as described previously.

Burrowing cycle

A second field study was conducted to monitor the burrowing activities of *H. scabra* in two areas of 100–150 m^2 at Kogu Veke (Fig. 1). The first area was chosen for its high concentration of individuals ≤100 mm and the second one for its extensive aggregation of larger specimens (>100 mm). Both areas had a minimum depth of 20 cm at low tide. In October 1997 and February 1998, the burrowing state of at least 25 individuals was recorded every 6 h over a 24-h cycle at each area. Upon sighting, the length of the sea cucumber (mouth to anus, contracted length) was assessed to the nearest 5 mm and its position (on the surface, semi-burrowed or burrowed) was determined as described by Mercier et al. (1999). The experiment was repeated twice in the same location, on a clear day and when it had been raining for over 6 h. Some observations were made during exceptionally high water temperature (≥30 °C) at the study areas.

Settlement and recruitment

Another field study was conducted to monitor newly-settled and early juvenile *H. scabra* <10 mm. In the laboratory, larvae settled preferentially on seagrass leaves and remained there until they reached ca. 10 mm (Mercier et al., 2000). Samples of seagrass leaves (mostly *Enhalus acoroides* and a few *Thalassia hemprichii*) were collected to determine the presence of newly-settled sea cucumbers in Kogu Veke (Fig. 1). Around 50–60 leaves were collected monthly 3–7 days before the full moon and 12–16 days after the full moon between April 1998 and April 1999, inclusive. These sampling dates were chosen to determine if there was a relationship between the lunar cycle and recruitment. The leaves were cut with scissors close to the sediment before being transferred gently to a plastic bag. Removal of leaves from water had been tested previously and found not to provoke the detachment of newly-settled sea cucumbers. Within a few minutes, 1000 ml of 99% ETOH was added to the bag. The sample was shaken gently every 15 min. for 1 h. The content of the bag was poured through an 80 μm sieve into sampling bottles. The leaves were washed thoroughly with fresh water to detach all remaining larvae from the leaves. The material was concentrated and stored in 99% ETOH before being observed under a light microscope. Identification of young *H. scabra* was performed using ossicle morphology (Massin et al., 2000). Each sea cucumber was measured (contracted length) under the microscope using a graduated ocular. The abundance of other organisms settled on the leaves was noted also.

Movement of juveniles on different substrata

A separate field study was conducted to examine the daily and long-term movements of juvenile *H. scabra* on different substratum types. Considering the relatively low abundance of juveniles found in the field, cultured juveniles were used for this investigation, similar to the ones used in a previous behavioural study (Mercier et al., 1999). The experimental site was a small bay located ca. 45 km west of Honiara on Guadalcanal (Fig. 1). The bay was chosen because it possessed the physico-chemical attributes of environments that supported populations of *H. scabra* in Solomon Islands and was accessible on a daily basis. In addition, no *H. scabra* occurred there, which made monitoring of the introduced specimens easier.

A total of 360 juveniles, equally divided in three size classes (>10–40 mm; >40–80 mm; >80–120 mm), was released in different habitats within the bay. The first habitat was characterised by a dense seagrass bed, the second had a uniform sandy bottom and the third was composed mainly of shells, crushed coral and dead coral pebbles. The three habitats were located within a 35 000 m^2 area bordered by mangroves. As for the first field study, each habitat was characterised using granulometry and OM content of the sediment, current velocity, temperature and salinity, using the techniques previously described. A total of 120 sea cucumbers (40 of each size-class) was released in each of the three habitats on 2 April 1998 around 18.00h. The exact release points in each habitat were marked by a plastic rod planted into the sediment. In each of the habitats, 10 juveniles of each size class were marked individually by scratching the dorsal surface prior to their release using a method similar to that of Reichenbach (1999). Preliminary experiments demonstrated that the mark remained visible for about 10 days before disappearing and did not induce any detectable behavioural change. The position of marked sea cucumbers was subsequently monitored daily for 7 days. The distance moved, and the angle with respect to the reference point, were noted to establish polar plots of movements (it is important to note that the daily recapture of sea cucumbers accounted only for the net displacement and did not show the exact extent of territory covered in a day). Only juveniles that were identified positively every day for 6 consecutive days were used for data analyses. Observations at the site were also made 1 week, 2 weeks, 1 month and 2 months after the initial release to recapture and measure as many *H. scabra* as possible.

Aggregative behaviour

Clusters of *H. scabra* were observed periodically in the field during the course of investigations at Kogu Veke and Kogu Halingi. We examined the aggregative behaviour of adults in five large 3000–4000 l outdoor tanks supplied by a flow-through system of natural seawater at a rate of ca. 360 l h^{-1}. Each tank was filled with ca. 30 cm of sand and contained eight adult sea cucumbers >250 mm. The position of the 40 sea cucumbers was recorded daily between

07.00h and 15.00h from February to May 1998. The number of individuals showing aggregative behaviour was established by noting pairs, trios and groups of more than three. Aggregation was considered to occur between individuals when they were <10 cm apart. The grouped animals were sexed on one occasion, using a syringe to collect gametes through the body wall, in order to verify if the aggregations were sex related. Cases of spawning were recorded. Similar observations of aggregative behaviour were carried out using smaller cultured juveniles distributed in five 140-l outdoor concrete tanks supplied with 300 l h^{-1} of seawater and containing ca. 10 cm of sand. The individuals used were categorised into size classes: <10 mm, 10–50 mm, >50–100 mm, >100–150 mm, and stocked into tanks in groups of 8–20 individuals depending on their size. Observations were undertaken as described above. All experiments were monitored under a natural day/night cycle.

Statistical analysis

One-way analysis of variance (ANOVA) was used to compare the densities and displacement of sea cucumbers. The effect of different sedimentary features (grain size, organic matter) on the spatial distribution, with depth as a co-variable, was tested using analysis of covariance (ANCOVA) (α=0.05). Homogeneity of variances was assessed using Cochran's test (α=0.05) and, where necessary, transformations, either natural log (x) or $\sqrt{x}$, were carried out to normalise data and homogenise variances. Movement and orientation by juvenile *H. scabra* were tested for randomness with the Rayleigh test. The directionality of movement by individuals was analysed via a second-order analysis, which allowed inferences to be made on the populations of individuals examined. In this case, the cumulative frequency distribution of the length of the mean vectors 'r' (the Raleigh statistic) of each individual's directions of movement was compared to the theoretical distribution of 'r' for the same sample size by the Kolmogorov-Smirnov goodness of fit test (Batschelet, 1981).

Results

Distribution of Holothuria scabra

Description of the habitat

The bay of Kogu Veke (Fig. 1) covered an area of ca. 12 000 m^2 in a semi-enclosed lagoon with no freshwater input except for rain. The site had a gentle slope <4°, and was characterised by *Enhalus acoroides* and *Thalassia hempricii* beds on sandy and/or muddy sediment, and by coarse coral and shell substrata (Fig. 2). The northern limit of the study site was bordered uniformly by an extensive mangrove swamp inundated at high tide for a distance of ca. 70 m. A small channel was carved by the tidal flux at the head of the bay. The subtidal area along the southern limit was characterised by the presence of numerous coral patch reefs. About 49% of the study site was exposed at low tide (excluding the mangrove area), while the deepest areas had a maximum depth of ca. 3 m (Fig. 2). The bay was protected from storms by its geographical location and limited fetch. Mixed tides generated currents and gyres as strong as 10–20 cm s^{-1} in some areas with an average of ca. 3–5 cm s^{-1} for the entire bay, depending on the tidal cycle (Fig. 2). On clear days, the pH varied between 8.4 and 8.8, dissolved oxygen was 5.9–6.4 mg l^{-1} and the horizontal water transparency varied from ca. 1 to 5.5 m.

Sediment in the low-energy section near the mangrove swamp, especially in the channel, was associated with the highest OM content (30–74%) and the smallest particles (0.03±0.02 mm in diameter) which formed a fine, unstable silt (Fig. 2). Sediments with between 10 and 30% OM surrounded the channel and portions of the exposed shore. Lesser OM content was found in or around the seagrass bed, in deeper waters and over the shell and coral substratum (particles of 5.1±0.9 to 12.6±3.2 mm in diameter). The intertidal zone was composed mainly of muddy sand (0.09±0.02 mm) fringed on both sides by shells and coral pebbles. Deeper locations were characterised by sandy substrata (0.36±0.08 to 0.78±0.11 mm in diameter) containing <5% OM (Fig. 2).

Water temperatures in Kogu Veke varied according to location and with the tidal cycle. At high tide, the temperature was higher near the mangrove swamp (around 27–29 °C) and decreased progressively to become a uniform 26 °C over the subtidal area. At low tide, the surface temperature reached ca. 37 °C near the upper channel area and progressively decreased to 25–27 °C over the subtidal area. These observations corresponded to the slack water tide interval on a sunny day. Salinity under the same conditions fluctuated from 35 near the mangrove swamp to 33 in the open bay. However, the salinity on a rainy day dropped down to 22–24 close to the channel and remained around 32–33 over the subtidal section of the

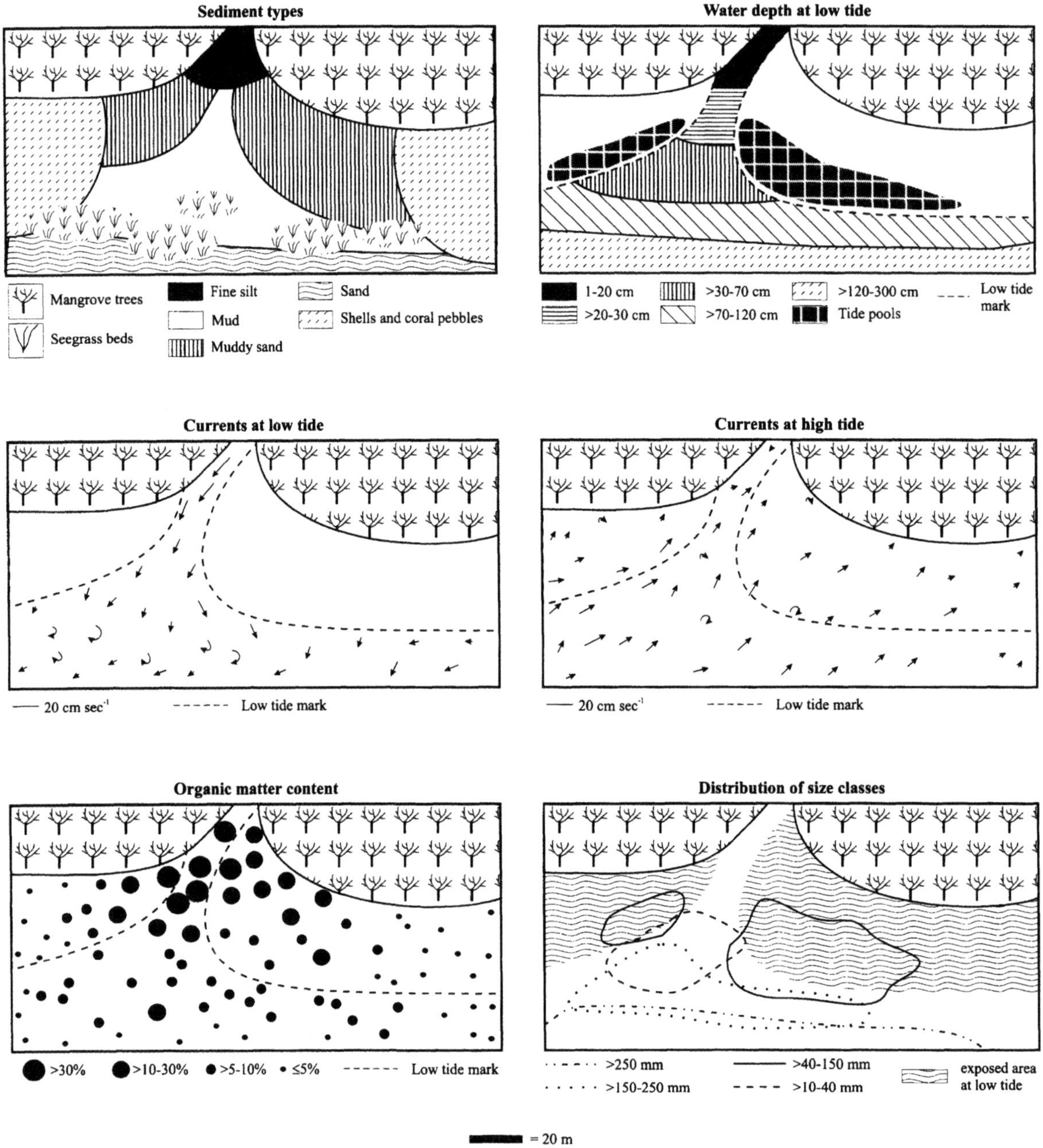

Figure 2. Schematic representation of Kogu Veke based on data from February 1998, illustrating substratum types, depth profiles (add ca. 45 cm for values at high tide), current conditions (ca. 30 min before the end of ebb tide and ca. 30 min before the end of flow tide), organic matter content and gross size-class distribution in the study site.

study site. Occasionally, and only for a short period of time, the salinity dropped to 10–15 in the upper intertidal zone and temperature increased at low tide to 36–37 °C in the entire bay under intense sun.

Spatial distribution and population structure

Totals of 90±4 and 134±6 *H. scabra* >100 mm were found in Kogu Veke in October 1997 and February 1998, respectively. These values were gathered by surveying the whole bay. Of these individuals, 28 females and 26 males were sexed successfully, giving a sex ratio that was not significantly different from 1:1 (χ^2, p>0.05). As for individuals <100 mm, 76 were recorded in October 1997 and 25 were observed in February 1998, using the quadrat method.

The total density of *H. scabra* >100 mm in Kogu Veke, from all substrata combined, was 0.75±0.03 ind. 100 m^{-2} in October 1997 and 1.12±0.05 ind. 100 m^{-2} in February 1998. However, the density of *H. scabra* estimated from the combined surveys was greater on mud (2.19±0.86 ind. 100 m^{-2}), muddy-sand (1.02±0.18 ind. 100 m^{-2}) and sand (0.86±0.08 ind. 100 m^{-2}) than on silt, coral pebbles and seagrass beds (<0.20 ind. 100 m^{-2}). Individuals <100 mm were concentrated mainly on muddy sand (34.75±29.03 ind. 100 m^{-2}) and mud (26.87±18.56 ind. 100 m^{-2}), although data were more variable and based on rough estimates from examination of ca. 3% of each substratum type. The highest density measured, pooling all size classes together, was 34 individuals on 10 m^2 of muddy substratum containing between 5 and 10% OM, at a depth of ca. 45 cm in an area that was never exposed at low tide (Fig. 2). Typically, individuals of all sizes avoided the upper channel area, which was covered with organically rich and fine silt, as well as the area at the margin of the mangrove swamp, which was exposed at low tide for 2–3 h longer than the lower intertidal zone. Areas of broken shell and coral pebbles were also sparsely populated (Fig. 2).

Holothuria scabra from Kogu Veke were distributed according to a depth gradient coupled with the effects of granulometry and OM content of the sediment (ANCOVA, p<0.01). The largest individuals, >250 mm, were located mainly in the deeper zone, >120 cm deep, on a sandy substratum (Figs 2 and 3). Animals ranging from 150 to 250 mm length were located mainly around seagrass beds, on the mud or muddy sand substrata in water 30–120 cm deep. Intermediate-sized individuals ranging from 40 to 150 mm were found on mud and muddy sand in the shallow water fringing the intertidal zone and in the intertidal zone itself (Figs 2 and 3). Individuals on the exposed portion of the substratum were always found burrowed at low tide and the small depression created by their burrowing retained water. No individuals were ever found exposed directly to the air. Finally, the smallest *H. scabra* (>10–40 mm) were usually observed on the mud and muddy sand substrata, sometimes inside the seagrass beds, in 20–120 cm of water. Thus, most *H. scabra* were on mud and muddy sand, although the largest size classes were located most commonly on sand (Fig. 3). Most individuals were found in areas of 5–10% OM (Fig. 4). The largest individuals occurred almost exclusively on those areas with <5% OM, whereas a few small and medium-sized individuals were localized on substrata with >10% of OM content and only four of them in areas of >30% OM content (Fig. 4).

The size-frequency distribution of *H. scabra* in Kogu Veke (Fig. 5) suggested multiple cohorts in the population with sizes varying from >10 mm to ca. 330 mm. Totals of 25 and 76 individuals ≤100 mm were observed in October 1997 and February 1998, respectively, while 90±4 and 134±4 individuals >100–330 mm were observed on the same dates. Conversely, only 45±5 individuals >100–330 mm were collected at Kogu Halingi on both sampling dates, indicating a population structure without recent recruitment (Fig. 5). Intermediate-sized individuals, 35±5 measuring between 120 and 280 mm, also dominated collections at Malmaragiri Inlet, with only three specimens <100 mm found in October 1997 and 7 in February 1998 (Fig. 5).

Burrowing cycle

The burrowing cycle of *H. scabra* in Kogu Veke varied according to environmental conditions. On a clear day, individuals ≤100 mm burrowed at sunrise and surfaced at sunset (Fig. 6). Individuals >100 mm presented almost the same but slightly dephased burrowing pattern, with individuals beginning to surface earlier in the middle of the afternoon. However, when it rained, 97% of individuals of both sizes classes remained burrowed all day. On two occasions, the water temperature increased to more than ca. 30 °C (with a peak of ca. 37 °C) and most individuals were found on the surface, remaining there until the temperature dropped below ca. 28–29 °C and did not follow their usual burrowing cycle.

Settlement and recruitment

Several *H. scabra* <10 mm were found on leaves of *E. acoroides* and *T. hemprichii* seagrass but never on other substrata (Fig. 7). The highest density of newly-settled and early juvenile *H. scabra*, corresponding to the smallest average size of 1.7±0.5 mm, were found typically before the full moon (Fig. 7). By contrast, the lowest densities, comprising individuals of 5.4±2.2 mm, were usually observed 12–16 days after the full moon. This pattern was repeated each month, except between September and November 1998. Also,

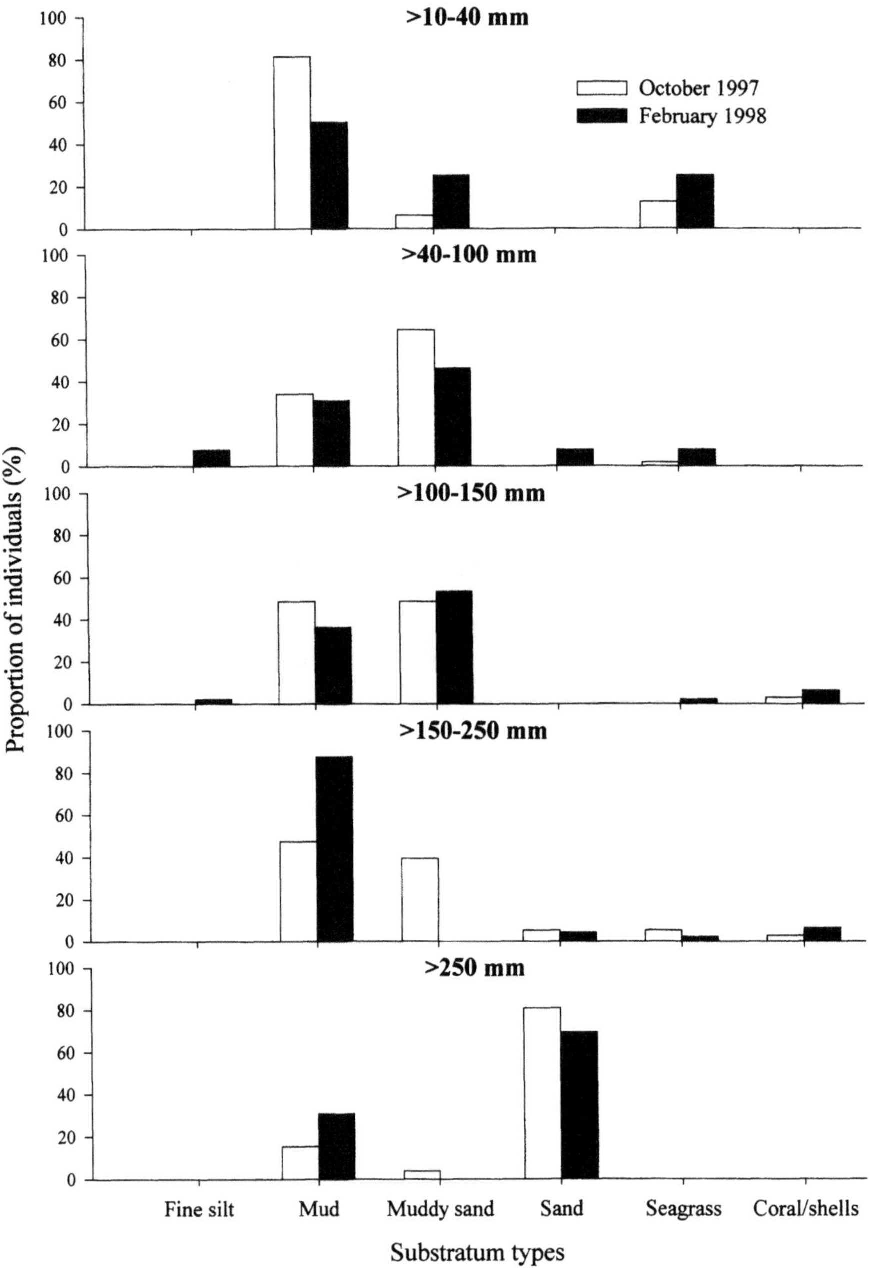

Figure 3. Length–frequency distribution of *Holothuria scabra* on the different substratum types of Kogu Veke in October 1997 and February 1998.

samples from December 1998 and January and February 1999 revealed very low or zero recruitment (Fig. 7). The occurrence of other fauna on the seagrass leaves, including numerous potential predators such as polychete worms, tectibranchs, crabs, gastropods, copepods, shrimps, flat worms and isopods, varied monthly in number and diversity. Maximum abundance of potential predators was correlated with peak recruitment of young *H. scabra* (Fig. 7). Newly-settled specimens of other species of sea cucumbers were also observed on the seagrass leaves, although it was impossible to identify exactly which species were involved. An average of 17±9 unidentified juveniles per sample was recorded.

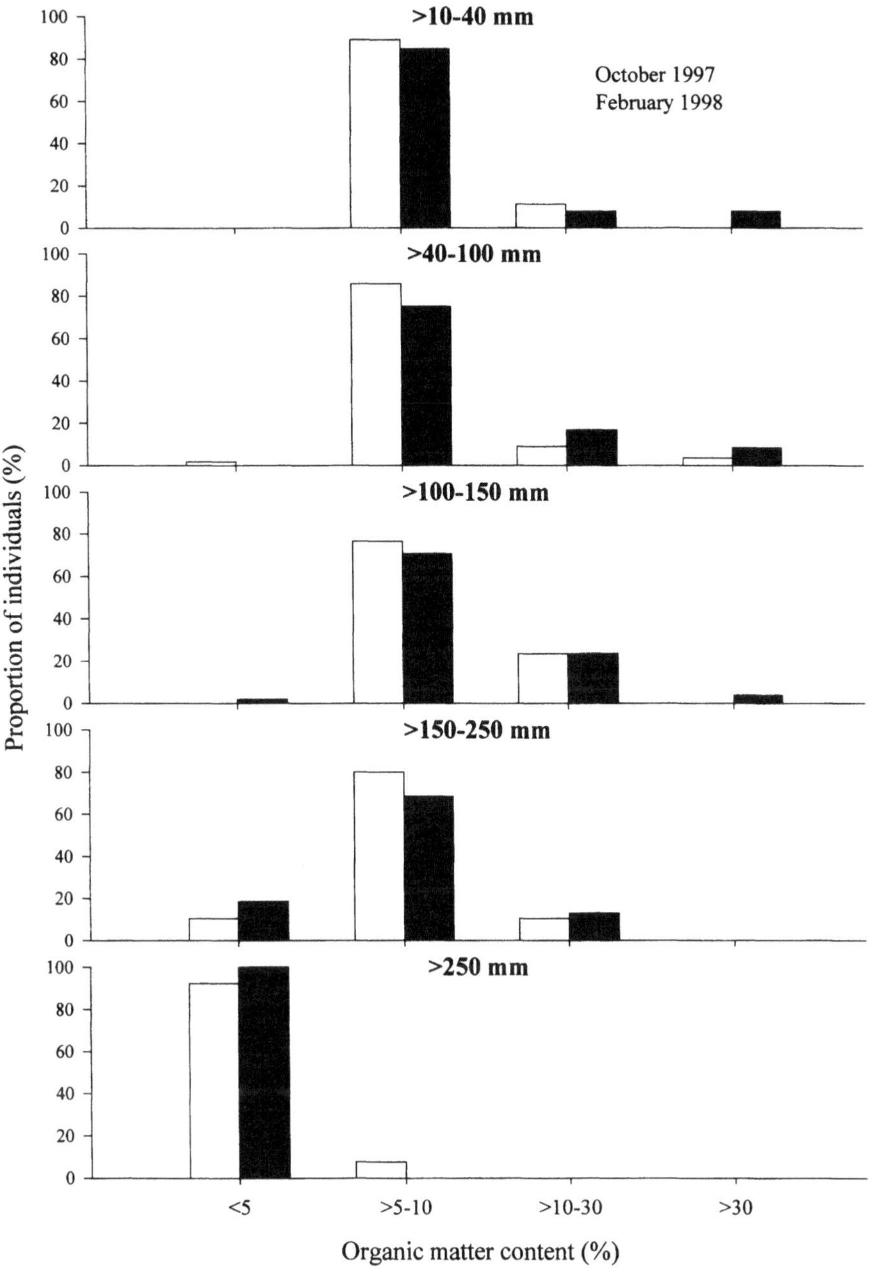

Figure 4. Length–frequency distribution of *Holothuria scabra* with respect to organic matter content in Kogu Veke in October 1997 and February 1998.

Movements of juveniles on different substrata

All three habitats surveyed on northwestern Guadalcanal were covered at low tide by 35–40 cm of water and had a slope <3°. They were bordered by a mangrove swamp and a fringing reef in a highly protected environment with maximum currents recorded of ca. 10 cm s^{-1}. Habitat 1 was composed of ca. 91% *E. acoroides* and *T. hemprichii* seagrass beds. The OM content of the sediment fluctuated between 7 and 15% and the substratum was muddy-sand. Habitat 2 was characterised by a sediment composed of shell and coral pebbles with <5% OM. Habitat 3 was characterised by a uniform sand substratum with a few coral

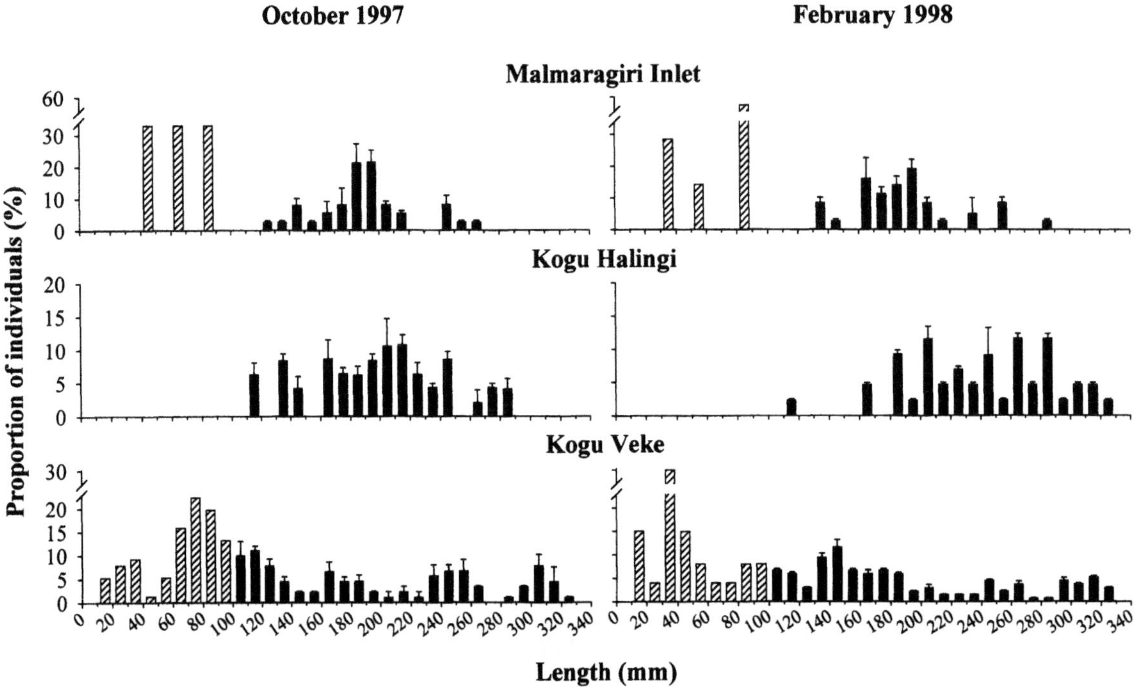

Figure 5. General population structure of *Holothuria scabra* in the three study sites in October 1997 and February 1998. For animals >100 mm, the sites were surveyed by three observers and the data are presented as mean±SE, n=3. For animals <100 mm, a quadrat method was used, hence the absence of error bars. Note the different y-axis scales.

boulders and an OM content fluctuating between 4 and 8%. Sporadic seagrass clumps were scattered among the habitats. Most sea cucumbers released in each of the three habitats remained resident, and the majority of marked individuals were recovered and measured every day. All size classes of young *H. scabra* released in the field in the three different habitats exhibited apparently random displacement (Rayleigh Test of directionality, $p<0.001$) changing direction every day (Fig. 8). Juvenile *H. scabra* tended to avoid being close to or under seagrass beds and were found mainly in clearings among the macrophytes.

The daily movement patterns for marked sea cucumbers are summarised in Figures 8 and 9. In general, the larger the sea cucumber, the greater distance travelled (One-Way ANOVA, $p<0.05$), except in the seagrass bed habitat (One-Way ANOVA, $p>0.05$). In all habitats, individual displacements remained relatively constant over the study period for all size classes. There was a significant difference in distance travelled when the three habitats were compared both between comparable size classes and across size classes (One-Way ANOVA, $p<0.05$). Regardless of size classes, the mix of shells and crushed coral resulted in the greatest displacement with ca. 80.2±9.1 cm d^{-1}. The seagrass bed habitat induced the second greatest movement with ca. 51.3±11.3 cm d^{-1} and the sand substratum induced the smallest movement with ca. 41.3±8.7 cm d^{-1} (Fig. 9). Movements with respect to OM content of the substratum in the three habitats were less conclusive and the highest mobility was recorded on the sediment of intermediate richness (One-Way ANOVA, $p>0.05$).

Growth differed amongst locations (χ^2, $p<0.01$) (Table 1). After 2 months, the largest sea cucumbers (239±57 mm) were found in the sandy habitat, having increased in size by 385%. Juveniles released in the seagrass bed were on average 203±61 mm, having increased by 327% and the smallest sea cucumbers recaptured at 156±43 mm were found among shells and crushed coral, with an average size increase of 252% (Table 1). No aggregation was observed in individuals <120 mm long, but larger individuals expressed an aggregative pattern similar to the one described below for adult sea cucumbers.

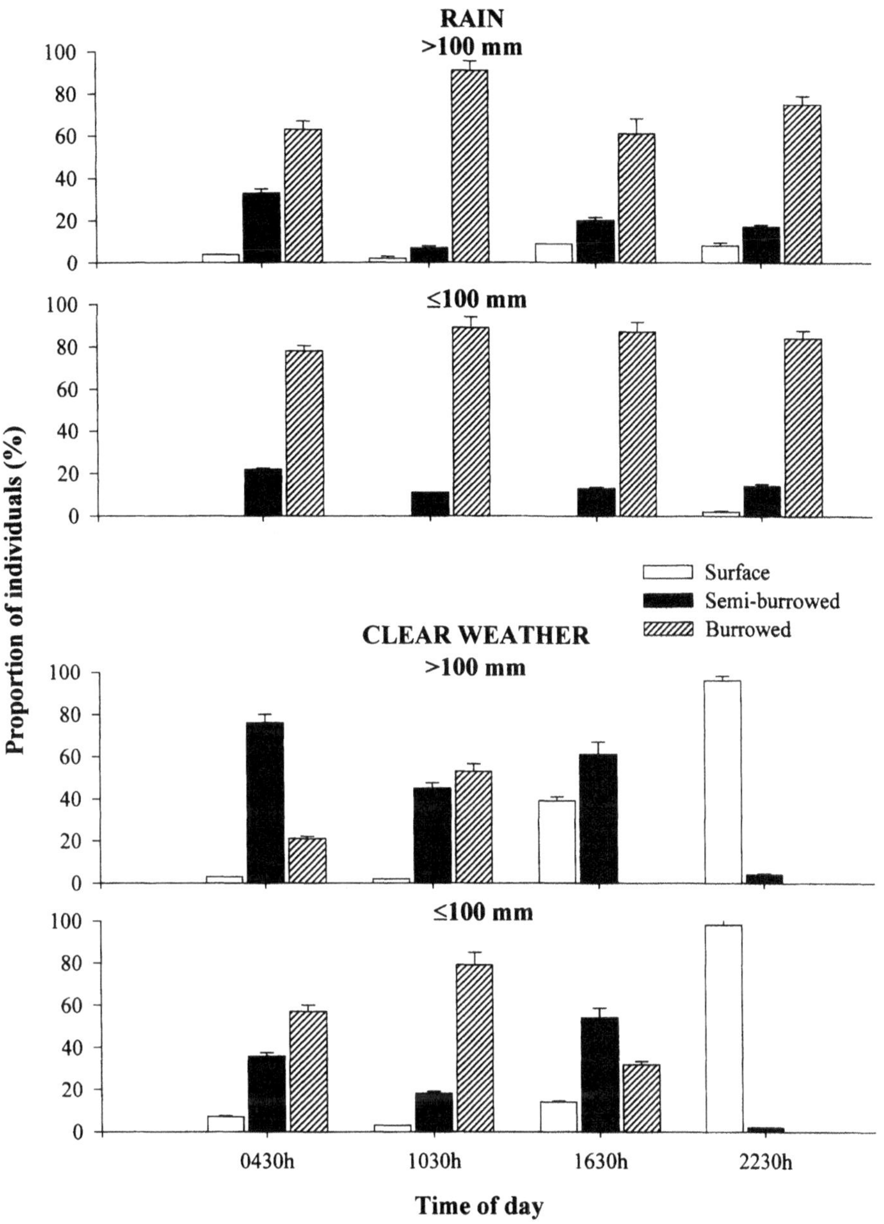

Figure 6. Daily burrowing cycle of two size classes of *Holothuria scabra* during rainy episodes and under clear weather conditions. Data were recorded in October 1997 and February 1998 (mean±SE, n=2).

Aggregative behaviour

Aggregation of *H. scabra* was monitored daily from February to May 1998 (Fig. 10). Most individuals remained away from each other during the absence of the moon, and started to form pairs, trios and larger groups progressively after the new moon. Aggregation peaked a few days before the full moon when >95% of the individuals participated. Subsequently, aggregation decreased until the next new moon. Most of the time, pair formations were more common than larger aggregations, with no progressive pattern. Male and female spawnings were observed in the tanks, typically around the full moon (Fig. 10). Spawning occurred both during the peak in aggregation and later on, when the individuals were more evenly distributed (Fig. 10). No aggregation was noted in individuals <110 mm in length.

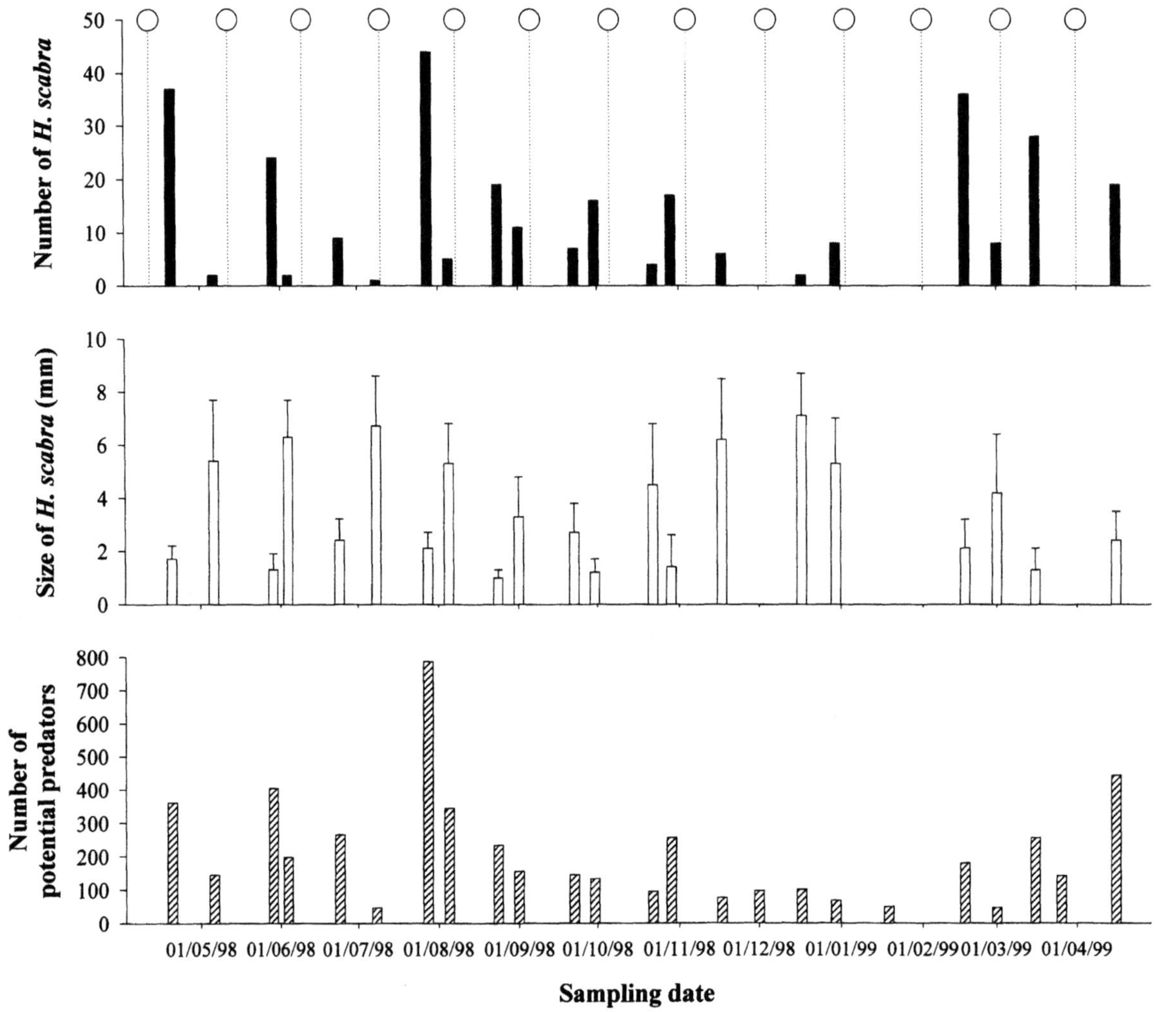

Figure 7. Number and size (mean±SE) of newly-settled *Holothuria scabra* found on seagrass leaves in Kogu Veke between April 1998 and April 1999. The abundance of other animals that could possibly prey on *H. scabra* is also shown. Data were collected 3–7 days before the full moon (shown as open circles) and 12–16 days after it.

Table 1. Mean length of *Holothuria scabra* juveniles prior to their release and 2 months later in the three different habitats of northwestern Guadalcanal (Mean±SE, *n*=29–38)

Habitats	Mean length (mm)		Estimated mean growth (mm d^{-1})
	Start	After two months	
Seagrass bed	65±26	203±61	2.3
Shells and coral pebbles	62±27	156±43	1.5
Sand	64±31	239±57	2.9

Discussion

The study revealed that the distribution of *H. scabra* on the soft substrata of Kogu Veke was heterogeneous. The location and abundance of the different size classes were correlated with the characteristics of the habitat such as depth, granulometry, richness of the substratum and presence of seagrass beds. Additional factors such as temperature, salinity, burrowing habits and aggregative behaviours influenced the finer distribution on daily to monthly scales. Monthly pulses in the recruitment of *H. scabra* to seagrass beds were greatest from February to August, roughly corresponding to the dry season. Field movements of juveniles >60 mm was estimated to be random, but the

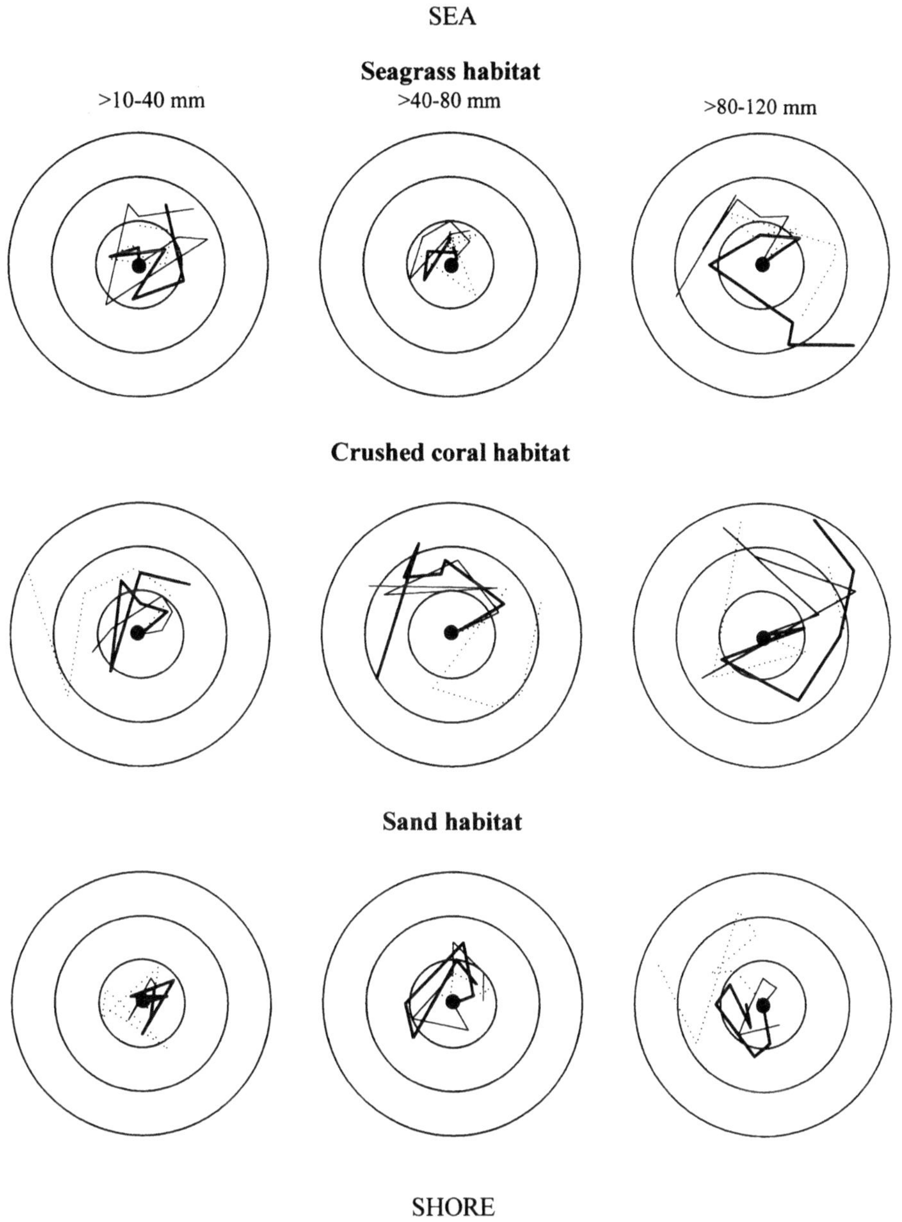

Figure 8. Movements of *Holothuria scabra* juveniles over 6 days. Data compiled from daily positions with respect to a central mark are presented for three individuals of each size-class in the three different habitats. The distance between the concentric circles is 50 cm.

net daily distance travelled and growth rates were a function of substratum type, suggesting that the young foraged in search of the most suitable feeding grounds.

Spatial distribution

Depth-dependent distributions of holothurians have been observed previously. Conand (1993) suggested that the occurrence of large *Stichopus variegatus* in deeper waters was a result of a size-related migration, whereas Bulteel et al. (1992) postulated that factors such as water movement, food availability and seagrass density could affect the distribution of *H. tubulosa*: small individuals favoured dense seagrass beds and larger specimens were found in more denuded areas, possibly because of an increased accessibility to food, resulting in a greater growth rate. This correlates well with the present study: *H. scabra* >10–

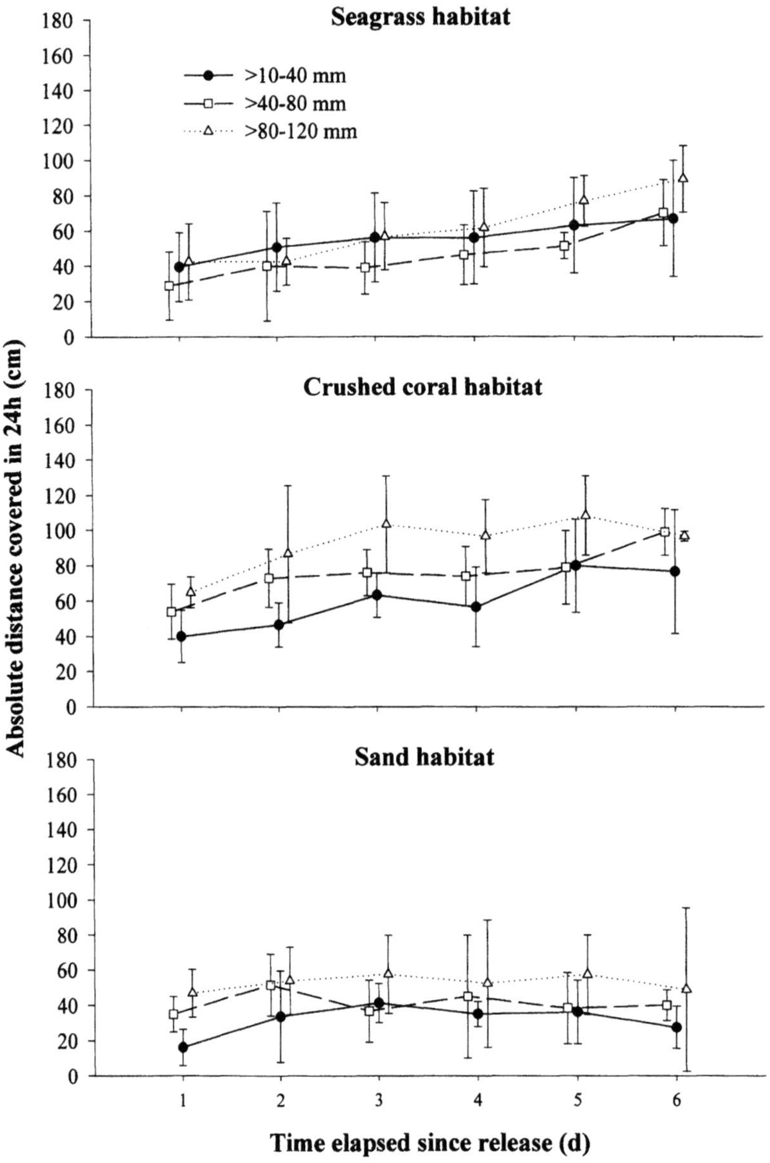

Figure 9. Daily displacement of *Holothuria scabra* juveniles of each size class over 1 week in three different habitats (mean ± SD; n= 7).

150 mm were often found close to seagrass beds, while larger individuals were not usually observed close to vegetation, most probably because the presence of dense root systems prevented their normal burrowing and feeding behaviours. The release of juvenile *H. scabra* in the Guadalcanal site supports this hypothesis since more limited movements and higher growth rates were recorded in sand habitats without dense seagrass beds. Furthermore, most individuals observed within the beds avoided seagrass shoots and moved toward clearer areas.

In contrast, Pomory et al. (1998) indicated that *H. floridiana* occurred more frequently in deeper water where thick seagrass patches were found, and suggested this distribution was related to resource acquisition, pointing out that seagrass beds favoured the deposition of particulate matter, making large proportions of smaller grain sizes available. While the importance of seagrass in the promotion of food sources for deposit feeders is likely, the difference in the distribution of *H. floridiana* and *H. scabra* may only be a question of accessibility, since *H. scabra* is a peri-

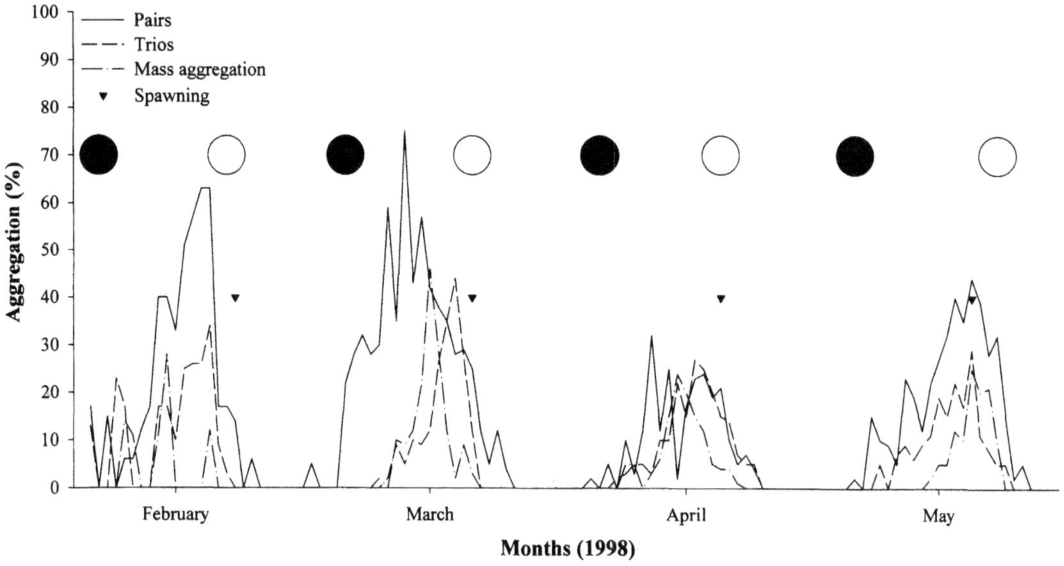

Figure 10. Aggregative behaviour (lines) and spawning (triangles) of adult *Holothuria scabra* in outdoor tanks between February and May 1998. Open circles indicate full moons and darkened circles represent new moons.

odic burrower. Furthermore, Long & Skewes (1997) observed that abundance of *H. scabra* in the Torres Strait, Papua New Guinea, increased with the percentage of seagrass cover, although they did not present fine scale data, making it impossible to determine whether the animals were located directly within the beds. While most adult *H. scabra* in the Solomon Islands were not found amongst seagrass, they were within a few meters of such habitats. On the other hand, newly-settled and young *H. scabra* <10 mm occurred exclusively on seagrass leaves. Our results thus suggest that seagrass beds represent the settlement substratum, offer shelter for newly-settled individuals, and supply food resources to nearby adults, by trapping detritus or through vegetal degradation.

Aside from the accessibility factor, the fact that larger *H. scabra* stay away from seagrass beds and prefer to inhabit deeper waters could be the result of their greater sensitivity to environmental variations. However, the possibility that selective harvest might have occurred from shallow areas within the bay cannot be ruled out. Choe (1963) proposed that local temperature gradients could explain the depth-related distribution in *Stichopus japonicus*. While permanent gradients are not likely to influence the size distribution of *H. scabra* in Solomon Islands, the increased sensitivity of larger specimens to varying water temperatures may force their migration toward slightly deeper and more stable areas. Previous studies by Mercier et al. (1999) showed that the burrowing cycle of *H. scabra* >40–140 mm was entrained by temperature variations, showing the sensitivity of juveniles to minor changes in temperature, and that smaller specimens were less affected by salinity fluctuations. Deeper distribution of large individuals associated with more stable environments was observed previously for the northern sea cucumber *Cucumaria frondosa* (Hamel & Mercier, 1996a) and the sea urchin *Strongylocentrotus droebachiensis* (Drouin et al., 1985).

In the present study, individuals exposed to high temperatures or low salinities modified their normal burrowing behaviour and feeding activity. They were found burrowed mostly during heavy rains, until the salinity returned to ambient conditions. Echinoderms have large areas of epidermal tissue exposed to the environment and lack differential organs for osmoregulation and are considered to be stenohaline (Binyon, 1966). Nevertheless, an increasing number of species are known to tolerate salinities below 20 (Binyon, 1966; Turner & Meyer, 1980). Our results showed that *H. scabra* can tolerate decreases in salinity but do not show any preference for brackish water. On the contrary, freshwater influence can cause a disturbance in their natural burrowing and feeding cycles (Mercier et al., 1999). Our data suggest that they favour shallow marine zones (<4 m depth) with high terrigenous input. This is consistent with the observations of Long & Skewes (1997) and Conand (1994), but differs from previous reports by James & James (1994) and James et al. (1994) who stipulated that this species preferred estuarine habitats and low salinities. We also observed that many *H. scabra* tended to remain on the surface for longer intervals during episodes of high water temperature, which can be explained by the fact that individuals >40 mm burrow in response to temperature variations and remain on the surface when temperature is above ca. 29 °C (Mercier et al., 1999), possibly to facilitate gas exchanges.

In general, *H. scabra* is found in areas which are sheltered, not greatly affected by wave and current action, and characterised by sediment that offers a compromise between stability and richness. Bakus (1973) indicated that numerous tropical holothurians were found in calm areas, which accumulate organic sediments. Similar distributions related to substratum type were found in *H. leucospilota* and *H. nobilis* (Shelley, 1981), and in *H. mexicana* and *Isostichopus badionotus* (Hammond, 1982). Conand & Chardy (1985) and Massin & Doumen (1986) also observed defined preferences in other tropical holothurians. We have found that *H. scabra* are associated with rich and relatively fine substrata (Mercier et al., 1999), but this apparent preference has clear boundaries. The avoidance of crushed shells and coral fragments is easy to understand, as they are more difficult to process for food and to burrow into. But *H. scabra* also avoid very rich substrata containing over 30% OM, which are usually extremely fine and possibly anoxic under the first few millimeters. The preferred substratum of *H. scabra* appears to combine a relatively high OM content with a grain size that allows easy burrowing while providing more stability than muddy silt. A compromise also exists between the presence of this preferred substratum and excessive exposure to air in the higher littoral zone of Kogu Veke. *Holothuria scabra*, especially specimens of small and intermediate sizes, respond by burrowing into the sediment when exposed during low tide.

Density, size frequency and recruitment

Between October 1997 and February 1998, the density of *H. scabra* >100 mm increased in Kogu Veke while that of small specimens decreased. This suggests that recruitment to the population takes place more commonly during the dry season. Approximately five *H. scabra* cohorts of various strengths appear to have been present in Kogu Veke. In contrast, less regular recruitment appears to take place at the other two sites as reflected in the irregular size–frequency distributions observed in Kogu Halingi and Malmaragiri Inlet. The reasons appear to be the influence of local harvesting for beche-de-mer which reduces the spawning biomass at these sites and possibly poor recruitment conditions (i.e. reduced water exchange or greater exposure to wind and stronger current flow; Hamel et al., 1999). In a brief report, Lokani et al. (1996) mentioned the occurrence of uni- and bimodal size distributions of *H. scabra* in Papua New Guinea, with respective mean sizes of 180 and 200 mm. Long & Skewes (1997) reported similar data with a conspicuous absence of individuals <100 mm. The data reflect the difficulty in finding small specimens, but probably also indicate a lack of recruitment as a result of overfishing as we have hypothesised for Kogu Halingi and Malmaragiri Inlet. Reported size-frequency distributions are very variable among other tropical holothurian species. For instance, Massin & Doumen (1986) noted uni- and bimodal length-frequency distributions of *H. atra* in Papua New Guinea, Seeto (1994) described a unimodal population structure of *H. atra* from Fiji, and Reichenbach (1999) observed uni-, bi- and multimodal weight class distributions of *H. fuscogilva* in the Maldives. Differences may well be owing to different reproductive strategies and migration patterns, but also to the difficulty involved in sampling very recent recruits <1 year old, which are rarely observed (Long & Skewes, 1997) on account of their cryptic nature and small size (Seeto, 1994). To our knowledge, distribution and growth of small recruits have been reported previously only for the temperate holothurians *Parastichopus californicus* (Cameron & Fankboner, 1989) and *C. frondosa* (Hamel & Mercier, 1996a).

It has been suggested that starfish recruitment is regulated by the proximity of the adult population to an adjacent nursery area, with juveniles gradually entering the adult population after they have attained a relatively large body size (Birkeland et al., 1971; Barker, 1977; Scheibling, 1980; Rumrill, 1989). Such a scenario was observed in the present study. In fact, a clear portrait can be drawn of *H. scabra* reproduction and recruitment when combining data from laboratory experiments and field surveys gathered during our present and previous studies of the species. First, the adults tend to aggregate increasingly toward the full moon. Spawning has been estimated to occur in the following days, as evidenced by direct observation in outdoor tanks and by recruitment data: large numbers of small recruits were found in the middle of the lunar cycle, whereas small numbers of larger recruits were observed just before the full moon. Moreover, we noted that juveniles remained approximately 4 weeks on seagrass leaves before moving onto underlying sand when reaching ca. 10 mm in length, (Mercier et al., 2000), a size at which they begin to show the typical burrowing behaviour (Mercier et al., 1999). This delay, which might even be shorter in nature, correlates well with the field data: monthly seagrass samples always showed the growth of new recruits with no overlapping of newly-settled cohorts. Thus, spawning seems to generate recruitment pulses on seagrass leaves roughly every month in Kogu Veke, which could account for the presence of a distinct cohort of juveniles >10–50 mm. A similar phenomenon has been observed previously in starfish populations (Barker & Nichols, 1983; Guillou & Guillaumin, 1985; Rumrill, 1989). Having attained a shelter size and acquired the ability to burrow and feed on sand, young *H. scabra* can move from the nursery seagrass beds to nearby mud-sand substrata, as observed with the starfish *Oreaster reticulatus* (Scheibling, 1980). Settlement on seagrass and subsequent migration to deeper habitats have also been suggested to occur in the sea cucumber *H. fuscogilva* (Reichenbach, 1999).

The possible influence of lunar cycles on holothurian reproduction has received little attention. Babcock et al. (1992) monitored the spawning events of several holothurians on the Great Barrier Reef and estimated that at least a few, including *Bohadschia argus*, *Euapta goddefroyi* and *S. chloronotus* typically spawned during or shortly after the full moon. These observations are consistent with our own data on *H. scabra*. In contrast, Engstrom (1980) reported that *H. scabra* had semiannual breeding cycles in Japanese waters, induced by dilution from rain, but differences associated with latitudinal factors are likely to be involved. In fact, Ong Che & Gomez (1985) observed *H. scabra* spawning several times a year at irregular intervals in the Philippines and Tuwo (1999) observed that *H. scabra* had two major spawning periods in Indonesia, but specified that mature gonadal tubules that released gametes could be observed throughout the year. Furthermore, based on gonadal observations, Ramofafia et al. (unpublished data) were able to detect up to 10–12 spawnings of *H. scabra* in Solomon Islands over a year. The monthly variation in the abundance of juvenile *H. scabra* recorded on seagrass leaves during the present survey could, therefore, be a result of partial spawning events. Alternatively, the presence of larval and juvenile predators on the leaves and the changing environmental conditions could play a role in the variable recruitment of *H. scabra*. The last hypothesis is supported by the fact that both *H. scabra* recruits and other fauna were less abundant on seagrass leaves during the wet season, between September and February.

Other factors affecting distribution

The well marked burrowing cycle of *H. scabra* drastically affects their vertical distribution at any given time. Individuals of various sizes have different periodicities in their cycle, and environmental factors such as salinity and temperature can modify their burrowing behaviours (Mercier et al., 1999). Additionally, aggregations appear to modify the distribution of adults on a monthly basis. Since no clusters were observed in individuals <120 mm, a size around which they reach sexual maturity (Mercier et al., 1999), and given the fact that maximum aggregation typically occurred just before spawning events, aggregative habits of *H. scabra* appear related to reproduction. Breeding aggregations have been observed in a number of asteroids (Ormond et al., 1973; Scheibling, 1980; Komatsu, 1983; Slattery & Bosch, 1993; Hamel & Mercier, 1995) and echinoids (Levitan et al., 1992; Young et al., 1992), and many authors suggest that aggregations minimize sperm dilution and increase fertilization success. Recently, Rodgers & Bingham (1996) observed aggregations of the sea cucumber *Cucumaria lubrica* and proposed that they were a result of their subtidal zonation in response to light. To our knowledge, aggregations related to reproduction have only been suggested to occur in deep-sea holothurians (Tyler et al., 1992), although inter-individual chemical communication during reproductive processes has been demonstrated in *C. frondosa* by Hamel & Mercier (1996b, 1999).

Movements of juveniles on different substrata

Juvenile *H. scabra* >10–120 mm moved a mean distance between 40 and 80 cm in a day, depending on substratum type. Larger mean daily movements have been reported previously for several species (Hammond, 1982; Da Silva et al., 1986). Nonetheless, results similar to our own have also been published; those of Hammond (1982) who measured daily movements of ca. 50 cm in *H. mexicana* and *I. badionotus* over 10–11 days, Rodgers & Bingham (1996) who noted a maximum displacement of 139 cm over 8 days (ca. 17 cm d^{-1}) in *C. lubrica*, and Reichenbach (1999) who estimated from bimonthly observations that *H. fuscogilva* moved about 2 m d^{-1}. The fact that our data are associated mostly with juvenile rather than adult holothurians certainly accounts for part of the variance, especially since a significant relationship between size and distance travelled was established in our study. Another factor is the difference in day lengths associated with the respective geographical locations, and the variable abundance of food sources in tropical versus temperate habitats, as proposed by Da Silva et al. (1986) and discussed below. It may be that a bias could also arise from the relocation of juveniles in our study. The fact remains that it would have been very difficult to record juvenile migrations in their true habitat owing to low densities.

Such manipulation was necessary to test the effect of different habitats, which allowed us to determine that the mobility of juveniles was related to substratum type and, to a second degree, food availability. The foraging activity was minimal on the optimal substratum of sand with intermediate OM content. Conversely, individuals of all sizes moved greater daily distances on pebbles and coral fragments, which do not allow them to burrow or feed as easily. Although seagrass beds offer the richest sediments, juveniles seem to avoid them, probably for the same reasons as adults, as discussed above in the distribution section. The growth patterns of juveniles in the different habitats support further the hypothesis that greater movements were associated with poorer food acquisition. Despite this, only a small number of juveniles dispersed outside the monitoring area, inferring that resources were not so limited that they drove animals from the study site. Undirectional movements, as observed in our study, are often associated with homogenous food resources (Levinton, 1972; Hammond, 1982; Da Silva et al., 1986). The patterns observed during this trial supported the observed distribution of *H. scabra* in Kogu Veke by showing that, upon leaving the nursery seagrass shoots, juveniles can progressively move away from them during their normal foraging activities and settle on muddy and/or sandy patches of intermediate richness where their growth is optimum.

Acknowledgements

We would like to thank Dr Claude Massin (Institut Royal des Sciences Naturelles of Belgium) and Dr. Tony Larkum (University of Sydney) for helping us with the identification of young sea cucumbers and seagrass species, respectively. We are grateful to Thomas Tabounaba, Maxwell Saurongo, Evizel Seymour, Christain Ramofafia, Idris Lane and the staff of ICLARM Coastal Aquaculture Centre and Nusa Tupe Station, for their precious help during the field and laboratory work. Thanks to Dr Johann Bell for his comments on the draft manuscript. This research was funded by the Canadian International Development Agency (CIDA) under the CGIAR-Canada Linkage Fund Programme. This is ICLARM contribution no 1543.

References

Babcock, R., C. Mundy, J. Keesing & L. Oliver, 1992. Predictable and unpredictable spawning events: *in situ* behavioural data from free-spawning coral reef invertebrates. Invert. Reprod. Develop. 22: 213–228.

Bakus, G. J., 1973. The biology and ecology of tropical holothurians. In Jones, O. A. & R. Endean (eds), Biology and Geology of Coral Reefs. Academic Press 2: 325–367.

Barker, M. F., 1977. Observations on the settlement of the brachiolaria larvae of *Stichaster australis* (Verrill) and *Coscinasterias calmaria* (Gray) (Echinodermata: Asteroidea) in the laboratory and on the shore. J. exp. mar. biol. Ecol. 30: 95–108.

Barker, M. F. & D. Nichols, 1983. Reproduction, recruitment and juvenile ecology of the starfish, *Asterias rubens* and *Marthasterias glacialis*. J. mar. biol. Ass. U.K. 63: 745–765.

Baskar, B. K., 1994. Some observations on the biology of the holothurian *Holothuria (Metriatyla) scabra*. Bull. Cent. Mar. Fish. Inst. 46: 39–43.

Batschelet, E., 1981. Circular Statistics in Biology. Academic Press Inc., New York: 371 pp.

Battaglene S. C. & J. D. Bell, 1999. Potential of the tropical Indo-Pacific sea cucumber, *Holothuria scabra*, for stock enhancement. In Howell, B. R., E. Moskness & T. Svasand (eds), Stock Enhancement and Sea Ranching. Bergen, Norway, Blackwell Science, Oxford: 478–490.

Battaglene, S. C., J. E. Seymour & C. Ramofafia, 1999. Survival and growth of cultured juvenile sea cucumbers *Holothuria scabra*. Aquaculture 178: 293–322.

Binyon, J., 1966. Salinity tolerance and ionic regulation. In: Boolootian, R. A. (ed.), Physiology of Echinodermata. Interscience Publishers, New York: 359–377.

Birkeland, C., F.-S. Chia & R. R. Strathmann, 1971. Development, substrate selection, delay of metamorphosis and growth in the seastar *Mediater aequalis*. Biol. Bull. 141: 99–108.

Bulteel, P., M. Jangoux & P. Coulon, 1992. Biometry, bathymetric distribution, and reproductive cycle of the holothuroid *Holothuria tubulosa* (Echinodermata) from Mediterranean seagrass beds. P.S.Z.N. I: Mar. Ecol. 13: 53–62.

Cameron, J. L. & P. V. Fankboner, 1989. Reproductive biology of the commercial sea cucumber *Parastichopus californicus* (Stimpson) (Echinodermata: Holothuroidea). II. Observations on the ecology of development, recruitment and the juvenile life stage. J. exp. mar. biol. Ecol. 127: 43–67.

Choe, S., 1963. Biology of the Japanese common sea cucumber *Stichopus japonicus*, Selenka. Pusan National University, Pusan: 226 pp.

Conand, C., 1993. Ecology and reproductive biology of *Stichopus variegatus* an Indo-Pacific coral reef sea cucumber (Echonodermata: Holothuroidea). Bull. mar. Sci. 52: 970–981.

Conand, C., 1994. Les holothuries, ressource halieutique des lagons. Rapport scientifique et techniques, Centre de Nouméa, Orstom 65: 86 pp.

Conand, C. & P. Chardy, 1985. Les holothuries aspidochirotes du lagon de Nouvelle-Calédonie sont-elles de bons indicateurs des structures récifales? Proc. Fifth Int. Coral Reef Congr. 5: 291–296.

Conand, C. & A. Tuwo, 1996. Commercial holothurians in South Sulawesi, Indonesia: fisheries and mariculture. SPC Beche-de-mer Inf. Bull. 8: 17–44.

Da Silva, J., J. L. Cameron & P. V. Fankboner, 1986. Movement and orientation patterns in the commercial sea cucumber *Parastichopus californicus* (Stimpson) (Holothuroidea: Aspidochirodita). Mar. Behav. Physiol. 12: 133–147.

Drouin, G., J. H. Himmelman & P. Béland, 1985. Impact of salinity fluctuations on echinoderm and mollusc populations. Can. J. Zool. 63: 1377–1387.

Engstrom, N. A., 1980. Reproductive cycles of *Holothuria floridiana, H. (H.) mexicana* and their hybrids (Echinodermata: Holothuroidea) in southern Florida. Int. J. Invert. Repro. 2: 237–244.

Guillou, M. & A. Guillaumin, 1985. Variations in the growth rate of *Asterias rubens* (L.) from west and south Brittany (France). In Keegan, B. F. & B. D. S. O'Connor (eds), Proc. Int. Echinod. Conf., Galway. A.A. Balkema, Rotterdam: 513–521.

Hamel, J.-F. & A. Mercier, 1995. Prespawning behavior, spawning and development of the brooding starfish *Leptasterias polaris*. Biol. Bull. 188: 32–45.

Hamel, J.-F. & A. Mercier, 1996a. Early development, settlement and growth of the sea cucumber *Cucumaria frondosa* (Holothuroidea: Echinodermata). Can. J. Fish. aquat. Sci. 53: 253–271.

Hamel, J.-F. & A. Mercier, 1996b. Evidence of chemical communication during the gametogenesis of holothurians. Ecology 77: 1600–1616.

Hamel, J.-F. & A. Mercier, 1999. Mucus as a mediator of gametogenic synchrony in the sea cucumber *Cucumaria frondosa* (Holothruoirea: Echinodermata). J. mar. biol. Ass. U.K. 79: 121–129.

Hamel, J.-F., P. K. L. Ng & A. Mercier, 1999. Life cycle of the pea crab *Pinnotheres halingi* sp. nov., an obligate symbiont of the sea cucumber *Holothuria scabra* Jaeger. Ophelia 50: 149–175.

Hammond, L. S., 1982. Patterns of feeding and activity in deposit-feeding holothurians and echinoids (Echinodermata) from a shallow back-reef lagoon, Discovery Bay, Jamaica. Bull. mar. Sci. 32: 549–571.

Holland, A., 1994. The status of global beche-de-mer fisheries with special reference to the Solomon Islands and the potentials of holothurian culture. MSc Thesis, University of Newcastle Upon Tyne, U.K.: 134 pp.

James, D. B. & P. S. B. R. James, 1994. A handbook on Indian sea-cucumbers. Central Marine Fisheries Research Institute Special publication 59, Cochin, India: 43 pp.

James D. B., A. D. Gandhi, N. Palaniswamy & J. X. Rodrigo, 1994. Hatchery techniques and culture of the sea-cucumber *Holothuria scabra*, Central Marine Fisheries Research Institute Special Publication 57, Cochin, India: 47 pp.

Kerr, A. M., E. M. Stoffel & R. S. Yoon, 1993. Abundance distribution of holothuroids (Echinodermata: Holothuroidea) on a windward and leeward fringing coral reef, Guam, Mariana Islands. Bull. mar. Sci. 52: 780–791.

Komatsu, M., 1983. Development of the sea-star, *Archaster typicus*, with a note on male-on-female superposition. Annot. Zool. JPN 56: 187–195.

Krishnan, S., 1968. Histochemical studies on reproductive and nutritional cycles of the holothurian, *Holothuria scabra*. Mar. Biol. 2: 54–65.

Lawrence, J. M., 1979. Numbers and biomass of the common holothuroids on the reef flat at Enewetak Atoll, Marshall Islands. Proc. European Echinoderms Colloquium 1: 201–204.

Levinton, J. S., 1972. Stability and trophic structure in deposit-feeding and suspension-feeding communities. Am. Nat. 106: 472–486.

Levitan, D. R., M. A. Sewell & F.-S. Chia, 1992. How distribution and abundance influence fertilization success in the sea urchin *Strongylocentrotus fransciscanus*. Ecology 73: 248–254.

Lokani, P., P. Polon & R. Lari, 1996.Management of beche-de-mer fhisheries in the western province of Papua New Guinea. SPC Beche-de-mer Inf. Bull. 8: 7–11.

Long, B. & T. Skewes, 1997. Distribution and abundance of beche-de-mer on Torres Strait reefs. South Pacific Commission Beche-de-mer Inf. Bull. 9: 17–22.

Massin, C., A. Mercier & J.-F. Hamel, 2000. Ossicle change in Holothuria scabra with a discussion of ossicle evolution within the Holothuriidae (Echinodermata). Acta Zool. 81: 77–91.

Massin, C. & C. Doumen, 1986. Distribution and feeding of epibenthic holothuroids on the reef flat of laing Island (Papua New Guinea). Mar. Ecol. Prog. Ser. 31: 185–195.

Mercier, A., S. C. Battaglene & J.-F. Hamel, 1999. Daily burrowing cycle and feeding activity of juvenile sea cucumber *Holothuria scabra* in response to environmental factors. J. exp. mar. Biol. Ecol. 239: 125–156.

Mercier, A., S. C. Battaglene & J.-F. Hamel, 2000. Settlement preferences and early migration of the tropical sea cucumber *Holothuria scabra*. J. exp. mar. Biol. Ecol. 249, 89–110.

Ong Che, R. G. & E. D. Gomez, 1985. Reproductive periodicity of *Holothuria scabra* Jaeger at Calatagan, Batangas, Phillipnes. Asian Mar. Biol. 2: 21–30.

Ormond, R. F. G., A. C. Campbell, S. H. Head, R. J. Moore, P. R. Rainbow & A. P. Saunders, 1973. Formation and breakdown of aggregations of the crown-of-thorns starfish, *Acanthaster planci* (L.) Nature 246: 167–169.

Pomory, C. M., T. W. Foret, S. K. Hill & J. M. Lawrence, 1998. Characteristics of a population of Holothuria floridiana from Florida. In Mooi, R. & M. Telford (eds), Echinoderms. Balkema, Rotterdam: 507–511.

Preston, G. L., 1990. Mass beche-de-mer production in Fiji. SPC Beche-de-mer Inf. Bull. 1: 4–5.

Reichenbach, N., 1999. Ecology and fishery biology of *Holothuria fuscogilva* (Echinodermata: Holothuroidea) in the Madives, Indian Ocean. Bull. mar. Sci. 64: 103–113.

Rodgers, S. A. & B. L. Bingham, 1996. Subtidal zonation of the holothurian *Cucumaria lubrica* (Clark). J. exp. mar. Biol. Ecol. 204: 113–129.

Rumrill, S. S., 1989. Population size-structure, juvenile growth, and breeding periodicity of the sea star *Asterina miniata* in Barkley Sound, British Columbia. Mar. Ecol. Prog. Ser. 56: 37–47.

Scheibling, R. E., 1980. Abundance, spatial distribution and size structure of populations of *Oreaster reticulatus* (Echinodermta: Asteroidea) on sand bottoms. Mar. Biol. 57: 107–119.

Seeto, J., 1994. The reproductive biology of the sea cucumber *Holothuria atra* Jaeger, 1833 (Echinodermata: Holothuroidea) in Laucala Bay, Fiji, with notes on its population structure and symbiotic associations. MSc Thesis, University of Otago, Dunedin, New Zealand.

Shelley, C. C., 1981. Aspects of the distribution, reproduction, growth and fishery potential of holothurians (beche-de-mer) in the Papuan coastal lagoon. MSc thesis, University of Papua New Guinea.

Shelley, C. C., 1985. The potential for re-introduction of a beche-de-mer fishery in Torres Strait. Torres Strait Fisheries Seminar, 1985, Port Moresby. Australian Gov. Pub. Service, Canberra.

Slattery, M. & I. Bosch, 1993. Mating behavior of a brooding Antarctic asteroid, *Neosmilaster georgianus*. Invert. Reprod. Devel. 24: 97–102.

Sloan, N. A. & B. Von Bodungen, 1980. Distribution and feeding of the sea cucumber *Isostichopus badionotus* in relation to shelter and sediment criteria of the Bermuda platform. Mar. Ecol. Prog. Ser. 2: 257–264.

Turner, R. L. & C. E. Meyer, 1980. Salinity tolerance of the brackish-water echinoderm *Ophiophragmus filograneus* (Ophiuroidea). Mar. Ecol. Prog. Ser. 2: 249–256.

Tuwo, A., 1999. Reproductive cycle of the holothurian *Holothuria scabra* in Saugi Island, Spermonde Archipelago, Southwest Sulawesi, Indonesia. SPC Beche-de-mer Inf. Bull. 11: 9–12.

Tyler, P. A., C. M. Young, D. S. M. Billett & L. A. Giles, 1992. Pairing behaviour, reproduction and diet in the deep sea holothurian genus *Parozira* (Holothuroidea: Synallactidae). J. mar. biol. Ass. U.K. 72: 447–462.

Uthicke, S., 1994. Distribution patterns and growth of two reef flat holothurians, *Holothuria atra* and *Stichopus chloronotus*. In David, B., A. Guille, J.-P. Féral & M. Roux (eds), Echinoderms Through Time. Balkema, Rotterdam: 569–576.

Webb, K. L., W. D. DuPaul & C. F. D'Elia, 1977. Biomass and nutrient flux measurements on *Holothuria atra* populations on windward reef flats at Enewetak, Marshall Islands. Proc. Third Int. Coral Reef Symp. 1: 409–416.

Wiedemeyer, W. L., 1994. Biology of small juveniles of the holothurian *Actinopyga echinites*: growth mortality and habitat preferences. Mar. Biol. 120: 81–93.

Young, C. M., P. A. Tyler, J. L. Cameron & S. G. Rumrill, 1992. Seasonal breeding aggregations in low-density populations of the bathyal echinoid *Stylocidaris lineata*. Mar. Biol. 113: 603–612.

Zhou, S. & T. C. Shirley, 1996. Habitat and depth distribution of the red sea cucumber *Parastichopus californicus* in southeats Alaska Bay. Alaska Fishery Res. Bull. 3: 123–131.

Hydrobiologia **440:** 101–109, 2000.
M.B. Jones, J.M.N. Azevedo, A.I. Neto, A.C. Costa & A.M. Frias Martins (eds), Island, Ocean and Deep-Sea Biology.

A 'minimum stress inflexion' in relation to environmental and biotic influences on the dynamics of subtidal encrusting communities?

Ben C. Maughan & David K. A. Barnes
Department of Zoology and Animal Ecology, University College Cork, Lee Maltings, Prospect Row, Cork, Ireland
E-mail: bmaughan@hotmail.com

Key words: encrusting community, competition, disturbance, sediment, stress, diversity

Abstract

The sedentary fauna of sublittoral boulders was studied at five sites with very different environmental conditions within Lough Hyne Marine Nature Reserve (51° 30′ N, 9° 18′ W). The degree of competition was assessed through the construction of competitive hierarchies built up from the results of nearly 3000 interactions between organisms. Communities at all sites were predominantly hierarchically organised with few stand-offs recorded. Density of interactions was correlated with total space occupation. Flow speed at all sites was logged over 24 h and this factor also correlated with space occupation. Degree of disturbance was assessed through the use of marked boulders at each site. Species diversity peaked where all factors were 'moderate' and was lowest at 'extreme' sites. The effects of these environmental variables and biotic factors in determining community composition are discussed, and a working model detailing a 'Minimum Stress Inflexion' is described. This represents an extension of the intermediate disturbance hypothesis in that it is proposed that disturbance in this system may serve to reduce sediment load as opposed to preventing space monopolisation in the classical model.

Introduction

The process by which species coexist and weaker competitors persist in the presence of stronger species (Hutchinson, 1959) has been of fundamental interest but still remains theoretical and debated in most instances. A wide variety of ideas has gained support and popularity at differing times and localities. These have included temporal environmental fluctuations (Hutchinson, 1961), niche differentiation (MacArthur & Levins, 1967; Knowlton & Jackson, 1994), variation in colonisation and recruitment (Paine, 1979; Underwood & Fairweather, 1989), variation in assemblage transitivity or networks vs hierarchies (Jackson, 1979), and disturbance mediated coexistence (Sousa, 1979). Over the last few decades, however, many studies have suffered from acute observer influences (Underwood & Fairweather, 1985; Benedetti-Cecchi et al., 1996), the stigma of departing from strong paradigms (Elner & Vadas, 1990) or a lack of comparative sites (McGuinness, 1990). A suggested ideal was if diversity could be compared over a wide range of environments within a study site along with potentially influencing environmental variables (Menge & Sutherland, 1987).

The marine nature reserve of Lough Hyne, in southwest Ireland encompasses a diverse set of environmental conditions within an area of less than 1 km^2. At the upper level of energetics, these include a complex wave-exposed oceanic headland and a turbulent channel with very high flow rates. There are sites with moderate or unidirectional flow and much of the remainder of the lough is subject to high water residence rates and, thus, high sedimentation (Bassindale et al., 1948). Previous work on boulder communities within the lough has described a diverse encrusting fauna (Lilly et al., 1953). That boulder shores are common around the globe and so convenient in portability and size range provision has made them an ideal substratum for investigation of various aspects of community dynamics.

In this study, we expand on the work of Lilly et al. (1953) and attempt to elucidate what role environmental factors and competitive processes play in influencing assemblage composition, complexity and diversity.

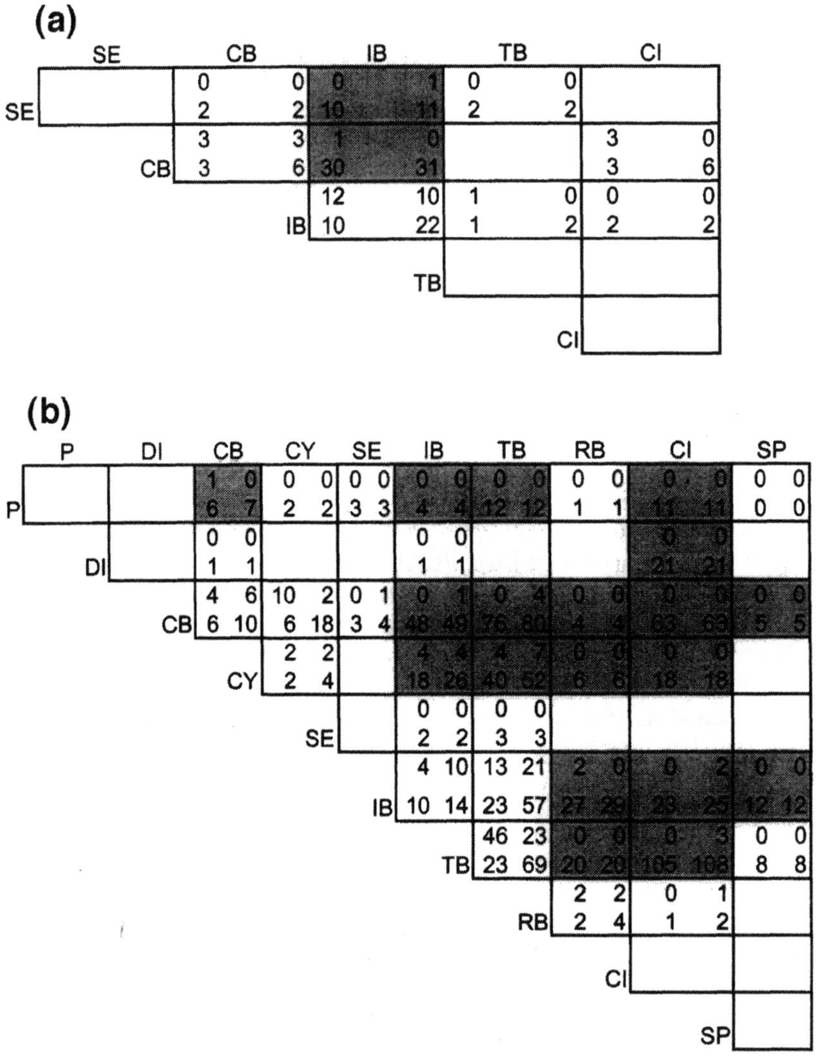

Figure 1 (a) and (b). Matrices of competitive interactions between functional groups at four sites (abbreviations as in Table 1). (a) Carrigathorna, (b) Rapids 3 m, (c) Rapids 6 m, (d) Whirlpool Cliff. Groups at the top of columns arranged in order of decreasing dominance (left to right). Bottom right number in each cell represents number of interactions observed between those two groups at that site, number in top left, bottom left and top right, respectively, represent number of standoffs, number of wins by group to the left of the row and number of wins by group at the top of column. Shaded cells represent significant winner in each case (chi square >3.84, d.f=1, $p<0.05$).

Materials and methods

Lough Hyne, Southwest Ireland (51° 30′ N, 9° 18′ W) is a small body of water (<1 km^2) connected to the Atlantic Ocean by a shallow (20 m wide) channel. This channel is subject to very high flow speeds on both incoming and outgoing tides. The incoming current is intercepted at Whirlpool cliff, a nearly vertical wall of rock which extends to 18 m below chart datum. Flow speeds here, though generally lower than the rapids, can be high on incoming tides. In the sheltered interior of the lough is Labhra Cliff, which extends to over 30 m depth. Because of the low levels of water movement here, all horizontal and partially inclined surfaces are covered in a layer of fine sediment. The North Basin of the lough is a very sheltered bowl up to 25 m deep and has a sediment floor. The final site chosen for this study was the highly exposed Atlantic coast outside the lough at Carrigathorna. All sites are well described in Bassindale et al. (1948) and Kitching & Ebling (1967).

In the summer of 1998, between 30 and 50 rocks (between 4 and 1100 cm^2) were sampled randomly

(c)

	P	CV	CB	M	CY	SE	IB	RC	TB	PC	RB	CI	SP	F
P			0 0 / 2 2	0 0 / 1 1	0 0 / 1 1	0 0 / 5 5	0 0 / 2 2	0 0 / 0 0	0 0 / 0 0	0 0 / 1 1	0 0 / 3 3	0 0 / 1 1	0 0 / 1 1	0 0 / 2 2
CV			0 0 / 5 5		0 0 / 1 1		0 0 / 9 9	0 0 / 1 1						
CB			9 5 / 5 14	0 1 / 1 2	6 3 / 4 13	0 2 / 15 17	1 1 / 29 31	0 0 / 5 5	0 0 / 12 12	1 0 / 10 11	0 0 / 11 11	1 0 / 9 10	0 0 / 12 12	0 0 / 2 2
M					0 0 / 1 1		0 1 / 3 4		0 0 / 2 2					
CY					1 0 / 0 1	0 0 / 3 3	3 2 / 7 12	1 0 / 3 4	0 1 / 2 3	0 0 / 3 3		0 0 / 3 3		
SE							0 6 / 6 12	0 0 / 1 1	0 0 / 2 2	0 1 / 2 3		0 0 / 2 2	0 0 / 6 6	0 0 / 1 1
IB							12 4 / 4 16	0 0 / 2 2	1 3 / 9 13	0 1 / 9 10	0 0 / 7 7	0 0 / 3 3	0 0 / 2 2	0 0 / 2 2
RC									0 3 / 0 3	4 8 / 16 28	0 1 / 0 1	0 4 / 1 5	0 0 / 5 5	0 0 / 1 1
TB										0 1 / 1 2	0 4 / 2 6	0 0 / 5 5	0 1 / 3 4	0 0 / 1 1
PC											0 1 / 0 1		0 0 / 3 3	0 0 / 1 1
RB											0 1 / 1 1			
CI														
SP														
F														

(d)

	SA	DI	CA	CV	P	M	CB	SE	CY	IB	TB	RC	PC	RB	SP	CI
SA							0 0 / 1 1		0 0 / 2 2	0 0 / 9 9						
DI			0 2 / 0 2				1 0 / 1 2	0 0 / 1 1	0 0 / 6 6	0 0 / 3 3	0 0 / 1 1		0 0 / 1 1	0 0 / 5 5	0 0 / 1 1	
CA							1 0 / 1 2	0 0 / 1 1	0 0 / 6 6		0 0 / 1 1			0 0 / 5 5	0 0 / 1 1	0 0 / 4 4
CV					5 0 / 1 6		1 0 / 9 10			1 1 / 8 10	0 0 / 1 1		0 0 / 4 4	0 0 / 6 6		0 0 / 14 14
P						0 0 / 2 2	4 18 / 24 46	0 0 / 28 28	0 2 / 38 40	4 10 / 25 39	0 0 / 6 6	0 0 / 1 1	0 0 / 6 6	0 0 / 13 13	0 0 / 8 8	0 0 / 73 73
M							1 3 / 4 8	0 0 / 1 1	3 0 / 8 11	1 1 / 6 8	0 0 / 2 2		0 0 / 3 3	0 1 / 5 6		1 0 / 2 3
CB							25 25 / 25 50	0 20 / 21 41	4 23 / 18 45	14 8 / 72 94	0 0 / 26 26	0 0 / 1 1	1 2 / 24 27	0 0 / 50 50	1 0 / 3 4	6 4 / 59 69
SE									0 3 / 4 7	0 3 / 31 34	0 0 / 6 6	0 0 / 1 1	0 0 / 9 9	0 0 / 27 27		0 0 / 15 15
CY									22 14 / 14 36	7 13 / 28 48	1 4 / 31 46	1 0 / 0 1	2 0 / 9 11	0 0 / 25 25	0 0 / 1 1	0 0 / 19 19
IB										23 32 / 32 55	2 6 / 43 51	0 0 / 1 1	1 4 / 21 26	0 4 / 63 67	0 0 / 3 3	1 5 / 33 39
TB											0 2 / 2 2	0 1 / 0 1	0 4 / 14 18	0 2 / 29 31	0 1 / 1 2	0 0 / 1 1
RC													0 2 / 1 3	0 1 / 0 1		
PC														0 17 / 5 22		0 5 / 7 12
RB																0 1 / 2 3
SP																0 1 / 0 1
CI																

Figure 1 (c) and (d).

Table 1. List of functional groups employed in this study, type species for each group and abbreviations

Functional group	Type species	Abbreviation
Solitary ascidian	*Ascidiella scabra*	SA
Colonial ascidian	*Botryllus schlosserii*	CA
Didemnid ascidian	*Didemnum maculosum*	DI
Corallimorpharia	*Corynactis viridis*	CV
Porifera	*Amphilectus fucorum*	P
Serpulidae	*Pomatoceros triqueter*	SE
Calcareous bryozoan	*Escharoides coccinea*	CB
Cyclostomes	*Tubulipora lilacea*	CY
Intermediate bryozoan	*Microporella ciliata*	IB
Thin bryozoan	*Haplopoma graniferum*	TB
Pink encrusting coralline	*Lithothamnion* sp.	PC
Red encrusting coralline	*Hildenbrandia* sp.	RC
Runner bryozoan	*Hippothoa divaricata*	RB
Cirripedia	*Verruca stroemia*	CI
Mollusca	*Anomia ephippium*	M
Foraminifera		F
Spirorbidae	*Spirorbis spirorbis*	SP

at each site from an area of approximately 4 m^2. Depths sampled ranged from intertidal to 18 m depth. Sampling was dictated to some extent by boulder availability and bathymetric profiles; depths sampled at each site are given in Table 2. This generally resulted in a total substratum surface area of approximately 2500 cm^2 sampled at each depth of each site.

All organisms were identified to species level where possible. Flora, other than *Lithothamnion* and *Hildenbrandia*, was very sparse and was not considered further. As most of the rocks were fairly flat, percentage cover of encrusting biota was estimated using a grid of dots on a transparent acetate sheet, generally resulting in at least 100 individual points being sampled. On each rock, the outcome of all interactions where competiton for space occurred was recorded. Outcomes were regarded as overgrowth if the growing edge of one individual (or colony) obscured any feeding apertures of the competitor (Rubin, 1982; Russ, 1982; Sebens, 1986). A standoff was recorded if the meeting between competitors resulted in redirected growth or cessation of growth. As more than 120 species were present, it was deemed necessary to condense these into functional groups for further analysis (Table 1). Pairwise interactions between these groups were analysed using contact matrices (Turner & Todd, 1994). These were constructed for sites where over 50 interactions were recorded; owing to the paucity of interactions at the North Basin and Labhra sites, interaction matrices were not constructed. Where the nature and outcome of interactions at different depths at the same site were similar, data were pooled to construct the matrices, i.e. Whirlpool 6, 12 and 18 m and Carrigathorna 6 and 12 m. Chi-square contingency tests were carried out on each pairwise interaction with the null hypothesis of equal probability of wins and losses between each pair of interacting groups (Nandakumar, 1995). Where one group outcompeted a second group a significant ($p<0.05$) number of times, the relevant box in the matrices was shaded grey (Fig. 1). Using the statistically significant interactions thus generated, competitive hierarchies could be drawn for each site (Nandakumar *op. cit.*). Disturbance of boulders at each site was quantified by repeated observations of sets of marked boulders of varying sizes at each site (Maughan, 2000). Flow rate for a spring tide at each site (June 1999) was monitored over 24 h of spring tides by the deployment of a Valeport logging flow meter (BFM 105) set to record every minute. Sedimentation at each site was assessed qualitatively. Assemblage growth rates at each site were assessed through a two year artificial substrata experiment (Maughan, 2000).

Results

Alpha diversity varied substantially between the five sites but peaked at 'moderate' sites of intermediate levels of flow and sedimentation (Table 3). Diversity was lowest at the least disturbed site (North Basin), but the next lowest alpha diversity value was at the most disturbed site. Total number of species was by far the greatest at Whirlpool Cliff (90) and lowest at North Basin (10). The Rapids and Labhra Cliff had similar numbers of species (70 and 67, respectively), though with few species in common. The exposed coastal site at Carrigathorna had fewer species than most sites within the lough (46). Of the total 114 taxa encountered, most were species though a few species complexes (e.g. spirorbid worms) were present but not separated.

Less than 10% (286) of the 2917 interactions recorded had tied outcomes (stand-offs), so overgrowth usually resulted. The number of interactions differed between sites, as did the number of significant winning groups (Fig. 1). The proportion of intra-group interactions (i.e. those involving two individuals/colonies from the same functional group) varied from 33% (Carrigathorna) to 9.9% (Whirlpool Cliff). The level of transitivity (Tanaka & Nandakumar, 1994) in all the matrices was high such that all could be described as hierarchies. As is typical of hierarchies, as opposed to networks (see Quinn, 1982; Nandakumar, 1995), the frequency of backloops (where a lower ranked competitor overgrows one of higher ranking) was generally low. There were, however, more frequent backloops at the Whirlpool site (8.5% of significant interactions between groups) relative to Carrigathorna, Rapids 3 m and Rapids 6 m (2.4%, 2.5% and 3.8% respectively).

Competitive ranking followed taxonomic or functional group (Table 1) significance rather than solitary or colonial form, although colonial competitors overgrew those of solitary form in the ratio 17:10. The ranking of functional groups was essentially similar in all of the competitive hierarchies (Fig. 2) though some groups were absent from some sites (e.g. ascidians from Carrigathorna and Rapids 6 m) leading to different taxa dominating. The hierarchy for Carrigathorna was very simple; there was little space occupied at this site, relatively low diversity and so few functional groups were common enough for many interactions (Fig. 2a). Serpulid worms were the dominant taxa here. The hierarchies for both Rapids sites (Figs 2b, c) were essentially similar in being dominated by sponges and in having barnacles, spirorbids and runner bryozoans near the base. The anthozoan, *Corynactis viridis*, was common at the 6 m site but not at the 3 m site and was a fairly dominant competitor. The Whirlpool hierarchy (Fig. 2d) was substantially more complicated than any of the others with both more functional groups and more connections. Ascidians, solitary and colonial, dominated the assemblage rather than sponges, but the two phyla did not meet so true assessment was not possible.

There was a significant correlation between area occupied and mean number of interactions cm^{-2} (Table 2, Pearsons correlation coefficient 0.898, $p<0.01$). There were also highly significant correlations between mean water flow rate and area occupied (0.938, $p<0.01$) and mean water flow rate and density of interactions (0.979, $p<0.01$). Total number of species at each site was unrelated to interaction density (0.211, $p>0.05$). Growth rate of the community as a whole was fastest at high flow sites (Table 3, see Maughan, 2000).

Flow rate varied enormously between sites, peaking at 260 cm/s in the Rapids, to approximately 100 cm/s at Whirlpool Cliff and, at the other extreme, was

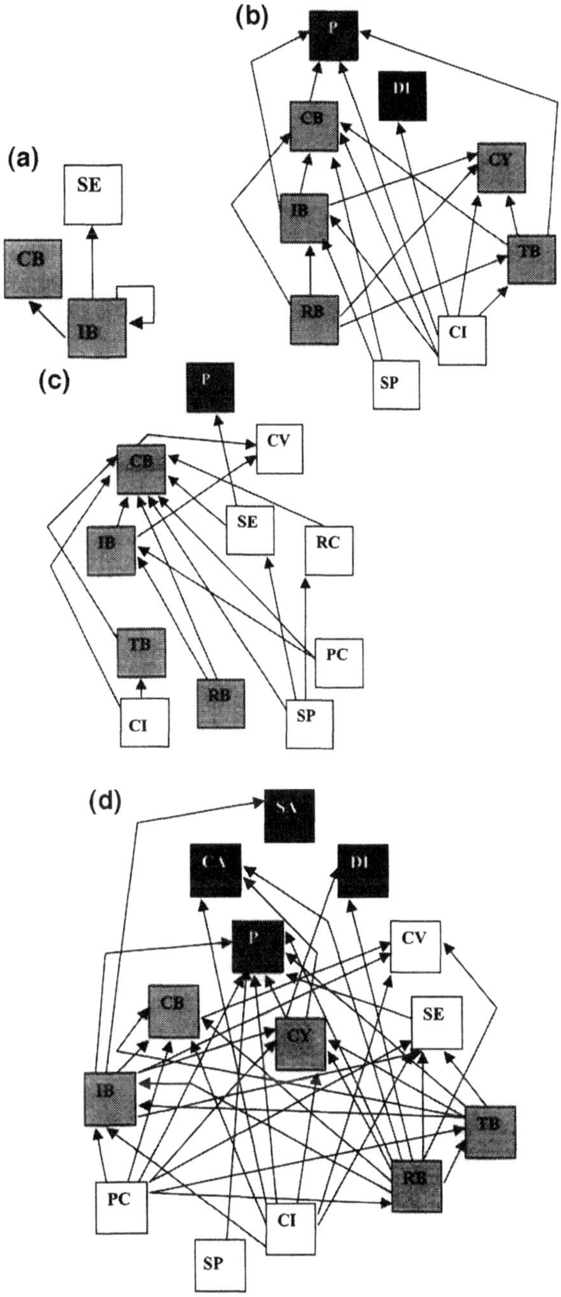

Figure 2. Hierarchies of functional groups observed at four sites (prepared from significant interactions detailed in Figure 1). Abbreviations as given in Table 1. Boxes are shaded as follows; ascidians-black, sponges-dark grey and bryozoans-light grey, all other groups are white.

almost negligible at the North Basin site. Disturbance was highest at the Carrigathorna site; waves of over 5 m height have been recorded in winter storms here (data from E. Murphy, Met. Éireann). Both intensity

Table 2. Density of interactions, percentage cover and flow conditions at each of the study sites

Site	Int. cm^{-2}	Mean % cover	Mean flow cms^{-1}	Max. flow cms^{-1}
Rapids 3 m	0.68	81.0	32	260
Rapids 6 m	0.16	36.5	7	150
Whirlpool 6 m	0.29	49.7	13	130
Whirlpool 12 m	0.14	39.0	9	100
Whirlpool 18 m	0.10	48.6	10	131
Carrigathorna 6 m	0.03	5.2	Turbulent	Turbulent
Carrigathorna 12 m	0.002	2.1	Turbulent	Turbulent
Labhra 12 m	0.014	4.9	$\ll 0.01$	0.02
Labhra 18 m	0.006	4.3	$\ll 0.01$	0.02
North Basin 18 m	0.000	2.3	0.00	0.00

Table 3. Environmental and community factors acting at each site in relation to species diversity. Size of circle is related to relative amount of that factor present for the first three factors listed and to speed of community growth for the last factor. Values in parentheses in alpha diversity column represent total number of species

Site	Disturbance	Sediment Load	Competition	Growth	Log alpha diversity
Whirlpool	●	●	●	●	12.6 (90)
Labhra	•	●	•	•	11.1 (67)
Rapids	●	•	●	●	8.5 (70)
Carrigathorna	●	•	•	•	7.0 (46)
North Basin	•	●	•	•	2.6 (10)

and frequency of disturbance decreased with distance from the Lough mouth to the North Basin, whilst sedimentation showed the converse pattern (Table 3).

Discussion

Competition

The encrusting assemblages on boulders around the shores of Lough Hyne were characterised by high levels of interspecific competition, low incidence of tied outcomes and hierarchical organisation. Such results are reasonably typical for the region (Rubin, 1982; Turner & Todd, 1994). Space occupation was highly correlated with interaction density across the range of sites studied, in accordance with the work of Nandakumar (1995). The number of interactions, interacting pairs of competitors and, therefore, complexity of hierarchies was very different between sites, being minimal at the sites of least and most water-born disturbance and maximal at Whirlpool (Fig. 2, Table 3). The percentage of backloops at the Whirlpool sites was much higher than for other sites. This may be an important factor in promoting and maintaining the high species diversity observed here through network effects preventing space monopolisation by competitive dominants (Jackson, 1979; Karlson & Jackson, 1981). Hierarchical complexity may be a better predictor of species diversity than absolute intensity of competition; boulders from the Rapids had very high interaction density but, as the community here was dominated by a few species, low hierarchical complexity.

Growth rate of a community has been implicated as a major factor in determining diversity (Huston, 1979). This variable would be closely linked with competition, as a higher growth rate of a community would be expected to result in greater average space occupation, which would lead to more competitive interactions. These predictions are supported by the data in Tables 2 and 3.

A model for community regulation in the study system

Menge & Sutherland (1987) have noted the dearth of community ecological studies in the literature where investigators deliberately chose sites spanning a broad range of physical conditions. They contended that most natural experiments looking at diversity suffered from an insufficiently broad set of local environmental gradients. Ideally, investigations purporting to examine influences along a given environmental gradient should include conditions from 'harsh' to 'benign', though quite what constitutes these has often attracted debate (e.g Pickett et al., 1989). Lough Hyne, however, has attracted marine biologists and ecological study for many decades precisely because of the wide environmental gradient span within such a small horizontal spatial scale (Table 3). The present work, utilising sites within this gradient, may then be regarded as unusually complete, avoiding the need for comparisons between different regions, workers, protocols or species pools.

Classical work on shore zonation (e.g. Connell, 1961; Paine, 1974; Menge & Sutherland, 1987) has implicated that biotic factors assume greater importance than physical factors in community regulation at the low end of the environmental stress continuum. This is a key requirement for the effects of the intermediate disturbance hypothesis to become apparent (Sousa, 1979). Competition, being one of the biotic factors discussed by Menge & Sutherland (1987), should be more intense at sites of least disturbance (i.e Labhra and North Basin). Both the low values for interaction density and percentage cover at these sites (Table 2) would seem to refute this prediction. Both number of species and alpha diversity were highest at Whirlpool Cliff. This, being the site with intermediate levels of disturbance, is in accordance with the intermediate disturbance hypothesis. The absence of any significant competitive dominants at the low end of the disturbance scale would seem to suggest, however,

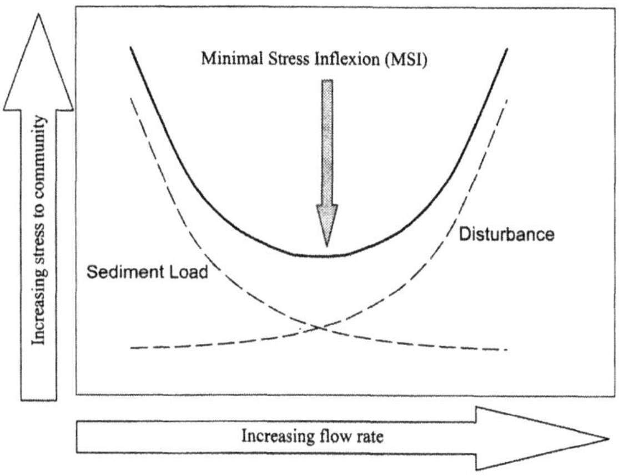

Figure 3. Proposed bi-directional stress gradient. Slope of disturbance and sediment loads need not necessarily be reflections of one another but for the sake of simplicity are shown here as such. The parabola represents the sum of the stress imposed on the community by these two factors.

that the disturbance occurring at Whirlpool Cliff does not act through preventing space monopolisation, but maybe through some other mechanism. We propose here an extension of the intermediate disturbance hypothesis in that there is a bi-directional stress gradient across the range of sites we have studied. Sediment load showed an opposite trend to disturbance (Table 3) and has often been shown to be a potential source of stress to encrusting fauna (Rogers, 1990; Umar et al., 1998). The interaction between these factors has been noted previously (Hiscock, 1983; Airoldi, 1998). Lilly et al. (1953), working on boulder communities at Lough Hyne, noted the importance of current in determining sediment load and, ultimately, species distribution. They did not, however, consider the effects of disturbance on the community as that force went largely unrecognised at the time.

The most benign environmental conditions, then, would occur where the sum of stresses imposed on the community by sediment load and disturbance is at a minimum (Fig. 3). The stress gradient proposed here is, however, for the encrusting community as a whole; sediment tolerant species will predominate on the left side of the parabola (Hiscock, 1983; Saiz-Salinas & Urdangarin, 1994) whilst opportunistic, disturbance tolerant species dominate the right hand side (Warner, 1984; Barnes & Clarke, 1999). The boundary between these two distinct faunal types occurs around the minimal stress inflexion (MSI) and a high species diversity is predicted here. The Whirlpool Cliff site, at Lough Hyne, was intermediate for both sediment load and disturbance frequency, and had the highest species diversity of all sites studied (Table 3). Competition and growth rate are also at intermediate levels here, which is in accordance with Huston's (1979) 'General hypothesis of species diversity'.

The relative gradient of both disturbance and sediment load parameters lead to interesting predictive spin-offs. Very large boulders or cliff faces will be disturbed very infrequently which would have the effect of pushing the MSI further to the right (i.e. to higher water movement). Disturbance, being very infrequent for such large pieces of substratum, would assume less importance in preventing competitive exclusion in this case. Large rocks in high flow speed may be dominated by better competitors rather than the predominantly opportunistic species on smaller rocks at such a site. This has been demonstrated by stabilisation of small rocks (McGuinness, 1987). In the present study, ultimately flow speed itself (extreme right hand side of the parabola; Figure 3) would start to become a major source of stress to the community (reviewed in Hiscock, 1983) and only highly tolerant species would persist. The prediction is supported by the fact that the sides of large boulders in the Rapids tend to be dominated by *Corynactis viridis, Metridium senile* and a few sponges such as *Hymenacidon perleve* (Lilly et al., 1953). All these species rank high in overgrowth capabilities (Fig. 2) and are characteristic of high flow sites (Hiscock, 1983). The tops of such rocks (i.e. in the highest flow), however, are virtually 100% covered in a thick crust of *Lithothamnion* sp. (Lilly et al., 1953).

Most systems are, regrettably, too complex to be reduced to a trade-off between two parameters and other minor environmental gradients may be present at Lough Hyne (Kitching & Ebling, 1967; Hiscock, 1983). The model developed here, however, may be usefully applied to the study system as we have shown. We believe the findings of this study may be of considerable use in interpretation of other work along diversity gradients or on the principles underlying diversity and community complexity. The model could be improved by predicting competition intensity, complexity and outcomes at other sites in Lough Hyne prior to investigation, and factoring in other parameters, such as predation, larval supply and incident radiation. Future work may examine faunal diversity on rocks of such a size that disturbance is minimal, even in very high flow conditions.

Acknowledgements

Thanks to all dive buddies who assisted with the project, especially A. Greenwood. Thanks also to D. O'Donnell (National Parks and Wildlife Service) for granting permission to undertake the study. Comments from two anonymous reviewers helped to improve the manuscript.

References

Airoldi, A., 1998. Roles of disturbance, sediment stress, and substratum retention on spatial dominance in algal turf. Ecology 79: 2759–2770.

Barnes, D. K. A. & A. Clarke, 1999. The ecology of an assemblage dominant: the encrusting bryozoan *Fenestrulina rugula.* Invert. Biol. 117: 331–340.

Bassindale, R., F. J. Ebling, J. A. Kitching & R. D. Purchon, 1948. The ecology of the Lough Ine rapids with special reference to water currents. I. Introduction and hydrography. J. Ecol. 36: 305–322.

Benedetti-Cecchi, L., L. Airoldi, M. Abbiati & F. Cinnelli, 1996. Estimating the abundance of benthic invertebrates: a comparison of procedures and variability between observers. Mar. Ecol. Prog. Ser. 138: 93–101.

Connell, J., 1961. Effect of competition, predation by *Thais lapillus*, and other factors on natural populations of the barnacle *Balanus balanoides*. Ecol. Monogr. 31: 61–104.

Elner, R. W. & R. L. Vadas,1990. Inference in ecology: the Sea urchin phenomenon in the northwestern Atlantic. Am. Nat. 136: 108–125.

Hiscock, K., 1983. Water movement. In Earll, R. & D. G. Erwin (eds), Sublittoral Ecology: The Ecology Of The Shallow Sublittoral Benthos. Clarendon Press, Oxford: 58–96.

Huston, M., 1979. A general hypothesis of species diversity. Am. Nat. 113: 81–101.

Hutchinson, G. E., 1959. Homage to Santa Rosalia, or why are there so many different kinds of animals? Am. Nat. 93: 145–159.

Hutchinson, G. E., 1961. The paradox of the plankton. Am. Nat. 95: 137–145.

Jackson, J. B. C., 1979. Overgrowth competition between encrusting cheilostome ectoprocts in a Jamaican cryptic reef environment. J. anim. Ecol. 48: 805–823.

Karlson, R. H. & J. B. C. Jackson, 1981. Competitive networks and community structure: a simulation study. Ecology 62: 670–678.

Kitching, J. A. & F. J. Ebling, 1967. Ecological studies at Lough Ine. Adv. ecol. Res. 4: 197–291.

Knowlton, N. & J. B. C. Jackson, 1994. New taxonomy and niche partitioning on coral reefs: Jack of all trades or master of some? Trend. Ecol. Evol. 9: 7–9.

Lilly, S. J., J. F. Sloane, R. Bassindale, F. J. Ebling & J. A. Kitching, 1953. The ecology of the Lough Ine rapids with special reference to water currents. IV. The sedentary fauna of sublittoral boulders. J. anim. Ecol. 22: 87–122.

MacArthur, R. H. & R. Levins, 1967. The limiting similarity, convergence and divergence of coexisting species. Am. Nat. 101: 377–385.

Maughan, B. C., 2000. Ecology of encrusting epifauna in Lough Hyne Marine Nature Reserve, Co. Cork, Ireland. Unpubl. Ph. D thesis, University College Cork, Ireland: 154 pp.

McGuinness, K., 1987. Disturbance and organisms on boulders. II. Causes of patterns in diversity and abundance. Oecologia 71: 420–430.
McGuinness, K., 1990. Physical variability, diversity gradients and the ecology of temperate and tropical reefs. Aust. J. Ecol. 15: 465–476.
Menge, B. A. & J. P. Sutherland, 1987. Community regulation: variation in disturbance, competition and predation in relation to environmental stress and recruitment. Am. Nat. 130: 730–757.
Nandakumar, K., 1995. Competitive interactions among sessile organisms in Tomioka Bay, South Japan: importance of light conditions on the panel surface. Mar. Biol. 121: 713–719.
Paine, R. T., 1979. Disaster, catastrophe and local persistence of the sea palm *Postelsia palmaeformis*. Science 205: 685–687.
Paine, R. T., 1974. Intertidal community structure: experimental studies on the relationship between a dominant competitor and its principal predator. Oecologia 15: 93–120.
Pickett, S. T. A., J. Kolasa, J. J. Armesto & S. L. Collins,1989. The ecological concept of disturbance and its expression at various hierarchical levels. Oikos 54: 129–136.
Quinn, J. F., 1982. Competitive hierarchies in marine benthic communities. Oecologia 53: 12–19.
Rogers, C. S., 1990. Responses of coral reefs and reef organisms to sedimentation. Mar. Ecol. Prog. Ser. 62: 185–202.
Rubin, J. A., 1982. The degree of intransitivity and its measurement in an assemblage of encrusting cheilostome bryozoa. J. exp. mar. Biol. Ecol. 60: 119–128.
Russ, G. R., 1982. Overgrowth in a marine epifaunal community: competitive hierarchies and competitive networks. Oecologia 53: 12–19.
Saiz-Salinas, J. I. & I. Urdangarin, 1994. Response of sublittoral hard substrate invertebrates to estuarine sedimentation in the outer harbour of Bilbao (N. Spain). Mar. Ecol. 15: 105–131.
Sebens, K. P., 1986. Spatial relationships among encrusting marine organisms in the New England subtidal zone. Ecol. Monogr. 56: 73–96.
Sousa, W. P., 1979. Disturbance in Marine intertidal boulder fields: the non-equilibrium maintenance of species diversity. Ecology 60: 1225–1239.
Tanaka, M. & K. Nandakumar, 1994. Measurement of the degree of intransitivity in a community of sessile organisms. J. exp. mar. Biol. Ecol.182: 85–95.
Turner, S. J. & C. D. Todd, 1994. Competition for space in encrusting bryozoan assemblages: The influence of encounter angle, site and year. J. mar. biol. Ass. U. K . 74: 603–622.
Umar, M. J., L. J. McCook & I. R. Price, 1998. Effects of sediment deposition on the seaweed *Sargassum* on a fringing coral reef. Coral Reefs 17: 169–177.
Underwood, A. J. & P. G. Fairweather, 1989. Supply-side ecology and benthic marine assemblages. Trends Ecol. Evol. 4: 16–20.
Underwood, A. J. & P. G. Fairweather, 1985. Intertidal communities: do they have different ecologies or different ecologists? Proc. Ecol. Soc. Aust. 14: 7–16.
Warner, G. F., 1984. Dynamic stability in two contrasting epibenthic communities. In Gibbs, P. E. (ed.), Proceedings of the 19th European Marine Biology Symposium. Cambridge University Press: 401–410.

Hydrobiologia **440:** 111–117, 2000.
M.B. Jones, J.M.N. Azevedo, A.I. Neto, A.C. Costa & A.M. Frias Martins (eds), Island, Ocean and Deep-Sea Biology.

Shell occupancy of the intertidal hermit crab *Clibanarius erythropus* (Decapoda, Diogenidae) on São Miguel (Azores)

Andrea Z. Botelho & Ana C. Costa*
Universidade dos Açores, Departamento de Biologia, Rua Mãe de Deus, 58, P-9500 Ponta Delgada, Portugal.
*Tel: +351-296653044. Fax: +351-296653455. E-mail: costa@alf.uac.pt (*Author for correspondence)*

Key words: hermit crabs, shell utilisation, Azores

Abstract

The intertidal hermit crab *Clibanarius erythropus* was collected at three sites on São Miguel (Azores) during low spring tides. Shells occupied were identified and measured. Crab sizes ranged from 1.78 to 13.67 mm (cephalothoracic shield length), with an average size of 4.40 ± 1.44 mm. Of the 19 different shells utilised, the most frequent were *Littorina striata* (23.8%), *Nassarius incrassatus* (22.5%) and *Mitra* sp. (22.0%). At Fenais da Luz, *L. striata* was most frequently occupied, while at Água de Alto it was *N. incrassatus* and, at Caloura, *Mitra* sp. shells were most frequently used. Shell selection appears to be determined by respective sizes of hermit crab and shell species. Small size-class crabs occupy more shell species than larger crabs. The smallest crab was found at Fenais da Luz occupying a small *Bittium* sp., whereas the largest crab was found at Caloura inhabiting *Stramonita haemastoma*.

Introduction

Hermit crabs occupy empty gastropod shells for protection against predators (Vance, 1972) and physical stress (Reese, 1969). The strong association between hermit crabs and their adopted shelters has influenced greatly almost all aspects of their biology (Hazlett, 1981). Occupied shells, a limiting resource to most populations (Fotheringham, 1976; Kellog, 1976), may vary in size, shape and condition. Hermit crabs have preferences for shells of particular size, weight, aperture width, volume and condition; also important for selection are the presence of epibionts and the shell species (Hazlett, 1981). In natural populations, shell selection is determined primarily by shell availability and, to a lesser extent, by preference, despite the active choice of shells shown in laboratory experiments (Kellogg, 1976; Bertness, 1980, 1981; Reddy & Biseswar, 1993).

The intertidal hermit crab *Clibanarius erythropus* (Latreille, 1818) is a common species of Mediterranean shores and is also present along the Atlantic coasts from Brittany to the Azores (Zariquiey Alvarez, 1968; Ingle, 1993). However, references to the general biology and ecology of this species are scarce (Gherardi, 1991). In the Azores, *C. erythropus* (known since the early naturalists' studies in the 19th C, e.g. Barrois, 1888) is abundant in low-shore pools on the rocky shores of São Miguel. The aim of this study was to examine the shell utilisation by *C. erythropus*, in S. Miguel.

Description of sites

The Azores archipelago (37° 40' N and 25° 31′ W) consists of nine volcanic islands (Fig. 1). São Miguel is the largest island located in the oriental group of the archipelago. The coastline of this island is irregular and the shoreline consists mainly of high cliffs falling directly to the sea, or rocky platforms. Boulder and cobble beaches, resulting from erosion of volcanic rocks, are also common. Most shores are subject to swell and semi-diurnal tides with a range of less than 2 m. Three populations of *Clibanarius erythropus* were studied for shell occupancy (Fig. 1). Fenais da Luz is a wave-exposed site on the north shore; Cerco da Caloura is a sheltered, gently sloping bay on the

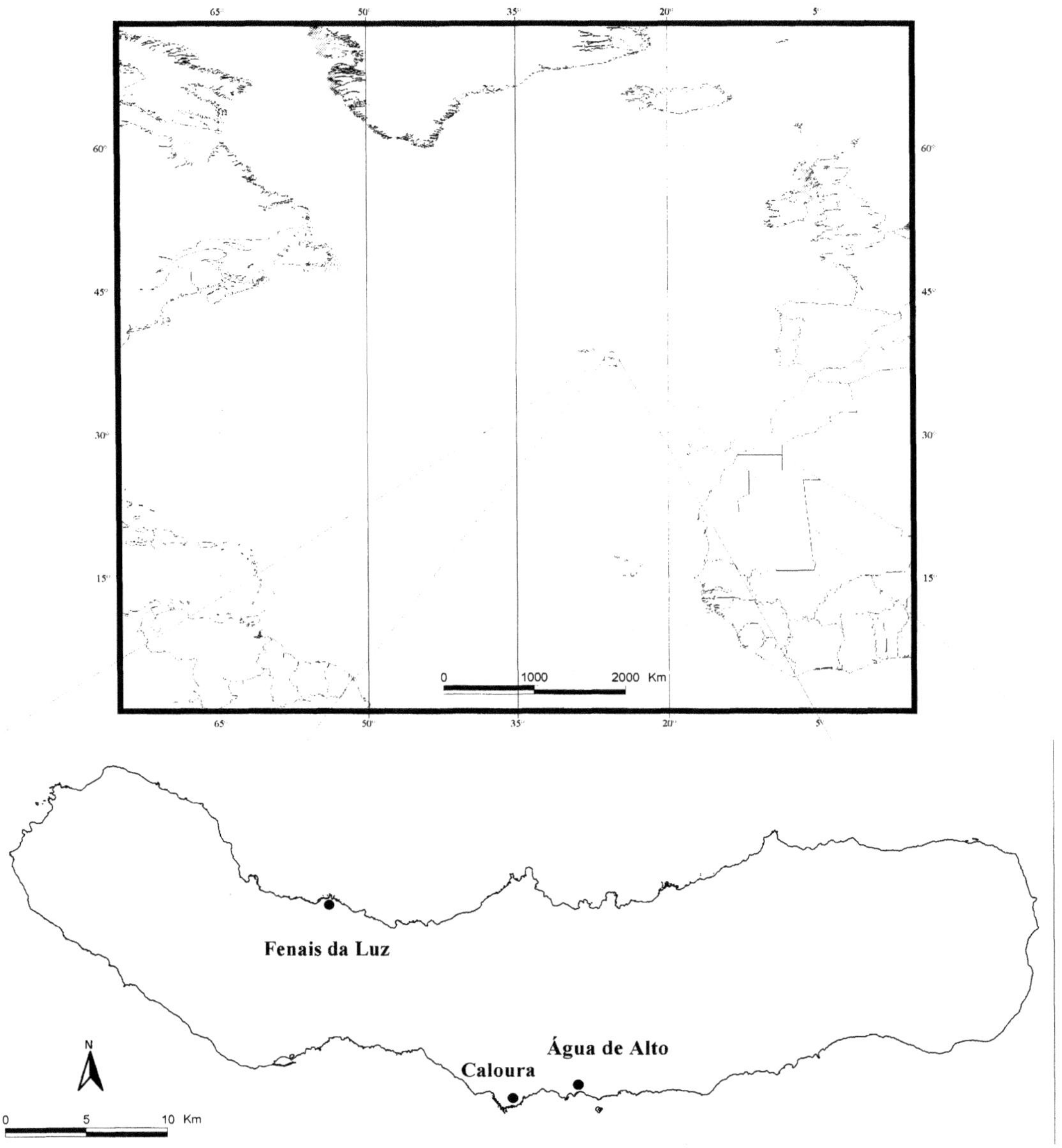

Figure 1. Location of the Azores and of the sampling sites in São Miguel island.

south coast, and Água de Alto is a rocky platform with boulders and rock pools backed by a sheer cliff, on the south coast.

Materials and methods

From each site, crabs were hand sampled during the low spring tides between November 1998 and June 1999. Eight hundred and forty nine crabs were examined. During crab collection, qualitative observations of living gastropods and of the availability of empty shells on each shore were made. In the laboratory, occupied shells were identified, measured (aperture length, shell length and shell width) with callipers (precision 0.01 mm) and weighed after removal of the crab. Cephalothoracic shield length (CL) of each crab was measured and specimens were grouped into 0.85 mm size classes calculated according to the highest and lowest sizes (e.g. 1.78–13.67 mm), following the method used by Manjón-Cabeza & García-Raso (1998).

Species composition of occupied shells at the three sites was compared with a Chi square (χ^2) test per-

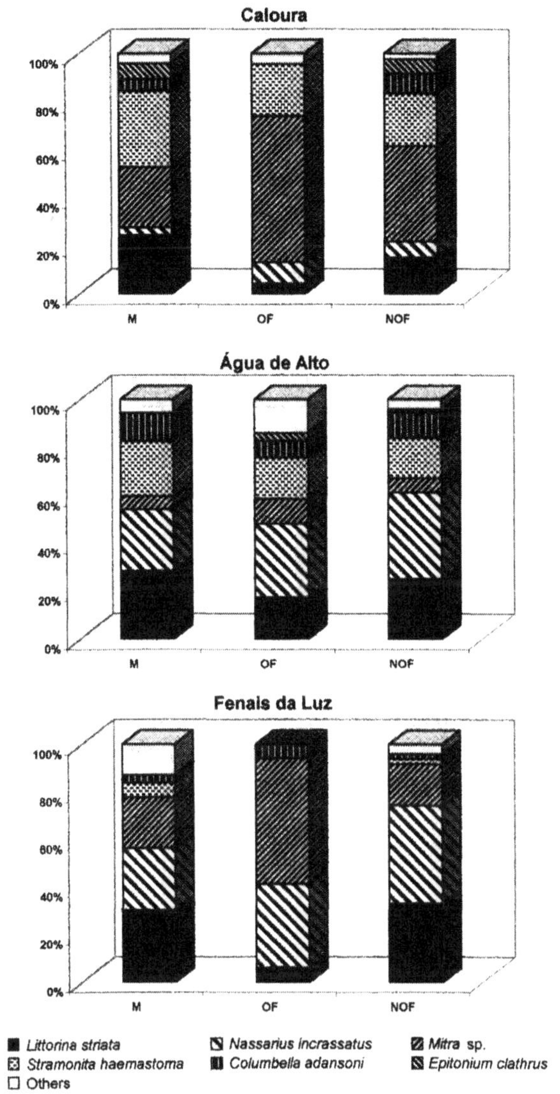

Figure 2. Frequency of the shell species occupied by each sex and ovigerous females, on the sites under study. M – male; OF – ovigerous females; NOF – non ovigerous females. Others – shell species present in low percentage (<1%).

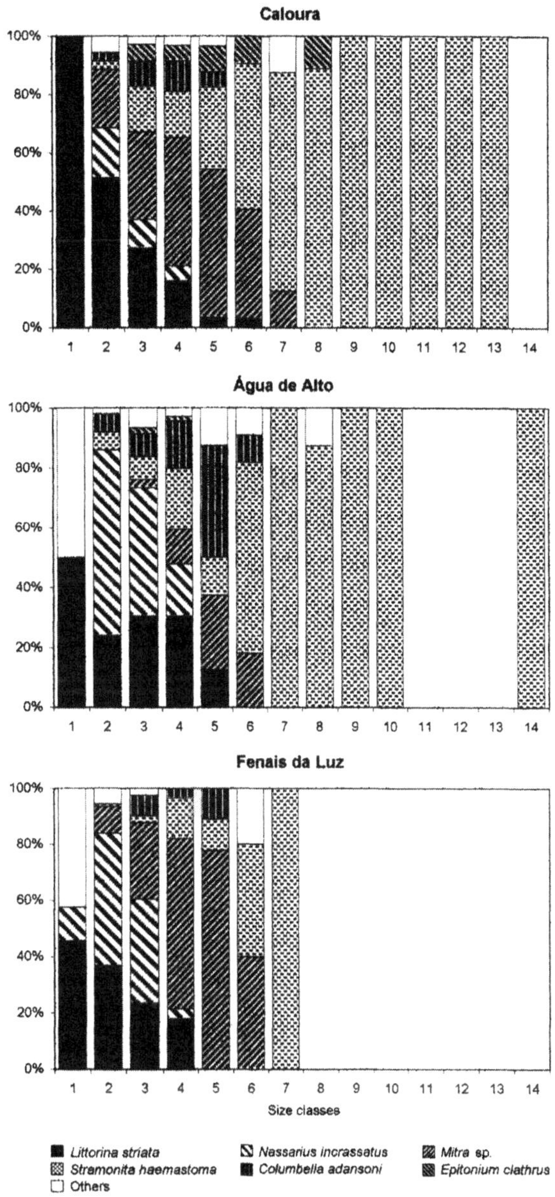

Figure 3. Shell species related to crab size classes.

formed on the seven most common shell species and an 8th category representing a pooled count for the remaining species. Differences in shell occupancy between the sexes, and between ovigerous and non ovigerous females, were tested by Spearman rank correlation (Zar, 1996).

To identify relationships between crab sizes and shell size variables, data were plotted, and correlation coefficients and linear regressions were calculated. Correlation between brood size of ovigerous females and size of the shell was determined. The relationship between shell species and hermit crab length (CL) was also subjected to statistical analysis (analysis of variance – ANOVA, Fry, 1996; Zar, 1996) against the null hypothesis that equal sizes of crabs would occupy different shell species. Twenty one individuals, occupying each of the six of the more common shell species, were chosen randomly for the ANOVA test, equal sample size, minimises type one error due to heterogeneity of variances (Green, 1979). Assumptions of ANOVA were checked applying Bartlett's test

Table 1. Frequencies of occupancy of shell species by *Clibanarius erythropus*

Species	Caloura	Água de Alto	Fenais da Luz	Total
Bittium sp.	0	0	1.5	0.5
Calliostoma sp.	0.3	1.1	0.4	0.6
Columbella adansoni	6.3	10.7	3.0	6.7
Epitonium clathrus	5.7	1.1	0	2.5
Fossarus ambiguus	0	0	1.5	0.5
Gibbula magus	0.6	0	0	0.2
Jujubinus sp.	0.3	0.7	0.4	0.5
Littorina striata	17.8	25.1	29.7	23.8
Melarhaphe neritoides	0.3	0	0	0.1
Mitra sp.	35.6	6.3	22.1	22.0
Natica sp	0.3	0	0	0.1
Nassarius incrassatus	5.7	32.5	32.3	22.5
Ocinebrina aciculata	0.6	2.6	0.8	1.3
Pedipes pedipes	0	0	0.8	0.2
Raphitoma sp.	0	0	0.4	0.1
Stramonita haemastoma	25.7	18.8	4.2	16.8
Tricolia pullus azorica	0	0	0.8	0.2
Vermetus sp.	0.3	1.1	0.4	0.6
None identified	0.3	0	0	0.1

for homogeneity of variance (Green, 1979; Fry, 1996; Zar, 1996). Raw data were transformed by $\log(x+1)$ to achieve normality and equal variance (Green, 1979; Zar, 1996). *Post hoc* multiple comparison was made using the Tukey's test. The significance level for all tests was $P<0.05$.

Results

Sizes of *Clibanarius erythropus* ranged from 1.78 to 13.67 mm (cephalothoracic shield length), with an average size of 4.40 ± 1.44 mm. Crabs occupied 19 species of gastropod shells; those most commonly occupied were *Littorina striata*, *Nassarius incrassatus*, *Mitra* sp. and *Stramonita haemastoma* (Table 1). Rare shell species (not present in Água de Alto) were usually occupied by males and never by ovigerous females.

Crabs occupied the same shells, but in significantly different proportions at each of the study sites (χ^2= 218.9, $P<0.05$). At Fenais da Luz, *L. striata* was the most common shell used while *N. incrassatus* was the most commonly inhabited shell at Água de Alto. At Caloura, *Mitra* sp. was the dominant shell used (Fig. 2). No empty shells utilised by *C. erythropus* was found at any shore.

The two sexes showed no significant difference in shell occupancy (Spearman rank correlation test: r_S=0.714, $P<0.05$). Although there was no significant difference between shells occupied by ovigerous and non ovigerous females (Spearman rank correlation test: r_S=0.855, $P<0.05$), 39% of ovigerous females were found in *Mitra* sp.

A direct relationship between size of hermit crab and host shell species was found (Fig. 3). While *S. haemastoma* was occupied by large crabs, smaller animals occupied a wide range of shell species. Significant differences were found in the size of crabs inhabiting different shell species (ANOVA, F=19.98; df=125; $P<0.05$). Crabs under 8.58 mm were found in 18 shell species, whereas those over 8.58 mm were found only in *S. haemastoma*. An *a posteriori* mean comparison test revealed that the bigger average size of crabs occupying *S. haemastoma* was responsible for the differences observed in the ANOVA results. The majority of small crabs occupied *L. striata* and *N. incrassatus* (Fig. 3). The largest crab was found at Caloura in a *S. haemastoma* shell, while the smallest was collected at Fenais da Luz occupying a small *Bittium* sp.

Shell sizes (both length and weight) occupied by hermit crabs were correlated with crab size. For the most commonly occupied shell species, the strongest relationship was found for *S. haemastoma* and *L. striata* (Fig. 4; Table 2). Brood size was also correlated with shell length for the shell species inhabited by ovigerous females (Pearson, r = 0.58).

Discussion

Present data are consistent with the observations made by Reddy & Biseswar (1993) on a congener species *Clibanarius virescens*, from South Africa. These authors reported that *C. virescens* occupied a similar number of shell species (23) and there was a close relationship between the crab size and the size of the occupied shell. In S. Miguel, *C. erythropus*, and its counterpart in Natal, displayed a shift in the shell species utilised during growth. Both species seem to occupy a wider variety of shell species in the size range of 3–4 mm than seen for larger individuals.

The availability of certain gastropod species influences the shell utilisation by hermit crabs in the natural habitat as they tend to be opportunistic with

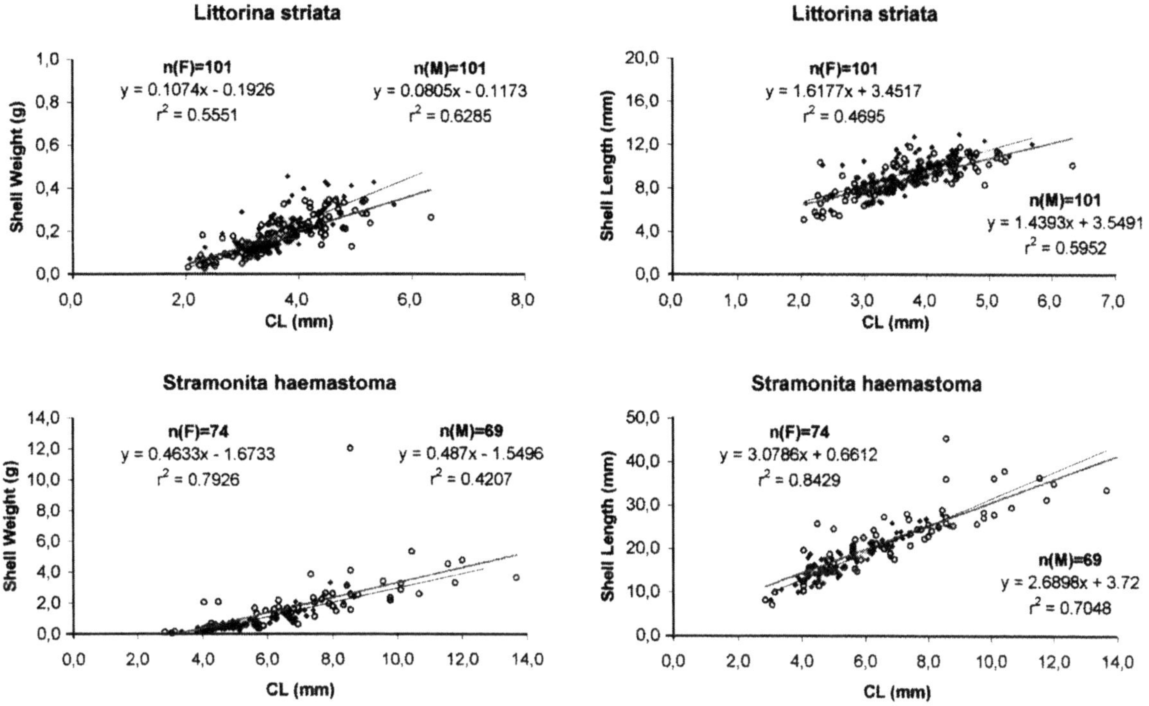

Figure 4. Morphometric relations between *C. erythropus* and two of the most common inhabited shells species, *Littorina striata* and *Stramonita haemastoma* (♦ Females; ○ Males), thicker line illustrates regression calculated for ♂.

regards to the shells they inhabit. Nevertheless, it is interesting to note that *C. virescens*, in contrast to *C. erythropus*, did not occupy *Littorina* sp. and *Mitra* sp., probably because of the availability of more suitable tropical gastropod shells. In southwest Britain, Southward & Southward (1977) found *C. erythropus* occupying mainly *Nucella lapillus*, whereas *Littorina littorea* and other species of shells accounted for a very small proportion of occupancy. Gherardi (1991) reported that a high number of shell species were occupied by this hermit species in Sardinia, where, amongst the most occupied shells, only two genera, *Columbella* and *Gibbula*, were common in our study.

Differences in shell occupation between the sampled populations can be related to the availability of empty shells at the specific site but can also be explained by the different sizes of crabs found in each location. For example, at Fenais da Luz, crab size was probably limited by the lack of large shells, and the abundance of *L. striata* and other small shell species (Fig. 3). It is known that crab growth is slowed by occupation of a small shell (Hazlett, 1981; Lancaster, 1988).

Availability of suitable empty shells in the intertidal is not only dependent on the mortality rate of gastropods but also on the type of mortality as this influences condition of the shells. Empty gastropod shells are usually very rare or absent in the intertidal (Fotheringham, 1976; Kellog, 1976; Bertness, 1980). This absence was also noted for the sites under study on the Azores, where crabs probably find shells subtidally. Physical processes (e.g. wave action) also remove or destroy shells and also influence availability (Gherardi, 1991).

Wilber (1990) reported that the hermit crab *Pagurus longicarpus* changed its preferences for shell species with size. The same probably happens with *C. erythropus*, as sizes of crabs inhabiting different shell species were different. It is, therefore, possible that larger animals choose *Stramonita haemastoma* shells, whereas smaller ones would occupy several shell species.

Differences in shell occupancy between sexes in *C. erythropus* were found by Gherardi (1991) who gave some behavioural explanations for this. Advantage of certain types of shells for females can be related to energy saving for reproductive process, whereas for males more importance might be giving to protection against predation (e.g. heavy *S. haemastoma* occupied by large males). Unbalanced sex ratio at certain size

Table 2. Pearson correlation test (r) and determination coefficients (r^2) obtained for linear regressions between shell parameters (length – SL (mm) and weight – SW (g)) and CL (mm) of the hermit crab (F – female; M – male), for the most abundant shell species

	Shell species	n		r^2		r	
		M	F	M	F	M	F
	Stramonita haemastoma	69	74	0.42	0.79	0.65	0.89
SW/CL	*Littorina striata*	101	101	0.63	0.56	0.79	0.75
	Mitra sp.	68	119	0.33	0.29	0.57	0.54
	Nassarius incrassatus	67	124	0.35	0.20	0.59	0.45
	Stramonita haemastoma	69	74	0.71	0.84	0.84	0.92
SL/CL	*Littorina striata*	101	101	0.60	0.47	0.77	0.69
	Mitra sp.	68	119	0.37	0.37	0.60	0.61
	Nassarius incrassatus	67	124	0.41	0.48	0.64	0.69

classes could also influence results (Gherardi, 1991). The narrow shell aperture of *Mitra* sp. might cause it to be a suitable shell for an ovigerous female as it not only provides enough space to accommodate a clutch, but also confers protection against desiccation.

For the most commonly occupied shell species, a good correlation was found between the sizes of shells occupied by hermit crabs and crab size as predicted by Hazlett (1981). However, linear regressions between shell measurements and crab size were not significant for most shell species; a well-fitted linear regression was found for *L. striata* and *S. haemastoma* (Fig. 4). Other host species are probably sub-optimal as there is a wide range of hermit size occupying the same shell size. Features of the shell, namely its architecture, could be responsible for this. The criteria by which hermit crabs select shells are not well understood, but it has been shown that shell configuration, aperture size, weight index (shell weight/crab weight) and volume index (shell volume/shell weight) are important for selection (Hazlett, 1981; Lancaster, 1988). Nevertheless, laboratory experiments are necessary to test differences in shell preferences of this species.

Crab size (CL) was better correlated with shell length than with shell weight. This latter relationship may be influenced by the frequent incrustation and damage of the shells.

From the correlation results of this study, it can be seen that ovigerous females in larger shells had bigger clutches. *Mitra* sp. seems to be preferred by ovigerous females and some of them will have larger broods than the size of the shell would predict. The measure chosen (length) as a descriptor of size may not be a good predictor for available internal space for clutch accommodation. The type of gastropod shell inhabited by a hermit crab influences both growth rate and clutch size (Bertness, 1981). Shell occupation affords egg protection, but the volume of eggs that can be protected is limited by shell volume. Shell volume is correlated with female size but a female hermit crab that obtains a larger shell may be able to carry a larger volume of well-protected eggs (Fotheringham, 1976; Hazlett, 1981; Lancaster, 1988).

More detailed studies to determine gastropod abundances and availability of shells at both intertidal and subtidal, and choice experiments in the laboratory, will elucidate the role of biotic/abiotic factors and behaviour shell occupancy by *Clibanarius erythropus*.

Acknowledgements

The authors thank João Brum for his assistance with fieldwork, Dr Sérgio Ávila for his expertise in gastropod identification and Dr Ian Tittley for text revision. Data analyses and preparation of this manuscript were supported financially by programme Estagiar-L, from Direcção Regional da Juventude Emprego e Formação Profissional (Azores).

References

Barrois, Th.,1888. Catologue des Crustacés Marins Recueillis aux Açores. Lille.

Bertness, M. D., 1980. Shell preference and utilization patterns in littoral hermit crabs of the bay of Panama. J. exp. Mar. Biol. Ecol. 48: 1–16.

Bertness, M. D., 1981. The influence of shell-type on hermit crab growth and clutch size. Crustaceana 40(2): 197–205.

Fotheringham, N., 1976. Hermit crab shells as a limiting resource (Decapoda, Paguridae). Crustaceana 31(2): 193–199.

Fry, J. C., 1996. In Fry, J. C. (ed.), Biological Data Analysis – A Practical Approach. Oxford University Press: 1–40.

Gherardi, F., 1991. Relative growth, population structure and shell-utilization of the hermit crab *Clibanarius erythropus* in the Mediterranean. Oebalia 17(N.S.): 181–196.

Green, R. H., 1979. Sampling design and statistical methods for environmental biologists. John Wilwey & Sons, New York.

Hazlett, B., 1981. The behavioral ecology of hermit crabs. Ann. Rev. Ecol. Syst. 12: 1–22.

Ingle, R., 1993. Hermit crabs of the Northeastern Atlantic Ocean and the Mediterranean Sea. Chapman & Hall: 495 pp.

Kellogg, C. W., 1976. Gastropod shells: a potentially limiting resource for hermit crabs. J. exp. Mar. Biol. Ecol. 22: 101–111.

Lancaster, I., 1988. *Pagurus bernhardus* (L.) – An introduction to the natural history of hermit crabs. Field Stud. 7: 189–238.

Manjón-Cabeza M. E. & J. E. García-Raso, 1998. Population structure and growth of the hermit crab *Diogenes pugilator* (Decapoda: Anomura: Diogenidae) from the northeastern atlantic. J. crust. Biol. 18(4): 753–762.

Reddy, T. & R. Biseswar, 1993. Patterns of shell utilization in two sympatric species of hermit crabs from the Natal coast (Decapoda, Anomura, Diogenidae). Crustaceana 65(1): 13–24.

Reese, E. S., 1969. Behavioral adaptations of intertidal hermit crabs. Am. Zool. 9: 343–355.

Southward, A. & E. Southward, 1977. Distribution and ecology of the hermit crabs *Clibanarius erythropus* in the western channel. J. mar. biol. Ass. U.K. 57: 441–452.

Vance, R. R., 1972. The role of shell adequancy in behavioral interactions involving hermit crabs. Ecology 53: 1075–1083.

Wilber, T. P. Jr. 1990. Influence of size, species and damage on shell selection by the hermit crab *Pagurus longicarpsus*. Mar. Biol. 104: 31–39.

Zar, J., 1996. Biostatistical Analysis. Prentice Hall Int. Upper Saddle River.

Zariquiey Alvarez, R., 1968. Crustáceos decápodes ibéricos. Invest. Pesq. Tomo 32. Barcelona.

Hydrobiologia **440:** 119–128, 2000.
M.B. Jones, J.M.N. Azevedo, A.I. Neto, A.C. Costa & A.M. Frias Martins (eds), Island, Ocean and Deep-Sea Biology.

Direct and indirect effects of *Littorina littorea* (L.) on barnacles growing on mussel beds in the Wadden Sea

Christian Buschbaum
Alfred-Wegener-Institut für Polar- und Meeresforschung, Wadden Sea Station Sylt, Hafenstraße 43, D-25992 List, Germany
Fax: +49-4651-956200. Tel: +49-4651-956135. E-mail: cbuschbaum@awi-bremerhaven.de

Key words: grazing, direct and indirect effects, barnacles, algae, mussel bed

Abstract

On the extensive sedimentary tidal flats of the Wadden Sea, beds of the blue mussel *Mytilus edulis* represent the only major hard substratum and attachment surface for sessile organisms. On this substratum, the barnacle *Semibalanus balanoides* is the most frequent epibiont. In summer 1998, it occurred on over 90% of the large mussels (>45 mm shell length) and the dry weight of barnacles reached 65% of mussel dry weight. However, the extent of barnacle overgrowth is not constant and differs widely between years. Periwinkles (*Littorina littorea*) may reach densities >2000 m^{-2} on intertidal mussel beds. Field experiments were conducted to test the effect of periwinkle grazing on barnacle densities. An experimental reduction of grazing and bulldozing pressure by periwinkles resulted in increased recruitment of barnacles, while barnacle numbers decreased with increasing snail density. The highest numbers of barnacles survived in the absence of *L. littorea*. However, a lack of periwinkle grazing activity also facilitated settlement of ephemeral algae which settled later in the year. Field experiments showed that the growth rate of barnacles decreased in the presence of these ephemeral algae. Thus, *L. littorea* may reduce initial barnacle settlement, but later may indirectly increase barnacle growth rate by reducing ephemeral algae. It is suggested that periwinkle density may be a key factor in the population dynamics of *S. balanoides* on intertidal mussel beds in the Wadden Sea.

Introduction

The Wadden Sea is one of the largest coastal soft bottom habitats in the world. In contrast to rocky shores, it is dominated by rather unstable sediments, and primary hard substrata for colonization of sessile epifauna and macroalgae are generally rare. However, epibenthic mussel beds of *Mytilus edulis* L. provide a widely available substratum for epibionts (aggregations of live blue mussels and shell fragments form three-dimensional structures within otherwise macroscopically unstructured mud and sandflats). This secondary hard substratum is utilized by a large number of sessile epifauna and algae (Sebens, 1982; Matsumasa & Nishihira, 1994; Albrecht, 1998).

On rocky shores, herbivorous grazers are very important in structuring patterns of distribution and abundance of algae (e.g. Lubchenco, 1978; Lubchenco & Gaines, 1981; Hawkins & Hartnoll, 1983; Janke, 1990; Vadas, 1992; Anderson & Underwood, 1997; Kim, 1997). Removal of algae by grazing activity causes direct and indirect effects on other sessile organisms (e.g. Petraitis, 1983, 1987; Bertness, 1984; Dungan, 1986, 1987; Vadas, 1992; Anderson & Underwood, 1997; Kim, 1997). For example, Van Tamelen (1987) showed that removal of algae by gastropod grazers indirectly led to higher recruitment of barnacles. On mussel beds in the Wadden Sea, grazing periwinkles, *Littorina littorea* (L.), achieve high densities of hundreds of snails per m^2 (Wilhelmsen & Reise, 1994). Their grazing pressure limits the distribution of ephemeral macroalgae which are generally rare in this habitat (Albrecht, 1998). Grazing and bulldozing by herbivorous snails was also found to be a mortality factor in young balanids (Connell, 1961; Dayton, 1971; Denley & Underwood, 1979; Hawkins, 1983; Underwood et al., 1983; Petraitis, 1983; Farrell, 1988; Miller & Carefoot, 1989). Nevertheless, intertidal mussel beds are often strongly overgrown by the barnacle *Semibalanus balanoides* L. The extent of barnacle epigrowth, however, shows strong temporal variation and differs widely between years (pers. obs.).

The aim of this study was to test the hypothesis that these activities of *L. littorea* would reduce survival of recently-settled barnacles on mussel beds in the Wadden Sea. In this case, inter-annual fluctuations in snail densities may be responsible for different barnacle settlement success. However, recently-settled barnacles may grow very fast (Bertness, 1984; Luther, 1987) and, within a short time period, attain a size refuge that protects them from grazing and bulldozing activity (Denley & Underwood, 1979). Then, the snails may actually have positive effects on the barnacles because their grazing activity suppresses algal epigrowth which may have disadvantageous effects for balanids (Crisp, 1960). Therefore, it was hypothesized that exclusion of snails would facilitate the development of ephemeral algae on mussel shell surfaces and that increasing algal epigrowth would reduce barnacle growth rate. Thus, the effect of periwinkle activity would then switch from a direct negative effect on early stages towards an indirect positive effect on the growth rate of balanids that had attained a size refuge from grazing and bulldozing pressure.

Material and methods

Study area

Investigations were carried out in Königshafen, a shallow tidal bay of about 4.5 km^2, north of the island of Sylt in the North Sea (55° 02′ N, 08° 26′ E) (Fig. 1). Sandy dunes form the northern and western boundary of the bight while it is limited by dikes in the south. The tidal water enters the bight through a tidal channel in an east-west direction. Due to aerial input of dune sand, the sediment is relatively coarse grained compared with other intertidal areas of the north German Wadden Sea. The area belongs to the cold temperate region with a mean annual water temperature of about 9 °C, a summer average of 15 °C, and a winter average of 4°C. The tides are semidiurnal and average tidal range is about 1.8 m. Salinity remains close to 30 psu. Further information about hydrography, geology, sediments and biota of the study area is given elsewhere (Reise, 1985; G. Austen, 1994; I. Austen, 1994; Bayerl & Higelke, 1994; Reise et al., 1994).

Intertidal mussel beds cover 1% of the intertidal area of Königshafen and are partly covered by the brown algae *Fucus vesiculosus* forma *mytili* (Nienburg), which lacks a holdfast and reproduces only vegetatively (Albrecht, 1998). In general, large mussels (>45 mm shell length) are located in the top layer of the mussel beds and are often heavily overgrown by balanids. Smaller and newly-settled mussels (shell length ≤45 mm) are situated underneath and are attached to the byssus threads of the large mussels. Finally, many small mussels stick in sediment which is finer than in the surrounding tidal flats (Albrecht, 1994; Okun, 1999) due to biodeposit accumulation (Dittmann, 1987).

In the northern Wadden Sea, the periwinkle *L. littorea* achieves high abundance in these mussel beds. Average densities are about 400 m^{-2} (Reise et al., 1994) but, in some years, they are far more abundant. Ephemeral algae are dominated by different *Enteromorpha* species. These are restricted to intertidal flats and rarely occur on mussel beds due to high periwinkle grazing pressure (Wilhelmsen & Reise, 1994; Albrecht, 1998). Growth of ephemeral algae starts mostly in the beginning of June after the settlement of *S. balanoides* in April and May (Luther, 1987).

Extent of barnacle overgrowth on intertidal mussel beds

From June 1998 to August 1999, the extent of *Semibalanus balanoides* was examined on an intertidal mussel bed located north of the tidal channel (Fig. 1). Species composition and abundance of mussels in this bed did not differ from other mussel beds in the Königshafen area. Samples were taken in June and September 1998, and in March, May and August 1999 using a box corer of 400 cm^2 (sampling depth of 25 cm). At each sampling, six replicates were taken. In the laboratory, samples were washed over a 1000 μm sieve, and mussels and *L. littorea* were counted. Afterwards, barnacle epigrowth was removed from the mussels and barnacles, and *M. edulis* were dried separately (3 days, 75 °C) to a constant weight. Dry weight of mussels and barnacle overgrowth was determined to the nearest 0.01 g. However, only *M. edulis* with a shell length >45 mm were examined because smaller mussels were not overgrown by barnacles. The commencement of barnacle settlement on mussels was noted.

Grazing experiments

The influence of periwinkles on the abundance of recently-settled barnacles on mussel shell surfaces was

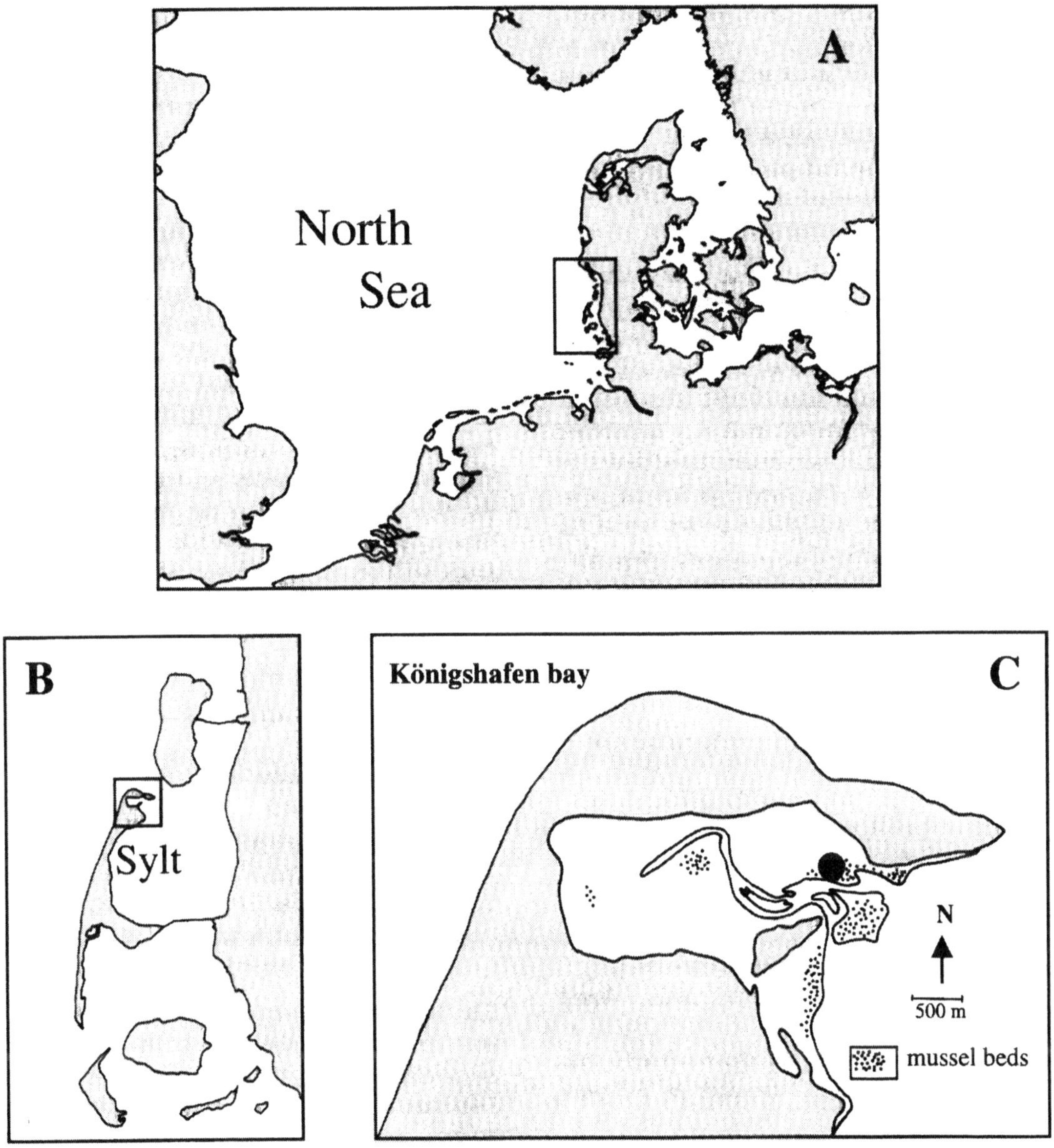

Figure 1. Location of the study area: the Königshafen bay (C) in the north of the island of Sylt (B) in the North Sea (A). Investigation site is marked by a black circle.

investigated by grazer exclusion experiments on the same intertidal bed of *Mytilus edulis* where barnacle overgrowth was determined. To keep snail densities constant, circular cages (25 cm high, 20 cm in diameter) with walls made of 6 × 6 mm^2 rust resistant wire netting were used. The upper and lower end of the cages were stabilized with a plastic ring (20 cm in diameter). Roofed cages were used to prevent snail migration. On 20 March, 1999 (i.e. before settlement of barnacles), the cages were fixed carefully to the mussel bed using three iron rods per cage (50 cm length; 6 mm in diameter). Care was taken to prevent periwinkles from passing underneath the cages. Caged mussels were cleaned of any epigrowth using a knife. The experiment included six treatments:

1. no snails in the cages;
2. reduced natural density of *L. littorea* (20 snails per cage corresponding to a density of approximately 650 m^{-2});
3. natural (ambient) snail density of March 1999 (50 snails per cage; ~1500 snails m^{-2}); and

4. three fold natural density (150 *L. littorea* per cage; 4500 snails per m^2).
   Two treatments served as a control of possible cage artefacts:
5. open cages with walls which left a 10 cm space above the ground so that snails could freely pass into and out of the treated area; and
6. untreated areas of the same size as cages.

Each treatment was replicated 6-fold with a random distribution on the mussel bed. On 13/14 May 1999 (after an experimental period of 8 weeks), when most barnacles had settled, barnacle epigrowth in the cages was determined. This was done using a clear sheet of pvc which was pierced by 100 randomly distributed holes each of 4 mm diameter, resembling the design used by Bertness (1984). The sheet had the same diameter as the cages, and was placed on the top of the cage after removing the roof and enclosed snails. Percent cover of the barnacle epigrowth on caged mussels was obtained by counting the holes with balanids visible underneath. The barnacle percent cover per cage was a rough estimate of barnacle abundance.

Periwinkle grazing effect on different barnacle size classes

A laboratory experiment was performed to test at which size newly-settled *S. balanoides* are protected against grazing pressure of *L. littorea*. Five size classes of barnacles were tested (each 6-fold replicated):

1. attached but not metamorphosed cypris larvae;
2. freshly metamorphosed barnacles with a basal shell diameter of 0.3–0.5 mm;
3. barnacles with basal shell diameter of about 1 mm;
4. barnacles with basal shell diameter of 2–3 mm; and
5. barnacles with a basal shell diameter of 5 mm.

For this experiment, overgrown mussels with a shell size of 50–60 mm were collected in the field and all barnacles were removed except for one size class. The number of the remaining barnacles was counted (10–15 barnacles per mussel) and each mussel was transferred to an aquarium (18 cm length, 12 cm width, 12 cm height) filled with a 2-cm layer of sandy sediment and a 10-cm layer of filtered seawater. Two specimens of *L. littorea* were added to each aquarium (containing a single mussel). A circular fence made of galvanized 6 $\times$ 6 mm^2 wire netting surrounded the mussel and the two periwinkles, preventing escape of the snails because they avoid contact with galvanized surfaces (Bertness, 1984). As a control, the same experiment was repeated without snails. The experiment started on 13 May, 1999. The number of barnacles remaining per mussel was counted after 24 h.

Exclusion of grazers

The effect of *L. littorea* on algal development and barnacle growth rates was tested by a further field experiment using the same cages as before (25 cm high, 20 cm in diameter). On 24 June, 1999, cages were fixed on a tidal flat with mussel clusters near the investigated mussel bed. Each cage contained seven mussels (shell size: 50–60 mm) with 10–15 barnacles per mussel and no algal epiphytes initially. The number of barnacles per cage was counted and the rostro-carinal length of the barnacle apertures was measured to the nearest 0.01 mm using a stereo microscope with ocular micrometer. At the beginning of the experiment, barnacle apertures had a length of 2.00–3.00 mm. There were three treatments, each replicated 6-fold:

1. cages with barnacle overgrown mussels and 20 snails added;
2. cages with barnacle overgrown mussels but without snails; and
3. cages in which barnacle overgrown mussels were cleaned of algal epiphytes by hand. This was done using a brush and a sponge at weekly intervals. All snails were removed from these cages at the beginning of the experiment.

The last treatment served as a control to test if it was really the cleaning activity of snails that influences algal development and barnacle growth. On 26 September 1999 (after an experimental period of 13 weeks), apertures of barnacles were measured and coverage by ephemeral algae of each mussel shell surface was estimated to the nearest 10%. To determine barnacle mortality, the number of live balanids was counted.

Statistical analysis

Results were calculated as mean values with standard deviation ($\bar{x}\pm$SD). All experiments were analysed by means of analysis of variance (ANOVA). Cochran's test was used to test for homoscedasticity of variances. Different levels within a significant experimental factor were analysed using Tukey's Honest-Significant-Difference (HSD) multiple com-

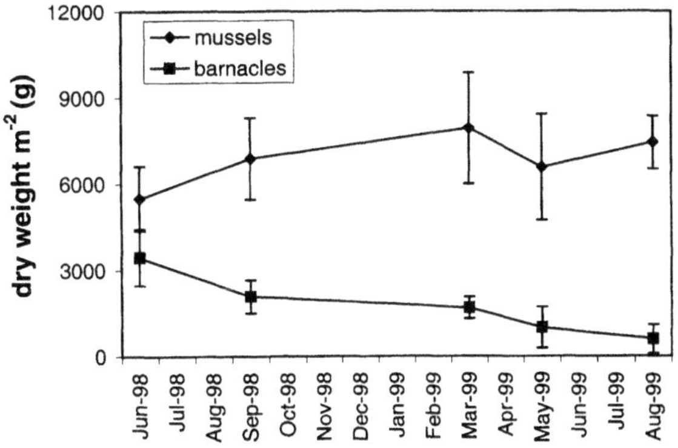

Figure 2. Dry weight of mussels and barnacle overgrowth (means of six replicates ± SD) over the investigation period from June 1998 to August 1999.

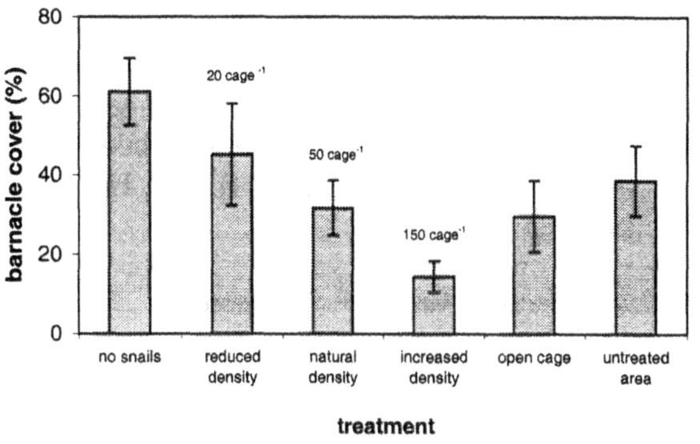

Figure 3. Mean percentage barnacle cover (± SD) (n = 6 cages) on blue mussels *Mytilus edulis* in treatments with different densities of *Littorina littorea* after an experimental period of 8 weeks.

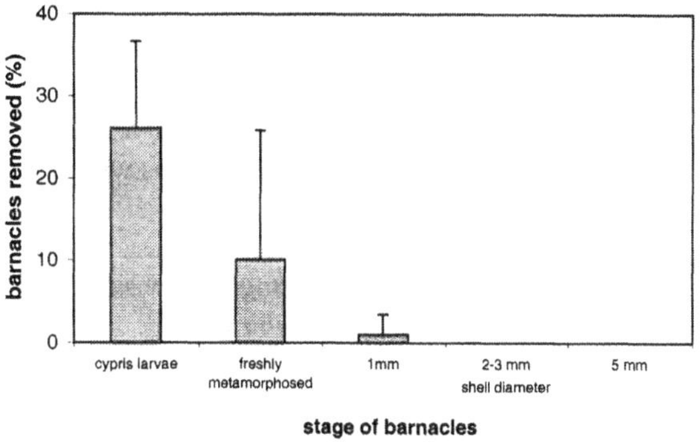

Figure 4. Percentage of barnacles (+ SD) (n = 6 aquaria) of different size classes removed by *Littorina littorea* after 24 hours.

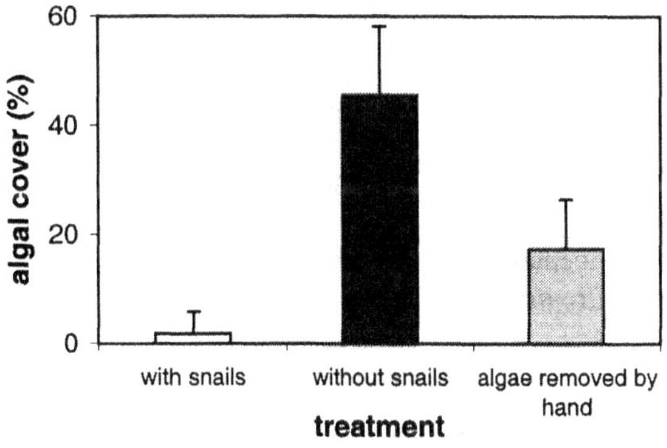

Figure 5. Mean percent of algal cover (+ SD) (n = 6 cages) on blue mussels *Mytilus edulis* in three different treatments.

parison test. Data of all experiments were homoscedastic in variance except for the estimation of algae percent cover in the grazing exclusion experiment. The data of this experiment were arcsine transformed to eliminate heterogeneity of variance. Effects were considered to be statistical significant, if p-value was <0.05.

Results

Extent of barnacle overgrowth

On the investigated mussel bed, overgrowth of *S. balanoides* was present throughout the 14 month investigation period. In June 1998, barnacles occurred on 91.9 (± 6.3)% of the large sized mussels (>45 mm) and barnacle dry weight reached 65.2 (± 21.5)% of mussel dry weight. However, the extent of barnacle overgrowth strongly decreased from 3445 (± 978) g m^{-2} dry weight at the beginning of the investigation period (June 1998) to 589 (± 496) g m^{-2} in August 1999 (Fig. 2). In contrast, mussel dry weight (mean over investigation period 6855 ± 924 g m^{-2}) and density [mean 461 (± 96) m^{-2}] did not show this tendency indicating that a fluctuation in available shell surface was not responsible for decreasing barnacle overgrowth. Mean density of *L. littorea* was 1240 (± 553) m^{-2} with highest densities in spring [1842 (± 362) m^{-2} in 1998 and 1739 (± 356) m^{-2} in 1999] shortly after barnacle settlement.

Grazing effects on newly-settled barnacles

Experimental manipulation of snail density resulted in a significant correlation between snail density and barnacle percent cover in the cages (one-way ANOVA, F = 34.027, df = 3, p < 0.0001) (Fig. 3). The highest percentage of barnacles became established in the absence of periwinkles, and increasing snail numbers coincided with decreasing barnacle percentage cover. The mean percentage of recently-settled barnacles differed significantly between the four experimental snail density levels (Tukey's test, $p < 0.05$). The barnacle epigrowth in open cages and the untreated areas did not differ from the cages with a natural density (50 snails cage^{-1}) of periwinkles (one-way ANOVA, $F = 1.963$, df = 2, $p = 0.175$). Thus, cage artefacts were not detected.

Grazing effects on barnacle size classes

Barnacle survival in the presence of periwinkles depended significantly on barnacle shell size (one-way ANOVA, $F = 10.404$, df = 4, $p < 0.0001$) (Fig. 4). In the laboratory, highest removal of barnacles occurred in attached but not metamorphosed cypris larvae and differed significantly from all other levels of size (Tukey's test, $p < 0.03$). Removal of metamorphosed balanids decreased with increasing shell size but differences were not significant (Tukey's test, $p > 0.05$). The presence of periwinkles did not affect barnacles > 2 mm shell size. In the control experiments without *L. littorea*, no barnacles disappeared.

Exclusion of grazers and cascading effects on algal development and barnacle growth rate

Algal overgrowth
After 13 weeks, the percentage cover of algae on mussels and attached balanids significantly differed over the three treatments (one-way ANOVA, $F = 183.477$, df = 2, $p < 0.0001$, followed by Tukey's test, $p < 0.001$) (Fig. 5). Algal development was highest in snail exclusion cages (45.56 ± 12.51%) and the lowest percentage cover occurred in the treatment with added periwinkles (1.85 ± 3.96%) (Fig. 6). Mussels cleaned of epigrowth by hand showed an intermediate growth of algae (17.41 ± 9.03%), indicating that experimental cleaning activity was less effective than that of *L. littorea*.

Growth rate of barnacles
After the experimental period of 13 weeks, mean growth of balanids in the presence of *L. littorea* was 1.62 (± 0.23) mm. This was significantly higher than growth of barnacles in the absence of snails (1.27 ± 0.19 mm) (one-way ANOVA, $F = 4.9854$, df = 2, $p = 0.0218$, followed by Tukey's test, $p = 0.0178$) (Fig. 7). Weekly removal of algae by hand resulted in intermediate barnacle growth of 1.48 (± 0.14) mm. Differences were not significant compared to the treatments with and without snails (Tukey's test, $p > 0.05$). No differences were found in barnacle mortality (one-way ANOVA, $F = 0.8498$, df = 2, $p = 0.4471$). The mortality rate ranged from 40 to 50% in all three treatments.

Discussion

Seasonal variation of barnacle epigrowth

The barnacle *Semibalanus balanoides* occurs frequently on epibenthic mussel beds in the northern Wadden Sea and, in some years, intertidal mussel beds may be completely covered by barnacle epigrowth (pers. obs.). In June 1998, the extent of barnacle overgrowth reached 65% of mussel dry weight, but decreased to only 8% of mussel dry weight by August 1999. The high density of balanids in 1998 resulted from a strong barnacle settlement in spring 1996 while settlement success was much lower in the following 3 years (pers. obs.). The decreasing abundance of *S. balanoides* from 1998 to 1999 indicates that the barnacle recruitment in these years was not high enough to balance mortality. Such annual variations in recruitment of *S. balanoides* are also reported from other areas (Kendall et al., 1985; Carroll, 1996).

In the present study, the cage experiments demonstrated a strong negative correlation between recruitment of *S. balanoides* on mussel beds and abundance of *L. littorea*. Barnacle percent cover decreased significantly with increasing snail densities in the cages. Abundance of periwinkles on intertidal mussel beds may vary strongly between years (Wilhelmsen & Reise, 1994; Nicolaysen, 1996; Fenske, 1997). This may be a consequence of variations in snail recruitment, varying predation intensity (e.g. by crabs; Scherer & Reise, 1981), or parasitic infestation (Lauckner, 1984). Interestingly, according to Buschbaum & Reise (1999), the barnacle cover on periwinkle shells had a negative effect on snail survival and fecundity, and is, therefore, regarded as a further important factor influencing the population dynamics of *L. littorea*. Although there are no long-term observations on the relationship of snail abundances to the extent of barnacle overgrowth on mussel beds, the strong interaction of snail density and barnacle epigrowth found in this investigation suggests that fluctuating snail densities may be an important factor in interannual variation of *S. balanoides* overgrowth on intertidal mussel beds in the Wadden Sea.

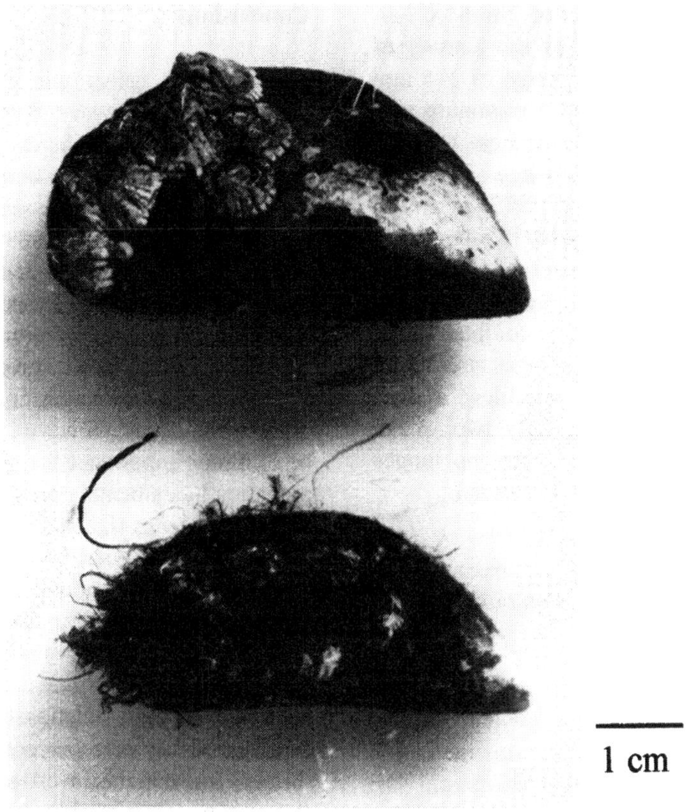

Figure 6. Extent of algal overgrowth on blue mussels *Mytilus edulis* in the cages with *Littorina littorea* (above) and in snail exclusion cages (below) after an experimental period of 13 weeks.

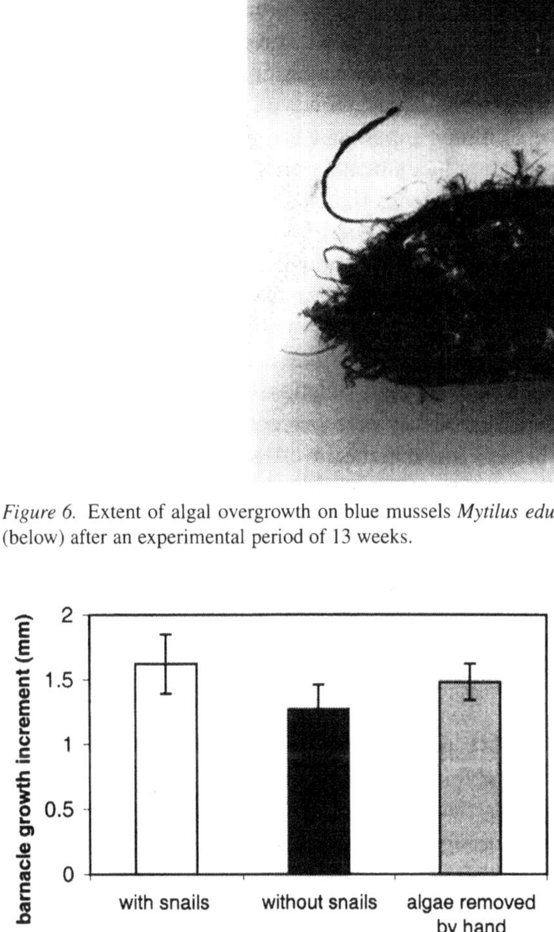

Figure 7. Mean barnacle growth increment ± SD ($n = 6$ cages) over an experimental period of 13 weeks. Growth rate was highest in the presence of *Littorina littorea*.

Negative effects of Littorina littorea *on survival of recently-settled barnacles*

Littorina littorea is a herbivorous snail, feeding mainly on early successional or ephemeral algae among which different *Enteromorpha* and *Ulva* species are preferred (Lubchenco, 1978; Watson & Norton, 1985; Imbrie et al., 1989; Norton et al., 1990; Wilhelmsen & Reise, 1994). Thus, periwinkle grazing activity reduces the development of ephemeral algae on intertidal mussel beds (Wilhelmsen & Reise, 1994; Albrecht, 1998). However, herbivorous snails may also have deleterious effects on sessile organisms outside their food spectrum (e.g. Dayton, 1971; Denley & Underwood, 1979; Menge et al., 1986; Petraitis, 1987; Farrell, 1988; Miller & Carefoot, 1989). These include the inadvertent dislodgement and consumption of settled larvae as the snails browse across the mussel shell surface (Denley & Underwood, 1979; Petraitis, 1987; Farrell, 1988; Miller & Carefoot, 1989), and the bulldozing of newly-settled larvae from their substratum by pushing and dislodging them with their foot, mouth, shell or tentacles (Dayton, 1971; Miller & Carefoot, 1989). This may explain how the grazing activity of the periwinkle *L. littorea* on mussel shell surfaces may have reduced survival of recently-settled barnacles in the present study.

Reaching a basal shell diameter of 2 mm, *S. balanoides* were no longer affected by the activity of *L. littorea*. This size is within the range of 2–4 mm estimated by Dayton (1971) to be a minimum size refuge for *Balanus glandula* (Darwin) from bulldozing by limpets, and is similar to the range of 3–4 mm estimated by Denley & Underwood (1979) as a minimum size refuge for *Tesseropora rosea* (Krauss) from mortality caused by *Cellana tramoserica* (Sowerby). Thus, on intertidal mussel beds, direct effects of *L. littorea* on barnacle epigrowth may be confined to the period of larval settlement and early stages after metamorphosis. *Semibalanus balanoides* achieve a shell diameter of 2 mm after approximately two weeks (Bertness, 1984), after which they were no longer vulnerable to the grazing activity of *L. littorea*.

Positive effects of periwinkle grazing on barnacle growth

The grazing activity by periwinkles prevented dense ephemeral algae epigrowth. Indirectly, this increased barnacle growth rate, because barnacle feeding was hindered by the presence of algae epiphytes. Direct observations showed that the opening of barnacle opercula was not hindered by algal epigrowth but that the feeding activity of the setose cirri was hampered. Algal epiphytes may also reduce the water current which may result in a restricted food supply for barnacles (Crisp, 1960). In addition, removal of gastropod grazers may lead to an accumulation of sediment (Bertness, 1984; Anderson & Underwood, 1997) which was also apparent in the treatments within snail-free cages in the present study. This sediment layer may impair living conditions for filter feeders and was supposed to be a mortality factor in balanids (Bertness, 1984). However, barnacles may survive for long periods without feeding (Barnes et al., 1963; Barnes & Barnes, 1967). This may be the reason why, in this study, barnacles covered by algal epigrowth did not show a higher mortality rate than uncovered *S. balanoides*. Nervertheless, the low growth might indicate a reduced fitness and cause a higher mortality in winter and a lower reproductive output in the next spring. Considering the direct negative effects of grazing on barnacle settlement and the indirect positive effects on barnacle growth, best living conditions for barnacles are expected to occur at intermediate periwinkle densities.

Conclusions

Many factors affect the population dynamics of *Semibalanus balanoides*. Barnacle settlement may be limited by planktonic larval supply (Bertness et al., 1992), while bioorganic films developed on the attachment surface influence settlement of barnacle cypris larvae (Maki et al., 1988; Keough & Raimondi, 1995). On rocky shores, Jernakoff (1985) and Jenkins et al. (1999) showed that newly-settled barnacles are hindered in early recruitment by algal epigrowth. As the barnacles grow, predation and competition for space become increasingly important. On intertidal mussel beds, however, these two factors may be of minor importance for barnacle density, because predation by epibenthic predators such as shore crabs *Carcinus maenas* (L.) and starfish *Asterias rubens* L. seem to be important only in the adjacent subtidal zone (Buschbaum, unpublished). Since *S. balanoides* settle earlier in the year than ephemeral algae, and other sessile invertebrates are not common, interspecific competition for space can be disregarded. Therefore, the negative relationship between *L. littorea* density and barnacle percent cover on mussel shell surfaces found in this investigation indicates that grazing activity of periwinkles is a very important factor in settlement and post-settlement processes structuring barnacle distribution on intertidal mussel beds in the Wadden Sea.

Acknowledgements

I would especially like to thank Hildegard and Richard Myland for their advice during all phases of my work. Bettina Saier, Dr Martin Thiel and two anonymous reviewers gave valuable comments on the manuscript. I am also grateful to Dr Werner Armonies and Prof. Karsten Reise for discussions, critical remarks and encouragement. This article is based in part on a doctoral study of C. Buschbaum in the Faculty of Biology, University of Hamburg, Germany.

References

Albrecht, A. S., 1994. Effects of *Fucus vesiculosus* covering intertidal mussel beds in the Wadden Sea. Helgoländer Meeresunters. 48: 243–256.

Albrecht, A. S., 1998. Soft bottom versus hard rock: community ecology of macroalgae on intertidal mussel beds in the Wadden Sea. J. exp. mar. Biol. Ecol. 229: 85–109.

Anderson, M. J. & A. J. Underwood, 1997. Effects of gastropod grazers on recruitment and succession of an estuarine assemblage: a multivariate and univariate approach. Oecologia 109: 442–453.

Austen, G., 1994. Hydrodynamics and particulate matter budget of Königshafen, southeastern North Sea. Helgoländer Meeresunters. 48: 183–200.

Austen, I., 1994. The surficial sediments of Königshafen variations over the past 50 years. Helgoländer Meeresunters. 48: 163–171.

Barnes, H. & M. Barnes, 1967. The effect of starvation and feeding on the production of egg masses in the boreo-arctic cirripede *Balanus balanoides* (L.). J. exp. mar. Biol. Ecol. 1: 1–6.

Barnes, H., M. Barnes & D.M. Finlayson, 1963. The metabolism during starvation of *Balanus balanoides*. J. mar. biol. Ass. U.K. 43: 213–223.

Bayerl, K. A. & B. Higelke, 1994. The development of northern Sylt during the last Holocene. Helgoländer Meeresunters. 48: 145–162.

Bertness, M. D., 1984. Habitat and community modification by an introduced herbivorous snail. Ecology 65: 370–381.

Bertness, M. D., S. D. Gaines, E. G. Stephens & P. O. Yund, 1992. Components of recruitment in populations of the acorn barnacle *Semibalanus balanoides* (Linnaeus). J. exp. mar. Biol. Ecol. 156: 199–215.

Buschbaum, C. & K. Reise, 1999. Effects of barnacle epibionts on the periwinkle *Littorina littorea* (L.). Helgol. Mar. Res. 53: 56–61.

Carroll, M. L., 1996. Barnacle population dynamics and recruitment regulation in southcentral Alska. J. exp. mar. Biol. Ecol. 199: 285–302.

Connell, J. H., 1961. Effects of competition, predation by *Thais lapillus*, and other factors on natural populations of the barnacle *Balanus balanoides*. Ecol. Monogr. 31: 61–104.

Crisp, D. J., 1960. Factors influencing growth-rate in *Balanus balanoides*. J. anim. Ecol. 29: 95–116.

Dayton, P. K., 1971. Competition, disturbance and community organizations: the provision and subsequent utilization of space in a rocky intertidal community. Ecol. Monogr. 41: 351–389.

Denley, E. J. & A. J. Underwood, 1979. Experiments on factors influencing settlement, survival and growth of two species of barnacles in New South Wales. J. exp. mar. Biol. Ecol. 36: 269–293.

Dittmann, S., 1987. Die Bedeutung der Biodeposite für die Benthosgemeinschaft der Wattsedimente. Unter besonderer Berücksichtigung der Miesmuschel *Mytilus edulis* L. PhD thesis, University of Göttingen, Germany, 182 pp.

Dungan, M. L., 1986. Three-way interactions: barnacles, limpets and algae in a sonoran desert rocky intertidal zone. Am. Nat. 127: 292–316.

Dungan, M. L., 1987. Indirect mutualism: complementary effects of grazing and predation in a rocky intertidal community. In Kerfoot, W. C. & A. Sih (eds), Predation: Direct and Indirect Impacts on Aquatic Communities. University Press of New England, Hanover, NH: 188–200.

Farrell, T. M., 1988. Community stability: effects of limpet removal and reintroduction in a rocky intertidal community. Oecologia 75: 190–197.

Fenske, C., 1997. The importance of intraspecific competition in a *Littorina littorea* population in the Wadden Sea. Hydrobiologia 355: 29–39.

Hawkins, S. J., 1983. Interactions of *Patella* and macroalgae with settling *Semibalanus balanoides*. J. exp. mar. Biol. Ecol. 71: 55–72.

Hawkins, S. J. & R. G. Hartnoll, 1983. Grazing of intertidal algae by marine invertebrates. Oceanogr. Mar. Biol. Annu. Rev. 21: 195–282.

Imbrie, D. W., S. J. Hawkins & C. R. McCrohan, 1989. The olfactory-gustatory basis of food preference in the herbivorous prosobranch, *Littorina littorea* (Linnaeus). J. Moll. Stud. 55: 217–225.

Janke, K., 1990. Biological interactions and their role in community structure in the rocky intertidal of Helgoland (German Bight, North Sea). Helgoländer Meeresunters. 44: 219–263.

Jenkins, S. R., T. A. Norton & S. J. Hawkins, 1999. Settlement and post-settlement interactions between *Semibalanus balanoides* (L.) (Crustacea: Cirripedia) and three species of fucoid canopy algae. J. exp. mar. Biol. Ecol. 236: 49–67.

Jernakoff, P., 1985. The effect of overgrowth by algae on the survival of the intertidal barnacle *Tesseropora rosea* Krauss. J. exp. mar. Biol. Ecol. 94: 89–97.

Kendall, M. A., R. S. Bowman, P. Williamson & J. R. Lewis, 1985. Annual variation in the recruitment of *Semibalanus balanoides* on the North Yorkshire coast. J. mar. biol. Ass. U.K. 65 (4): 1009–1030.

Keough, M. J. & P. T. Raimondi, 1995. Responses of settling invertebrate larvae to bioorganic films: effects of different types of films. J. exp. mar. Biol. Ecol. 185: 235–253.

Kim, J. H., 1997. The role of herbivory, and direct and indirect interactions, in algal succession. J. exp. mar. Biol. Ecol. 217: 119–135.

Lauckner, G., 1984. Impact of trematode parasitism on the fauna of a North Sea tidal flat. Helgoländer Meeresunters. 37: 185–199.

Lubchenco, J., 1978. Plant species diversity in a marine intertidal community: Importance of herbivore food preference and algal competitive abilities. Am. Nat. 112: 23–39.

Lubchenco, J. & S. D. Gaines, 1981. A unified approach to marine plant-herbivore interactions. I. Populations and communities. Ann. Rev. Ecol. Syst. 12: 405–437.

Luther, G., 1987. Seepocken der deutschen Küstengewässer. Helgoländer Meeresunters. 41: 1–43.

Maki, J. S., D. Ritschoff, J. D. Costlow & R. Mitchell, 1988. Inhibition of attachment of larval barnacles, *Balanus amphitrite*, by bacterial surface films. Mar. Biol. 97: 199–206.

Matsumasa, M. & M. Nishihira, 1994. Habitat structure provided by *Mytilus edulis* and the modification by sessile inhabitants. Bull. Mar. Biol. Stn. Asamushi 19: 51–63.

Menge, B. A., J. Lubchenco & L. R. Ashkenas, 1986. Experimental separation of effects of consumers on sessile prey in the low zone of a rocky shore in the Bay of Panama: direct and indirect consequences of food web complexity. J. exp. mar. Biol. Ecol. 100: 225–269.

Miller, K. M. & T. H. Carefoot, 1989. The role of spatial and size refuges in the interaction between juvenile barnacles and grazing limpets. J. exp. mar. Biol. Ecol. 134: 157–174.

Nicolaysen, M., 1996. Die Bedeutung von Seegraswiesen (*Zostera noltii*) für die Strandschnecke (*Littorina littorea* L.) im Nordsylter Wattenmeer. MSc thesis, University of Kiel, Germany, 67 pp.

Norton, T. A., S. J. Hawkins, N. L. Manley, G. A. Williams & D. C. Watson, 1990. Scraping a living: a review of littorinid grazing. Hydrobiologia 193: 117–138.

Okun, N., 1999. Einfluß der Sedimentation auf die Miesmuschel (*Mytilus edulis* L.). MSc thesis, University of Hamburg, Germany, 68 pp.

Petraitis, P. S., 1983. Grazing patterns of the periwinkle and their effect on sessile intertidal organisms. Ecology 64: 522–533.

Petraitis, P. S., 1987. Factors organizing rocky intertidal communities of New England: herbivory and predation in sheltered bays. J. exp. mar. Biol. Ecol. 109: 117–136.

Reise, K., 1985. Tidal flat ecology. Springer, Berlin, 191 pp.

Reise, K., E. Herre & M. Sturm, 1994. Biomass and abundance of macrofauna in intertidal sediments of Königshafen in the northern Wadden Sea. Helgoländer Meeresunters. 48: 201–215.

Scherer, B. & K. Reise, 1981. Significant predation on micro- and macrobenthos by the crab *Carcinus maenas* L. in the Wadden Sea. Kieler Meeresforsch., Sonderh. 5: 490–500.

Sebens, K. P., 1982. Competition for space: growth rate, reproductive output, and escape in size. Am. Nat. 120: 189–197.

Underwood, A. J., E. J. Denley & M. J. Moran, 1983. Experimental analyses of the structure and dynamics of mid-shore rocky intertidal communities in New South Wales. Oecologia 56: 202–219.

Vadas, R. L., 1992. Littorinid grazing and algal patch dynamics. In Grahame, J., Mill, P. J., Reid, D. G. (eds), Proceedings of the Third International Symposium on Littorinid Biology 197–209.

Van Tamelen, P. G., 1987. Early successional mechanisms in the rocky intertidal: the role of direct and indirect interactions. J. exp. mar. Biol. Ecol. 112: 39–48.

Watson, D. C. & T. A. Norton, 1985. Dietary preferences of the common periwinkle *Littorina littorea* (L.). J. exp. mar. Biol. Ecol. 88: 193–211.

Wilhelmsen, U. & K. Reise, 1994. Grazing on green algae by the periwinkle *Littorina littorea* in the Wadden Sea. Helgoländer Meeresunters. 48: 233–242.

Hydrobiologia **440:** 129–135, 2000.
M.B. Jones, J.M.N. Azevedo, A.I. Neto, A.C. Costa & A.M. Frias Martins (eds), Island, Ocean and Deep-Sea Biology.

Age and growth, reproduction and diet of a sublittoral population of the rock goby *Gobius paganellus* (Teleostei, Gobiidae)

José Manuel N. Azevedo & Ana Maria V. Simas
Departamento de Biologia, Universidade dos Açores, Rua Mãe de Deus, 9500 Ponta Delgada, Portugal
Tel. + 351 296 653044. E-mail: azevedo@notes.uac.pt

Key words: fish biology, length–weight relationship, length–frequency distributions, modal analysis, growth rate, life span

Abstract

Basic biological information for a sublittoral population of the rock goby *Gobius paganellus* Linnaeus, 1758 is presented based on a 2-year study involving 1680 specimens. The length–weight relationship was given by TW = 0.0089 * $TL^{3.163}$ (where TW= total weight in g; TL= total length in mm). Age at length data were inferred by modal analysis of the monthly length–frequency distributions. The parameters of the fitted Von Bertalanffy growth equation (with seasonal component, birth date on the 1st of January) were L_∞ = 13.8 cm; K = 0.73 yr^{-1}; t_0 = −0.22 yr; C = 0.95; W = 0.07. This growth rate is much higher than that described for northern Europe populations (where K is about 0.3 yr^{-1}) and is probably associated with a shorter life span. Macroscopic examination of the gonads, and analysis of the monthly values of the gonadosomatic index, indicated that reproduction occurs in winter and early spring, with a maximum in February and March, when water temperatures are lowest. Individuals become sexually mature around 6–7 cm TL, a size that can be reached in less than 1 year. By contrast, individuals of this goby in the British Isles mature in their second or third year. Stomach contents were mainly small benthic invertebrates, predominantly crustaceans.

Introduction

The rock goby *Gobius paganellus* Linnaeus, 1758 is a relatively large goby found along the northwest Atlantic coastline, from western Scotland to Senegal, and on the Atlantic islands of the Azores, Madeira and the Canaries (Miller, 1986). It is the only species of *Gobius* from the Azores (Miller, 1984) and is considered mainly as an intertidal species (Miller, 1986). The two major studies of its biology (Miller, 1961; Dunne, 1978) used intertidal animals as a matter of convenience, however, Dunne (1978) recognized that the species is common subtidally "to a depth of approximately five metres". In the Azores, *G. paganellus* is found in the rocky subtidal down to 15 m (Patzner et al., 1992), where it is one of the most abundant benthic species, together with *Tripterygion delaisi*, *Parablennius incognitus* and *P. ruber* (Azevedo, 2000). This paper describes various aspects of the biology of a sublittoral population of *G. paganellus* at São Miguel (Azores).

Materials and methods

Between December 1993 and November 1995, 1680 specimens were captured at 3–5 m depth on a rocky reef near Ponta Delgada harbour. One to three samples were taken each month by SCUBA diving, using rotenone (3–67 animals per month, average 23, standard deviation 18). Each specimen was measured (total length, TL, to the nearest mm below), weighed (total wet weight, TW, in g), fixed in formalin and transferred to 70% ethanol after 2–4 weeks. Later, a monthly sub-sample of up to five specimens for each 1 cm length class was selected. Each of these was sexed by external examination of the shape of the genital papilla (following Miller, 1961) and by macroscopic observation of the gonads, which were removed and weighed (gonad weight, GW, in g). Gonad observation and weighing was not generally possible for smaller specimens (<5 cm TL). A gonosomatic index (GSI) was calculated as GSI=GW/(TW−GW)*100. Stomach contents, from 2 to 6 specimens in each monthly

Table 1. Length-frequency distribution of male and female *Gobius paganellus*. Asteriscs indicate the probability of the deviations from a 1:1 sex ratio, calculated from the binomial distribution (* $p<0.05$; *** $p<0.001$)

TL (cm)	Females	Males
2		1
3	8	3
4	19	10*
5	42	49
6	20	65
7	143	99***
8	134	121
9	165	147*
10	99	126*
11	26	92***
12	1	13***
13	1	4
N	690	726
Average	8.0	8.5
St. deviation	1.7	1.9
t	5.058	$p<0.001$

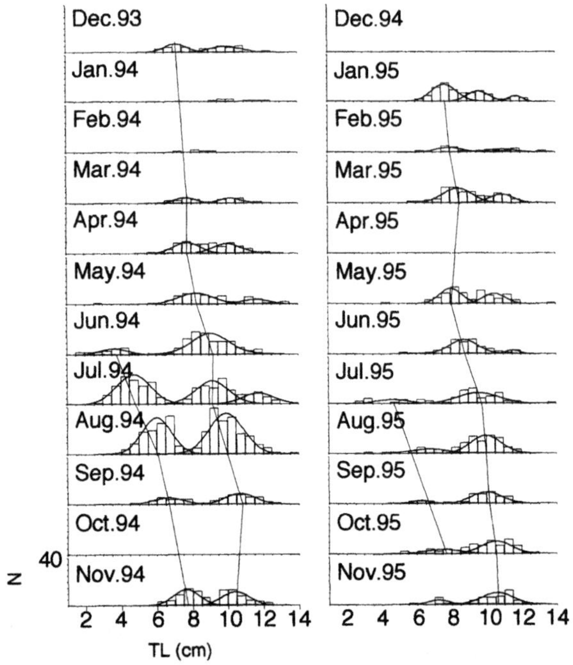

Figure 1. Monthly length–frequency distributions of *Gobius paganellus*, with superimposed normal curves from modal analysis.

sample selected to cover the entire size range present, were removed and placed in 70% ethanol. They were sorted under a dissecting microscope and identified to the lowest possible taxonomic rank. Results are presented as percentage occurrence in the stomachs (i.e. number of stomachs with item over total number of stomach contents examined).

Age at length data were deduced from modal analysis of the monthly length–frequency distributions and used to calculate the parameters of the Von Bertalanffy seasonalized growth equation, following the methods described in Sparre & Venema (1992) and using the FiSAT software (Gayanillo et al., 1994). The 1st of January was chosen as the date of birth. Age-length data for this species were read from graphs in Miller (1961) and Dunne (1978), and converted to total length. The age–length points obtained for these populations were used to determine their Von Bertalanffy parameters in the same way as data from the present study.

Results

Age and growth

Table 1 shows the length–frequency distributions of sexed individuals. There is a statistically significant difference in average total length between males and females due, mainly, to the higher proportion of males in the largest length classes. The length–weight relationship, calculated separately for each sex, was found to be statistically identical. The relationship for the entire sample is:

$$TW = 0.0089 * TL^{3,163} \qquad (n = 1152; r = 0.995).$$

The monthly length–frequency distributions, and respective modal analysis, are given in Figure 1, both sexes were considered together. Several well-defined modes can be seen and a displacement to the right can be followed between successive months. These modes were assumed to correspond to distinct age groups, or cohorts. The 1994 cohort was the most distinct and can be traced from its beginning in June 1994 up to the end of the sampling period. The 1993 cohort was deduced by comparison and its displacement could be followed until November 1994. Small specimens were also captured in the summer of 1995 but in much smaller numbers than the previous year, making it difficult to follow its modal progression. Based on these premises, an age–length key was computed (Table 2). This key contains enough age–length points to allow the adjustment of a von Bertalanffy growth curve with

Table 2. Age–length key for the Azorean population of *Gobius paganellus*, based on modal analysis data

Date	1993				1994				1995			
	Age (years)	TL (cm) Average	St.dv.	*n*	Age (years)	TL (cm) Average	St.dv.	*n*	Age (years)	TL (cm) Average	St.dv.	*n*
9/12/93	0.94	7.1	0.5	18								
10/3/94	1.19	7.7	0.6	12								
12/4/94	1.28	7.8	0.8	39								
11/5/94	1.36	8.3	1	50								
8/6/94	1.43	9	1.2	103	0.43	3.4	0.9	21				
19/7/94	1.55	9.3	0.9	86	0.55	4.7	0.9	112				
18/8/94	1.63	10	1	171	0.63	6	0.8	118				
24/9/94	1.73	10.7	0.9	40	0.73	6.7	0.8	22				
9/11/94	1.85	10.4	0.8	46	0.85	7.7	0.7	48				
18/1/95					1.05	7.7	0.6	50				
18/2/95					1.13	7.8	0.6	11				
13/3/95					1.19	8.4	1	59				
5/5/95					1.51	8	0.7	40				
5/6/95					1.42	8.8	0.9	55				
10/7/95					1.52	9.6	1.2	55	0.52	4.6	1.5	14
16/8/95					1.62	10	0.9	68				
26/9/95					1.73	10.1	0.8	41				
26/10/95					1.82	10.6	1	51	0.82	7.7	1.4	26
18/1/95					1.88	10.6	0.8	39				

a seasonal component. The parameters obtained were:

$$L_\infty = 12.77 \text{ cm}; K = 0.89 \text{ yr}^{-1};$$
$$t_0 = -0.16 \text{ yr}; C = 0.99; W = 0.06$$

The L_∞ obtained is smaller than the length of the largest specimen captured (13.8 cm). The latter is also the value obtained by the Powell–Wetherall method (see Sparre & Venema, 1992). When this value of $L\infty$ is made invariable, the new equation parameters become:

$$L_\infty = 13.8 \text{ cm}; K = 0.73 \text{ yr}^{-1};$$
$$t_0 = -0.22 \text{ yr}; C = 0.95; W = 0.07$$

This growth curve, and the points used for its computation, are show in Figure 2. This figure also shows the growth curves calculated from the data of Miller (1961) and Dunne (1978), whose parameters are, respectively:

$$L_\infty = 12.00 \text{ cm}; K = 0.31 \text{ yr}^{-1};$$
$$t_0 = -0.32 \text{ yr}; C = 1; W = 1$$
$$L_\infty = 14.15 \text{ cm}; K = 0.27 \text{ yr}^{-1};$$
$$t_0 = -0.03 \text{ yr}; C = 0.86; W = 0.85$$

The Azores population has a much higher growth rate, measured by K. In their first year, Azorean rock gobies attain an average size double that of their northern conspecifics. Values within 10% of L_∞ are attained in just 3 years in the Azores, compared with 7–8 years in the British Isles populations.

Reproduction

Monthly changes in gonadosomatic index for each sex are shown in Figure 3. Minimum values were observed between July and November, and maximum values between January and May. Females with ripe gonads (oocytes extruding under light pressure on the abdomen) were observed mainly in February and March, suggesting that reproduction occurs in winter and early spring.

The relationship between length and the gonadosomatic index is given in Figure 4. Well developed gonads were found only in specimens with total length equal to or greater than 6 cm (males) or 7 cm (females). These, presumably, represent the size at first sexual maturation for this population. According to the growth equation calculated above, males and females can reach sexual maturity in the reproductive season immediately following the one in each they were born.

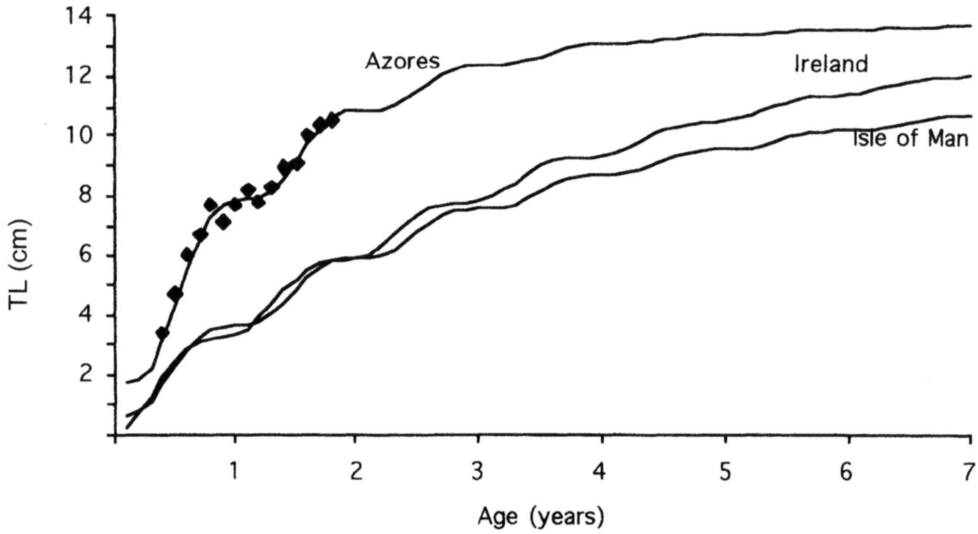

Figure 2. Growth curves for different populations of *Gobius paganellus:* Azores (present study), Ireland (Dunne, 1978) and Isle of Man (Miller, 1961). Points show age–length values obtained by modal analysis. See text for curve parameters.

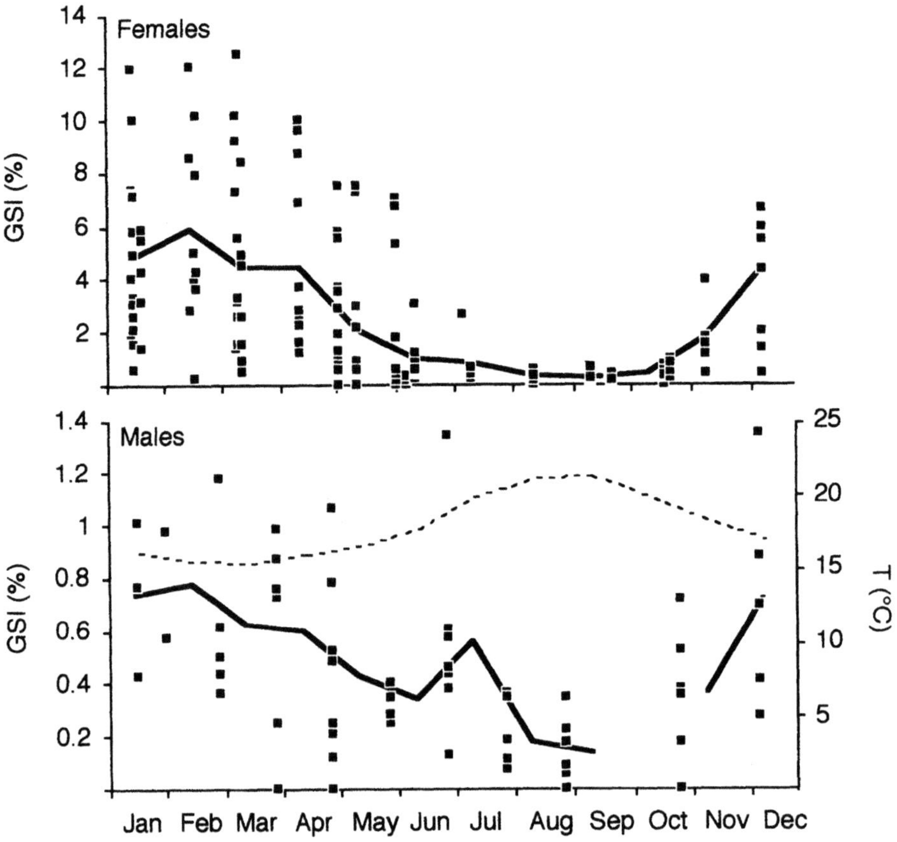

Figure 3. Monthly values of the gonadosomatic index for each sex of *Gobius paganellus*. The solid line connects the monthly averages. Note different scales on the GSI axes of each graph.

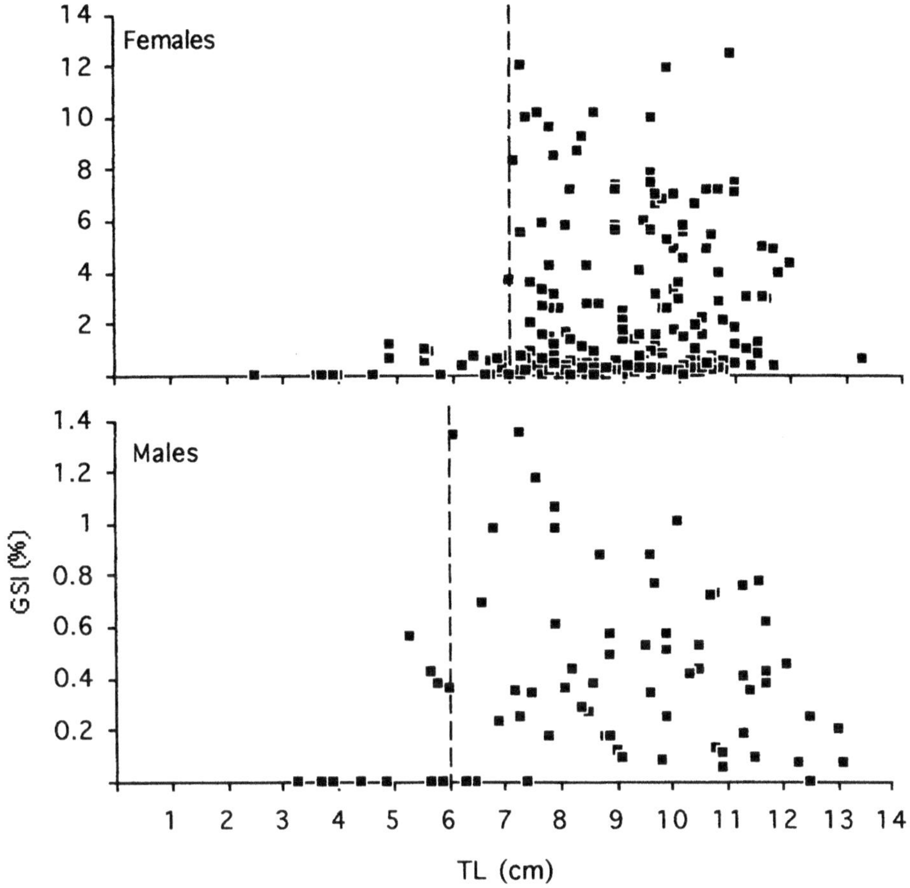

Figure 4. Length–GSI relationship for both sexes of the studied population of *Gobius paganellus*. Broken line mark presumptive size at first sexual maturation. Note different scales on the GSI axes of each graph.

Food

Table 3 shows the percentage occurrence of different food items in rock goby stomachs. Small benthic invertebrates accounted for the bulk of diet. Dominant items were isopods and amphipods, although gastropods, polychaetes and crabs were observed frequently.

Discussion

Separate growth rates for each sex could not be assessed by the methods used in the present study, but the higher proportion of males in the largest length classes suggests that males grow faster than females. For the closely related *G. niger* from the Adriatic Sea, Fabi & Giannetti (1985) also found a statistically significant dominance of males in the largest size classes; using otolith data, they showed that males grew faster than females. Miller (1961) and Dunne (1978) found no sexual differences in growth rate for *G. paganellus*. The former author, however, noted the disappearance of the larger members of each age group, suggesting that selective mortality or migration to the subtidal could account for this. This disappearance could mask the effects of differences in growth rate between sexes.

The main difference between the growth curve deduced in this study from those of more northern locations (Miller, 1961; Dunne, 1978) is the higher value of K. This implies a faster growth rate for the Azores population. Arruda et al. (1993) reported age-size values intermediate between these two extremes for a population of *G. paganellus* from the Ria de Aveiro, near the centre of the west coast of mainland Portugal. In this latter work, the average length of 0+ and 1+ specimens was given as 6.2 and 7.3 cm, respectively. These compare with sizes at the end of the first and second years of 3.3–3.7 and 5.9 for the British populations, and 7.9 and 10.9 for the Azores. There is an inverse correlation between growth rate and latitude, suggesting that temperature might be the

Table 3. Stomach contents (percentage occurrence) of different populations of *Gobius paganellus*. Items occurring in more than half the stomachs analysed are shown in bold

			Present study	Gibson (1972)	Dunne (1978)	Costa (1988)
Algae			3	16	37	
Foraminifera				1		
Cnidaria	Hydrozoa			1		
Annelida	Polychaeta		11	7	30	71
Mollusca	Bivalvia		8			
Mollusca	Gastropoda		28	15	31	
Arthropoda	Chelicerata	Pycnogonida		8		
		Acarina		9		
	Crustacea	Ostracoda		38	27	
		Copepoda	3	**82**	34	
		Cirripedia		1		
		Cumacea		5		
		Euphausiacea	3			
		Decapoda			15	
		Decapoda, Dendrobranchia	3			
		Decapoda Brachyura	14	32		19
		Mysidacea				14
		Isopoda	**69**	25		24
		Amphipoda	**100**	**69**	**67**	33
	Insecta	Coleoptera		1		
		Diptera		8	42	1
Echinodermata	Ophiuroidea		11	11		1
		N=	36	170	329	21

main factor involved. This was proposed originally by Dunne (1978) to explain differences between Carna and the Isle of Man populations. It must be pointed out, however, that other factors may be involved. In particular, the sublittoral habitat of the present population may be more benign in providing opportunities for feeding, unrestricted by tidal cycles, compared to the tide-pool habitat studied by the other authors.

Most individuals of the Azores population probably reproduce in their first year. This contrasts with the observations of Miller (1961) and Dunne (1978), who report a size at first maturation which corresponds to an age of 2–3 years. The eggs of *G. paganellus*, like those of most gobiids (Ruple, 1984), are demersal, with adhesive filaments. They are placed by the female under rocks or empty bivalve shells where they are guarded by the male until hatching (Miller, 1961). The larvae remain close to the bottom until metamorphoses (Russel, 1930), after which they assume the cryptobenthic behaviour of the adults. In the Plymouth area, the time span from spawning to absorption of the vitelline sac by the larvae is 30 days (Lebour, 1919). There is very little information regarding the following phases of larval development and none about the effect of temperature on development. In the Isle of Man, spawning takes place from mid-April to mid-June. Post-metamorphic juveniles (9.5–10.5 mm in length) are in rock pools by early August (Miller, 1961), almost 4 months after the beginning of the reproductive season. In the present study, the first recruits (TL between 2.8 and 4 cm) were captured in May, in both years. Even assuming that spawning starts in January, these data are consistent with a faster larval development in the Azores. This accelerated life history of Azorean individuals may mean a shorter life span. The oldest fishes observed by Dunne (1978) belonged to the 7+ age group, but Miller (1961) reported specimens older than 10 years. No direct age determination was attempted in the present study but, if the longevity estimation measure proposed by Taylor (1962, $A_{0.95}$, the time an individual takes to reach 95% of L_{∞}) is used, the Azores population scores only 3.9, considerably lower than that of the Isle of Man (9.3) or Carna (11.1).

In the Azores, the diet of *G. paganellus* is based on small benthic invertebrates, generally similar to that of other populations. Crustaceans were the main stomach items in Ireland (Dunne, 1978) and in northern France (Gibson, 1972). Costa (1988), on the other hand, stated that the most frequent item in the population from the Tagus Estuary was polychaetes. This may be due to the peculiarities of this estuarine habitat, since the remaining areas were fully marine. Algal fragments have a frequency in the stomach contents of between 16% (Costa, 1988) and 37% (Dunne, 1978). This latter author suggested that algal material is ingested accidentally, together with the animal constituents of the diet. This view is substantiated by the present study, since algal fragments were found only in one of the 36 stomach contents analysed.

References

Arruda, L., J. M. N. Azevedo & A. I. Neto, 1993. Abundance, age-structure and growth, and reproduction of gobies (Pisces; Gobiidae) in the Ria de Aveiro Lagoon (Portugal). Estuar. coast. Shelf Sci. 37: 509–523.

Azevedo, J. M. N., 2000. Now you see me, now you don't: a case study of the effect of the sampling method on the perceived structure ichthyological communities. Arquipélago, Life Mar. Sci. Supp. 2(a): 133–143.

Costa, M., 1988. Écologie alimentaire des poissons de l'estuarie du Tage. Cybium 12 (4): 301–320.

Dunne, J., 1978. The biology of the rock goby, *Gobius paganellus* L., at Carna. Proc. r. Irish Acad. 78B (12): 179–191.

Fabi, G. & G. Giannetti, 1985. Growth parameters of the black goby (*Gobius niger* L.) in the Adriatic Sea, based on otholiths reading. Rapp. Comm. int. Mer Médit. 29(8): 87–90.

Gayanillo, F. C. Jr., P. Sparre & D. Pauly, 1994. The FAO-ICLARM stock assessment tools (FiSAT) user's guide. FAO, Rome: 186 pp.

Gibson, R. N., 1972. The vertical distribution and feeding relationships of intertidal fish on the Atlantic coast of France. J. anim. Ecol. 41: 189–207.

Lebour, M. V., 1919. The young of the Gobiidae from the neighbourhood of Plymouth. J. mar. biol. Ass. U.K. 12: 48–80.

Miller, P., 1984. The gobiid fishes of temperate Macaronesia (Eastern Atlantic). J. Zool., Lond. 204 (3): 363–412.

Miller, P., 1986. Gobiidae. In Whitehead, P. J. P., M.-L. Beauchot, J.-C. Hureau, J. Nielsen & E. Tortonese (eds), Fishes of the Northeastern Atlantic and the Mediterranean. UNESCO, Paris: 1019–1085.

Miller, P. J., 1961. Age, growth, and reproduction of the rock goby, *Gobius paganellus* L., in the Isle of Man. J. mar. biol. Ass. U.K. 41: 737–769.

Patzner, R., R. Santos, P. Ré & R. Nash, 1992. Littoral fishes of the Azores: an annotated checklist of fishes observed during the 'Expedition Azores 1989'. Arquipélago, Life Earth Sci. 10: 101–111.

Ruple, D., 1984. Gobioidei: development. In Moser, H. G. (ed.), Ontogeny and Systematics of Fishes. American Society of Ichthyologists and Herpetologists, Special Volume nr. 1: 582–587.

Russel, F., 1930. The vertical distribution of marine macroplankton. IX. The distribution of the pelagic young of teleostean fishes in the Plymouth area. J. mar. biol. ass. U. K. 15: 829–850.

Sparre, P. & S. Venema, 1992. Introduction to tropical fish stock assessment. Part 1. Manual. FAO Fisheries Technical Paper 306/1. FAO, Rome: 376 pp.

Taylor, C., 1962. Growth equations with metabolic parameters. J. Con. perm. int. Explor. Mer. 27: 270–286.

Hydrobiologia **440:** 137–144, 2000.
M.B. Jones, J.M.N. Azevedo, A.I. Neto, A.C. Costa & A.M. Frias Martins (eds), Island, Ocean and Deep-Sea Biology.

Luminous capabilities and life-history traits of the polychromatic ophiuroid *Amphipholis squamata* in two isolated populations from Oliveri-Tindari lagoon system, Sicily

S. Dupont[1] & J. Mallefet
Laboratoire de Physiologie animale, Université catholique de Louvain, B-1348 Louvain-la-Neuve, Belgium [1]*Present address: 5 Place Croix du Sud, B-1348 Louvain-la-Neuve, Belgium*
Tel: +32-10473472. Fax: +32-10473476. E-Mail: dupont@bani.ucl.ac.be

Key words: bioluminescence, population dynamics, colour variety, microevolution, ophiuroid

Abstract

Amphipholis squamata is a small ophiuroid with a worldwide distribution except for Polar regions. Global populations of *A. squamata* are morphologically uniform but inter- and intrapopulational variability occurs at levels from the population to the genotype. The species is polychromatic and two colour varieties (dark-brown and spotted) from two adjacent populations (separated from each other for only 20 years) in a lagoonal system (Sicily) were compared using bioluminescence and life-history traits as indicators of variability. The dark-brown variety was dominant in Lago Nuovo and the spotted one was dominant in Porto Vecchio. Important variations were observed within populations, however, there were significant differences in life-history traits and luminous capabilities between the dark-brown and spotted varieties. The dark-brown variety invested more energy in somatic growth than reproduction (1 cohort each year and growth rate of 0.279 mm each month), whereas energy was devoted mainly to reproduction in the spotted variety (4 cohorts each year and growth rate of 0.171 mm each month). The chronology of the life cycles was completely different between the two varieties. Fertilization occurred mainly in spring for the spotted but only in autumn for the dark-brown. Recruitment occurred mainly in autumn for spotted and in spring for dark-brown individuals. Finally, significant differences were observed for maximal intensity of emitted light and kinetic parameters between the two varieties.

Introduction

Amphipholis squamata (Delle Chiaje, 1828) has a worldwide distribution being absent only in polar regions. It occupies a wide range of habitats from the intertidal zone down to 1330 m depth, often occurring in dense local populations (Gage et al., 1983; Jones & Smaldon, 1989). Global populations of *A. squamata* are morphologically uniform (Hyman, 1955) but, surprisingly, several works report loose inter- and intrapopulational variability at all levels studied from the population to genotype (DeBremaeker et al., 1994; Mallefet et al., 1994; Deheyn et al., 1997; Dupont & Mallefet, 1999; Poulin et al., 1999; Sponer et al., 1999). The species displays various colours of the disc and arms (Tortonese, 1932; Binaux & Bocquet, 1971), and all levels of intraspecific variation are linked to this polychromatism. Preliminary results on genetic variation, as revealed by RAPDs (Random Amplified Polymorphic DNA), indicate that genetic structure is homogenous for each colour variety within a population while it shows interpopulational variations for the same variety, suggesting that polychromatism is a good indicator of variability in genotypes only at intrapopulationnal level (Dupont et al., 1999). Within populations of *A. squamata* at Langrune-sur-mer (Deheyn et al., 1997) and Porto Vecchio (Dupont, 1998), each colour variety exhibits its own capacity to produce light. Individuals of the beige and black variety, two of the six colour varieties described in Langrune-sur-mer, differ in the nature of their cholinergic receptor and some of their neuromodulators (DeBremaeker et al., 1994; Mallefet et al., 1994). Assuming that polychromatism is an indicator of genetic speciation (Tuttle & Lindhal, 1980), these observations suggest that rapid speciation could explain the differences between the various *A. squamata* colour varieties. The species might be a complex of sibling species, each morph (genetically differentiated colour variety) becoming a new species.

The present work aims to test this hypothesis by comparing two colour varieties from two neighbouring populations (separated from each other 20 years ago) using bioluminescence and, for the first time, life-history traits as indicators of variability.

Description of sites studied

Two populations of *A. squamata* from the lagoon system of Oliveri-Tindari (Sicily) were studied (Fig. 1). The lagoons were created in 1967 by a sediment deposit. Lago Nuovo has been isolated from Porto Vecchio and the open sea since 1980, and 10 years later Porto Vecchio was isolated from the sea (Crisafi et al., 1981; Costa et al., 1996). Both lagoons have similar abiotic (temperature, salinity, oxygen concentration, pH, etc.) and biotic conditions, although the Lago Nuovo ecosystem has more faunistic and floristic diversity (Crisafi et al., 1987). Ophiuroids were collected from the green alga *Chaetomorpha mediterranea* (Kützing) Kützing from a shallow *Cymodocea nodosa* (L.) Del meadow; high density aggregations were observed, with up to 2000 individual brittle stars in 1 l of seaweed (Dupont & Mallefet, 1999).

Materials and methods

Sampling

Specimens of *A. squamata* were collected by hand in the phanerogam meadows. Sediment, algae and plants containing ophiuroids were sampled to isolate all ophiuroid size classes. Animals were transferred to the laboratory (University of Messina) where they were kept in aerated lagoonal seawater. Between 250 and 500 specimens were sampled randomly every 3 months between June 1997 and September 1999 for Porto Vecchio, and between June 1998 and September 1999 for Lago Nuovo.

Population dynamics

After anaesthesia by immersion in 3.5% w/w $MgCl_2$ in artificial seawater (ASW) for 3 min, individuals were examined under a binocular microscope and the following were noted: (i) the colour pattern of disc and arms; and (ii) the disc diameter (the distance between the distal part of radial shields and the aboral limit of the opposite interradius) and arm length. As adult individuals (disc diameter > 1.45 mm; Dupont & Mallefet, 1999) are able to breed both larvae and juveniles in 10 genital bursae located in the disc (Fell, 1946), the brooding condition was determined by dissecting the bursae. The number of larvae and/or juveniles was counted and the disc diameter of juveniles was measured.

The population dynamics of the spotted colour variety has been described for Porto Vecchio (Dupont & Mallefet, 1999). In this study, a similar analysis was performed for the dark-brown variety of the Lago Nuovo lagoon. Since the different cohorts present a Gaussian distribution, analysis of size–frequency distributions was performed for the different colouration patterns using the method described by Bhattacharya (1967) to estimate number, proportion, mean and standard error of Gaussian components (Alvà, 1996). Chi-square statistic was used to test the stability of frequency of each variety through time (Fienberg, 1977).

Luminous capabilities

The arms are the only luminescence body parts of *A. squamata* (Brehm & Morin, 1977) and each of the five arms from a single individual produces light of the same intensity (Mallefet et al., 1992). Measurement of light capabilities was carried out in a dark room. Arms of anaesthetised animals were removed from the disc using a fine scalpel and placed in a small chamber containing 200 μl of ASW with the following composition (in mm): 400.4 NaCl; 9.6 KCl; 52.3 $MgCl_2$; 9.9 $CaCl_2$; 27.7 Na_2SO_4; 20 Tris HCl buffer; pH 8.3. KCl is often used to measure bioluminescent capabilities of marine organisms (Herring, 1974) and KCl 200 mM triggers maximal light emission in *A. squamata* (Mallefet et al., 1992). Each isolated arm was stimulated by injection in the chamber of 200 μl of 400 mM KCl solution in ASW where NaCl had been replaced by KCl to obtain a final KCl concentration of 200 mM. Light responses were measured using a photomultiplier phototube (PM 270 D, EMI), the electric signal was amplified by a Radiometer (type IL 7000, International Light) before being recorded (Minigor 510). Each week, the system was calibrated using a tritium-phosphore light source (Beta Light Source, Saunder & Roe, Nuclear enterprise). Four parameters were used to character-

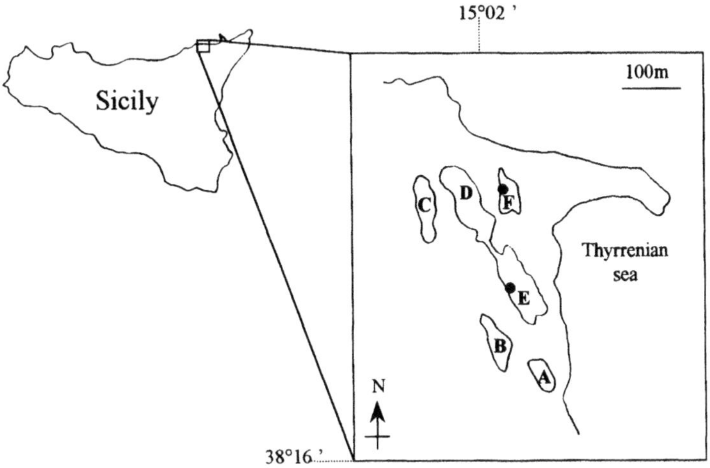

Figure 1. Oliveri-Tindari lagoons: Lago Marinello (A), Lago Megolo (B), Lago Verde (C), Fondo Porto (D), Porto Vecchio (E), Lago Nuovo (F); points of sampling (●).

ize the monophasic light response of *A. squamata* to KCl stimulation (Mallefet et al., 1992), the maximal intensity of light expressed in megaquanta per seconds (Lmax in Mq.s^{-1}) and three kinetic parameters expressed in seconds: (1) the time elapsing between the application of the KCl stimulus and the beginning of the light production (Latency time, LT); (2) the time between the beginning of the light production and the light maximum production (Tlmax), and (3) the time to reach half extinction (T1/2ext). For both lagoons, data from all the sampling periods were pooled. A canonical discriminant analysis was used to separate colour varieties, a classification variable, using luminous parameters as quantitative variables (Kshirsagar, 1972). Each mean value is expressed with its standard error of mean (mean ± SEM); analysis of variance (ANOVA) and *t*-tests were used to determine the significance of the observed differences between the groups. Simple linear regression models were used to test the relation type between variables. All statistical methods used (ANOVA, *t*-test, linear regression and canonical discriminant analysis) are designed under the assumption that the data are normally distributed. Tests were used to check that the data were a random sample from a normal distribution. If the sample size was less than or equal to 2000, the Shapiro–Wilk (1965) statistic (*W*) was used; if the sample size was greater than 2000, the Kolmogorov statistic (*D*) was used (Stephens, 1974). When data were not normally distributed or when heteroscedasticity occurred, a logarithmic transformation of data was performed as indicated by Sokal & Rohlf (1995). Analyses were performed using Statistical Analysis System (SAS institute).

Results

Colouration patterns

Three colour patterns (varieties) based on disc and arm colouration were observed: (1) the beige variety exhibited homogenous beige to yellowish arms and disc; (2) the dark-brown variety had a brown disc with few black spots and grey homogenous arms; (3) the spotted variety had a similar disc but with greenish arms with some white segments.

Although the density values of *Amphipholis squamata* in the two lagoons were similar (mean of 1230 and 1296 individuals in a 1 l of green seaweed in Lago Nuovo and Porto Vecchio respectively), the colour varieties were present in different proportions in the two populations (Fig. 2). In Porto Vecchio, two varieties were present and spotted individuals were more abundant (94% of the population) than the dark-brown ones (6%). In Lago Nuovo, the dark-brown variety was dominant (82%), followed by the spotted (10%) and the beige (8%) variety. Chi-square statistics indicate that these proportions of each variety were constant from one season to another (p=0.416 for Lago Nuovo; p=0.125 for Porto Vecchio). As a consequence of its rarity, the beige variety was not used for analysis of the population dynamics.

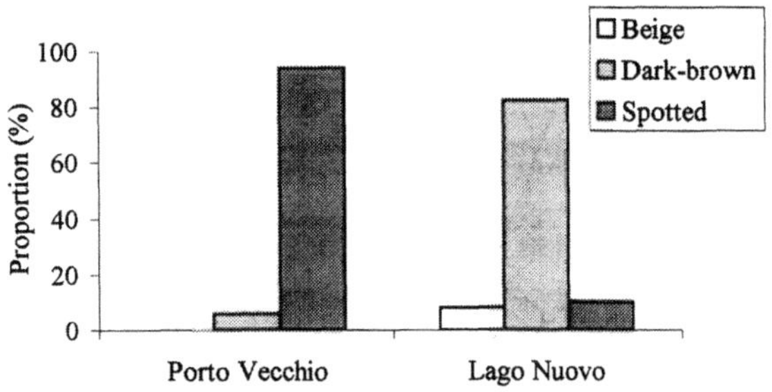

Figure 2. Proportion of the three colour varieties (beige, dark-brown and spotted) in the lagoons Porto Vecchio and Lago Nuovo from Oliveri-Tindari.

Population dynamics of the dark-brown variety

Population dynamics of the dark-brown variety was studied in the Lago Nuovo lagoon (Fig. 3). The Bhattacharya analysis indicated that, in September and December, the population comprised one cohort (mean size of 1.67±0.07 mm and 2.44±0.19 mm, respectively). A new cohort appeared in March (mean size 0.82±0.11 mm) and dominated (70.8%) the population, with the older cohort forming only 29.2% of the population (mean size 3.34±0.11 mm). In June, the older cohort disappeared and the population was unimodal (mean size 1.6±0.09 mm). These results suggest that the longevity of each cohort is 1 year with recruitment occurring in autumn (in March in 1999 and probably in early June in 1998) and mortality of the oldest individuals occurring mainly in spring.

Individuals ranged in size from 0.75 to 4.75 mm with an average >1.7 mm (Table 1). Simple linear regression between time and disc diameter showed a constant growth rate of the disc of 0.279 mm per month for both studied cohorts (Fig. 4). Brooding duration and individual longevity were extrapolated from this rate as follows: (a) the smallest releasable individual had a disc diameter of 0.75 mm corresponding to 2.68 months; (b) the largest brooded juvenile had a disc diameter of 1.05 mm corresponding to 3.76 months; (c) the oldest adult had a discal diameter of 4.75 mm corresponding to 17 months. These data indicate that *A. squamata* has a brooding duration of 2–4 months and a maximal longevity of 17 months.

The number of brooded juveniles increased with disc diameter (Fig. 5). Other observations confirmed that the recruitment occurred in spring. For example, the proportion of brooding individuals decreased from more than 50% to less than 1% between March and June (Fig. 6) and the quantity of brooded juveniles was lower, falling to 1 in June (Table 2). Fertilization, and the appearance of brooded eggs and/or larvae, occurred in autumn when the proportion of brooding individuals increased from zero to more than 50% (Fig. 6). The growth from eggs to larvae, and from larvae to juvenile, occurred in winter since the proportion of brooding individuals was constant (Fig. 6), the number of brooded larvae and/or juveniles increased from 3.81 to 8.61 (Table 2), and the disc diameter of brooded juveniles increased from 0.44 (±0.04) to 0.54 (±0.02) mm. These results suggest that fertilization occurs in autumn and recruitment in spring.

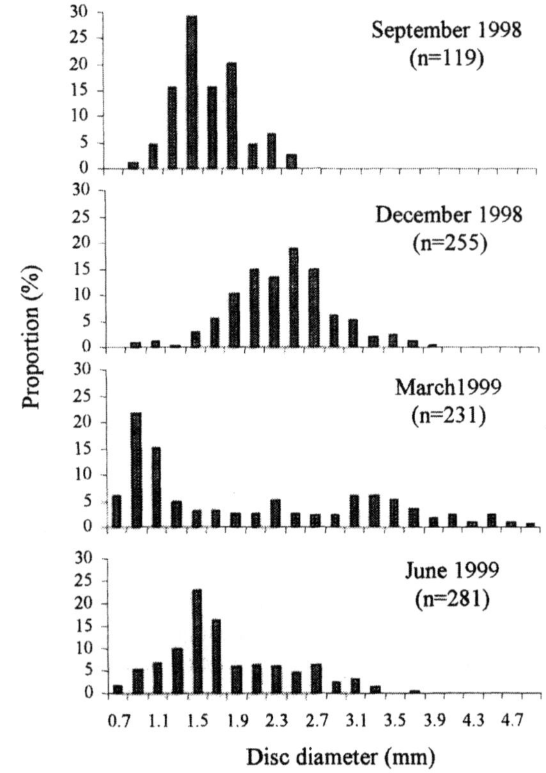

Figure 3. Seasonal size-frequency distribution of disc diameter of the dark-brown variety in Lago Nuovo.

Table 1. Disc diameter ranges (mean size ± SEM), maximum, minimum and sample size of dark-brown variety in Lago Nuovo at each collecting period

	Mean	Minimum	Maximum	Sample size
September 1998	1.72±0.03	1	2.6	119
December 1998	2.42±0.03	1	4	255
March 1999	2.02±0.11	0.75	4.75	231
June 1999	1.72±0.04	0.75	3.7	281

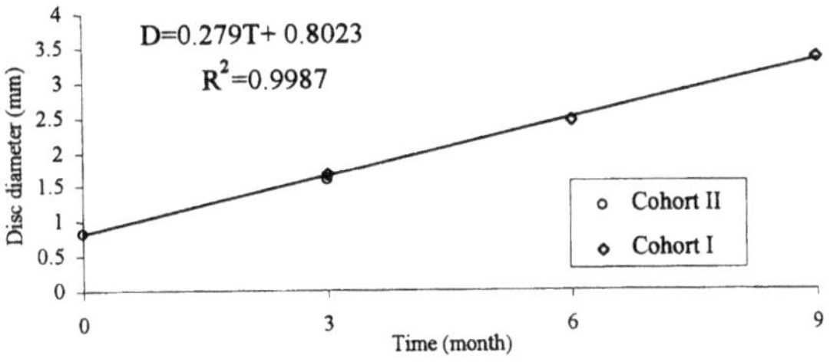

Figure 4. Linear regression between time (T in month) and disc diameter (D in mm, 0 corresponding to the time when juveniles were released) of two cohorts of the dark-brown variety from Lago Nuovo.

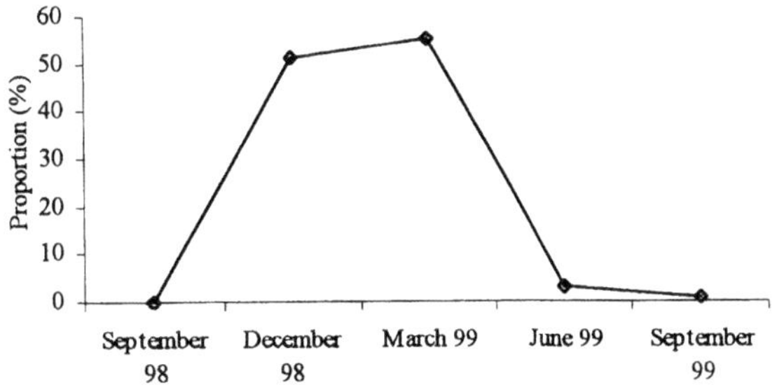

Figure 6. Proportion (%) of brooding individuals of dark-brown variety from Lago Nuovo at each season.

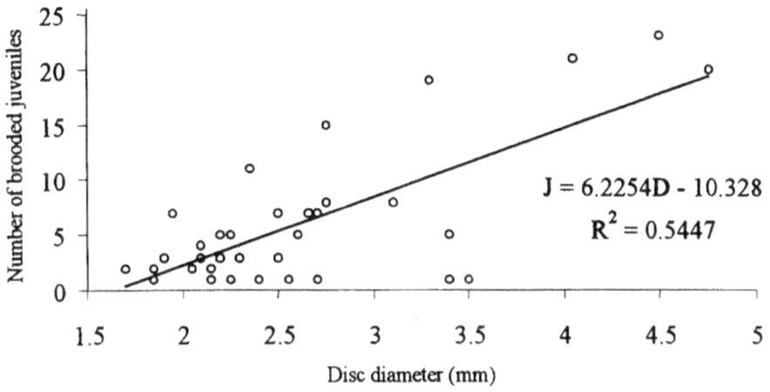

Figure 5. Linear regression between disc diameter (D in mm) and number of brooded juveniles (J) of the dark-brown variety in Lago Nuovo.

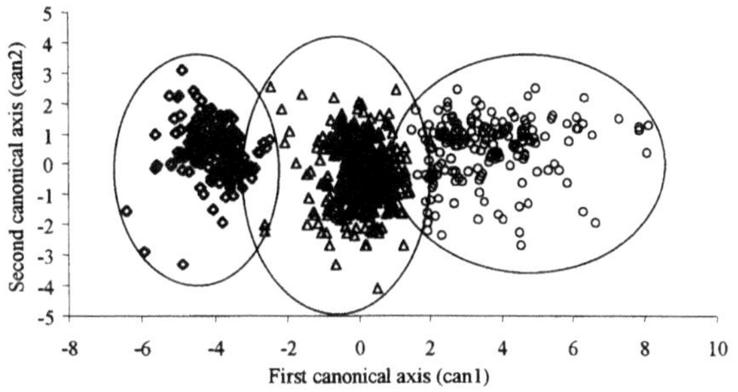

Figure 7. Canonical discriminant analysis used on colour variety of Lago Nuovo: ◇, beige variety; △, Dark-brown variety and °, spotted variety. ($can1=0.99*Lmax-0.57*LT+0.09*TLmax-0.80*T1/2$ and $can2=0.05*Lmax+0.23*LT+0.99*TLmax+0.27*T1/2$).

Luminous capabilities

A canonical discriminant analysis was used to summarize between-class variations from colour variety (a classification variable) and luminous parameters as quantitative variables. In the population of Lago Nuovo, the results indicated that variations of luminous capabilities could be related to polychromatism since each colour variety was well separated (significant distances between varieties, $p>0.0001$) from the others and 96.6% of the variations could be explained by the first canonical axis (Fig. 7). In Porto Vecchio, the two varieties were well separated by a canonical discriminant analysis since only one canonical axis could explain all the variability of luminous parameters between the colour varieties. Four parameters were estimated and compared for the three colour patterns, and each variety produced light with intensity and kinetic parameters (excepted Tlmax) significantly different from other varieties (ANOVA, $p<0.01$; Fig. 8). For example, spotted individuals were able to produce more intense light than the dark-brown ones and three times more intense light than the beige ones.

Table 2. Number of brooded larvae and/or juveniles per brooding ophiuroid of dark-brown variety (±standart error of mean, minimum, maximum and sample size) at each collecting period (in situ) in Lago Nuovo

	Mean	Minimum	Maximum	Sample size
December 1998	3.81±0.52	1	7	132
March 1999	8.61±1.8	1	23	125
June 1999	1±0	1	1	1

Discussion

For the first time, two geographically close, but isolated, populations of *Amphipholis squamata* have been compared. Although the populations had similar densities, they showed significant differences in the frequency of colour varieties. The dark-brown variety was dominant in Lago Nuovo while the spotted one was dominant in Porto Vecchio. These differences cannot be a consequence of differences in the biotope, since the ophiuroids were located in the same habitat (green seaweeds), and both lagoons have similar biotic and abiotic conditions. As the two populations have been completely isolated from each other for 20

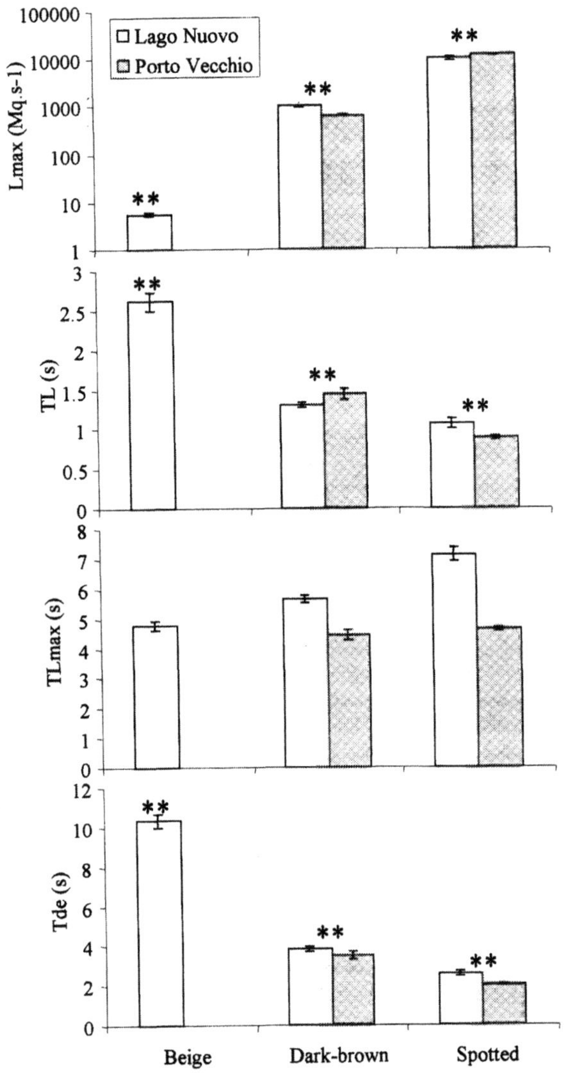

Figure 8. Luminous capabilities of the three varieties (Beige, dark-brown and spotted) from Lago Nuovo and Porto Vecchio: maximal intensity of light (Lmax in Mq.s^{-1}); Latency time (LT in s); Time to reach maximal intensity (Tlmax in s); Time to reach half-intensity (T1/2 in s). (mean±SEM; ** = $p<0.01$, statistically different from each of the three other varieties after logarithmic transformation).

years, gene flow has ceased and, therefore, the random foundation effect and/or natural selection might be responsible for the observed variations in variety frequencies. Furthermore, observations on *A. squamata* from these lagoons suggest that differential parasitism (e.g. the orthonectidae *Rhopalura ophiocomae* specifically castrates the spotted variety) and differential predation by crustaceans (e.g. the shrimp *Palaemon elegans* eats the spotted variety more rapidly) occurs between these two varieties. These parameters could induce rapid changes in the proportions of colour varieties as suggest by our unpublished observation in Porto Vecchio where an infestation by *R. ophiocomae* induced an important modification in variety frequencies in less than 6 months (September 1999).

Polychromatism, a common phenomenon amongst echinoderms, is a sign of speciation (Tuttle & Lindahl, 1980). In *A. squamata*, intraspecific variations are linked to the colour of disc and arms since the genetic structure, polychromatism, luminescence capabilities and control mechanisms are linked (DeBremaeker et al., 1994; Mallefet et al., 1994; Deheyn et al., 1997; Dupont & Mallefet, 1999; Poulin et al., 1999; Sponer et al., 1999). Moreover, bioluminescence is a good indicator of genetic variability (Dupont et al., 1999). Since large differences were observed in the luminous capabilities between colour varieties, each studied variety producing light with its own capabilities, we postulate that these three colour varieties (beige, dark-brown and spotted) correspond to three genetically differentiated morphs.

In this work, an original approach was used to study the differences between these colour morphs. We compared the population dynamics of the colour varieties co-existing in the same biotope. For practical reasons (abundance), the beige variety was not used in this part of the analysis. The life-history traits of the dark-brown variety are summarized in Table 3. The population comprised 1 or 2 cohorts, reproduction occurred in autumn and larvae grew in the genital bursae for 2–4 months. Juveniles (disc diameter between 0.75 and 1.05 mm) were released in spring and became sexually mature within 2 months. The life span of a sexually mature adult is up to 11 months since mortality of the oldest individuals occurred mainly in spring. Therefore, longivity of this colour variety is about 15–17 months. Dark-brown *A. squamata* reach a relatively large size (maximal disc diameter of 4.75 mm) and disc growth is constant within a cohort (0.28 mm per month).

Comparison of these characteristics with those of the spotted one identified differences (Table 3). The number of brooded juveniles per adult of the spotted variety is smaller than in the dark-brown, where up to 23 juveniles were recorded. These results suggest that the fecundity is lower for spotted than dark-brown ophiuroids. This lower fecundity is compensated by several periods of reproduction for the spotted variety since a new cohort appears at each season while only one new cohort appears each year for the dark-

Table 3. Comparison of life-history characteristics of two colour varieties from Oliveri-Tindari lagoon system

	Dark-brown (This work)	Spotted (Dupont & Mallefet, 1999)
Free animal disc diameter (mm)	0.75–4.75	0.45–3.45
Reproduction	Autumn	All year
Number of brooded juveniles per adult		
-maximal	23	10
-monthly mean (max)	8.61	3.82
Brooding duration (month)	2–4	2–6
Recruitment	Spring	Each season (mainly in Autumn)
Juvenile size at recruitment (mm)	0.75–1.05	0.75–0.99
Growth rate (mm.month^{-1})	0.279	0.171
Longevity (month)	15–17	15–20
Mortality	Spring	Winter and Summer

brown variety. The chronology of the life cycle is completely different for the two varieties. Fertilization occurs mainly in spring for the spotted and only in autumn for the dark-brown form. Recruitment occurs mainly in autumn for spotted individuals but only in spring for the dark-brown. Mortality is more important in winter and summer for the spotted variety, and in spring for the dark-brown. The growth rate is 1.6 times greater for the dark-brown variety than for the spotted one. Dark-brown individuals can reach a larger size even if longevity, juvenile size recruitment and brooding duration are similar to the spotted type. These results demonstrate that the dark-brown variety invests more energy in somatic growth rather than in reproduction compared to the spotted one (that invests more energy in reproduction).

The life history of *A. squamata* has been studied in several places throughout the world (France, Scotland, England, United States and Italy) and significant differences have been observed between distant populations. Authors suggest that variations in environmental conditions, such as temperature, might be an important factor influencing population dynamics (see Dupont & Mallefet, 1999 for review). For the first time, significant differences in population dynamics have been observed between two varieties of *A. squamata* co-existing in the same biotope. These differences, including those observed for luminous capabilities, cannot be explained by differences in biotic and/or abiotic environment and are probably of genetic origin.

This work supports the concept that *A. squamata* is a complex of sibling species, each colour morph becoming a new species since colour varieties of Oliveri-Tindari present different life-history traits and extremely different luminous capabilities (maximal intensity and kinetic parameters) which cannot be explained as an effect of biotic or abiotic environment. Moreover, our results provide clues on the mechanisms inducing speciation in this species. Hermaphroditism, selfing, aggregation behaviour and low genetic flow might favour the isolation of individuals (Poulin et al., 1999). Since isolation promotes speciation (Baguette & Lebrun, 1990; Blondel, 1995), allopatric speciation (the evolution of new species when geographically isolated from the parent species) could partially explain intraspecific variations linked to polychromatism. Nevertheless, we observed two varieties, co-existing in the same population, with different reproductive periods (spotted mainly in spring and dark-brown only in autumn). Therefore, individuals from one colour variety are reproductively isolated from individuals of other varieties. We propose that sympatric speciation (the evolution of a new species within the same geographical region as the parent species) and/or allopatric speciation followed by dispersion could explain these intrapopulational differences.

Acknowledgements

We thank F. Bieswal, N. Donato, M-T. Costanzo and Prof. U. Brancato for field access and assistance; Prof. A. Donato and Prof. L. Salpietro for providing facilities at the Dipartimento di Biologia Animale ed

Ecologia Marina of the University of Messina (Sicily, Italy); Prof. M.C. Thorndyke, G. Delva and anonymous referees for criticism of the manuscript; I. Cogneau for technical assistance. This work was supported by research exchange program between FNRS (Belgium) and CNR (Italy) and CGRI (Belgium) and ministery of foreign affair (Italy). Research supported by FRIA grant to S. Dupont. J. Mallefet is Research Associate of the National Fund for Scientific Research (FNRS, Belgium).

References

Alvà, V., 1996. Reproduction, développement, incubation et dynamique de population de l'ophiure *Amphipholis squamata* (Echinodermata) en baie de Seine. PhD thesis, Université Libre de Bruxelles, Belgium.

Baguette, M. & P. Lebrun, 1990. Isolement des populations et conservation des espèces. Rev. Quest. Sci. 161(3): 319–332.

Bhattacharya, C. G., 1967. A simple method of resolution of a distribution into Gaussian components. Biometrics 23: 116–135.

Binaux, R. & C. Bocquet, 1971. Sur le polychromatisme de l'ophiure *Amphipholis squamata* (Delle Chiaje). C.R. Acad. Sci. Paris 273: 1618–1619.

Blondel, J., 1995. Biogéographie. Approche écologique et évolutive. Masson, Paris: 297 pp.

Brehm, P. & J. G. Morin, 1977. Localization and characterization of luminescent cells in *Ophiopsila californica* and *Amphipholis squamata* (Echinodermata: Ophiuroidea). Biol. Bull. 152: 12–25.

Costa, F., S. Giacobbe & L. Salpietro, 1996. Taxonomic and ecological notes on *Buenia affinis* Iljin 1930 (Pisces: Gobiidae). Oebalia 22: 15–23.

Crisafi, E., F. Azzaro, D. Caridi, P. De Francesco, L. Genovese, M. Giacobbe, S. Giacobbe, S. Greco, R. LaFerla, M. Leonardi, G. Maimone, G. Pandolfino, F. Perdichizzi, P. Rinelli & R. Zaccone, 1987. Indagine interdisciplinare sul sistema degli stagni salmastri costieri di Oliveri-Tindari (Messina). I. Risultati delli campagne stagionali di 1987. Rapporti di Istituto Sperimentale Tallassographico, Messina: 142 pp.

Crisafi, E., S. Giacobbe & M. Leonardi, 1981. Nuovo ricerche idrobiologiche nell'area lagunare di Oliveri-Tindari (Messina). I. Morfologia dei bacini e caratteristiche fisico-chimiche delle acque e du sedimenti. Mem. Biol. Mar. Oceanol. 11(4): 139–186.

De Bremaeker, N., J. Mallefet & F. Baguet, 1994. Luminescence control of *Amphipholis squamata* (Ophiuroidea): nature of cholinergic receptor. In David, B., A. Guille, J. P. Féral & M. Roux (eds), Echinoderms Through Time. A.A. Balkema, Rotterdam: 405–407.

Deheyn, D., J. Mallefet & M. Jangoux, 1997. Intraspecific variations of bioluminescence in a polychromatic population of *Amphipholis squamata* (Echinodermata: Ophiuroidea). J. mar. biol. Ass. U.K. 77: 1213–1222.

Delle Chiaje, S., 1828. Memorie sulla storia e anatomia degli animali senza vertebre del Regno di Napoli. 3: 231 pp.

Dupont, S., 1998. Characterization of bioluminescence variations in an isolated population of *Amphipholis squamata* (Ophiuroidea, Echinodermata) from Oliveri-Tindari's lagoon (Sicily). Master thesis, Université catholique de Louvain, Belgium.

Dupont, S., S. Chaufer, E. Poulin, J.-P. Féral & J. Mallefet, 1999. Is there a link between morphological, physiological and genetic variability of the ophiuroid *Amphipholis squamata*. In Ter Maat A. & R. C. Peters (eds), Proceeding of the 6th Benelux Congress of Zoology. Utrecht University, Utrecht: 47 pp.

Dupont, S. & J. Mallefet, 1999. Population dynamics of *Amphipholis squamata* (Echinodermata: Ophiuroidea) in the Oliveri-Tindari's lagoon system, Sicily. In Candia Carnevali M. D. & F. Bonasoro (eds), Echinoderm Research 1998. A.A. Balkema, Rotterdam: 65–70.

Dupont, S., G. Vandemeulebroecke & J. Mallefet, 1999. Effect of bioluminescence on predation in the polychromatic ophiuroid *Amphipholis squamata*. In Ter Maat & R. C. Peters (eds), Proceeding of the 6th Benelux Congress of Zoology. Utrecht University, Utrecht: 46 pp.

Fell, H. B., 1946. The embryology of the viviparous ophiuroid *Amphipholis squamata* Delle Chiaje. Trans r. Soc. N.Z. 75: 419–464.

Fienberg, S. E., 1977. The Analysis of Cross-Classified Data. MIT Press, Cambridge.

Gage, J. D., M. Pearson, A. M. Clarck, G. L. J. Paterson & P. A. Tyler, 1983. Echinoderms of Rockall Through and Adjacent Area. 1. Crinoidea, Asteroidea and Ophiuroidea. Bull. British Mus. 45: 263–308.

Herring, P. J., 1974. New observations on the bioluminescence of echinoderms. J. Zool., Lond. 38: 127–159.

Hyman, L. H., 1955. The Invertebrates, 4. Echinodermata. McGraw-Hill Book Co. Inc., New York: 631 pp.

Jones, M. B. & G. Smaldon, 1989. Aspects of biology of a population of the cosmopolitan brittle star *Amphipholis squamata* (Echinodermata) from the Firth of Forth, Scotland. J. Nat. Hist, Lond. 23: 613–625.

Kshirsagar, A. M., 1972. Multivariate Analysis. Marcel Dekker, Inc, New York.

Mallefet, J., N. De Bremaeker & F. Baguet, 1994. Neuromodulatory effects of GABA, S1 and S2 FMRF-amide and antho-Rfamide on *Amphipholis squamata* (Echinodermata) luminescence. In Campbell, A., L. Kricka & P. Stanley (eds), Bioluminescence and Chemiluminescence, Fundamentals and Applied Aspects. John Wiley & Sons, Chichester: 109–112.

Mallefet, J., P. Vanhoutte & F. Baguet, 1992. Study of *Amphipholis squamata* luminescence. In Scalera-Liaci L. & C. Canicatti (eds), Echinoderm Research 1991. A.A. Balkema, Rotterdam: 125–130.

Poulin, E., J. P. Féral, F. Mireia, L. Cornudella & V. Alvà, 1999. Selfing and outcrossing in the brood protecting ophiuroid *Amphipholis squamata*. In Candia Carnevali M. D. & F. Bonasoro (eds), Echinoderm Research 1998. A.A. Balkema, Rotterdam: 147–150.

Shapiro, S. S. & M. B. Wilk, 1965. An analysis of variance test for normality (complete samples). Biometrika 52: 591–611.

Sokal, R. R. & F. J. Rohlf, 1995. Biometry. Freeman & Co, New York: 887 pp.

Sponer, R., M. S. Roy & P. V. Mladenov, 1999. Molecular evolution of *Amphipholis squamata* (Echinodermata: Ophiuroidea): hign genetic differences among populations. In Candia Carnevali M. D. & F. Bonasoro (eds), Echinoderm Research 1998. A.A. Balkema, Rotterdam: 405–408.

Stephen, M. A., 1974. EDF statistics for googness of fit and some comparisons. J. am. Stat. Ass. 69: 730–737.

Tortonese, E., 1932. Osservazioni sul colore di alcuni Echinodermi. Natura (Milano) 23: 160–164.

Tuttle R. D. & R. Lindahl, 1980. Genetic variability in three co-occuring forms of the starfish genus *Othilia* (=*Echinaster*). Experientia 36: 923–925.

Hydrobiologia **440:** 145–152, 2000.
M.B. Jones, J.M.N. Azevedo, A.I. Neto, A.C. Costa & A.M. Frias Martins (eds), Island, Ocean and Deep-Sea Biology.

Physical control of plankton population abundance and dynamics in intertidal rock pools

M. P. Johnson
Division of Biodiversity and Ecology, School of Biological Sciences, University of Southampton, Biomedical Sciences Building, Bassett Crescent East, Southampton SO16 7PX, U.K.
Present address: School of Biology and Biochemistry, The Queen's University of Belfast, 97 Lisburn Road, Belfast BT9 7BL, Northern Ireland, U.K.
E-mail: m.johnson@qub.ac.uk

Key words: intertidal, rock pool plankton, model, plankton, spring-neap, tidal cycle

Abstract

Little is known about the population structure and dynamics of plankton of intertidal rock pools. A numerical model was developed for rock pool plankton with growth limited by both tidal washout and the stress associated with adverse conditions in high-shore pools. This model predicts that a stress tolerant species will tend to have maximum population densities in high-shore pools and that populations will fluctuate in opposite phase to the spring-neap tidal cycle. Conversely, where a species is susceptible to stress in high-shore pools, the maximum population density is likely to occur lower on the shore, and numbers in upper shore pools will cycle in phase with the spring-neap cycle. These two alternative predictions were sufficient to classify the dynamics of the most abundant species in time series taken from rock pools in the Isle of Man. The dinoflagellate *Oxyrrhis marina* followed the predictions of the stress tolerant model. In comparison, the spatiotemporal patterns of other taxa, including a ciliate, a dinoflagellate and cryptophytes, suggested stress-susceptible life histories.

Introduction

Rock pools are ubiquitous yet relatively understudied components of rocky shores (Metaxas & Scheibling, 1993; Underwood & Skilleter, 1996). There is considerably more information in the literature on the communities of the emergent substrata than there is on the inhabitants of pools. This observation is particularly true for the populations of planktonic organisms found in rock pools (but see Faure-Fremiet, 1948; Dethier, 1980; Metaxas & Scheibling, 1996; Powlik, 1998). A central question about rock pool plankton is whether their presence represents a resident community or merely reflects the species found in adjacent coastal waters. Studies of rock pool communities, in general, have found species which are rarely or never recorded from open waters (Dethier, 1980; Jonsson, 1994). Often, such rock pool species have physiological or behavioural adaptations which can aid population persistence in individual pools (Blackwell & Gilmour, 1991; Jonsson, 1994). Rock pool plankton may, therefore, be considered to live in 'islands' of favourable habitat. As pools are easy to sample, rock pool plankton may represent a convenient model system to study wider ecological concepts such as community assembly, spatial population dynamics and local extinction (Pfister, 1998).

Only a few studies have described spatial and temporal patterns in the population density of rock pool plankton. Communities in rock pools have been classified partly on the basis of height on the shore (Droop, 1953; Metaxas & Lewis, 1992), however, a more extensive study by Metaxas & Scheibling (1996) suggested that pool height had no significant effect on the plankton abundance in pools. Temporal patterns have been compared between separate dates, but no studies have sampled at timescales equivalent to the growth responses of plankton. This is an important omission as it is difficult to interpret the effects of variable local conditions on plankton populations without taking regular samples at close to daily intervals (Sephton & Harris, 1984; Harris & Trimbee, 1986).

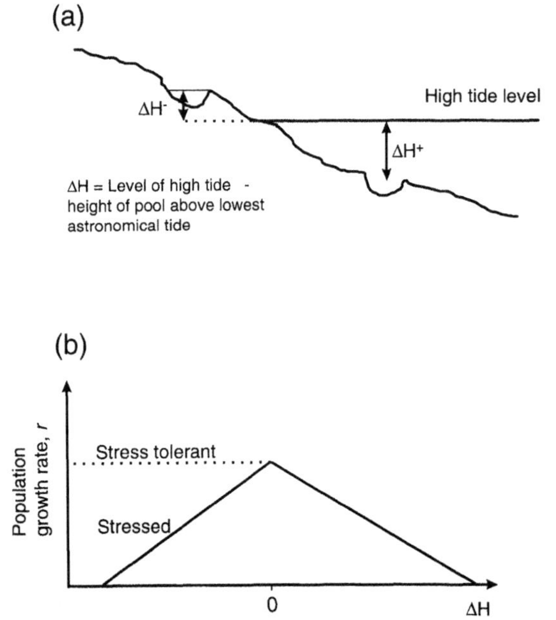

Figure 1. (a) The effects of flushing and isolation on population growth rates are defined using the difference between the height of high tide and the height of each pool (ΔH). (b) Negative values of ΔH have either no effect on population growth rate if the species is stress tolerant or reflect a reduction in the growth rate of stress susceptible species.

The study presented in this paper addresses some of the issues overlooked in earlier descriptions of rock pool plankton communities. Samples were taken every other day throughout two spring-neap tidal cycles to track population responses to changes in tidal height. A numerical model was developed to provide a framework for interpreting the data collected. The basis for this model was that plankton population growth rates could be reduced by both the stress of extreme conditions in upper shore pools and the loss of cells due to washout during high tides. Different predictions about planktonic abundances are made depending on whether a species is considered susceptible to stress or stress tolerant.

Materials and methods

A model of plankton population growth in rock pools

A numerical model of plankton populations suitable for use in spreadsheets can be derived from a well-known (Begon et al., 1996) difference equation for a single population:

$$N_{t+1} = \frac{rN_t}{(1 + aN_t)}, \tag{1}$$

where N_t and N_{t+1} are the population sizes at time t and $t+1$, respectively, r is the intrinsic population growth rate while a represents the population response to crowding and, therefore, sets the local carrying capacity. The changes in loss rate due to flushing and any possible effects of extreme conditions in upper shore pools can be included by making r a function of the difference between the level of high tide and height of individual pools (ΔH, Fig. 1):

$$r = r - d\Delta H^{+} \text{ or } r = r + s\Delta H^{-}, \tag{2}$$

where d relates to loss rates due to flushing when ΔH is positive and s describes the slow down of growth under environmental stress when ΔH is negative. The anomaly of negative growth rates is avoided by defining the minimum value of r to be zero. The reasoning behind this dependence of population growth rate on the relative heights of high-tide and individual pools is straightforward. Increasingly positive values of ΔH imply that a pool spends more of each tidal cycle submerged. This will reduce the net population growth rate as cells are flushed from pools during submergence and net growth can only occur during the period that pools are above sea level. Pools isolated above the high-tide mark have dramatic variations in temperature and salinity. A pool at the level of high-water springs will be isolated for 14 days at a time, whereas a pool at the level of high-water neaps will only be isolated for a few hours each day. These differences in isolation period are assumed to relate to the probability of extreme conditions occurring in any individual pool. Two alternatives are proposed for plankton in pools isolated above the level of any particular high tide. A species can be unaffected by the extreme conditions (stress tolerant) and hence the growth rate will not change with the period of isolation. In comparison, the growth rates of stress susceptible species are assumed to be related to the period of isolation of individual pools.

Model predictions can be derived for individual pools at different heights on the shore by defining a series of daily values of ΔH for each pool and substituting the appropriate values into Equations 1 and 2. Stress tolerant species are more abundant in upper shore pools as tidal washout is the only factor limiting population growth (Fig. 2). In comparison, stress

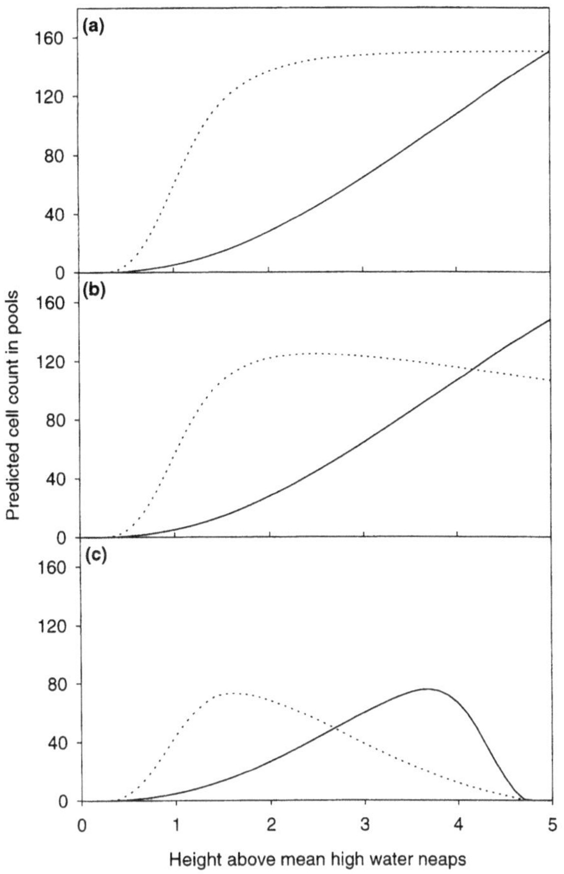

Figure 2. Predictions of population size for plankton in pools at different heights on a shore on spring (solid line) and neap (dotted line) tides. Species modelled are (a) stress tolerant, $s = 0$; (b) stress susceptible, $s = 0.1$; (c) stress susceptible, $s = 0.5$. The difference between high water neaps and high water springs was an arbitrary 5 units, other parameters used: $d = 0.5$, $r = 2.5$, $a = 0.01$.

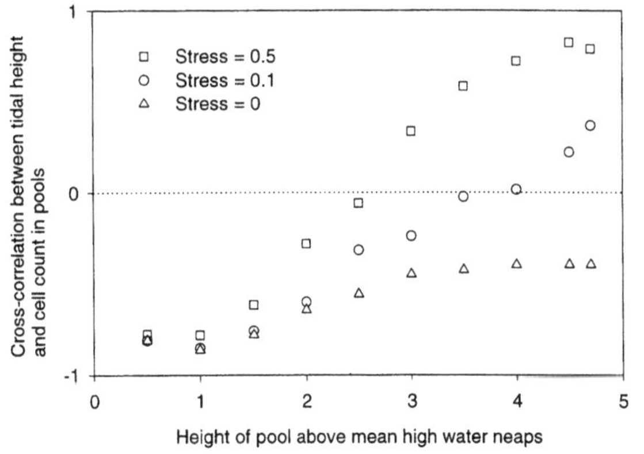

Figure 3. Cross-correlations (non parametric) between predicted cell counts and level of high tide. Simulations used a 14-day spring–neap cycle with the same model parameters as Figure 2.

susceptible species can be less abundant or completely absent from upper shore pools, depending on the extent to which isolation in pools limits the population growth rate. If tidal heights vary with the spring-neap cycle, the predictions of the model are also dynamic. For stress tolerant species, counts from different pools are correlated with decreases in expected cell number occurring on spring tides (Fig. 3). Stress susceptible species, however, have more complex dynamics. In pools low on the shore, tidal washout is still the most important limit on population growth so populations fall when tidal heights are at their maximum (spring tides). In comparison, population growth in upper shore pools is limited by stressful conditions, so cell counts fall during neap tides when pools are isolated for longer periods of time. The model, therefore, makes different predictions for stress tolerant and stress susceptible species. For stress tolerant species, cell counts from individual pools will have strong positive correlations with height of the pool above chart datum and negative temporal correlations with tidal height during the spring–neap cycle. Correlations of stress susceptible species counts with pool height will be inconsistent, ranging from negative to positive and including the possibility of mid-shore abundance peaks. The relationships of cell counts with tidal height during spring–neap cycles will also be complex, with negative correlations low on the shore and positive correlations higher on the shore.

As an alternative to population growth in rock pools, plankton may simply be washed in by successive tides. If populations do not decay when pools are isolated above the high-tide mark, there will be no relationship between tidal height and population size, and no temporal correlations between cell counts and tidal height. Perhaps more likely is the case where cells suffer greater mortality as a result of isolation within a pool. This will cause high-shore pools to have lower population densities than low-shore pools. All temporal correlations of cell counts with tidal height will, however, be positive, reflecting the fact that populations can only be replenished by tidal supply. Hence, the model describing population growth within pools (Equations 1 and 2) makes predictions which are distinct from the alternatives expected when populations do not grow or decay in pools.

Data collection and analysis

A total of 88 pools was mapped in a 5 m wide transect perpendicular to the shoreline at Port St. Mary, Isle of

Man (4° 44′ W, 54° 4′ N). The pools consisted of shallow depressions in a gently sloping limestone shore (see Southward, 1953, for a general description of the area). The tidal range for this shore is approximately 4.2 m at neaps and 8.4 m during springs. The average height of high water on neap tides is approximately 7.3 m above chart datum. Pools were generally small and depths ranged from 1.5 to 21.5 cm and areas from 0.04 to 2.1 m^2. A Cowley level was used to estimate the relative heights of pools on the shore. These measurements were converted to heights above chart datum by recording the position of the high-tide mark on a number of calm days and using these records as reference levels. The heights of pools sampled ranged from 7.29 to 9.62 m above chart datum and, therefore, covered the entire range of high-tide levels between springs and neaps.

All the pools were surveyed on two separate dates (4 December 1997 and 22 July 1998). Temporal patterns were investigated by sampling every two days over a period of 31 days during July and August (1998). To reduce the number of samples counted, the time series was taken from nine of the mapped pools. Sampling was stratified so that three pools were chosen randomly from the top, middle and bottom of the transect. In each of the groups, pool heights ranged from 7.34 to 7.36 m (bottom pools), from 8.46 to 8.6 m (middle pools) and from 9.43 to 9.53 m above chart datum (pools at the top of the transect).

Water samples were taken using a 125 ml wide mouth polypropylene bottle placed horizontally into each pool. It was possible to fill bottles quickly in all but the shallowest pools, and with minimum disturbance of benthic cells and detritus. Shallow pools (less than 5 cm depth) were sampled by pouring the contents of several 'half-fills' of the sample bottle into a 125 ml collection bottle. Plankton samples were preserved immediately after collection by adding a few drops of Lugol's iodine (Parsons et al., 1984). Preserved samples were counted using the inverted microscope method after overnight settling in 25 ml (winter samples) or 10 ml (summer samples) chambers (Hasle, 1976). The entire base plate of the chamber was counted unless the cell number exceeded 1000 per sample, in which case a representative area was counted. Before estimation of correlations between cell counts and the average height of high tide, time series were differenced to remove long-term trends (Chatfield, 1996).

Table 1. Non parametric correlations (Spearman's r_s) between plankton counts in pools and the pool height above chart datum (ns = not significant, * $p < 0.05$, ** $p < 0.01$, *** $p < 0.001$)

Plankton taxa	Winter	*p*	Summer	*p*
Oxyrrhis marina	0.616	***	0.416	***
Gymnodinium spp.	0.260	*	−0.320	**
Cryptophytes	−0.689	***	0.133	ns
Strombidium oculatum	0.371	***	−0.688	***

Results

Various taxa were recorded from pools including copepods, ostracods, armoured dinoflagellates (typically *Ceratium* and *Dinophysis* spp.), chlorophytes, gymnostome and hypotrich ciliates, scuticociliates and various diatoms. However, only four unicellular plankton taxa occurred in significant numbers across a range of pools and were, therefore, common enough to test for general patterns of distribution and temporal abundance. These taxa were the dinoflagellate *Oxyrrhis marina* Dujardin, an unidentified *Gymnodinium* species, cryptophytes and the ciliate *Strombidium oculatum* Gruber.

Data collected in 88 pools showed clear positive relationships between *Oxyrrhis* population density and pool height for both winter and summer samples (Fig. 4, Table 1). The pattern was not so clear for other plankton taxa. *Gymnodinium* spp. was less common in upper shore pools from the summer sample. The relationship between *Gymnodinium* counts and shore heights in the winter sample was positive but there also appears to be a peak in cell counts at about 8.5 m above chart datum. Cryptophytes were more common in lower shore pools during winter. *Strombidium* switched from a positive relationship with shore height to rarity in upper shore pools between the sampling dates.

Examples of the time series from a single pool are given in Figure 5. *Oxyrrhis* counts seem to peak around the time of neap tides. Evaporation during the period that pools were isolated above the high-tide level caused the salinity of mid- and upper-shore pools to rise. The sharp fall in cell counts on day 215 was associated with a rainstorm when the salinity fell from 65.2 to 34.6. Points on this date were ignored in subsequent time series analysis to remove confounding effects of the rainstorm on any patterns associated with the spring-neap cycle. Abundance peaks for the

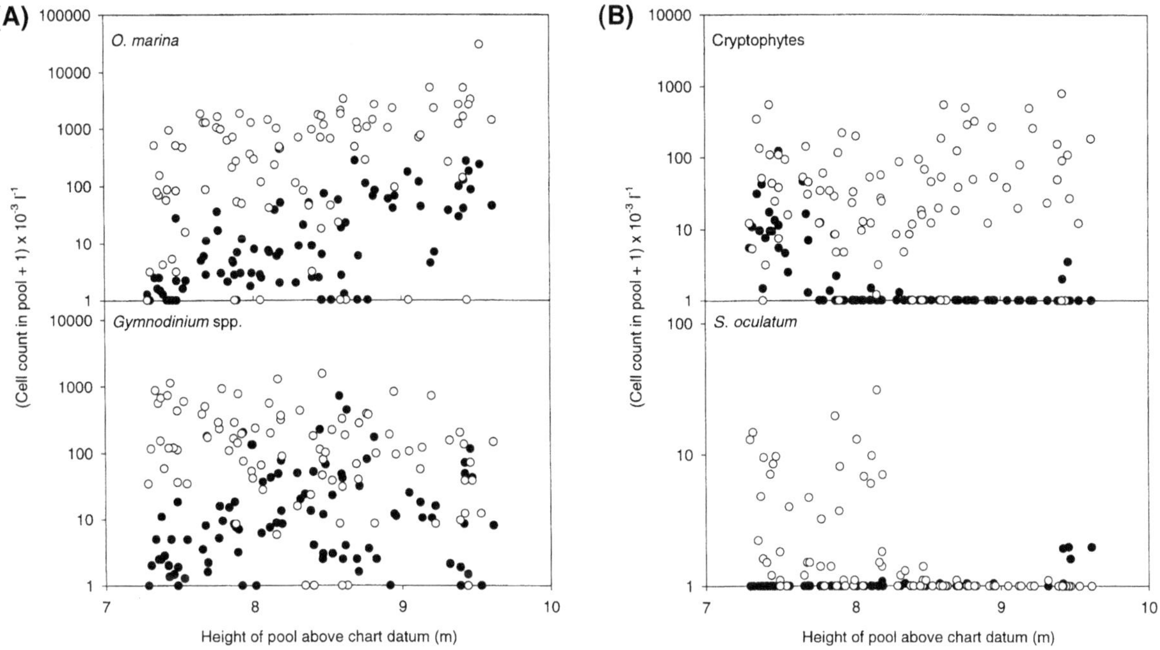

Figure 4. Cell counts from pools at Port St. Mary, Isle of Man. Filled circles are from the winter samples and open circles are summer data.

other taxa appear to be out of phase with the peaks in *Oxyrrhis* counts. Summary data from time series in different pools reveal further variation between the dynamics of different taxa at different heights on the shore (Fig. 6). *Oxyrrhis* counts always fell during spring tides, regardless of pool height on the shore. In comparison, *Gymnodinium* spp., cryptophytes and *Strombidium* had negative correlations with high-tide height low on the shore but positive correlations in mid-shore pools. *Gymnodinium* spp. counts were also correlated with tidal height in an upper shore pool whereas cryptophyte counts had no strong relationships with the spring-neap cycle at the top of the shore. However, the overall trend in cryptophyte cross-correlations with shore height was positive ($r_s = 0.964$, $p < 0.001$).

Discussion

The model developed in this paper seems to offer an explanation for the observed patterns of cell counts for abundant planktonic species in intertidal rockpools. The temporal and spatial patterns recorded suggest that the four taxa studied exist as stress tolerant or stress susceptible pool residents rather than open water species. *Oxyrrhis marina* appears to be a stress tolerant species as cell counts increase with pool height on the shore and decrease during spring tides. The other taxa appear to conform to the predictions for stress susceptible species. The closest fit to model predictions for a stress susceptible species is with *Gymnodinium* spp. Cell counts of this species were lower in upper shore pools than elsewhere and there was a switch from population minima on spring tides to population minima on neap tides with increasing pool height on the shore.

Gymnodinium spp, *Strombidium* and *Oxyrrhis* were also found to be dominant protist taxa in intertidal rock pools sampled in north-west France (Jonnson, 1994). Difficulties in establishing cultures of *Strombidium* (Jonsson, 1994) support the hypothesis that this species is susceptible to environmental stress. Cultures of *Gymnodinium* spp. and *Strombidium*, based on cells isolated from pools on the Isle of Man, were also difficult to maintain. In comparison, *Oxyrrhis* is considered to be a robust species with wide environmental tolerances and is, therefore, suitable for culture (Droop, 1970). The precise nature of the 'stress' which reduces population growth rates in high-shore pools is unclear. The counts in pools can be corrected to remove the effect of shore height using a linear regression. Residuals from this regression indicate pools with unexpectedly high or low species counts of a species. However, there were no signific-

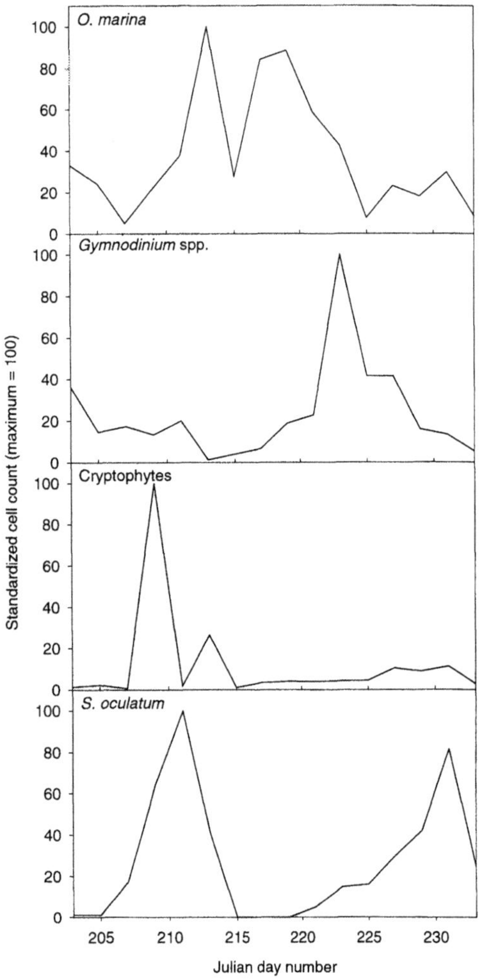

Figure 5. Time series of counts from a single pool at Port St Mary during summer 1998. Neap tides occurred on Julian days 214 and 228.

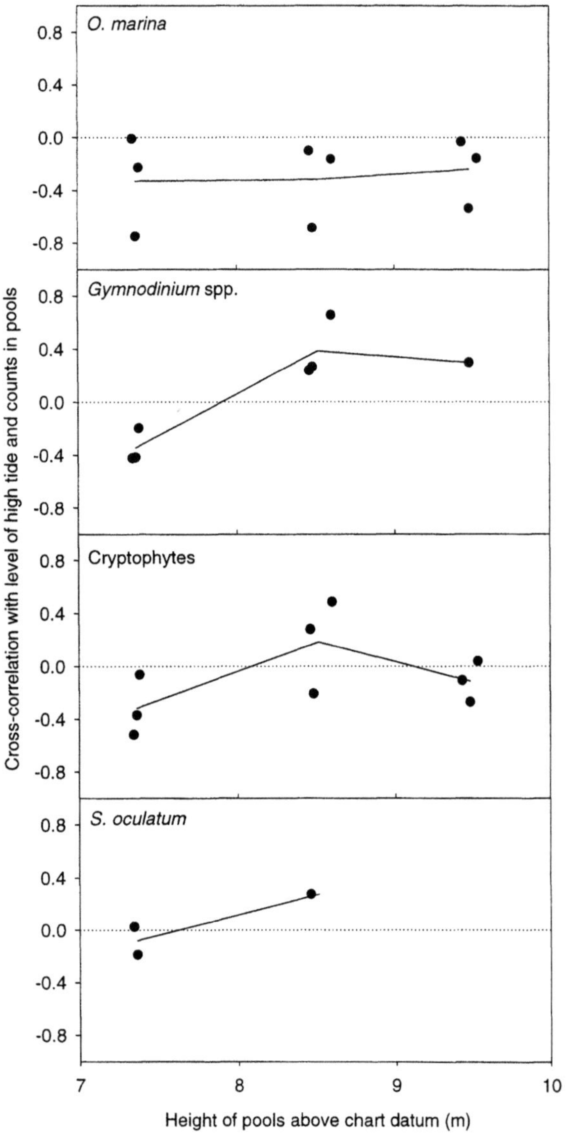

Figure 6. Cross-correlations (non parametric) between the time series of counts in different pools and the level of high tide at Port St. Mary in summer 1998. Lines are drawn between the means of groups of pools at the bottom, middle and top of the transect. To avoid spurious relationships, cross-correlations were only calculated for pools where at least 80% of each time series had non-zero counts.

ant negative correlations between the residual counts for different species. This suggests that environmental stresses, rather than species interactions such as competition or predation, were responsible for the reduced growth rates of stress susceptible species in high-shore pools. The high-shore copepod *Tigriopus brevicornis* (Müller) was not, however, included in this analysis and may, therefore, be acting to reduce the population growth rates of stress susceptible species.

Results from the current study appear to contradict the conclusions of a previous study by Metaxas & Scheibling (1996). Samples taken over a 15 month period from a shore in Nova Scotia, Canada, did not show any pattern of plankton abundance with pool elevation (Metaxas & Scheibling, 1996). Two important methodological points relate to any comparisons of the work presented here with the study of Metaxas & Scheibling (1996). The Canadian study used samples collected from four pools in each of three arbitrarily defined zones (1.2–2.3 m, 2.5–3.0 m and 2.8–4.5 m above chart datum) and tested for differences in sample counts between zones. Unfortunately, this approach assumes that the zones have *a priori* some meaningful interpretation. However, there seems no need to propose the existence of zones in the face of a continuous gradient of pool height. It could be argued that the choice of zones was merely a way of partitioning a range of pool heights. In the face of continuous variation in the treatment variable, however, analysis by partitioning a gradient may lead to various misinterpretations (Dawkins, 1983; Perry, 1986; Cousens, 1988). In particular, an analysis comparing means may suggest no treatment effect when a plot of means against treatment effect shows a clear trend. Obviously in such cases, the treatment effects must have some logical order or scale. This condition is satisfied when pools have a measurable height on the shore. The second difference between the current study and that of Metaxas & Scheibling (1996) concerns the sampling frequency. Whereas samples were taken every 2 days over two spring-neap tidal cycles from Port St. Mary, Metaxas & Scheibling (*op. cit.*) sampled approximately monthly and ignored any potential role of the spring-neap cycle. As is clear from Figure 2, the pattern of cell counts in pools at different shore heights is predicted to vary within the spring-neap cycle. At certain times, there may be very little variation in plankton counts between pools at different heights – even when the system is totally controlled by the flushing and isolation stresses on individual pools. This is consistent with the effect of zones varying considerably between sampling dates in Metaxas & Scheibling (*op. cit.*). These methodological considerations, therefore, indicate that pool height may be more important for certain rock pool plankton than was concluded by Metaxas & Scheibling (*op. cit.*). This conclusion was reached by taking a dynamic view of population growth and by using methods appropriate to describing pattern along continuous gradients. Of course, populations will still be influenced by the physical characteristics of individual pools and the occurrence of chance events – as can be seen from the scatter around the central trends in Figure 4.

The current study emphasises the importance of sampling at time scales close to the population growth responses of the organisms involved when testing hypotheses about patterns of abundance. The large difference between spring and neap tidal levels in the Isle of Man emphasised the role of spring-neap physical forcing of the system. On other shores, the influence of storms may surpass the influence of the spring-neap cycle. In such cases, a more stochastic model, which structures the plankton populations by time since last storm event, may be required.

Acknowledgements

Work was funded by a Natural Environmental Research Council fellowship. Thanks to David Montagnes for general discussion and access to laboratory and microscope facilities.

References

Begon, M., J. L. Harper & C. R. Townsend, 1996. Ecology. Individuals, Populations and Communities. Blackwell Science, Oxford: 1068 pp.

Blackwell, J. R. & D. J. Gilmour, 1991. Stress tolerance of the tidal pool chlorophyte *Chlorococcum submarinum*. Brit. J. Phycol. 26: 141–147.

Chatfield, C., 1996. The Analysis Of Time Series. An Introduction. 5th Edition. Chapman and Hall, London: 283 pp.

Cousens, R., 1988. Misinterpretations of results in weed research through inappropriate use of statistics. Weed Res. 28: 281–289.

Dawkins, H. C., 1983. Multiple comparisons misused: why so frequently in response curve studies? Biometrics 39: 789–790.

Dethier, M. N., 1980. Tidepools as refuges: predation and the limits of the harpacticoid copepod *Tigriopus californicus* (Baker). J. exp. mar. Biol. Ecol. 42: 99–111.

Droop, M. R., 1953. On the ecology of flagellates from some brackish and fresh water rockpools of Finland. Acta bot. fenn. 51: 3–52.

Droop, M. R., 1970. Nutritional investigation of phagotrophic Protozoa under axenic conditions. Helgoländer Meeresunters. 20: 272–277.

Faure-Fremiet, E., 1948. The ecology of some infusorian communities of intertidal pools. J. anim. Ecol.17: 127–130.

Harris, G. P. & A. M. Trimbee, 1986. Phytoplankton population dynamics of a small reservoir: physical/biological coupling and the time scales of community change. J. Plankton Res. 8: 1011–1025.

Hasle, G. R., 1976. The inverted microscope method. In Sournia, A. (ed.), Phytoplankton Manual. Monographs on Oceanographic Methodology 6. UNESCO, Paris: 88–96.

Jonsson, P. R., 1994. Tidal rhythm of cyst formation in the rock pool ciliate *Strombidium oculatum* Gruber (Ciliophora, Oligotrichida): a description of the functional biology and an analysis of the tidal synchronization of encystment. J. exp. mar. Biol. Ecol. 175: 77–103.

Metaxas, A. & A. G. Lewis, 1992. Diatom communities of tidepools: the effect of intertidal height. Bot. Mar. 35: 1–10.

Metaxas, A. & R. E. Scheibling, 1993. Community structure and organization of tidepools. Mar. Ecol. Prog. Ser. 98: 187–198.

Metaxas, A. & R. E. Scheibling, 1996. Spatial heterogeneity of phytoplankton assemblages in tidepools: effects of abiotic and biotic factors. Mar. Ecol. Prog. Ser. 130: 179–199.

Parsons, T. R., Y. Maita & C. M. Lalli, 1984. A Manual of Chemical and Biological Methods for Seawater Analysis. Pergamon Press, Oxford: 173 pp.

Perry, J. N., 1986. Multiple comparison procedures: a dissenting view. J. econ. Entomol. 79: 1149–1155.

Pfister, C. A., 1998. Extinction, colonization and species occupancy in tidepool fishes. Oecologia 114: 118–126.

Powlik, J. J., 1998. Seasonal abundance and population flux of *Tigriopus californicus* (Copepoda: Harpacticoida) in Barkley Sound, British Columbia. J. mar. biol. Ass. U.K 78: 467–481.

Sephton, D. H. & G. P. Harris, 1984. Physical variability and phytoplankton communities: VI. Day to day changes in primary productivity and species abundance. Arch. Hydrobiol. 102: 155–175.

Southward, A. J., 1953. The ecology of some rocky shores in the South of the Isle of Man. Proc. Transact. Liverpool Biol. Soc. 59: 1–50.

Underwood, A. J. & G. A. Skilleter, 1996. Effects of patch size on the structure of assemblages in rock pools. J. exp. mar. Biol. Ecol. 197: 63–90.

Hydrobiologia **440:** 153–157, 2000.
M.B. Jones, J.M.N. Azevedo, A.I. Neto, A.C. Costa & A.M. Frias Martins (eds), Island, Ocean and Deep-Sea Biology.

New records of brown algae (Phaeophyta) from the Azores

M. I. Parente[1], R. L. Fletcher[2] & A. I. Neto[1,*]
[1]*Secção de Biologia Marinha, Departamento de Biologia, Universidade dos Açores, Rua da Mãe de Deus, 9500 Ponta Delgada, Azores, Portugal*
Tel: +351-296653044. Fax: +351-296653155. E-mail: aneto@notes.uac.pt
[2]*University of Portsmouth, Institute of Marine Sciences, Ferry Road, Eastney, Portsmouth, Hampshire PO4 9LY, U.K.*
*Fax: +351-296653155. E-mail: aneto@notes.uac.pt (*Author for correspondence)*

Key words: Phaeophyta, Atlantic Ocean, the Azores, new records

Abstract

The following five species of microscopic tuft-forming/encrusting brown algae (Phaeophyta) are newly recorded for the Island of São Miguel (Azores): *Nemoderma tingitana* Schousboe ex Bornet, *Pseudolithoderma roscoffense* Loiseaux (Lithodermataceae), *Hecatonema terminalis* (Kützing) Kylin (Punctariaceae), *Compsonema saxicolum* (Kuckuck) Kuckuck, and *Microspongium gelatinosum* Reinke (Scytosiphonaceae). The species are described, and information is presented on their ecology and distribution around the island.

Introduction

Relatively little attention has been given to studies of the smaller, tuft-forming and crustose brown algae (Phaeophyta) of the Azores. Neto (1994) listed 70 species of Phaeophyta from the Azores, including six tufted and one crustose species, the cosmopolitan *Ralfsia verrucosa*, first listed for the islands of Pico and Faial by Tittley & Neto (1994). To provide better knowledge of the marine brown algal flora of the Azores, with particular respect to the smaller microscopic species, field work was undertaken on the island of São Miguel. Particular attention was given to searching for crustose species of brown algae. As a result of this survey, five new records of small brown algae are given, along with preliminary data on their distribution, seasonality and ecology.

Materials and methods

From May 1998 to May 1999, seven intertidal sites on the island of São Miguel (Fig. 1) were sampled at monthly intervals. Study sites were surveyed, and collections of stones, shells, chiseled portions of bedrock, and trapped plastic and wood substrata with obvious brown spots, crusts or small tufts were made. In the laboratory, algae were identified using a combination of stereoscopic and compound microscopes. Where appropriate, measurements were made of cells and other structures using a calibrated micrometer eye piece. After staining in either fast green or crystal violet, permanent slides were made of all algae collected. A reference collection was made, and each specimen was given an herbarium code number and stored in 5% buffered formaldehyde seawater solution. A representative collection of colour slides was made using an OLYMPUS-PM 10-35 AD-1 microphotographic system. Camera lucida drawings of all diagnosing features were made using a Leitz Laborlux microscope with a drawing tube. All collections are numbered and deposited at the University of Azores.

Results

The following five new records of small brown algae were recorded during the survey and are decribed below: *Pseudolithoderma roscoffense* and *Nemoderma tingitana* (Lithodermataceae); *Compsonema saxicolum* and *Microspongium gelatinosum* (Scytosiphonaceae); *Hecatonema terminalis* (Punctariaceae)

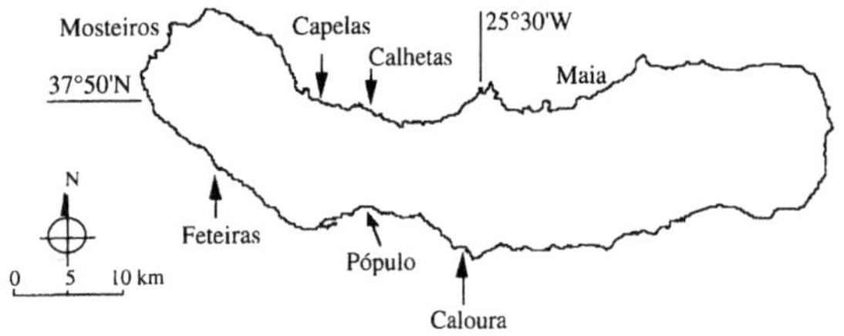

Figure 1. Location of the seven study sites on the island of São Miguel (Azores).

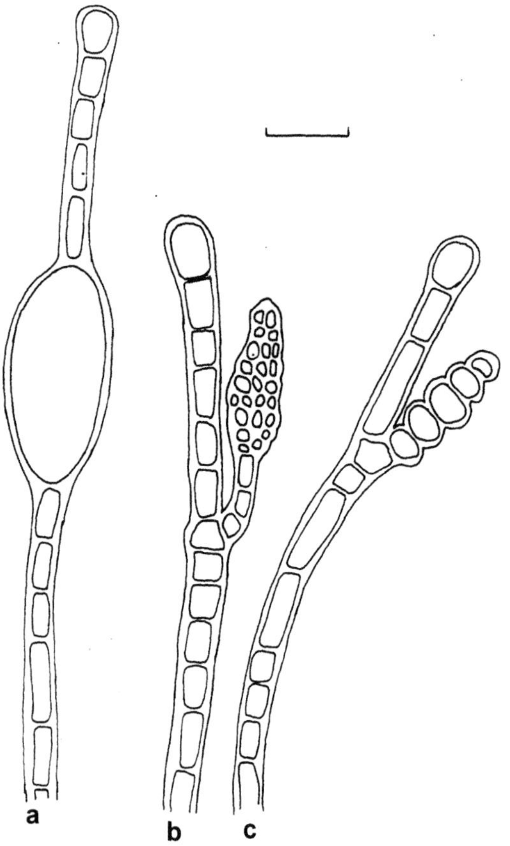

Figure 2. Nemoderma tingitana: (a) sporangia arising from intercalary cells; (b) male and (c) female gametangia arising laterally. Bar = 25 μm.

(see Fletcher, 1987 for taxonomic grouping). Under each species, the herbarium code numbers are listed.

Descriptions of plants

Lithodermataceae

Nemoderma tingitana Schousboe ex Bornet (Fig. 2) SMG-98-470; SMG-98-713; SMG-98-1043a; SMG-98-1352; SMG-98-1394; SMG-98-1527; SMG-98-1597; SMG-99-230; SMG-99-452; SMG-99-792.

Epilithic, more rarely epizoic, forming thin, smooth, slightly spongy, yellow crusts, often confluent and irregularly spreading to one square metre, and closely adherent and firmly attached to the substratum, usually without obvious rhizoids. In squash preparations, plants reveal a discoid base of which all cells give rise to erect, monosiphonous unbranched filaments up to 46 cells (1006 μm) in length, easily separable under light pressure. Central cells of erect filaments commonly rectangular, measuring 8–21 × 8–29 μm; upper cells more quadrate, elongate or slightly pyriform, measuring 11–18 × 10–26 μm (Fig. 2a). Cells light coloured with several scattered discoid plastids with several pyrenoids. Hairs abundant, grouped and emerging from depressions. In the older crusts, the lower cells contain few or no plastids and appear slightly colourless. Both unilocular sporangia and gametangia were observed. Unilocular sporangia (17–30 × 40–67 μm) develop singly from intercalary cells in the middle to upper parts of the erect filaments and are ovate (Fig. 2a). Male and female gametangia, found on the same plant, are borne laterally at about the same level and also in the upper part of the filaments (Fig. 2b). They form small, elongate clusters of compartments, female gametangia distinctly larger (25–55 × 13–20 μm) than male (40–45 × 13–20 μm).

Plants, observed throughout the year at all the sites except Mosteiros, covered large stones, bedrock and other hard substrata. in the mid to low-tide region of exposed shores.

Pseudolithoderma roscoffense Loiseaux (Fig. 3) SMG-98-573; SMG-98-616; SMG-98-752; SMG-98-1080; SMG-98-1524; SMG-98-1620; SMG-98-1623; SMG-99-41; SMG-99-446; SMG-99-760; SMG-99-802.

Epilithic, forming thin to moderately thick crusts, light brown to black. Adhering firmly to the substratum, plants are irregular and confluent, 5 cm or more in extent. Thallus comprising a discoid base, giving rise to erect, strongly united monosiphonous filaments up to 22 cells (195 μm) in length (Fig. 3a), enclosed by a thick cuticle. Cells of the erect filaments measure 8–26 × 8–18 μm and each contains 3–6 or more discoid plastids without obvious pyrenoids. In surface view, apical cells are polygonal, measuring 7–17 μm in diameter (Fig. 3b) and contain several, peripherally placed, discoid plastids; hairs not observed. The rarely observed plurilocular sporangia are terminally positioned in slightly raised mucilaginous

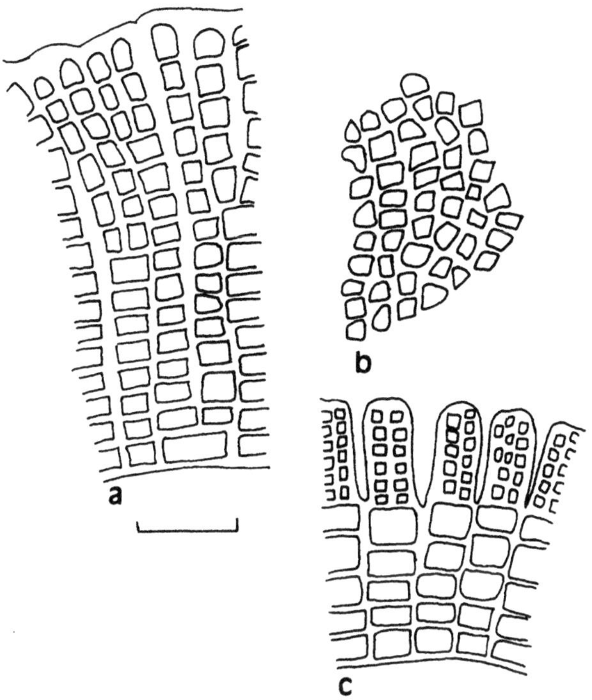

Figure 3. *Pseudolithoderma roscoffense*: (a) squash preparation of vegetative crust; (b) surface view; (c) squash preparation of fertile crust with terminal plurilocular sporangia. Bar = 50 μm.

sori (Fig. 3c). They are biseriate, rarely multiseriate, and are without accompanying paraphyses. Usually, they are covered with a thick cuticle. In vertical section, they appear clavate to slightly pyriform, up to 4–6 compartments high and measure 26–52 × 10–23 μm. Unilocular sporangia unknown.

Plants common throughout the year at all localities except Mosteiros and Calhetas, occurring on rocks and stones at low-tide level, often in pools.

Scytosiphonaceae
Compsonema saxicolum (Kuckuck) Kuckuck (Fig. 4)
SMG-98-1133; SMG-98-1396; SMG-98-1460; SMG-98-1461; SMG-98-1463; SMG-98-1464; SMG-98-1523; SMG-98-1525; SMG-98-1621; SMG-99-40.

Epilithic, tuft-forming, light yellow-brown and spreading to approximately 1–2 mm in diameter. Tufts comprise a basal layer of branched, outwardly spreading and loosely associated filaments, from which erect filaments arise (Fig. 4a). Cells of the basal layer mainly rectangular, sometimes quadrate measuring 10–16 × 16–30 μm, and frequently biseriate. Erect filaments linear, up to 44 cells (492 μm) long, monosiphonous, unbranched, loosely associated in a

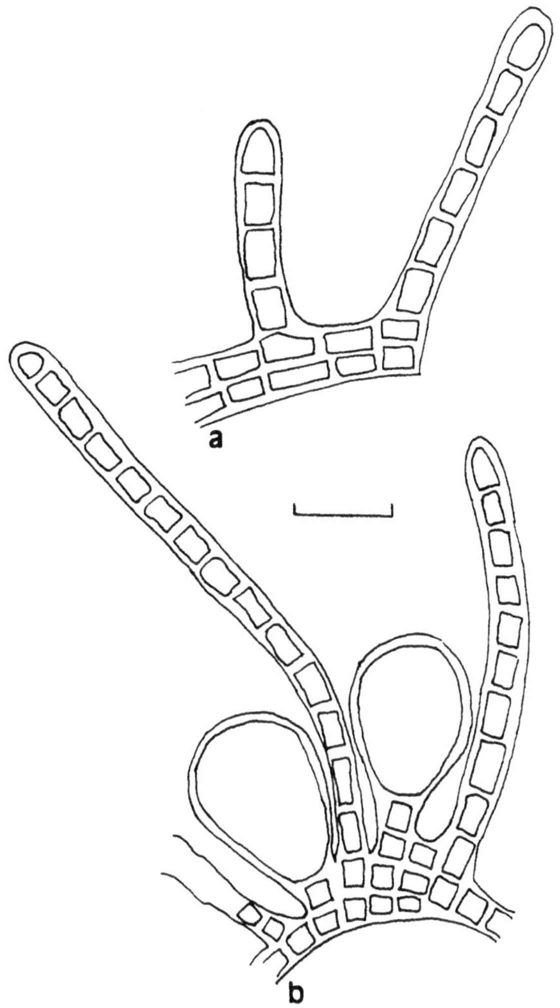

Figure 4. Squash preparation of *Compsonema saxicolum*: (a) vegetative thalli showing erect filaments arising from a distromatic base; (b) fertile thalli showing unilocular sporangia. Bar = 50 μm.

gelatinous matrix and easily separable under pressure, comprising cells mainly quadrate (8 × 13 μm), sometimes rectangular (10–13 × 16–26 μm). Cells with a single, plate-like plastid with a conspicuous pyrenoid. Longitudinal divisions sometimes present in the central region of some filaments (Fig. 4b). Hairs common, arising from the basal layer or from terminal cells of the erect filaments. Unilocular sporangia common, pyriform or elongate pyriform, sessile or stalked on basal layer (Fig. 4b), or in some cases, terminal or lateral on erect filaments. Plurilocular sporangia unknown.

Plants uncommon and only recorded at Maia, Caloura and Feteiras; occurring throughout the year on rocks at low-tide level or in pools.

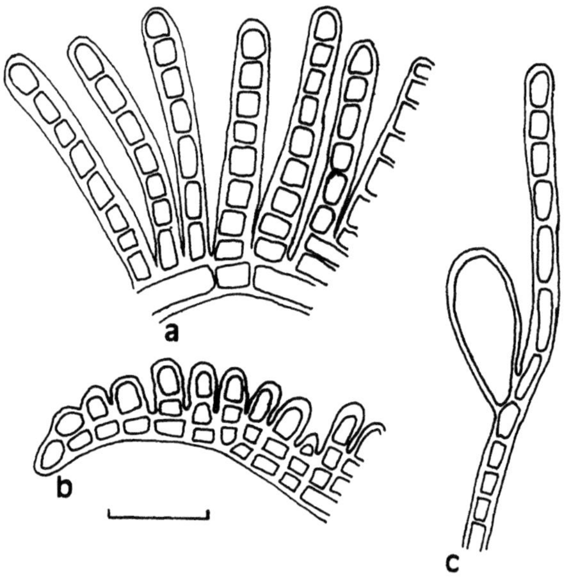

Figure 5. Squash preparation of *Microspongium gelatinosum*: (a) vegetative thalli showing loosely adjoined erect filaments; (b) crust edge showing marginal apical cell; (c) a fertile thalli showing unilocular sporangia arising at base of paraphyses. Bar = 50μm.

Microspongium gelatinosum Reinke (Fig. 5) SMG-98-573b; SMG-98-617; SMG-98-620; SMG-98-754; SMG-98-811; SMG-98-829; SMG-98-880; SMG-98-947; SMG-98-1043a; SMG-98-1249.

Epilithic, crustose to slightly pulvinate, spongy in texture, dark-brown to black, circular or more commonly irregularly spreading over several centimeters, firmly attached to the substratum, usually without obvious rhizoids. Crusts with a monostromatic base, giving rise to erect, usually unbranched filaments up to 12 cells (140 μm) long, coherent but easily separable under pressure (Figs 5a, b). Basal cells, commonly rectangular, measuring 13–30 × 10–13 μm, central cells of erect filaments also rectangular but slightly smaller, measuring 10–23 × 8–16 μm; upper cells either quadrate, elongate or slightly pyriform, measuring 10–44 × 8–16 μm. Cells with a single, parietal, plate-like plastid, located in the upper cell region, with one pyrenoid. Unilocular sporangia, in extensive, spongy sori on crust surface, elongate-pyriform or elongate-cylindrical, measuring 60–133 × 18–42 μm, commonly sessile (Fig. 5c). Linear paraphyses up to10 cells. Plurilocular sporangia not observed.

Plants common at all sites except Calhetas and Mosteiros, on rocks and in pools at low-tide level; usually absent during the winter.

Punctariaceae

Hecatonema terminalis (Kützing) Kylin (Fig. 6) SMG-98-370; SMG-98-521; SMG-98-572; SMG-98-751; SMG-98-809; SMG-98-810; SMG-98-1622; SMG-99-186; SMG-99-334; SMG-99-615.

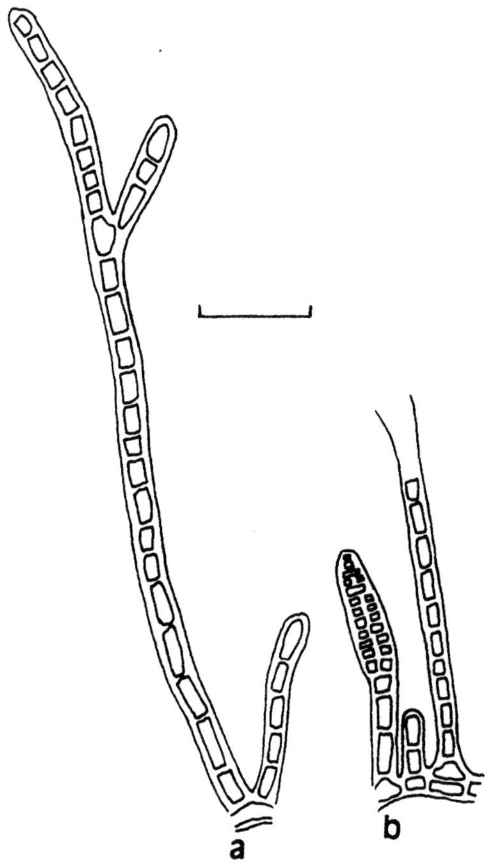

Figure 6. Squash preparation of *Hecatonema terminalis*: (a) erect filaments arising from basal layer; (b) fertile thalli showing erect filaments arising from a monostromatic/distromatic base, with plurilocular sporangia. Bar = 50 μm.

Epilithic, forming small, microscopic, light-brown tufts, discrete or more commonly confluent, spreading to approximately 2–4 cm in diameter. In squash preparations, plants reveal a pseudodiscoid base, comprising cells of variable shape, measuring 7–13 × 16–23 μm. Basal cells giving rise to erect, unbranched filaments, up to 40 cells (963 μm long), linear and loosely associated in tufts (Fig. 6a). Cells of filaments are mainly rectangular, measuring 10–13 × 17–27 μm, occasionally with longitudinal divisions, each containing 1–3 plate-like plastids, with several pyrenoids; hairs common, arising from the basal layer or terminal on erect filaments. Plurilocular sporangia common, arising from the basal layer on stalks or terminal on short erect filaments (Fig. 6b), ovate-lanceolate, biseriate or more commonly multiseriate,

20 × 60 μm, up to 29 loculi long. Unilocular sporangia unknown.

Plants common at all sites except Calhetas, occurring throughout the year on stones in pools or exposed at low-tide level.

Discussion

The five new records increase the marine brown algal flora of the Azores to 75 species. Many more collections made on the island of São Miguel await examination and it is likely that the total number of brown algae present will considerably exceed this as more have been recorded on the mainland coast. For example, Ardré (1970) recorded 98 species for Portuguese coasts. Order Ectocarpales is unrepresentative in the list of Azorean species yet is recorded for the mainland coast. It is likely that more collections of large, well known 'host' algae such as *Sargassum* spp. and *Cystoseira* spp. will increase considerably the species list. None of the new records is likely to represent recent additions to the flora of the Azores, but have been overlooked until now because of their small size and /or encrusting habit. It is perhaps only surprising that *Nemoderma tingitana* has not been reported previously, in view of its abundance and the extent of spread of the yellow crusts observed in some localities.

The present paper has extended the distribution of these small algae in the North Atlantic. The Azores now represent the southern limit of distribution for *Pseudolithoderma roscoffense*, *Compsonema saxicolum* and *Microspongium gelatinosum*, and the northern limit for *Nemoderma tingitana*. Note, however, that*Pseudolithoderma roscoffense*, only reported previously for northern France (Loiseaux, 1968) and the British Isles (Fletcher, 1987; Fletcher & Maggs, 1985), may well be conspecific with *Lithoderma adriaticum* (Fletcher, 1987; Guiry, 1997) which has been reported widely in the Mediterranean Sea. The presence of these species in the Azores confirms the reports of the Azorian flora representing a mixture of cold and warm water species (Neto, 1997).

The discovery of both *Compsonema saxicolum* and *Microspongium gelatinosum* was particularly interesting as both algae have been reported as phases in the life histories of members of the order Scytisphonales (Fletcher, 1987) which is well represented by four species *(Colpomenia sinuosa, Hydroclathrus clathratus, Scytosiphon lomentaria* and *Endarachne binghamiae*) in the Azores (Neto, 1994). Culture studies are presently underway to establish any possible life-history connections between these new records and the erect Scytosiphonales, and will be the subject of a later communication. Also interesting was the first report of *Hecatonema terminalis* in the Azores, although an unidentified *Hecatonema* sp. was reported by Neto (1994) in her checklist. Material described as *Ectocarpus terminalis* for the Canary islands by Børgesen (1926) differs from the present material by being more luxuriant in its growth, with erect filaments and plurilocular sporangia reaching much greater dimensions. It is possible that *Hecatonema terminalis* is conspecific with *Hecatonema maculans* (Collins) Sauvageau in which case the former would have priority (Wynne, 1998).

Acknowledgments

We thank the following members of the Universidade dos Açores: Sandra Monteiro for technical assistance in the field and for help with the culture work, Rui Sousa for his help in the field and Marisa Toste for helping with the drawings. We also thank the anonymous referees whose comments improved the manuscript.

References

Ardré, F., 1970. Contribution à l'étude des algues marines du Portugal. I – La Flore. Portugaliae Acta Biológica. (B) 10: 137–555.

Børgesen, F., 1926. Marine algae from the Canary Islands especially from Tenerife and Gran Canaria. II Phaeophyceae. K. danske Vidensk. Selsk. Sur. (Biol.) 6 (2): 1–112.

Fletcher, R. L., 1987. Seaweeds of the British Isles. Vol. 3. Fucophyceae (Phaeophyceae) Part I. British Museum (Natural History), London: ix + 359 pp.

Fletcher, R. L. & C. A. Maggs, 1985. Two crustose marine brown algae new to Ireland. Ir. Nat. J. 21: 523–526.

Guiry, M. D., 1997. Benthic red, brown and green algae. In Howson, C. M. & B. E. Picton (eds), The Species Directory of the Marine Fauna and Flora of the British Isles and Surrounding Areas. Ulster. Museum and the Marine Conservation Society: 341–367.

Loiseaux, S., 1968. Recherches sur les cycles de développement des Myrionématacées (Phéophycées). III Tribu des Ralfsiées. IV Conclusions generales. Rev. gén. Bot. 75: 295–318.

Neto, A. I., 1994. Checklist of the benthic marine macroalgae of the Azores. Arquipélago, Life Mar. Sci., 12A: 15–34.

Neto, A. I., 1997. Studies on algal communities of São Miguel, Azores. Tese de Doutoramento. Universidade dos Açores, Ponta Delgada: x+309 pp.

Tittley, I. & A. I. Neto, 1994. Benthic marine algae (seaweeds) recorded from Faial and Pico. Arquipélago, Life Mar. Sci. 12A: 1–13.

Wynne, M., 1998. A checklist of benthic marine algae of the tropical and subtropical western Atlantic: first revision. Nova Hedwigia Beiheft 116: J. Cramer, Berlin, Stuttgart: iii + 155 pp.

The Open Ocean

Invited Keynote Address:

Quantification of diel vertical migration by micronektonic taxa in the northeast Atlantic

M.V. Angel[1] & P.R. Pugh

[1] *Southampton Oceanography Centre*
Empress Dock
Southampton SO14 3ZH
United Kingdom

Hydrobiologia **440:** 161–179, 2000.
M.B. Jones, J.M.N. Azevedo, A.I. Neto, A.C. Costa & A.M. Frias Martins (eds), Island, Ocean and Deep-Sea Biology.

Quantification of diel vertical migration by micronektonic taxa in the northeast Atlantic

M.V. Angel & P.R. Pugh
Southampton Oceanography Centre, Empress Dock, Southhampton SO14 3ZH, U.K.

Key words: diel vertical migration, micronekton, northeast Atlantic, biomass fluxes

Abstract

Analyses of day/night changes in the bathymetric distribution of micronektonic biomass at 16 stations in the northeastern Atlantic, sampled between 1978 and 1994, provided quantitative estimates of the organic carbon fluxes associated with diel vertical migration of individual taxa of micronekton. Gelatinous taxa contributed 50–80% of the integrated standing crop by volume but, apart from tunicates, contributed relatively little to the active migratory fluxes when expressed in terms of carbon. Total micronektonic migratory fluxes into the upper 200 m ranged from 12.5 to 58 mgC per m^2. At 15 stations, fish and pteropods provided 50–80% of the fluxes into the upper 100, 200 and 400 m. At one station, tunicates (pyrosomes) contributed substantially. Wherever tunicates or the medusa *Pelagia* were swarming, migrations by other taxa appear to be suppressed. The mean proportions of the stock (in terms of biomass) of each of the dominant migratory taxa entering and leaving the upper 100 m were 23% for tunicates, 18% for fish, 22% for pteropods, 8% for decapod crustaceans and 23% for euphausiids. The maximum proportions for these five taxa were 90%, 60%, 75%, 25% and 75%, respectively. Similar estimates of the mean fluxes into and out of the upper 400 m were generally higher: 19% for tunicates, 39% for fish, 28% for pteropods, 49% for decapods and 55% for euphausiids; the respective maxima were 99%, 74%, 99%, 72% and 91%. It is estimated for fish that if these migrations occur throughout the year, they will result in an active carbon export (both POC and DOC) from the wind-mixed layer and immediate sub-thermocline depths of about 500% of the mean annual standing stock. If this estimate can be extended to other taxa, then the material fluxes resulting from these active migrations will be quantitatively similar to those resulting from the deposition of phytodetritus at temperate latitudes.

Introduction

Diel vertical migration (DVM) is characteristic behaviour of many pelagic taxa (Longhurst, 1976). The usual diel migration pattern is for populations to occupy deeper strata during the day and to migrate up towards the surface at dusk as light levels begin to fall. The majority of individuals feed actively during the night but not during the day. Hence, they migrate down at dawn with full stomachs and return to near the surface at dusk with empty stomachs (e.g. Foxton & Roe, 1974; Atkinson et al., 1996; Hopkins et al., 1996). Generally, the stimulus for these migrations is considered to be rapid alterations in the rate of change of light intensity (Ringelberg, 1964; Longhurst, 1976) and their function to be a mechanism for minimising exposure to visual predation (Zaret & Suffern, 1976; Iwasa, 1982; Vuorinen, 1987; Frost, 1988). In some freshwater plankton, it has been demonstrated experimentally that the migratory behaviour is triggered by chemical exudates from predators (Lampert, 1989; Ringelberg, 1991). In the North Sea, recent data imply that interannual variations in the intensity of diel vertical migration are correlated to fluctuations in stocks of planktivorous fish (Hays et al., 1996).

The range of the migrations, and the numbers and species composition of the migrators, must also influence ecological processes in the upper water column. The oscillations in the composition of the assemblages between day and night must be having substantial repercussions on grazing and predation pressure, and on intraspecific interactions. Those organisms, which have gut clearance rates slower than the duration of the migratory phase, will carry organic matter deep into the water column (Vinogradov, 1968; Angel, 1985, 1989; Longhurst & Harrison, 1988, 1989). Such transport may account for sediment trap observations of flux maxima at midwater depths of around 500 m. In earlier studies, these maxima were considered to result either from lateral advection (Bishop et al., 1980) or to be an artifact resulting from the presence of 'swimmers' (Lee et al., 1988). Based on sediment trap data, and the observed changes in chemical composition of the sedimenting particles, Walsh et al. (1988) concluded that for bioactive transport – "packaging and transport of material by zooplankton and nekton is the most likely cause of the widely observed mid-water column flux maximum". It may also serve to account for the peak in fluxes of snow aggregates observed between 0800 and 1200h at subthermocline depths in the northeast Atlantic (Lampitt et al., 1993a).

The contribution by the migrants to particulate fluxes is supplemented further by fluxes of dissolved organic and nitrogenous materials released as a result of their metabolic activity (Small et al., 1983; Dam et al., 1995: Steinberg et al., 2000). Since both carbon and nitrogen are predominantly assimilated by the migrants during the shallow phase of their migration cycle, any subsequent excretion of respiratory carbon dioxide (Longhurst et al., 1990), and loss of dissolved organic matter (DOM) and dissolved nitrogenous organic matter (DON) during the day-time phase of their migration deep within the water column, will result in an export flux (Longhurst & Harrison, 1988; Longhurst et al., 1989). Likewise, the death of any migrant at deeper daytime depths as a result of predation, disease or senescence will also result in a net downward transfer of material. Quantitatively, the individual daily migratory fluxes may be relatively small, but when integrated over a full year, may make a significant contribution to carbon and nutrient budgets. Thus, if the migrants contribute only 2–3% of the export flux at the base of the thermocline, this would still be equivalent to the amount of organic carbon that reaches the sea-floor. There is also evidence for asynchronous migrations from passive traps (Harding et al., 1986), with the downward migrants having fuller guts than those migrating up. The fluxes resulting from these active migrations may make a substantial contribution to the material fluxes reaching the sea-floor, because the vertical velocity of the 'gut-flux' is much faster than the sedimentary flux. So the material is being transported rapidly out of the warm upper layers of the water column, where microbial metabolism is more active, into cooler waters where microbial degradation is likely to be slower. It is also 'packaged' into larger aggregates (faecal pellets) that will sink faster and have a shorter residence time in midwater, hence, it will be less likely to be intercepted by detrivores during its downward passage.

Table 1. Station positions and dates, shown in the latitudinal order used in all the figures

Station	Position	Date	Maximum sampling depth	
			By day	By night
9791	50° N, 14° W	May, 1978	1000 m	1000 m
53304	49°15′ N, 12° 50′ W	April, 1994	1000 m	1000 m
53305	49°12′ N, 12° 17′ W	April, 1994	1000 m	1000 m
53318	49°30′ N, 11° 30′ W	April, 1994	500 m	500 m
11794	47° N, 20° W	June/July, 1988	1000 m	1000 m
11056	46° N, 14° W	April, 1984	500 m	800 m
11040	39° N, 13° W	March/April, 1984	1000 m	800 m
9801	42° N, 17° W	May, 1978	1000 m	1000 m
10379	35° N, 33° W	June, 1981	1000 m	1000 m
10382	33° N, 32° W	June, 1981	1000 m	1000 m
10376	33° N, 33° W	May, 1981	1000 m	1000 m
10378	32° N, 34° W	June, 1981	1000 m	1000 m
10380	30° N, 34° W	June, 1981	1000 m	1000 m
10228	33° N, 32° W	November, 1980	1000 m	1000 m
10233	32° N, 32° W	November, 1980	1000 m	1000 m
10222	30° N, 30° W	October, 1980	800 m	1000 m

In this study, changes in biomass profiles between day and night have been used to derive quantitative estimates of the fluxes associated with the migrations (Allison & Wishner, 1986; Angel, 1989a; Roe et al., 1990; Angel & Hargreaves 1992). Longhurst (1976) estimated that up to 25% of zooplankton and micronekton biomass in the upper 1 km of an oceanic water column could be entering the epipelagic zone at night (i.e. the euphotic and seasonal thermocline zones). Around the Azores Front, Angel (1989b) showed that at some stations at night, the 50% level for the total micronektonic biomass in the upper 1000 m was shoaling by >350 m. Angel & Hargreaves (1992) compared changes in the total biovolumes in day and night samples from a series of stations along 20° W but made no distinction between gelatinous and non-gelatinous taxa. Here, night-time increases in the biomass (expressed in terms of carbon) of the main micronektonic taxa sampled at a series of stations occupied in the northeast Atlantic in 1978–1991 have been used quantitatively to estimate the vertical migrations.

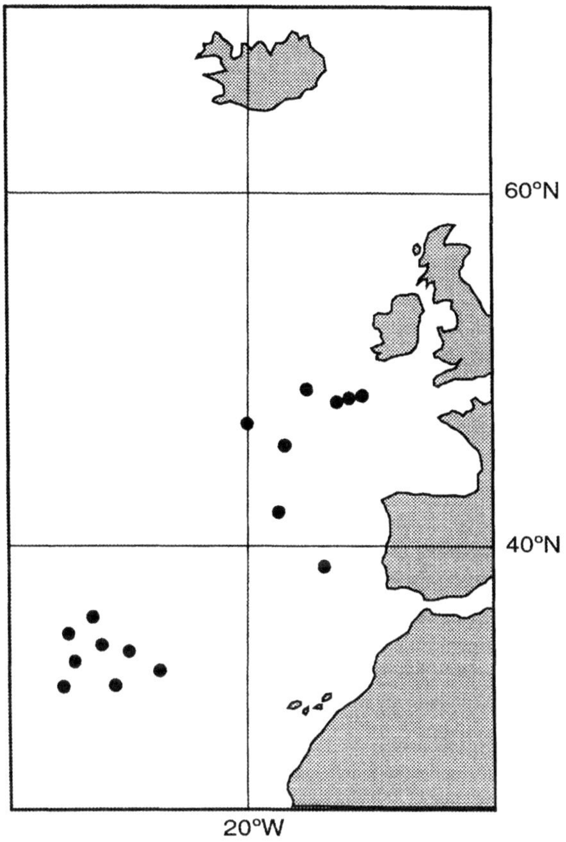

Figure 1. Map showing the locations of the stations analysed in this study.

Materials and methods

The data are derived from 16 series of macroplankton and micronekton profiles observed during a series of sampling programmes conducted in the northeast Atlantic by the Institute of Oceanographic Sciences (now part of the Southampton Oceanography Centre) (Table 1, Fig. 1). The latitudes straddle the major faunistic boundary at around 40° N that marks the southern limit of deep-winter mixing and marked seasonality in the production cycle (see Campbell & Aarup, 1992; Longhurst, 1995). Most stations were occupied during spring or early summer, but three were conducted in late autumn as part of a major study of the Azores Front (Irwin et al., 1983; Gould, 1985; Fasham et al., 1985; Domanski, 1986; Angel, 1989b; Angel & Hargreaves, 1992); these provide limited information on seasonal variations in the migratory fluxes.

At each station, bathymetrically-stratified series of samples were collected by day and by night using an RMT(1+8)M sampler (Roe & Shale, 1979). Most series consisted of contiguous, but non-overlapping horizontal tows at 100 m depth intervals from the surface to depths of 1000 m. Some of the series were, however, curtailed by bad weather (see Table 1). One station (53 318) was situated over the continental slope where the sounding was only 600 m. The RMT(1+8)M is a multiple opening/closing Rectangular Midwater Trawl enabling collection of three sequential pairs of samples. Each pair consists of a macroplankton sample collected in a nominally 1 m^2 net (RMT1) which has a mesh size of 0.32 mm, and a micronekton sample in a nominally 8 m^2 net (RMT8) with a mesh size of 4.5 mm. The duration of each haul was usually 1 h, and the continuous read-out from a flow meter mounted on the net was used to keep *in situ* towing speeds as close as possible to 2 knots (ca. 1 $m.s^{-1}$). During a typical tow, about 2 650 m^3 of water were filtered by the RMT1 and about 30 500 m^3 by the RMT8, assuming 100% filtering efficiency.

The samples were fixed initially in 5% formalin seawater and, for each sample, the preservative was renewed after about 1 day, when routinely the biovolume of the total sample was measured. In the laboratory, the total displacement volume of each sample was re-measured prior to being sorted into the major taxonomic groups. The biovolume of each group sorted was then measured. The replicated measurements of displacement volumes of some of the samples showed measurement errors amounted to 5–10%. There were also inter-sample errors stemming from the varying amounts of interstitial water trapped within the sample. Although more of the interstitial water could have been removed, any marginal improvements in the precision of the biomass estimates would have been more than offset by the increased damage to the specimens. Differentiating between the contributions to the fluxes by gelatinous and non-gelatinous groups has proved to be important. The errors associated with the sampling and the subsequent measurements are large, but will generally lead to underestimates of the quantities involved in the migratory fluxes. For example, no allowance was made for the shrinkage of gelatinous organisms that occurs after preservation. All groups were not sorted for all the series, hence, in some cases, the total flux estimates are incomplete.

Two parameters have been used to quantify the migrations. Firstly, night/day differences in displacement volumes integrated over three ranges 0–100 m, 0–200 m and 0–400 m have been calculated for each station assuming that the quantitative estimates from the nets were not subject to sampling errors or variations resulting from patchiness. Secondly, estimates of the carbon fluxes resulting from the migrations based on the integrated displacement volumes of each taxon were calculated, with a distinction being made between the contributions of gelatinous and non-gelatinous taxa.

The biovolume data for each haul were standardised to volume per 1000 m^3 of water filtered, and then converted into biovolume per m^2 within each sample range; these could then be summed to give the total volume from the surface to any selected depth within the sampling range. The daytime and night-time standing-crops were compared, assuming that the sources of error described above were small. However, in some cases, this was clearly not the case. For example, patchiness was probably the cause of the apparently large reverse migration by amphipods (the vast majority of which were *Themisto compressa*) at station 11794.

The data presented here are estimates of carbon (mgC m^{-2}) based on conversions of biovolume to organic carbon. For non-gelatinous plankton, the conversion used was that of Wiebe et al. (1975) and subsequently modified by Wiebe (1988):

$$\log_{10}V = 0.820.\log_{10}C - 1.434,$$

where C is the biomass in terms of organic carbon per m^3 of water filtered and V is the displacement volume (ml/m^3). For the gelatinous organisms, a displacement volume of 1 ml was assumed to contain 3.649 mgC; this value was derived by Pugh et al. (1997) from the data of Larson (1986). The taxa considered to be gelatinous were medusae, Siphonophora, Tunicata (thaliaceans and pyrosomes) and Chaetognatha. The taxa considered to be non-gelatinous were fish, Decapoda, Euphausiacea, Mysidacea, Amphipoda and Pteropoda.

Results

Figure 2 shows histograms of micronekton (RMT 8 catches) integrated biomass in the total water column sampled (generally 900–1000 m, but see Table 1 for the exceptions) expressed as gC m^{-2} by day and by night. At 5 of the 16 stations, the integrated catches were larger by day than by night. The five exceptional stations were: 1. At 47° N (station 11794) where the upper 200 m of the water column was dominated by a massive swarm of the hyperiid amphipod *Themisto compressa*. This amphipod appeared to be excluding (or consuming) other taxa from the upper water column. Patchy swarms of *T. compressa* are a regular seasonal feature of the temperate northeast Atlantic (e.g. Lampitt et al., 1993b). At night, there was a substantial reduction in the catches of this species integrated throughout the total water column. This was probably a result of patchiness but, in addition, if a large proportion of the population had migrated to within 5 m of the surface, it would not have been sampled by the nets, and hence the migratory flux would be grossly underestimated. 2. At 42° N, large numbers of salps were taken in the near-surface daytime catches, but very few were caught at night. Salps form dense, but highly heterogeneous swarms, so this was a patchiness problem. 3. At 35° N during summer and also at 33° N (4) and 30°N (5) in the winter in the vicinity of the Azores Front, large numbers of large *Pyrosoma* were caught by day at depths of 500–700 m. However, at night, when they could be seen in very large numbers at the surface (both in the lights of the ship and in the dark because of their intense bioluminescence), the near-surface nets collected relatively few. The trawl, as noted above, is ineffective in sampling the topmost 5 m of the water column in the wake of the ship.

The larger than usual disparities between day and night at the 46° N station resulted from less of the water column being sampled by day (to 500 m by day and 800 m at night, Table 1) as a result of bad weather. At 39° N, however, the daytime sampling extended down to 1000 m compared to only 800 m at night so the disparity is exacerbated. At least some of this disparity is explained by the exceedingly large swarms of pteropods in the upper 200 m at night that were largely absent from the daytime samples. At 49° N, the sounding at the shallowest slope station was only 600 m. Diel migrants appeared to have been migrating up from the benthopelagic zone where they are known to accumulate (Wishner, 1980; Hargreaves et al., 1984; Angel, 1990), which is likely to have greatly enhanced the night-time catches. The other highly disparate catch was at the 33° N station (10 376) which was targeted at the Azores Front itself; one series may have been more successful in sampling the frontal system than the other. What is clearly noticeable in Figure 1 is the quite sharp quantitative decline in total standing crop from ~100–150 mgC m^{-2} to ~50 mgC m^{-2} that occurred south of 39° N.

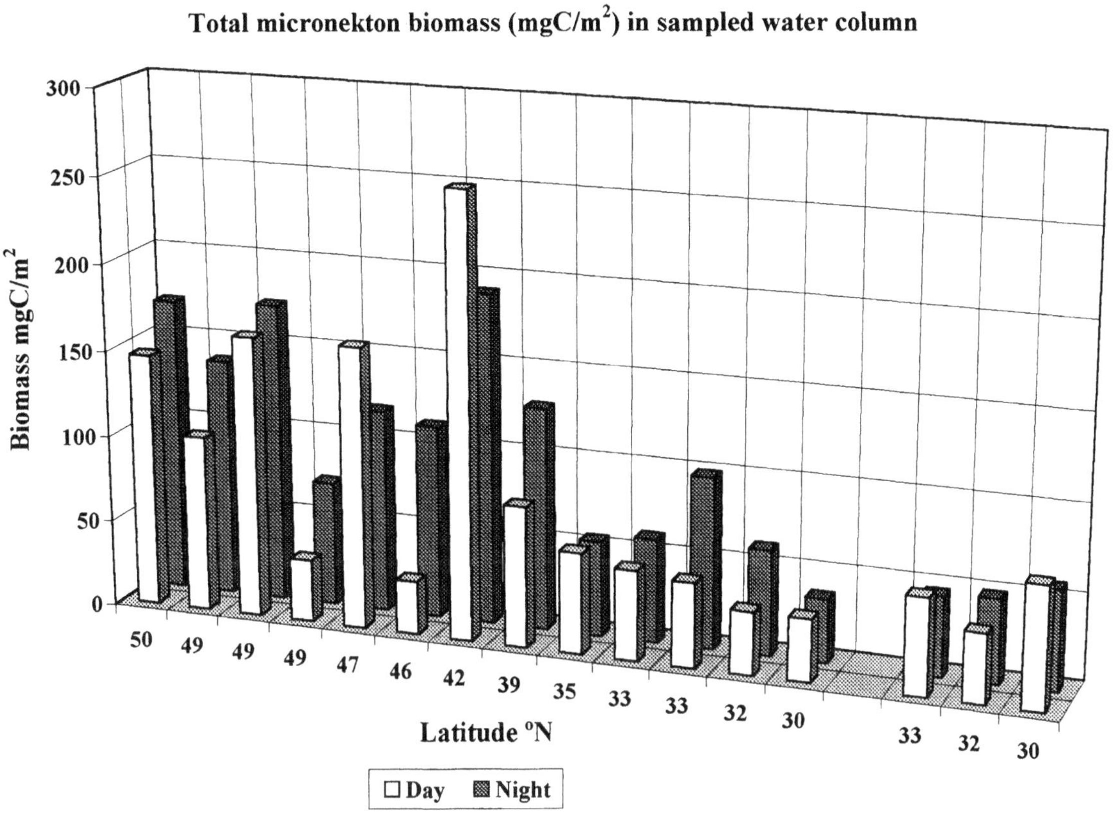

Figure 2. Day and night standing crops (mgC/m^2) of micronekton integrated over the total water column sampled (mostly 900 or 1000 m but see Table 1 for exceptions) in latitudinal order as shown in Table 1. The wintertime Azores Front station values are separated from the summertime values.

Individual taxa

Amphipoda (Fig. 3)

Figure 3 shows histograms of the night minus day standing crops (expressed in units of carbon) which are assumed to represent the carbon fluxes contributed by the diel vertical migrations of amphipods into the upper 100, 200 and 400 m of the water column. The backrow of histograms are plots of the integrated standing crop at each station (the largest value whether day or night is plotted). Generally, the standing crops of amphipods were quite small <6 mgC m^{-2}, but at 47° N and 42° N, the hyperiid *Themisto compressa* was the dominant component of micronekton biomass in the upper 200–300 m. It was also abundant at 39° N. At 47° N, there was a small positive flux (5% of the standing crop) into the upper 100 m, but the fluxes into the deeper two strata appeared to be negative. These negative fluxes, as discussed above, are mostly artifacts resulting from either patchiness, or the migration of a substantial proportion from daytime depths of 200–400 m, into the upper 5 m at night where they would have been under-sampled. It is unlikely that *T. compressa* was undertaking a reverse migration (i.e. migrating down at night). The basis for this conclusion is that there was no evidence for this behaviour in the composition of the hauls. Moreover, sampling undertaken near 50° N in 1991 during the BOFS programme (Savidge et al., 1992) showed that *T. compressa* was undertaking diel migrations of the normal pattern. However, during a storm, the migrations were disrupted (Lampitt et al., 1993b). At both 39° N and 42° N, where there were high standing crops of *T. compressa* ~60% of its biomass was migrating into the top 200 m.

Cephalopoda (not illustrated)

The standing crop of Cephalopoda seldom exceeded 1 mgC m^{-2}, and at five stations too few specimens were caught for any analysis to be meaningful. At 47° N, where reasonable numbers were caught, the bulk of the population migrated up into the top 200 m, but

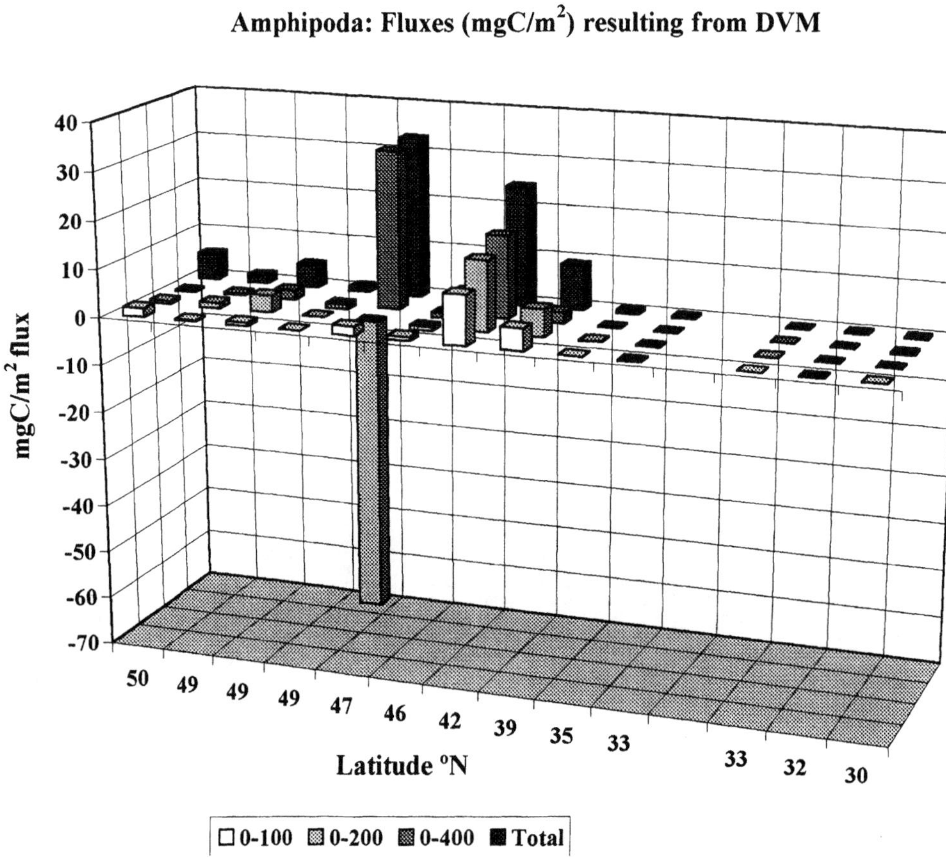

Figure 3. Amphipoda: migratory fluxes (mgC/m^2) into the upper 100, 200 and 400 m of the water column, together with the total night-time integrated standing crop in the water column sampled, in the same order as shown in Figure 2.

still generated a small carbon flux. While sampling at several of the stations at lower latitude, large numbers of ommastrephid squid were seen around the ship at night feeding on myctophid fish attracted in by the lights. Migratory fluxes by such large, actively swimming species remain undetected by samplers of the size used in this study. At those stations where there were adequate catches, mean fluxes increased from 17% of the total population sampled into the top 100 m to 55% into the top 400 m. However, none of the catches of this active group of animals was likely to be representative.

Chaetognatha (Fig. 4)

At the majority of stations, the total water column standing crop of chaetognaths, in terms of carbon, was <6 mgC m^{-2}. However, at 50° N, large populations were sampled at depths of 100–400 m and 600–800 m, and the total standing crop attained was 45 mgC m^{-2}. At all stations, the diel migratory fluxes were <0.3 mgC m^{-2}. At seven stations, the fluxes were apparently negative and the possibility of some species undertaking reverse migrations cannot be ruled out. The maximum percentages of the standing crop moving up into the upper 100 m, 200 m and 400 m strata examined were 5, 13 and 22%, respectively.

Ctenophora (not illustrated)

Only one species, *Beroë* sp., was sampled regularly. Since this species is a specialist predator of other ctenophores, its occurrence indicates that other species were present but not sampled. The great fragility of many ctenophores is well documented (Harbison et al., 1978), so the other species probably disintegrated in the net and were extruded. At only three stations were the numbers of *Beroë* worth analysing. It was undertaking vertical migrations at 42° N and 46° N, where its standing crop was high (7.5 and 1.7 mgC m^{-2}, respectively), but was mostly confined to depths >400 m. Thus, no estimate of the contribution by cten-

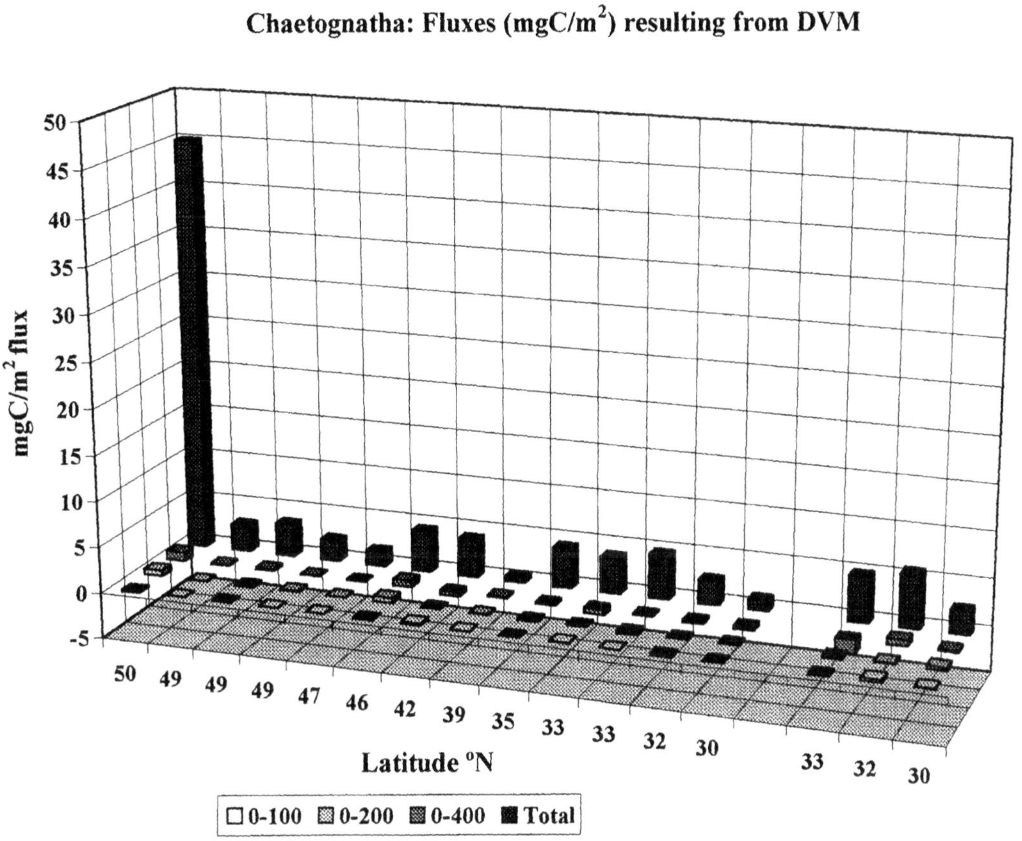

Figure 4. Chaetognatha: migratory fluxes (mgC/m^2) into the upper 100, 200 and 400 m of the water column, together with the total nighttime integrated standing crop in the water column sampled, in the same order as shown in Figure 2.

ophores to migratory fluxes could be made on the basis of these data, but there was no indication that the ctenophores were a major contributor to migratory fluxes. However, there is no denying the potential importance of the group for other aspects of water column ecology.

Decapoda (Fig. 5)

Total standing crop of decapods ranged only from 4.9 to 19.0 mgC m^{-2}. This group is one of the most conspicuous of the diel migrators. Their standing crops generally increased polewards but were particularly high in the vicinity of the Azores Front at 33° N (Domanski, 1986). Decapod biomass varied little between the late autumn and early summer in the vicinity of the Azores Front. At all stations, except the summer 30°N station in Western Atlantic Water near the Azores Front, there were detectable upward fluxes into the top 100 m at night, but only at five stations did these fluxes exceed 1 mgC m^{-2}; the highest being 3.6 mgC m^{-2}(ca.27% of the standing crop) at 42° N. Upwards fluxes into the upper 200 m exceeded 2 mgC m^{-2} at 10 stations; the maximum of 4.2 mgC m^{-2} again being at 42° N. However, the migratory fluxes into the upper 400 m were much larger (2 – 7.8 mgC m^{-2}). A substantial proportion of the total decapod populations was involved in these migrations, the maximum being 71% at 35° N (where the upper 1000 m was sampled). Thus, diel vertical migration is an important element in decapod ecology. However, since most of the decapod migrants remained below the seasonal thermocline, they will contribute relatively little to export fluxes from the euphotic zone, but may be playing an important role in subthermocline fluxes. The mean percentage of the population migrating into the upper 100 m was 8% (maximum 27%), and into the upper 200 m and 400 m increased to 22% (44)% and 49% (71)%, respectively.

Euphausiacea (Fig. 6)

Generally, euphausiid standing crops were relatively small, but ranged over more than an order of mag-

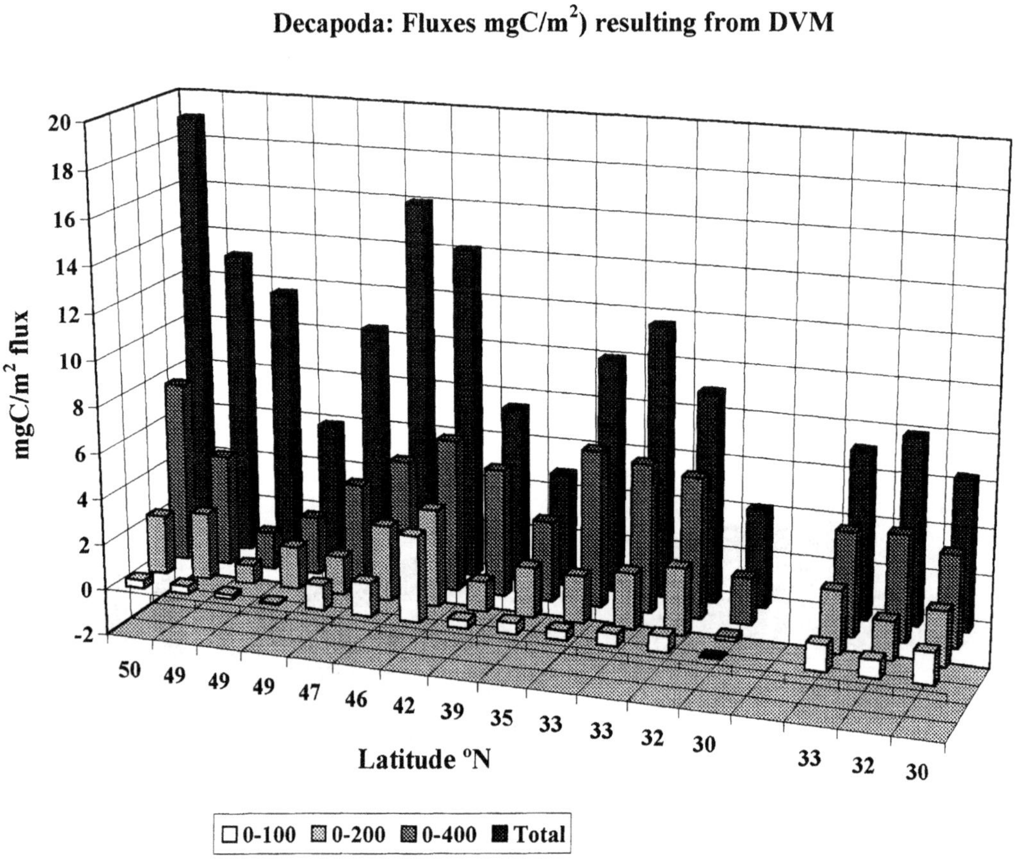

Figure 5. Decapoda: migratory fluxes (mgC/m^2) into the upper 100, 200 and 400 m of the water column, together with the total nighttime integrated standing crop in the water column sampled, in the same order as shown in Figure 2.

nitude from 2.5 to 38.3 mgC m^{-2}. The highest biomasses (i.e. >7 mgC m^{-2}) were at middle latitude stations from 39° to 47° N. There seemed to be a positive correlation between the size of the standing crop and the percentage of the population undertaking diel migrations into the upper 100 m (>75% at 42° N). The maximum migration fluxes (>30 mgC m^{-2}) into the upper 200 and 400 m of the water column were observed at 39° N. At the three near-slope stations at 49° N, increasing proximity to the slope depressed both the percentage of the population migrating and the total flux. As for decapods, positive fluxes into the deeper strata were observed at all stations. Mean (maximum) percentages of the euphausiid population migrating into the upper 100, 200, and 400 m were 23 (64)%, 52 (88)% and 55 (92)%, respectively.

Fish (Fig. 7)

Standing crops of fish, like those of decapod crustaceans, varied over a limited range from 11.9 (at 46° N) to 66.0 mgC m^{-2} (at the Azores Front, 33° N). The highest fluxes, and the highest percentages of the total population migrating up into the upper 100 m at night, were at 42° N and at the Azores Front station at 33° N. The highest percentage of migrants was at the shallowest slope station at 49° N (sounding 600 m), but one of the lowest percentages was the next station on the cross slope transect where the sounding was 1000 m. This was a station at which Pteropoda occurred in very high abundances. One hypothesis to explain these sharp contrasts over such short distances is that the exceptional localised abundance of food deep within the water column may have reduced the need for the fish to migrate. At most stations, migration fluxes increased down to 400 m. The lowest percentage of migrants in the total fish assemblage was in Western Atlantic Water at 30° N, which coincided with the lowest observed standing crop. The mean (maximum) percentages of the total water column population migrating up into

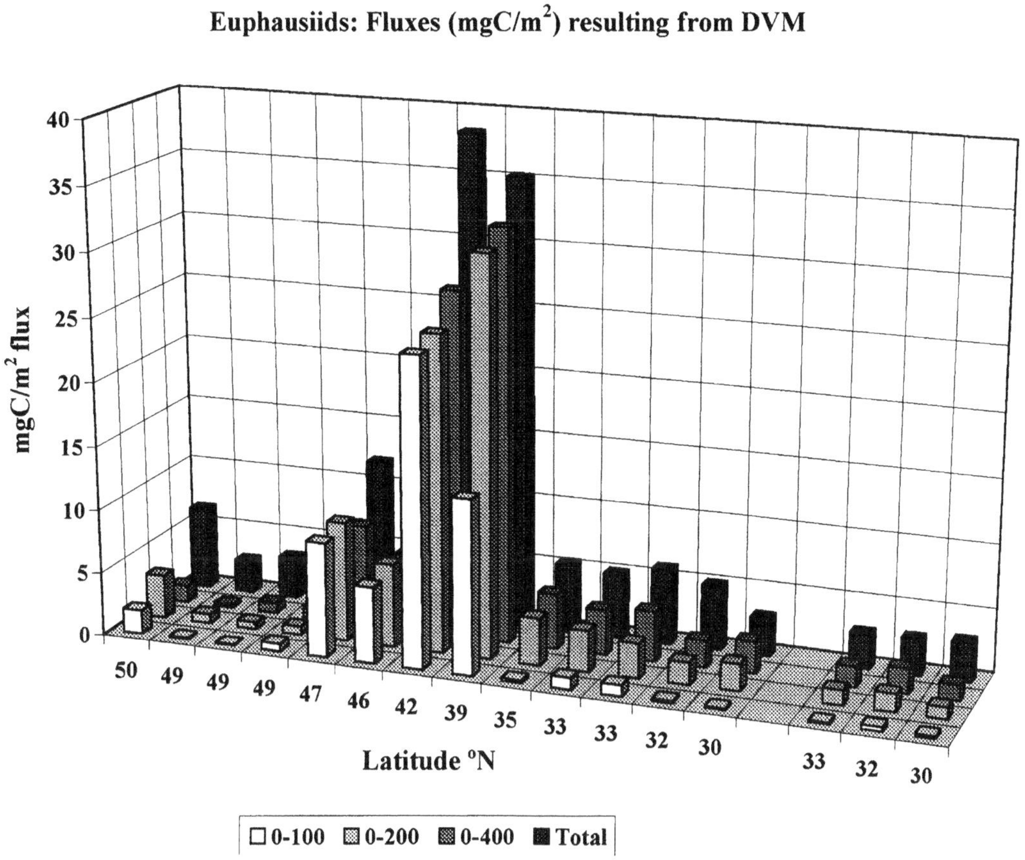

Figure 6. Euphausiacea: migratory fluxes (mgC/m^2) into the upper 100, 200 and 400 m of the water column, together with the total nighttime integrated standing crop in the water column sampled, in the same order as shown in Figure 2.

the upper 100, 200 and 400 m, respectively, were 18(59)%, 29(71)% and 39(74)%.

Medusae (not illustrated)

The standing crops and migrations of medusae were analysed at the seven more northerly stations only. Standing crops at two stations (50° N and 42° N) exceeded 10 mgC m^{-2} and at each the medusae appeared to be undergoing reverse migrations. At 42° N, there was a large near-surface swarm of *Pelagia*, and such large swarms may inhibit upward migrations by other taxa. The mean percentages of the population migrating up did not exceed 2%, a value well within the range of likely observational error.

Mysidacea (not illustrated)

The standing crops of mysids (predominantly a single species *Eucopia unguiculata*) ranged from 0.4 to 5.6 mgC m^{-2} at northern stations (>39° N) and from 0.2 to 2.4 mgC m^{-2} at southern stations. The species was almost totally restricted to depths >400 m and so was not contributing to fluxes in the upper water column. However, data from repeated 48 h of sampling collected at 44° N, 13° W showed that *E. unguiculata* was undertaking diel migrations to at least 250 m during the spring bloom (Roe et al., 1984). Similarly, recent sampling in the vicinity of the Almeria-Oran Front in the Mediterranean in January (1997), showed that *E. unguiculata* species was a component of the population occupying daytime depths of <500 m and was undertaking diel migrations (Howell, pers. comm.). Hence, for most of the year, mysids make little or no contribution to migratory fluxes in the northeast Atlantic but may make a small contribution around the timc of any scasonal bloom.

Pteropoda (Fig. 8)

Seasonally, pteropods form dense swarms and their contribution to the standing crops in these samples was highly erratic. They made substantial (>15 mgC m^{-2})

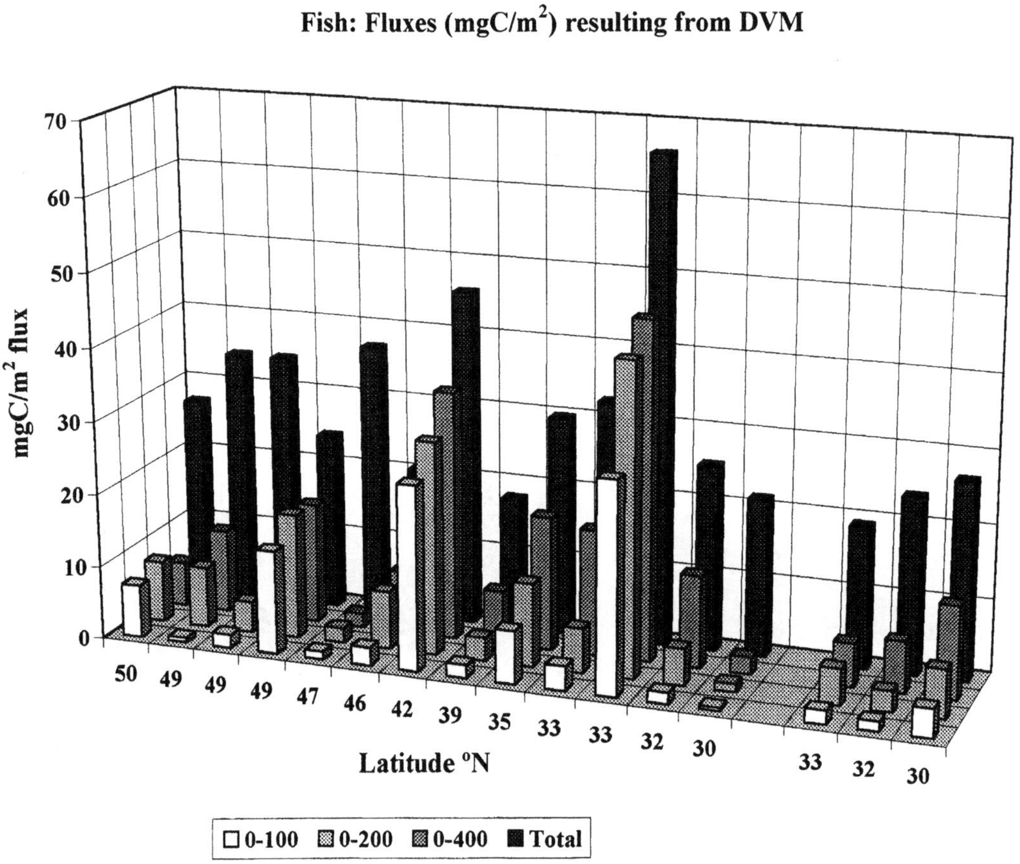

Figure 7. Fish: migratory fluxes (mgC/m^2) into the upper 100, 200 and 400 m of the water column, together with the total nighttime integrated standing crop in the water column sampled, in the same order as shown in Figure 2.

contributions to the total water column biomass at 39° N, 46° N and two of the near-slope stations at 49° N. In contrast, at all the more southerly stations (<39° N) their contributions were trivial. Whenever they were abundant, they made substantial contributions to the migratory fluxes. At 39° N and 46° N, their migratory fluxes into the upper 100 m amounted to 26 and 13 mgC m^{-2}, respectively. At the two outer slope stations at 49° N, their migratory fluxes into the upper 200 m were 28–38 mgC m^{-2}, but there was little migration into the upper 100 m. Ten days after the samples used in this analysis were collected at 39° N, the position was revisited. Already, the massive swarm of pteropods originally observed had disappeared, either because it had been advected out of the region or the population had collapsed. A similar situation prevailed at 42° N where a swarm of pteropods was sampled at 300–500 m by day, but only a relatively small biomass was taken at night mostly in the upper 100 m. This serves to highlight the high variability of the contribution to migratory fluxes that may be made by such swarming species. The mean (maximum) percentages of the populations migrating up into the upper 100, 200 and 400 m, respectively, were 22(82)%, 31(98)% and 28(98)%.

Tunicata (thaliaceans and pyrosomes) (Fig. 9)

These gelatinous taxa often contributed substantially to total water column biovolumes but, in terms of carbon, their contributions to the total standing crop were quite small. At the southern stations in the vicinity of the Azores Front, the dominant species was *Pyrosoma atlantica*. In both seasons sampled, the whole of its population of large specimens was undertaking migrations from daytime depths of 500–700 m up into the upper 100 m at night. At the more northerly stations, *P. atlantica* was replaced by patchy swarms of the aggregate stages of *Salpa* and *Iasis* at some of the stations. These salps also undertook diel vertical migrations but not as extensive as the *P. atlantica* mi-

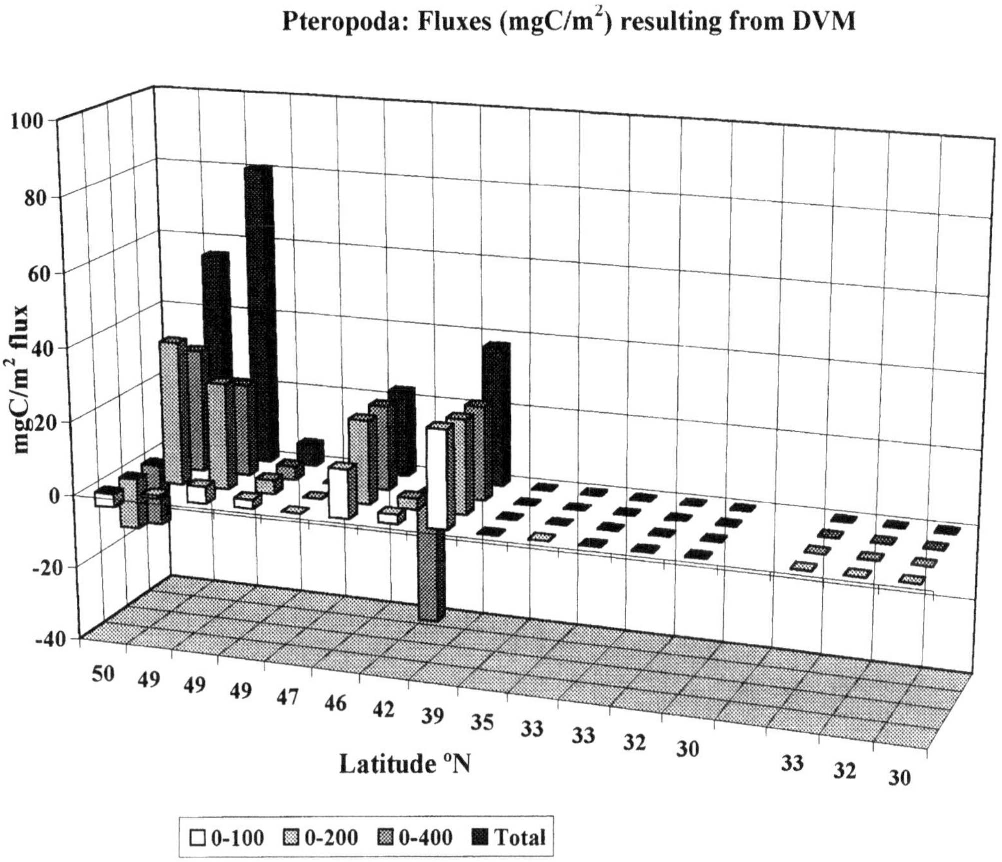

Figure 8. Pteropoda: migratory fluxes (mgC/m^2) into the upper 100, 200 and 400 m of the water column, together with the total nighttime integrated standing crop in the water column sampled, in the same order as shown in Figure 2.

grations, nor those of up to 2000 m reported for *Salpa aspera* in the northwest Atlantic (Wiebe et al., 1978). The migratory fluxes by *P. atlantica* exceeded 10 mgC m^{-2} into the upper 100 m at two of the Azores Front stations, and those by salps at some of the more northerly stations were around 3–4 mgC m^{-2}. As with the pteropods, when these thaliaceans and pyrosomes are abundant, they can contribute substantially to fluxes, but their occurrences tend to be patchy in both time and space.

Siphonophora (Fig. 10)

The standing crop of siphonophores diminished south of 39° N. Their main centre of abundance is usually at depths of 200–400 m where they are usually dominant in terms of biovolume (Pugh, 1991; Pugh et al., 1997). Their maximum contribution to migratory fluxes into the upper 100 m was around 3 mgC/m^{-2} at two of the Azores Front summertime stations where 45–50% of the population was migrating into the upper 100 and 200 m. Otherwise, their only major contribution was an apparent flux of 17.5 mgC m^{-2} into the upper 400 m involving 57% of the population at the 50° N station. There were several apparently negative fluxes into the upper 100, 200 and 400 m amounting to ~3–4 mgC m^{-2}, but only two of these negative fluxes (both into the upper 400 m) involved >10% of the total siphonophoran population. Mean population movements into all three depth bands were close to 10%.

Discussion

The approach adopted here suffers from four important sources of error. Firstly, quantities of some gelatinous organisms may have been substantially underestimated because many of the more fragile siphonophores, tunicates (particularly some salp species) and ctenophores disintegrate in nets, and are extruded

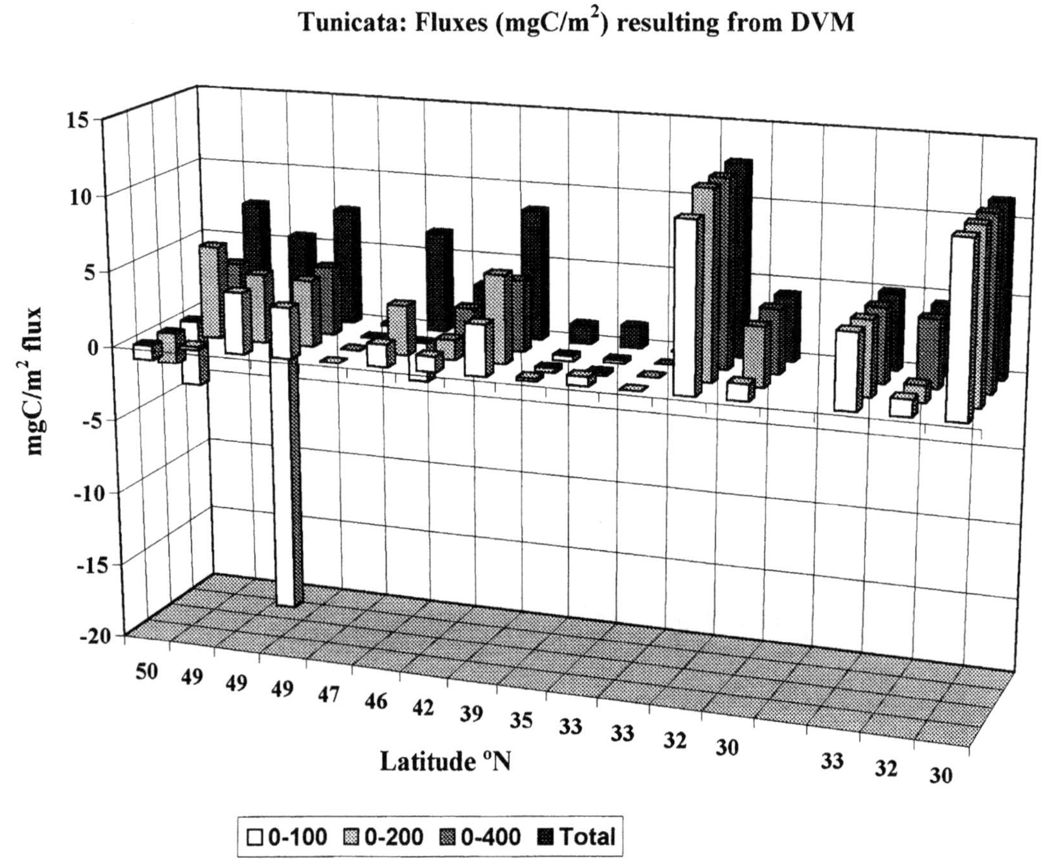

Figure 9. Tunicata: migratory fluxes (mgC/m^2) into the upper 100, 200 and 400 m of the water column, together with the total nighttime integrated standing crop in the water column sampled, in the same order as shown in Figure 2.

through the meshes (e.g. Harbison et al., 1978; Pugh, 1989). Such extrusion effects will be unaffected by the day/night cycle. Secondly, avoidance of the nets by the organisms tends to be greater by day than by night. This contributes to the integrated biomass estimate for the daytime total water column sampled being lower and this will tend to result in migratory fluxes being over-estimated. The taxonomic groups that usually demonstrate the greatest day/night disparity are euphausiids (up to 40%), fish (10–20%) and decapod crustaceans (10–20%).

The third source of error arises if the vertical migrations are ranging to depths deeper than the lowest range of the sampling. Some species of fish and decapods are known to migrate from daytime depths of 1600–1700 m and 1200–1300 m, respectively, in the central oligotrophic gyre around the Azores Current region (Angel, 1989b; Domanski, 1986). However, in repeat samples from a depth of 1000 m in more productive temperate waters at 42° N, 17° W, Angel et al. (1982) found that only one species of myctophid fish, *Notoscopelus elongatus kroyeri*, showed variations in abundance throughout 48 h of sequential sampling, and which were consistent with it undertaking diel vertical migrations from that depth synchronised with the light cycle. So at night, some of the shallow catches will have included some specimens, which may have migrated from depths deeper than the lower limit of daytime sampling. Over the slope, in particular, where the nets could not be fished very close to the sea-bed, the night-time catches may have been supplemented with diel migrants that are known to aggregate close to the sea-bed during the day (Wishner, 1980; Angel & Hargreaves, 1984; Angel, 1990). However, Pugh (unpublished data) has shown that several siphonophore species are unable to adapt their vertical ranges to shoaling or deepening bathymetry.

The fourth source of error is the high spatial heterogeneity shown by pelagic communities in general, which are known to show both passive and active re-

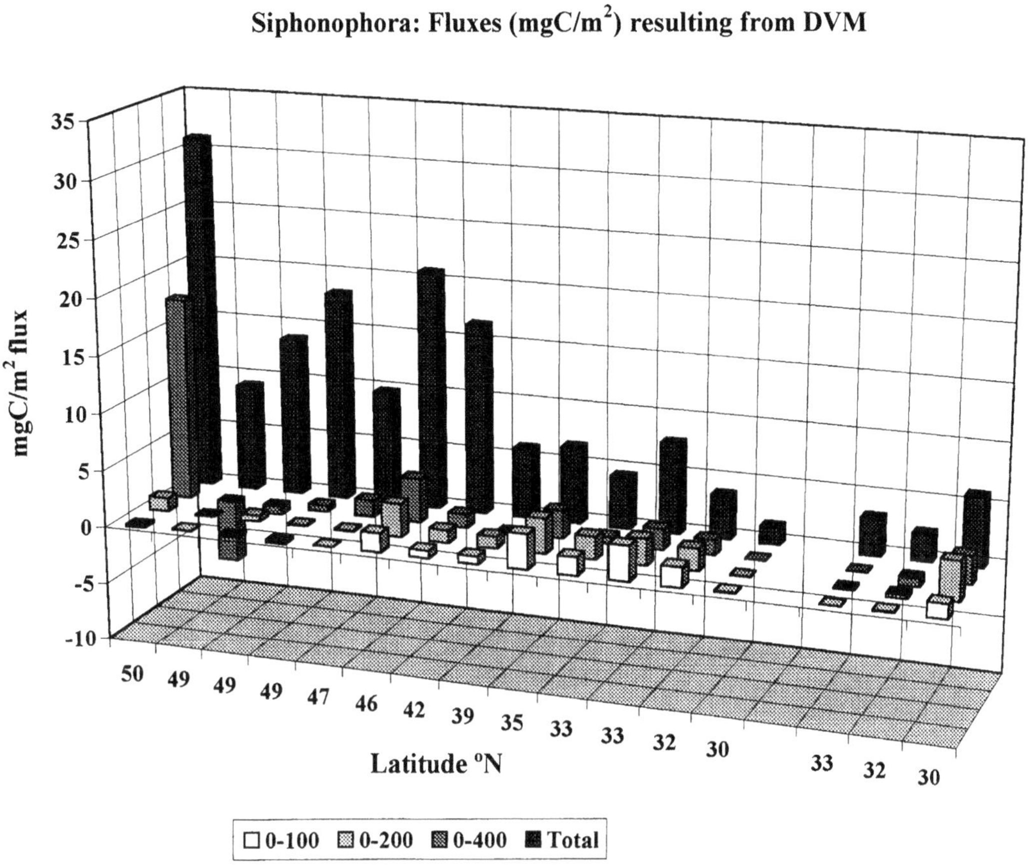

Figure 10. Siphonophora: migratory fluxes (mgC/m^2) into the upper 100, 200 and 400 m of the water column, together with the total nighttime integrated standing crop in the water column sampled, in the same order as shown in Figure 2.

sponses to meso-scale eddies. Passive aggregation can occur at local convergences. But aggregation may be active, as postulated to explain the occurrence of much higher abundances of fish observed within the Azores Front than a few tens of kilometres on either side of it (Angel, 1989b). Aggregations can also build up as a result of very rapid reproduction and growth. Some taxa have exceptionally high potential rates of growth and reproduction. For example, Heron (1972) attributed the rapid build up (within a few days) of massive swarms of a *Thalia* species to its growth being more closely correlated to the second power of body length than to the usual third power. In other words, this species grows at a rate related to some function of area rather than body volume, and so it has the ability to respond exceptionally rapidly to localised enhancement of productivity. Swarming is a common phenomenon in several common pelagic taxa, notably thaliaceans, medusae and pteropods. Consequently, abundances of these groups may fluctuate by an order of magnitude over quite small spatial scales and hence between successive samples. So by chance, a large swarm might have been sampled during the day but not by night (or vice versa); this will greatly distort our estimates of migratory fluxes.

Bearing these sources of error in mind, the data presented here are probably qualitatively reasonable but quantitatively questionable (may be accurate to within a factor of two or three), the consistency of many of the trends gives a measure of confidence. Until a technology capable of surveying large volumes of water is developed and validated (e.g. using bioacoustics), net-derived data remain the only basis on which the active migratory component of the biological pump can be assessed.

Biovolume, despite its widespread use in literature reports, is an unsatisfactory parameter for assessing the importance of these migratory fluxes. Gelatinous taxa in open ocean environments are often volumetrically dominant in the upper few hundred metres, but

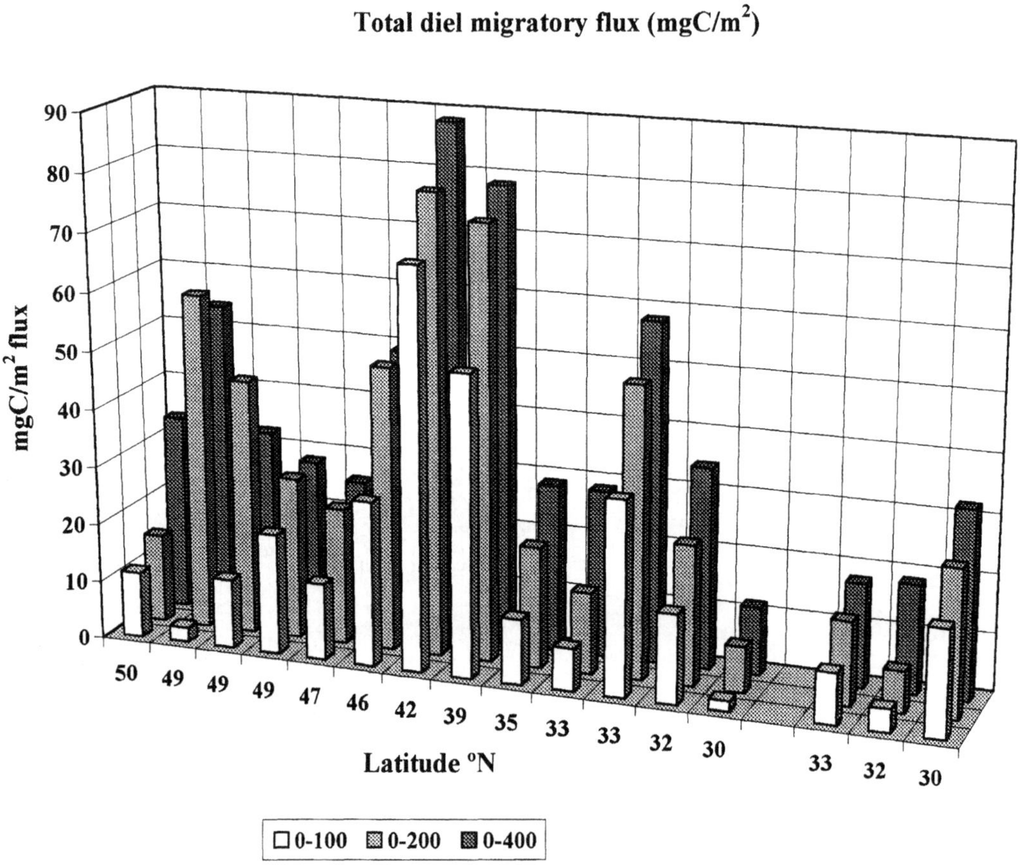

Figure 11. Total micronektonic migratory nighttime fluxes (mgC/m^2) into the upper 100, 200 and 400 m at all stations sampled. The fluxes values were calculated by summing the flux contributions by each of the dominant taxa.

in terms of organic carbon, their quantitative importance is lower. Since metabolic rates are more closely attuned to the organic content of the organisms rather than to their physical bulk, estimates based on carbon content are likely to provide a more precise estimate of the active fluxes resulting from the migrations. Even then, this assumes that all active fluxes are associated with migrations that are synchronised with the light cycle. If non-synchronised movements are widespread (Pearre, 1979), the fluxes resulting from active transport may be even larger than estimated here. In addition, the intensity of acoustic back-scatter is better related to carbon content than biovolume.

The percentage of the total micronektonic biomass (in terms of carbon) involved in diel migrations into the upper 100 m ranged from 1.9% to 40.3%; from 8.8% to 58.0% into the upper 200 m; and from 19.7% to 61.4% into the upper 400 m. There appear to be systematic changes in these percentages related to latitude. The percentages migrating can be expected to vary seasonally, particularly at the higher latitudes where many species undertake both ontogenetic and seasonal migrations, but the sampling coverage was inadequate to investigate such variations.

The fluxes into the upper 100 m integrated over all taxa at each station ranged from 1.8 to 68 mgC m^{-2}, and peaked at 39° and 42° N. The integrated flux at the Azores Front was high, but was low on either side of it but particularly on the western Atlantic Water side. There was also a low flux into the upper 100 m at the outermost slope station off the Goban Spur at 49° N, which was in marked contrast with the high flux into the upper 200 m at the same station. Integrated fluxes into the upper 200 m ranged from 7.8 to 78.5 mgC m^{-2}. Again, the maximum values were at middle latitudes at 39° and 42° N, closely followed by the high flux at the outermost station off the Goban Spur. Again, the lowest fluxes were associated with the waters either side of the Azores Front especially in winter; the flux at 50° N was also low. Integrated

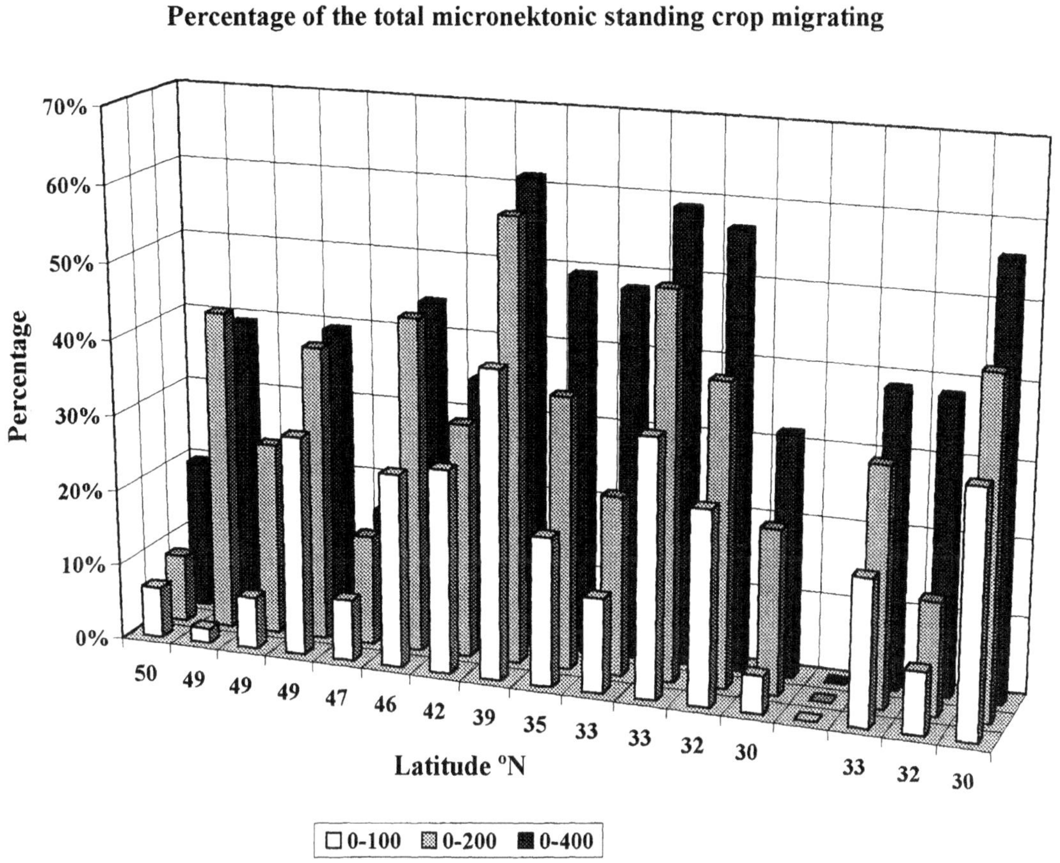

Figure 12. Percentages of the total micronekton standing crop engaged in vertical migration into the upper 100, 200 and 400 m.

fluxes into the upper 400 m ranged from 11.5 to 88 mgC m^{-2} and were, with one exception, higher than those into the upper 200 m.

Inter-taxa correlations in fluxes

A correlation analysis of the fluxes into the upper 100 m showed that the fluxes of amphipods, decapods, euphausiids, fish and pteropods were correlated positively at the 5% level with the total flux. However, there were few inter-taxa correlations. Flux estimates resulting from migrations by amphipods correlated positively with those generated by decapods and euphausiids, and those of decapods correlated with those of euphausiids and chaetognaths. Fluxes of euphausiids and pteropods into the upper 200 m were correlated positively with the total fluxes. However, the only significant correlation (at the 5% level) between individual taxa was a negative correlation between fluxes resulting from the migrations by amphipods and cephalopods, but this is almost certainly an artifact resulting from the incomplete data set for the cephalopods. Generally, the number of correlations is small and within the number that statistically can happen by chance. In other words, these data provide no evidence for the migrations of the individual taxa being inter-related.

Fluxes up into the upper 400 m resulting from the migrations by decapods, euphausiids, fish and pteropods were correlated significantly with the total fluxes. However, the only positive correlation between pairs of taxa were between fish and amphipods, and pteropods and euphausiids; again there was a negative correlation between cephalopods and amphipods. These results suggest that the carbon fluxes associated with the migrations of the various taxa are independent of each other, the few positive correlations being no more than would be expected by chance.

Taxa contributing most to active fluxes

The taxa making the major contributions to the migratory fluxes varied with latitude. At seven stations, fish contributed the most to the fluxes into the upper 100 m; pteropods and tunicates were the major contributors at four stations, and euphausiids at one. Substantial contributions were made by fish at another two stations, euphausiids at a further three, tunicates at a further two. Generally, active carbon fluxes into the upper 100 m were dominated by four taxa: fish, pteropods, euphausiids and tunicates. Fluxes into the upper 200 m were dominated by these same taxa with very little change in their relative contribution. Fluxes into the upper 400 m were supplemented by decapod crustaceans whose fluxes ranging from 1.55 to 7.78 mgC m^{-2}, a relatively small range compared with most other taxa. At one station, they were the largest contributor and the second or third largest at eight other stations.

In previous sampling in the region and elsewhere, tunicates and pteropods have proved to be very patchily distributed in both time and space. Populations of both groups are highly responsive to the spring bloom in the north Atlantic (e.g. Vane & Colebrook, 1962; Hunt, 1968). Fish, on the other hand, are longer lived and less spatially patchy. Many fish species are migratory and many undertake ontogenetic migrations. So migratory fluxes can be expected to vary seasonally.

During May and June 1991, during a Lagrangian study centred at about 48° N, 17° 30′ W, the amphipod *Themisto compressa* was numerically the dominant component of the micronekton (Lampitt et al., 1993). At night, the ratios between the biomasses of fish:decapods:euphausiids:amphipods at 50–25 m were 1:10:100:450. Generally, during the day, populations of *T. compressa* were found at depths ranging from 100 to 400 m. However, one set of samples, collected immediately after a storm while wave conditions were still rough, showed that *T. compressa* had not migrated down and was still in very large numbers in the wind-mixed layer. Was this observation exceptional, or was it representative of a regular response by this (and maybe other species) to weather and rough seas? These amphipods were not only consuming marine snow but also producing faecal pellets, which resembled marine snow sampled deeper down (270 m) and contained viable cyanobacteria. Lampitt et al. (1993) estimated the gut flux at the time to be 0.43 mgC m^{-2}, which amounted to about 0.4% of the passive vertical flux of dry mass in floating sediment traps. Generally, in the stations reported here, amphipods were unlikely to be contributing more than a small percentage of the total active (migratory) flux.

Estimates of diel migrants' contribution to export

Longhurst & Harrison (1989) suggested that 20–50% of zooplankton vertically migrate, and Roman et al. (1993) reported that at night the zooplankton biomass nearly doubled in the upper 160 m of the water column near Bermuda. The possibility of diel vertical migration contributing to the transmission of food down through the water column was first postulated by Vinogradov (1968), who described the concept as "a ladder of migrations". Evidence for migrations contributing to the vertical fluxes of both organic and inorganic material has been reported subsequently in a number of publications (Angel, 1985, 1989; Longhurst & Harrison, 1988; Longhurst et al., 1990; Small & Ellis 1992; Dam et al., 1995; Steinberg et al., 2000). For gut fluxes to be substantial, feeding activity needs to be concentrated during the shallow phase of the migrations and gut clearance rates need to be long enough for the downward migrations to have been completed before evacuation occurs. This is well documented in indices of gut fullness in fishes (Merrett & Roe, 1974; Baird et al., 1975; Roe & Badcock, 1984; Gorelova & Kobylyansky, 1985; Hopkins et al., 1996).

Comparisons of the gut fullness of micronektonic and macroplanktonic species caught during the upward phase of migrations at dusk, and the downward phase at dawn (in Spring 1990; see Savidge et al., 1992), confirmed that, in most, their guts were full (Howell pers.com.). Similar observations of periodic feeding in mesoplankton have been published (e.g. Atkinson et al., 1996). However, reported gut clearance rates in copepods seem to be too rapid for them to be effective transporters of faecal material, even so they will still be contributing to active fluxes of carbon released either as respiratory carbon dioxide or as DOC – dissolved organic carbon (Dagg, 1976; Small et al., 1983). In *Calanus pacificus*, 17–19% of ingested carbon is released as DOC (Copping & Lorenzen, 1980). The migrants will also be exporting nitrogen from the euphotic zone, which is taken up during feeding near surface at night and then released by day at depth mostly as inorganic nitrogen (ammonia) but also as dissolved organic nitrogen (urea and amino acids) (e.g. Ikeda & Skoldal, 1989), and this probably makes a small, but still significant, contribution to the nutrient fluxes (Longhurst & Harrison, 1988; Longhurst et al., 1989; Steinberg, et al., 2000).

It is worth attempting a tentative carbon export budget based on some of these estimates. In the Gulf of Mexico, Hopkins et al. (1996) estimated that the midwater standing stock of fishes of 133 kgC km^{-2} (i.e. 133 mgC m^{-2}) produces 0.6–0.9 kgC km^{-2} d^{-1} of faeces. This represents a daily carbon transfer into faeces of 0.45–0.7% of the standing crop, which represents an annual transfer of about 150–250% of the fish standing crop. Since our data imply that 50% of the fish biomass is migrating, the fish gut flux is 75–125% of the mean standing crop of fish. Assuming that the daily ration of the fishes is 5% of the standing stock and that, like copepods, 20% of the carbon ingested is released as DOC. Then, a mean standing crop of fish of 100 mgC m^{-2} produces an annual flux of DOC of 365 mgC m^{-2}, half of which will be released during the day at greater depths (i.e. 180%). Estimating mortality fluxes is even more problematical. But, if it is assumed that fish have annual life cycles, then the mean annual fish production will approximate the standing crop. So if half of the mortality occurs at the deeper daytime depths, then the annual mortality flux would be of the order of 50%. Thus, the total export flux will be equivalent to 100% (gut) plus 180% (DOC) and 50% (mortality) (i.e. of the order of 300% of the average standing crop). If other taxa contribute commensurately, then the export flux of migrating micronekton will be of the order of 300% of the total annual mean migrating population which, based on these samples, will be ~300–450 mgC m^{-2} at latitudes >40° N and 150 mgC m^{-2} at latitudes <40° N; this is comparable to the annual input of carbon of 70–390 mgC m^{-2} estimated to be have reached an abyssal benthic community in the form of phytodetritus at 47° N, 20° W (Pfannkuche, 1993). While these estimates are highly speculative, and overlook seasonal variations in migration activity, they do serve to highlight the potential importance of these active fluxes, which so far with a few notable exceptions, has been ignored by those developing carbon and nutrient budgets for the deep ocean.

References

Allison, S. K. & K. F. Wishner, 1986. Spatial and temporal patterns of zooplankton biomass across the Gulf Stream. Mar. Ecol. Prog. Ser. 31: 233–244.

Angel, M. V., 1985. Vertical migrations in the oceanic realm: possible causes and probable effects. In Rankin, M. A. (ed.), Migration, Mechanisms and Adaptive Significance. Contributions in Marine Science, Texas, 27: supplement: 45–70.

Angel, M. V., 1989a. Does mesopelagic biology affect the vertical flux? In Berger, W. H., V. S. Smetacek & G. Wefer (eds), Productivity of the Oceans: Past and Present. Proceedings of Dahlem Workshop on...). John Wiley: 155–173.

Angel, M. V., 1989b. Vertical profiles of pelagic communities in the vicinity of the Azores Front and their implications to deep ocean ecology. Prog. Oceanogr. 22: 1–46.

Angel, M. V., 1990. Life in the benthic boundary layer: connections to the mid-water and sea floor. Phil. Trans. r. Soc., Lond. A 331: 15–28.

Angel, M. V. & P. M. Hargreaves, 1992. Large scale patterns in the distributions of planktonic and micronektonic biomass in the N.E.Atlantic. ICES J. mar. Sci. 49: 403–411.

Angel, M. V., P. Hargreaves, P. Kirkpatrick & P. Domanski, 1982. Low variability in planktonic and micronektonic populations at 1000 m depth in the vicinity of 42° N 17° W; Evidence against diel vertical migratory behaviour in the majority of species. Biol. Oceanogr. 1: 287–319.

Atkinson, A., P. Ward & E. J. Murphy, 1996. Diel periodicity of subantarctic copepods: relationships between vertical migration, gut fullness and gut evacuation rate. J. Plankton Res. 18: 1387–1406.

Baird, R. C., T. L. Hopkins & D. F. Wilson, 1975. Diet and feeding chronology of *Diaphus taaningi* (Myctophidae) in the Cariaco Trench. Copeia 1975 (2): 356–365.

Bishop, J. K. B., R. W. Collier, D. R. Ketten & J. M. Edmond, 1980. The chemistry, biology and vertical flux of particulate matter from the upper 1500 m of the Panama Basin. Deep-Sea Res. 27: 615–640.

Campbell, J. W. & T. Aarup, 1992. New production in the North Atlantic derived from seasonal patterns of surface chlorophyll. Deep-Sea Res. 39: 1669–1694.

Copping, A. E. & C. J. Lorenzen, 1980. Carbon budget of a marine phytoplankton-herbivore system with carbon-14 as a tracer. Limnol. Oceanogr. 25: 873–882.

Dagg, M. J., 1976. Complete carbon and nitrogen budgets for the carnivorous amphipod, *Calliopius laeviusculus* (Kröyer). Int. Rev. ges. Hydrobiol. 61: 297–357.

Dam, H. G., M. R. Roman & M. J. Youngbluth, 1995. Downward export of respiratory carbon and dissolved inorganic nitrogen by diel-migrant mesozooplankton at the JGOFS Bermuda time-series station. Deep-Sea Res. 42: 1187–1197.

Domanski, P. A., 1986. The Azores Front: A zoogeographical boundary? UNESCO Tech. Pap. Mar. Sci. 49: 73–83.

Frost, B. W., 1988. Variability and possible adaptive significance of diel migration in *Calanus pacificus*, a planktonic marine copepod. Bull. mar. Sci. 43: 675–694.

Gorelova, T. A. & S. G. Kobylyansky, 1985. Feeding of deep-sea fishes of the Family Bathylagidae. J. Ichthyol. 32: 264–274.

Gould, W. J., 1985. The physical oceanography of the Azores Front. Prog. Oceanogr. 14: 167–190.

Fasham, M. J. R., T. Platt, B. Irwin and K. Jones, 1985. Factors affecting the spatial pattern of the deep chlorophyll maximum in the region of the Azores Front. Progr. Oceanogr. 14: 129–165.

Foxton, P. & H. S. J. Roe, 1974. Observations on the nocturnal feeding of some mesopelagic decapod Crustacea. Mar. Biol. 28: 37–49.

Harbison, G. R, L. P. Madin & N. R. Swanberg, 1978. On the natural history and distribution of oceanic ctenophores. Deep-Sea Res. 25: 233–256.

Harding, G. C., W. P. Vass, B. T. Hargrave & S. Pearre Jr., 1986. Diel vertical movements and feeding activity of zooplankton in

St George's Bay N.S., using net tows and a newly developed passive trap. Can. J. Fish. aquat. Sci. 43: 952–967.

Hargreaves, P. M., C. J. Ellis & M. V. Angel, 1984. An assessment of biological processes close to the sea bed in a slope region and its significance to the assessment of sea bed disposal of radioactive waste. IOS Report 185: 121 pp.

Hays, G. C., A. J. Warner & D. Lefevre, 1996. Long term changes in the diel vertical migration behaviour of zooplankton. Mar. Ecol. Prog. Ser. 141: 149–159.

Heron, A. C., 1972. Population ecology of a colonizing species: the pelagic tunicata Thalia democratica. 1. Individual growth rate and generation time. Oecologia 10: 269–293.

Hopkins, T. L., T. T. Sutton & T. M. Lancraft, 1996. The trophic structure and predation impact of a low latitude midwater fish assemblage. Prog. Oceanogr. 38: 95–153.

Hunt. H. G., 1968. Continuous plankton records: Contribution towards a plankton atlas of the North Atlantic and North Sea. Part XI: The seasonal and annual distributions of Tahliacea. Bull. mar. Ecol. 6: 225–249.

Ikeda, T. & H. R. Skjodal, 1989. Metabolism and elemental composition of zooplankton from the Barents Sea during early Arctic summer. Mar. Biol. 100: 173–183.

Irwin, B., T. Platt, P. Lindley, M. J. R. Fasham & K. Jones, 1983. Phytoplankton productivity in the vicinity of a front, SW of the Azores during May 1981. Can. Data Rep. Fish. aquat. Sci. 400: 1–100.

Iwasa, Y., 1982. Vertical migration of zooplankton: a game between predator and prey. Am. Nat. 20: 171–180.

Lampert, W., 1989. The adaptive significance of diel vertical migration of zooplankton. Funct. Ecol. 3, 21–27.

Lampitt, R. S., W. R. Hillier & P. G Challenor, 1993a. Seasonal and diel variation in the open ocean concentration of marine snow aggregates. Nature, London, 362: 737–739.

Lampitt, R. S., K. F. Wishner, C. M. Turley & M. V. Angel, 1993b. Marine snow studies in the Northeast Atlantic Ocean: distribution composition and role as a food source for migrating plankton. Mar. Biol. 116: 689–702.

Larson, R. J., 1986. Water content, organic content and carbon and nitrogen composition of medusae from the Northeast Pacific. J. exp mar. Biol. Ecol. 99: 107–120.

Lee, C., S. G. Wakeham & J. I. Hedges, 1988. The measurement of oceanic particle flux – are 'swimmers' a problem? Oceanography, November: 34–36.

Longhurst, A. R., 1976. Vertical migration. In Cushing, D. H. & J. J. Walsh (eds), The Ecology of the Seas. W. B. Saunders: 116–137.

Longhurst, A. R., 1995. Seasonal cycles of pelagic production and consumption. Progr. Oceanogr. 36: 77–168.

Longhurst, A. R. & W. G. Harrison, 1988. Vertical nitrogen flux from the oceanic photic zone by diel migrant zooplankton and nekton. Deep-Sea Res. 35: 881–889.

Longhurst, A. R. & W. G. Harrison, 1989. The biological pump: Profiles of plankton production and consumption in the upper ocean. Prog. Oceanogr. 22: 47–123.

Longhurst, A. R., A. W. Bedo, W. G. Harrison, E. J. H. Head, E. P. Irwin & C. Morales, 1989. NFLUX: a test of vertical migration flux by diel migrant biota. Deep-Sea Res. 36: 1705–1719.

Longhurst, A. R., A. W. Bedo, W. G. Harrison, E. J. H. Head & D. D. Sameoto, 1990. Vertical flux of respiratory carbon by diel migrant biota. Deep-Sea Res. 37: 685–694.

Merrett, N. R. & H. S. J. Roe, 1974. Patterns and selectivity in the feeding of certain mesopelagic fishes. Mar. Biol. 28: 115–126.

Oceanographic Laboratory, Edinburgh, 1973. Continuous plankton records; a plankton atlas of the North Atlantic and the North Sea. Bull. mar. Ecol. 7: 1–174.

Pearre, S., 1979. Problems of detection and interpretation of vertical migration. J. Plankton Res. 1: 29–44.

Pfannkuche, O., 1993. Benthic response to the sedimentation of particulate organic matter at the BIOTRANS station, 47° N, 20° W. Deep-Sea Res. II 40: 135–149.

Pugh, P. R., 1989. Gelatinous zooplankton – the forgotten fauna. Prog. Underwater Sci. 14: 67–78.

Pugh, P. R., 1991. Co-occurrence of hippopodid siphonophores and their potential prey. Hydrobiologia 216/217: 327–334.

Pugh, P. R., F. Pages & B. Boorman, 1997. Vertical distribution and abundance of pelagic cnidarians in the eastern Weddell Sea, Antarctica. J. mar. biol. Ass. U.K. 77: 341–360.

Ringelberg, J., 1964. The positively phototactic reaction of *Daphnia magna* Straus: a contribution to the understanding of diurnal vertical migration. Neth. J. Sea Res. 2: 319–406.

Ringelberg, J., 1991. Enhancement of the phototactic reaction in *Daphnia hyalina* by a chemical mediated by juvenile perch (*Perca fluviatilis*). J. Plankton Res. 13: 17–25.

Roe, H. S. J. & J. Badcock, 1984. The diel migrations and distributions within a mesopelagic community in the northeast Atlantic. 5. Vertical migrations and feeding of fish. Progr. Oceanogr. 13: 389–424.

Roe, H. S. J. & D. M. Shale, 1979. A new multiple rectangular trawl (RMT 1+8M) and some modifications to the Institute of Oceanographic Sciences' RMT 1+8. Mar. Biol. 50: 283–288.

Roe, H. S. J., A. De C. Baker, R. M. Carson, R. Wild & D. M. Shale, 1980. Behaviour of the Institute of Oceanographic Sciences' Rectangular Midwater Trawls: theoretical aspects and experimental observations. Mar. Biol. 56: 247–259.

Roe, H. S. J., D. S. M. Billett & R. S. Lampitt, 1990. Benthic/midwater interactions on the Madeira Abyssal Plain; evidence for biological transport pathways. Progr. Oceanogr. 24: 127–140.

Roe, H. S. J., M. V. Angel, J. Badcock, P. Domanski, P. T. James, P. R. Pugh & M. T. Thurston, 1984. The diurnal migrations and distributions of a mesopelagic community in the North-east Atlantic. Progr. Oceanogr. 13: 245–511.

Roman, M. R., H. G. Dam, A. L. Gauzens & J. M. Napp, 1993. Zooplankton biomass and grazing at the JGOFS Sargasso Sea time series station. Deep-Sea Res. 1: 40: 883–901.

Savidge, G., D. R. Turner, P. H. Burkill, A. J. Watson, M. V. Angel, R. D. Pingree, H. Leech & K. J. Richards, 1992. The BOFS 1990 Spring Bloom Experiment: Temporal evolution and spatial variability of the hydrographic field. Prog. Oceanogr. 29: 235–281.

Small, L. F. & S. G. Ellis, 1992. Fecal carbon production by copepods in the Santa Monica Basin; the effects of body size and carnivorous feeding. Prog. Oceanogr. 30: 197–221.

Small, L. F., S. W. Fowler, S. A. Moore & J. Larosa, 1983. Dissolved and fecal pellet carbon and nitrogen release by zooplankton in tropical waters. Deep-Sea Res. 10: 1199–1220.

Steinberg, D. K. , C. A. Carlson, N. R. Bates, S. A. Goldthwait, L. P. Madin & A. F. Michaels, 2000. Zooplankton vertical migration and the active transport of dissolved organic and inorganic carbon in the Sargasso Sea. Deep-Sea Res. I, 47: 137–158.

Vane, F. R. & J. M. Colebrook, 1962. Continuous Plankton Records: contribution towards a plankton atlas of the North-east Atlantic and the North Sea. Part vi: The seasonal and annual distributions of the Gastropoda. Bull. Mar. Ecol. 5: 247–253.

Vinogradov, M. E., 1968. Vertical Distribution of the Oceanic Zooplankton. Nauka, Moscow: 339 pp (translation: Israel Program for Scientific Translation, Jerusalem, 1970).

Vuorinen, I., 1987. Vertical migration of *Eurytemora* (Crustacea, Copepoda), a compromise between the risk of predation and decreased fecundity. J. Plankton Res. 9: 1037–1046.

Walsh, J., J. Dymond & R. Collier, 1988. Rates of recycling of biogenic components of settling particles in the ocean derived from sediment trap experiments. Deep-Sea Res. 35: 43–58.

Wiebe P. H., S. Boyd. & J. L. Cox, 1975. Relationships between zooplankton displacement volume, wet weight, dry weight and carbon. Fish. Bull. 73: 777–786

Wiebe, P. H., 1988. Functional regression equations for zooplankton displacement volumes, wet weight, dry weight, and carbon: a correction. Fish. Bull. 86: 833–835.

Wiebe, P. H., L. P. Madin, L. R. Haury, G. R. Harbison & L. M. Philbin, 1978. Diel vertical migration by *Salpa aspera*: potential for large-scale particulate organic matter transport to the deep-sea. Mar. Biol. 53: 249–256.

Wishner, K. F., 1980. The biomass of deep-sea benthopelagic plankton. Deep-Sea Res. 27: 203–216.

Zaret, T. M. & J. S. Suffern, 1976. Vertical migration in zooplankton as a predator avoidance mechanism. Limnol. Oceanogr. 21: 804–813.

Hydrobiologia **440:** 181–189, 2000.
M.B. Jones, J.M.N. Azevedo, A.I. Neto, A.C. Costa & A.M. Frias Martins (eds), Island, Ocean and Deep-Sea Biology.

Colonization of floating seaweed by pelagic and subtidal benthic animals in southwestern Iceland

Agnar Ingólfsson
Institute of Biology, University of Iceland, Grensasvegur 12, IS-108 Reykjavík, Iceland
E-mail: agnaring@hi.is

Key words: floating seaweed, colonizers, macrofauna, life histories, Iceland

Abstract

Coastal floating seaweed, originating from the intertidal, is colonized by various pelagic and subtidal benthic animals. At southwestern Iceland, common macrofaunal colonizers include the fish *Cyclopterus lumpus* L., the harpacticoid *Parathalestris croni* (Krøyer), the amphipods *Calliopius laeviusculus* (Krøyer), *Gammarellus angulosus* (Rathke), *Dexamine th*ea Boeck and *Ischyrocerus anguipes* (Krøyer), and the isopod *Idotea baltica* Pallas. The colonization patterns of these species were studied by collecting samples from experimental and natural clumps of floating seaweed at approximately monthly intervals for ca. 18 months. The occurrence of colonizers was highly seasonal with maximum numbers between April and September. For most species, numbers were correlated with weight of clumps. The relationship of density (numbers/100 g algae) to weight of clump and distance from shore varied between species. The algal species composition of clumps appeared to influence some colonizers. Only juvenile *C. lumpus* were found, while all *P. croni* were sexually mature; juveniles predominated in other species. The different colonization patterns of the species can be explained partly by their biological traits. The clumps are vital for the breeding of *P. croni* and are used for feeding by *C. lumpus* fry. Other species possibly colonize floating algae accidentally, being programmed to seek attached vegetation. Floating clumps may, nevertheless, serve to disperse these species.

Introduction

Seaweed detached from the intertidal or shallow subtidal and floating at the surface of the sea is inhabited by a number of animal species. The latter includes animals originally living on the seaweed *in situ* and a number of species, both fish and invertebrates, that quickly colonize the floating clumps (e.g. Druce & Kingsford, 1995; Kingsford, 1995; Ingólfsson, 1995, 1998). For some of the colonizing species, the floating seaweed is clearly essential (Ingólfsson & Ólafsson, 1997), while, in most cases, the adaptive significance of seeking floating seaweed is not readily apparent. The aim of this study was to obtain seasonal and life-history data of the floating seaweed colonized to help evaluate the significance of this colonizing behaviour.

Data are presented on the following macrofaunal species known to be common in floating seaweed off the coasts of Iceland (Ingólfsson, 1995, 1998; Ingólfsson & Ólafsson 1997): the fish *Cyclopterus lumpus* L., the harpacticoid *Parathalestris croni* (Krøyer), the amphipods *Calliopius laeviusculus* (Krøyer), *Gammarellus angulosus* (Rathke), *Dexamine th*ea Boeck, and *Ischyrocerus anguipes* (Krøyer), and the isopod *Idotea baltica* Pallas.

Materials and methods

At intervals of 3–7 weeks from June 1995 to January 1997, floating seaweed clumps were collected in the Bay of Faxaflói, southwestern Iceland. On each of the 16 sampling occasions, a fixed transect (Fig. 1) was sailed by the rescue vessel, *Henry Hálfdánarson*, as described by Ingólfsson (1998). The transects went from the shore to a point approximately 31 km from nearest point on land. Seaweed clumps estimated to be larger than about 10 g wet weight were scooped up

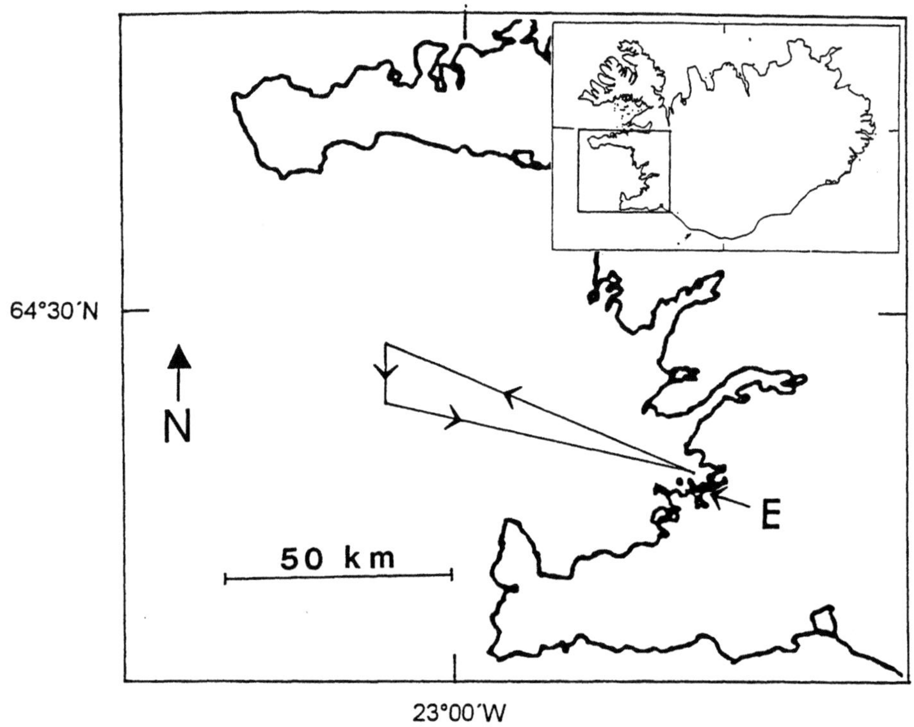

Figure 1. Approximate transect route in the Bay of Faxaflói, southwestern Iceland and the site of the Eidsvik experiments (E).

with a plankton net (diameter 44 cm, mesh size 0.1 mm); the net also caught planktonic animals in the immediate vicinity of the clumps. The clumps were placed in neutral 5% formaldehyde. An attempt was made to collect at least 35 clumps on each trip, but scarcity of clumps sometimes prevented this. A total of 372 clumps, ranging in wet weight from 6 to 6140 g (mean 354 g), was collected.

On 18 occasions, at approximately monthly intervals from May 1995 to November 1996, branches of the brown seaweed *Ascophyllum nodosum*, weighing 200–500 g (wet weight), were cut from the upper part of the shore at low tide at Bygggardar, near Reykjavik, southwestern Iceland. Each time, 16 branches were placed in plastic bags without any treatment and transported to Eidsvik (Figure 1) where they were carefully floated and tethered two to a buoy with 4 and 8 m nylon string, respectively. Control samples lacked the seven animal species under study here when floated, although numerous intertidal animals were present. The buoys were between 300 and 600 m from the shore at a depth of 7–14 m below mean level of low spring tides. Some kelp (*Laminaria* spp.) grew on the seabed. The branches were retrieved with a plankton net after approximately 10 days. The total number of branches retrieved was 261.

In the laboratory, the algae were washed over a 0.5 mm sieve and wet weighed. The animals retained by the sieve were placed in 70% ethanol, identified to species (if possible), and counted. Lumpsucker fry were measured (total length) to the nearest mm, while the peracarids were measured to the nearest 0.01 mm. Copepods (*Parathalestris croni*) were sexed by the form of the first antenna and the presence/absence of an egg sac was noted. Attempts were made to sex all the peracarids, the unsexable animals being regarded as juveniles. Males were recognized by their penises, while the presence of oostegites was the chief character used in defining females. The developmental stage of the oostegites and the presence of eggs or young in the brood pouch was noted.

Pearson correlations were used to relate numbers and densities of animals to weight of algal clumps or their constituents, and their distance from nearest shore, using Bonferroni probabilities. Calculations used SYSTAT® 7.0 software, which was also employed in drawing the figures.

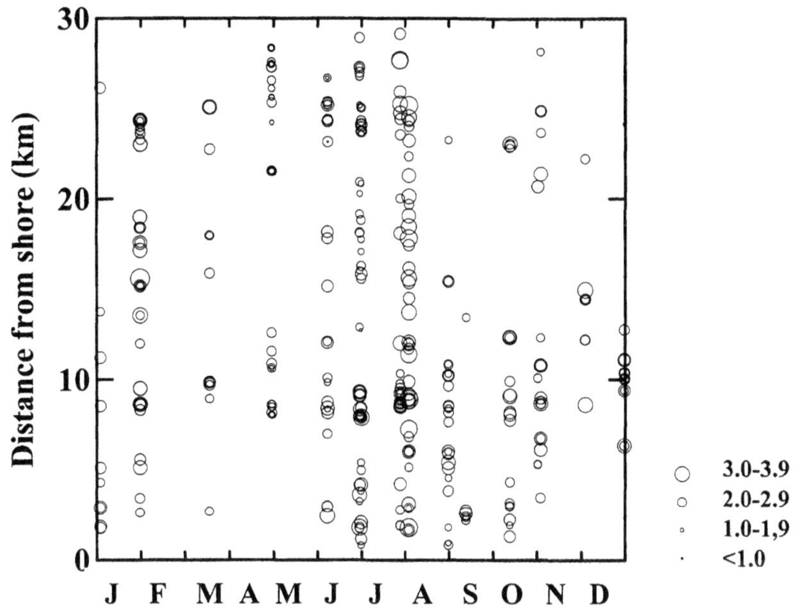

Figure 2. The location (distance from shore) and weight (g) of clumps collected on 16 occasions on the Faxaflói transect from June 1995 to January 1997. The size of the symbols reflect the log-transformed weights of the clumps.

Results

Floating seaweed were obtained on the Faxaflói transect on each sampling trip, but the clumps were scarce on some dates, especially in autumn and winter. Thus, less than 20 clumps were collected on the following dates (number of clumps in parenthesis): 30 October 1995 (18), 29 November 1995 (8), 28 December 1995 (16), 10 September 1996 (7), 28 October 1996 (6) and 3 January 1997 (10). However, on some dates in autumn and winter, clumps were common. The weight of clumps did not differ markedly with season, while the distribution of clumps along the transect was quite variable (Fig. 2).

The occurrence of all colonizing species, both from the Faxaflói survey and from the Eidsvik experiment, was highly seasonal. Maximum densities were recorded during summer and animals were virtually absent during the middle of the winter (Figs 3 and 4). In the Faxaflói survey, the sharpest seasonal peaks were shown by *Calliopius laeviusculus* (August) and *Parathalestris croni* (early May–late June) (Fig. 3), while the peaks were less distinct for *Cyclopterus lumpus, Gammarellus angulosus* and, especially, *Idotea baltica* which occurred commonly but rather irregularly from April to November. In the Eidsvik experiment, the peaks were clear for *Dexamine thea* (June–August), *C. laeviusculus* (May–September) and *Ischyrocerus anguipes* (June–November), while less distinct, as in the Faxaflói survey, for *C. lumpus* and *G. angulosus* (Fig. 4).

In summer (April–September), *Calliopius laeviusculus* was the most abundant species in Eidsvik, followed by *Dexamine thea* and *Ischyrocerus anguipes*, while *Idotea baltica* and *Parathalestris croni* were rare (Table 1). In the Faxaflói survey, *P. croni* was present in the highest densities, followed by *Cyclopterus lumpus* and *Gammarellus angulosus. Dexamine thea* was rare and few individuals of *I. anguipes* were recorded (Table 1). The animals were almost always highly overdispersed amongst clumps (calculated on numbers/100 g algae) with the following exceptions: *G. angulosus* at Eidsvik did not show a significant deviation from random dispersion, while the dispersion of *C. lumpus* at Eidsvik deviated significantly from random towards regular (s^2/x = 0.51, t=−4.31, N= 156, P<0.001).

The experimental *Ascophyllum nodosum* fronds used in Eidsvik had a weight range of 200–500 g, and there was no correlation between numbers of individuals of the species considered here and weight of fronds. The *A. nodosum* fronds often caught pieces of other floating algae while tethered and the numbers of *G. angulosus*, but not those of other species, correlated with the weight of these adventitious algae (r = 0.35, N = 156, P <0.001). The main constituents of the

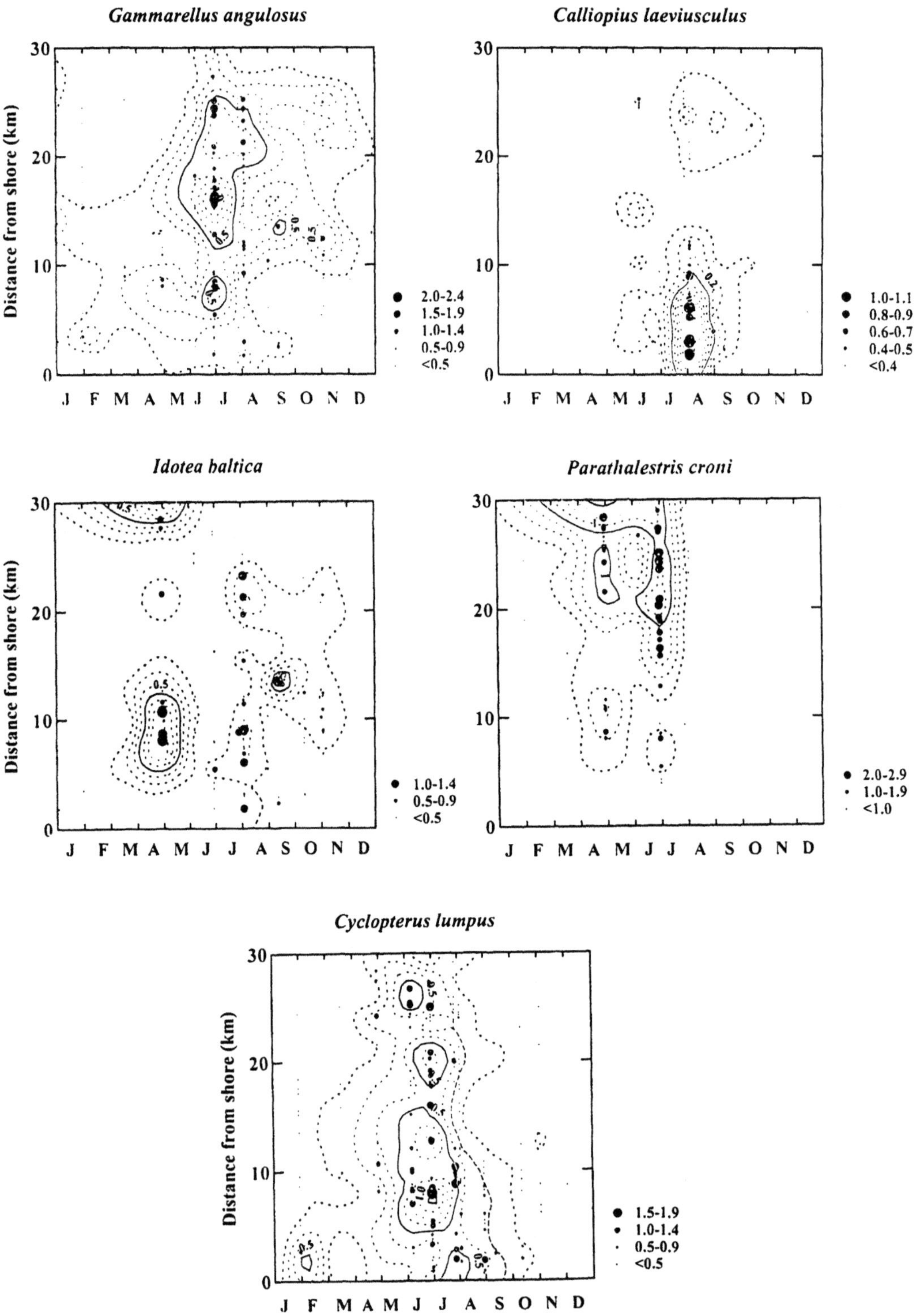

Figure 3. Densities of common macrofaunal colonizers of floating seaweed clumps on the Faxaflói transect, according to time of year and distance from nearest shore. The analyses are based on log-transformed data and sizes of symbols indicate densities as log(1+number/100 g algae). Contours are drawn by SYSTAT®, using default settings. Note that zero values are not shown.

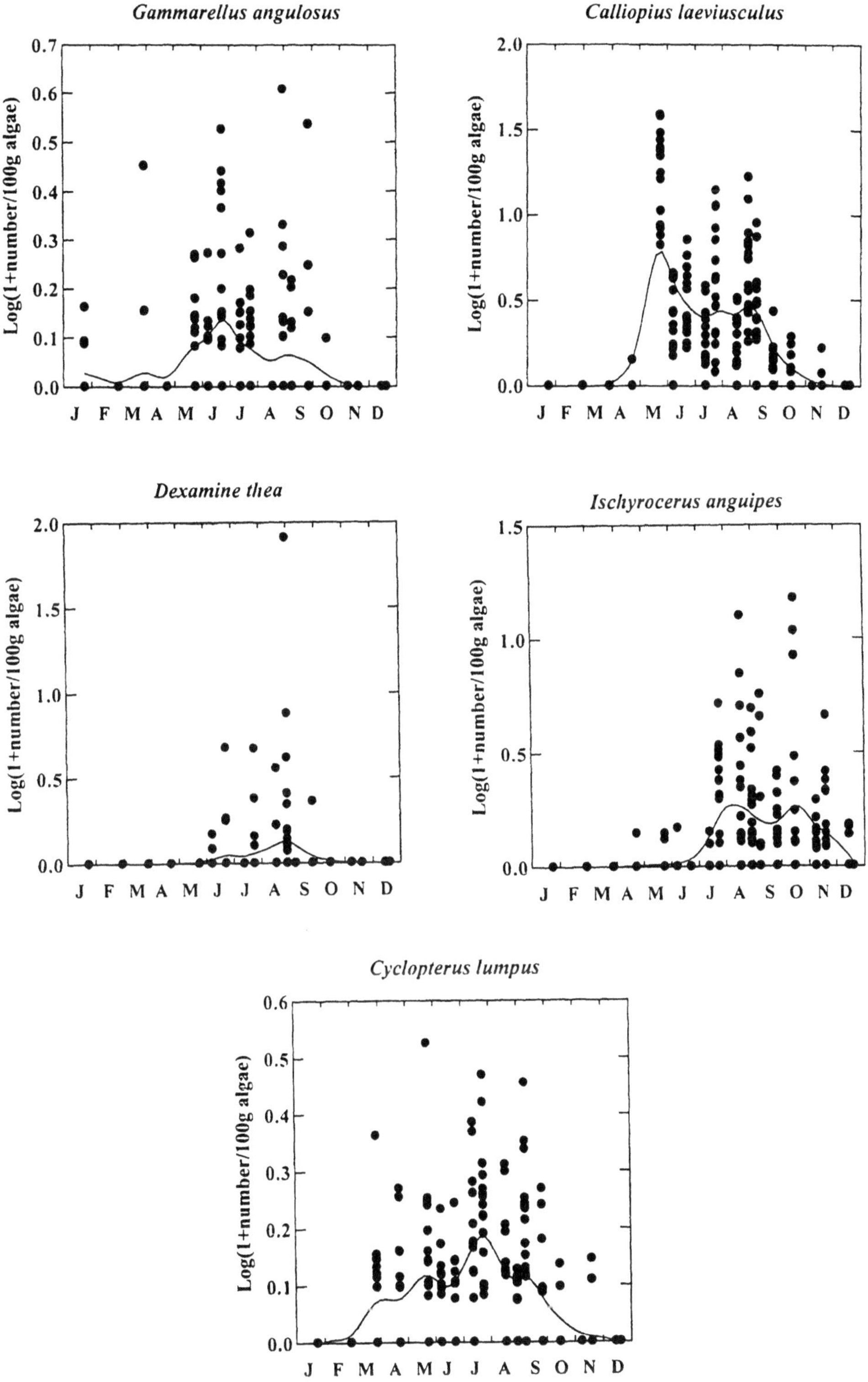

Figure 4. Log-transformed densities of common macrofaunal colonizers of *Ascophyllum nodosum* experimental fronds at Eidsvik, according to time of year. Fronds were kept afloat for approximately 10 days tethered to buoys. The lines are fitted by distance weighted least squares by SYSTAT®(tension=0.7).

Table 1. The average density (numbers pre 100 g algae, wet weight) with standard errors during April–September of some colonizers of floating experimental fronds at Eidsvik and of natural clumps collected on the Faxaflói transect.

	Eidsvík (N=156)	Faxaflói (N=235)
Parathalestris croni	0.01±0.01	9.38±1.68
Calliopius laeviusculus	3.59±0.53	0.49±0.11
Gammarellus angulosus	0.24±0.04	3.32±0.65
Dexamine thea	0.71±0.52	0.12±0.09
Ischyrocerus anguipes	0.58±0.11	0.01±0.00
Idotea baltica	0.04±0.01	1.12±0.21
Cyclopterus lumpus	0.37±0.04	3.47±0.47

372 clumps collected in Faxaflói were: *Ascophyllum nodosum* (54% of total weight of algae in clumps), *Fucus vesiculosus* (24%), *Polysiphonia lanosa* (an epiphyte on *A. nodosum*, 3%), *Fucus distichus* (3%) and *Fucus spiralis* (1.5%). Clumps composed of two or more separate fronds and/or algal species were common. There was no correlation between the weight of clumps or their main individual algal constituents and distance from nearest shore.

During summer (April–September), all common species in the Faxaflói samples showed a highly significant positive correlation between numbers and weight of clumps, with the exception of *Parathalestris croni* (Table 2). When numbers were correlated with the weight of individual algal constituents of clumps in partial correlations, keeping total weight of clumps constant, it was found that *Calliopius laeviusculum* correlated positively with weight of *Ascophyllum nodosum* (r=0.43, N=235, P<0.01), while *Cyclopterus lumpus*, *Gammarellus angulosus* and *Dexamine thea* correlated positively with *Polysiphonia lanosa* [(r=0.24 (P<0.05), 0.55 (P<0.01), 0.28 (P<0.05), respectively], in spite of this alga being a relatively minor constituents of clumps. The density (number/100 g algae) of *C. laeviusculus* was positively correlated with weight of clumps, while this correlation was significantly negative for *Cyclopterus lumpus* and *P. croni*. The density of *C. laeviusculus* was highly negatively correlated with distance from the nearest shore, while *P. croni* showed a highly positive correlation (Table 2, Fig. 3). This is in accord with the difference in average densities of these two species in Eidsvik (samples <1 km to shore) and Faxaflói (average distance from shore >10 km). In Faxaflói, *C. lumpus* also showed a negative correlation with distance from shore in spite of its considerably greater density in Faxaflói than Eidsvik (Tables 1 and 2). Although *Idotea baltica* was virtually absent from Eidsvik, there was no correlation with distance from shore for this species in Faxaflói.

All *Parathalestris croni* obtained were adults, the great majority of which were females, often ovigerous (Table 3). The proportion of ovigerous females did not change markedly from April to July. Although very few animals were found in clumps in January – March, some of these were ovigerous females. All *Cyclopterus lumpus* obtained were juveniles, almost all <35 mm in total length and 1 year or less in age. Newly-hatched fry with the yolk sac still present were often abundant. The majority of individuals of the common peracarids were juveniles (Table 3), although sexually mature animals were found for all species. This included some ovigerous females in all species except *Gammarellus angulosus*. The highest proportion of sexually mature animals was found in *Ischyrocerus anguipes* and *Calliopius laeviusculus* in Eidsvik samples, while very few sexually mature *Dexamine thea* from Eidsvik and *C. laeviusculus* and *G. angulosus* from Faxaflói were recorded.

Although the proportion of mature *Idotea baltica* from Faxaflói was almost 20%, there were virtually no ovigerous females amongst them. The data are too few to allow detailed analysis of seasonal changes in proportions of juvenile and mature peracarids in samples, but juveniles were usually especially dominant in samples from late summer (July, sometimes June) to autumn (September–October).

Discussion

The results of the Faxaflói survey are in agreement with earlier studies that showed that *Parathalestris croni* seeks floating clumps for breeding purposes, a substratum being essential for the naupliar larvae (Ingólfsson & Ólafsson, 1997). This species is apparently planktonic for the greater part of the year and then shows no or little affinity for floating seaweed. It colonized mainly off-shore clumps and was very rare in clumps <5 km from shore, suggesting an off-shore planktonic habit. Weight of algal clumps had no effect on numbers of *P. croni* present for all species. Presumably, large clumps are more easily located by colonizers than small ones but, in the case of *P. croni*, other qualities of clumps, in addition to distance from

Table 2. Pearson correlations between number or density (number/100 g algae) of common colonizers and weight of clumps and distance from nearest shore, of clumps collected on the Faxaflói transect, April–September (N=235). Probabilities are Bonferroni.

	Number *vs.* weight	Density *vs.* weight	Density *vs.* distance
Parathalestris croni	NS	−−	+++
Calliopius laeviusculus	+++	++	−−−
Gammarellus angulosus	+++	NS	NS
Idotea baltica	+++	NS	NS
Cyclopterus lumpus	+++	−	−

Symbols:
NS = not significant; + or − = P<0.05 (positive or negative); ++ or −− = P<0.01; + + + or − − − = P<0.001.

Table 3. Life-cycle stages of six crustaceans retrieved from experimental fronds in Eidsvik and from algal clumps collected on the Faxaflói transect.

Species	Stage	Eidsvik		Faxaflói	
		Nos.	%	Nos.	%
Parathalestris	Juveniles			0	0
croni	Females total			1737	98.7
	Ovigerous females			709	40.3
	Males			22	1.3
Calliopius	Juveniles	970	76.1	451	96.6
laeviusculus	Females total	152	11.9	10	2.1
	Ovigerous females	34	2.7	2	0.4
	Males	152	11.9	6	1.3
Gammarellus	Juveniles	115	82.7	1746	99.9
angulosus	Females total	8	5.8	0	0
	Ovigerous females	0	0	0	0
	Males	16	11.5	1	0.1
Dexamine	Juveniles	153	94.4	78	85.7
thea	Females total	5	3.1	10	11.0
	Ovigerous females	2	1.2	2	2.2
	Males	4	2.5	3	3.3
Ischyrocerus	Juveniles	314	68.9		
anguipes	Females total	44	9.6		
	Ovigerous females	16	3.5		
	Males	98	21.5		
Idotea	Juveniles			915	81.8
baltica	Females total			107	9.6
	Ovigerous females			2	0.2
	Males			97	8.7

shore (uncorrelated with weight of clumps), might be of greater importance. Previous studies suggested that the species showed special affinities to the alga *Ascophyllum nodosum* and its epiphyte *Polysiphonia lanosa* (Ingólfsson & Ólafsson, 1997). The present, more extensive data set, however, does not support this conclusion, although both algae correlated positively with numbers of other colonizers.

All other colonizing species considered here colonizing Faxaflói clumps showed a positive correlation between numbers and weight of clumps. The negative correlation between the density of *Cyclopterus lumpus* and clump weight in Faxaflói was, however, unexpected. These results, taken in conjunction with the underdispersion of fry in Eidsvik fronds, might indicate some form of territorialism (i.e. fry excluding other individuals from their immediate vicinity). One possibility is that this might result from cannibalism, common in floating algae. The fry also feed on a variety of other animals in the clumps, and it seems likely that floating algae are an important habitat for the species during the first year, while older fry, although pelagic, are not found associated with floating seaweed (A. Ingólfsson & B. K. Kristjánsson, pers. obs.). Although the fry can also be found among attached algae during their first year, the rapid colonization of floating fronds from the time of hatching (Ingólfsson, 1998) must result from a widespread habit of swimming from the bottom towards the surface, presumably an adaptive trait.

The five peracarid crustaceans considered here are usually regarded as subtidal, benthic and mostly phytal in habit, occasionally extending into the low intertidal (Naylor, 1955; Lincoln, 1979). In late summer, *Calliopius laeviusculus* may swarm "at the surface and may then be found in the plankton far out to sea" (Steele & Steele, 1973), although typically found clinging to the algae at the edge of water. Locke & Corey (1989) cite other examples of planktonic records of the species which they found to be strongly associated with floating seaweed in the Bay of Fundy, Canada. In Faxaflói, the species has been caught occasionally in plankton samples, but there was no evidence of swarming (Ingólfsson, 1998). Animals colonizing seaweed are mostly juveniles but sexually mature animals, including ovigerous females, are also involved. Clearly this behaviour is highly seasonal, at least in Iceland, being confined to the summer months, but apparantly not directly linked to reproduction. *Calliopius laeviusculus* is unique among the species studied here in showing a positive relationship between density and clump weight, similar to that seen in many intertidal species found in floating algae (Ingólfsson, 1998). This could result from greater attractiveness of larger clumps in comparison with smaller ones, while the explanation suggested for intertidal animals, i.e. a decrease in probability of dropping off with increasing clump size (Ingólfsson, 1998), seems difficult to apply to a good swimmer such as *C. laeviusculus*. *Gammarellus angulosus*, and/or the closely similar *G. homari*, has also been recorded in planktonic samples (Vader, 1975; Locke & Corey, 1989; Ingólfsson, 1998) and Tully & Ó Céidigh (1987) suggested that the species may be permanently pelagic. This is hardly supported by the available evidence (Steele & Steele, 1972; Vader, 1975). In Iceland, its occurrence in floating seaweed is less strictly seasonal than that of other amphipod species studied here, although there is a clear difference between summer and winter. Juveniles tend to be very dominant and there was no instance of any ovigerous female from floating seaweed. *Idotea baltica* is sometimes common in planktonic (neustonic) samples, especially juveniles (Tully & Ó Céidigh, 1986) and its association with floating seaweed is well known (Tully & Ó Céidigh, 1986; Locke & Corey, 1989; Davenport & Rees, 1993; Ingólfsson, 1995). The planktonic, free-swimming behaviour of *C. laeviusculus*, *Gammarellus* spp. and *I. baltica* might be a dispersal mechanism (cf. Tully & Ó Céidigh, 1986) and not necessarily directed towards drifting seaweed, although this seems likely. Another possibility is that the colonization of floating seaweed is an accidental consequence of the behaviour of seeking attached vegetation. Floating seaweed, when encountered, may, in any case, greatly enhance dispersal over distances far beyond the swimming abilities of the animals (cf. Locke & Corey, 1989).

Dexamine thea and *Ischyrocerus anguipes* have not to my knowledge been recorded elsewhere as planktonic or associated with floating seaweed. *Ischyrocerus anguipes* is a tube-building species (Lincoln, 1979), the males of which are sometimes known to swim around in search of mates (Tully & Ó Céidigh, 1987). It may be significant that this species showed the highest proportion of males of all amphipods considered, although juveniles still predominated.

Acknowledgements

I thank Bjarni Kr. Kristjánsson and Magnea Karlsdóttir who were responsible for carrying out the fieldwork of this study. Bjarni also worked a large number of samples in the laboratory, while the remaining laboratory work, including the ageing and sexing of the crustaceans, was mainly in the capable hands of Magnea with invaluable assistance from María Björk Steinarsdóttir. Jörundur Svavarsson read a draft of this manuscript and made numerous helpful suggestions. Two anonymous reviewers improved this paper substantially. This study was supported by grants from Lydveldissjodur and the University of Iceland Research Fund.

References

Davenport, J. & E. I. S. Rees, 1993. Observations on neuston and floating weed patches in the Irish Sea. Estuar. coast. mar. Sci. 36: 395–411.

Druce, B. E. & M. J. Kingsford, 1995. An experimental investigation on the fishes associated with drifting objects in coastal waters of temperate Australia. Bull. mar. Sci. 57: 378–389.

Ingólfsson, A., 1995. Floating clumps of seaweed around Iceland: natural microcosms and means of dispersal for fauna. Mar. Biol. 122: 13–21.

Ingólfsson, A., 1998. Dynamics of macrofaunal communities of floating seaweed clumps off western Iceland: a study of patches on the surface of the sea. J. exp. mar. Biol. Ecol. 231: 119–137.

Ingólfsson, A. & E. Ólafsson, 1997. Vital role of drift algae in the life history of the pelagic harpacticoid *Parathalestris croni* in the northern North Atlantic. J. Plankton Res. 19: 15–27.

Kingsford, M. J., 1995. Drift algae: a contribution to near-shore habitat complexity in the pelagic environment and an attractant for fish. Mar. Ecol. Prog. Ser. 116: 297–301.

Lincoln, R. J., 1979. British Marine Amphipoda: Gammaridea. British Museum (Natural History), London: 658 pp.

Locke, A. & S. Corey, 1989. Amphipods, isopods and surface currents: a case for passive dispersal in the Bay of Fundy, Canada. J. Plankton Res. 11: 419–430.

Naylor, E., 1955. The ecological distribution of British species of *Idotea* (Isopoda). J. anim. Ecol. 24: 255–269.

Steele, D. H. & V. J. Steele, 1972. Some aspects of the biology of *Calliopius laeviusculus* (Krøyer)(Crustacea, Amphipoda) in the northwestern Atlantic. Can. J. Zool. 51: 723–728.

Steele, D. H. & V. J. Steele, 1973. Biology of *Gammarellus angulosus* (Crustacea, Amphipoda) in the Northwestern Atlantic. J. Fish. Res. Bd Canada 29: 1337–1340.

Tully, O. & P. Ó Ceidigh, 1986. The ecology of *Idotea* species (Isopoda) and *Gammarus locusta* (Amphipoda) on surface driftweed in Galway Bay (west of Ireland). J. mar. biol. Ass. U. K. 66: 931–942.

Tully, O. & P. Ó Ceidigh, 1987. Investigations of the plankton of the west cost of Ireland.-VIII. The neustonic phase and vertical migratory behaviour of the benthic peracaridea in Galway Bay. Proc. R. Ir. Acad. 87B: 43–64.

Vader, W., 1975. Het vlokreeften-geslacht Gammarellus in Nederland. De Levende Natuur 78: 111–114.

Hydrobiologia **440:** 191–198, 2000.
M.B. Jones, J.M.N. Azevedo, A.I. Neto, A.C. Costa & A.M. Frias Martins (eds), Island, Ocean and Deep-Sea Biology.

Semper's (zoanthid) larvae: pelagic life, parentage and other problems

J. S. Ryland[1], Samantha de Putron[1,2], R. S. Scheltema[3], P. J. Chimonides[4] & D. G. Zhadan[5]
[1]*University of Wales Swansea, Swansea SA2 8PP, U.K.*
[2]*Bermuda Biological Station for Research, St Georges, Bermuda*
[3]*Woods Hole Oceanographic Institution, Woods Hole, MA 02543, U.S.A.*
[4]*Natural History Museum, London SW7 5BD, U.K.*
[5]*Department of Invertebrate Zoology, Moscow State University, Moscow 119899, Russia*

Key words: larval types, distributions, sea temperature, zooxanthellae, deep sea, *Epizoanthus*

Abstract

Semper's larvae were obtained from <300 out of ~1800 plankton tows taken in the world's oceans (1964–1993). Zoanthellae (larvae of Sphenopidae) occurred at 217 stations and zoanthinae (larvae of Zoanthidae) at 86, the two larval types showing distributions clearly delimited by a minimum sea temperature (~22 °C for zoanthellae, ~18 °C for zoanthinae; a statistically significant difference, $P<0.001$). Length of formalin-fixed zoanthellae was ~2–8.6 mm and of zoanthinae ~1.5–5.9 mm. Endodermal zooxanthellae were present in 9/24 zoanthinae but in no zoanthellae (of 19). Three larvae contained an endo-commensal/parasitic amphipod. Septa were externally visible in larger zoanthinae and were counted in transverse sections of other larvae, a majority of which (both kinds) had 12 septa, the normal maximum. The pattern was brachycnemic in 40/43 larvae and anomalous (but non-macrocnemic) in three. If macrocnemic genera reproduce by Semper's larvae, they should have been represented in such a large sample. The distribution of adult *Epizoanthus* was examined: many species are deep sea (recorded down to ~5000 m) but shallow-water species are relatively plentiful in, for example, the Adriatic and North Seas. No Semper's larva has ever been recorded from either. Some *Parazoanthus* species also occur in shallow water, especially associated with western Atlantic reef sponges. If they produce Semper's larvae, these have never been found. It is probable that macrocnemic zoanthids settle from planulae that do not develop into recognizable zoanthellae or zoanthinae.

Introduction

Semper's larvae are named from their discoverer (Semper, 1867), who identified them only as belonging to Anthozoa. He found two different kinds: the first was a 6 mm, almost cylindrical planula with a long, midventral ciliary band extending from the blastopore much of the distance to the anterior pole; the second was smaller, ovoid, with an equatorial ciliary band. Semper (1867) suggested that the second larva might be an earlier stage of the first. His 'first' larva was obtained from the Moçambique Current, off the Cape of Good Hope (42° S) and off the south coast of Java. Two further specimens were obtained south of the Cape Verde islands by the German Plankton Expedition, one of which was fully described by van Beneden (1890), who later (1897) named it *Zoanthella*. Van Beneden (1897) also described a specimen of Semper's 'second' larva, taken from the Guinea Current, naming it *Zoanthina*, and similar larvae, from off Beaufort, North Carolina, were examined by McMurrich (1891). Both authors (Van Beneden 1890, 1897; McMurrich, 1891) recognized Semper's larvae as zoanthids. Van Beneden's terms have remained in use as the names of larval types but not of genera.

Zoanthids (order Zoanthidea) are distinguished and classified according to the arrangement of the internal septa (Fig. 1). Appreciation of this is necessary to follow the arguments in this paper. In all Anthozoa, septa which reach the actinopharynx or (below it) bear a thickened 'filament' are termed 'perfect'; septa which neither reach the actinopharynx nor bear a filament are 'imperfect'. Polyps and larvae display biradial symmetry and septa which constitute mirror

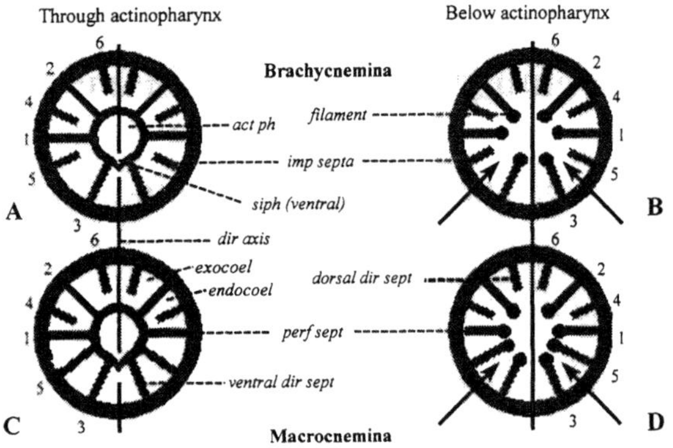

Figure 1. Septa and the classification of zoanthids. Diagrams are cross-sections through larvae or embryos at the 12-septa stage of development (at which the diagnostic patterns first become apparent). The numbers 1–6 indicate the sequence of septal formation, the state of the fifth couple (imperfect or perfect) being diagnostic for Brachycnemina and Macrocnemina, respectively; further septa are added in the proliferation chambers (ventro-lateral exocoels), marked by the solid arrows, on each side of the ventral directive septa. The upper diagrams (A, B) represent Brachycnemina, the lower pair (C, D) Macrocnemina; the left-hand diagrams (A, C) are cut through the actinopharynx, the right-hand pair (B, D) below the actinopharynx; perfect septa bear filaments (shown black) at this level. The solid vertical lines mark the directive axes: left and right sides are mirror images, opposite septa being termed couples. Spaces between directive couples, and ventral to perfect septa (except for the fifth couple in Macrocnemina) are endocoels; spaces ventral to imperfect septa (and to the perfect fifth couple in Macrocnemina) are exocoels. Act ph, actinopharynx; dir axis = directive axis; dir sept = directive septum; imp sept = imperfect septum/septa; perf sept = perfect septum/septa; siph = siphonoglyph.

images comprise 'couples'. The couples which border the axis of symmetry ('directive axis') are 'directive' septa. In zoanthids, the dorsal directives are imperfect and the ventral directives perfect; the remainder, with the exception of the fifth couple (counting from the dorsal directives), are alternately perfect and imperfect. As established by Erdmann (1885), the fifth couple may be either 'macrocnemic' (perfect: suborder Macrocnemina Haddon & Shackleton, 1891) or 'brachycnemic' (imperfect: suborder Brachycnemina Haddon & Shackleton, 1891). Septa first appear in the developing larvae; small larvae contain six septa, larger ones contain 12. McMurrich (1891) and Van Beneden (1897) established that the arrangement of internal septa in their larvae corresponded to the brachycnemic pattern, but disagreed somewhat over the sequence of their development.

Menon (1902, 1914, 1926) settled the developmental sequence of septa in the larvae (numerals 1–6 in Fig. 1) and established which genera reproduce with a zoanthella (*Sphenopus, Protopalythoa*), and which with a zoanthina (*Isaurus, Zoanthus*). Kimura et al. (1972) later added *Palythoa* to the former group. These are all brachycnemic genera but form two natural groupings on another ground: the former (with a zoanthella) incorporate sand into their mesogloea, the latter (with a zoanthina) do not. Sand incorporation was recognized by Gray (1867) as a significant character, but in recent years – despite the correlation with larval type – both groups have been inappropriately united within the family Zoanthidae. Here, Sphenopidae Hertwig (1882) is reinstated to distinguish the former.

Carlgren (1923) sectioned *Isozoanthus giganteus* and found that it brooded embryos which clearly displayed the macrocnemic pattern, with the 5th couple perfect at the 12-septa stage. Studied species in the macrocnemic genera *Epizoanthus* (Muirhead et al., 1986) and *Parazoanthus* (Ryland, unpub.), on the other hand, have proven oviparous, although no corresponding macrocnemic larvae have ever been found. Zoanthid reproduction generally has been reviewed by Ryland (1997).

Important results have accrued from Scheltema's (1968, 1971, 1989) studies of the occurrence of larvae, including zoanthellae and zoanthinae, of shoal-water benthic invertebrates in the oceanic plankton. First for the Atlantic (Scheltema, 1968) and then for the Pacific (1989), Scheltema showed that the larvae ranged east–west right across the oceans (in the Pacific, the only larval group in his study not to show significant reduction east of longitude 160° W), but that they were restricted to warm low latitude waters. There are now many more samples, especially from the Atlantic, than when the earlier papers were written and the entire collections of Semper's larvae obtained between 1952 and 1993 form the basis of the present paper. Our objectives are: (1) to analyse and compare the distribution patterns of these larvae in the surface waters of the world's oceans; (2) to study in them the occurrence and distribution of zooxanthellae; and (3) to investigate why, roughly a century after Van Beneden's and Menon's first papers (Van Beneden, 1890; Menon, 1902), larvae of zoanthids having macrocnemic septation are – as at the end of the nineteenth century (Carlgren, 1906) – still unknown.

Materials and methods

Nearly 1800 successful plankton samples were taken on almost 60 cruises between 1952 and 1993. Collec-

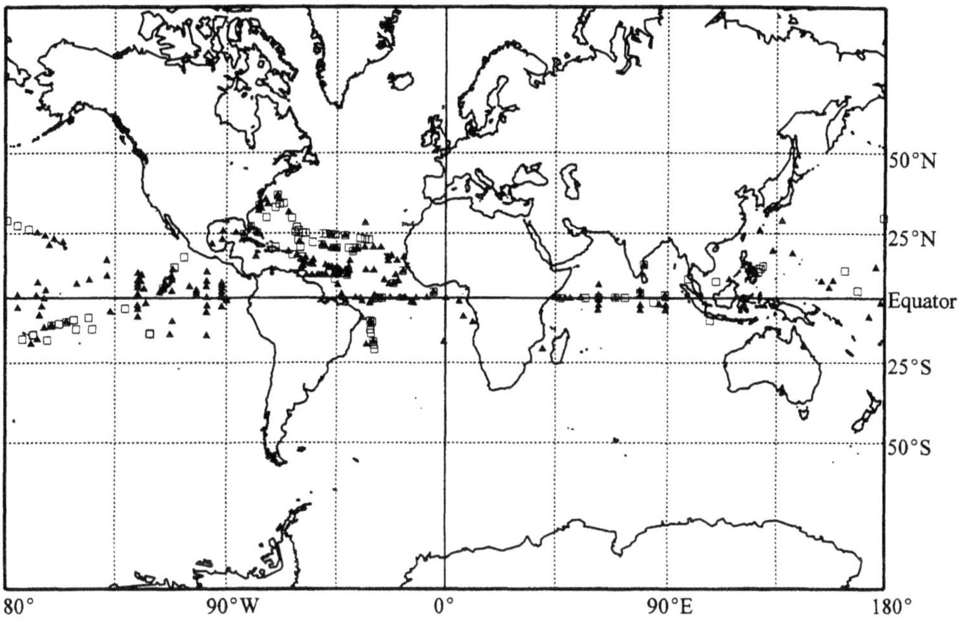

Figure 2. Occurrences of Semper's larvae in the world's oceans, including the few known from earlier literature (▲ zoanthella, □ zoanthina).

tion and sample sorting methods have been described elsewhere (Scheltema, 1968, 1971, 1986, 1989). Exact procedures varied slightly between vessels of different operators, but sea surface temperature and depth were generally recorded by Woods Hole Oceanographic Institution (WHOI) ships. Tows were oblique and generally sampled from a depth of 100–150 m to the surface. Larvae for sectioning were drawn, or photographed using dark field illumination under a Wild M20 Makroskop, and measured with a micrometer eyepiece. Sections were cut at 7 or 8 μm (some at WHOI at 5 and 10 μm), and stained with haematoxylin and eosin or Mallory's triple. Since many of the specimens had been in preservative for decades, and were expected to be either soft or brittle, a double embedding technique was used at Swansea. Specimens were dehydrated in ethanol and then soaked for 24 h through two changes of necoloidine solution before being cleared in chloroform. Embedding in molten paraffin wax was repeated in four changes to ensure complete penetration before cooling the block ready for sectioning. Sections were examined primarily to determine the numbers, types and arrangement of septa, and the absence/presence and distribution of zooxanthellae, but the amount of yolk in the coelenteron, the state of the cell layers, and the thickness of the mesogloea proved highly variable and were recorded.

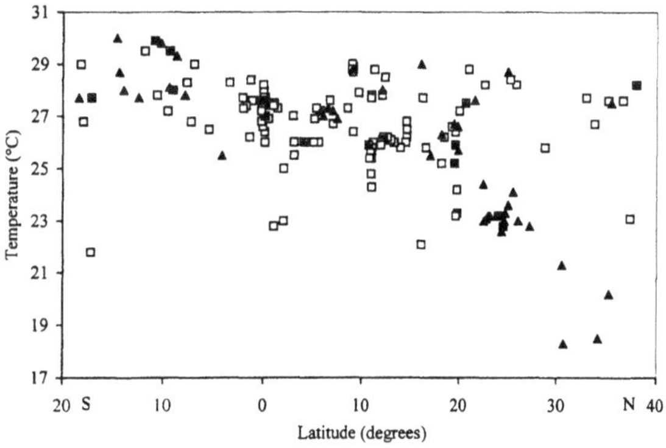

Figure 3. Temperature distribution of Semper's larvae (▲ zoanthella, □ zoanthina).

Cruise and station details were entered into a database. A few additional records have been obtained from the literature. Collection coordinates (latitude and longitude converted to decimal degrees) have been used for map compilation and distributions have been correlated with sea surface temperature when available.

The principal macrocnemic genera are *Epizoanthus* and *Parazoanthus*. The former is better known because, though both tend to be commensal, adult *Epizoanthus* colonies are associated with molluscs (especially living neogastropods); with pagurid crabs, with which they form carcinoecia, replacing the ori-

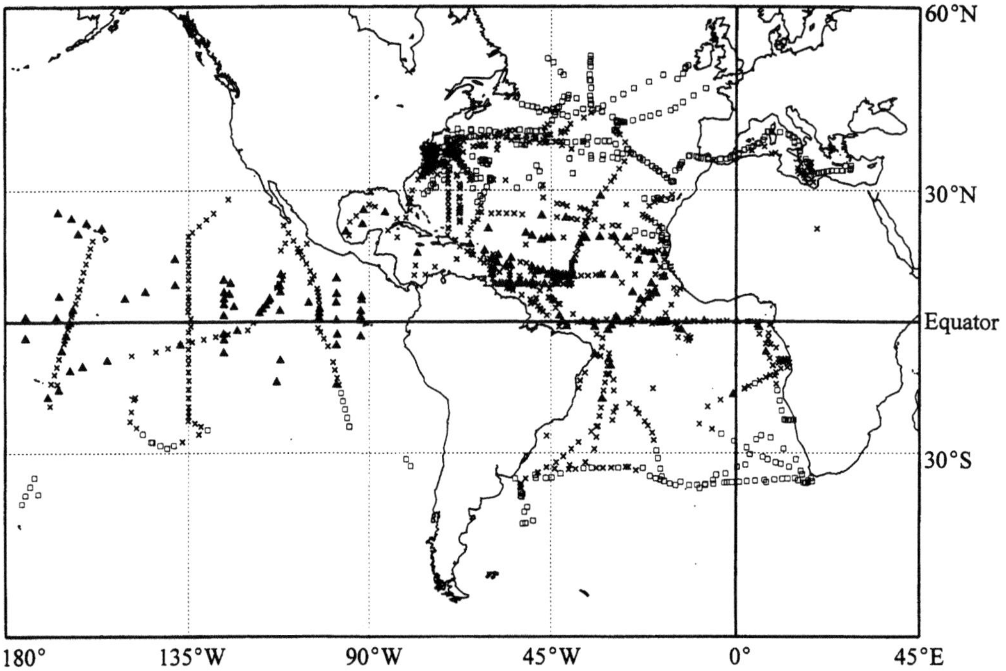

Figure 4. Atlantic and eastern Pacific records of zoanthella larvae, together with negative stations (▲ zoanthellae, × negative stations with temperature ≥22 °C, □ negative stations with temperature <22 °C).

ginal gastropod shell (Carlgren, 1923; Muirhead et al., 1986); or with the rope-like stems of hyalonematid glass sponges (Schulze, 1887), all of which tend to ensure that specimens are retained in collections. We have collated and analysed records of the geographic and bathymetric distribution of *Epizoanthus*, partly from the literature but mainly from specimens or information sent earlier to A. Muirhead, or to JSR, and from the substantial collections in the Moscow State Museum, so that these largely new data may be compared with the geographic distributions of Semper's larvae.

Results

Zoanthellae were obtained at 210 stations and zoanthinae at 79, generally as single larvae but occasionally in greater numbers (rarely >10) (Fig. 2). The distributions of both larval types were ocean wide but only within warm, essentially tropical waters. There are, however, differences in temperature ranges of the two: zoanthellae were never found at <22 °C (Figs 3 and 4), whereas substantial numbers of zoanthinae occurred at 18–22 °C (Figs 3 and 5), with a clearly more northerly distribution in the Atlantic (Fig. 2). The summed occurrences by temperature, tested by the Kolmogorov-Smirnov 2-sample test (Sokal & Rohlf, 1995), were highly significantly different ($P < 0.001$).

Zoanthellae were commonly >6 (maximum 8.6) mm preserved length, zoanthinae 3–4 (maximum 5.9) mm. Forty three larvae (19 zoanthellae, 24 zoanthinae) were sectioned, though state of preservation, plane of sectioning, and the difficulty in orientating extremely small larvae, made it impossible to obtain all the sought information from every larva. The condition of the mesogloea and cell layers varied in a manner not correlated with size or presence/absence of zooxanthellae. The mesogloea was generally quite thick, and corrugated in some septal pairs; this condition was associated with a thick ectoderm and often with yolk between the septa. In other larvae, the mesogloea was membranous, the ectoderm sparse (in the region of the ciliary band) or missing, and the coelenteron 'empty'. All transversely-sectioned larvae having 12 septa conformed to the brachycnemic pattern (only septal couples 2, 4 and 6 on each side of the directive axis either reaching the actinopharynx or bearing a filament at the centripetal end), with the exception of two that were anomalous. In all, there were three anomalous arrangements: one zoanthina was asymmetrical, with seven perfect and five imperfect

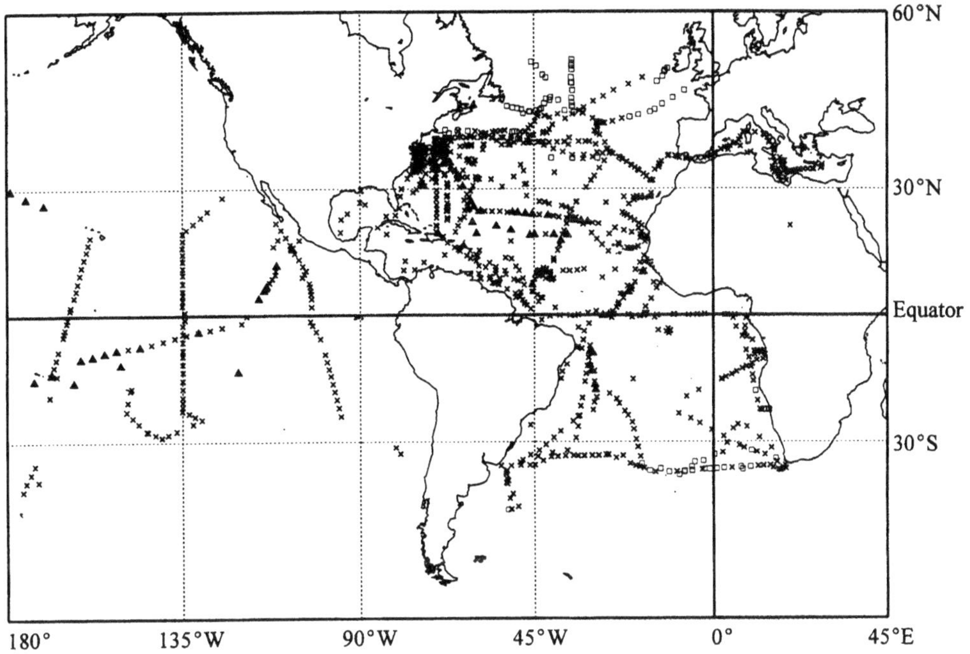

Figure 5. Atlantic and eastern Pacific records of zoanthina larvae, together with negative stations (▲ zoanthinae, × negative stations with temperature ≥18 °C, □ negative stations with temperature <18 °C).

septa; one zoanthina had all 12 septa, except the dorsal directives, perfect; and one zoanthella had 16 septa, 8 of them – including the 1st couple (dorsal directives) – perfect. None had a macrocnemic arrangement and none included smaller, less-developed larvae in the coelenteron (cf. Van Beneden, 1897). Zooxanthellae were present in the endoderm of nine zoanthinae but in none of the zoanthellae. Three larvae, all zoanthinae, contained a small endo-commensal/parasitic amphipod.

We have established that, unlike Semper's larvae, adult *Epizoanthus* are distributed from the equator to high latitudes (Fig. 6) and also from the intertidal (e.g. *E. minutus* in Florida (JSR unpublished)) to great depths (~5000 m) (Fig. 7). While many of the larger species are bathyal or abyssal, there are some which are plentiful in well-studied, shallow seas (e.g. *E. papillosus* in the North Sea, *E. couchii* around western European coasts, and *E. arenaceus* in the Adriatic), in depths <200 m.

Discussion

That the oceanic distribution of zoanthellae is highly correlated with sea surface temperature was established early in these studies (Scheltema, 1971). What is curious about the present results is the extent to which the temperature limits of zoanthellae and zoanthinae differ. Perhaps a physiological response to temperature limits survival and, therefore, modifies the ocean distribution of larvae, determined initially by their areas of origin and the currents by which they are transported. Nothing known about the geographical distributions and reproductive periods of *Palythoa/Protopalythoa* and *Zoanthus* adults (the genera most likely to provide zoanthellae and zoanthinae, respectively) explains this difference. In marginal areas generally (southern Florida, KwaZulu-Natal and southern Queensland), the two groups occur in similar abundances and to similar limits; *Protopalythoa* occurs in both the Canary Islands and Madeira (Wirtz, 1995) but *Zoanthus* does not. It, therefore, would be expected *a priori* that zoanthellae and zoanthinae should occur mixed in broadly constant (though, of course, not necessarily equal) proportions; but this was not so. While some kind of artifact might be suspected, the difference is statistically too significant ($P<0.001$) to ignore. The fact that the Pacific was relatively undersampled, and that different areas of ocean were visited at different times of year, does not, *per se*, explain certain areas being dominated by zoanthellae and others by zoanthinae.

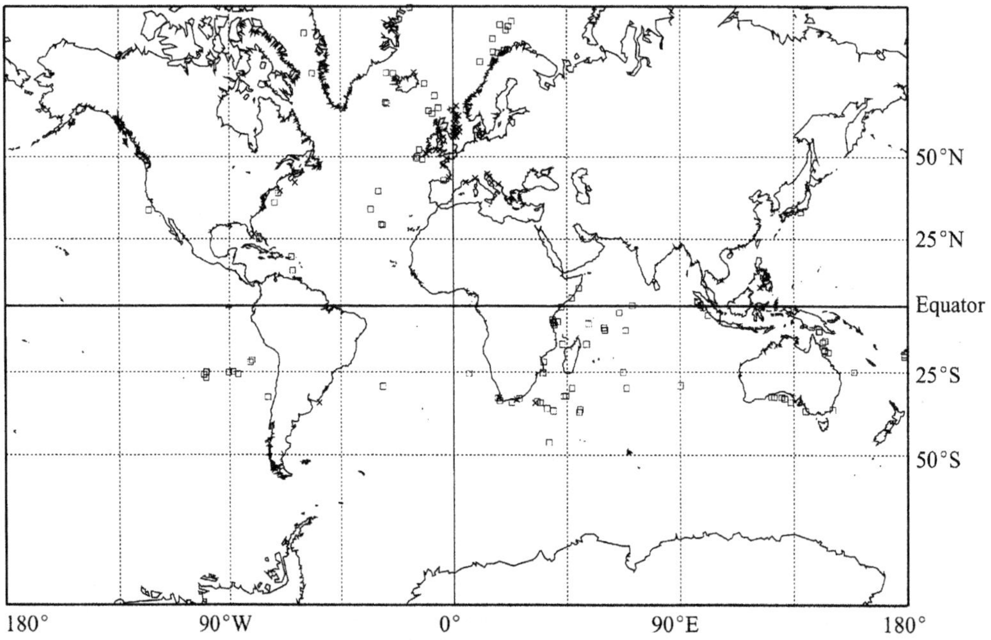

Figure 6. Geographic distribution of the genus *Epizoanthus*. Data are mainly from unpublished sources (see text) with additional records from Schulze (1887) and Carlgren (1923) (☐ stations with depth <200 m, × depth ≥200 m).

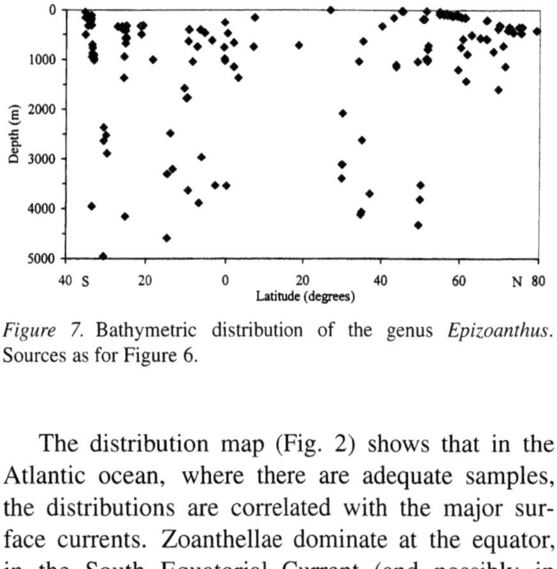

Figure 7. Bathymetric distribution of the genus *Epizoanthus*. Sources as for Figure 6.

The distribution map (Fig. 2) shows that in the Atlantic ocean, where there are adequate samples, the distributions are correlated with the major surface currents. Zoanthellae dominate at the equator, in the South Equatorial Current (and possibly in the eastwardly-flowing Equatorial Undercurrent below 50 m depth), through the Caribbean Sea into the Gulf of Mexico, into the Florida Current and also, during the summer months, in the seasonally flowing Equatorial Counter Current as far northeast as the Cape Verde Islands. From ~20° N, however, zoanthinae prevail in the North Equatorial Current, the Antilles Current, and the outer Gulf Stream east of Cape Hatteras. No larvae, even zoanthinae, occurred in the 20–24 °C band of Gulf Stream water that flows easterly to the Azores, where the water temperature attenuates rapidly and seasonal differences are quite marked.

In the Pacific, there were fewer larvae, particularly where temperature was also recorded, but clusters of zoanthinae occurred to the west of the Hawaiian chain and around French Polynesia (Fig. 2). The most striking feature of the Pacific distributions is the absence of an eastern Pacific barrier (see Scheltema, 1989), with a notable abundance of zoanthellae east of the Line Islands, despite there being few sources of larvae east of longitude 180°. Many teleplanic larvae of a wide variety of invertebrates retain the ability to metamorphose (Scheltema, unpublished) but Semper's larvae held over several months have never done so. This may be because the requisite cues were absent, but it could also be because the larvae have lost competence. Some of the sectioned larvae appeared in good condition, others appeared 'wasted'. In a recent study on competence and longevity in alyconacean larvae, Zaslow & Benayahu (1998) found that both zooxanthellate and non-zooxanthellate larvae could survive 7–22 weeks, according to species, but lost competence to metamorphose at 7–11 weeks. In particular, zooxanthellate species, presumably aided by translocated products from photosynthesis, survived for several weeks after they had lost competence. It is, therefore, impossible to decide whether teleplanic dispersal of Semper's larvae is adaptive or not (i.e. whether larvae carried in the Equatorial Currents or Countercurrents remain competent to settle after an Atlantic crossing). Unfortunately, the unsatisfactory state of adult zoanthid taxonomy renders meaningful comparison of east and west Atlantic faunas impossible, though the distinctive *Isaurus tuberculatus* (with a zoanthina) has a pan-tropical distribution (Muirhead & Ryland, 1985).

It was hoped that, by sectioning a substantial number of larvae, one or more displaying macrocnemic septation would be discovered. This did not happen: all larvae have been brachycnemic or (a few) anomalous. The occurrence of a larva which was brachycnemic on one side of the directive axis and macrocnemic on the other is not surprising, the condition being well-known in adults of brachycnemic genera. Adults with a symmetrical excess of perfect septa must, on the other hand, be unusual and certainly none with perfect dorsal directives has been recorded previously. The absence of macrocnemic larvae, despite the known occurrence of several species of *Epizoanthus* and *Parazoanthus* in shallow tropical waters (e.g. West, 1979; Crocker & Reiswig, 1981), suggests that the fertilized eggs of such zoanthids do not develop into planktonic Semper's larvae.

Our other significant morphological findings concern zooxanthellae. About half the zoanthinae had zooxanthellae in their endoderm, perhaps acquiring them in early development whilst still inshore. It is, in general, not known how larvae and/or adults acquire their zooxanthellae (Ryland, 1997). However, the absence of them from all zoanthellae was unexpected in view of the occurrence of maternal transmission in one species of *Protopalythoa* well known in the western tropical South Pacific (Ryland & Babcock, 1991; Burnett et al., 1997) (though the only sampling in this area was a chain of stations (barren of Semper's larvae) between Fiji and New Zealand). Since the largest zoanthid eggs measure ~0.5 mm diameter, and many are much smaller (Ryland, 1997), Semper's larvae grow considerably, with or without zooxanthellae, providing strong circumstantial evidence that they engulf food through the blastopore/mouth. However, the undamaged small crustaceans present in three zoanthinae appeared not to have been eaten but to be living commensals or endo-parasites. Amphipods have not been recorded as internal commensals of adult zoanthids though they occur in the coelenteron of actinians (Vader, 1983).

Following the absence of macrocnemic larvae in the samples, we surveyed the spatial distribution of adult *Epizoanthus* for any light it might shed on their larvae. As anticipated, both latitudinal and bathymetric ranges are extensive, in marked contrast to those of brachycnemic genera. Although *Isozoanthus giganteus* broods embryos (Carlgren, 1923), at least some species of *Epizoanthus* and *Parazoanthus* are known to be broadcast spawners (Muirhead et al., 1986; Ryland, unpublished observations, Bermuda, 1999). Whilst the larvae of bathyal and abyssal species may never leave the deep sea (Gage & Tyler, 1991), not all non-tropical zoanthids are restricted to great depths. Recent surveys have shown *Epizoanthus papillosus* (=*incrustatus*) to be abundant in parts of the North and Irish Seas (J.E. Lancaster, J. Ellis, pers. comms) and *E. arenaceus* is similarly common in the northern Adriatic (Pax & Müller, 1962; M. Stachowitsch, pers. com.). In both seas, the depths of occurrence are much less than 200 m. Although the North and Adriatic Sea areas were not sampled during our surveys, the former has been intensively studied in the past (Carlgren, 1906; Thorson, 1946; Widersten, 1968). For the Adriatic, after many years of study, Pax & Müller (1962) concluded that *Epizoanthus* has no true pelagic stage. Likewise, in northern waters inhabited by *Epizoanthus* (Fig. 7), no Semper's larvae have been recorded from multi-depth plankton studies (see Murray & Hjort, 1912). While negative evidence inevitably leaves an element of doubt, Semper's larvae are large and distinctive and it seems probable that, should they occur in the above localities, they would have been recognized by now. The firm conclusion derived from the combined surveys of Semper's larvae and of *Epizoanthus* adults, therefore, is that macrocnemic zoanthids, as suggested earlier by Menon (1926) and Pax & Müller (1962), do not reproduce by means of a pelagic zoanthella or zoanthina but by some other, presumably less well differentiated, form of planula which possibly (as the latter authors suggested) never leaves the vicinity of the sea bed. Settlement may even occur before the septa have differentiated, as in the anemone *Aiptasia* (Widersten, 1968). Without septa, such a planula could not be referred to any particular cnidarian taxon.

Acknowledgements

The first author acknowledges a grant from the Leverhulme Trust which enabled him to visit Woods Hole Oceanographic Institution. We thank the following for specimens of, or data on, *Epizoanthus*: P. Arnold and M. Pichon (*Cidaris* cruises), P. Arnold and A. Birtles (Cleveland Bay), P.A. Tyler and C.M. Young (Bahamas, Tongue of the Ocean), S. Beaulieu, J. Ellis, S. Jennings, B. Picton and J. Turner. We also acknowledge the constructive comments of two anonymous referees.

References

Babcock, R. C. & J. S. Ryland, 1990. Larval development of a tropical zoanthid (*Protopalythoa* sp.). Invertebr. Reprod. Dev. 17: 229–236.

Burnett, W. J., J. A. H. Benzie, J. A. Beardmore & J. S. Ryland, 1997. Great Barrier Reef and Torres Straits zoanthids (Cnidaria, Hexacorallia): evolution, systematics and a key to species. Coral Reefs 16: 55–68.

Carlgren, O., 1906. Die Actinien-larven. Nordisches Plankton 11: 65–89.

Carlgren, O., 1923. Ceriantharia und Zoantharia der Deutschen Tiefsee-Expedition. Wiss. Ergebn. dt. Tiefsee-Exped. 'Valdivia' 19(7): 12–89.

Crocker, L. A. & H. M. Reiswig, 1981. Host specificity in sponge-encrusting Zoanthidea (Anthozoa: Zoantharia) of Barbados, West Indies. Mar. Biol. (Berl.) 65: 231–236.

Erdmann, A., 1885. Ueber einige neue Zoantheen. Ein Beitrag zur anatomischen und systematischen Kenntnis der Actinien. Jena. Z. Naturw. 19: 430–488.

Gage, J. & P. A. Tyler, 1991. Deep-sea Biology: a Natural History of Organisms at the Deep-sea Floor. Cambridge, Cambridge University Press: 504 pp.

Gray, J. E., 1867. Notes on Zoanthinae, with the descriptions of some new genera. Proc. zool. Soc. Lond. 15: 233–240.

Haddon, A. C. & A. M. Shackleton, 1891. Revision of the British Actiniae, Part II. The Zoantheae. Sci. Trans. r. Dublin Soc. 4: 609–672.

Hertwig, R., 1882. Report on the Actiniaria dredged by H.M.S. 'Challenger' during the years 1873–1876. Rep. Sci. Res. 'Challenger' Exped., Zool. 6(15): 1–134.

Kimura, S., Y. Hashimoto & K. Yamazato, 1972. Toxicity of the zoanthid *Palythoa tuberculosa*. Toxicon 10: 611–617.

McMurrich, J. P., 1891. Contributions on the morphology of the Actinozoa. III. The phylogeny of the Actinozoa. J. Morphol. 5: 125–164.

Menon, K. R., 1902. Notes on Semper's larvae. Proc. Cambr. Phil. Soc., Biol. Sci. 11: 407–417.

Menon, K. R., 1914. Preliminary note on the metamorphosis of *Zoanthella*. Rec. Ind. Mus. 10: 212–213.

Menon, K. R., 1926. On the adults of *Zoanthella* and *Zoanthina*. Rec. Ind. Mus. 28: 61–64.

Muirhead, A. & J. S. Ryland, 1985. A review of the genus *Isaurus* Gray, 1828 (Zoanthidea), including new records from Fiji. J. Nat. Hist. 19: 323–335.

Muirhead, A., P. A. Tyler & M. H. Thurston, 1986. Reproductive biology and growth of the genus *Epizoanthus* (Zoanthidea) from the north-east Atlantic. J. mar. biol. Ass. U.K. 66: 131–143.

Murray, J. & J. Hjort, 1912. The depths of the ocean. London, Macmillan: 821 pp.

Pax, F. & I. Müller, 1962. Die Anthozoenfauna der Adria. Fauna et Flora Adriatica. Split, Institut za Oceanografiju i Ribastvo. No. 3: 343 pp.

Ryland, J. S., 1997. Reproduction in Zoanthidea. Invertebr. Reprod. Dev. 31: 177–188.

Ryland, J. S. & R. C. Babcock, 1991. Annual cycle of gametogenesis and spawning in a tropical zoanthid, *Protopalythoa* sp. In Williams, R. B. P. F. S. Cornelius, R. G. Hughes & E. A. Robson (eds), Coelenterate Biology: Recent Research on Cnidaria and Ctenophora. Proceedings of the Fifth International Conference on Coelenterate Biology. Hydrobiologia 216/217: 117–123.

Scheltema, R. S., 1968. Dispersal of larvae by equatorial ocean currents and its importance to the zoogeography of shoal-water tropical species. Nature (Lond.) 217(5134): 1159–1162.

Scheltema, R. S., 1971. The dispersal of the larvae of shoal-water benthic invertebrate species over long distances by ocean currents. In Crisp, D. J. (ed.), Proceedings of the Fourth European Marine Biology Symposium. Cambridge, Cambridge University Press: 7–28.

Scheltema, R. S., 1986. Long-distance dispersal by planktonic larvae of shoal-water benthic invertebrates among Central Pacific islands. Bull. mar. Sci. 39: 241–256.

Scheltema, R. S., 1989. The eastern Pacific as a barrier to the dispersal of planktonic larvae of benthic invertebrates. In Klekowski, R. Z., E. Styczynska-Jurewicz & L. Falkowski (eds), Proceedings of the 21st European Marine Biology Symposium (1986). Wroclaw, etc., Polish Academy of Sciences: 435–445.

Schulze, F. E., 1887. Report on the Hexactinellida. Rep. Sci. Res. 'Challenger' Exped., Zool. 21: 1–513.

Semper, C., 1867. Ueber einige tropische Larvenformen. Zt. wiss. Zool. 17: 407–428.

Sokal, R. R. & F. J. Rohlf, 1995. Biometry. San Francisco, W.H. Freeman: 887 pp.

Thorson, G., 1946, Reproduction and larval development of Danish marine bottom invertebrates. Medd. Komm. Havundersøg., Kbh., Ser. Plank.,Bd. 4, No. 1: 523 pp.

Vader, W., 1983. Associations between amphipods (Crustacea: Amphipoda) and sea anemones (Anthozoa, Actiniaria). Mem. Austr. Mus. 18: 141–153.

Van Beneden, E., 1890. Les Anthozoaires pélagiques recueillis par M. le professeur Hensen dans son expédition du plankton. Communication préliminaire. I. – Une larve voisine de la larve de Semper. Archs Biol., Liège 10: 485–521.

Van Beneden, E., 1897. Les Anthozoaires de la 'Plankton-Expedition'. Ergebn. Plankton-Exped. Humboldt Stift. 2 K(e): 485–521.

West, D. A., 1979. Symbiotic zoanthids (Anthozoa: Cnidaria) of Puerto Rico. Bull. mar. Sci. 29: 253–271.

Widersten, B., 1968. On the morphology and development in some cnidarian larvae. Zool. Bidr. Upps. 37: 139–179.

Wirtz, P., 1995. One vascular plant and ten invertebrate species new to the marine flora and fauna of Madeira. Arquipelago. Cienc. Biol. Mar. Mar. Biol. Sci. (13A): 119–123.

Zaslow, R. B.-D. & Y. Benayahu, 1998. Competence and longevity in planulae of several species of soft corals. Mar. Ecol. Prog. Ser. 163: 235–243.

Invited Keynote Address:

A review of the distribution of hydrothermal vent communities along the northern Mid-Atlantic Ridge: dispersal vs. environmental controls

D. Desbruyères[1], A. Almeida, M. Biscoito,
T. Comtet, A. Khripounoff, N. Le Bris,
P.-M. Sarradin & M. Segonzac

[1]*Département "Environnement Profond"*
Centre de Brest de l'IFREMER
B.P. 70
F-29280 Plouzané
France

Hydrobiologia **440:** 201–216, 2000.
M.B. Jones, J.M.N. Azevedo, A.I. Neto, A.C. Costa & A.M. Frias Martins (eds), Island, Ocean and Deep-Sea Biology.

A review of the distribution of hydrothermal vent communities along the northern Mid-Atlantic Ridge: dispersal vs. environmental controls

D. Desbruyères[1], A. Almeida[2], M. Biscoito[3], T. Comtet[1], A. Khripounoff[1], N. Le Bris[1], P. M. Sarradin[1] & M. Segonzac[1]
[1]*Département 'Environnement Profond', Centre de Brest de l'IFREMER, B.P. 70, 29280 Plouzané, France*
[2]*Laboratório Marítimo da Guia, Universidade de Lisboa, Estrada do Guincho, 2750 Cascais, Portugal*
[3]*IMAR-Museu Municipal do Funchal, Rua da Mouraria, 31 – 9004-546 Funchal, Madeira, Portugal*

Key words: deep sea, hydrothermal vents, ecology, Mid-Atlantic Ridge, Azores Triple Junction

Abstract

Until 1985, seven vent fields were described from the Mid-Atlantic Ridge (MAR). An eighth field, Mount Saldanha (36° N), discovered in 1998, showed unusual geological and biological settings. Vent sites on the MAR exhibit varied environmental conditions, resulting from depth variation of the axis and associated physical parameters, and different source rocks. These could be considered as first order (i.e. most dominant) factors affecting the composition of vent communities on the MAR, in contrast to the East Pacific Rise (EPR) where geographical isolation appears to be a major determinant of faunal differences. In this paper, the geological setting and vent fluid composition of the fields are considered together with their community composition to tentatively ascertain the order of a hierarchy between dispersal and environmental control. The deepest fields (>3000 m) are rather stable systems. The shallower fields, especially Rainbow and Menez Gwen, present some evidence of instability in time and space. The variability in fluid composition is related to phase separation processes (boiling/distillation of subsurface vent fluids) and to the nature of the basement rocks. Depending on depth, phase separation produces gas-enriched and metal-depleted fluids (Menez Gwen, Lucky Strike) or metal-enriched brines (Rainbow, TAG). In addition, high methane content characterises the fluids formed in ultramafic rocks (Rainbow, Logatchev) compared to basaltic rocks. The discrepancy in mineral particulate fluxes at Lucky Strike and Menez Gwen, on one hand, and TAG and Rainbow, on the other, is correlated to the predominance of the vapour or brine phase. The semi-quantitative description of the faunal composition of the different vent fields displays a continuum from *Rimicaris*-dominated to *Bathymodiolus*-dominated assemblages. Rather than geographic or bathymetric zonation, this gradation appears to be related to the metal content of the fluids. In addition, the penetration of non vent species into the vent environment increases with decreasing hydrostatic pressure and/or metal content in the fluids. Similarity analysis between vent communities shows that similarity is strongest between Menez Gwen and Lucky Strike (the shallowest fields), less significant between these sites and Rainbow, and weakest for Snake Pit. The inverse relationship between filter feeding organisms and metal concentration in vent fluids could result from a hindrance of mussel bed development by particulate or toxic metal fluxes, and has to be further investigated. Conversely, high metal and particulate content would less affect the more mobile *Rimicaris* populations. Considering specific similarities of endemic fauna between the four best known hydrothermal vents, the distance between vent fields appears to be a first order parameter. Nevertheless, within the proximity of the Azores Triple Junction area, and in the absence of geographical discontinuity, the similarity between fields stays rather low suggesting faunal islands that have distinct composition and habitat requirements.

Introduction

Light penetrates only ~300 m into the ocean and the deep-sea environment is devoid of local photosynthetic production, which usually fuels the ecosystems of the Earth. Deep-sea benthic communities are fuelled by surface primary production, which is partially mineralised during its descent to the deep-sea floor. Given this limitation of energy inputs to the deep sea, the biomass is generally low (i.e. <1 g wet weight

/m^2) and the deep-sea environment is usually considered as a desert (life in the deep-sea habitat is scarce because its food is scarce). Conversely, deep-sea environments are generally known for the high diversity found amongst the mostly small-bodied invertebrates inhabiting the sediments.

In contrast to the rest of the deep sea, the hydrothermal vent assemblages, discovered first at the Galápagos Spreading Centre (GSC) at 2500 m water depth, are spectacular in their biomass, distribution and composition (Hessler & Smithey, 1983). Since 1977, faunal assemblages, similar to those discovered at GSC, have been sampled at other locations on the East Pacific Rise (EPR) (Lutz & Kennish, 1993), Juan de Fuca Ridge (Tunnicliffe, 1988), Mid-Atlantic Ridge (MAR) (Van Dover, 1995), and the spreading centres of Western and South-western Pacific Back Arc Basins (Hessler et al., 1987; Desbruyères et al., 1994). The deep MAR exploration began in 1985 and is, presently, restricted to a section delimited by Vema Fracture Zone (11° N) to the south and the Azores Plateau (38° N) to the north, with the exception of few indications of activity (only dense beds of sponges observed) at the Iceland periphery (see Van Dover, 1995 for review). Between these two latitudes, seven vent fields have been localised and studied (Fig. 1, Table 1) with variable sampling effort ranging from a few dives (e.g. Logatchev and Broken Spur vent fields) to long series of diving cruises (e.g. Snake Pit and Lucky Strike vent fields). This discrepancy in data makes comparison difficult between vent field communities.

'Hydrothermal communities' thriving on chemoautotrophic bacterial production are strictly linked to reduced fluid emission. Clustered around vents, they are roughly distributed on the mesoscale in a linear arrangement along the ridge axis. Because active hydrothermalism occupies only a small portion of the ridge axis at any one time, that portion of the hydrothermal habitat that can support life is discontinuous in space. Distribution patterns are affected greatly by the shape and size of the underlying magma chamber, and related spreading rates (Cann et al., 1994). Discontinuity in vent emission rate has been observed directly (Haymon et al., 1993; Desbruyères, 1998) or inferred from radiochronological data (Lalou et al., 1989, 1990) in different ridge dynamic contexts.

When studying the faunal composition of vent communities, the most basic pattern to emerge is an inverse relationship between faunal similarity and distance between vent fields (Hessler & Lonsdale, 1991). The increase in number and size of ridge axis discontinuities, as well as increasing distance between vent fields, will result in faunal differences partially because of the geographic isolation of breeding groups and genetic pools. Several works dealing with hot vent biogeography, and mainly based on dispersion in a linear fragmented habitat, have been published recently using plate tectonic history (Tunnicliffe, 1988, 1991, 1996; Hessler & Lonsdale, 1991; Van Dover, 1995; Tunnicliffe & Fowler, 1996; Tunnicliffe et al., 1996, 1998). This model only applies to communities having fairly homogenous habitats, uniform bathymetry, unique source rocks and linear heat sources, such as observed on fast spreading ridges such as the EPR. The situation is much more complex for the slow spreading Mid-Atlantic Ridge where the local conditions affected by changing source rocks, and depth variations of the axis (proximity of hot spot), might be first order factors affecting the composition of vent communities.

This paper aims to describe the patterns of the vent distribution and habitat diversity on the MAR between 11° N and 38° N. It aims also to compare community composition to determine the relative importance of dispersal vs. environmental controls. The characteristics of the studied areas, with depth ranging from 800 to 3800 m and source rocks varying from basaltic to ultramafic rocks, are of importance in structuring vent communities.

Cruises

The work presented here is based chiefly on the most recent dives made at the Azores Triple Junction during cruises DIVA 2 (1994), MARVEL (1997), PICO & SALDANHA (1998) with the manned submersible Nautile. Nevertheless, complementary data were also collected during former cruises of the Nautile (HYDROSNAKE, MICROSMOKE) and Alvin (MAR'93), on different MAR vent fields.

Distribution of hot hydrothermalism along Mid-Atlantic Ridge between 11° and 38° N (Fig. 1)

Present knowledge of the vent distribution relies on a complex sequence of (1) swath bathymetry and side scan sonar, (2) hydrocast sampling, CTD and transmissometer-nephelometer measurements, (3) dredging of sulphide and source rocks, (4) camera deep-towing, and (5) submersible operations. These methods have allowed the study of hydrothermal fields that have been primarily localised by the detection

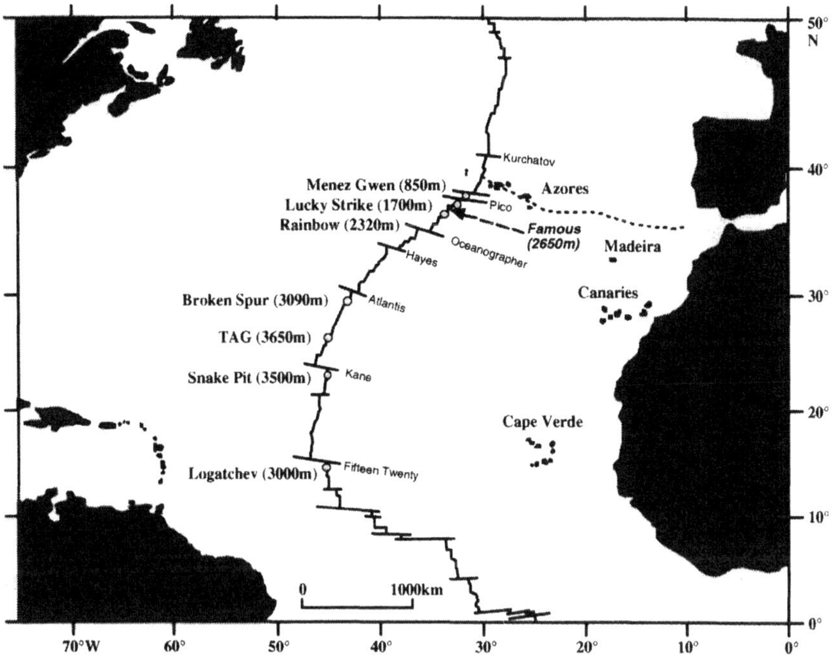

Figure 1. Location map of the eight active hydrothermal vent fields known to date on the Mid-Atlantic Ridge between Equator and the Azores Archipelago. The major Transform Faults, which constitute the principal isolation structures, are also shown.

of hydrothermal tracers (manganese, hydrogen, methane) in seawater. Experience and theory predict that at slow-spreading centres, the style of volcanism and hydrothermalism provides for relatively stable but isolated vent systems (Cann et al., 1994). TDM[1] CH_4 anomalies and bathymetry data on two sections of the Mid-Atlantic Ridge (German & Parson, 1998) inferred that the frequency of venting is at least one vent field every ~110 km between 12° N and 26° N, and one vent field every 25–30 km along axis between 36° and 38°N. The distribution of individual sites among the vent fields is likely to differ between MAR and EPR, being less numerous in the first case but larger in size and more stable in time (Cann et al., 1994). The above estimation, relying on major tracer anomaly, pays no attention to small-sized phenomena such as low temperature diffuse venting.

In fact, some conspicuous vent fields (e.g. Menez Gwen) were found in areas without significant hydrothermal signals in the water column because of the special fluid pattern encountered. A new kind of hydrothermal vent, known as a methane vent, was recently observed on the MAR. Its occurrence and biological importance are still unknown and deserves further investigation. This type of vent was found after strong anomalies of methane were detected (Charlou et al., 1997) at the southern end of the FAMOUS segment (36 °30′ N). This area was later surveyed and sampled during the FLORES and MARVEL diving cruises, but no visible hydrothermal manifestation or typical hydrothermal vent fauna was then discovered. During the Portuguese-French SALDANHA cruise in 1998, the area was revisited and six dives with the Nautile were made. A submarine hill of about 700 m altitude, on the southern tip of the FAMOUS segment (Mount Saldanha, 36° 33′ N, 33° 26′ W and 2300 m depth), was explored thoroughly and finally, at its summit, a discharge of clear, warm, gas-bearing fluid was discovered. This discharge originates from small orifices through the sediment, covering an area of at least 50 m^2 (Barriga et al., 1998). The summit and flanks of Mount Saldanha are covered with very fine sediment. The orifices through which the vent fluid escapes contain metallic oxides and sulphates (Barriga, 1999). Such important methane anomalies were detected also in the southern end of the Lucky Strike segment but, up to the present time, no submersible dive has taken place there.

The biological setting of this new active site is completely different from those found at Rainbow, Lucky Strike and Menez Gwen fields. The most striking feature is the absence of benthic vent macrofauna

[1] Total dissolved manganese.

Table 1. Location of vent fields observed on the M. A. R. between 11° N and 38° N

Vent field	Latitude N	Longitude W	Depth	References
Logatchev	14° 45′	44° 58.7′	2930 – 3020	Gebruk et al. (1997b)
Snake Pit	23° 23′	44° 56.1′	3 480	Segonzac (1992)
T.A.G.	26° 08′	44° 49.6′	3635 – 3670	Rona et al. (1993)
				Van Dover (1995)
Broken Spur	29° 10′	43°1 0.4′	3050 – 3875	Van Dover (1995)
				Murton et al. (1995)
Rainbow	36° 13′	33° 54.1′	2260 – 2350	Fouquet et al. (1997)
Mount Saldanha	36° 33′	33° 28′	2300	
Lucky Strike	37° 17′	32° 16.3′	1620 – 1730	Langmuir et al. (1993)
				Van Dover et al. (1996)
Menez Gwen	37° 51′	31° 31.2′	840 – 865	Fouquet et al. (1997)
				Fouquet et al. (1999)

in the active site. In some orifices, thin whitish filaments were observed undulating with the fluid. These could well be methanotrophic bacteria, but this remains to be confirmed. Another interesting feature is the presence of large concentrations of hairy, round tetractinellid sponges, possibly close to the genus *Cinachyra*, in the flanks and near the summit of Mount Saldanha. Further away, large concentrations of sessile invertebrates, in particular hexactinellid sponges, hydroids, gorgonians, alcyonarians and stalked crinoids, can be found on the edge of basaltic crests where currents are important. The mutual influence of enrichment by chemoautotrophic local production and/or topographical effect is unknown and deserves further study in the near future.

Habitat heterogeneity

At a regional scale, vent habitats among MAR fields differ mainly in their fluid chemistry (including particulate content) and mineral deposit pattern. These groups of factors are deeply influenced by water depth and the geology of source rocks and hydrothermal deposits.

Geological settings

According to Fouquet (1997), the size and morphology of the sulphide deposits are controlled by the permeability of the seafloor, the temporal stability of the discharge, the existence of geological caps over the system and the water depth which affects the phase separation. The Atlantic deep sites (Logatchev, Snake Pit, TAG, Broken Spur) are mainly massive sulphide mounds with focused discharges of high temperature fluid venting from chimneys or fractures [for reviews, see Van Dover (1995) and Gebruk et al. (1997a)]. They are rather stable systems. For example, radiochronological dating (Lalou & Brichet, 1981; Lalou et al., 1990, 1993) of sulphide documented several hydrothermal episodes at the same place over a period of 26 000 years at TAG. Conversely, Rainbow vent field comprises more than 30 groups of active small sulphide chimneys spread over 15 000 m^2. Because of the great number of inactive edifices observed in the same area, and the changes in individual smokers at an interval of 1 year, the discharge in this vent field is likely to be unstable in time and place, a character often attributed to fast-spreading ridges. Lucky Strike vent field is clustered around a lava lake (caldeira) (Fouquet et al., 1995). Twenty seven vent sites have been observed in the border of this caldeira. At the scale of individual mounds comprising clusters of a few vent openings, the presence of impermeable cap rocks (slabs) prevents the mixing and dilution of fluid in ambient seawater. Hydrothermal solutions are conductively cooled and precipitate sulphides before their emission on the seafloor (Fouquet, 1997). This subsurface circulation could be a favourable zone of thermophilic biological production, and flocs of bacterial mats have been observed, emerging from cracks and warm vent openings. In Menez Gwen vent field, barite and anhydrite dominate the surface precipitates and hydrothermal chimneys are situated at the top of sulphate mounds. This shallow system can be affected by explosive volcanic activity (Fouquet et al., 1999) on an area of several square kilometres as disclosed by the distribution of volcanic ejecta on the bottom (ash, sand and lapilli). According to Fouquet et al. (1994), the Menez Gwen is, geologically speaking, very young (few decades?): chimneys are very small, growing directly on fresh pillow lava.

Hydrothermal fluids and mixing zone chemical characteristics

Deep-sea hydrothermal communities live in the interfacial zone where hydrothermal fluid mixes turbulently with bottom seawater. This fluctuating environment provides a cyclical access to reduced chemical species (H_2S, CH_4) provided by vent fluid and seawater oxidised compounds (especially O_2), both of which are required for chemolithoautotrophic bacterial primary production (Fisher, 1990; Tunnicliffe, 1991; Childress & Fisher, 1992). In other respects, this environment is enriched in potentially toxic species (sulphide and heavy metals) to which the organisms are exposed.

The chemical characteristics of habitats can be derived, in the first instance, from the dilution of a pure vent fluid in seawater. They also depend on abiotic and biotic processes occurring in the subsurface and within the mixing zone (Johnson et al., 1988; Sarradin et al., 1999). Characterisation of the medium surrounding the communities thus provides the more relevant way to compare the different habitats. However, data related to the mixing zone are characterised by large spatial and temporal variability, and are limited to a few hydrothermal fields. End-member fluid concentrations, extrapolated from hot fluids collected on smokers, provide a more complete and consistent database on which a first differentiation can be established.

Hydrothermal fluids are seawater which has been chemically altered during its passage through the oceanic crust at elevated temperature and pressure. Two main processes control fluid chemistry (Von Damm et al., 1995). At high temperature and pressure, saline solutions can undergo phase separation processes which produce metal-rich brines and a gas-enriched vapour phase. Fluid chemistry is also controlled by seawater-rock interaction undergone at high temperature and pressure. In fact, vent fluids often issue as a mixture of the vapour phase, brine and (altered) seawater.

Chemical composition of these vapour and brine contributions depends on the temperature and pressure conditions met during the hydrothermal circulation (Edmond et al., 1995; Douville et al., 1999) and the nature of the underlying rocks. In particular, fluid emissions occurring in ultramafic rock (Rainbow, Logatchev), as distinguished from basaltic systems are characterised by high methane content resulting from serpentinization (Donval et al. 1997).

The concentration of most of the major chemical components in vent fluid end-members is summarised in Table 2. These parameters can be considered as characteristics of the processes controlling the hydrothermal emissions, highlighting differences between sites. Furthermore, some of these species, if not all, are directly or indirectly related to biological activity. H_2S, CO_2, CH_4 being the source of carbon and energy for chemoautotrophic bacteria are essential to the functioning of the ecosystem. pH is expected to play a role in biotic and abiotic processes occurring in the mixing zone (Scott et al., 1994; Toulmond et al., 1994; Goffredi et al., 1997). Iron and manganese, the most abundant transition metals in the MAR hydrothermal fluids (James et al., 1995b; Douville et al., 1997), as in other MOR fluids (Von Damm, 1995), can be regarded as major sources of mineral particles. Iron and manganese enrichments are also usually associated with enrichment of other transition metals, potentially toxic to organisms. Zinc and copper are the next most concentrated transition metals in proportions ranging between 1 and 10% of the iron contents (Douville et al., *op. cit.*). Several other transition metals (e.g. Cd, Pb, Ag, Sb) have been also found in MAR fluids at concentrations ranging from a few nanomoles per kilogram to a few tens of nanomoles per kilogram of fluid (Douville et al., 1997).

Snake Pit, TAG and Lucky Strike vent fluids are well documented with data coming from several diving cruises, enabling inter-annual and intra field comparison. For the other fields, especially Rainbow, Broken Spur and Logatchev, the data are scarcer and do not necessarily account for the potential variability within a field.

The Menez Gwen fluids exhibit a uniform chemical composition. Chloride depletion and CH_4 enrichment indicates a dominant contribution of the vapour phase resulting from phase separation (Douville et al., 1999). This gas enrichment is consistent with the fact that Menez Gwen is considered a young site at the beginning of its activity. Menez Gwen is located in a basaltic environment, methane is produced by outgassing of carbon from the mantle and is related to the carbon-enriched character of basalt (Charlou et al., 1997). In addition, the low pH and low Fe and Si concentrations are consistent with a short duration of

Table 2. End-member temperature (°C) and concentrations of chemical species (mmol kg^{-1}) for the different MAR vent fields (MG: Menez Gwen (Douville et al., 1997, 1999; Fouquet et al., submitted; Charlou et al., 2000); LS: Lucky Strike (Von Damm et al., 1998; Langmuir et al., 1997; Donval et al., 1997; Douville et al., 1997; Fouquet et al., submitted); Rb: Rainbow (Charlou et al., 1997; Donval et al., 1997; Douville et al., 1997, 1999; Charlou et al., 2000, submitted); BS: Broken Spur (James et al., 1995a; Fouquet et al., submitted); T.A.G.: Trans Atlantic Geotraverse (Charlou & Donval, 1993; Von Damm, 1995; Edmonds et al., 1996; Rudnicki & Elderfield, 1992; Charlou et al., 1996; Gamo et al., 1996); SP: Snake Pit (Von Damm, 1995; Campbell et al., 1988; Jean-Baptiste et al., 1991; Edmond et al., 1995; Rudnicki and Elderfield, 1992); Lg: Logatchev (Charlou et al., 1997, submitted)

Site	MG	LS	Rb	BS	T.A.G.	SP	Lg
T	265–284	152–333	360–365	356–364	270–363	335–356	>353
pH	4.2–4.8	3.5–4.9	2.8–3.1		2.5–3.4	3.7–3.9	<3.3
Si	8.2–11.2	9.1–17.5	6.9–8.0		18–22	18–20	7–8.2
Cl^-	360–400	410–540	>750	469	633–675	550-563	515–522
CO_2	17–20	8.9–28	<16		2.9–4.1	10.63	
H_2S	1.5–2	1.4–3.3	1–2.5	9.3	2.5–6.7	2.7–6.1	<1
CH_4	1.35–2.63	0.5–0.97	2.2–2.5	0.065	0.14–0.62	0.046–0.062	2.1
Fe	0.002–0.018	0.13–0.86	24	1.68–2.16	1.64–5.45	1.8–2.56	2.50
Mn	0.068	0.45	2.25	0.26	1	0.49	0.33

fluid-rock interaction linked to a shallow circulation system induced by the Azores Hot Spot (Douville et al., 1999).

A high variability in fluid composition between vents was observed at Lucky Strike and two sources of fluids were hypothesised (Von Damm et al., 1998). Lucky Strike vent fluids have chlorinities equal or depleted with respect to ambient seawater and are thought to have undergone phase separation (Von Damm et al., *op. cit.*). Lucky Strike is associated usually with Menez Gwen when comparing vent fluid characteristics. Both are basalt-hosted sites, strongly affected by recent volcanic activity and shallow circulation systems. The vapour phase enrichment is, however, more moderated at Lucky Strike than at Menez Gwen. The unusual aspect of Lucky Strike fluid chemistry is the relatively high pH and low concentration of Fe, Mn and Zn. This was attributed to reaction with an altered substrate, depleted in these elements (Von Damm et al., *op. cit.*).

Rainbow fluids are uniform in composition and influenced by phase separation, which generates stable Cl-rich fluids indicating a high brine contribution (Douville et al., 1997). Compared to the Menez Gwen field, little evidence of recent volcanic activity or hydrothermal activity was found at Rainbow. Hydrothermal circulation is considered to be tectonically controlled and activated by exothermic serpentinization processes (Douville et al., 1997). Serpentinization is the cause of high methane concentration in fluids. Transition metal (Fe, Mn, Cu) content in Rainbow fluids is the highest observed in the MAR hydrothermal area. This metal enrichment is likely to be related to the formation of Cl-complexes at the high temperature reached by these fluids (Douville et al., 1999).

Broken Spur is located in a basaltic substratum. Hydrothermal activity within this field was shown to be controlled by a combination of recent volcanic activity and tectonic activity (Murton et al., 1995). Fluid depletion in chloride (–14%) is attributed to supercritical phase separation.

Sampled TAG fluids are inferred to be from the same source of fluid (Von Damm et al., 1998). As for Rainbow, high chlorinity and metal content are attributed to phase separation (Douville et al., 1997). Hydrothermal circulation through the basaltic layer carries CO_2 and CH_4 of magmatic origin (Charlou et al., 1996). The Snake Pit fluids have almost the same composition and temperature as fluids collected in the TAG area, suggesting that the factors controlling the chemistry of the major elements are similar (Jean-Baptiste et al., 1991).

Like Rainbow, Logatchev is located in an ultramafic environment. Composition of hot fluids confirms that serpentinization is a major process producing high methane flux (Charlou et al., 1997).

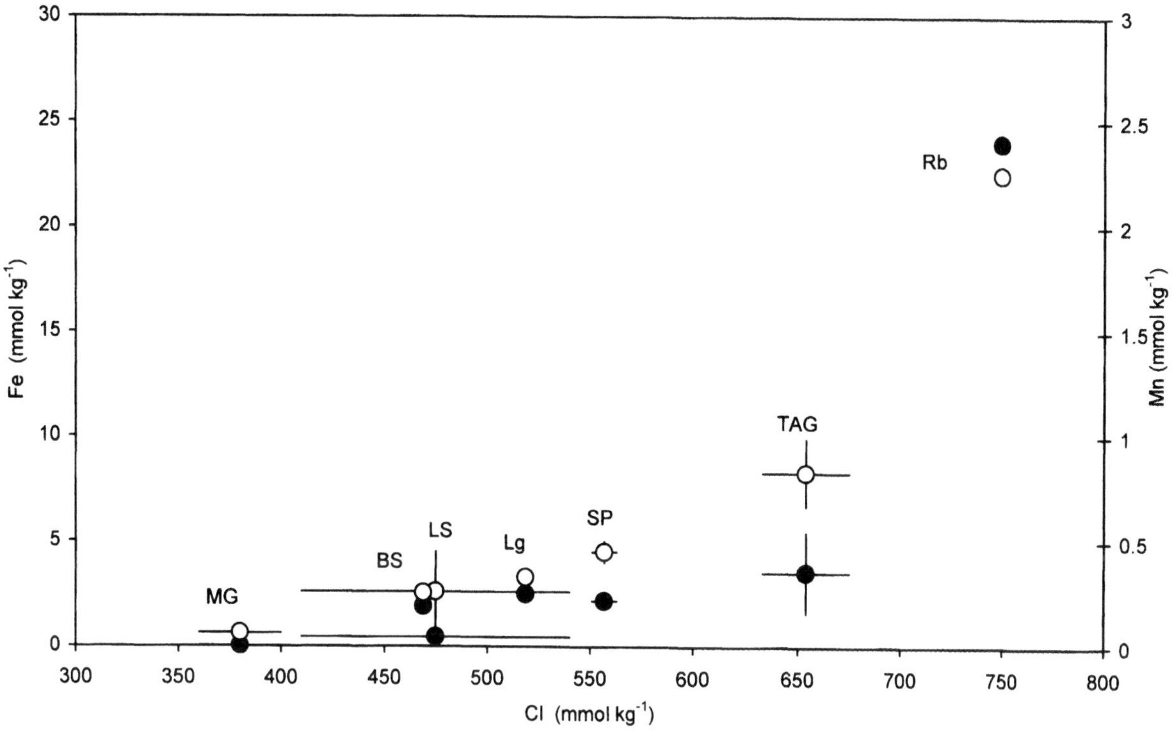

Figure 2. Iron (black circles) and manganese (open circles) concentrations versus chloride concentration in the end-member fluids of the MAR hydrothermal fields (when more than one data set is available, error bars indicate the range of end-member concentrations determined for one site). Chloride concentration in seawater is about 550 mmol kg^{-1}. End-member chloride concentrations above this value indicates a high brine contribution, and fluid depleted in chloride compared to seawater should be enriched in the vapour phase.

Comparison of end-member chemical composition between fields

The above observations, and characteristics listed in Table 2, enable the sites to be distinguished with regards to their end-member fluid composition. Figure 2 shows that manganese and iron covary with chloride as expected from metal enrichment in the brine following phase separation processes. The opposite trend predicted for CH_4 content is less noticeable. Rainbow differs clearly from the other sites with a high methane content associated with the highest chlorinity and metal content. Methane enrichment in the fluids of Logatchev vents, which is also hosted in an ultramafic environment, was also highlighted (Charlou et al., 1997).

As shown in Figure 3, CH_4 enrichment (Log atchev, Rainbow, TAG) corresponds to H_2S depletion in the fluids and this inverse relation is verified for all the studied MAR vent fields.

Particle fluxes

The flux of mineral and organic particles in the vicinity of vents is a major ecological factor both in terms of nutrition and toxicity. Mineral particulate flux can alter the feeding activity of filter feeding and deposit feeding animals and, when accumulating on the bottom, may change edaphic properties and affect larval or post-larval settlement. In the studied areas, fluid aspect ranges from clear shimmering fluids at Menez Gwen to black particle loaded fluids at Rainbow and TAG. Quantification (Table 3) of particulate flux was made using sediment traps positioned with the submersible in the close vicinity of vents. A sediment trap moored off the vent fields gave the pelagic input reference.

The total particle fluxes measured with sediment traps (Khripounoff & Albéric, 1991) are of the same magnitude at the two deepest sites studied. At the shallow fields (Menez Gwen and Lucky Strike), total fluxes are ten times smaller than those of TAG and Rainbow. The same trends were observed for the particulate sulfur contribution. The particulate organic carbon flux was of the same order of magnitude (9–48 mg m^{-2} d^{-1}) in all the vent fields studied.

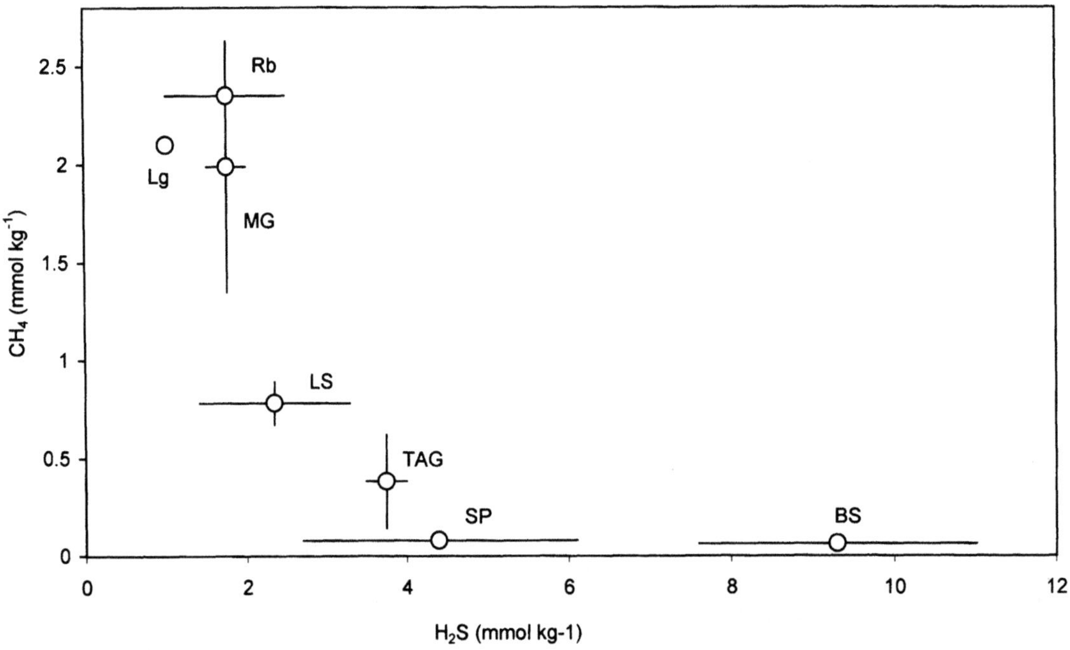

Figure 3. Methane concentration versus chloride concentration in the end-member fluids of the MAR hydrothermal fields.

Table 3. Particle fluxes (raw data not corrected for current headings) measured by sediment traps at four hydrothermal vent fields on the Mid-Atlantic Ridge

Site	Location from the vent	Total flux mg m^{-2} d^{-1}	Organic C flux mg m^{-2} d^{-1}	Inorganic C flux mg m^{-2} d^{-1}	Total Nitrogen flux mg m^{-2} d^{-1}	Total S flux mg m^{-2} d^{-1}
TAG	2 m northward	9024.7	48.8	not detected	4.46	2733.7
	pelagic reference	11.3	0.49	0.95	0.05	0.2
Rainbow	2 m Southwest	6890	9.3	not detected	0.9	1171
	Pelagic reference	11.2	0.62	1.32	0.07	0.04
Lucky Strike	2 m southward	264.5	12.06	9.4	2.0	29.0
	Pelagic reference	7.7	0.55	0.75	0.09	0.07
Menez Gwen	2 m southward	704	11.26	not detected	0.85	126.7
	Pelagic reference	28.5	2.51	2.18	0.41	0.11

Table 4. Schematic description of the hydrothermal vent communities from the known hydrothermal fields on MAR based on published observations and personal data. Numbers (ranging from 1 to 5) between square brackets are visual estimation of the distribution and dominance of megafauna: [1], rare; [2], rare / patchy; [3], common, [4] dominant/patchy, [5] Exclusive

	Dominant species on chimneys walls	Dominant species on chimneys bases	Dominant species on surrounding blocs within active area	Dominant accompanying species	Dominant carnivorous	Dominant peripheral species
Logatchev (Gebruk et al., 1997a; Gebruk et al., 1997b)	*Bathymodiolus* cf. *puteoserpensis* [3] + *Branchipolynoe seepensis* [3] *Rimicaris exoculata* [2]	*Bathymodiolus* cf. *puteoserpensis* [3] + *Branchipolynoe seepensis* [3]		*Ophioctenella acies* [3] *Mirocaris keldyshi* [3]	*Segonzacia* sp.	Actinaria gen. sp. [2]
Snake Pit (Segonzac, 1992; Segonzac et al., 1993)	*Rimicaris exoculata* [4]	*Chorocaris chacei* [2] *Alvinocaris markensis* [1]	*Bathymodiolus puteoserpensis* [1] + *Branchipolynoe seepensis* [1] *Ophioctenella acies* [1]	*Pseudorimula midatlantica* *Peltospira* sp.	*Segonzacia mesatlantica* [2] *Pachychara thermophilum* [2]	*Parasicyonis ingolfi* [2] *Candelabrum serpentarii* [1] *Chaetopterus* sp. [1]
TAG (Gebruk et al., 1997a; Van Dover, 1995)	*Rimicaris exoculata* [4]	*Chorocaris chacei* [2] *Alvinocaris markensis* [1]	*Ophioctenella acies* [1]	*Stygiopontus pectinatus*	*Segonzacia mesatlantica* *Pachychara thermophilum* [1]	*Chaetopterus* sp. *Parasicyonis* ? sp. [3] *Munidopsis* sp.
Broken Spur (Murton et al., 1995; Gebruk et al., 1997a)	*Rimicaris exoculata* [2]	*Mirocaris fortunata* [1]	Spionid polychaetes [2] *Bathymodiolus puteoserpensis* [1] & *Bathymodiolus azoricus* *Branchipolynoe seepensis* [1] *Ophioctenella acies* [2]		*Segonzacia mesatlantica* [2]	*Chaetopterus* sp. Actininaria gen. sp.
Rainbow (Desbruyères et al., subm)	*Rimicaris exoculata* [3] *Mirocaris fortunata* [2]	*Mirocaris fortunata* [2] *Chorocaris chacei* [1]	*Bathymodiolus azoricus* [2] + *Branchipolynoe seepensis* [1]	*Amatys lutzi*	*Segonzacia mesatlantica* [2]	*Chaetopterus* sp. [1] *Prionospio* sp. [2]
Lucky Strike (Desbruyères et al., subm) (Van Dover et al., 1996)	*Bathymodiolus azoricus* [5] + *Branchipolynoe seepensis* [5] *Chorocaris chacei* [2] *Mirocaris fortunata* [3]	*Bathymodiolus azoricus* [5] + *Branchipolynoe seepensis* [5] *Mirocaris fortunata* [3]	*Bathymodiolus azoricus* (young specimens) [3] + *Branchipolynoe seepensis* [3]	*Amathys lutzi* *Lepetodrilus* n. sp.	*Segonzacia mesatlantica* [2]	*Candelabrum phrygium* [2]
Menez Gwen (Desbruyères et al. subm) (Colaço et al., 1998)		*Bathymodiolus azoricus* [3–5] according to the vent sites *Mirocaris fortunata* [2–3] *Chorocaris chacei* [1]	*Bathymodiolus azoricus* [3–5] according to the vent sites *Mirocaris fortunata* [2–3] *Chorocaris chacei* [1]	*Protolyra valvatoidae*	*Chaceon affinis* [2] *Segonzacia mesatlantica* [1]	*Eudendrium* sp. *Grammaria abietina*

MAR hydrothermal vent communities

Faunal composition of the hydrothermal vent communities on MAR was described previously (Grassle 1986; Moskale, 1992; Segonzac, 1992; Van Dover 1995; Murton et al., 1995; Van Dover et al., 1996; Saldanha et al., 1996; Gebruk et al., 1997a; Desbruyères et al., in press). Several phylogenetic analyses, based on molecular techniques, have been made on the dominant species along the MAR and between ridges in order to resolve relationships (Comtet, 1998; Shank et al., 1998a, 1999; Chevaldonné et al., 1998; Jollivet et al., 1998). Nevertheless, part of the taxonomic work is still in progress and the following faunal description could be subject to future variations in taxonomic content, at least at the level of accompanying species; for an updated species list see Desbruyères et al. (submitted). Based on a 'landscape approach', used previously in MAR hydrothermal vent fields by Gebruk et al. (1997a), Table 4 gives a global scheme of the dominant species in vent communities along the ridge axis from Logatchev to Menez Gwen vent fields. A continuum from *Rimicaris*-dominated assemblages to *Bathymodiolus*-dominated assemblages comes out when considering the whole range of variations within MAR hydrothermal vent communities. The MAR vent communities, ordered according to their faunal composition, range from TAG to Menez Gwen assemblages: TAG ↔ Snake Pit ↔ Rainbow ↔ Broken Spur ↔ Logatchev ↔ Lucky Strike ↔ Menez Gwen. Compared to other vent field assemblages, Broken Spur and Logatchev appear to be substantially impoverished in terms of abundance. The two dominant taxa are associated with autotrophic bacteria: *Bathymodiolus* species (Mytilidae) are sessile organisms with endosymbionts belonging to sulphide-oxidising chemoautotrophs and methanotroph bacteria (Cavanaugh et al., 1992; Fiala-Médioni et al., 1996). Nevertheless, although anchored by byssal filaments, *Bathymodiolus* is capable of migration along sulphide gradients and is also able to grow on methane alone as demonstrated by Dando et al. (submitted). Although symbiotrophic, *Bathymodiolus* spp. still behave as filter feeding organisms, requiring a constant flow of water through their gills. *Rimicaris exoculata* (Bresiliidae) is a mobile swimming species and its source of nutrition is still debated, but probably (Segonzac et al., 1993; Polz et al., 1998; Rieley et al., 1999) largely reliant on bacterial epibionts living inside the branchial chamber and also on gut microflora. Based on stable isotope analysis, Colaço et al. (submitted) recognise, among all the MAR vent fields considered in this study, an obvious dichotomy of the trophic web based on either *Bathymodiolus* spp. beds being significantly depleted in ^{13}C or *Rimicaris* swarms being substantially enriched in ^{13}C. In a first analysis, the gradation between the two assemblages, based on dominant populations among vent communities, cannot be related to geographic distances or to an obvious bathymetric zonation. When considering the relationship between the chemical environment and the characteristics of the faunal assemblages, a high level of metals and high chlorinity of the fluid is associated with *Rimicaris*-dominated communities, while low levels are related to *Bathymodiolus*-dominated communities. In a first attempt, when considering our data on mineral particle fluxes, among the four studied vent fields, *Rimicaris*-dominated assemblages are characteristic of high mineral particle fluxes while *Bathymodiolus*-dominated assemblages are characteristic of fluids depleted in mineral particles. The very high production of inorganic particles at TAG and Rainbow can not only affect the filtration of the mussels, but also juvenile growth as demonstrated previously in *Mercenaria mercenaria* (Bricelj et al., 1984).

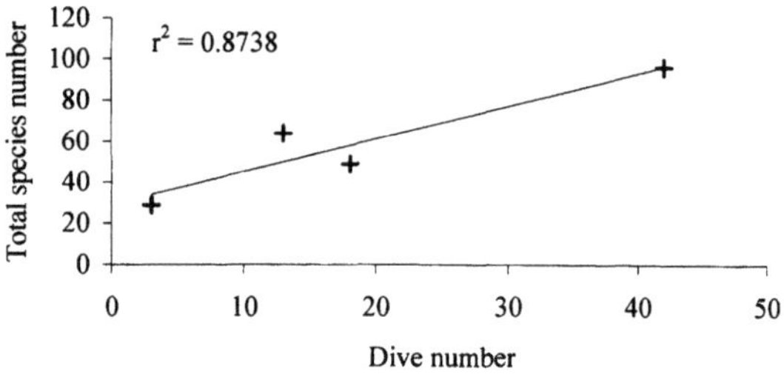

Figure 4. Number of species collected on the four studied vent fields against the number of dives on each field. The total number of species ranges from 29 for Rainbow to 96 for Lucky Strike.

MAR geographic patterns in hydrothermal vent fauna

The taxonomic work on MAR hydrothermal vents, on which the present biogeographic analysis relies, is still in progress both for classical taxonomy and molecular phylogenetic analyses, and may still evolve in the future. Our biogeographical study is based on the analysis of a presence/absence matrix referring to the theory that the distribution of taxa represents information on its history over an evolutionary time scale. The species list has been made from that published in Desbruyères & Segonzac (1997), and updated from recent papers and from unpublished results kindly communicated by taxonomists. Species have been separated into 'hydrothermal' and 'penetrating'. A species has

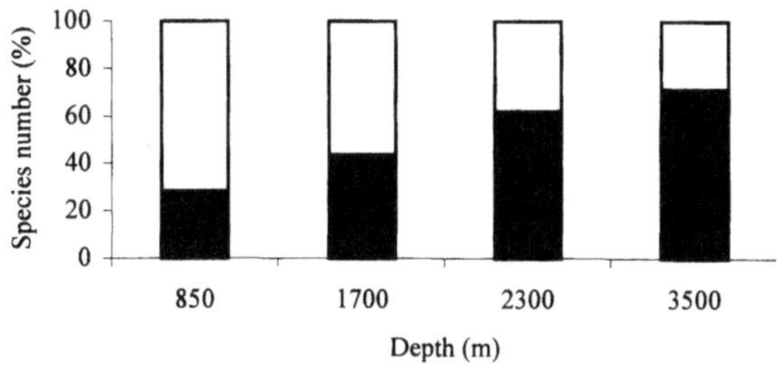

Figure 5. Relative numbers of hydrothermal (black) and penetrating (white) species against depth inside the hydrothermal vent influence area.

been considered 'hydrothermal' if it is known only from hydrothermal vent environments and if it lives close to fluid emissions. Thus, each new species found in active areas has been considered as 'hydrothermal'. Conversely, each species known to occur in the normal deep-sea habitats has been considered as 'penetrating'. This led us to distinguish 67 'hydrothermal' and 105 'penetrating' species among those sampled on MAR in this environment. An important bias in such a study is caused by the differences in sampling effort in each vent field. The number of species increases with the number of dives, which is used as an estimation of the sampling effort (Fig. 4). This trend is amplified by the fact that the poorly dense vent fields deserve less sampling effort due to scientific profitability of dives. This is the reason why we considered herein only the four best sampled fields (Menez Gwen, Lucky Strike, Rainbow and Snake Pit). Even with such care, our results give only a provisional image of what we know at the time of the analysis.

Relation between depth and faunal composition

As expected, the deepest sites are dominated by 'hydrothermal' species whereas 'penetrating' species are the main component of the shallowest field (i.e. Menez Gwen), where 'hydrothermal' species represent only 28% of the community (Fig. 5).

The penetration of non-hydrothermal fauna inside the vent habitat may be a consequence of phase separation caused by the decrease of hydrostatic pressure leading to a non-metal loaded fluid as revealed by the low metal/chlorinity content (see Fig. 2) rather than a direct bathymetric effect. This explanation, based on the low toxicity of metal-depleted hydrothermal fluids, was proposed for the first time by Valdès & Bouchet (1998) for the occurrence of a nudibranch (*Dendronotus*) in the Lucky Strike vent field.

Similarity analysis

In order to analyse species distribution between the four best-known vent fields, we used the Sorensen similarity coefficient:

$$I_s = \frac{2a}{2a + b + c},$$

where a is the number of species found in the two vent fields, b the number of species found only in the first vent field, and c the number of species found only in the second vent field. Results are given in Table 5.

Similarity values are very low when only the 'penetrating' species are considered. This means that these deep-sea species are different from one field to another. As they are those found in the common deep-sea environment, the observed differences are likely to be due to differences in the zonation patterns between the four fields, but may also result from edaphic characteristics of the peripheral areas (soft bottom vs. hard bottom).

Similarities between hydrothermal vent communities are higher when only the 67 'hydrothermal' species are considered. Their significance level has been tested using a Monte Carlo procedure (Sokal & Rohlf, 1980). As recommended by Jackson & Somers (1989), we performed 10 000 random permutations of the initial matrix of presence/absence data. The similarity coefficient was calculated after each permutation. The probability that similarity is due to chance, is then the ratio between the number of times the recalculated coefficient is superior to the observed one (i.e. when the similarity by chance is greater than the observed one) and the number of permutations. Results are given in Table 6.

These results show that the hydrothermal fauna in Snake Pit is clearly different from that of the three other fields ($p>0.6$ for each of the three similarities). The most significant similarity between faunas ($p<0.01$) is observed for the closest fields (Lucky Strike and Menez Gwen, separated by 49 nautical miles), whereas the similarities between these fields and Rainbow are less significant (p value ranging from 0.05 to 0.1). These values have been used to draw a similarity tree using the simple linkage clustering method (Fig. 6a). When considering the total fauna, similarity trees are different, grouping Rainbow and Snake Pit on one hand, Lucky Strike and Menez Gwen on the other hand, although the similarities are less significant (Fig. 6b). The 'penetrating' species composition is likely to be controlled by depth whereas

Table 5. Sorensen similarity coefficients and geographic distances: above diagonal: 'hydrothermal' species and geographic distances in nautical miles; under diagonal: 'penetrating' species and total species

	Snake Pit	Rainbow	Lucky Strike	Menez Gwen
Snake Pit	1	0.3396 / 199	0.3896 / 300	0.3396 / 349
Rainbow	0.1600 / .02821	1	0.4667 / 101	0.4444 / 151
Lucky Strike	0.0588 / 0.2345	0.0615 / 0.2560	1	0.5333 / 49
Menez Gwen	0.0667 / 0.1947	0.0702 / 0.2151	0.2800 / 0.3750	1

Table 6. Probability that the similarity is due to chance, calculated after 10 000 permutations of the presence/absence data matrix

	Snake Pit	Rainbow	Lucky Strike	Menez Gwen
Snake Pit	–			
Rainbow	0.6977	–		
Lucky Strike	0.9999	0.1015	-	
Menez Gwen	0.6808	0.0509	0.0072	–

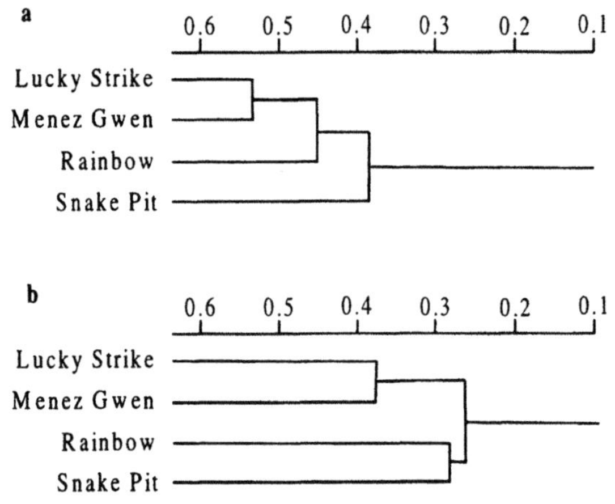

Figure 6. Similarity trees based on the Sorensen similarity coefficient. These trees were drawn using the simple linkage clustering method (a) hydothermal species, (b) all species.

'hydrothermal' species composition is controlled by the chemistry of the fluid.

Discussion and conclusion

The occurrence of hot hydrothermal venting on MAR between 12° N and 38° N, as revealed by chemical tracer detection, is more important than previously thought, mainly in the northernmost part where the frequency of occurrence is close to that described previously on fast-spreading ridges. A new kind of warm methane venting was observed on the Famous segment (Saldanha vent site) and deserves further studies. This new hydrothermal site is not colonised by endemic hydrothermal species, but by filamentous bacterial mats and tetractinellid sponges (*Cynachira*). Hexactinellid sponges (Euplectellidae) were also observed at the periphery of active vents in the North Fiji Basin (Desbruyères et al., 1994) and methanotrophic *Cladorhiza* (Desmospongiae) were abundant at Barbados cold seeps at c.a. 5000 m (Vacelet & Boury-Esnault, 1995; Olu et al., 1997). Although the distribution of the phenomenon is unknown, several clues lead us to think that many methane vent fields could be present in the vicinity of the Azores Triple Junction. Nevertheless, because of its faunal composition, it seems unlikely that this new kind of methane venting could serve as stepping stones between hot vent fields.

A continuum between *Rimicaris*-dominated and *Bathymodiolus*-dominated assemblages emerges from our 'landscape approach' based on semi-quantitative data. Bathymetric zonation or geographic distance between vent fields cannot explain this continuum. Most likely it is related to the metallic content (dissolved and particulate) of the end-member fluid. The development of mussel beds could be hindered by iron oxide precipitation and/or heavy mineral particle fluxes being deleterious both to filter feeding behaviour and larval settlement. Among the seven vent fields considered, only Broken Spur does not follow this trend, possibly not only due to an incomplete series of chemical samples within a heterogeneous vent field (James et al., 1995a). Nevertheless, no relation is obvious among studied MAR vent fields with respect to abundance of organisms in communities and Fe/H_2S ratio as suggested previously by Von Damm (1995). Our hypothesis of an inverse relationship between filter feeding organisms and mineral particles has to be tested experimentally. Conversely, *Rimicaris* populations are able to thrive in an environment with high concentration of metals, possibily not only because of their ability to escape from contamination, but also to detoxify the environment through symbiotic relationships. Tyler & Young (1999) assume that young individuals and berried females of *Rimicaris* may live away from the vent influence or in less toxic parts of the habitat (Shank et al., 1998b). The absence of swarms of *Rimicaris* in shallowest vent fields remains enigmatic as only few individuals were sampled at the Lucky Strike vent field. This absence may result from interspecific competition with mussels for access to habitat, which for both, it is the porous parts of the chimney walls according to Segonzac et al. (1993). The end member fluid chemistry provides a first estimate of the major ecological structuring factors at the scale of the ridge. Nevertheless, further work on fluid chemistry, at the level of population, by *in-situ* analysis is still needed as well as measurements of chemical fluxes (dissolved and particulate) at the very proximity of organisms to assess the microscale factors involved.

Specific similarities between the four best known vent fields give a complex image, which results from dispersal and recruitment parameters. Larval biology, hydrodynamic patterns and distance between vent fields influence dispersal. Propagule availability and local environmental conditions influence recruitment. Examination of the species list demonstrates an increase in non-endemic species ('penetrating') within the vent habitat at the approach of the Azores Plateau where depth decrease from 2300 m in the Rainbow vent field to 840 m in the Menez Gwen vent field. This increase results probably from a decrease in fluid toxicity as revealed by metallic content and sulphide concentration. When considering the cladogram of specific similarities of endemic fauna between the four best known hydrothermal vents, the influence of the distance between vent fields appears to be a first order parameter. Transform faults may represent major barriers to dispersal of vent species along the ridge. Between 14° 45′ N and 37° 51′ N, only three major transform faults (Kane, Oceanographer and Famous) break the ridge ranging from 145 to 20 km in length (Van Dover, 1995). Within the Azores Triple Junction area, the geographic proximity, and the absence of conspicuous discontinuities between vent fields, allows the exchange of propagules of the vent endemic species when bathymetric variations, induced by the proximity of the hot spot, introduce zonation differences among non-endemic assemblages. Nevertheless, when considering presence/absence matrix, the faunal similarity between the different vent fields remains relatively low ($\leq$30% of hydrothermal species only are shared between two vent fields). One cannot consider MAR vent communities, or even the Azores Triple Junction communities, as a single biogeographical entity but as a series of distinct faunal islands having different composition and habitat requirements.

Acknowledgements

We thank the crews of N/O 'Nadir', 'L'Atalante' and manned submersible 'Nautile' for invaluable technical assistance with deployment and retrieval of experimental arrays and the collection of specimens. We are indebted to the taxonomists and chemists who informed us of the content of their pending manuscripts (J.-L Charlou, R. Von Cosel, A. Warén and P. Bouchet), and G. Beaugrand (SAHFOS, Plymouth, U.K.) for his help in the analysis of similarity. We are grateful to C. Mével, A. Fiala-Médioni, C. Van Dover, D. Prieur, F. Barriga and Y. Fouquet chief scientists of the cruises 'Hydrosnake', 'MAR'93', 'Microsmoke' and 'Saldanha' respectively. This research has been undertaken in the framework of the AMORES project. We acknowledge the support from the European Commission's Marine Science and Technology Program (MAST III) under contract MAS3-CT95-0040. The cruises Diva 2, Marvel and Pico were supported by IFREMER.

References

Barriga, F., Y. Fouquet, A. Almeida, M. Biscoito, J.-L. Charlou, R. Costa, A. Dias, A. Marques, J. M. Miranda, K. Olu, F. Porteiro & M. G. Queiroz, 1998. Discovery of the Saldanha hydrothermal field on the FAMOUS segment of the M.A.R.(36° 30′ N)., Eos Trans. Am. Geophys. Union: 837 pp.

Barriga, F. J. A. S., 1999. Actividade hidrotermal no fundo do mar dos Açores: Estado da Arte, Colóquio/Ciências, 23: 44–59.

Bricelj, V. M., R. E. Malouf & C. De Quillfeldt, 1984. Growth of juvenile *Mercenaria mercenaria* and the effect of resuspended bottom sediments, Mar. Biol. 84: 167–173.

Campbell, A. C., M. R. Palmer, G. P. Klinkhammer, T. S. Bowers, J. M. Edmond, J. R. Lawrence, J. F. Casey, G. Thompson, S. Humphris, P. Rona & J. A. Karson, 1988. Chemistry of hot springs on the Mid-Atlantic Ridge, Nature 335: 514–519.

Cann, J., C. Van Dover, C. Walker, P. Dando & B. Murton, 1994. Diversity of vent ecosystems (DOVE). Bridge 4: 1–39.

Cavanaugh, C. M., C. O. Wirsen & H. W. Jannasch, 1992. Evidence for methylotrophic symbionts in a hydrothermal vent mussel (Bivalvia: Mytilidae) from the Mid-Atlantic Ridge. Appl. Environ. Microbiol. 58: 3799–3803.

Charlou, J.-L., J.-P. Donval, E. Douville, J. Knoery, Y. Fouquet, H. Bougault, P. Jean-Baptiste, M. Stievenard & S. Michel, 1997. High methane flux between 15° N and the Azores Triple Junction, Mid-Atlantic Ridge. Eos Trans. Am. Geophys. Union 78 (46): F831.

Charlou, J. L. & J. P. Donval, 1993. Hydrothermal methane venting between 12° N and 26° N along the Mid-Atlantic Ridge. J. Geophys. Res., B, 98: 9625–9642.

Charlou, J. L., J. P. Donval, P. Jean-Baptiste, A. Dapoigny & P. A. Rona, 1996. Gases and helium isotopes in high temperature solutions sampled before and after ODP leg 158 drilling at TAG hydrothermal field (26° N, MAR). Geophys. Res. Lett. 23 (23): 3491–3794.

Charlou, J.-L., J.-P. Donval, E. Douville, P. Jean-Baptiste, J. Radford-Knoery, Y. Fouquet, A. Dapoigny & M. Stievenard, 2000. Compared geochemical signatures and the evolution of Menez Gwen (37° 50′ N) and Lucky Strike (37° 17′ N) hydrothermal fluids, south of the Azores Triple Junction on the Mid-Atlantic Ridge. Chem. Geol. 171: 49–75.

Charlou, J.-L., J.-P. Donval, N. Holm, E. Douville, J. Radford-Knoery, Y. Fouquet & C. German, submitted. Hydrogen: energy source for abiotic hydrocarbons and prebiotic molecule synthesis at the Rainbow serpentinized hydrothermal field (36° 14′ N, M.A.R.), Geochim. Cosmochim. Acta.

Chevaldonné, P., D. Jollivet, R. A. Feldman, D. Desbruyères, R. A. Lutz & R. C. Vriejenhoek, 1998. Commensal scale-worms of the genus *Branchipolynoe* (Polychaeta: Polynoidae) at deep sea hydrothermal vents and cold-seeps. Cah. Biol. Mar. 39: 347–350.

Childress, J. J. & C. R. Fisher, 1992. The biology of hydrothermal vent animals: Physiology, biochemistry, and autotrophic symbioses. Oceanogr. Mar. Biol. Annu. Rev. 30 (4044): 337–441.

Colaço, A., F. Dehairs & D. Desbruyères. Nutritional relations of deep sea hydrothermal fields at the Mid-Atlantic Ridge: a stable isotope approach, Deep-Sea Res. submitted.

Colaço, A., D. Desbruyères, T. Comtet & A. M. Alayse, 1998. Ecology of the Menez Gwen hydrothermal vent field (Mid-Atlantic Ridge/Azores Triple Junction). Cah. Biol. Mar. 39: 237–240.

Comtet, T., 1998. Structure des populations, reproduction, croissance et phylogéographie des Mytilidae des champs hydrothermaux Lucky Strike et Menez Gwen (37° 17′ N et 37° 50′ N sur la dorsale médio-atlantique), Doctorat d'Université, thèse à l'Université de Bretagne Occidentale, Brest, France.

Dando, P., A. Fiala-Médioni, R. Stöhr, P.-M. Sarradin & A.-M. Alayse-Danet, 1998. Behaviour of vent mussels with methanotrophic and sulphur-oxidising symbionts. Nature submitted.

Desbruyères, D., 1998. Temporal variations in the vent communities on the East Pacific Rise and Galapagos Spreading Centre: a review of present knowledge. Cah. Biol. Mar. 39: 241–244.

Desbruyères, D., A. M. Alayse-Danet, S. Ohta & Scientific parties of STARMER and BIOLAU cruises, 1994. Deep sea hydrothermal communities in Southern Pacific back-arc basins (the North Fiji and Lau Basins): composition, microdistribution and food-web. Mar. Geol. 116: 227–242.

Desbruyères, D., M. Biscoito, J.-C. Caprais, A. Colaço, T. Comtet, P. Crassous, Y. Fouquet, A. Khripounoff, N. Le Bris, K. Olu, R. Riso, P.-M. Sarradin, M. Segonzac & A. Vangriesheim, in press. Variations in deep sea hydrothermal vent communities on the Mid-Atlantic Ridge when approaching the Azores Plateau. Deep-Sea Res. II

Desbruyères, D. & M. Segonzac, 1997. Handbook of Deep-sea Hydrothermal Vent Fauna. 279 pp. IFREMER, Brest:

Donval, J.-P., J.-L. Charlou, E. Douville, J. Knoery, Y. Fouquet, E. Poncevera, P. Jean-Baptiste, M. Stievenard & C. German, 1997. High H_2 and CH_4 content in hydrothermal fluids from rainbow site newly sampled at 36° 14′ N on the AMAR segment, Mid-Atlantic Ridge (diving Flores cruise, July 1997). Comparison with other MAR sites. Eos Trans. Am. Geophys. Union 78 (46): 832.

Douville, E., P. Bienvenu, J. L. Charlou, J.-P. Donval, Y. Fouquet, P. Appriou & T. Gamo, 1999. Yttrium and rare earth elements in fluids from various deep sea hydrothermal systems. Geochim. Cosmochim. Acta. 63(5): 627–643.

Douville, E., J.-L. Charlou, J.-P. Donval, D. Hureau & P. Appriou, 1999. Le comportement de l'arsenic (As) et de l'antimoine (Sb) dans les fluides provenant de différents systèmes hydrothermaux océaniques. C. R. Acad. Sci. Paris, Sér. II. 328: 97–104.

Douville, E., J.-L. Charlou, J.-P. Donval, J. Knoery, Y. Fouquet, P. Bienvenu & P. Appriou, 1997. Trace elements in fluids from the new Rainbow hydrothermal field (36° 14′ N, MAR): a comparison with other Mid-Atlantic fluids. Eos Trans. Am. Geophys. Union. 78 (46): 832.

Edmond, J. M., A. C. Campbell, M. R. Palmer, G. P. Klinkhammer, C. R. German, H. N. Edmonds, H. Elderfield, G. Thompson & P. Rona, 1995. Time series studies of vent fluids from the TAG and MARK sites (1986, 1990) Mid-Atlantic ridge: a new solution chemistry model and a mechanism for Cu/Zn zonation in massive sulphide orebodies. In Parson, C.L.W.L.M. & D.R. Dixon (eds), Hydrothermal Vents and Processes. The Geological Society of London, London: 77–86.

Edmonds, H. N., C. R. German, D. R. H. Green, Y. Huh, T. Gamo & J. M. Edmond, 1996. Continuation of the hydrothermal fluid chemistry time series at TAG, and the effects of ODP drilling. Geophys. Res. Lett. 23 (23): 3487–3489.

Fiala-Médioni, A., C. Cavanaugh, P. Dando & C. Van Dover, 1996. Symbiotic mussels from the Mid-Atlantic ridge: adaptations to trophic resources. J. Conf. Abstr. 1 (2): 788, 1996.

Fisher, C. R., 1990. Chemoautotrophic and methanotrophic symbioses in marine invertebrates. Rev. Aquat. Sci. 2: 399–613.

Fouquet, Y., 1997. Where are the large hydrothermal sulphide deposits in the oceans? Phil. Trans. r. Soc. Lond. A 355: 427–441.

Fouquet, Y., J.-L. Charlou, I. Costa, J.-P. Donval, J. Radford-Knoery, H. Pellé, H. Ondréas, N. Lourenço, M. Segonzac & M. K. Tivey, 1994. A detailed study of the Lucky Strike Hydrothermal site and discovery of a new hydrothermal site: Menez Gwen; preliminary results of the Diva 1 Cruise (5–29 May, 1994). InterRidge News 3 (2): 14–17.

Fouquet, Y., J.-L. Charlou, J.-P. Donval, J. Radford-Knoery, H. Ondréas, P. Cambon, H. Bougault, J. Etoubleau, F. J. A. S. Barriga, I. M. A. Costa, N. Lourenço & M. K. Tivey, submitted. Hydrothermal processes on shallow volcanic segments: Mid Atlantic Ridge near the Azores Triple Junction. J. Geophys. Res. B.

Fouquet, Y., J.-L. Charlou, H. Ondréas, J. Radford-Knoery, J.-P. Donval, E. Douville, R. Apprioual, P. Cambon, H. Pellé, J.-Y. Landuré, A. Normand, E. Ponsevera, C. German, L. Parson, F. Barriga, I. Costa, J. Relvas & A. Ribeiro, 1997. Discovery and first submersible investigations on the Rainbow hydrothermal field on the MAR (36° 14 N). Eos Trans. Am. Geophys. Union 78 (46): 832.

Fouquet, Y., J. P. Eissen, H. Ondréas, F. Barriga, R. Batiza & L. Danyushevsky, 1999. Extensive volcanoclastic deposits at the Mid-Atlantic Ridge axis: results of deep-water basaltic explosive volcanic activity? Terra Nova 10: 280–286.

Fouquet, Y., H. Ondréas, J.-L. Charlou, J.-P. Donval, J. Radford-Knoery, I. Costa, N. Lourenço & M. K. Tivey, 1995. Atlantic lava lakes and hot vents. Nature 377: 201.

Galkin, S. V. & L. I. Moskalev, 1990. Hydrothermal fauna of the Mid-Atlantic Ridge. Oceanology 30 (5): 624–627.

Gamo, T., H. Chiba, H. Masuda, H. N. Edmonds, K. Fujioka, Y. Kodama, H. Nanba & Y. Sano, 1996. Chemical characteristics of hydrothermal fluids from the TAG mound of the Mid-Atlantic ridge in August 1994: implications for spatial and temporal variability of hydrothermal activity. Geophys. Res. Lett. 23 (23): 3483–3486.

Gebruk, A. V., S. V. Galkin, A. L. Vereshchaka, L. I. Moskalev & A. J. Southward, 1997a. Ecology and biogeography of the hydrothermal vent fauna of the Mid-Atlantic Ridge. Adv. Mar. Biol. 32: 93–144.

Gebruk, A. V., L. I. Moskalev, P. Chevaldonné, S. M. Sudarikov & P. Chernyaev, 1997b. Hydrothermal vent fauna of the Logatchev area (14° 45′ N, MAR): preliminary results from first Mir and Nautile dives in 1995. InterRidge News 6 (2): 1–8.

German, C. R. & L. M. Parson, 1998. Distributions of hydrothermal activity along the Mid-Atlantic Ridge: interplay of magmatic and tectonic controls. Earth Planet. Sci. Lett 160: 327–341.

Goffredi, S. K., J. J. Childress, N. T. Desaulniers & F. H. Lallier, 1997. Sulfide acquisition by the vent worm *Riftia pachyptila* apprears to be *via* uptake of HS-, rather than H2S. J. Exp. Biol. 200: 2609–2616.

Grassle, J. F., 1986. Animals at Mid-Atlantic Ridge Hydrothermal vents. Eos Trans. Am. Geophys. Union. 67 (44): 1022.

Haymon, R. M., D. J. Fornari, K. L. Von Damm, M. D. Lilley, M. R. Perfit, J. M. Edmond, W. C. Shanks, R. A. Lutz, J. M. Grebmeier, S. Carbotte, D. Wright, E. S. McLaughlin, M., N. Beedle & E. Olson, 1993. Volcanic eruption of the mid-ocean ridge along the East Pacific Rise crest at 9° 45–52′ N: Direct submersible observations of seafloor phenomena associated with an eruption event in April, 1991. Earth Planet. Sci. Lett. 119: 85–101.

Hessler, R. R., S. C. France & M. A. Boudrias, 1987. Hydrothermal vent communities of the Mariana back-arc basin. Eos Trans. Am. Geophys. Union 68 (44): 1531.

Hessler, R. R. & P. F. Lonsdale, 1991. Biogeography of Mariana Trough hydrothermal vent communities, Deep-Sea Res. 38: 185–199.

Hessler, R. R. & W. M. Smithey, Jr., 1983. The distribution and community structure of megafauna at the Galapagos Rift hydrothermal vents. In Rona, P. A., K. Boström, L. Laubier & K. L. Smith, Jr. (eds), Hydrothermal Processes at Seafloor Spreading Centers. Plenum Press, New York: 735–770.

Jackson, D. A. & K. M. Somers, 1989. Are probability estimates from the permutation model of Mantel's test stable?. Can. J. Zool. 67: 766–769.

James, R. H., H. Elderfield & M. R. Palmer, 1995a. The chemistry of hydrothermal fluids from the Broken Spur site, 29° N Mid-Atlantic Ridge, Geochim. Cosmochim. Acta 59 (4): 651–659.

James, R. H., H. Elderfield, M. D. Rudnicki, C. R. German, M. R. Palmer, C. Chin, M. J. Greaves, E. Gurvich, G. P. Klinkhammer, E. Ludford, R. A. Mills, J. Thomson & A. C. Williams, 1995b. Hydrothermal plumes at Broken Spur, 29° N Mid-Atlantic ridge: chemical and physical characteristics. In Parson, C. L. W. L. M. & D. R. Dixon (eds), Hydrothermal Vents and Processes. The Geological Society of London, London: 97–109.

Jean-Baptiste, P., J. L. Charlou, M. Stievenard, J. P. Donval, H. Bougault & C. Mevel, 1991. Helium and methane measurements in hydrothermal fluids from the mid-Atlantic ridge: the Snake Pit site at 23° N. Earth Planet. Sci. Lett. 106: 17–28.

Johnson, K. S., J. J. Childress, R. R. Hessler, A.C.M. Sakamoto & C. L. Beehler, 1988. Chemical and biological interactions in the Rose Garden eastern Pacific ocean hydrothermal vent field Galapagos spreading center. Deep-Sea Res. A (Ocean. Res. Paper) 35 (10-11): 1723–1744.

Jollivet, D., T. Comtet, P. Chevaldonné, S. Hourdez, D. Desbruyères & D. Dixon, 1998. Unexpected relationship between dispersal strategies and speciation within the association *Bathymodiolus* (Bivalvia) *Branchipolynoe* (Polychaeta) inferred from the rDNA neutral ITS2 marker. Cah. Biol. Mar. 39: 359–362.

Khripounoff, A. & P. Albéric, 1991. Settling of particles in a hydrothermal vent field (East Pacific Rise 13° N) measured with sediment traps. Deep-Sea Res. 38: 729–744.

Lalou, C. & E. Brichet, 1981. Possibilités de datation des dépôts de sulfures métalliques hydrothermaux sous-marins par les descendants à vie courte de l'Uranium et du Thorium. C. r. Acad. Sci. Paris, Sér. II 293: 821–824.

Lalou, C., E. Brichet & J. Lange, 1989. Fossil hydrothermal sulfide deposits at the Galapagos Spreading Centre near 85 ° 00 West: geological setting, mineralogy and chronology. Oceanol. Acta 12 (1): 1–8.

Lalou, C., J. L. Reyss, E. Brichet, M. Arnold, G. Thompson, Y. Fouquet & P. A. Rona, 1993. New age data for Mid-Atlantic Ridge hydrothermal sites: TAG and Snakepit chronology revisited. J. Geophys. Res. B 98: 9705–9713.

Lalou, C., G. Thompson, M. Arnold, E. Brichet, E. Druffel & P. A. Rona, 1990. Geochronology of TAG and Snakepit hydrothermal fields, Mid-Atlantic Ridge: Witness to a long and complex hydrothermal history. Earth Planet. Sci. Lett. 97 (1704): 113–128.

Langmuir, C., S. Humphris, D. Fornari, C. Van Dover, K. Von Damm, M. K. Tivey, D. Colodner, J. L. Charlou, D. Desonie, C. Wilson, Y. Fouquet, G. Klinkhammer & H. Bougault, 1997. Hydrothermal vents near a mantle hot spot: the Lucky Strike vent field at 37° N on the Mid-Atlantic Ridge. Earth Planet. Sci. Lett. 148: 69–91.

Langmuir, C. H., D. Fornari, D. Colodner, J.-L. Charlou, I. Costa, D. Desbruyères, D. Desonie, T. Emerson, A. Fiala-Médioni, Y. Fouquet, S. Humphris, L. Saldanha, R. Sours-Page, M. Thatcher, M. Tivey, C. Van Dover, K. Von Damm, K. Wiese & C. Wilson, 1993. Geological setting and characteristics of the Lucky Strike vent field at 37° 17′ N on the Mid-Atlantic Ridge. Eos Trans. Am. Geophys. Union 74 (43): 99 pp.

Lutz, R. & M. Kennish, 1993. Ecology of Deep sea Hydrothermal Vent Communities: a Review. Rev. Geophys. 31(3) (4053): 211–242.

Murton, B. J., C. Van Dover & E. Southward, 1995. Geological setting and ecology of the Broken Spur hydrothermal vent field: 29° 10′ N. In Parson, C. L. W. L. M. & D. R. Dixon (eds), Hydrothermal Vents and Processes. The Geological Society of London, London: 33–41.

Olu, K., S. Lance, M. Sibuet, P. Henry, A. Fiala-Médioni & A. Dinet, 1997. Cold seep communities as indicators of fluid expulsion patterns through mud volcanoes seaward of the Barbados accretionary prism. Deep-Sea Res. (I Oceanogr. Res. Pap.) 44 (5): 811–841.

Polz, M. F., J. J. Robinson, C. M. Cavanaugh & C. L. Van Dover, 1998. Trophic ecology of massive shrimp aggregations at a Mid-Atlantic Ridge hydrothermal vent site. Limnol. Oceanogr. 43 (7): 1631–1638.

Rieley, G., C. L. Van Dover, D. B. Hedrick & G. Eglinton, 1999. Trophic ecology of *Rimicaris exoculata:* a combined lipid abundance/stable isotope approach. Mar. Biol. 133: 495–499.

Rona, P. A., M. D. Hannington, C. V. Raman, G. Thompson, M. K. Tivey, S. E. Humphris, C. Lalou & S. Petersen, 1993. Active and relict sea-floor hydrothermal mineralization at the TAG hydrothermal field, Mid-Atlantic Ridge. Econ. Geol. 88: 1989–2017.

Rudnicki, M. D. & H. Elderfield, 1992. Helium, radon and manganese at the TAG and Snakepit hydrothermal vent fields, 26° N and 23° N, Mid-Atlantic Ridge. Earth Planet. Sci. Lett. 113: 307–321.

Saldanha, L., M. Biscoito & D. Desbruyères, 1996. The Azorean deep sea hydrothermal ecosystem: its recent discovery. In Deep sea and shallow water extreme habitats: Affinities and adaptation. Biosyst. Ecol. Ser. 11: 383–388.

Sarradin, P. M., N. Le Bris, D. Birot & J. C. Caprais, 1999. Laboratory adaptation of the methylene blue method to flow injection analysis: towards *in situ* sulfide analysis in hydrothermal seawater. Anal. Commun. 36: 157–160.

Scott, K. M., C. R. Fisher, J. S. Vodenichar, E. R. Nix & E. Minnich, 1994. Inorganic carbon and temperature requirements for autotrophic carbon fixation by the chemoautotrophic symbionts of the giant hydrothermal vent tube worm *Riftia pachyptila*. Physiol. Zool. 67 (3): 617–638.

Segonzac, M., 1992. Les peuplements associés à l'hydrothermalisme océanique du Snake Pit (dorsale médio-atlantique; 23° N, 3480 m): composition et microdistribution de la mégafaune. C. R. Acad. Sci. Paris, Sér. III 314: 593–600.

Segonzac, M., M. De Saint Laurent & B. Casanova, 1993. L'énigme du comportement trophique des crevettes Alvinocarididae des sites hydrothermaux de la dorsale médio-atlantique. Cah. Biol. Mar. 34: 535–571.

Shank, T. M., M. B. Black, K. M. Halanych, R. A. Lutz & R. C. Vrijenhoek, 1999. Miocene radiation of deep sea hydrothermal vent shrimp (Caridea: Bresiliidae): evidence from mitochondrial Cytochrome oxidase subunit I. Mol. Phylogen. Evol. 13 (2): 244–254.

Shank, T. M., D. J. Fornari, K. L. Von Damm, M. D. Lilley, R. M. Haymon & R. A. Lutz, 1998a. Temporal and spatial patterns of biological community development at nascent deep sea hydrothermal vent (9° 50′ N, East Pacific Rise). Deep-Sea Res. II 45: 465–515.

Shank, T. M., R. A. Lutz & R. C. Vrijenhoek, 1998b. Molecular systematics of shrimp (Decapoda: Bresiliidae) from deep sea hydrothermal vents, I: Enigmatic 'small orange' shrimp from the Mid-Atlantic Ridge are juvenile *Rimicaris exoculata*. Mol. Mar. Biol. Biotechnol. 7 (2): 88–96.

Sokal, R. R. & F. J. Rohlf, *Biometry*, 1980. 859 pp., W. H. Freeman and Company, New York: 859 pp.

Toulmond, A., F. H. Lallier, J. De Frescheville, J. J. Childress, R. Lee, N. K. Sanders & D. Desbruyères, 1994. Unusual carbon dioxide-combining properties of body fluids in the hydrothermal vent tubeworm *Riftia pachyptila*. Deep-Sea Res. (I Oceanogr. Res. Pap.) 41 (10): 1447–1456.

Tunnicliffe, V., 1988. Biogeography and evolution of hydrothermal-vent fauna in the eastern Pacific Ocean. Proc. R. Soc. Lond. B 233: 347–366.

Tunnicliffe, V., 1991. The biology of hydrothermal vents: Ecology and evolution. Oceanogr. Mar. Biol. Annu. Rev. 29: 319–407.

Tunnicliffe, V., 1996. Paleobiology – an alternative viewpoint. J. Geosci. Education 44: 251.

Tunnicliffe, V. & C. M. R. Fowler, 1996. Influence of sea-floor spreading on the global hydrothermal vent fauna. Nature: 531–533.

Tunnicliffe, V., C. M. R. Fowler & A. G. McArthur, 1996. Plate tectonic history and hot vent biogeography. In MacLeod C. J., P. A. Tyler & C. L. Walker (eds), Tectonic, Magmatic, Hydrothermal and Biological Segmentation of Mid-Ocean Ridges. Geological Society: 225–238.

Tunnicliffe, V., A. G. McArthur & D. McHugh, 1998. A biogeographical perspective of the deep sea hydrothermal vent fauna. Adv. Mar. Biol. 34: 254–404.

Tyler, P. A. & C. M. Young, 1999. Reproduction and dispersal at vents and cold seeps. J. Mar. Biol. Ass. U.K. 79: 193–208.

Vacelet, J. & N. Boury-Esnault, 1995. Carnivorous sponges. Nature 373: 333–335.

Valdès, A. & P. Bouchet, 1998. Naked in toxic fluids: a nudibranch mollusc from hydrothermal vents. Deep-Sea Res. (II Top. Stud. Oceanogr.) 45: 319–327.

Van Dover, C. L., 1995. Ecology of Mid-Atlantic Ridge hydrothermal vents. In Parson, C. L. W. L. M. & D. R. Dixon (eds), Hydrothermal Vents and Processes. Geological Society Special Publication, London: 257–294.

Van Dover, C. L., D. Desbruyères, M. Segonzac, T. Comtet, L. Saldanha, A. Fiala-Médioni & C. Langmuir, 1996. Biology of the Lucky Strike hydrothermal field. Deep-Sea Res. 43 (9): 1509–1529.

Von Damm, K. L., 1995. Controls on the chemistry and temporal variability of seafloor hydrothermal fluids. In Humphris, R. A., Z. D. E. L. S. Mullineaux & R. E. Thompson (eds), Seafloor Hydrothermal Systems. American Geophysical Union: 222–247.

Von Damm, K. L., A. M. Bray, L. G. Buttermore & S.E. Oosting, 1998. The geochemical controls on vent fluids from the Lucky Strike vent field, Mid-Atlantic Ridge. Earth Planet. Sci. Lett 160: 521–536.

Hydrobiologia **440**: 217–225, 2000.
M.B. Jones, J.M.N. Azevedo, A.I. Neto, A.C. Costa & A.M. Frias Martins (eds), Island, Ocean and Deep-Sea Biology.

The effect of water depth on bacterial numbers, thymidine incorporation rates and C:N ratios in northeast Atlantic surficial sediments

J. L. Dixon & C. M. Turley*
*Plymouth Marine Laboratory, Prospect Place, The Hoe, Plymouth, PL1 3DH, U.K. (*Author for correspondence)*

Key words: bacteria numbers, [^{3}H]-thymidine incorporation, deep sea, surficial sediment, phytodetritus, northeast Atlantic

Abstract

The effect of water depth on bacterial biomass and their ability to synthesise DNA, by measuring their rate of [^{3}H]-thymidine incorporation, was investigated in the northeast Atlantic at three sites of varying water depth (1100–3580 m) and sediment characteristics. Thymidine incorporation rates (y) in surficial sediments varied between 0.028 and 1.44 pmol h^{-1} g^{-1} and showed an exponential relationship with depth (x) according to the equation $y = 2.05e^{-0.0011x}$ (r=0.9830 for n=7, P<0.001). However, this relationship failed when a layer of phytodetritus was found overlying the surface sediment and [^{3}H]-thymidine incorporation rates increased by 80–339%. In contrast, bacterial numbers varied between 1.09 and 11.96 $\times$ 10^8 cells g^{-1} (dry weight) and showed no significant relationships with water depth or sediment POC/TN content. Significant exponential relationships were also found between water depth (x) and the POC (y_1) and total nitrogen (TN, y_2) content of surficial sediments according to the following equations: where $y_1 = 719e^{-0.0003x}$ (r=0.8700 for n=9, P<0.01) and $y_2 = 76e^{-0.0002x}$(r=0.7582 for n=9 P<0.02). These relationships were irrespective of the presence or absence of an overlying layer of phytodetritus. This suggests that the POC and TN content of these surficial deep sea sediments is directly related to the flux of material through the water column, which significantly impacts bacterial production.

Introduction

The deep-sea benthos has only relatively recently been recognised as a community that depends upon the supply of relatively fresh, often aggregated particulate organic matter (POM) sedimenting from the surface oceans to the deep sea bed (Rice, 1983; Rice et al., 1986; Gooday & Turley, 1990 and all references therein; Pfannkuche, 1993). Indeed, the presence and persistence of life itself on the ocean floor can be viewed as a response to such organic inputs (Lochte & Turley, 1988; Thiel et al., 1988/1989; Gooday & Turley, 1990). By virtue of the seasonal and ephemeral nature of POM delivery at depths (Lampitt, 1985), the deep sea is thus a food-limited environment in which the abundance and biomass of benthic organisms is related directly to the amount of food (POM) reaching the sediment surface (Sibuet et al., 1984; Sibuet, 1987; Tseytlin, 1987).

The benthic boundary layer (BBL), which in the deep ocean extends from the sediment surface to several tens of metres above the bottom, is a region of enhanced mixing and exchange between benthic and pelagic realms (Richards, 1990). It is considered to be a bioreactive interface in which material sedimenting from above is biologically transformed over a wide range of time scales, eventually forming the sedimentary record. The seasonal flux of labile phytodetritus through the BBL to the surface sediment stimulates benthic biological activity (Gooday & Turley, 1990), leading to the consumption of oxygen. Much of the resulting activity is attributable to bacteria and protozoans, rather then to metazoans. Indeed, bacteria contribute the greatest biomass in the top 10 cm of sediment (Lochte, 1992).

In general, the flux of particulate organic carbon (POC) through the water column decreases exponentially with depth in the northeast Atlantic (see review by Turley et al., 1995) so supply to sediments at different depths (should or) may vary accordingly.

However, the flux in this region exhibits a strong seasonal signal (Newton et al., 1994; Turley &

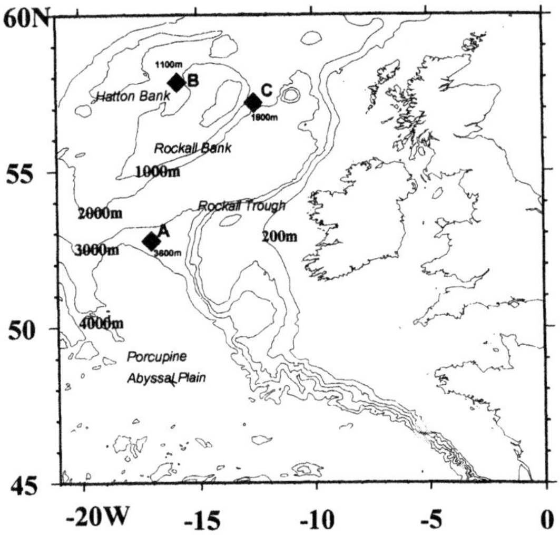

Figure 1. Location of BENBO sampling stations A–C in the northeast Atlantic.

Mackie, 1995) in quantity and composition. This study, therefore, aimed to investigate the effects of water depth on the numbers and production of bacteria, and the carbon and nitrogen composition of surficial sediments in the northeast Atlantic. In addition, the European Committee for Ocean and Polar Sciences (ECOPS) recently identified 'Variability in the deep sea Environment' as one of its five Grand Challenges (Suess & Le Pichon, 1994). We, therefore, present the preliminary results from a study examining the spatial and temporal variations in sediment composition, bacterial numbers and production between three predetermined stations in the northeast Atlantic.

Materials and methods

Study sites

Three areas of seabed, around the Rockall Trough region in the northeast Atlantic, were chosen as the field study area (Fig. 1). The sites were chosen on the basis of their primary characteristics which were examined during previous studies and surveys. Station A (59.918° N, 16.917° W) was located in the mouth of the Rockall Trough in 3580 m of water, characterised by low currents. Station B (57.425° N, 15.683° W) was in the Hatton-Rockall basin in 1100 m of water charcterised by a high POC flux and low currents. Station C (57.100° N 12.515° W) was a current-affected site in the Flank of Feni Drift in 1920 m of water. Three scientific cruises (CD107, August 1997; CD111, May 1998; CD113, July 1998) were conducted in this area as part of the NERC thematic program on the deep ocean benthic boundary layer (BENBO).

Sampling

Relatively undisturbed surficial sediment and overlying water was collected from the three sampling sites in the northeast Atlantic using a multiple corer (Barnett et al., 1984). On retrieval to the surface, cores were taken immediately to a constant temperature laboratory held at *in situ* temperature. The top 0–2 cm sections of all sediment cores were removed using a sterilised slicer and pooled in a sterilised container. The sediment was then slurried 1:1 with filtered (0.2 μm pore size) seawater overlying the sediment cores, inoculated with [^{3}H]-thymidine and re-pressurised for the rate process measurements, as described below. Using a slurry ensures homogeneity in labelling, sub sampling (in time-course experiments) and replication. However, this process may physically disrupt the bacteria's micro-environment, resulting in changes to the measured processes. But this is likely to be less significant in surficial deep-sea sediment samples where there is often resuspension and resettling of the surface layer (Lampitt et al., 2000), and the sediments are aerobic to an approximately depth of 10 cm. The alternative sampling method is via direct injection into undisturbed whole cores (Epstein, 1997). However, this method also suffers from methodological constraints such as diffusion, replication and control problems; not withstanding the technical difficulties involved with processing large numbers of samples under pressure (Deming, 1997).

The effects of depressurisation during sample retrieval is also a problem and one which has received little attention. Although it is known that laboratory cultures, which have received short depressurisation periods, do not suffer from irreversible damage (Deming, 1997). However, it is unknown whether whole microbial community rate processes, such as those measured in this study, represent 'real' *in situ* rates or activities. For instance, there could be inhibition of deep-sea bacteria or stimulation of surface-originating bacteria by depressurisation during sampling, despite their immediate re-pressurisation onboard ship (Deming, 1997). In addition, the proportion of surface-originating bacteria, and the effects of changing pressure in deep-sea sediments, is also unknown.

Table 1. Variation in water depth, *in situ* water temperature (10 mab) and *in situ* pressure for the BENBO stations A–C used in the incubation experiments

	Station B Hatton-Rockall Basin	Station C Flank of Feni Drift	Station A Mouth of Rockall Trough
Depth (m)	1100	1920	3580
In-situ temperature (°C)			
August 1997	6.3	3.1	2.4
May 1998	5.5	3.0	2.5
July 1998	5.0	3.0	2.0
In-situ pressure (atms)			
August 1997	111	193	357
May 1998	110	192	358
July 1998	109	194	357

The *in situ* temperatures were measured at each station by a CTD vertical profile and were within 10 m above the seabed (mab) (Table 1).

Measurement of bacterial DNA synthesis

The incorporation of [^{3}H]-thymidine (specific activities ranging between 76 and 79 Ci mmol^{-1} in aqueous solution, Amersham) into DNA of sediment-associated bacteria was determined as follows: sediment slurry (four replicates of 0.5 ml 1:1 slurry, as detailed above) obtained at each station during the May 1998 and July 1998 cruises was put into 5 ml sterile polyethylene bags (1 of the replicates served as a control and was killed with 27.5 μl ice-cold 100% trichloroacetic acid (TCA), 5% final concentration), 50 μl of [^{3}H]-thymidine working solution added (giving a final concentration of 100 nM) and incubated in 1 l pressure vessels (Turley & Lochte, 1990) at *in situ* temperature and pressure (Table 1). Samples were terminated as for the controls after 3 h (this was within the period of linear incorporation) and stored frozen until later analysis in the laboratory.

During the August 1997 survey, incubations were carried out at 1 atmosphere in 1.7 ml sterile micro test tubes at *in situ* temperature. Resulting [^{3}H]-thymidine incorporation rates were corrected subsequently to rates occurring at *in situ* pressure from experiments examining the effect of pressure on [^{3}H]-thymidine incorporation rates for each station by applying the following correction factors: Station A × 1.042, Station B × 0.793 and Station C × 1.43.

Samples were defrosted and 0.5 ml slurry aliquots were pipetted into sterile micro test tubes. Samples were washed three times (to remove unincorporated [^{3}H]-thymidine) with 0.5 ml ice-cold TCA (5%), whirled (for 5 s), centrifuged (5 min at >3000 rpm) and the supernatant discarded. Following this procedure, total macromolecules were extracted with 0.5 ml of an alkaline extractant containing 0.3 N NaOH, 25 mm EDTA and 0.1% SDS for 12 h at 30 °C (Findlay, 1993). In addition, DNA (10 mg ml^{-1}, final concentration) and cold thymidine (5 mg ml^{-1}, final concentration) were also added to optimise macromolecular recovery, i.e. thymidine minimises control values by reducing non-specific binding of [^{3}H]-thymidine, and DNA promotes the precipitation of low concentrations of radiolabelled nucleic acids (Thorn & Ventullo, 1988 and references therein). Following total macromolecule extraction from the sediment, the DNA and protein were separated from the RNA, residual unincorporated [^{3}H]TdR and catabolised products (i.e. ^3H_2O etc.) by acid precipitation, whirled (for 5 s) and stored on ice for 30 min. The pellet was then centrifuged for 15 min (15 000 rpm) at 5 °C and the supernatant discarded. The pellet was washed a further three times with 1 ml of ice-cold 5% TCA, whirled and centrifuged as described previously. The DNA was hydrolysed with 1 ml TCA (5%) at 100 °C for 30 min (the remaining pellet containing protein). The resulting supernatant (approximately 1 ml) was pipetted into a scintillation vial and 2.8 ml of scintillation liquid (Optiphase Hisafe 3) added. The radioactivity incorporated into the DNA was then counted on a Wallac 1409 liquid scintillation counter. Counting efficiency was corrected for quenching by internal standardisation assuming that the efficiency of counting the spike was equal to that of the sample.

Sediment analysis

Particulate organic carbon (POC) and total nitrogen (TN) in the surficial sediments were analysed by using a Europa Roboprep CN Analyser. The POC was determined after removing carbonate carbon by acidification with sulphurous acid *in situ* with aluminium sample cups (Verardo et al., 1990). The dry-weight content of the sediments was determined by weighing additional subsamples (0.5–2.5 ml) dried at 60 °C for 24 h.

Enumeration of bacteria

Bacterial numbers were determined by direct counts with epifluorescence microscopy of Acridine Orange stained cells. However, it was first necessary to break up the sediment particles to disperse the cells and ensure an even distribution on the filter (Turley & Mackie, 1995). A total sonification time of 80 s on ice (16 × 5 s, pulse 5 s on, pulse 5 s off) was found to produce optimum cell numbers without cell break-up. A Vibra Cell (Sonics & Materials Inc.) 20 kHz 400 Watt ultrasonic disintegrator with a stepped microtip (3 mm diameter) was used to dislodge the bacterial cells from the sediment particles.

Sediment slurry (1:1 as detailed previously) samples were diluted (×10 with 2% particle-free SEM grade gluteraldehyde) and incubated at room temperature for 20 min, whilst shaking, to help disperse the bacteria. Samples were then pre-cooled on ice and sonicated as detailed above, before further dilution of c. ×2000 fold. Sample volumes (1–2 ml as necessary to count between 15 and 30 cells per quadrat, Turley, 1993) were then filtered and prepared as detailed in Turley & Hughes (1992). In addition, 1 ml of citrate buffer (0.1 m Na-citrate) was used to rinse the samples after staining to reduce background fluorescence. Samples were viewed under an Olympus BH2 epifluorescence microscope with UV excitation (340–380 nm) at a final magnification of ×1250. Three replicate counts were carried out for each sample and >300 cells counted from each replicate. Although cell loss has been reported to occur during the storage of preserved seawater samples (Turley & Hughes, 1992, 1994), this does not seem to be a problem in deep sea sediment-associated bacteria (K. Lochte, pers. com.). All analyses have been standardised to dry sediment weight.

Results and discussion

Spatial and temporal variations

The rates of [^{3}H]-thymidine incorporation range between 0.028 and 1.44 pmol h^{-1} g^{-1} and show both spatial and temporal variations (Fig. 2a). The rates at the deeper station A are between 10 and 31 times lower than those at stations B and C, which suggests that the bacterial production at station A is relatively limited. The data do not show any consistent temporal trends for the stations. However, stations B and C both showed maximum rates of [^{3}H]-thymidine uptake during August 1997 compared to May and July 1998 (39–49% and 144–339% higher, respectively).

Bacterial numbers ranged between 1.09 and 11.96 × 10^8 cells g^{-1} (dry weight, Fig. 2b) and fall within the range of bacterial densities given in Deming & Yager (1992), and references therein, for the surface layer (0–1 cm) of marine sediments between 200 and 6000 m water depth. Bacterial numbers followed similar trends both in space and time to [^{3}H]-thymidine incorporation rates with the following exceptions: firstly, the concentration of bacteria associated with surface sediments at station A was only between 1.7 and 9.0 times less than those at stations B and C, a lower figure than one might have envisaged based on [^{3}H]-thymidine incorporation rates. Secondly, bacterial numbers at station B during July 1998 were lower than anticipated from incorporation rates, i.e. despite [^{3}H]-thymidine incorporation rates approximately 10 times higher than station A, bacterial numbers were actually 15% lower.

Variation of [^{3}H]-thymidine incorporation rates and bacterial numbers with water depth

In order to investigate what factors were causing and sustaining the spatial variations observed in [^{3}H]-thymidine incorporation rates and bacterial numbers, both variables were plotted as a function of water depth (Fig. 3a, b, respectively). The following statistically significant exponential relationship was found between water depth (x) and [^{3}H]-thymidine incorporation rate (y), excluding the surficial sediment samples found with overlying phytodetritus, i.e. station C during August 1997 & July 1998:

$$y = 2.05e^{-0.0011x},$$

where n=7, r=0.9830 (P<0.001). This relationship suggests that surficial sediment-associated bacterial production, calculated from [^{3}H]-thymidine incorporation rates, could be predicted from a knowledge of the water depth, provided that the system had not received any recently added energy stimulus such as phytodetritus. Interestingly, the presence of an overlying phytodetrital layer resulted in increased bacterial [^{3}H]-thymidine incorporation rates of between 80 and 339% in the underlying sediments. This implies that these sediments have been 'stimulated' by the arrival of phytodetritus through either enrichment, presumably of organic carbon, via. diffusion, bioturbation or resuspension by bottom currents and subsequent incorporation of phytodetrital material into sediments (Lampitt, 1985; Turley & Mackie, 1995). Although the above significant relationship between [^{3}H]-thymidine incorporation rates and water depth is apparent, it is suggested, however, that it manifests as a consequence of changes in the concentration of organic carbon of

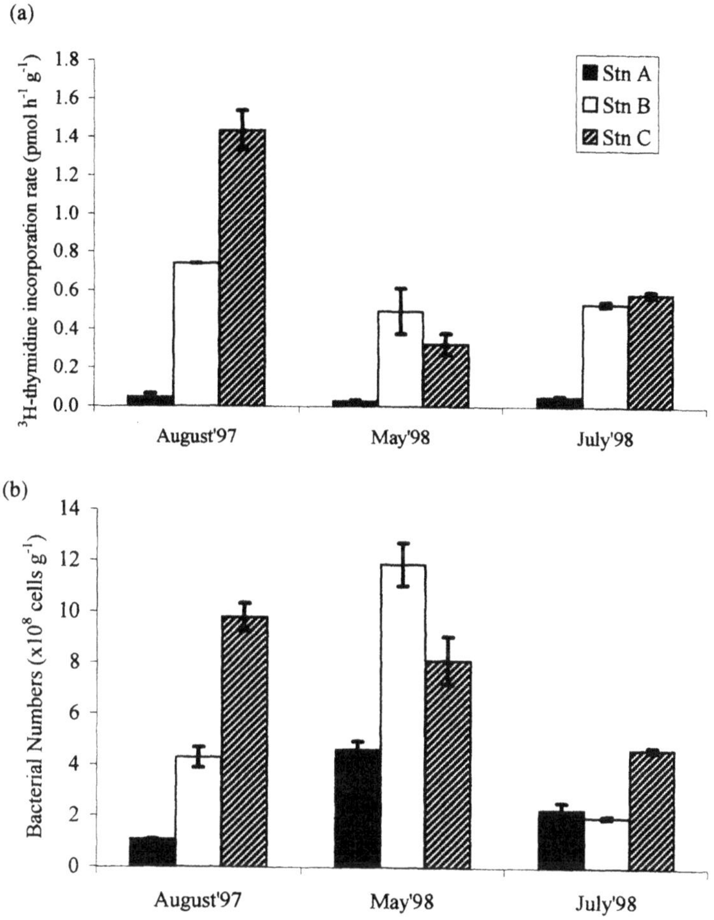

Figure 2. Variation in (a) [^{3}H]-thymidine incorporation rates and (b) bacteria numbers in surficial sediments both among cruises and stations. Error bars represent 1 standard deviation based on three replicate samples.

the surficial sediments and substrate quality (Touratier et al., 1999) with depth. The supply of POM to the sea bed depends on the magnitude of the surface productivity, the vertical flux of biogenic particles out of the surface waters (the biological pump), and the rate of particle solubilisation during descent (Turley & Mackie, 1995; Turley et al., 1995). Changes in the magnitude of all these processes will influence the surficial sediment bacterial numbers and production. Other potential sources of energy to deep-sea sediment bacteria may include labile organic matter sorbed to particle surfaces (Keil et al., 1994) and lateral advection of particles by near bottom currents (Thomsen, 1999).

This depth relationship clearly does not exist for bacterial numbers (Fig. 3b). However, it is notable that the numbers of bacteria associated with the phytodetrital material were approximately 6 times greater than those associated with the underlying surficial sediments, and are in agreement with the range found on other phytodetrital material sampled in the northeast Atlantic quoted in Lochte & Turley (1988) and Thiel et al. (1988/1989).

Particulate organic carbon and nitrogen content of surficial sediments: variation with water depth

The analysis of POC and TN of surficial sediments from stations A–C (Table 2) showed that the organic carbon and nitrogen content, on a percentage basis, of the shallower stations B and C are approximately 2 and 1.5 times greater than that observed at the deep station A, respectively. This may indicate an enhanced nutritional status of the surficial sediments at stations B and C compared to A. There were no systematic changes in POC, TN and C:N among cruises (Table 2), however,

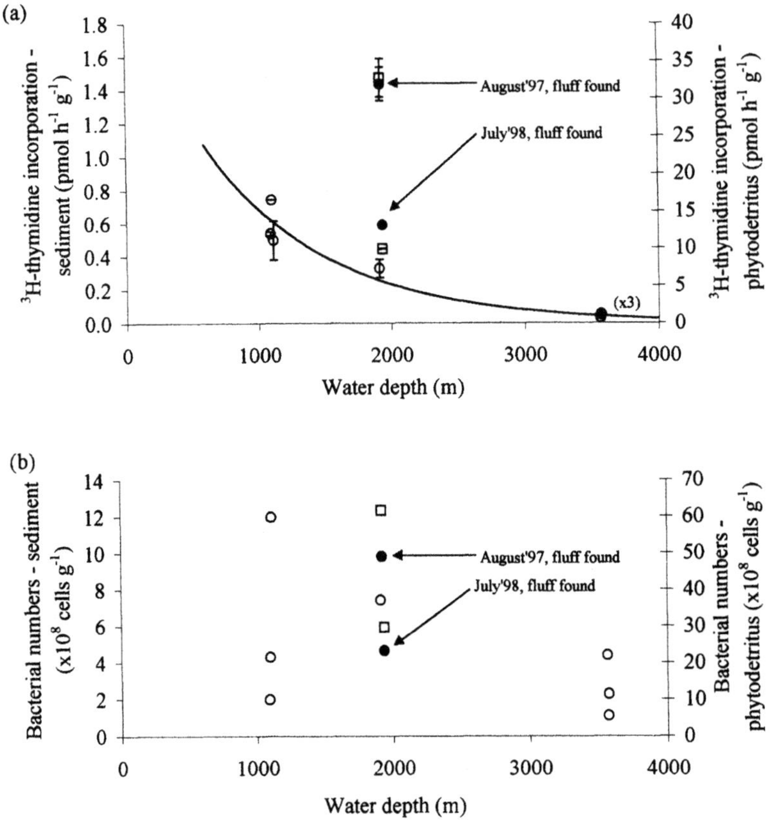

Figure 3. Variation in (a) [^{3}H]-thymidine incorporation rates and (b) bacteria numbers with water depth, where (○) represents surficial sediments, (•) represents sediment samples with an overlying phytodetrital layer and (□) denotes the actual phytodetritus material, plotted on the secondary *y*-axis. Note the scale differences. Error bars represent 1 standard deviation based on three replicate samples.

for both POC and TN station C > station B > station A. In addition, at station C during August 1997 and July 1998, a 1–3 cm layer of phytodetrital 'fluff' material was observed overlying the sediments. The POC and TN content of this material was between 56–62% and 40–70%, respectively, higher that the underlying sediment which indicates the relatively enhanced nutritional value of this material. Likewise, the rates of incorporation of [^{3}H]-thymidine were between 17 and 23 times greater for the phytodetritus compared to the underlying sediment.

The concentration of POC, TN and C:N ratio of surficial sediments were plotted as a function of water depth (Figs 4a, b, c respectively). All three variables showed a statistically significant exponential relationship with depth (Table 3). This suggests that the sediment POC and TN content of these surficial deep-sea sediments is directly related to the flux of material through the water column. The aforementioned water depth relationships with POC and TN also indicate that the deep station A is substrate limited relative to stations B and C. This could explain the lower bacterial [^{3}H]-thymidine incorporation rates measured at station A (Fig. 2a). In addition, Fukuda et al. (1998) found that the C:N ratio of natural bacterial assemblages from oceanic and coastal waters ranged between 5.9 and 6.8. This compares well to the lower C:N range found at the deep Station A of between 6.7 and 7.3 (Table 2), and further suggests that Station A may be limited by substrate quality. Indeed, the lowest C:N ratio of 6.7 occurred during May 1998, concurrent with the lowest rates of [^{3}H]-thymidine incorporation of 0.028 ± 0.005 pmol h^{-1} g^{-1} (Fig. 2a).

The consequence of an exponentially decreasing C:N ratio with depth means that the surficial sediments at station A are comparatively richer in nitrogen relative to carbon relative to those at stations B and C, i.e. the concentration of POC was 48% less at station A compared to station B, whereas the TN concentration was only 40% less at station A compared to B. This

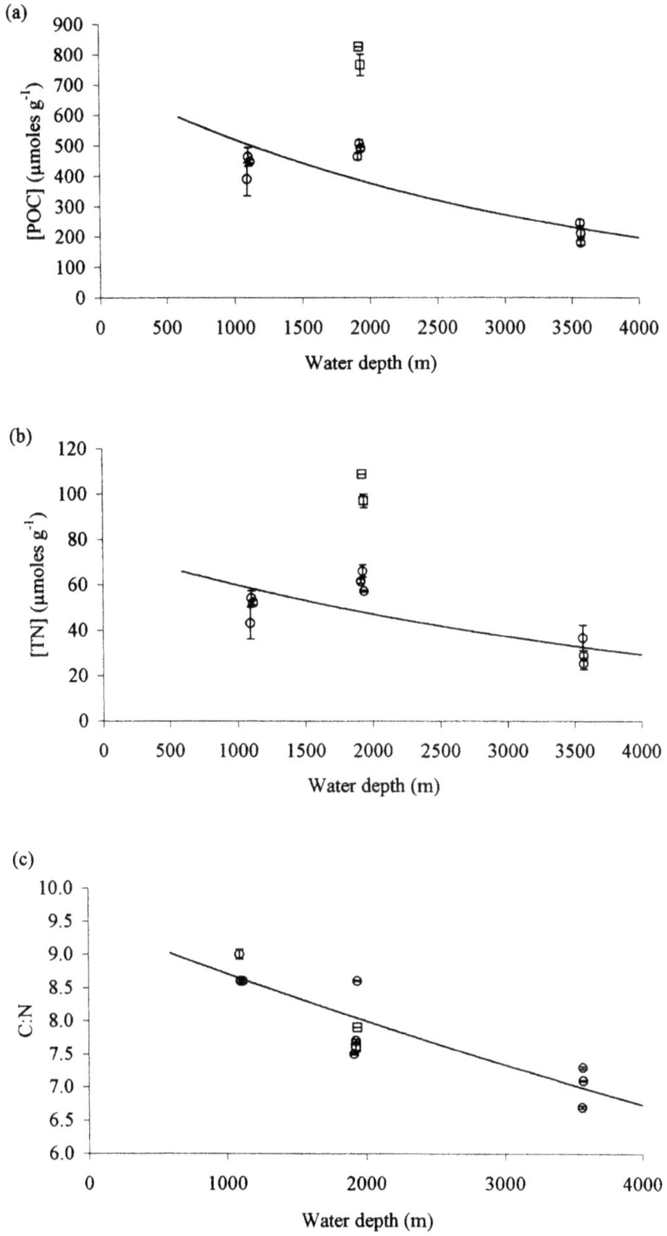

Figure 4. Variation in concentration of (a) particulate organic carbon, (b) total nitrogen and (c) carbon:nitrogen ratio in surficial sediments with water depth, where (○) represents surficial sediment and (□) phytodetrital samples. Error bars represent 1 standard deviation based on three replicate samples.

could either be due to preferential remineralisation or utilisation of the POC during the descent through the water column or after arrival on the sea bed. This also suggests that carbon is the more limiting resource rather than nitrogen, bacteria being able to utilise dissolved organic and/or inorganic nitrogen in the pore waters. The results of Goldman et al. (1987) led them to speculate that the C:N ratio of substrates utilised by bacteria in marine waters was >10:1. If this is the case, then the C:N ratios of the surficial deep-sea sediments of this study (Table 2) are all relatively deficient in carbon in comparison.

Table 2. Variation in POC, TN and C:N ratio of the surficial sediment characteristics at BENBO stations A–C during the three cruises. The numbers in brackets refer to phytodetrital 'fluff' material when it was found overlying the surficial sediments

		Station		
		B	C	A
POC (%)				
	August 1997	0.556	0.608 (0.990)	0.254
	May 1998	0.538	0.556	0.295
	July 1998	0.468	0.588 (0.918)	0.216
TN (%)				
	August 1997	0.076	0.093 (0.130)	0.035
	May 1998	0.073	0.086	0.051
	July 1998	0.060	0.080 (0.136)	0.036
C:N				
	August 1997	8.6	7.7 (7.6)	7.3
	May 1998	8.6	7.5	6.7
	July 1998	9.0	8.6 (7.9)	7.1

Table 3. Regression equations and statistics for relationships between surficial sediment POC, TN and C:N ratio with water depth (x), where r is the correlation coefficient, n is number of observations and P is the probability level

	Formula x = water depth (m)	r	n	P
POC (μmoles g^{-1})	$y=719e^{-0.0003x}$	0.8700	9	<0.01
TN (μmoles g^{-1})	$y=76e^{-0.0002x}$	0.7582	9	<0.02
C:N	$y=9.5e^{-0.00009x}$	0.8948	9	<0.01

Conclusion

These bacterial production measurements represent one of the few deep-sea surficial sediment datasets published in the literature (refer to Deming & Yager, 1992; Deming & Baross, 1993). We conclude that thymidine incorporation rates in surficial sediments of the northeast Atlantic decreased significantly and exponentially with depth, and assuming no recent phytodetritus stimulus, this relationship could be used to predict bacterial production. Whilst in comparison, bacterial numbers showed no significant relationships with water depth or sediment POC content. This may be because [^{3}H]-thymidine incubations are short-term rates measured at specific times, whereas bacterial numbers reflect longer term changes such as growth, respiration or mortality through grazing or cell lysis.

Acknowledgements

Our thanks to Estelle Le Pape and to the Officers and crew of the RRS Charles Darwin This work was supported by a NERC, BENBO Thematic Special Topics Award No GST/02/1761 and is BENBO publication number 10. Our indebted thanks also go to Dr Hilary Kennedy and Dr Stathis Papadimitriou for carrying out the POC and TN analysis of the BENBO sediment samples.

References

Barnett, P. R., J.Watson & D. Connelly, 1984. A multiple corer for taking virtually undisturbed samples from the shelf, bathyal and abyssal sediments. Oceanol. Acta 7: 399–408.

Deming, J. W., 1997. Unusual or extreme high pressure marine environments. In Hurst, C. J., G. R. Knudsen, M. J. McInerney, L. D. Stetzenbach & M. V. Walter (eds), Manula of Environmental Microbiology. ASM Press, Washingotn D.C: 366–376.

Deming, J. W. & J. A. Baross, 1993. The early diagenesis of organic matter: Bacterial activity. In Engel, M. H. & S. A. Macko (eds), Organic Geochemistry. Plenum Press 6, New York: 119–144.

Deming, J. W. & P. L. Yager, 1992. Natural bacterial assemblages in deep sea sediments: Towards a global view. In Rowe, G. T. & V. Pariente (eds), Deep Sea Food Chains and the Global Carbon Cycle. Kluwer Academic Publishers, Dordrecht, The Netherlands: 11–27.

Epstein, S. S., 1997. Microbial food webs in marine sediments. I. Trophic interactions and grazing rates in two tidal flat communities. Microbiol. Ecol 34: 188–189.

Findlay, S., 1993. Thymidine incorporation into DNA as an estimate of sediment bacterial production. In Kemp, P. F., B. F. Sherr, E. B. Sherr & J. J. Cole (eds), Handbook of Methods in Aquatic Microbial Ecology. CRC Press, Inc., Boca Raton, Florida: 505–508.

Fukuda, R., H. Ogawa., T. Nagata & I. Koike, 1998. Direct determination of carbon and nitrogen contents of natural bacterial assemblages in marine environments. Apl. envir. Microbiol. 64: 3352–3358.

Goldman, J. C., D. A. Caron & M. R. Dennett, 1987. Regulation of gross efficiency and ammonium regeneration in bacteria by substrate C:N ratio. Limnol. Oceanogr. 32: 1239–1252.

Gooday, A. J. & C. M. Turley, 1990. Responses by benthic organisms to inputs of organic material to the ocean floor: a review. Phil. Trans. r. Soc., Lond. A. 331: 119–138.

Keil, R. G., D. B Montlucon, F. G. Prahl & J. I. Hedges, 1994 Sorptive preservation of labile organic matter in marine sediments. Nature 370: 549–552.

Lampitt, R. S., 1985. Evidence for the seasonal deposition of detritus to the deep sea floor and its subsequent resuspension. Deep Sea Res. 32: 885–897.

Lampitt, R. S., P. P Netwon, T. D. Jickells, T. D. Thompson & P. King, 2000. Near-bottom particle flux in the abyssal Northeast Atlantic. Deep Sea Res. 47: 9–11.

Lochte, K., 1992. Bacterial standing stock and consumption of organic carbon in the benthic boundary layer of the abyssal North Atlantic. In Rowe, G. T. & V. Pariente (eds), Deep Sea Food Chains and the Global Carbon Cycle. NATO-ASI Series C Volume 360, Mathematical and Physical Sciences, Kluwer Academic Publishers, Dordrecht/Boston/London: 1–10.

Lochte, K. & C. M. Turley, 1988. Bacteria and cyanobacteria associated with phytodetritus in the deep sea. Nature 333: 67–69.

Newton, P. P., R. S. Lampitt, T. D. Jickells, P. King & C. Boule, 1994. Temporal and spatial variability of biogenic particle fluxes during the JGOFS northeast Atlantic process study at 47° N, 20° W. Deep Sea Res. I. 41: 1617–1642.

Pfannkuche, O., 1993. Benthic response to the sedimentation of particulate organic matter at the BIOTRANS station 47° N, 20° W. Deep Sea Res. II. 40: 135–150.

Rice, A. L., 1983. Thomas Henry Huxley and the strange case of *Bathybius haeckelii*: a possible alternative explanation. Arch. nat. Hist. II. 2: 169–180.

Rice, A. L., D. S. M. Billet, J. Fry, A. W. G. Fry, R. S. Lampitt, R. F. C. Mantoura & R. J. Morris, 1986. Seasonal deposition of phytodetritus to the deep sea floor. Proc. r. Soc. Edin. B. 88: 265–279.

Richards, K. J., 1990. Physical processes in the benthic boundary layer. Phil. Trans. r. Soc., Lond. A. 331: 3–13.

Sibuet, M., 1987. Structures des peuplements benthique en relation avec les conditions trophique en milieu abyssal dans l'ocJan Atlantique. ThPse de Doctorat d'Etat Js Sciences Naturelles. UniversitJ Pierre et Marie Curie, France.

Sibuet, M., C. Monniot, D. DesbruyPres, A. Dinet, A. Khripounoff, A. Rowe & M. Segonzac, 1984. Peuplements benthique et caractJristiques trophique du milieu dans la plaine abyssale de Demerara dans l'Ocean Atlantique. Oceanol. Acta 7: 345–358.

Suess, E. & X. Le Pichon, 1994. The deep sea floor as a changing environment. Report of the ECOPS Euroconference, Kiel, GEOMAR publication: 48 pp.

Thiel, H., O. Pfannkuche, G. Schriever, K. Lochte, A. J. Gooday, C. H. Hemleben, R. F. C. Mantoura, C. M. Turley, J. W. Patching & F. Riemann, 1988/89. Phytodetritus on the deep sea floor in a central oceanic region of the northeast Atlantic. Biol. Oceanogr. 6: 203–239.

Thomsen, L., 1999. Processes in the benthic boundary layer at continental margins and their implications for the benthic carbon cycle. J. Sea Res. 41: 73–86.

Thorn, P. M. & R. M. Ventullo, 1988. Measurement of bacterial growth rates in subsurface sediments using the incorporation of tritiated thymidine into DNA. Microbiol. Ecol. 16: 3–16.

Touratier, F., L. Legendre & A. VJzina, 1999. Model of bacterial growth influenced by substrate C:N ratio and concentration. Aquat. Microbiol. Ecol. 19: 105–118.

Tseytlin, V. B., 1987. Detritus flux to the ocean bed and benthic biomass. Oceanologia 27: 98–101.

Turley, C. M., 1993. Direct estimates of bacterial numbers in seawater samples without incurring cell loss due to sample storage. In Kemp, P. F., B. F. Sherr, E. B. Sherr & J. J. Cole (eds), Handbook of Methods in Aquatic Microbial Ecology. CRC Press, Inc., Boca Raton, Florida: 143–147.

Turley, C. M. & D. J. Hughes, 1994. The effect of storage temperature on bacterial cell loss in preserved seawater samples. J. mar. biol. Ass. U.K. 74: 259–262.

Turley, C. M. & D.J. Hughes, 1992. Effects of storage on direct estimates of bacterial numbers of preserved seawater samples. Deep Sea Res. 39: 375–394.

Turley, C. M. & K. Lochte, 1990. Microbial response to the input of fresh detritus to the deep sea bed. Palaeogeography, Palaeoclimatology, Palaeoecology (Global and Planetary change section) 89: 3–23.

Turley, C. M. & P. J. Mackie, 1995. Bacterial and cyanobacterial flux to the deep NE Atlantic on sedimenting particles. Deep Sea Res. I. 42: 1453–1474.

Turley, C. M., K. Lochte & R. S. Lampitt, 1995. Transformations of biogenic particles during sedimentation in the northeastern Atlantic. Phil. Trans. r. Soc., Lond. B. 348: 179–189.

Verardo, D. J., P. N. Froelich & A. McIntyre, 1990. Determination of organic carbon and nitrogen in marine sediments using the Carlo Erba NA-1500 Analyser. Deep Sea Res. II. 37: 157–165.

Hydrobiologia **440:** 227–238, 2000.
M.B. Jones, J.M.N. Azevedo, A.I. Neto, A.C. Costa & A.M. Frias Martins (eds), Island, Ocean and Deep-Sea Biology.

Distribution of live benthic Foraminifera at three oceanographically dissimilar sites in the northeast Atlantic: preliminary results

J.A. Hughes[1], A.J. Gooday[1] & J.W. Murray[2]
[1]*George Deacon Division for Ocean Processes, Southampton Oceanography Centre, Southampton SO14 3ZH, U.K.*
[2]*School of Ocean and Earth Science, Southampton Oceanography Centre, Southampton SO14 3ZH, U.K.*

Key words: Rockall, deep sea, Foraminifera, seasonality

Abstract

Live (i.e. rose Bengal stained) benthic foraminiferal assemblages in the 0–1 cm layer of multiple core samples were examined at three contrasting sites in the northeast Atlantic as part of the Natural Environment Research Council Deep Ocean Benthic Boundary Layer (BENBO) Thematic Programme. Sites A (3600 m water depth) and C (1900 m) were located in the Rockall Trough while Site B (1100 m) was in the Hatton–Rockall Basin. Wet-sorting sediment residues (>125 μm fraction) revealed more abundant (98–190 individuals/10 cm^2) and diverse (71–99 species /27 cm^2) assemblages than found previously in studies employing dry-sorting techniques. At all three sites, the assemblages were numerically dominated by delicate soft-bodied and agglutinated species, most of which are undescribed. Calcareous taxa formed a relatively small proportion of the assemblages (23% of individuals at Site B, 13% at Site C and 7% at Site A). Conversely, the agglutinated hormosinaceans (mainly *Reophax* species) became more prominent with increasing depth and accounted for 8% of the assemblage at Site B, 33% at Site C and 51% at Site A. Foraminifera represented 67–81% of the fauna in the samples and greatly outnumbered the metazoans. The assemblage at Site A has similarities with abyssal assemblages, while those at Sites B and C contain characteristic bathyal species. At Site B, *Uvigerina peregrina* and *Melonis barleeanum* are abundant, suggesting that there are high organic carbon inputs at this site. Reduced species diversity was observed at Site A following the spring phytodetritus bloom due to the presence of large numbers of juvenile *Hoeglundina elegans*, although it is not clear whether this species is responding to phytodetritus inputs.

Introduction

Bathyal continental margins, which occupy 17.8% of the World Ocean area (Atlas of the Oceans, 1980, cited in Zezina, 1997) are highly variable regions of the deep sea due to differences in water depth, primary productivity, current activity, the topography of the sea floor, sediment characteristics, lateral sediment transport processes, and the physical and chemical nature of the water masses. The northeast Atlantic continental margin is of increasing commercial interest as technological and economic developments make deep-sea fisheries and hydrocarbon extraction economically viable, while the use of the deep sea as a location for waste disposal is an increasingly contentious issue (Rice & Owen, 1999). Understanding the biological processes involved in carbon cycling on continental slopes may also be important due to their potential role as a sink for excess atmospheric carbon dioxide, via the export of shelf primary productivity (Walsh et al., 1991; Van Weering et al., 1998). Despite the obvious importance of oceanic margins, however, relatively little is known about the indigenous biological assemblages.

Despite being a focus of research since the start of oceanography (Deacon, 1971; Currie, 1986), many of the processes occurring in the deep northeast Atlantic are still poorly understood (McCave, 1986). Three contrasting sites in and around the Rockall Trough were studied as part of the Deep Ocean Benthic Boundary Layer (BENBO) Programme, funded by the U.K. Natural Environment Research Council (Black, 1999) (Fig. 1). One of the main aims of this programme was to determine the effects of phytodetritus

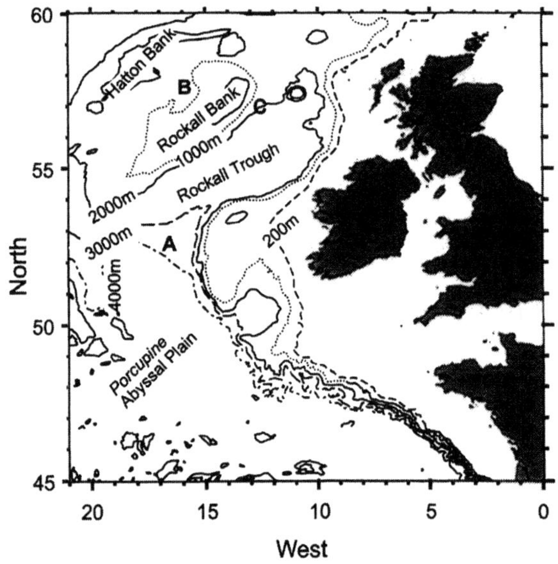

Figure 1. The northeast Atlantic, indicating the location of Sites A, B and C.

deposition on processes in the benthic boundary layer. The primary productivity of the overlying waters at all three sites is similar although the sites are located at different water depths, so that they receive differing organic carbon inputs. The seasonal deposition of phytodetritus is now well documented at bathyal and abyssal depths in the northeast Atlantic (Billett et al., 1983; Rice et al., 1986; Thiel et al., 1989) where it may form an important food resource for deep-sea benthic organisms (Gooday & Turley, 1990). In particular, certain species of benthic foraminifera have been shown to respond rapidly to seasonal phytodetritus deposition (Gooday, 1988, 1993, 1996; Gooday & Lambshead, 1989; Gooday & Rathburn, 1999). Here, we examine the distribution of foraminiferal assemblages at the three BENBO stations and present provisional results on temporal variability at one of the stations (Site A at 3600 m).

The study area

Three sites were examined, two in the Rockall Trough (Sites A and C) and one in the Hatton-Rockall Basin (Site B) (Fig. 1). The Rockall Trough is an elongate basin with the continental slopes of Scotland and Ireland to the east, and the Rockall Bank to the west (Fig. 1). The Trough is closed by a chain of banks to the north which connect the Rockall Bank to the Scottish continental shelf by way of the Wyville–Thomson Ridge (minimum depth 550 m), while to the south it opens onto the Porcupine Abyssal Plain at depths of 3000–4000 m. Site A (3600 m), located at the south of the Trough, is a tranquil, relatively oligotrophic site, while Site C (1900 m), located on the Feni Drift on the western side of the Trough, is thought to experience both high current velocities and organic carbon inputs. The Hatton–Rockall Basin is a depression on the centre of the Rockall Plateau, flanked to the northwest by Hatton Bank and to the southeast by Rockall Bank (Fig. 1). The basin is closed to the northeast by George Bligh Bank and is open to the southwest. Site B (1100 m) is located on the east side of the Basin in an area of assumed low current activity and high organic carbon inputs.

Hydrography

Water enters the Rockall Trough from the south and south-west, and forms two main water masses (reviewed by Ellett et al., 1986). The upper water (0–1200 m) derives from the Eastern North Atlantic Water (ENAW; 35.4–35.0 psu at 12–4 °C; Harvey, 1982). During late winter, the upper water is mixed to depths of around 600 m, to a maximum of 1000 m during severe winters (Meincke, 1986). The northeast Atlantic exhibits temporal (seasonal to decadal) variability of upper water mass properties and circulation features, with successions of salinity anomalies (Belkin et al., 1998) and changes in convective activity related to atmospheric patterns such as the north Atlantic Oscillation (Dickson et al., 1996). In the present study, only Site B (1100 m) will be directly influenced by poleward moving ENAW. This site may also be influenced by an extension of the north Atlantic current which sweeps sediment northeastwards off the Hatton Bank into the Hatton–Rockall Basin (Ruddiman, 1972). Also, water that cools over the Rockall Plateau during winter (Lee & Ellett, 1965) may sink into the Hatton–Rockall Basin, and flow southwest out of the trough, transporting sediments (Poore & Bergren, 1975).

Below 1200 m, Labrador Sea Water (LSW) is trapped in the Trough where it circulates cyclonically. This water is centred upon a salinity minimum (34.94–34.96 psu, 3.4–3.7 °C) and oxygen maximum (normally 6.0 ml l^{-1}) between 1600 and 1900 m (Ellett et al., 1986) which impinges upon Site C (1900 m). LSW circulation is probably responsible for the relatively high current velocities (mean 16 cm/s at 1750 m) observed at SMBA Station F, located close to Site C (Booth, 1983). This circulation may have been responsible for the deposition of the Feni Drift (Holliday et al., 2000) an extensive deposit of fine-grained sediment formed during the Plio-Pleistocene and previously thought to have been formed by the periodic flow of deep Arctic water over the Wyville–Thomson Ridge (Jones et al., 1970). Recent evidence suggests that the effect of this overflow on the hydrography of the Rockall Trough is limited. In the deeper regions of the Trough, where Site A (3600 m) is located, there appears to be a system of deep boundary currents which bring Lower Deep Water ($<$34.94 psu, $<$3 °C) into the Trough (Holliday et al., 2000). This water is characterised by high silicate contents, reflecting the diluted influence of Antarctic Bottom Water (Schmitz & McCartney, 1993).

Organic carbon inputs

Direct measurements of annual primary production are not available for the individual BENBO sites, although the productivity in the region appears to lie within the range of 60–90 g m^{-2} yr^{-1} (Berger et al., 1987; Berger, 1989). Thomson (pers. com.) found lower sedimentation rates and total organic carbon concentrations at Site B than at Site C, suggesting that additional sediment deposition is occurring at the latter station, although this may contain quite refractory organic carbon. This is consistent with the hypothesis that the Feni Drift is an area where active sediment deposition is still occurring.

The 1998 spring bloom appeared to peak sometime in June in the study region (Bett et al., 1998), preceding cruise CD113. During cruise CD113, a few millimetres to one centimetre of 'fluffy' phytodetritus was found at Site C, none was found at Site B and only trace amounts at Site A (Black, 1998).

Sediments

Below 700 m depth, sediments in the Rockall Trough are composed of an increasing proportion of fine calcareous particles of pelagic origin (Gage, 1986). Sediments at sites A and B are biogenic calcareous oozes with a $CaCO_3$ content in surface sediments of approximately 80%. The sediment at Site C is a marl with a $CaCO_3$ content of around 54% (Thomson, pers. com.).

Previous work on benthic foraminifera in the northeast Atlantic

Relatively few foraminiferal studies have been carried out in the BENBO study area. Early accounts of deep-water foraminifera in the area were published by Parker & Jones (1865), Carpenter et al. (1870), Thomson (1873), Brady (1884) and Jones (1994). Recent studies have been carried out by Murray & Taplin (1984), Weston & Murray (1984), Murray et al. (1986), Jones (1986), Lukashina (1983, 1988a, b), and Gadyukov & Lukashina (1988). Several studies have examined the live foraminiferal assemblages in deep-sea areas close to the BENBO sites. Gooday (1988, 1996, 1999), Gooday et al. (1995, 1998) and Smart & Gooday (1997) examined samples from the Porcupine Abyssal Plain. Gooday (1986a, b), Gooday & Lambshead (1989) and Wright (1889, 1891) examined samples from the Porcupine Seabight, while Weston (1985) compared live and dead assemblages in the Porcupine Seabight with those in the Western Approaches.

Materials and methods

Samples were obtained during three cruises on the R.R.S. *Charles Darwin*: CD107 (August, 1997), CD111 (May, 1998) and CD113 (July, 1998) (Table 1). All samples were taken using a Barnett–Watson multiple corer (Barnett et al., 1984), which takes virtually undisturbed core samples of 59 mm internal diameter (27.3 cm^2 surface area). After carefully removing the overlying water, cores were sectioned in 0.5 cm slices to 2 cm sediment depth and then in 1 cm sections to 10–15 cm depth. Samples were fixed in 10% formalin buffered with sodium borax. The results of the 0–0.5 cm and 0.5–1 cm sections are presented here. From Site A, results are presented from all three cruises, while from Sites B and C results are presented only from CD107 (Table 1).

In the laboratory, the disaggregated sediment of individual core slices was allowed to settle in a measuring cylinder for at least three days and then the volume was taken. The sediments were wet sieved over 150 μm and 125 μm sieves using filtered tap water, and the residues stained overnight in rose Bengal solution (1 g of rose Bengal and 10 g of phenol dissolved in 1 l of filtered tap water). The sample residues ($>$150 μm, 125–150 μm) were wet-sorted under a stereo-microscope. When samples took a long time to sort, they were occasionally (approximately weekly) re-stained. Individual specimens were picked out using a glass Pasteur pipette and stored in glycerol on a glass cavity slide (proteinaceous and

Table 1. Depth, date and location details of the multiple core samples examined during this study

Site	Cruise	Depth (m)	Date	Sample	Latitude (N)	Longitude (W)
A	CD107	3569	29/08/97	10704#3	52° 55.07′	16° 55.02′
B	CD107	1100	22/08/97	10702#2	57° 25.57′	15° 41.04′
C	CD107	1926	24/08/97	10703#2	57° 06.03′	12° 30.83′
A	CD111	3576	07/05/98	54404#10	52° 54.11′	16° 55.27′
A	CD113	3567	13/07/98	54704#8	52° 54.31′	16° 55.30′

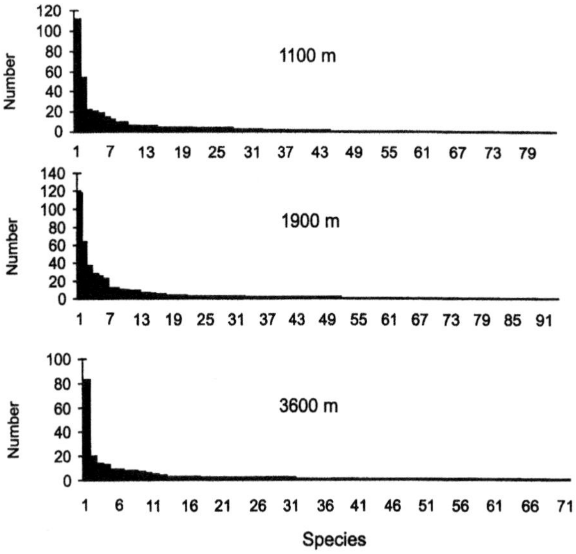

Figure 2. Ranked species dominance at Sites A, B and C (densities / 27 cm^2), August 1997.

Table 2. Number of species, number of individuals, Shannon-Wiener Index ($\log_{10}$), Fisher α index and Simpsons Diversity for the entire assemblages (all data/27 cm^2, except where indicated otherwise)

	Site				
	A	B	C	A	A
Date:-	August	August	August	May	July
Number of Species (0–0.5 cm)	52	64	56	74	49
Number of Species (0.5–1 cm)	33	35	55	35	27
Total Number of Species	71	81	93	90	63
Number of Individuals (0–0.5 cm)	189	184	303	387	278
Number of Individuals (0.5–1 cm)	80	260	215	101	74
Total Number of Individuals	269	444	518	488	352
Number of Individuals/10 cm^2	98	162	190	178	129
Shannon Weiner ($\log_{10}$)	1.39	1.38	1.45	1.48	1.34
Fisher α	31.5	29.0	33	32.4	21.8
Simpsons Diversity (1/D)	9.08	9.8	12.18	13.9	11.01
Total Number of Metazoans	61	214	163	106	58
Number of Metazoans/ 10 cm^2	22	78	60	39	21

agglutinated forms incorporating mineral grains), or dry on a micropalaeontological slide (calcareous species and agglutinated forms incorporating calcareous material) or in small vials in 10% formalin (tubular fragments, komokiaceans and large agglutinated specimens). All metazoans were retained in glycerol on glass cavity slides. Fragments of tubular taxa, such as *Rhizammina algaeformis* and komokiaceans, were not included in the results. Statistical analyses of the results were carried out using the program 'Biodiversity Pro' to calculate the following indices: The Shannon-Wiener function ($\log_{10}$) (Shannon & Weaver, 1963), Fisher α index (Fisher et al., 1943) and Simpson's Index (Simpson, 1949). Diversity is also represented graphically using rarefaction curves (Hurlbert, 1971).

Potential problems associated with the rose Bengal staining technique have been discussed by various authors (Bernhard, 1988; Corliss & Emerson, 1990; Murosky & Snyder, 1994; Murray & Bowser, 2000). The possibility that dead individuals may retain undegraded protoplasm for long periods after death is considered unlikely due to the seasonal population fluctuations that have been observed previously in species from continental margin sites (Gooday & Rathburn, 1999). The results presented here should be interpreted as reflecting individuals which were live at the time of collection or had died very recently (Corliss & Emerson, 1990).

Results

Faunal characteristics (August, 1997)

The assemblages at Sites A, B and C contained 269, 444 and 518 individuals/27 cm^2, respectively (= 98, 162 and 190 individuals/10 cm^2, respectively) (Table 2). At Site B, the majority of individuals was in the 0.5–1 cm section (59%), while at Sites A and C the

Table 3. Number of species, number of individuals, Shannon-Wiener Index ($\log_{10}$), Fisher α index and Simpsons Diversity for calcareous species only (all data/27 cm^2, except where indicated otherwise)

	Site				
	A	B	C	A	A
Date:-	August	August	August	May	July
Number of Species (0-0.5 cm)	7	11	16	19	13
Number of Species (0.5-1 cm)	5	11	10	6	5
Total Number of Species	11	17	19	20	16
Number of Individuals (0-0.5 cm)	10	36	47	45	93
Number of Individuals (0.5-1 cm)	9	100	21	14	10
Total Number of Individuals	19	136	68	59	103
Number of Individuals/10 cm^2	7	50	25	22	38
Shannon Weiner ($\log_{10}$)	1.00	0.89	0.96	1.11	0.49
Fisher α	10.09	5.13	8.75	10.65	5.33
Simpsons Diversity (1/D)	17.1	4.34	5.48	10.56	1.71

highest numbers (70% and 58%, respectively) were in the 0–0.5 cm section. The proportions of live specimens found in the 125–150 μm fraction were 25% (Site A), 54% (Site B) and 49% (Site C).

A total of 71 species was recognised at Site A, 81 at Site B and 93 at Site C, with more being found in the 0–0.5 cm sections than in the 0.5–1 cm sections (Table 2). At all sites, a few species were common and most were rare (Fig. 2), with Rank-1 dominance values of 31%, 25% and 23% at Sites A, B and C, respectively. This is reflected in the species diversity values, measured using the Shannon–Wiener Information Function and the Fisher α and Simpson indices, which were similar at the three sites (Table 2). Species richness, represented graphically using rarefaction, is also fairly uniform at all three sites (Fig. 3). Many of the species found were undescribed, although the proportion varied between taxa. Only two described species, *Nodellum membranacea* Rhumbler and *Vanhoeffenella gaussi* Rhumbler, were found amongst approximately 80 nominal soft-shelled monothalamous species, mainly allogromiids and saccamminids. This proportion increased for *Lagenammina* species (around 30% of which were previously described), hormosinaceans (around 40%) and was approximately 60% for other multichambered agglutinated taxa. At least 90% of the rotaliid species were previously described.

When only calcareous foraminifera are considered (Table 3), lower numbers of both species and individuals were observed. Calcareous foraminifera accounted for 15%, 21% and 20% of species, and 7%, 31% and 13% of specimmens at Sites A, B and C, respectively (Fig. 3; Table 4). Diversity indices were correspondingly reduced compared to the entire assemblages (Table 3). Simpson's diversity for the calcareous foraminiferans at Site A was higher than for the entire assemblage, possibly due to the inaccuracy of calculating diversity indices using low abundances. The lower diversity of the calcareous assemblage, when compared to the entire assemblage, was also shown clearly for Sites B and C using rarefaction (Fig. 3).

While calcareous foraminifera became scarcer with increasing water depth, hormosinaceans (mainly *Reophax* species) became more numerous, accounting for over 50% of the live assemblage at Site A (Table 4). Similarly, *Lagenammina* species were much more numerous at Site A than at the two shallower sites. At Site B, six of the ten most abundant species were calcareous, at Site C only two were calcareous, while at Site A all the top ten species were agglutinated (Table 5). Conversely, only one *Reophax* species appeared in the top ten at Site B, while there were four at Site C and five at Site A. The most abundant species at Sites B and C was '*Saccammina*' sp. 3. This small species was responsible for the high densities retained on 125 μm meshes (Table 2) and for the high percentage of 'soft-walled unilocular forms' (Table 4) observed at both these sites. This species was also found in low numbers at Site A.

Temporal variability at Site A

A total of 143 species and 1109 individuals occurred in the three samples from Site A. Cores from August, May and July contained 269, 488 and 352 individuals/27 cm^2, respectively (= 98, 178 and 129 individuals/10 cm^2, respectively) (Table 2). Almost 80% of specimens were found in the 0–0.5 cm section of the May and July samples, confirming what was observed during August 1997. The numbers of species observed were 71 in August, 90 in May and 63 in July. Calcareous species accounted for 22% and 25% of species, and 12% and 29% of individuals in the May and July samples, respectively. The high densities of calcareous foraminifera in the July sample was due to the presence of large numbers of *Hoeglundina elegans* (d'Orbigny) (Table 3). The majority of the *H. elegans* individuals were small juveniles, 195–220 μm in length and usually (91%) consisting of four chambers. This species was either absent or present in very low numbers in the samples from August and May. Species

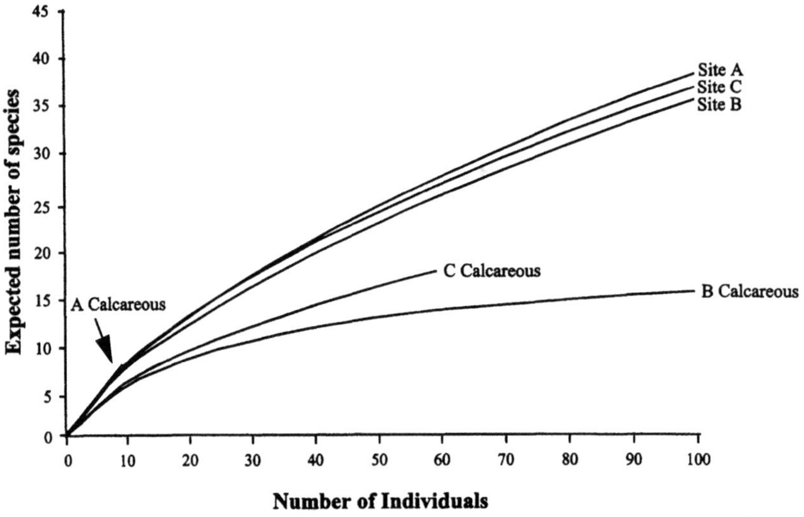

Figure 3. Rarefaction curves, showing the Expected Number of Species against Number of Individuals for entire assemblages (Sites A, B and C) and the calcareous assemblages (A calcareous, B calcareous and C calcareous) for samples taken during August 1997.

Table 4. Gross taxonomic composition, expressed as percentages, of the foraminiferal assemblages at Sites A, B and C, August, 1997 (0.0–1.0 cm sediment depth)

	Site A		Site B		Site C	
	Species	Individuals	Species	Individuals	Species	Individuals
Allogromids	11.3	8.2	12.3	3.6	8.6	4.2
Soft-walled unilocular forms	21.1	6.7	38.3	37.8	39.8	39.4
Lagenammina	15.5	22.3	6.2	14.9	9.7	7.3
Multiocular agglutinated forms:						
(a) Hormosinacea	22.5	50.6	13.6	7.7	17.2	32.8
(b) Others	14.1	5.2	8.6	5.4	4.3	3.1
Calcareous forms:						
(a) Miliolina	4.2	1.5	6.2	5.4	8.6	2.1
(b) Lagenina	1.4	0.4	1.2	0.2	2.2	0.6
(a) Rotalina	9.9	5.2	13.6	25.0	9.7	10.4
Total Numbers	71	269	83	444	93	518

diversity in the three cores from Site A were broadly comparable, with the sample from May the most diverse, as measured using all three indices (Table 2). The July sample showed the lowest H(S) and Alpha-index values, while the August sample had the lowest value for Simpson's Diversity (1/D).

Metazoans

In August 1997, metazoan densities ranged from 61/27 cm^2 (Site A) to 214/27 cm^2 (Site B) (Table 2). At Site A, densities ranged from 58/27 cm^2 (July) to 106/27 cm^2 (May) with a mean of 75 $\pm$ 27/27 cm^2. Metazoans were represented by a wide range of taxa. Polychaetes were the most abundant macrofaunal group, particularly at Site B, with juvenile bivalves, aplacophorans, echinoderms, cumaceans, tanaids and ostracods also present in the samples. Meiofaunal nematodes numerically dominated the metazoan assemblages at all sites, with harpacticoid copepods the second most abund-

Table 5. The 10 dominant foraminiferal species in each of the samples (abundances > 125 μm; 0.0–1.0 cm sediment depth)

August 1997 (Site A)		**Site B**		**Site C**	
Reophax sp. A	83	'*Saccamina*' sp. 3	112	'*Saccamina*' sp. 3	118
Lagenammina sp. 1	20	*Globigerina*-fragment *Lagenammina*	54	*Reophax* sp.	O.64
Lagenammina sp. 3.	14	*Uvigerina peregrina*	52	*Reophax* sp. M/N	37
Nodellum membranacea	10	*Reophax* sp. M/N	15	*Reophax* sp. T	28
Reophax sp. F	9	*Nonionella iridea*	15	*Nanionella iridea*	26
Lagenammina sp. 2.	8	*Cibicidoides* sp.	10	*Saccaminid* sp. E.	23
Globigerina-fragment *Lagenammina*	8	*Adercotryma glomerata*	10	*Lagenammina* sp. 12	
Reophax sp. B.	8	*Quinqueculina* sp. B.	7	*Gavinulinopsis lobatulus*	12
Large-glob. *Reophax* sp. 2	8	*Melonis barleeanum*	7	*Nodellum membranacea*	11
Reophax scorpiurus	6	*Karriella bradyi*	6	*Reophax* sp. U	10
May, 1998 (Site A)		**July, 1998 (Site A)**			
Reophax sp. A	95	*Hoeglundina elegans*	78		
Lagenammina sp. 1	57	*Reophax* sp. A	47		
Globigerina-fragment *Lagenammina*	43	*Lagenammina* sp. 1.	46		
Nodellum membranacea	41	*Nodellum membranacea*	16		
Lagenammina sp. 3	26	*Reophax* sp. F.	16		
Hoeglundina elegans	12	*Lagenammina* sp. 3	11		
Oridorsalis umbonata	10	*Reophax scorpiurus*	11		
Cibicidoides sp. 1	9	*Adercotryma* sp.	11		
'*Saccamina*' sp. 3	8	*Lagenammina* sp. 2	8		
Reophax sp. E.	7	*Globigerina*-fragment *Lagenammina*	8		

ant group. Metazoan densities were appreciably lower than foraminiferal densities in all samples.

Discussion

Limitations of the data set

The majority of foraminiferal studies are based on hard-shelled specimens which are of interest to palaeontologists as they have a high fossilisation potential. These taxa can be picked easily from dry sample residues. For biological investigations, however, it may be important to examine the 'entire' live (i.e. rose Bengal stained) foraminiferal assemblage which includes delicate soft-bodied and agglutinated forms destroyed by drying (Brodniewicz, 1965). These poorly known species, which are abundant in the deep sea (e.g., Gooday et al., 1998), can be examined only by wet-sorting, the method employed for the present study. Unfortunately, this technique is very time consuming and has limited the amount of material that could be examined to one sample per site.

We examined the >125 μm size fraction, which is often used in deep-sea foraminiferal studies (e.g. Douglas & Woodruff, 1981). Certain species are absent or inadequately represented in this size fraction (Schroder et al, 1987), including small species such as *Epistominella exigua* (Brady) and *Alabaminella weddellensis* (Parr) which have been shown to respond quickly to newly-delivered phytodetritus following the spring bloom (Gooday, 1993). Gooday et al. (1998) observed that, in the abyssal northeast Atlantic, a quarter or more of species were confined to the 63–125 μm fraction, although the greatest increase in diversity measures generally occurred >125 μm.

A final limitation is that we examined the foraminifera in the top 1 cm of sediment layers only and some infaunal taxa may live deeper in oxygenated sediments (Corliss, 1991). In areas with contrasting organic matter fluxes to the seafloor, however, the near-surface dwelling foraminifera should reflect the environmental differences more clearly than deeper infaunal species (Gooday, 1996).

Comparison with previous studies

Soft-bodied species were an important component of foraminiferal assemblages at all three BENBO sites. The relative proportion of these taxa increased with increasing water depth, while the proportion of the much better known hard-shelled taxa decreased. In common with many deep-sea assemblages, most of the soft-shelled species found during this study are undescribed, although the majority of calcareous species are well known (Gooday et al., 1998).

The general pattern of diversity, with a few common species and a large number of rare ones, is typical for deep-sea foraminiferal assemblages from well-oxygenated settings (Douglas & Woodruff, 1981). Comparisons with previous studies are hampered by the use of different methodologies, although the standing stocks observed here for the entire assemblages are much higher than those reported by most other workers. For example, Weston (1985) found 2–4 stained individuals 10 cm^{-2} (>125 μm) at depths greater than 1000 m from the Porcupine Seabight and Western approaches (255–1260 m), while Seiler (1975) found 5–24 stained individuals 10 cm^{-3} (>63 μm) at depths greater than 1000 m off the coast of Portugal. Similarly, our diversity values are higher than those obtained using conventional techniques due to the larger number of individuals and species observed. Murray (1991) observed that, in the deep sea off the Atlantic seaboard of Europe and Africa, Fisher α diversity values ranged from 9 to 22 (>13 off Portugal and 4–11 off northwest Africa). Comparable values for the entire assemblages at the three BENBO sites were α 29–33, while for the calcareous component it was α 5–10.

The only appropriate comparison is between our results and those of Gooday (1986a, 1996) who used almost identical methods. The faunas at Sites B and C are similar to the bathyal assemblage described by Gooday (1986a, b; Gooday & Lambshead, 1989) from the Porcupine Seabight (1330 m). In wet-sorted samples (>150 μm, 0–1 cm), Gooday (1986a) found 156–396 live individuals/10 cm^2, which is comparable to the results presented here. The diversity of this assemblage (>45 μm) was α 40–63, H(S) 3.17 – 4.09, which is higher than those found in the present study due to the finer sieve size used. *Nonionella iridea* Heron-Allen and Earland, the most abundant calcareous species at Site C, was the most abundant species found on the Porcupine Seabight (Gooday, 1986a), while *Uvigerina peregrina* Cushman and *Melonis barleeanum* (Williamson), both of which were found in relatively high densities at Site B, are shallow infaunal species characteristically found on continental slopes (Murray, 1991) in areas where there is persistently high organic carbon inputs (Fariduddin & Loubere, 1997). While the faunas at both Sites B and C show similarities to bathyal assemblages, the Site A assemblage appears to have more in common with an abyssal assemblage. At the Porcupine Abyssal Plain (PAP), Gooday (1996) found 314 individuals/10 cm^2 (>63 μm, 0–1 cm), with 48–54% of individuals >125 μm and an average of 145 species per sample. These results are lower than those reported in the present study. Diversity values at the PAP were higher than in the present study, probably due to the finer mesh used, with α 53–61 and H(S) 5.72–6.28.

Foraminifera represented 67–81% of the fauna (>125 μm) in the samples examined, greatly outnumbering the metazoans. This is consistent with the idea that benthic foraminifera form an increasingly important component of benthic communities with increasing water depth (e.g. Tietjen, 1971; Alongi & Pichon, 1988; Levin, 1991; Gooday, 1994). They have been shown to account for 50% or more of the eukaryotic biomass in the deep sea (Gooday et al. 1992).

Ecological controls on deep-sea benthic foraminifera

Mackensen et al. (1995) identified four primary, but interdependent, environmental factors which influence the generation and distribution of foraminiferal assemblages: (1) lateral advection and bottom water advection; (2) primary productivity and organic carbon flux rates; (3) bottom water carbonate corrosiveness; and (4) the energetic state of the benthic boundary layer. Following the work of Belanger & Streeter (1980), there has been increased emphasis on the importance of organic carbon inputs on the distribution and abundance of benthic foraminifera which, in most open ocean settings, is derived from the euphotic zone. In marginal environments, terrigenous inputs may be of varying importance. The relationship of live and total benthic foraminiferal assemblages to sediment organic carbon content has been examined along several continental margins (Mackensen et al., 1985; Corliss & Chen, 1988; Corliss & Emerson, 1990). Previous workers have observed decreasing foraminiferal densities with increasing depth down continental slopes (e.g. Lutze, 1980; Weston 1985), presumably reflecting a decrease in the amount of organic material reaching the sea bed with increasing depth, although absolute values of abundance appear to vary between locations (Weston, 1985).

Seasonal processes

Evidence for the importance of seasonal processes in the deep sea has emerged in recent years. One important seasonal signal arises from the deposition of aggregated phytodetritus on the sea floor following the spring bloom. This process is now well documented at bathyal and abyssal depths in the northeast Atlantic (Billett et al., 1983; Rice et al., 1986; Thiel et al, 1989; Gooday & Turley, 1990) where it may form an important food resource for deep-sea benthic organisms. Seasonally-pulsed inputs are also believed to lead to seasonal reproduction and growth in certain invertebrates (Gage & Tyler, 1991), and it has been suggested that the life histories of some deep-sea benthic foraminiferal species are linked to pulsed inputs (Gooday, 1993, 1994). These species are all small, thin-walled forms, referred to as 'phytodetritus species', which are thought to undergo rapid seasonal growth at both the individual and population levels (Gooday & Turley, 1990), characteristics which are believed to reflect their opportunistic nature (Gooday, 1993; Alve, 1994).

At Site A, however, the only species to clearly increase in abundance following the spring bloom (July 1998) was *Hoeglundina elegans*. This species has been reported from a wide depth range in the northeast Atlantic (200–3800 m; Ganssen & Lutze, 1982; Lukashina, 1983; Lutze & Colbourne, 1984; Murray, 1991). It differs from typical phytodetritus species in having a relatively large test with a thick aragonitic wall. Large numbers of juvenile specimens (195–220 μm) were present in one core (Table 3), although the species was much less common in the other two replicates from Site A. This suggests that a reproductive event had occurred, presumably triggered by a localised organic enrichment. Although *H. elegans* may be exploiting deposited phytodetritus, only small numbers were observed at this site during the July sampling cruise.

Conclusions

Live benthic foraminifera (>125 μm) were examined at three sites in the northeast Atlantic: Sites A (3600 m) and C (1900 m) were located in the Rockall Trough while Site B (1100 m) was in the Hatton-Rockall Basin. Wet-sorting sediment residues revealed high densities, with 269, 444 and 518 individuals/27 cm^2 at Sites A, B and C, respectively (0–1 cm sediment depth), and high numbers of species; 71 at Site A, 81 at Site B and 93 at Site C.

At all three stations, the assemblages were characterised by a few common species together with numerous rare ones. This was reflected in high diversity values, with Shannon–Weiner ($\log_{10}$) values ranging from 1.34 to 1.38 and Fisher α values ranging from 29 to 33.

The assemblages were dominated by delicate soft-bodied and agglutinated taxa which have been overlooked in previous studies employing dry-sorting techniques. Calcareous taxa formed a relatively small proportion of the assemblages, accounting for 23% of the individuals at Site B, 13% at Site C and only 7% at Site A. Hormosinaceans (mainly *Reophax* spp.) became more numerous with increasing water depth.

The majority of species found were undescribed, although the proportion of undescribed species varied between groups. The soft-bodied allogromiids and saccamminids were the least well known, with only 2% of the species previously described, while at least 90% of the calcareous species were described.

The faunas at Sites B and C are typical of bathyal assemblages, while those at Site A had more in common with an abyssal assemblage. The high densities of *Uvigerina peregrina* and *Melonis barleeanum* found at Site B suggest high organic carbon inputs at this location.

Reduced diversity was observed at Site A following the spring phytodetritus bloom due to the presence of large numbers of juvenile *Hoeglundina elegans*. This species differs from typical 'phytodetritus species' in that it produces a large, thick aragonitic test. It is not clear whether *H. elegans* was responding to phytodetritus deposition as only small amounts of phytodetritus were observed in cores recovered on the post-bloom cruise.

Acknowledgements

The authors would like to thank Dr John Thomson for informative discussions concerning the sediment geochemistry of the BENBO sites and Penny Holliday for advice on the hydrography. We would like to acknowledge the help of Dr Kevin Black, co-ordinator of the BENBO project, the Captains and crew of the RRS *Charles Darwin* on cruises CD 107, 111 and 113, and the many scientists who helped with the sampling. This is Publication Number 13 of the Thematic Research Programme BENBO, carried out under award

No. GST021749 from the U.K. Natural Environment Research Council. DEEPSEAS publication No. 42.

References

Alongi, D. M. & M. Pichon, 1988. Bathyal meiobenthos of the western Coral Sea: distribution and abundance in relation to microbial standing stocks and environmental factors. Deep Sea Res. 35: 491–503.

Alve, E., 1994. Opportunistic features of the foraminifer *Stainforthia fusiformis* (Williamson): evidence from Frierfjord, Norway. J. micropalaeontol. 13: 24.

Barnett, P. R. O., J. Watson & D. Connelly, 1984. A multiple core for taking virtually undisturbed samples from shelf, bathyal and abyssal sediment. Oceanol. Acta 7: 399–408.

Belanger, P. E. & S. S. Streeter, 1980. Distribution and ecology of benthic Foraminifera in the Norwegian-Greenland Sea. Mar. Micropaleontol. 5: 401–428.

Belkin, I. M., S. Levitus, J. Antonov & S. V. Malmberg, 1998. Great Salinity Anomalies in the North Atlantic. Prog. Oceanogr. 41: 1–68.

Berger, W. H., 1989. Global Maps of Ocean Productivity. In: Berger, W. H., V. S. Smetacek & G. Wefer (eds), Productivity of the Ocean: Present and Past. John Wiley & Sons Limited, New York: 429–455.

Berger, W. H., K. Fischer, C. Lai & G. Wu, 1987. Ocean productivity and organic carbon flux. Part 1. Overviews and maps of primary production, University of California, San Diego, SIO Reference 87–30.

Bernhard, J. M., 1988. Postmortem vital staining in benthic Foraminifera: duration and importance in population and distributional studies. J. foramin. Res. 18: 143–146.

Bett, B. J. et al., 1998. RRS Charles Darwin Cruise 112C, 19 May–24 June 1998. Atlantic Margin Environmental Survey: seabed survey of deep-water sreas (17th round tranches) to the north and west of Scotland. Southampton Oceanography Centre Cruise Report, No. 25: 171 pp.

Billett, D. S. M., R. S. Lampitt, A. L. Rice & R. F. C. Mantoura, 1983. Seasonal sedimentation of phytoplankton to the deep-sea benthos. Nature 302: 520–522.

Black, K., 1999. Investigating the deep ocean benthic boundary layer – from Huxley to BENBO. Ocean Challenge 9: 18–23.

Black, K. S., 1998. Cruise Report, RRS Charles Darwin Cruise 113, 28th June–22 July, 1998. Dunstaffange Marine Laboratory: 71 pp

Booth, D. A., 1983. Non Tidal Currents in the Rockall Trough. SMBA Marine Physics Group Report: 24 pp.

Brady, H. B., 1884. Report of the Foraminifera dredged by H.M.S. Challenger, during the years 1873–1876. Rept. Scientific Results Explor. Voyage H.M.S. Challenger, Zoology 9: 1–814 115pls.

Brodniewicz, I., 1965. Recent and some Holocene Foraminifera of the Southern Baltic Sea. Acta Palaeontologica Polonica 10: 131–236.

Carpenter, W. B., J. G. Jeffreys & C. W. Thomson, 1870. Preliminary report on the scientific exploration of the deep-sea in H.M.S. Porcupine, during the summer of 1869. Proc. r. Soc., 19: 146–221.

Corliss, B. H., 1991. Morphology and microhabitat preferences of benthic Foraminifera from the northwest Atlantic Ocean. Mar. Micropaleontol. 17: 195–236.

Corliss, B. H. & C. Chen, 1988. Morphotype patterns of Norwegian Sea deep-sea benthic Foraminifera and ecological implications. Geology 16: 716–719.

Corliss, B. H. & S. Emerson, 1990. Distribution of Rose Bengal stained deep-sea benthic Foraminifera from the Nova Scotian continental margin and Gulf of Maine. Deep Sea Res. 37: 381–400.

Currie, R. I., 1986. Introduction: early investigations in the Rockall Channel. Proc. r. Soc., Edinb. 88B: 1–3.

Deacon, M., 1971. Scientists and the Sea 1690–1900: a Study of Marine Science. Academic Press, London: 445 pp.

Dickson, R., J. Lazier, J. Meincke, P. Rhines & J. Swift, 1996. Long-term co-ordinated changes in the convective activity of the North Atlantic. Prog. Oceanogr. 38: 241–295.

Douglas, R. & F. Woodruff, 1981. Deep-sea Foraminifera. In Emiliani, C. (ed.), The Sea. Volume 7. John Wiley & Sons: 1233–1327.

Ellett, D. J., A. Edwards & R. Bowers, 1986. The hydrography of the Rockall Channel – an overview. Proc. r. Soc., Edinb. 88B: 61–81.

Fariduddin, M. & P. Loubere, 1997. The surface ocean productivity response of deeper water benthic Foraminifera in the Atlantic Ocean. Mar. micropaleontol. 32: 289–310.

Fisher, R. A., A. S. Corbet & C. B. Williams, 1943. The relationship between the number of species and number of individuals in a random sample of an animal population. J. anim. Ecol. 12: 42–58.

Gadyukov, A. A. & N. P. Lukashina, 1988. Distribution Patterns of Present-Day Benthic Foraminifera of the North Atlantic and Norwegian Sea as Indicated by Factor Analysis. Oceanology 28: 344–347.

Gage, J. D., 1986. The benthic fauna of the Rockall Trough: regional distribution and bathymetric zonation. Proc. r. Soc., Edinb. 88B: 159–174.

Gage, J. D. & P. A. Tyler, 1991. Deep Sea Biology. A Natural History of Organisms at the Deep-Sea Floor. Cambridge University Press, Cambridge: 504 pp.

Ganssen, G. & G. Lutze, 1982. The aragonite compensation depth at the northeastern Atlantic continental margin. 'Meteor' Forsch.-Ergebnisse 36: 57–59.

Gooday, A. J., 1986a. Meiofaunal Foraminiferans from the bathyal Porcupine Seabight (Northeast Atlantic): size structure, standing stock, taxonomic composition, species diversity and vertical distribution in the sediment. Deep Sea Res. 33: 1345–1373.

Gooday, A. J., 1986b. Soft-shelled Foraminifera in meiofauna samples from the bathyal northeast Atlantic. Sarsia 71: 275–287.

Gooday, A. J., 1988. A response by benthic Foraminifera to the deposition of phytodetritus in the deep sea. Nature 332: 70–73.

Gooday, A. J., 1993. Deep-sea benthic Foraminiferal species which exploit phytodetritus: Characteristic features and controls on distribution. Mar. micropaleontol. 22: 187–205.

Gooday, A. J., 1994. The Biology of Deep-Sea Foraminifera: A Review of Some Advances and Their Applications in Paleoceanography. Palaios 9: 14–31.

Gooday, A. J., 1996. Epifaunal and shallow infaunal Foraminiferal communities at three abyssal NE Atlantic sites subject to different phytodetritus input regimes. Deep Sea Res. 43: 1395–1431.

Gooday, A. J., 1999. Biodiversity of Foraminifera and other protists in the deep-sea: scales and patterns. Belg. J. Zool 129: 61–80.

Gooday, A. J. & P. J. D. Lambshead, 1989. Influence of seasonally deposited phytodetritus on benthic Foraminiferal populations in the bathyal northeast Atlantic: the species response. Mar. Ecol. Prog. Ser. 58: 53–67.

Gooday, A. J. & A. E. Rathburn, 1999. Temporal variability in living deep-sea benthic Foraminifera: a review. Earth Sci. Rev. 46: 187–212.

Gooday, A. J. & C. M. Turley, 1990. Responses of benthic organisms to inputs of organic material to the ocean floor: a review. Phil. Trans. r. Soc., Lond. 331: 119–138.

Gooday, A. J., B. J. Bett, R. Shires & P. J. D. Lambshead, 1998. Deep-Sea benthic Foraminiferal species diversity in the NE Atlantic and NW Arabian Sea: a synthesis. Deep Sea Res. II 45: 165–201.

Gooday, A. J., M. Carstens & H. Thiel, 1995. Micro-and Nano-Foraminifera from Abyssal Northeast Atlantic Sediments: A Preliminary Report. Int. Revue ges. Hydrobiol. 80: 361–383.

Gooday, A. J., L. A. Levin, P. Linke & T. Heeger, 1992. The role of benthic Foraminifera in deep-sea food webs and carbon cycling. In Rowe, G. T. & V. Pariente (eds), Deep-Sea Food Chains and the Global Carbon Cycle. Kluwer Academic Publishers, Dordrecht: 63–91.

Harvey, J., 1982. Theta-S relationships and water masses in the eastern North Atlantic. Deep Sea Res. 29: 1021–1033.

Holliday, N. P., R. T. Pollard, J. F. Read & H. Leach. 2000. Water Mass Properties and Fluxes in the Rockall Trough; 1975 – 1998. Deep Sea Res. 47: 1302–1332

Hurlbert, S. M., 1971. The nonconcept of species diversity: a critique and alternative parameters. Ecology 52: 577–586.

Jones, E. J. W., M. Ewing, J. I. Ewing & S. L. Eittreim, 1970. Influences of Norwegian Sea overflow water on sedimentation in the northern North Atlantic and Labrador Sea. J. Geophys. Res. 75: 1655–1680.

Jones, R. W., 1986. Distribution of morphogroups of Recent agglutinating Foraminifera in the Rockall Trough – a synopsis. Proc. r. Soc., Edinb. 88B: 55–58.

Jones, R. W., 1994. The Challenger Foraminifera. Oxford University Press, Oxford: 149 pp.

Lee, A. & D. Ellett, 1965. On the contribution of overflow water from the Norwegian Sea to the hydrographic structure of the North Atlantic Ocean. Deep Sea Res. 12: 129–142.

Levin, L. A., S. E. Childers & C. R. Smith, 1991. Epibenthic, agglutinating Foraminiferans in the Santa Catalina Basin and their response to disturbance. Deep Sea Res. 38: 465–483.

Lukashina, N. P., 1983. The Distribution of Recent Bottom-Dwelling Foraminifera in the Northeast Atlantic. Oceanology 23: 72–75.

Lukashina, N. P., 1988a. Benthic Foraminifera Communities and Water Masses of the North Atlantic and the Norway-Greenland Basin. Oceanology 28: 612–617.

Lukashina, N. P., 1988b. Distribution Patterns of Benthic Foraminifera in the North Atlantic. Oceanology 28: 492–497.

Lutze, G. F., 1980. Depth distribution of benthic Foraminifera on the continental margin off NW Africa. 'Meteor' Forsch.-Ergebnisse 32: 31–80.

Mackensen, A., G. Schmiedl, J. Harloff & M. Giese, 1995. Deep-sea Foraminifera in the South Atlantic Ocean: ecology and assemblage generation. Micropaleontology 41: 342–358.

Mackensen, A., H. P. Sejrup & E. Jansen, 1985. The distribution of living benthic Foraminifera on the continental slope and rise off Southwest Norway. Mar. Micropaleontol. 9: 275–306.

McCave, I. N., 1986. Epilogue: recommendations of the Workshop Session. Proc. r. Soc., Edinb. 88B: 291–298.

Meincke, J., 1986. Convection in the oceanic waters west of Britain. Proc. r. Soc., Edinb. 88B: 127–139.

Murosky, M. W. & S. W. Snyder, 1994. Vertical distribution of stained benthic Foraminifera in sediments of Southern Onslow Bay, North Carolina continental shelf. J. foramin. Res. 24: 158–170.

Murray, J. W., 1991. Ecology and Palaeoecology of Benthic Foraminifera. Longmann Scientific and Technical, New York: 397 pp.

Murray, J. W. & S. S. Bowser, 2000. Mortality, protoplasm decay rate, and reliability of staining techniques to recognize 'living' Foraminifera: a review. J. foramin. Res. 30: 66–70.

Murray, J. W. & C. M. Taplin, 1984. Larger agglutinated Foraminifera from the Faeroe Channel and Rockall Trough collected by W.B. Carpenter. J. Micropalaeontol. 3: 59–62.

Murray, J. W., J. F. Weston, C. A. Haddon & A. D. J. Powell, 1986. Miocene to Recent bottom water masses of the Northeast Atlantic: an analysis of benthic Foraminifera. Geological Society Special Publication 21: 219–230.

Parker, W. K. & T. R. Jones. 1865. On some Foraminifera of the North Atlantic and Arctic Oceans including Davis Strait and Baffins Bay. Phil. Trans r. Soc., Lond. 155: 325–441.

Pearce, G. J., 1980. Benthonic Foraminifera from the continental slope and their fossil counterparts. PhD thesis, University of Aberystwyth: 716 pp.

Poore, R. Z. & W. A. Berggren, 1975. Late Cenozoic plankton Foraminiferal biostatigraphy and paleoclimatology of Hatton-Rockall Basin: DSDP Site 116. J. foramin. Res. 5: 270–293.

Rice, A. L., D. S. M. Billett, J. Fry, A. W. G. John, R. S. Lampitt, R. F. C. Mantoura & R. J. Morris, 1986. Seasonal deposition of phytodetritus to the deep-sea floor. Proc. r. Soc., Edinb. 88B: 265–279.

Rice, A. L. & P. Owen., 1999. Decommissioning the Brent Spar. E & FN Spoon, London: 182 pp.

Ruddiman, W. F., 1972. Sediment redistribution on the Reykjanes Ridge. Geol. Soc. Am. Bull. 83: 2039–2062.

Schmitz, W. J. & M. S. McCartney, 1993. On the North Atlantic Circulation. Rev. Geophys. 31: 29–49.

Schroder, C. J., B. D. Scott & F. S. Medioli, 1987. Can smaller benthic Foraminifera be ignored in paleoenvironmental analysis? J. foramin. Res. 17: 101–105.

Shannon, C. E. & W. Weaver, 1963. The Mathematical Theory of Communication. University of Illinois Press, Urbana: 117pp.

Simpson, E. H., 1949. Measurement of diversity. Nature 163: 688.

Smart, C. W. & A. J. Gooday, 1997. Recent benthic Foraminifera in the abyssal Northeast Atlantic Ocean: Relation to phytodetrital inputs. J. foramin. Res. 27: 85–92.

Thiel, H., O. Pfannkuche, G. Schriever, K. Lochte, A. J. Gooday, C. H. Hemleben, R. F. G. Mantoura, C. M. Turley, J. W. Patching & F. Riemann, 1989. Phytodetritus on the Deep-Sea Floor in a Central Oceanic Region of the Northeast Atlantic. Biol. Oceanogr. 6: 203–239.

Thomson, C. W., 1873. The Depths of the Sea. MacMillan and Co., London: 527 pp.

Tietjen, J. H., 1971. Ecology and distribution of deep-sea meiobenthos off North Carolina. Deep Sea Res. 18: 941–957.

Van Weering, T. C. E., I. N. McCave & I. R. Hall, 1998. Ocean Margin Exchange (OMEX I) benthic processes study. Prog. Oceanogr. 42: 1–4.

Walsh, J. J., 1991. Importance of continental margins in the marine biogeochemical cycling of carbon and nitrogen. Nature 350: 53–55.

Weston, J. F., 1985. Comparison between Recent benthic Foraminiferal faunas at the Porcupine Seabight and Western Approaches Continental Slope. J. Micropalaeontol. 4: 165–183.

Weston, J. F. & J. W. Murray, 1984. Benthic Foraminifera as deep-sea water-mass indicators. Benthos '83; 2nd International

Symposium on Benthic Foraminifera (Pau, April, 1983). Rev. Paleobiol., Vol. Spec. 2: 605–610.
Wright, J., 1889. Report on the deep sea trawling cruise off the southwest coast of Ireland: Foraminifera. Ann. Mag. Nat. His. Series 6: 4.
Wright, J., 1891. Report on the Foraminifera obtained off the southwest of Ireland during the cruise of the 'Flying Falcon,' 1888. Proc. R. Ir. Acad. 3: 460–502.
Zezina, O. N., 1997. Biogeography of the Bathyal Zone. Adv. mar. Biol. 32: 389–426.

Hydrobiologia **440:** 239–247, 2000.
M.B. Jones, J.M.N. Azevedo, A.I. Neto, A.C. Costa & A.M. Frias Martins (eds), Island, Ocean and Deep-Sea Biology.

Foraminiferan (Protozoa) epizoites on deep-water arcturids (Crustacea, Isopoda)

Jörundur Svavarsson & Steinunn H. Ólafsdóttir
Institute of Biology, University of Iceland, Grensásvegur 12, 108 Reykjavík, Iceland and Sandgerði Marine Centre, Garðvegur 1, 245 Sandgerði, Iceland
Fax: +354-5254069. E-mail: jorundur@hi.is

Key words: epizoites, deep water, *Cibicides* spp., isopods, Arcturidae

Abstract

Foraminiferan (Protozoa) epizoites were examined on two deep-water isopods, *Neastacilla* sp. and *Pleuroprion hystrix* (Valvifera, Arcturidae), from the north Atlantic Ocean and the Nordic Seas. Most foraminiferans belonged to genus *Cibicides* and occurred on 27% of *Neastacilla* and 23% of *Pleuroprion hystrix.* The pattern of foraminiferan epibiosis was similar in both species, with a concentration on the dorsal body and on the posterior pereopods. The arcturids showed an increase in number of foraminiferans with increased body size, suggesting that available time span for settling on the younger stages (mancas) was relatively short. There was a trend towards larger numbers of foraminiferans occurring on larger arcturid species, suggesting that available space is the factor that determines the total number of foraminiferans for epibenthic arcturid isopod species.

Introduction

Crustaceans carry numerous epizoites, including bacteria, protozoans, algae and multicellular organisms (Becker & Wahl, 1996). Amongst the isopod crustaceans, various epizoites have been reported (Glynn, 1970; Jones & Kahn, 1970; Svavarsson & Davíðsdóttir, 1994; Key & Barnes, 1999). The behaviour of the host seems to be the most important factor in shaping epibiosis among crustaceans (Becker & Wahl, 1996), including isopods (Svavarsson & Davíðsdóttir, 1994, 1995).

Recent studies have shown that foraminiferans (Protozoa) occur frequently as epibionts on isopods (Moore, 1985; Svavarsson & Davíðsdóttir, 1994). The foraminiferans that do so belong almost entirely to the genus *Cibicides.* (Svavarsson & Davíðsdóttir, 1994, 1995; Svavarsson & Ólafsdóttir, 1999). These foraminiferans are common on a variety of hard substrata (rocks, many animals) and seem to prefer an elevated substratum (Lutze & Thiel, 1989). On isopods, they are most frequently found on epibenthic ones, which rarely or never clean themselves (Svavarsson & Davíðsdóttir, 1994), and can be quite abundant on individual species (up to ≈94 foraminiferans on the large arcturid *Arcturus baffini*). The isopods are colonised either by fairly large, vagrant foraminiferans or by foraminiferan larvae settling from the water column (Svavarsson & Davíðsdóttir, 1995). For many isopod species, the presence of foraminiferans does not seem to have any effects. On the anthuridean isopod *Calathura brachiata*, foraminiferans settled on the mouthparts and on appendages used for grasping prey while the isopod was at rest (Svavarsson & Ólafsdóttir, 1999). It has been suggested that epibiont overgrowth may even favour the isopod, providing camouflage (Moore, 1985). Being vagrant, the foraminiferans can find a new substratum while the isopods moults.

Little is known of the effect of isopod size, growth rate and sexual dimorphism on the pattern of epibiosis. Among some species (e.g. *Arcturus baffini*), there is an increase in the number of foraminiferans with increased size of the isopod host, while in other species (e.g. *Calathura brachiata*), there is no relationship between the size of the isopod host and the number of the foraminiferans (Svavarsson & Davíðsdóttir, 1995; Svavarsson & Ólafsdóttir, 1999), suggesting that the growth rate and/or intermoult time-span of the isopod may be important for the development of foraminiferan epibiosis. Svavarsson & Davíðsdóttir

(1994) found no significant differences between the mean sizes of a variety of isopod species with and without foraminiferan epizoites, indicating that the host size alone may not be important.

The aim of this paper was to describe the pattern of foraminiferan epibiosis on two small (<9 mm) arcturid (Crustacea, Isopoda, Valvifera) species. We examined the role of size and shape in the development of foraminiferan epibiosis, and compared these two small species with the large (often <40 mm) arcturid *Arcturus baffini*. We ask the question: is the pattern of foraminiferan epibiosis similar among differently sized host species having similar morphology? Furthermore, we ask: do similarly sized and shaped animals from different geographical locations have comparable epibiosis?

Materials and methods

Isopods were provided from the international BIOICE (Benthic Invertebrates of Icelandic waters; 1991) project. Samples were collected at 15 stations in Icelandic waters during cruises (1991–1994) with the research vessels '*Håkon Mosby*', '*Bjarni Sæmundsson*' and '*Magnus Heinason*' (Table 1). Samples were taken with a modified Rothlisberg-Pearcy epibenthic sled (RP sled) (Rothlisberg & Pearcy, 1976; Brattegard & Fosså, 1991) and a Sneli sled (Sneli, 1998). The RP sled was towed over the bottom and collected animals lifted by the turbulence in front of the sled. The sled net had a 0.5 mm mesh in the cod-end. A closing mechanism ensured that the sled was open only on the bottom and so reduced loss of material on retrieval. The RP samples were decanted over a 0.5 mm sieve, and the sediment and remaining animals sieved over a series of sieves (8, 4, 2, 1, 0.5 mm). The Sneli sled samples were sieved gently through 3 mm and 0.5 mm sieves. Samples were preserved in 10% neutralized formalin and transferred to 70% ethanol. Isopods were sorted at the Sandgerði Marine Centre, Sandgerði.

The species were collected at different geographical locations (Table 1). *Neastacilla* was collected in the most northern part of the North Atlantic Ocean, while *Pleuroprion* was collected in the most southern part of the Nordic Seas (Greenland, Iceland and Norwegian Seas). A total of 601 specimens of *Neastacilla* sp. and 91 specimens of *Pleuroprion hystrix* were studied under a dissection microscope using 12.5–80× magnification. Each part of the body, except the pleopods (enclosed by the uropods), of each individual was studied carefully to mark the position of the foraminiferans prior to their removal for measurement and identification. Isopods were staged, sexed and their body length measured. The isopod developmental stages were identified based on the development of the seventh pair of legs (absent in manca 1, small rudimentary in manca 2, large but rudimentary in manca 3), the number and occurrence of setae on the first antenna (only distally on precopulatory male, distally and ventrally on copulatory male), and the presence or absence of oostegites. Female *Neastacilla* with many long spines were considered of different developmental stages (preparatory II and brooding II) from those with a few and short spines (preparatory I and brooding I).

The species

The bodies of the two species, *Neastacilla* sp. and *Pleuroprion hystrix*, differ in structure. *Neastacilla* is fairly slender (brooding female body approximately 11 times longer than wide; copulatory male body approximately 17 times longer than wide) (Fig. 1A), while *Pleuroprion* is rather robust (brooding female body 4.6 times longer than wide; copulatory male body approximately nine times longer than wide) (Fig. 1B). *Neastacilla* and *Pleuroprion* sieve food particles from the water column and pereopods 1–4 are modified for that purpose. For each species, pereopods 5–7 are modified to attach to the bottom or to benthic organisms such as hydrozoans or sponges. The second antenna is rather long, nearly as long as the body of adults. In *Neastacilla*, sexual dimorphism is pronounced and is seen mainly in the structure of the fourth body segment, which is more slender and has fewer spines in males than females. Females may be spinier than males. In *Pleuroprion*, sexual dimorphism is modest. The female body is broader than the male body and is covered almost totally with spines dorsally and laterally. Males have fewer spines.

Results

Sizes and size increment of Neastacilla *and* Pleuroprion

Each species apparently passes through relatively few developmental stages before reaching maturity (Table 2). There is an indication that *Neastacilla* has

Table 1. Location of sampling sites for *Neastacilla* sp. and *Pleuroprion hystrix*

Species Station	Latitude N°	Longitude W°	Depth (m)	Temp. °C	Salinity	Sediment	Sampling gear
Neastacilla sp.							
BIOICE 2257	63° 14.61′	26° 29.14′	1209	4.1	34.9	Clay and shell sand	RP
BIOICE 2299	63° 00.10′	22° 39.61′	775	5.5	35	Sand	RP
BIOICE 2406	62° 59.20′	21° 47.00′	934	4.6	34.98	–	RP
BIOICE 2691	64° 26.10′	28° 14.20′	1162	3.7	34.92	Sand w/silt, gravel	Sneli
BIOICE 2697	64° 10.20′	27° 43.10′	1042	4.2	34.93	–	RP
BIOICE 2704	63° 50.50′	27° 42.80′	1295	3.8	34.92	–	RP
BIOICE 2707	63° 55.30′	28° 16.80′	1407	3.7	34.91	Silty sand	RP
Pleuroprion hystrix							
BIOICE 2004	65° 20.81′	13° 19.63′	177	2.8	–	Soft sediment	RP
BIOICE 2067	66° 08.60′	17° 36.42′	196	3.8	34.9	–	RP
BIOICE 2075	67° 11.65′	17° 32.04′	563	−0.4	34.7	Silty sand, gravel, stones	Sneli
BIOICE 2090	67° 13.34′	17° 48.94′	539	−0.4	34.9	–	RP
BIOICE 2126	66° 59.51′	18° 49.82′	208	2.7	34.9	Sandy gravel, few stones	Sneli
BIOICE 2317	64° 17.00′	9° 03.00′	996	–	–	–	RP
BIOICE 2324	63° 45.00′	10° 11.00′	554	–	–	Silt, stones	Sneli
BIOICE 2328	63° 20.00′	10° 57.00′	430	–	–	Sandy silt, gravel, stones	RP

two preparatory and brooding stages, while only a single stage was seen for *Pleuroprion*. Both species show a large size increment between the manca stages. In *Neastacilla* sp., the size increment between the first stages was 33% (manca 1–2), 29% (manca 2–3), 20% (manca 3-precopulatory males) and 14% (manca 3-preparatory females), whilst in *Pleuroprion* the size increment between the first stages was 23% (manca 1–2), 17% (manca 2–3), 11% (manca 3-precopulatory males) and 24% (manca 3-preparatory females).

Table 2. Sizes of different developmental stages of *Neastacilla* sp. and *Pleuroprion hystrix*

Stage	*Neastacilla* sp.		*P. hystrix*	
	Mean (mm)	Range (mm)	Mean (mm)	Range (mm)
Manca 1	2.3	2.1–2.5	2.8	2.4–3.1
Manca 2	3.4	3.1–4.1	3.6	3.0–4.1
Manca 3	4.8	3.9–5.5	4.4	3.9–4.9
Preparatory female I	5.6	5.0–6.7	5.8	4.7–6.9
Precopulatory male	6.0	4.9–6.7	4.9	4.8–5.0
Brooding female I	6.6	5.0–7.9	7.2	5.7–9.0
Copulatory male	6.7	5.7–8.1	5.2	4.3–6.0
Preparatory female II	8.5	8.2–8.9		
Brooding female II	7.0	6.3–7.5		

Neastacilla *sp.*

Foraminiferans occurred on 162 individuals (27%) of the 601 *Neastacilla* sp. examined (Table 3). In all, 410 foraminiferans were found, including *Cibicides refulgens* de Montfort, *C. lobatulus* Walker and Jacob, *C. wuellerstorfi* Walker and Jacob, and *Cibicides* spp. (juveniles). In addition, some specimens of the agglutinating family Ammodiscidae, one foraminiferan (family Rosalinidae) and some unidentified agglutinating foraminiferans were found. Other epizoites were mostly the suctorian ciliates *Ephelota* sp. and few other unidentified ciliates. Most *Cibicides* were small, with a diameter of less than 0.4 mm (Fig. 2A). The frequency of isopods with foraminiferans varied considerably between locations (8–67%) (Table 3). Mostly (67.9% of cases), one or two foraminiferan specimens occurred on the isopods, but up to 22 foraminiferan specimens were found on a single individual of *Neastacilla*. The number of foraminiferans, and their frequency, increased with the size and the development of the isopods (Fig. 3). Similar patterns were found for male and females. Only 2–7% of the mancas had foraminiferans and these were relatively small (Table 4). Later developmental stages had higher

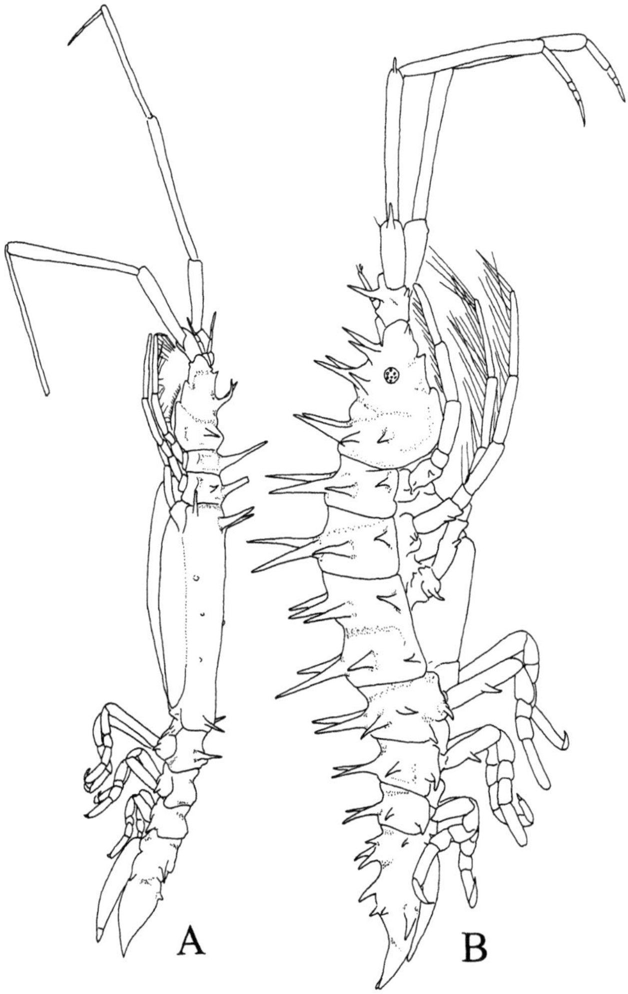

Figure 1. Line drawing of female *Neastacilla* sp. (A) (brooding female I, 7.2 mm body length) and *Pleuroprion hystrix* (B) (8.0 mm body length).

frequency of foraminiferans and there was a tendency towards larger size range in larger isopod specimens, although the foraminiferans were all quite small.

Foraminiferans were located on every part of the body (Table 5) and on all, except the first and third pairs, pereopods (Table 6). In each sex, the fourth pereonite is much longer than other pereonites, being more than one third (40–42%) of the total body length. There were significantly more foraminiferans on the head and first three pereonites (about 22% of the body length) of the females than on pereonite 4 (about 40% of the body length), pereonites 5–7, the pleotelson and the uropod (about 38% of the body length) (Table 7). Pereonite 4 had significantly higher numbers of foraminiferans than pereonites 5-7 and the pleotelson. Males had significantly higher number of foraminiferans on the head and pereonites 1–3 (about 20% of the body length) than on pereonite 4 (about 42% of the body), pereonites 5–7, the pleotelson and the uropod (about 38% of the body length). The posterior pereonites and the pleotelson of the males, however, had significantly higher number of foraminiferans than pereonite 4, unlike the females (Table 7). Foraminiferans were mainly found on the posterior pereopods (4 to 7) (Table 6), and on all articles, except the dactylus. Males and females had similar numbers of foraminiferans (males 1.04:1 females) on the body. Males, however, had significantly (χ^2=12.35, df =1, P<0.001) higher number of foraminiferans on pereonites 5–7, the pleotelson and uropod than the females.

Table 3. Number of *Neastacilla* sp. and *Pleuroprion hystrix* examined at each location, number of isopods with foraminiferans and number of isopods holding different numbers of foraminiferans

Species Station	No. of isopods	Isopods with forams No	Isopods with forams (%)	Total no. of foraminiferans on isopods	No. of isopods holding 0, 1,....,>10 foraminiferans 0	1	2	3	4	5	6	7	8	9	≥10
Neastacilla sp.															
BIOICE 2257	27	5	19	7	22	3	2	0	0	0	0	0	0	0	0
BIOICE 2299	27	18	67	94	9	1	5	2	2	1	2	2	2	0	1
BIOICE 2406	105	48	46	137	57	20	10	5	4	3	1	0	2	2	1
BIOICE 2691	5	1	20	1	4	1	0	0	0	0	0	0	0	0	0
BIOICE 2697	324	80	25	159	244	43	15	11	5	5	0	0	1	0	0
BIOICE 2704	106	9	8	10	97	8	1	0	0	0	0	0	0	0	0
BIOICE 2707	7	1	14	2	6	0	1	0	0	0	0	0	0	0	0
Total	601	162	(27)	410	439	76	34	18	11	9	3	2	5	2	2
Pleuroprion hystrix															
BIOICE 2004	1	1	(100)	2	0	0	1	0	0	0	0	0	0		
BIOICE 2067	8	1	13	2	7	0	1	0	0	0	0	0	0		
BIOICE 2075	1	0	(0)	0	1	0	0	0	0	0	0	0	0		
BIOICE 2090	4	1	25	1	3	1	0	0	0	0	0	0	0		
BIOICE 2126	6	1	17	6	5	0	0	0	0	0	1	0	0		
BIOICE 2317	55	11	20	21	44	6	2	2	0	1	0	0	0		
BIOICE 2324	1	0	(0)	0	1	0	0	0	0	0	0	0	0		
BIOICE 2328	15	6	40	19	9	2	1	1	1	0	0	0	1		
Total	91	21	(23)	51	70	9	5	3	1	1	1	0	1		

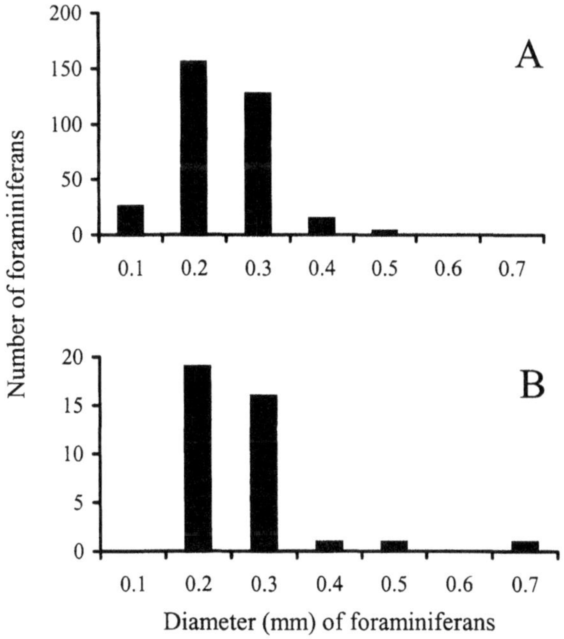

Figure 2. Size–frequency histogram of foraminiferans on *Neastacilla* (A) and on *Pleuroprion hystrix* (B).

The long second antenna of *Neastacilla* represents about 30% of the total surface of brooding females and about 36% of the surface of copulatory males (excluding pereopods). The density of the foraminiferans was significantly higher on the body than on the second antenna for both males (χ^2=68.48, df=1, P<0.001) and females (χ^2=28.68, df=1, P<0.001). Males had about 1.08 times more foraminiferans on the second antenna than the females. The difference was, however, not significant (χ^2=2.52, df=1, P>0.05). Nearly all foraminiferans (32 of 33) found on the second antenna occurred on the antennal peduncle. Most of the foraminiferans were found on the fourth article of the peduncle (the second most distal one; nine individuals), only two occurred on the second article and seven foraminiferans were observed on the third article. The foraminiferans appeared to be located randomly on dorsal, ventral and lateral surfaces of the second antenna.

Table 4. Number and size range of foraminiferans on different developmental stages of *Neastacilla* sp. and *Pleuroprion hystrix*

Species	Developmental stage	Number of isopods	Isopods with forams number	Isopods with forams %	Foraminiferans Number	Foraminiferans Mean size (mm)	Foraminiferans Size range (mm)
Neastacilla sp.							
	Manca 1	57	4	7	4	0.14	0.08–0.24
	Manca 2	83	2	2	2	0.16	0.08–0.23
	Manca 3	81	4	5	8	0.22	0.08–0.38
	Preparatory I female	48	7	15	7	0.14	0.05–0.24
	Precopulatory male	77	20	26	30	0.16	0.07–0.28
	Brooding I female	101	55	54	135	0.2	0.06–0.48
	Copulatory male	132	60	45	182	0.17	0.05–0.38
	Preparatory II female	3	1	(33)	1	(0.18)	
	Brooding II female	19	9	47	41	0.21	0.07–0.44
	Total	601	162	27	410		
Pleuroprion hystrix							
	Manca 1	10	0	0	0		
	Manca 2	12	0	p0	0		
	Manca 3	11	2	18	2	0.14	0.11–0.16
	Preparatory female	8	1	13	1	(0.2)	
	Precopulatory male	2	0	(0)	0		
	Copulatory male	34	10	28	30	0.2	0.13–0.29
	Brooding female	14	8	57	18	0.24	0.14–0.63
	Total	91	21	23	51		

Pleuroprion hystrix

A total of 51 foraminiferans was found on 21 (23%) of the 91 individuals studied (Table 3). These were identified as *Cibicides refulgens* and *Cibicides* spp. (juveniles). Five specimens of *Hemisphaerammina* sp., four specimens of Ammodiscidae spp. and one individual of the family Miliolidae were also found. Other epizoites were mostly the suctorian ciliates *Ephelota* sp., and a few unidentified ciliates and hydrozoans. Most *Cibicides* (93%) were small, with a diameter of less than 0.3 mm (Fig. 2B). Frequency of isopods with foraminiferans ranged from 13–40% (Table 3). Mostly, one or two foraminiferans occurred on the isopods. One individual held up to eight specimens of foraminiferans (Table 3). The number and the frequency of foraminiferans increased with the size of the isopods (Figs 3C and D). Foraminiferans occurred rarely on small *P. hystrix* and only two specimens were found on the manca stages (only manca 3) (Table 4). As for *Neastacilla*, the later developmental stages had a higher frequency of foraminiferans. The patterns of epibiosis differed between females and males, possibly due to different size range of the sexes (Figs 3C and D). Females had a larger size range than males (4.7–9 mm and 4.3–6 mm, respectively). The size range of foraminiferans indicates that their size increases as the isopod develops (Table 4).

Foraminiferans were located on most parts of the body except on pereopods 1–4, pereonite 3 and the uropod (Tables 5 and 6). There was no significant difference in the distribution of the foraminiferans on the second antenna and the body ($\chi^2 = 3.08$, df =1, $P>0.05$); the second antenna is approximately 20% of the total surface. There was no significant difference of the number of foraminiferans on the body and the pereopods of females and males ($\chi^2 = 0.028$, df =1, $P>0.05$).

Discussion

Foraminiferan epizoites were relatively common on each of the arcturid isopod species studied, supporting

Table 5. Distribution of foraminiferans on the body and on the second antenna of *Neastacilla* sp. and *Pleuroprion hystrix*

Species	Second antenna	Head	Pereonite 1	2	3	4	5	6	7	Pleo-telson	Uropod	Total
Neastacilla sp.												
Females:												
Dorsal	7	50	13	5	6	14	3	1	1	2	1	103
Lateral	6		12	9	5	16	1	1	2	1		53
Ventral	6					5	2			1		14
Total	19	50	25	14	11	35	6	2	3	4	1	170
Males												
Dorsal	5	48	13	18	3	6	6	1		11	4	115
Lateral	6		13	2	2	12	6	1	3	2		47
Ventral	3	1				11	6	4		1		26
Total	14	49	26	20	5	29	18	6	3	14	4	188
Pleuroprion hystrix												
Dorsal	3	4	6	1		4	1	1		3		23
Lateral	1	–	–	3		3	2	2	1	4		16
Ventral					4					4		
Total	4	4	6	4	0	11	3	3	1	7	0	43

Table 6. Distribution of foraminiferans on the pereopods of *Neastacilla* sp. and *Pleuroprion hystrix*

Species	Article	Pereopod 1	2	3	4	5	6	7	Total
Neastacilla sp.									
	Basis				1	11	3	1	16
	Ischium		1			2	1	1	5
	Merus		1		2				3
	Carpus		1		3		3	3	10
	Propodus					1	1	2	4
	Dactylus	–	–	–	–				0
	Total	0	3	0	6	14	8	7	38
Pleuroprion hystrix									
	Basis					3	1	2	6
	Ischium								0
	Merus						1		0
	Carpus								0
	Propodus							1	1
	Dactylus	–	–	–	–				0
	Total	0	0	0	0	3	2	3	8

previous studies on arcturids (Moore, 1985; Svavarsson & Davíðsdóttir, 1995). As in earlier studies on foraminiferan epizoites on isopods (Svavarsson & Davíðsdóttir, 1994), *Cibicides* species were the most commonly encountered foraminiferan epizoites. The general pattern of foraminiferan distribution is presumably shaped by the habitat preferences of the foraminiferans, which prefer an elevated substratum (Lutze & Thiel, 1989), and the behaviour of the arcturid isopods. All species presumably cling to hydrozoans or other benthic organisms with their posterior pereopods, elevating the anterior part of their bodies into the water column. Little is known of why species of other foraminiferan genera and families are rare as epizoites on isopod crustaceans. Many foraminiferan species occur commonly on other marine organisms, such as on bivalves and bryozoans (Haward & Haynes, 1976; Zampi et al., 1997) or even on sponges (Lutze & Thiel, 1989). Frequent moulting of crustaceans, and their feeding and moving activity, may possibly be too disturbing for some of the foraminiferan species, but other aspects, such as antifouling agents (Harrington & Armstrong, 1999), may also play an role.

There were similar general patterns in foraminiferan epibiosis in each species studied here des-

Table 7. Chi-square comparisons of frequency of foraminiferans on *Neastacilla* sp.

Sex	Comparison		Chi-square	df	*P*
Females					
	head and pereonites 1–3 vs. pereonite 4	h, p1–3>p4	87.82	1	<0.001
	head and pereonite 1–3 vs. pereonites 5–7, peleotelson and uropod	h, p1–3>p5–7, pl, up	122.59	1	<0.001
	pereonite 4 vs. pereonites 5–7 and pleotelson and uropod	p4>p5–7, pl, up	6.14	1	<0.05
Males					
	head and pereonites 1–3 vs. pereonite 4	h, p1–3>p4	120.93	1	<0.001
	head and pereonite 1–3 vs. pereonites 5–7, peleotelson and uropod	h, p1–3>p5–7, pl, up	76.32	1	<0.001
	pereonite 4 vs. pereonites 5–7 and pleotelson and uropod	p4<p5–7, pl, up	5.26	1	<0.05

pite being collected from different geographical areas. The pattern observed resembled that seen in the large arcturids *Astacilla longicornis* and *Arcturus baffini* (Moore, 1985; Svavarsson & Davíðsdóttir, 1995). Here, foraminiferans were frequent on the body and on the posterior pereopods, while few occurred on the slender setose first three or four pairs of pereopods, used to filter particles from the water column.

Significant difference between the arcturid species (*Neastacilla* sp., *Pleuroprion hystrix*, *Astacilla longicornis*, *Arcturus baffini*) was seen in the maximum number of reported foraminiferans. While *Neastacilla* and *Pleuroprion* had 22 and eight foraminiferans per isopod, respectively, *Arcturus baffini* has been reported with about 94 specimens (Svavarsson & Davíðsdóttir, 1995) and *Astacilla longicornis* up to 46 specimens (Moore, 1985). Comparing average sizes of isopod species with and without foraminiferans, Svavarsson & Davíðsdóttir (1994) did not find any significant differences, suggesting that size alone did not determine the presence/absence of foraminiferan epibionts. Among closely-related species, however, there may be a relationship between isopod size and the maximum/average number of epizoic foraminiferans. Species with similar morphology and ecology offer similar habitats for the foraminiferans, but larger isopods species supply more space for foraminiferans than small ones.

Unlike the anthuridean isopod, *Calathura brachiata*, where no relationship between the body size and the number of attached foraminiferans occurs (Svavarsson & Ólafsdóttir, 1999), the frequency and number of foraminiferans increased with the size of *Neastacilla* and *Pleuroprion hystrix*. This has been shown also for *Arcturus baffini* (Svavarsson & Davíðsdóttir, 1995). The difference may be explained by the different growth patterns of the species. For *Neastacilla* and *Pleuroprion*, a large increment occurs between the first manca stages. These species presumably attain full length in a fairly short time, while their adult stages (preparatory, precopulatory, brooding and copulatory stages) have presumably longer duration. *Calathura brachiata*, which presumably feeds on benthic crustaceans, may, however, spend a similar time span on each developmental stage, allowing similar densities of foraminiferans on the different developmental stages (Svavarsson & Ólafsdóttir, 1999).

The size range of the foraminiferans on both *Neastacilla* and *Pleuroprion* resembled those seen on small asellotan and larval gnathiid isopods (Svavarsson & Davíðsdóttir, 1994). Somewhat larger size range of foraminiferans with increased development of the isopods indicates more available time for the foraminiferans on the isopods. The foraminiferans on the youngest stages (manca 1–3) of *Neastacilla* were fairly small indicating that they settled from the water column as larvae. However, in the large, antenna-brooding arcturid *Arcturus baffini*, relatively large (up to 0.8 mm) foraminiferans migrate from the mother's second antenna onto the first two manca stages clinging to the second antenna of the mother (Svavarsson & Davíðsdóttir, 1995). Svavarsson & Davíðsdóttir (1995) suggested that antennal brooding in *Arcturus baffini* affected the number of foraminiferans on the second antenna of the mother. The frequency of foraminiferans on the body of the females of *A. baffini* was considerably higher (3.7–4 times) than on the males, whilst the frequency of the foraminiferans on the second antenna was not as high (2.1 times). In *Neastacilla*, in which antennal brooding has not been reported, the number of foraminiferans was similar in males and females on the body and the second an-

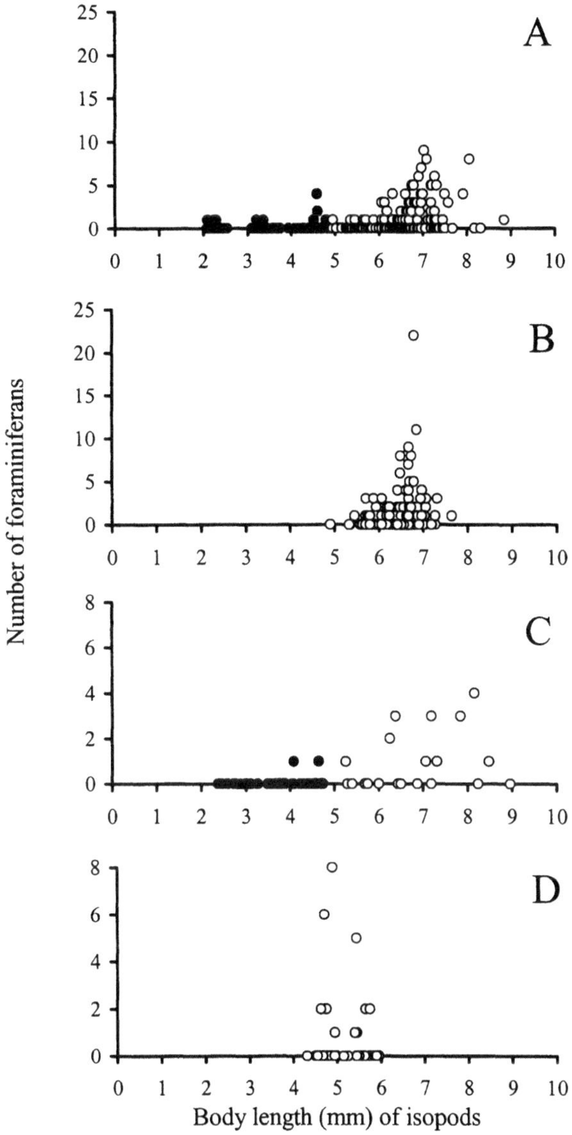

Figure 3. Number of foraminiferans on females (open circles) and mancas (closed circles) of *Neastacilla* (A), on *Neastacilla* males (B), *Pleuroprion* females (open circles) and mancas (closed circles) (C), and on *Pleuroprion* males (D).

tenna, supporting the suggestion that antennal brooding may affect foraminiferan density.

There were small differences in the pattern of epibiosis on the body of the males and females of *Neastacilla*. Males had fewest foraminiferans on pereonite 4, while the females had fewest foraminiferans on pereonites 5–7 and the pleotelson. This latter pattern is presumably a reflection of the pronounced sexual dimorphism in *Neastacilla*, seen in the very slender fourth segment of the males.

Acknowledgements

We thank Dr Torleiv Brattegard for his hospitality on board RV "*Håkon Mosby*", MRI, Reykjavík for the use of RV "*Bjarni Sæmundsson*" and the Fiskirannsóknastofa, Torshavn, for the use of RV "*Magnus Heinason*". We thank Sigrún Haraldsdóttir and the staff of the Sandgerði Marine Centre for sorting the isopods, and the Editorial Committee of the EMBS Proceedings for linguistic help. The paper was improved by the constructive comments of the referees. The Science Foundation of the University of Iceland and Lýðveldissjóður supported the study.

References

Becker, K. & M. Wahl, 1996. Behaviour patterns as natural antifouling mechanisms of tropical marine crabs. J. exp. mar. Biol. Ecol. 203: 245–258.

Brattegard, T. & J. H. Fosså, 1991. Replicability of an epibenthic sampler. J. mar. biol. Ass. U.K. 71: 153–166.

Glynn, P.W., 1970. Growth of algal epiphytes on a tropical marine isopod. J. exp. mar. Biol. Ecol. 5: 88–93.

Harrington, J. M. & P. B. Armstrong, 1999. A cuticular secretion of the horseshoe crab, *Limulus polyphemus*: a potential anti-fouling agent. Biol. Bull. 197: 274–275.

Haward, N. J. B. & J. R. Haynes, 1976. *Chlamys opercularis* (Linnaeus) as a mobile substrate for Foraminifera. J. Foraminiferal Res. 6: 30–38.

Jones, M. B. & M. A. Kahn, 1970. The occurrence of *Conidophrys* species (Protozoa, Ciliata) on members of the *Jaera albifrons* Leach group. Acta Protozool. 8: 149–153.

Key, M. M. Jr. & D. K. A. Barnes, 1999. Bryozoan colonization of the marine isopod *Glyptonotus antarcticus* at Signy Island, Antarctica. Polar Biol. 21: 48–55.

Lutze, G. F. & H. Thiel, 1989. Epibenthic foraminifera from elevated microhabitats: *Cibicidoides wuellerstorfi* and *Planulina ariminensis*. J. Foraminiferal Res. 19: 153–158.

Moore, P. G., 1985. *Cibicides lobatulus* (Protozoa: Foraminifera) epizoic on *Astacilla longicornis* (Crustacea: Isopoda) in the North Sea. J. nat. Hist., Lond. 19: 129–133.

Rothlisberg, P. C. & W. G. Pearcy, 1976. An epibenthic sampler used to study the ontogeny of vertical migration of *Pandalus jordani* (Decapoda, Caridea). Fish. Bull. 74: 994–997.

Sneli, J. A., 1998. A simple benthic sledge for shallow and deep-sea sampling. Sarsia 83: 69–72.

Svavarsson, J. & B. Davíðsdóttir, 1994. Foraminiferan (Protozoa) epizoites on Arctic isopods (Crustacea) as indicators of isopod behaviour? Mar. Biol. 118: 239–246.

Svavarsson, J. & B. Davíðsdóttir, 1995. *Cibicides* spp. (Protozoa, Foraminifera) as epizoites on the Arctic antenna-brooding *Arcturus baffini* (Crustacea, Isopoda, Valvifera). Polar Biol. 15: 569–574.

Svavarsson, J. & S. H. Ólafsdóttir, 1999. Foraminiferan (Protozoa) epizoites on the anthuridean (Crustacea, Isopoda) *Calathura brachiata* (Stimpson). J. nat. Hist., Lond. 33: 1423–1432.

Zampi, M., S. Benocci & S. Focardi, 1997. Epibiont foraminifera of *Sertella frigida* (Waters) (Bryozoa, Cheilostomata) from Terranova Bay, Ross Sea, Antarctica. Polar Biol. 17: 363–370.

Hydrobiologia **440:** 249–260, 2000.
M.B. Jones, J.M.N. Azevedo, A.I. Neto, A.C. Costa & A.M. Frias Martins (eds), Island, Ocean and Deep-Sea Biology.

Resting stages in a submarine canyon: a component of shallow–deep-sea coupling?

Luigi Della Tommasa[1], Genuario Belmonte[1], Alberto Palanques[2], Pere Puig[2] &
Ferdinando Boero[1]
[1]*Dipartimento di Biologia, Stazione di Biologia Marina, CoNISMa, Università degli Studi di Lecce, 73100 Lecce, Italy*
E-mail: boero@unile.it
[2]*Institut de Ciències del Mar (CSIC), Plaça del Mar s/n, 08039 Barcelona, Spain*

Key words: submarine canyon, resting stages, shallow–deep-sea coupling, life cycle

Abstract

The ecological importance of resting stages in shallow waters is well known, but their presence in the deep sea is practically unrecorded. Samples of sinking particles were collected from April 1993 to May 1994 in and around the Foix Canyon (northwest Mediterranean Sea) using PPS3 sediment traps located between −600 m and −1180 m. Dead and viable organisms were collected, and inorganic empty shells constituted most of the biologically-derived matter. Resting stages, considered as POM, had a flux of up to 70 000 items m^{-2} d^{-1}. They were the second most abundant fraction of total POM after tintinnids (mainly represented by empty, chitinous loricas), and first of the viable POM fraction. Most remained unidentified, but 58 morphotypes were referable to coastal species of Dinophyta, Tintinnina and Calanoida. Resting stages were rare in samples collected from the open slope adjacent to the canyon. These preliminary data suggest an important role of submarine canyons in concentrating POM and transferring it from shallow to deep-sea habitats. Due to their resistance to degradation processes, resting stages are probably the only POM component that can return to shallow areas by upwelling currents occurring in the canyon.

Introduction

The ecological importance of resting stages in neritic areas is widely investigated (Matsuoka, 1985; Ishikawa & Taniguchi, 1994; Belmonte et al., 1995; Sonnemann & Hill, 1997; Montresor et al., 1998). Lindley (1990) and Marcus (1995) stated that the presence of resting eggs of Calanoida, Cladocera and Rotifera in marine shelf sediments was inversely correlated with distance from the coastline and/or depth. The importance of cysts in deep waters is mostly unknown. Only Dale & Dale (1992) have recorded a flux of organic-walled and calcareous dinoflagellate cysts into the deep sea at a rate of up to several thousands items m^{-2} d^{-1}. Lack of interest in the patterns and processes related to cyst-mediated plankton-benthos dynamics is due probably to the tendency to consider plankton ecology as limited to the water column (Boero et al., 1996; Marcus & Boero, 1998). This is particularly true for deep waters, where only biogeochemical cycles are recognised as important components of benthic-pelagic coupling.

In marine canyons, physical oceanographic models suggest that water circulation is characterised by both downwelling and upwelling events (Hickey, 1995; Puig et al., 2000). Both hydrographic and satellite data indicate a role of submarine canyons off the northeast Spanish coast in the modification of the surface flow (Masó et al., 1990). Moreover, submarine canyons play an important role in biological events and are often underestimated areas of high species diversity. They attract bottlenose whales and small odontocetes (Gowans & Whitehead, 1995; Whitehead et al., 1997), and host many species of decapods (Cartes & Sarda, 1993; Maynou et al., 1996), nematodes (Soetaert & Heip, 1995), micronektonic crustaceans (Macquart-Moulin & Patriti, 1996), fish assemblages (Stefanescu et al., 1994) and hydromedusae (Gili et al., 1998, 1999).

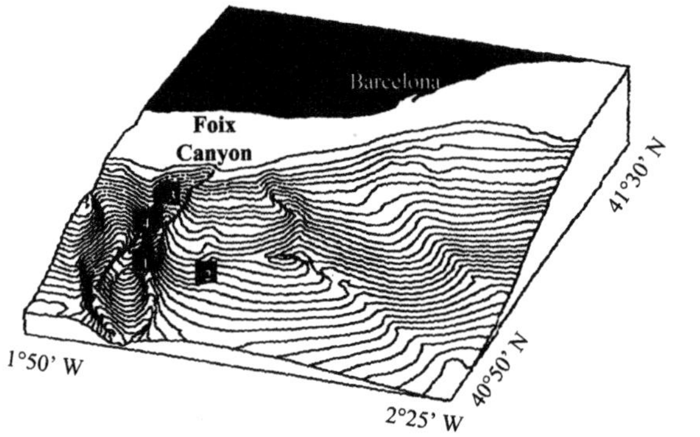

Figure 1. Foix Canyon topography and sediment trap locations.

This work investigates the flow of resting stages in a marine canyon, and discusses its possible role in benthic-pelagic (and deep-shallow sea) coupling.

Materials and methods

Sinking particles were collected in and around the Foix submarine Canyon off the Catalan coast near Barcelona (Spain) (Fig. 1), by a PPS3 sediment trap system with a 0.125 m^2collecting area for each trap. Collected particles were preserved automatically in 5% formalin solution to avoid organic matter degradation. A rotatory collector, electronically controlled by a timer/motor assembly, alternated six cups at intervals of 15/16 days (Heussner et al., 1990). From April 1993 to May 1994, samples were collected at four sites, three inside the canyon and one in the adjacent open slope, at different depths and heights from the bottom (see Puig & Palanques, 1998 for further details).

Eight samples were chosen for this study (Table 1): five (one for each season from April 1993 to May 1994) from the deepest station in the canyon (Station 1) and three (one from each other site) collected in the summer, when resting stages were most abundant at Station 1. Samples were sonicated for 3 mins by an ultrasound device (Branson Sonifier 2200), sieved at four different mesh sizes (250 μm, 125 μm, 63 μm and 20 μm), and diluted in 30 ml of 4% formalin solution. Two whole-sieved fractions (>250 μm and 250–125 μm) and 10% aliquots of the others (125–63 μm and 63–20 μm) were analysed for each sample. Aliquot size was achieved by a preliminary statistical analysis optimising subsampling according to Bros & Cowell (1987). Resting stage separation and identification were carried out with an inverted microscope (Zeiss Axiovert S 100) connected with a photo-camera. Resting stage (or 'cyst', according to Belmonte et al., 1995) abundance was compared with that of remaining biogenic particles (grouped according to nine categories: silicoflagellates, dinoflagellates, foraminiferans, radiolarians, tintinnids, Oligotrichina, metazoans, faecal pellets, others). The number of observed items was related with both collecting area and time, obtaining the estimated flux (items m^{-2} d^{-1}). All the biogenic material was separated qualitatively into two fractions: inorganic empty shells (of silicoflagellates, radiolarians and foraminiferans) and particulate organic matter (POM), to which cysts were assigned.

Cyst description was based on morphological traits (shape, size, surface features and content colour) which were compared with available literature information. Unidentified morphotypes were roughly separated, according to their size, into 'protistan' types (diameter up to 40 μm), *incertae sedis* (diameter from 41 to 70 μm) and 'metazoan' types (minimum diameter 70 μm).

Results

Among the five samples collected at the deepest canyon station (Site 1), the total flux of biogenic particles with a diameter >20 μm reached a maximum (1 779 490 items m^{-2} d^{-1}) in spring 1993 (sample ST01C11) and a minimum (632 990 items m^{-2} d^{-1}) in summer (sample ST06C11) (Fig. 2) when the highest flux of cysts was recorded. In summer, all four sampling sites combined (samples ST06C11, ST04C12, ST06C13, ST06C14), the total flux of biogenic particles (632 990 items m^{-2} d^{-1}) was highest at the deepest canyon site (sample ST06C11) and minimum (211 258 items m^{-2} d^{-1}) at 500 m above the canyon bottom (sample ST06C14).

The most abundant biogenic particles were silicoflagellates (recorded only as inorganic shells in all samples: up to 1 528 050 items m^{-2} d^{-1} in ST01C11 sample, see Fig. 2). The shells of the silicoflagellate *Dictyocha fibula* were always the most abundant items. Tintinnids were the second more abundant group, with a maximum (511 500 items m^{-2} d^{-1}) recorded in winter (sample ST05C21). Resting stages ranked third, with a peak (73 810 items m^{-2} d^{-1}) in the summer (sample ST06C11) and a minimum (2857 items m^{-2} d^{-1}) in the same season in the open slope adjacent to the canyon (sample ST04C12). The remaining categories (dinoflagellates, foraminiferans,

Table 1. General features of observed samples

Sample	Station	Sampling depth (m)	Distance from the bottom (m)	Sampling period
ST01C11	1	−1180	30	15/04/93–01/05/93
ST06C11	1	−1180	30	01/07/93–15/07/93
ST12C11	1	−1180	30	01/10/93–15/10/93
ST05C21	1	−1180	30	01/01/94–15/01/94
ST12C21	1	−1180	30	15/04/94–30/04/94
ST04C12	2	−980	30	01/06/93–15/06/93
ST06C13	3	−600	30	01/07/93–15/07/93
ST06C14	4	−710	500	01/07/93–15/07/93

radiolarians, Oligotrichina, metazoans, faecal pellets, and 'others') were a minority of the isolated particles (Fig. 2). As a whole, they ranged between 7536 items m^{-2} d^{-1} in the open slope adjacent to the canyon (summer: sample ST04C12) and 165 990 items m^{-2} d^{-1} (autumn: sample ST12C11), when a peak of faecal pellets occurred.

Most biogenic particles were inorganic empty shells of silicoflagellates, radiolarians and foraminiferans. In the canyon-adjacent open slope (sample ST04C12), the organic component (POM) reached its highest value (32.63% of total flux) during autumn (sample ST05C21), and lowest (2.6% of total flux) in summer (Fig. 3). In general, tintinnids (only 0.3% of this category in the form of viable items, the remaining as empty loricas) were the most abundant POM fraction, followed by cysts (only 23% of them in the form of empty shells, the remaining as viable forms) (Fig. 4).

Cysts

Cysts were represented by 58 morphotypes: 35 protists, 12 Metazoa and 11 *incertae sedis* due to taxonomic uncertainty. Most protistan cysts were unidentified (19 morphotypes); 10 morphotypes (nine Dinophyta and one Tintinnina) were identified to genus level and six to class level (Polyhymenophorea and Chrysophyceae). Among the metazoan cysts, seven morphotypes were calanoid copepods (the remaining ones were unidentified).

Due to difficulties in identification of the dinoflagellate *Scrippsiella* at species level on the basis of cyst morphology (Montresor et al., 1998), different morphotypes were considered under the group *Scrippsiella*-like. All recognised resting stages, except the pelagic dinoflagellate *Thoracosphaera albatrosiana*, were coastal species and most have been recorded from recent sediment samples (Matsuoka, 1988; Lewis, 1991; Sonneman & Hill, 1997; Belmonte, 1998; Montresor et al., 1998). *Thoracosphaera albatrosiana* is known only as a cyst and is recorded in great numbers from deep-sea sediment traps both in the Pacific and the Atlantic Oceans (Dale & Dale, 1992).

The total flux of protistan and *incertae sedis* cysts was seven times higher than the total flux of metazoan resting stages. The flux of empty protistan cysts was a seventh of that of viable ones, while the flux of empty metazoan cysts was eight times higher than that of viable ones. The most abundant cyst types were *T. albatrosiana* among protists, and 'Centropagiidae egg 1', among metazoa (see Appendix 1 for detailed description of cyst morphologies).

Discussion

Puig & Palanques (1998) stated that the Foix Canyon receives particles not only from the overlaying water column, but also from the shelf waters, whereas on the adjacent open slope, sinking particles come mainly from the overlying water column, with a small contribution of particles transferred from the shelf. Our data, obtained from the flux of biogenic sediment particles, support this model (Puig & Palanques, 1998). The flux of POM recorded in the open slope (10 557 items m^{-2} d^{-1}) was many times lower than the average (185 775 items m^{-2} d^{-1}) of all samples from inside the canyon. This suggests that the Foix Canyon accumulates POM enriched by organic biogenic sediment coming from the continental shelf.

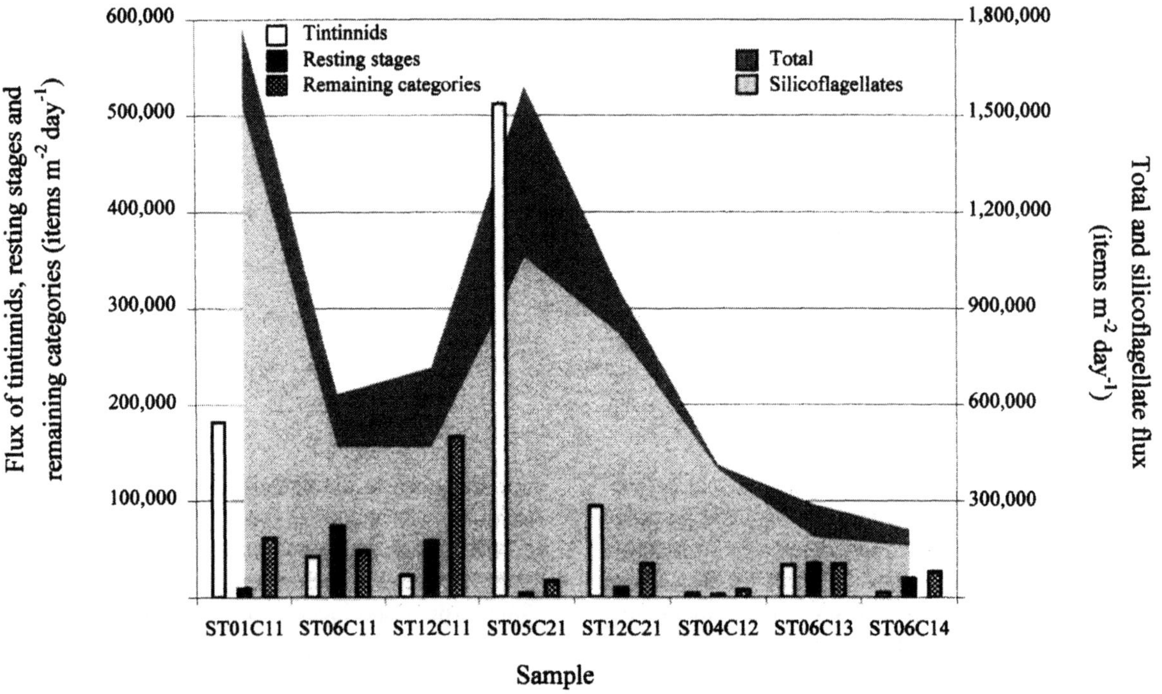

Figure 2. Estimated flux (n° items m^{-2} d^{-1}) of the most abundant categories (see Table 1 for sample description).

POM is the main source of biomass and energy for the deep sea; it includes large food falls, consisting of animal carcasses, along with terrigenous and coastal plant debris, as well as fine particulate organic matter mostly from planktonic animals, including faecal pellets, moults, and phytoplankton (Gage & Tyler, 1991). Resting stages have not been considered previously as a POM component. In the Foix Canyon, resting stages represent the second fine-POM fraction (even the most abundant fraction in sample ST06C11) and could play an important role in shallow–deep-sea coupling. Boero (1994), Boero et al. (1996), Belmonte & Rossi (1998) and Marcus & Boero (1998) suggested that cyst-mediated dynamics call for a re-appraisal of the classical separation of plankton and benthos. This has been stressed also by Pati et al. (1999) who showed that cysts (never considered by meiobenthologists) constituted 52% of the meiobenthos of a coastal lagoon in terms of number of individuals and 30% in terms of biomass!

Marcus & Boero (1998) hypothesised that cysts should represent a fundamental biological link, via submarine canyons, in shelf-slope and shallow-deep sea coupling. The preliminary data from the Foix Canyon corroborate their suggestion. All the 23 identified morphotypes in the present study, with the exception of *Thoracosphaera albatrosiana* (see Dale & Dale, 1992), belonged to coastal taxa. *Scrippsiella* spp. and *Calciodinellum operosum*, recorded from deep sea by Dale & Dale (1992), are dinoflagellate cysts, typically recorded in coastal areas by several authors (Lewis, 1991; Belmonte et al., 1995; Montresor et al., 1997, 1998). The occurrence of cysts of coastal species in the canyon suggests transport from neritic areas by downwelling phenomena. The flux of cysts inside the canyon varied from 4050 items m^{-2} d^{-1} to 73 810 items m^{-2} d^{-1}. The lowest flux was recorded in the canyon-adjacent slope (2857 items m^{-2} d^{-1}).

In the Atlantic and Pacific Oceans, Dale & Dale (1992) reported a flux of dinoflagellate cysts ranging between 401 cysts m^{-2} d^{-1} and 14 584 cysts m^{-2} d^{-1} from −389 m to −5582 m. In coastal areas, the sedimentation flux of resting stages is very variable. In the Mar Piccolo of Taranto, Rubino et al. (1996) recorded a sinking rate of about 23 000 resting stages m^{-2} d^{-1} (March 1996). In the Gulf of Naples, Montresor et al. (1998) found an average cyst flux of 280 000 cysts m^{-2} d^{-1}. Therefore, the resting stage flux recorded in the Foix Canyon appears more similar to fluxes recorded in coastal areas than to those of the deep sea.

The Foix Canyon, thus, acts as an accumulator of resting stages. Further investigations are needed to

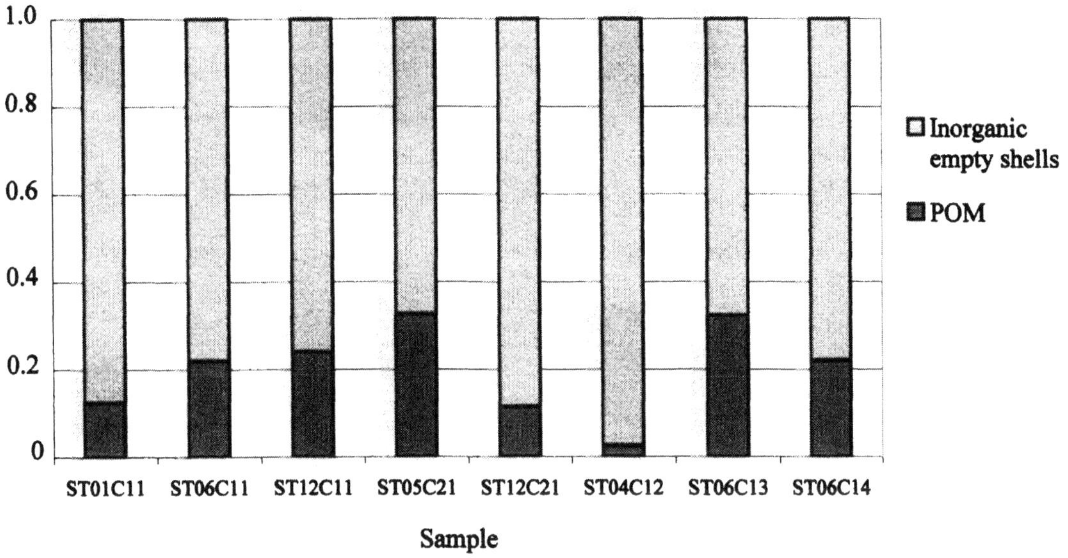

Figure 3. Composition of biogenic sediments (see Table 1 for sample description).

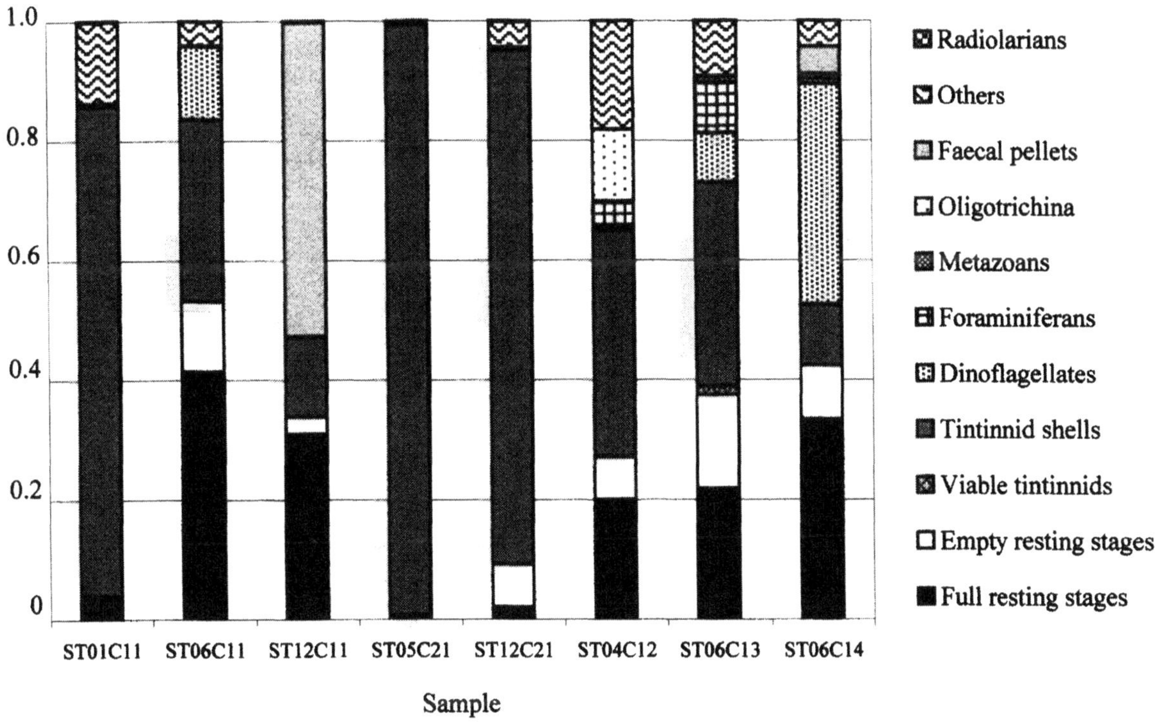

Figure 4. Relative abundance of POM components (see Table 1 for sample description).

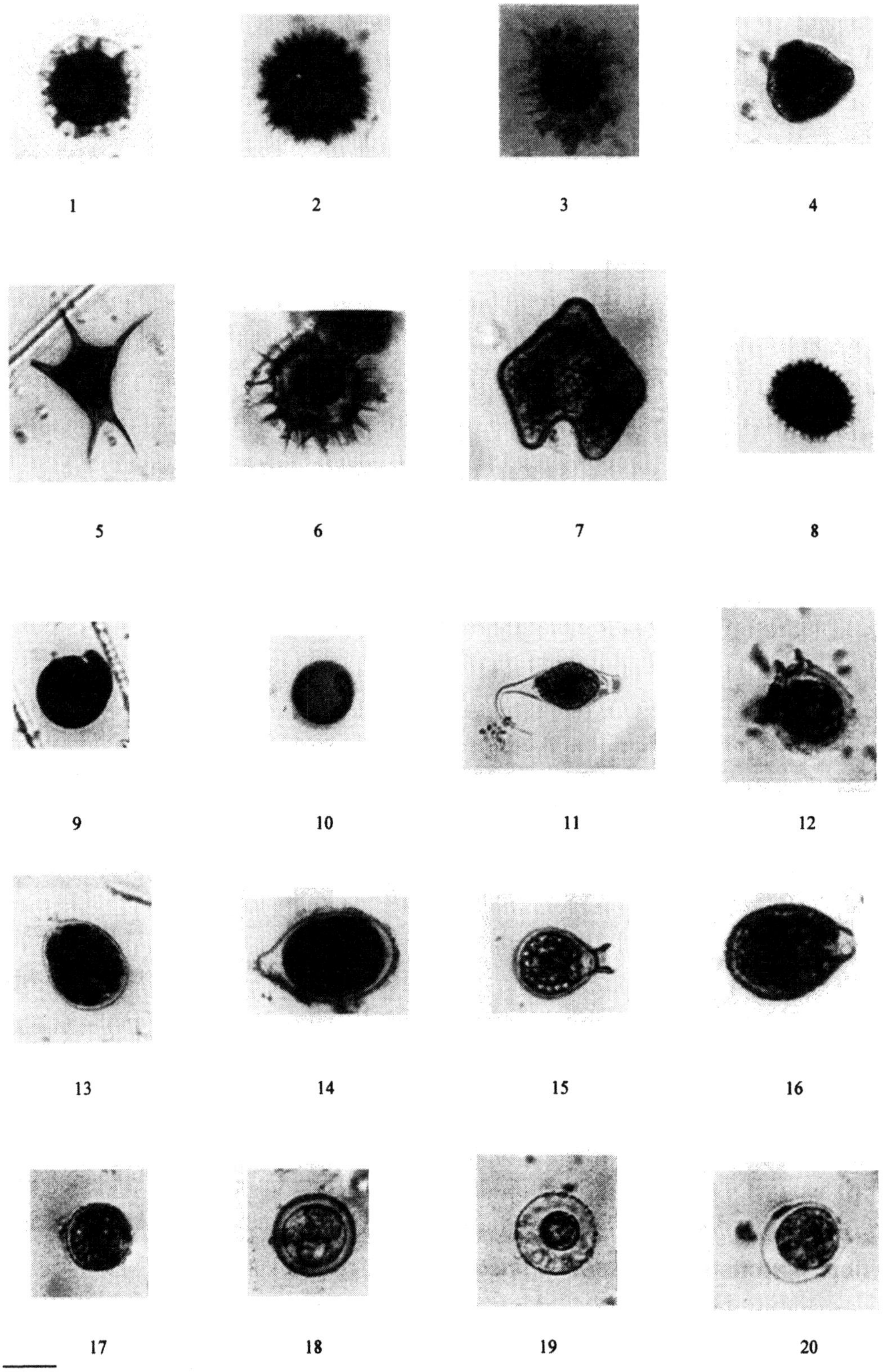

Plate 1. Observed resting stage morphotypes (see Appendix 1 for morphological descriptions). Scale bar 20 μm, except 14 (50 μm).

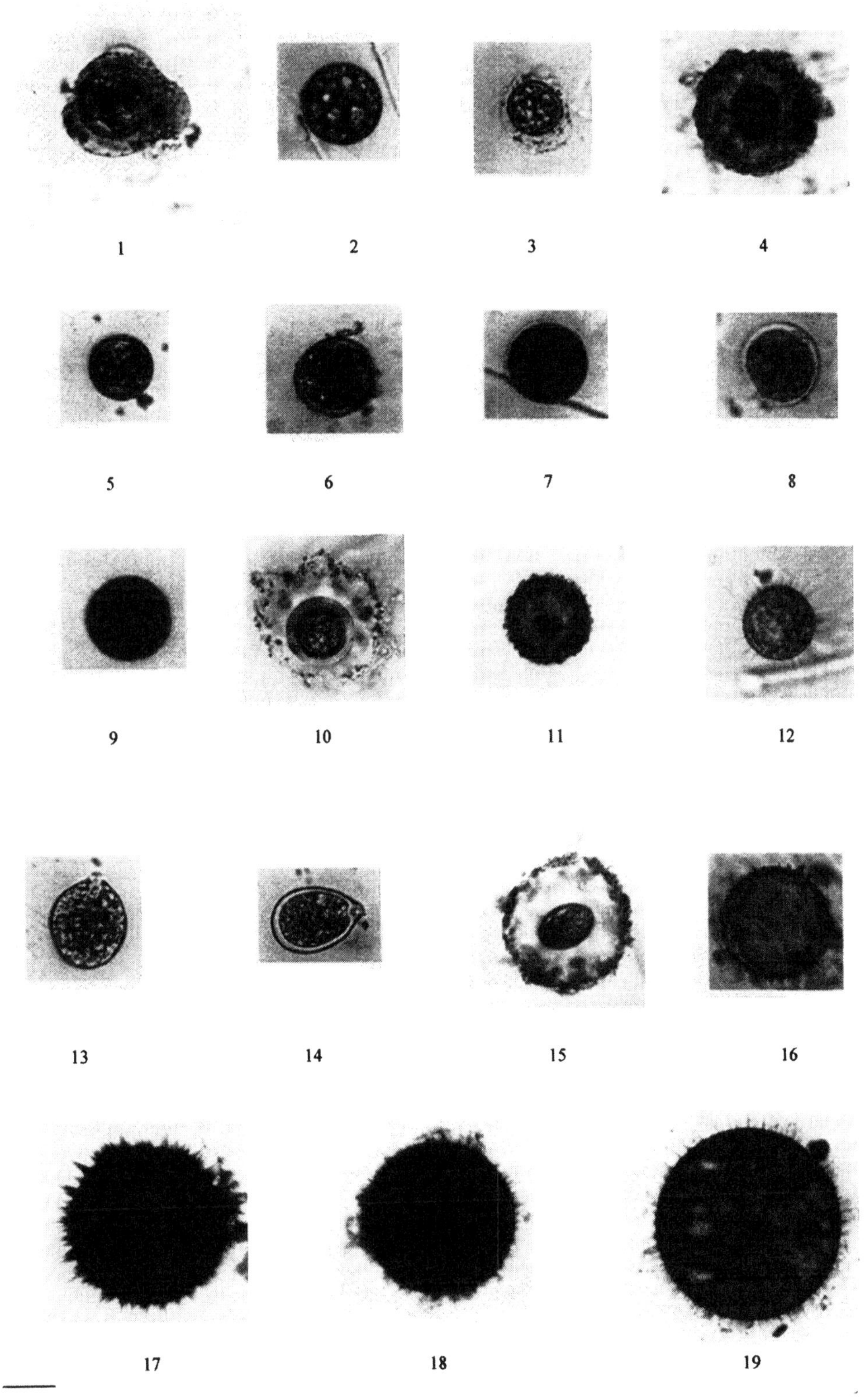

Plate 2. Observed resting stage morphotypes (see Appendix 1 for morphological descriptions). Scale bar: 20 μm Figures 1–14; 50 μm Figures 15–19.

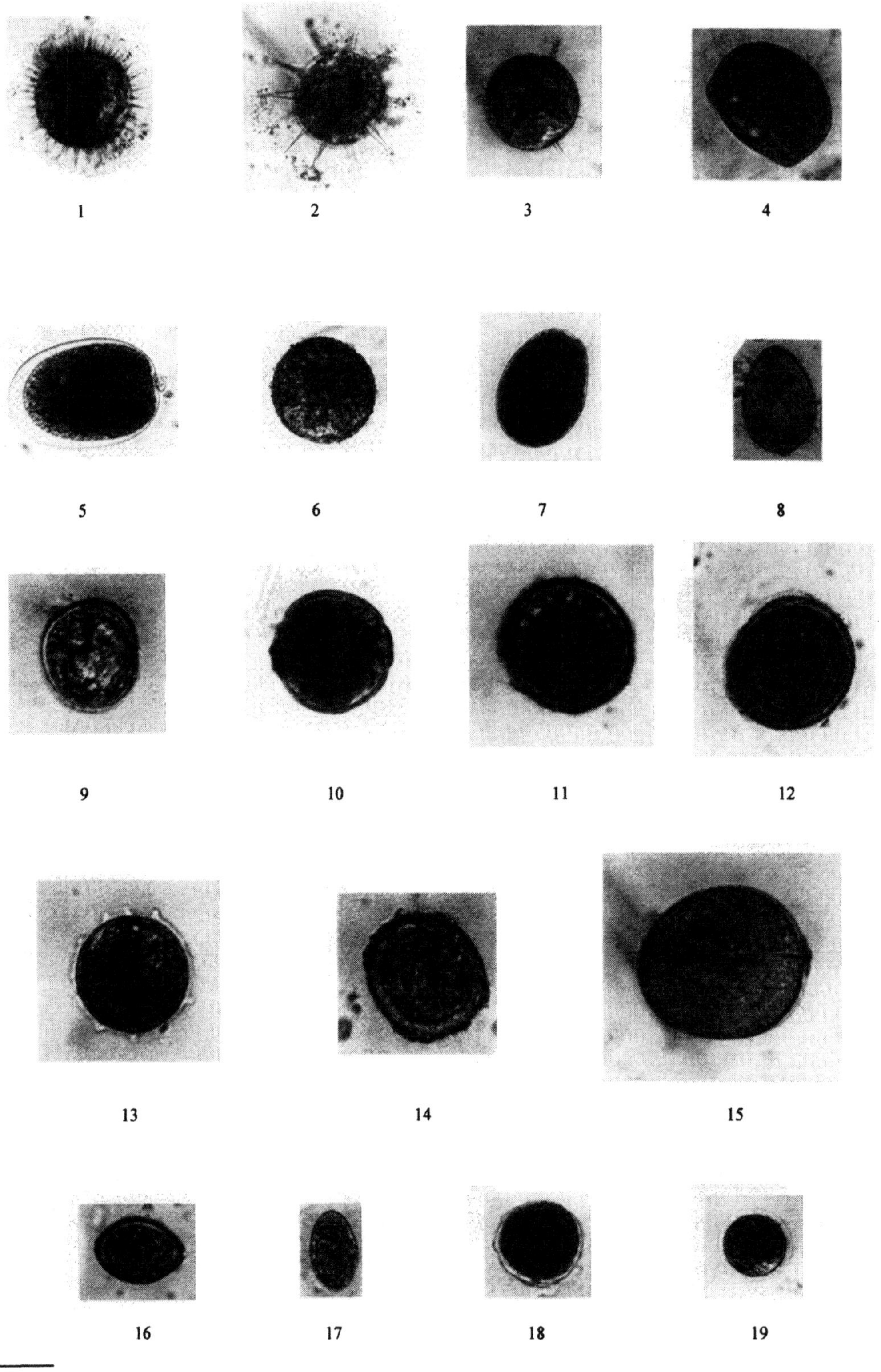

Plate 3. Observed resting stage morphotypes (see Appendix 1 for morphological descriptions). Scale bar: 50 μm Figures 1–8, 16–19; 20 μm Figures 9–15.

verify whether it refuels coastal planktonic populations via upwelling. During upwelling events, significant quantities of water coming from the deep bottom are pumped out of the canyon walls toward the coasts (Peffley & O'Brien, 1976; Klinck, 1995; Allen, 1996). If the canyon head is located sufficiently close to the coast, the canyon-upwelled water reaches the euphotic zone, becoming readily available to the biota (Denman & Powell, 1984; Hickey, 1995). As suggested by Marcus & Boero (1998), upwelled waters might affect coastal planktonic populations by not only supplying dissolved nutrients, but also recruiting propagules in the form of cysts. If this biological coupling between shallow and deep sea is confirmed, the functioning of coastal waters would be intimately linked with that of offshore ones through canyon-driven propagule circulation.

Many authors invoked the need to establish a complete model to better understand the functioning of marine systems (Denman & Powell, 1984). Hickey (1995) called for a synergy between physical and biological oceanography to link topographical features or physical mechanism to biological effects. On the other hand, Boero et al. (1996) and Marcus & Boero (1998) proposed biological cycles as a necessary complement to biogeochemical cycles. The study of resting stage dynamics in submarine canyons might be a further step in this direction.

Acknowledgements

We thank J. M. Gili (Institut de Ciències del Mar, CSIC, Barcelona) for his stimulating encouragement to start the study of submarine canyon sediments. This work partly fulfils the requirements for the PhD training of L. Della Tommasa. Funds from the programme "Biodiversità e interazioni nelle comunità marine" (COFIN) of the Ministero dell'Università e Ricerca Scientifica e Tecnologica. This research was supported by project AMB92-0251-CO2-01 funded by the *Comisión Interministerial de Ciencia y Tecnología* and by projects MAS2-CT93-0053 and MAS3-CT95-0037 funded by the EEC.

References

Allen, S. E., 1996. Topographically generated, subinertial flows within a finite length canyon. J. Phys. Oceanogr. 26 (8): 1608–1632.

Belmonte, G., 1998. The egg morphology of 7 Acartiidae species: a preliminary survey of the ootaxonomy of calanoids. J. mar. Syst. 15: 35–39.

Belmonte, G. & V. Rossi, 1998. Resurrection and time travelling: diapause in crustaceans (and others). Trends Ecol. Evol. 13 (1): 4–5.

Belmonte, G., P. Castello, M. R. Piccinni, S. Quarta, F. Rubino, S. Geraci & F. Boero, 1995. Resting stages in marine sediments off the italian coast. In Eleftheriou, A., A. D. Ansell & C. J. Smith (eds), Biology and Ecology of Shallow Coastal Waters. 28. E.M.B.S., Iraklio, Crete. International Symposium Series: 53–58.

Boero, F., 1994. Fluctuations and variations in coastal marine environments. P.S.Z.N.I: Mar. Ecol. 15 (1): 3–25.

Boero, F., G. Belmonte, G. Fanelli, S. Piraino & F. Rubino, 1996. The continuity of living matter and the discontinuities of its constituents: do plankton and benthos really exist? Trends Ecol. Evol. 11 (4): 177–180.

Bolch, C. J. & G. M. Hallegraeff, 1990. Dinoflagellate cysts in recent marine sediments from Tasmania, Australia. Bot. mar. 33: 173–192.

Bros, W. E. & B. C. Cowell, 1987. A technique for optimizing sample size replication. J. exp. mar. Biol. Ecol. 114: 63–71.

Cartes, J. E. & F. Sarda, 1993. Zonation of deep-sea decapod fauna in the Catalan Sea (Western Mediterranean). Mar. Ecol. Prog. Ser. 94 (1): 27–34.

Dale, B. & A. L. Dale, 1992. Dinoflagellate contribution to the deep sea. Ocean Biocenosis Series n° 5. Woods Hole Oceanographic Institution, Woods Hole: 77 pp.

Denman, K. L. & T. M. Powell, 1984. Effects of physical processes on planktonic ecosystems in the coastal ocean. Oceanography and Marine Biology: an Annual Review 22: 125–168.

Duff, K. E., B. A. Zeeb & J. P. Smol, 1995. Atlas of chrysophycean cysts. Kluwer Academic Publishers, Dordrecht, The Netherlands: 189 pp.

Gage, J. D. & P. A. Tyler, 1991. Deep-sea Biology: a Natural History of Organisms at the Deep-sea Floor. Cambridge University Press, Cambridge: 504 pp.

Gili, J. M., J. Bouillon, F. Pages, A. Palanques & P. Puig, 1999. Submarine canyons as habitats of prolific plankton population: three new deep-sea Hydroidomedusae in the western Mediterranean. Zool. J. linn. Soc., 125 (3): 313–329.

Gili, J. M., J. Bouillon, F. Pages, A. Palanques, P. Puig, & S. Heussner, 1998. Origin and biogeography of the deep-water Mediterranean hydromedusae including the description of two new species collected in submarine canyons of Northwestern Mediterranean. Sci. mar. 62 (1–2): 113–134.

Gowans, S. & H. Whitehead, 1995. Distribution and habitat partitioning by small odontocetes in the Gully, a submarine canyon on the Scotian shelf. Can. J. Zool. 73 (9): 1599–1608.

Heussner, S., C. Ratti & J. Carbonne, 1990. The PPS3 time-series sediment trap and the trap sample processing techniques used during the ECOMARGE experiment. Contin. Shelf Res. 10 (9-11): 945–958.

Hickey, B. M., 1995. Coastal submarine canyons. In Müller, P. & D. Henderson (eds), Topographic Effects in the Ocean. SOEST Special Publication, University of Hawaii, Manoa: 95–110.

Ishikawa, A. & A. Taniguchi, 1994. The role of cysts on population dynamics of *Scrippsiella* spp. (Dinophyceae) in Onagawa Bay, northeast Japan. Mar. Biol. 119: 39–44.

Klinck, J. M., 1995. Circulation near submarine canyon: a modelling study. J. geophys. Res. 101 (C1): 1211–1223.

Kokinos, J. P. & D. M. Anderson, 1995. Morphological development of resting cysts in cultures of the marine dinoflagellate

Lingulodinium polyedrum (= *L. machaerophorum*). Palynology 19: 143–166.

Lewis, J., 1991. Cyst-theca relationship in *Scrippsiella* (Dinophyceae) and related Orthoperidinioid genera. Bot. mar. 34: 91–106.

Lewis, J. & R. Hallett, 1997. *Lingulodinium polyedrum* (*Gonyaulax polyedra*) a blooming dinoflagellate. Oceanography and Marine Biology: an Annual Review 35: 97–161.

Lindley, J. A., 1990. Distribution of overwintering copepod eggs in sea-bed sediment around southern Britain. Mar. Biol. 104: 209–217.

Macquart-Moulin, C. & G. Patriti, 1996. Accumulation of migratory micronekton crustaceans over the upper slope and submarine canyons of the Northwestern Mediterranean. Deep-Sea Res. I 43 (5): 579–601.

Marcus, N. M., 1995. Seasonal study of planktonic copepods and their benthic resting eggs in northern California coastal waters. Mar. Biol. 123: 459–465.

Marcus, N. H. & F. Boero, 1998. Minireview: the importance of benthic-pelagic coupling and the forgotten role of life cycles in coastal aquatic systems. Limnol. Oceanogr. 43 (5): 763–768.

Masò, M., P. E. La Violette & J. Tintoré, 1990. Coastal flow modification by submarine canyons along the NE Spanish coast. Sci. mar. 54 (4): 343–348.

Matsuoka, K., 1985. Organic-walled dinoflagellate cysts from surface sediments of Nagasaki Bay and Senzaki Bay, West Japan. Bull. Faculty of Liberal Arts, Nagasaki Univ., (Natural Science) 25(2): 21–115.

Matsuoka, K., 1988. Cyst-theca relationships in the diplopsalid group (Peridiniales, Dinophyceae). Rev. Paleont. Palynol. 56: 95–122.

Maynou, F., G. Y. Conan, J. E. Cartes, J. B. Company & F. Sarda, 1996. Spatial structure and seasonality of decapod crustacean population on the Northwestern Mediterranean slope. Limnol. Oceanogr. 41 (1): 113–125.

Montresor, M., D. Janofske & H. Willems, 1997. The cyst-theca relationship in *Calciodinellum operosum* emend. (Peridiniales, Dinophyceae) and a new approach for the study of calcareous cysts. J. Phycol. 33: 122–131.

Montresor, M., A. Zingone & D. Marino, 1993. The calcareous resting cyst of *Pentapharsodinium tyrrhenicum* comb. nov. (Dinophyceae). J. Phycol. 29: 223–230.

Montresor, M., A. Zingone & D. Sarno, 1998. Dinoflagellate cyst production at a coastal Mediterranean site. J. Plankton Res. 20 (12): 2291–2312.

Pati, A. C., G. Belmonte, V. U. Ceccherelli & F. Boero, 1999. The inactive temporary component: an unexplored fraction of meiobenthos. Mar. Biol. 134: 419–427.

Peffley, M. B. & J. J. O'Brien, 1976. A three-dimensional simulation of coastal upwelling off Oregon. J. Phys. Oceanogr. 6: 164–180.

Puig, P. & A. Palanques, 1998. Temporal variability and composition of setting particle fluxes on the Barcelona continental margin (Northwestern Mediterranean). J. mar. Res. 56 (3): 639–654.

Puig, P., A. Palanques, J. Guillen & E. Garcia-Ladona, 2000. Deep slope currents and suspended particle fluxes in and around the Foix submarine canyon (NORTHWEST Mediterranean). Deep-Sea Res. I 47 (3): 343–366.

Reid, P. C. & A. W. G. John, 1978. Tintinnid cysts. J. mar. biol. Ass. U.K. 58: 551–557.

Rubino, F., O. D. Saracino, G. Fanelli, G. Belmonte & F. Boero, 1996. Plankton dynamics in the Mar Piccolo of Taranto: a pilot plan. Giornale Botanico Italiano 130: 1032–1036.

Soetaert, K. & C. Heip, 1995. Nematode assemblage of deep-sea and shelf break sites in the North Atlantic and Mediterranean Sea. Mar. Ecol. Prog. Ser. 125 (1–3): 171–183.

Sonneman, J. A. & D. R. A. Hill, 1997. A taxonomic survey of cyst-producing dinoflagellates from recent sediments of Victorian coastal waters, Australia. Bot. mar. 40: 149–177.

Stefanescu, C., B. Morales Nin & E. Massuti, 1994. Fish assemblage on the slope in the Catalan Sea (western Mediterranean): influence of a submarine canyon. J. mar. biol. Ass. U.K. 74 (3): 499–512.

Wall, D. & B. Dale, 1968. Modern dinoflagellate cysts and evolution of the peridiniales. Micropaleontology 14 (3): 265–304.

Whitehead, H., S. Gowans, A. Faucher & S. W. McCarrey, 1997. Population analysis of northern bottlenose whales in the Gully, Nova Scotia. Marine Mammal Science 13 (2): 173–185.

Appendix 1. Cyst morphological descriptions

Plate 1

Dinophyta cysts

1.1. *Calciodinellum operosum* Deflandre. Diameter 40 μm, spherical, wall calcareous, surface crested with a clear paratabulation on the outer layer, brown. References: Dale & Dale (1992: 25, Pl. 1.1 Figure 13); Montresor et al. (1997); Montresor et al. (1998: 2297, Figure 2f).

1.2. cf. *Diplopelta parva* (Abé, 1941) Matsuoka. Synonym: *Dissodium parvum* Abé. Diameter 32 μm, processes 6 μm, spherical, with many short spines, brownish. References: Bolch & Hallegraeff (1990: 183, Figures 31a–c); Matsuoka (1988: 102, Pl. 1 Figures G–K).

1.3. *Lingulodinium polyedrum* (Stein) Dodge. Paleontological taxon: *Lingulodinium machaerophorum* (Deflandre & Cookson) Wall. Synonym: *Gonyaulax polyedra* Stein. Diameter 26 μm, processes 6–8 μm, spherical, with both bent and upright long conical spines, brown. References: Wall & Dale (1968: 294, Pl. 1 Figure 18); Kokinos & Anderson (1995); Lewis & Hallett (1997).

1.4. *Pentapharsodinium tyrrhenicum* (Balech) Montresor et al. Paleontological taxon: *Calcicarpinum bivalvum* Versteegh. Synonym: *Peridinium tyrrhenicum* Balech. Diameter 30 μm, wall calcareous, with distinctive ridges conferring a broadly triangular-shape, brown. References: Montresor et al. (1993: 223, Figures 1 and 4); Sonneman & Hill (1997: 157, Figures 12a–b); Montresor al. (1998: 2296, Figure 2d).

Appendix 1. contd.

1.5. *Protoperidinium compressum* (Abé) Balech. Paleontological taxon: *Stelladinium reidii* Bradford; *Stelladinium stellatum* (Wall) Reid. Synonym: *Peridinium compressum* Abé; *P. stellatum* Wall & Dale. Length 20 μm, width 16 μm, processes 10–14 μm, stellate, with one apical, two antapical and two lateral processes, light brown. References: Bolch & Hallegraeff (1990: 181, Figure 22); Sonneman & Hill (1997: 163, Figure 19).

1.6. *Protoperidinium* cf. *nudum* (Meunier) Balech. Paleontological taxon: *Selenopemphix quanta* (Bradford) Matsuoka. Synonym: *Peridinium* cf. *nudum* (Meunier) Wall & Dale. Length 32 μm, width 28 μm, spines 6–8 μm, oval, with many conical spines, brown. References: Wall & Dale (1968: 301, Pl. 4 Figure 5); Sonneman & Hill (1997: 163–164, Figures 22a–b).

1.7. *Protoperidinium oblongum* (Aurillius) Parke & Dodge. Paleontological taxon: *Votadinium calvum* Reid. Synonym: *Peridinium divergens* Ehrenb. var. *oblongum* Aurivillius; *P. oblongum* (Aurivillius) Lebours; *P. oceanicum* Schiller. Length 60 μm, width 40 μm, peridinoid, dorso-ventrally compressed, with rounded apices and without spines, light brown. References: Wall & Dale (1968: 295, Pl. 1 Figure 23); Bolch & Hallegraeff (1990: 181, Figure 19a); Sonnneman & Hill (1997: 164, Figure 26a).

1.8. *Scrippsiella*-like. Length 25 μm–38 μm, width 22–31 μm, spherical or oval, calcareous, thickly provided with short squat or conical spines, body red to brown. References: Lewis, (1991); Montresor et al. (1998).

1.9 *Thoracosphaera albatrosiana* Kampaner. Diameter 25 μm, spherical, calcareous, surface smooth, archeopyl shaped as described by Dale & Dale (1992), brown. References: Dale & Dale (1992: 25, Pl. 1.1 Figure 14).

Chrysophycean cysts

1.10. Stomatocyst type 1. Diameter 20 μm, spherical, surface smooth, wall thick, apical pore with cylindrical collar. Similar to stomatocyst 151 Zeeb & Smol. Reference: Duff et al. (1995: 56, Figure 43).

Tintinnid cysts

1.11. cf. *Fusopsis* sp.. Length 20 μm, width 18 μm, flask-shaped, surface smooth, apically-capped, antapically-tailed, greenish. Reference: Reid & John (1978: 552, Figure 1f).

Polyhymenophorea (Protozoa, Ciliophora) flask-shaped cysts

1.12. Type: Cil un1. Length 28 μm, width 25 μm, envelope thickness 3 μm, flask-shaped, surface smooth, external envelope gelatinous, neck conical with a large pore, light brown.

1.13 Type: Cil un2. Length 44 μm, width 30 μm, flask-shaped, surface smooth, protoplasm granulous, apical process, brownish.

1.14. Type: Cil un3. Length 125 μm, width 78 μm, flask-shaped, surface smooth, wall thick, brown.

1.15. Type: Cil un4. Length 38-40 μm, width 25 μm, flask-shaped, surface smooth, neck conical with a large pore, greenish.

1.16. Type: Cil un5. Length 50 μm, width 34 μm, flask-shaped, surface smooth, wall double, apical prominence, brownish.

Unidentified protistan cysts

1.17. Type: Prot un1. Diameter 28-47 μm, spherical, surface smooth, wall thin, protoplasm granulous, greenish or brownish.

1.18 Type: Prot un2. Diameter 25–34 μm, spherical, surface smooth, wall thin, with gelatinous crown, yellowish.

1.19. Type: Prot un3. Diameter 14 μm, envelope 8 μm thick, spherical, surface smooth, wall thin, with thick gelatinous crown, greenish.

1.20. Type: Prot un4. Diameter 22 μm, envelope 2–10 μm thick, spherical, surface smooth, wall thin, with gelatinous crown and apical pore, greenish.

Plate 2

2.1. Type: Prot un5. Diameter 33 μm, spherical surface smooth, wall thin, with irregular mucilaginous layer, brown.

2.2. Type: Prot un6. Diameter 33 μm, spherical, surface smooth, wall thin, protoplasm with lipidic masses, greenish.

2.3. Type: Prot un7. Diameter 18 μm, spherical, surface smooth, wall thin, with mucilaginous crown, protoplasm with lipidic masses, yellowish.

2.4. Type: Prot un8. Diameter 20 μm, envelope 14 μm thick, spherical, surface smooth, wall thin, wrapped by mucilaginous crown, brownish.

2.5 Type: Prot un9. Diameter 22 μm, spherical, surface smooth, wall thick, protoplasm granulous, light brown.

Appendix 1. contd.

2.6.	Type: Prot un10. Diameter 25–38 μm, spherical, surface smooth, wall thick, with apical pore, brownish.
2.7.	Type: Prot un11. Diameter 32 μm, spherical, surface smooth, wall thick, protoplasm granulous, brown.
2.8.	Type: Prot un12. Diameter 25–37 μm, spherical, surface smooth, wall thick, with apical pore, yellowish.
2.9.	Type: Prot un13. Diameter 32–38 μm, spherical, surface smooth, wall thick, black.
2.10.	Type: Prot un14. Diameter 28 μm, gelatinous envelope 5 μm thick, spherical, surface smooth, wall thick, with gelatinous crown, outer mucilage, yellowish.
2.11.	Type: Prot un15. Diameter 35 μm, spherical, with many short and pointed processes, wall thick, red body in the protoplasm, brownish.
2.12.	Type: Prot un16. Diameter 25 μm; spines 3 μm, spherical, with many short and thin spines, yellowish.
2.13.	Type: Prot un17. Length 34 -55 μm, width 30–48 μm, oval, cyst, surface smooth, wall thin, with apical duct and lipid assemblages in the protoplasm, yellowish.
2.14.	Type: Prot un18. Length 32–52 μm, width 25–34 μm, oval, surface smooth, wall thick, with apical prominence, yellowish.
2.15.	Type: Prot un19. Length 38 μm, width 27 μm, envelope 32 μm thick, elliptical, surface smooth, wall thin, wrapped by mucilaginous crown, brownish.
Metazoan cysts	
2.16.	Calanoida egg1. Diameter 78–88 μm, spherical, with many short spines, empty.
2.17.	Calanoida egg2. Diameter 125 μm, spines 16–20 μm, spherical, with many conical spines, black.
2.18.	Calanoida egg3. Diameter 125 μm, spines 5 μm, spherical, with many short spines, brown.
2.19.	Calanoida egg4. Diameter 160 μm, spines 5–8 μm, spherical, with many short and thin spines, brownish.
Plate 3	
3.1.	Acartiidae egg1. Diameter 58 μm, spines 6-10 μm, spherical, with many pointed and branched conical spines, light brown.
3.2.	Centropagiidae egg 1. Diameter 48 μm, spines 15–20 μm, spherical, with long, pointed and branched conical spines, light brown.
3.3.	Centropagiidae egg 2. Diameter 60 μm, processes 16 μm, spherical, with long, thin and bent spines, light brown.
Unidentified metazoan resting stages	
3.4.	Type: Egg un1. Diameter 90 μm, cap-shaped, surface smooth, protoplasm with lipidic masses, brown.
3.5.	Type: Egg un3. Length 125 μm, width 88 μm, oval, surface smooth, wall thin, with gelatinous crown and apical pore, brown.
3.6.	Type: Egg un4. Diameter 82 μm, spherical, surface rough, with many squat processes, light brown.
3.7.	Type: Egg un5. length 90 μm, width 65 μm, oval, surface smooth, wall thin, with apical pore, brown.
3.8.	Type: Egg un6. Length 85 μm, width 58 μm, oval, surface smooth, wall thin, with apical duct, brownish.
Incertae sedis	
3.9.	Type: Inc un1. Diameter 45 μm, spherical, surface smooth, wall thick, yellowish.
3.10.	Type: Inc un2. Diameter 55 μm, spherical, surface smooth, wall thick, with amphipolar prominences, brown.
3.11	Type: Inc un3. Diameter 52 μm, spherical, surface smooth, wall double, protoplasm with lipidic masses, brownish.
3.12.	Type: Inc un4. Diameter 55 μm, spherical, surface irregular, wall double, light brown.
3.13.	Type: Inc un5. Diameter 47 μm, spherical, with short pointed processes, greenish.
3.14	Type: Inc un6. Diameter 52 μm, spherical, surface irregular, wall double, brown.
3.15.	Type: Inc un7. Length 40–65 μm, width 34–56 μm, round-oval, surface smooth, wall thin, with apical pore, light brown.
3.16.	Type: Inc un8. Length 70 μm, width 48 μm, oval, surface smooth, wall double, with apical pore, protoplasm with lipidic masses, brown.
3.17.	Type: Inc un9. Length 55 μm, width 36 μm, elliptical, surface smooth, wall thin, protoplasm granulous, greenish.
3.18.	Type: Inc un10. Diameter 65 μm, spherical, surface smooth, wall thin, with irregular gelatinous crown, brown.
3.19.	Type: Inc un11. Diameter 48 μm, spherical, surface smooth, wall double, with septed gelatinous crown, protoplasm with lipidic masses, brownish.

Hydrobiologia **440:** 261–271, 2000.
M.B. Jones, J.M.N. Azevedo, A.I. Neto, A.C. Costa & A.M. Frias Martins (eds), Island, Ocean and Deep-Sea Biology.

Patterns in deep-sea macrobenthos at the continental margin: standing crop, diversity and faunal change on the continental slope off Scotland

John D. Gage[1], Peter A. Lamont[1], Kerstin Kroeger[1], Gordon L.J. Paterson[2] & José Luis Gonzalez Vecino[1]
[1]*Scottish Association for Marine Science, P.O. Box 3, Oban, Argyll PA34 4AD, U.K.*
E-mail: jdg@dml.ac.uk
[2]*Zoology Department, Natural History Museum, Cromwell Road, London SW7 5BD, U.K.*

Key words: macrobenthos, deep sea, bathyal, community structure, downslope pattern

Abstract

Depth-related patterns of macrobenthic community structure and composition have been studied from box-core samples from the Scottish continental slope where deep-sea trawling and oil exploration are becoming increasingly important. There is a strong pattern of declining biomass and faunal abundance with increasing depth, but results also indicate reduced biomass and numbers of macrobenthos in the shallowest samples from just below the shelf edge where there are coarse sediments and a regime of strong bottom currents. There is also reduced species diversity at the shallowest stations, probably caused by hydrodynamic disturbance, but no clear mid-slope peak in species diversity as described from the northwest Atlantic. Taxonomic composition of the macrobenthic community shows most change between about 1000 and 1200 m, expressed as a major dichotomy in multivariate analysis by cluster analysis and ordination. It also shows up as a step-like increase in the rate of accumulation of new macrofaunal species. This corresponds to a change in hydrodynamic regime, from a seabed rich in suspension- and interface-feeding epifauna, to one where biogenic traces from large, burrowing deposit feeders are well developed, and visible epifauna rare in seabed photographs. It also corresponds to the depth zone where earlier study of megafaunal echinoderms in trawl and epibenthic sled samples also shows a clear peak in across-slope rate of change in faunal composition.

Introduction

In contrast to the huge area and relatively uniform conditions on the abyssal plains of the great ocean basins, the depth gradient at the continental margins represents a zone of rapid environmental change. Classically, three depth-related trends have been thought to be associated with the benthic fauna along this gradient (Carney et al., 1983): (1) an exponential decline in standing crop with increasing depth amongst meiofaunal to megafaunal size classes; (2) pronounced depth related changes in benthic faunal composition; and more arguably, (3) a species diversity maximum at mid-slope depths.

Such patterns represent expressions of complex biological systems that are still poorly understood in the deep sea (taken here as depths below about 200 m, or the shelf/slope break). Yet the upper continental slope, just beyond the shelf edge in the North Atlantic, is experiencing increasing industrial-scale interventions by man. This comes from fishermen targeting non-quota stocks using powerful trawlers in ever-deeper water to replace income lost by over fishing of traditional fishing grounds, and, more recently, as government-regulated oil/gas exploration moves into depths beyond the continental shelf. Although fishing impacts on the wider deep-water ecosystem remain largely uninvestigated, environmental impact assessment (EIA) has long been a statutory requirement for oil/gas prospecting. A major challenge is to provide better predictive power to EIA in an area where our understanding of the benthic biological ecosystem is more uncertain than in shallow water. Although we cannot ignore possible latitudinal gradients in faunal composition and community structure (Rex et al., 1993, 1997), at the regional scale such variability be-

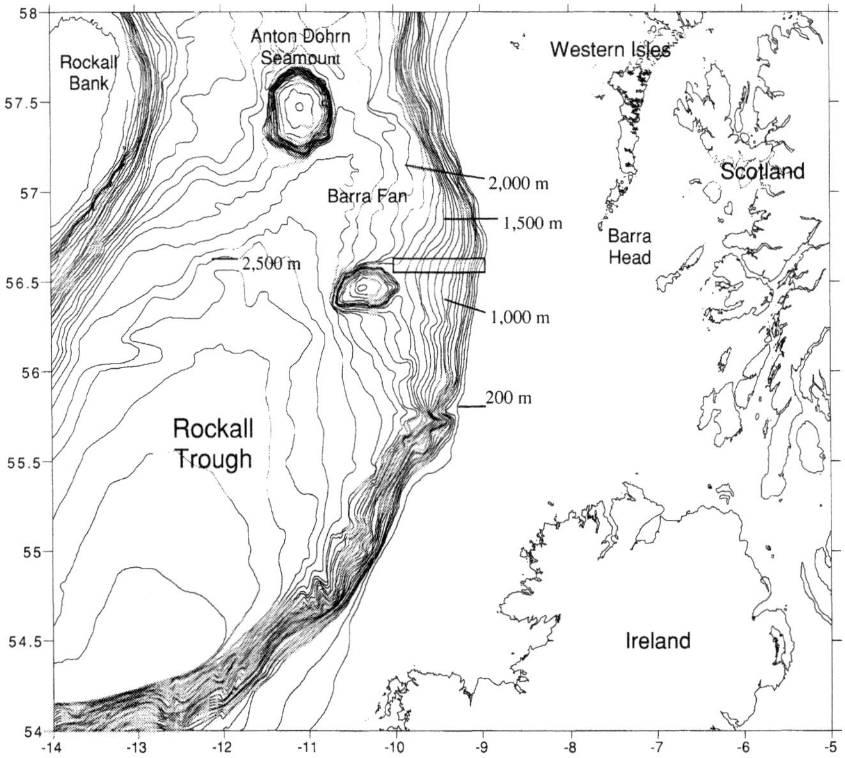

Figure 1. Hebridean slope area showing area of sampling stations (rectangular box) and bathymetry (contours at 100 m intervals from 200 m depth).

comes less important than depth in determining the nature of benthic life at the deep-sea bed. Our traditional approach is to work bathymetric transects, where depth is used as a master variable to which other parameters are associated, just as terrestrial ecologists relate variability to altitude as a major structuring force on land. Benthic biology has importance in these studies not only because these populations are consumers and mediators in re-cycling and relocating particulate organic carbon, but also because the usually limited motility of benthic fauna means they act as good indicators of change in environmental conditions.

We report here on depth-related patterns of the macrobenthos on a section of the continental slope west of Barra Head at the southermost point of the Hebridean islands off Scotland. This slope area has been regarded as a pristine environment, but has been trawled for deep-sea fish for a number of years and is now actively being prospected for oil.

Material and methods

The area of the box-coring stations is shown in Figure 1. Positional co-ordinates of the samples are listed in Table 1. Samples were taken using either the Mk 1 USNEL box corer (Hessler & Jumars, 1974) or, from 1982 onwards, the Mk 2 design incorporating 'Sandia' modifications (Gage & Tyler, 1991). This device samples a 50 × 50-cm area of seabed totalling 0.25 m^2 down to a maximum depth of 50 cm. In practice, few samples achieved a penetration greater than 30 cm, while some from the coarser, sandy sediments near the shelf break sampled a much shallower depth. The water overlying the sediment surface, which may include relatively motile taxa such as peracarid crustaceans, was drained from the core box after its removal from the box corer on deck, and any fauna present screened off using the same 0.42 mm square mesh sieve used later in washing the sediment. Water overlying the core had drained through the sediment layer in the core box in the samples from around 400 m depth, and it is, therefore, possible that fauna, particularly lighter bodied forms such as peracarids, had been lost

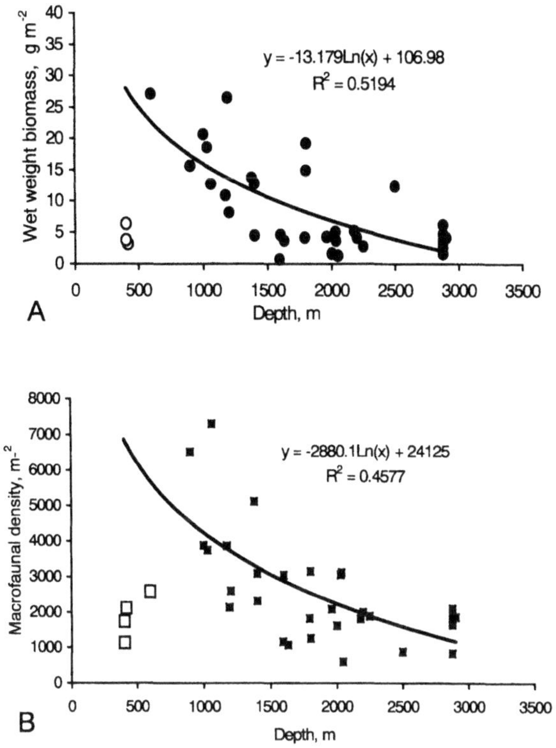

Figure 2. (A) Wet-weight macrobenthic metazoan biomass plotted against depth. The stations around 400 m depth (open circles) were excluded from the logarithmic regression fitted to the remaining data (filled circles); (B) numerical abundances of the total metazoan macrobenthos plotted against depth. The stations around 400 m depth (open squares) were excluded from the logarithmic regression fitted to the remaining data (filled squares).

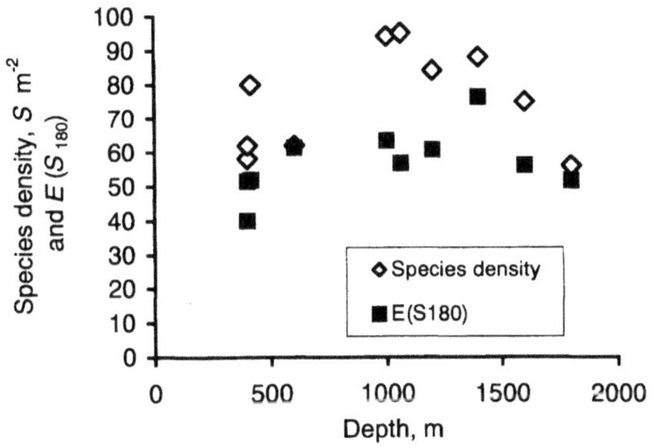

Figure 3. Numbers of species in each box core (standardised to S m^{-2}) and Expected Species Diversity for a sample size of 180 individuals, $E(S_{180})$, plotted against depth.

Table 1. Positional coordinates and other details for the box-core samples used for plotting depth-related pattern in biomass and macrofaunal abundance. The 10 samples shown in **bold** were those where a taxonomically harmonised data set of species-level data were available for the diversity and community analysis

Station	Date	Latitude	Longitude	Depth, m
SBC 46	06.09.75	55° 04′ N	12° 06′ W	2875
SBC 47	06.09.75	55° 03′ N	12° 03′ W	2875
SBC 48	07.09.75	55° 04′ N	12° 04′ W	2875
SBC 49	07.09.75	55° 03′ N	12° 05′ W	2875
SBC 50	07.09.75	55° 05′ N	12° 02′ W	2875
SBC 51	07.09.75	55° 03′ N	12° 04′ W	2875
SBC 58	21.06.76	54° 41′ N	12° 17′ W	2900
SBC 60	23.06.76	56° 35′ N	11° 03′ W	2500
SBC 63	**25.06.76**	**56° 37′ N**	**09° 49′ W**	**1800**
SBC 64	**26.06.76**	**56° 38′ N**	**09° 29′ W**	**1400**
SBC 65	**26.06.76**	**56° 39′ N**	**09° 40′ W**	**1600**
SBC 66	**26.06.76**	**56° 39′ N**	**09° 23′ W**	**1200**
SBC 67	**26.06.76**	**56° 39′ N**	**09° 13′ W**	**1000**
SBC 68	01.07.76	58° 42′ N	09° 43′ W	1800
SBC 209	**02.08.82**	**56° 41′ N**	**09° 00′ W**	**416**
SBC 210	**02.08.82**	**56° 40′ N**	**09° 00′ W**	**401**
SBC 211	**02.08.82**	**56° 41′ N**	**09° 00′ W**	**402**
SBC 212	**02.08.82**	**56° 43′ N**	**09° 00′ W**	**601**
SBC 213	02.08.82	56° 45′ N	09° 06′ W	841
SBC 214	**02.08.82**	**56° 45′ N**	**09° 11′ W**	**1062**
SBC 215	03.08.82	57° 02′ N	09° 47′ W	2001
SBC 237	24.07.83	57° 06′ N	09° 22′ W	1172
SBC 238	24.07.83	57° 05′ N	09° 21′ W	1028
SBC 275	06.08.83	56° 13′ N	10° 06′ W	1961
SBC 276	06.08.83	56° 14′ N	09° 51′ W	1792
SBC 277	06.08.83	56° 15′ N	09° 44′ W	1593
SBC 278	06.08.83	56° 15′ N	09° 46′ W	1631
SBC 279	06.08.83	56° 17′ N	09° 32′ W	1378
SBC 280	06.08.83	56° 18′ N	09° 25′ W	1188
SBC 281	06.08.83	56° 20′ N	09° 16′ W	1005
SBC 430	08.08.92	57° 00′ N	09° 55′ W	2003
SBC 431	08.08.92	57° 00′ N	09° 55′ W	2012
SBC 432	09.08.92	57° 00′ N	09° 59′ W	2016
SBC 433	10.08.92	57° 00′ N	09° 57′ W	2006
SBC 435	11.08.92	57° 00′ N	09° 58′ W	2032
SBC 440	16.08.92	57° 01′ N	09° 26′ W	1634
SBC 562	25.07.95	56° 45′ N	09° 25′ W	1540
SBC 568	28.07.95	57° 00′ N	09° 59′ W	2032
SBC 577	03.08.95	56° 31′ N	09° 18′ W	1009

from the sample before washing. Box samples were either washed in their entirety or only the top 20 cm layer was utilised (SBC 63 and 64 only). Occasion-

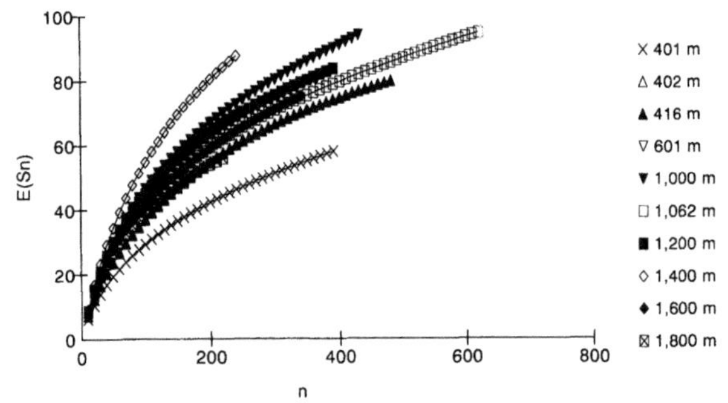

Figure 4. Rarefaction curves for the 10 Group 1 box-core samples.

ally, large-bodied benthic animals were encountered in the samples. These would qualify as megafauna under the usual definition of organisms easily resolved by a seabed camera if exposed on the sediment surface. They were not included in the comparison of biomass as they would have increased the total sample biomass by a large amount, and add greatly to already considerable variability in the data. Although excluded in this way in the present work, this is not to say that burrowed megafauna are unimportant overall in benthic standing crop, but merely a consequence of difficulty in collecting megafauna in sufficient numbers using a box corer sampling just 0.25 m^2 of seabed.

Samples were obtained, usually opportunistically, on research cruises of R.R.S. *Challenger* and R.R.S *Charles Darwin* to the Rockall Trough area on various dates from 1976 to 1995 (Table 1). Because of this, it was not possible to obtain replicates at individual stations except those at around 400 m. This means that it was not possible to estimate statistical sampling error at the other stations (we were reluctant to estimate error from the three replicates from 400 m because of uncertainty over sampling efficiency). It is difficult to guess how important such variability may be at the area of individual stations. However, Gage (1977) concluded from a series of six box-core replicates taken at mid-basin depth of around 2900 m depth in the southern Rockall Trough that single, good box-core samples provided reasonable estimates of community parameters over the area covered by the station (about 7 km^2 of relatively level seabed).

In some box-core samples, a small portion was removed for sedimentological or geochemical analysis. Where this occurred adjustment of the biomass/species abundance data was made to allow comparison with other samples. Of the samples listed in Table 1, all were utilised in assembling a data set for a depth-related comparison of macrofaunal biomass, but only a subset of 10 (referred to as Group 1) taken from the depth range of 400–1800 m on the Barra Fan, and the upper slope area just above (Fig. 1), are used here in analysis of community structure. Samples were washed by elutriation using filtered sea water on board ship. Washed samples were fixed immediately in borax-buffered sea-water formalin and later transferred to a solution of 90% ethanol to which 4% propylene glycol as a precaution against samples irrecoverably drying out. 'Macrofauna' (here taken to include all metazoans retained by a 0.42 mm sieve, excluding 'meiofaunal' taxa – Nematoda, Copepoda and Ostracoda) was sorted under a binocular microscope after a solution of 1% Rose Bengal had been added to the sample to make the organisms more easily detected amongst the sediment particles. For estimating biomass, sorted samples were bulked, blotted damp dry and weighed using an analytical balance. Samples used in analysis of macrofaunal community structure were sorted to the level of putative species. Many of these have been identified by taxonomic specialists, but these data will not be given here. Analysis of uni- and multivariate statistics of community structure were undertaken using the PC software packages BioDiversity Pro and Primer. Terminology and definitions of faunal diversity in Magurran (1988) are followed in the analysis of community structure applied here, and this reference may be consulted for a fuller explanation for all the univariate statistics of diversity applied here.

Results

Values of wet-weight biomass plotted against sample depth are given in Figure 2A for all samples listed in Table 1. Note that to increase the depth range, plotted data are included from depths to 2900 m depth in the mid Rockall Trough from where a series of replicated box core samples mentioned above (see Gage, 1977) are available. Total abundances of macrobenthos, standardised to 1 m^2, are plotted in Figure 2B.

A total of 511 putative species, excluding peracarid Crustacea, among 25 major groups has been identified amongst the 10 box-core samples for which species level data are available. Data for many peracarid groups are not yet available to a sufficiently high taxonomic resolution. Values for various univariate measures of habitat, 'local' or 'sample' diversity (α diversity *sensu* Magurran, 1988) applied to the single box-core samples were calculated. Species richness

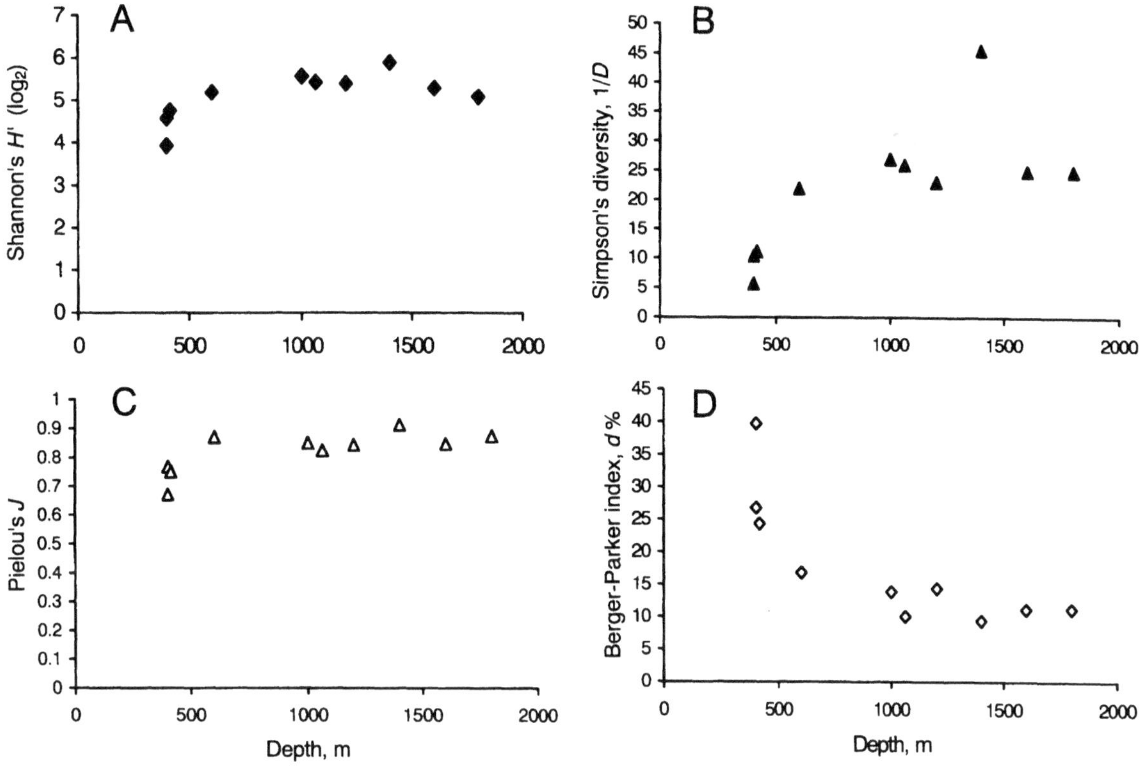

Figure 5. Alpha diversity statistics versus depth for the Group 1 samples. (A) Shannon's diversity H'; (B) Simpson's diversity, as $1/D$;. (C) evenness as Pielou's J; (D) Berger-Parker index, as $d\%$,, of sample dominance.

(the number or variety of species present) is expressed here in two ways: (1) as the number of species present per m^2, a plot of this as species density in relation to depth being given in Figure 3; and (2) numerical species richness which is the number of species per specified number of individuals or biomass. The method of rarefaction (Sanders, 1968), as modified by Hurlbert (1971), generates curves of expected numbers of species, $E(S_n)$, among increasing values of n individuals up to those found in the total sample. This provides a good method of comparing numerical species richness over the depth gradient among samples with differing total numerical abundances (Fig. 4). Expected species diversity for $n = 180$, $E(S_{180})$, is plotted in Figure 3, this value of n individuals representing the highest value for which it is possible to calculate a common knot size for $E(S_n)$ from the Group 1 box-core samples. A high common value of n minimises effects of differences in evenness which have greatest potential effect on the expected species diversity at low numbers values of n (Gage & May, 1994). The concept of species diversity incorporates both species richness and their relative abundance in the sample (evenness,

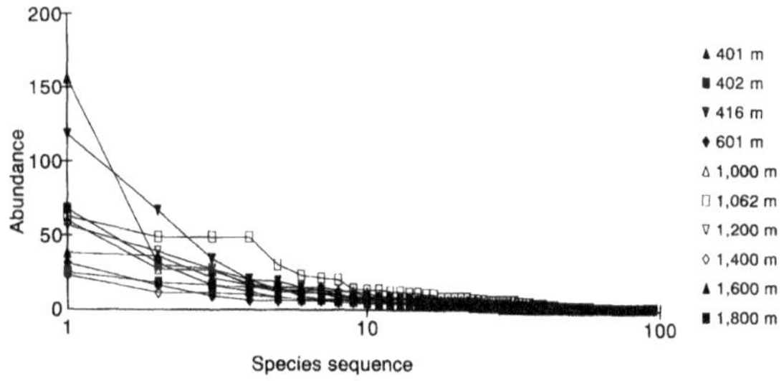

Figure 6. Ranked species abundances plotted for the fauna in the Group 1 box cores.

or dominance diversity). Shannon's diversity index H' embodies both aspects of diversity and is calculated as the sum of the logarithm of the proportion of the total abundance made up by each species. Following convention in marine biological studies, $\log_2$ is used in this calculation. Figure 5A shows values of H' plotted against depth. Simpson's diversity index, D, is heavily weighted to abundance of the commonest species and is the probability that any two randomly drawn individuals are the same species (Fig. 5B). It is calculated as the sum of the squared proportions of the total made

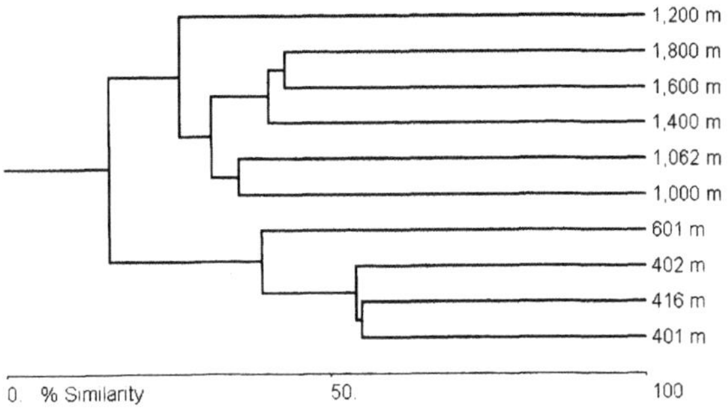

Figure 7. Dendrogram representation (single linkage clustering) of Bray–Curtis similarities among the untransformed species abundances in the Group 1 box cores.

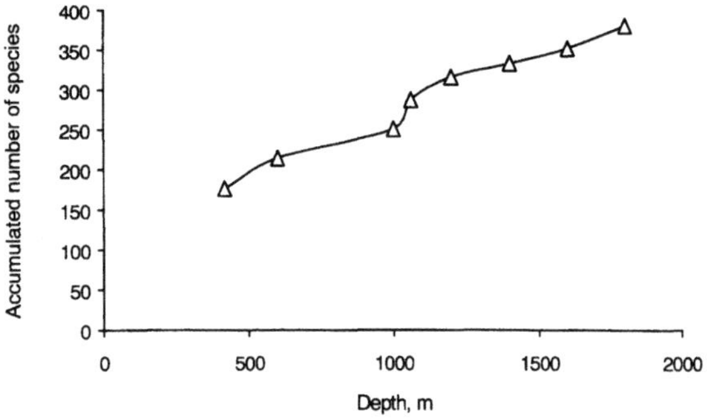

Figure 9. Downslope accumulation of macrobenthic species occurrences in the Group 1 box cores.

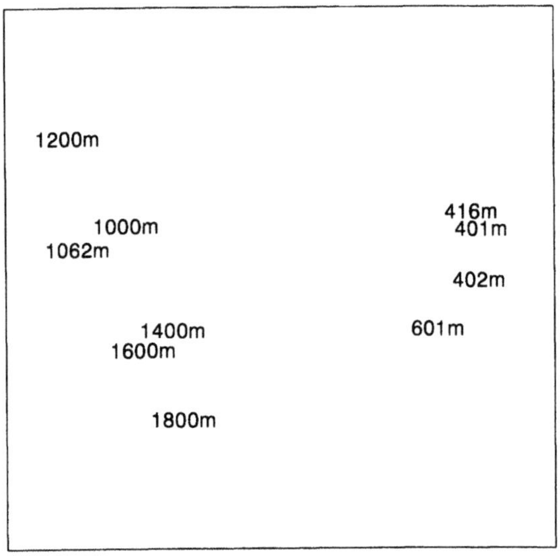

Figure 8. Ordination of untransformed species abundances in the Group 1 box cores by means of multidimensional scaling; stress = 0.04.

up by individual species. Values of Pielou's J (derived from Shannon's H' where J is the ratio between H' and maximum possible diversity, H_{max}, where all the species are equally abundant) are plotted against depth in Figure 5C, while the Berger-Parker index (simply the proportional abundance of the most abundant species) is plotted against depth in Figure 5D. A plot of ranked abundances of species in the samples is given in Figure 6.

The degree of change in species between habitats, or along an environmental gradient is usually referred to as β diversity. Multivariate analysis was applied to the total data to see what trends emerge in relation to depth. A dendrogram representation using single-link clustering (Lance & Williams, 1967; Gunn & Everitt, 1981) of Bray-Curtis similarities (Bray & Curtis, 1957), based on untransformed species abundances, is given in Figure 7. Ordination avoids the chance anomalies in the agglomerative process when clustering the smallest units, this sometimes obscurs true affinities. The technique of non-metric multidimensional scaling, a method of ordination which utilises the ranked orderings of pair-wise similarities (the same as those used to calculate a Bray–Curtis classification), was also applied. The result, presented in Figure 8, has a stress factor of 0.04 indicating excellent representation of the samples.

Finally, β diversity may also be expressed as the rate of turnover of species along an environmental gradient. It is possible to measure this rate of turnover simply as the rate at which new species accumulate in samples arranged along the depth gradient. Figure 9 shows a cumulative curve of species occurrences where an increased rate of accumulation, apparent as a step-like change in the upward slope of the curve at about 1000 m depth, indicates an abrupt change in composition of the benthic community.

Discussion

Downslope pattern in benthic biomass and abundance

The world-wide quantitative sampling undertaken mainly by Russian workers in the 1950s and 1960s has clearly shown a trend towards exponential decrease in biomass with increasing depth. A comprehensive list of references for this work is given by Rowe (1983). In the global context, the envelope of variability (see Fig. 7 in Rowe, 1983) is wide, being influenced by variables such as terrestrial and coastal run-off. Nevertheless, the pattern at individual sites is coherent

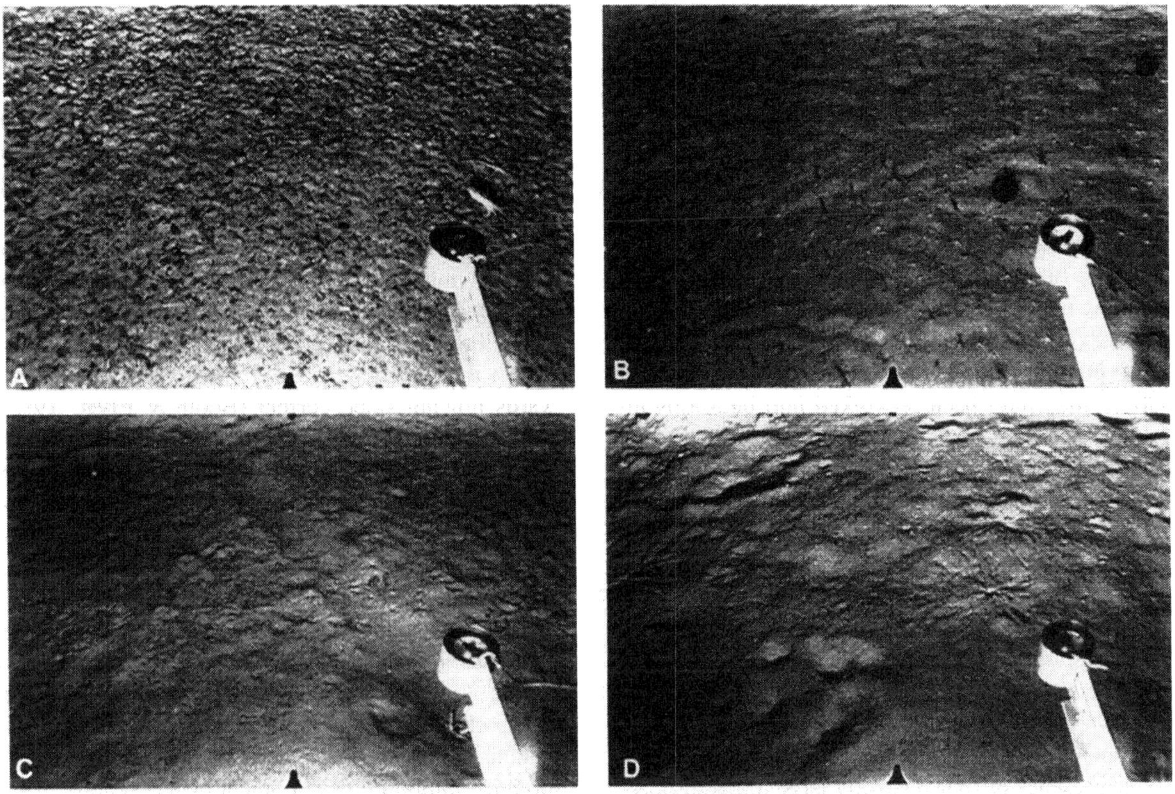

Figure 10. Seabed photographs taken as part of the NERC Shelf Edge Study showing changing benthic environment with increasing depth on the Hebridean continental slope west off Barra Head in the area of the box core samples. The view is oblique and covers about 1.2 m width by about 2.5 m from front to back. (A) 300 m, showing a fairly uniform field of pebble gravel and sand. Although there appears to be no evidence of hydrodynamic activity in this photograph, in others taken nearby ripple bedforms are visible along with evidence of advection of sand-sized sediment particles. A small octopus (probably *Eledone cirrhosa*) and a large brittle star (probably a species of *Ophiura*) are visible just above the compass at middle right. (B) 700 m, showing a muddy sand bed smoothed and marked by extensive biogenic activity. The tyre-tread like tracks are made by the large surface deposit feeding spatangoid, *Spatangus rashi,* one of which is seen just to the left of the top of the compass. Dense populations of the smal white brittle star, *Ophiocten gracilis*, and a cerianthid anemone with its tentacle crown oriented to face upstream into the current. In other photographs from this depth, there is evidence of slight bed-modification by oscillatory benthic currents moving along the slope contours. A sea star, probably a predatory species, is visible at top left. (C) 1,000 m, showing the muddy bed smoothed and marked by biogenic activity including some burrows and surface traces, such as the two circular patterns at middle left, made by burrowed deposit feeders (at other positions nearby at this depth, bedforms indicate occasional energetic events creating small cusp-shaped ripples by oscillating flows). A burrow opening visible just below the compass is occupied by a decapod crustacean, possibly a galatheid. The shallow furrow-like mark extending diagonally from top left to bottom right is probably an old trawl mark. (D) 1500 m showing a soft muddy bottom with numerous biogenic surface traces indicating it is extensively reworked by burrowed deposit feeders (none of the photographs taken from this site indicates hydrodynamic activity). The radial pattern surrounding a central hole visible to the left of the top of the compass is an example of this activity made by a polychaete or echiuran worm.

and usually somewhat similar to other sites within the same general area, for example those in the northwest Atlantic (see Rex, 1983). Although the absence of replication among most of the samples means sampling error is unknown, the apparently exponential decrease in macrobenthic standing crop in the present samples shown when plotted against increasing depth (Fig. 2A) supports this expectation only among samples taken at depths greater than about 600 m. The slope of the non-linear regression fitted to the data excluding the 400 m samples fits easily within the global range plotted by Rowe (1983). Values from the shallowest samples at around 400 m depth are particularly lower than expected from the curve extrapolated to the shelf edge. Although there is a possibility that these box-core samples may have lost fauna along with the water overlying the core water, both the similarity of values in the three samples suggests any bias was consistent, while the marked reduction in standing crop would seem to reflect a real reduction in benthic fauna at this depth. The benthic environment visible in seabed photographs taken at this depth level in the SES programme (LOIS Shelf-Edge Study CD-ROM, 1999), in the close vicinity of most of the samples, indicates vigorous hydrodynamic conditions that would easily be capable of moving coarse sediment particles as well as washing any fine sediment. The scale of physical disturbance transmitted as eddy energy from storms at the surface, along with tidal flow, will directly limit the development of the infauna at this site through a regime of physical disturbance. The seabed photographs taken below the shelf edge down to about 500–600 m (Fig. 10A) show only coarse grained sediments present, such as cobbles and sands which above 300 m often have been re-worked into bedforms indicative of strong currents. Usually, few epifauna are visible except for small, attached colonial organisms such as sponges, hydroids, bryozoans and tubeworms encrusting stones. There are only very sparse indications, from burrows or other traces, of large, burrowed fauna.

When the total abundance of the macrobenthic biota is plotted (Fig. 2B) a similar pattern is evident differing only with respect to the sample from 601 m. The relatively low numerical abundance in relation to biomass in this sample is caused, however, partly by the presence of four relatively large onuphid polychaetes whose combined biomass (2.2 g) formed a significant part of the total of 6.8 g (in other samples, single large organisms of megafaunal size had been removed before weighing, but these onuphids were considered only marginally large enough to fall in this category). Seabed photographs a little deeper than this station at about 700 m depth show a rich epifauna. Overall, the pattern is one where faunal biomass and density peaks in the upper bathyal at a zone just below the shelf edge where the seabed photographs show the bottom is affected by considerable hydrodynamic disturbance (Fig. 10A). Classically, bottom mixing caused by surface storms and tidal flow moderates sufficiently with increasing depth so that by the maximum depth of winter mixing at this latitude in the northeast Atlantic (Levitus, 1982), fine particles accumulate on the seabed. This corresponds to the 'mud line' concept of John Murray (see Murray & Hjort, 1912). A modern view of mixing processes affecting the sediments at the continental margin is given by Huthnance (1995) who describes other sources of turbulence such as internal tides and waves. But these are much less important than wind and tidal mixing on the relatively gentle declivity of the shelf edge area considered here (J.M. Huthnance, pers. comm.). On the Hebridean margin off Barra Head, wind-driven currents at the seabed just below the shelf break through most of the year are not likely to be as strong as tidal currents (typically 0.1–0.2 m s^{-1}), although wind-driven currents may increase this expectation in winter. Overall levels of turbulent kinetic energy are comparable with those on the adjacent shelf (Huthnance, 1986). But whether benthic diversity in the current-swept sediments on the outer shelf is similar to that at 400 m is as yet unknown, although Gray (1994) claims such communities maybe as diverse as those in the deep sea.

Below the area of predominantly tidal and wind-driven mixing, the seabed is affected by the northwards flowing slope current (Booth & Ellett, 1983; Huthnance, 1986). The slope current (typically 0.1–0.2 m s^{-1} near the bottom in the area of the slope sampled) augments any remaining tidal and wind-driven currents. The evidence from seabed photographs suggests this flow may help smooth the bottom, although most of the sediment displacement may actually be biogenic (Fig. 10B), but in places ripple bedforms generated by currents are seen. The flow regime is certainly strong enough to re-suspend and transport light detrital particles utilisable as food by benthic fauna, but insufficient to cause physical disturbance inhibiting development of the benthic macrofauna. Furthermore, the presence in early summer after the spring bloom of large quantities of flocculent detritus at the bottom on the continental slope has been noted by several investigators from seabed photographs (Billett et al., 1983; Hecker, 1990; Bett, pers. comm.), indicating that such material sedimenting from phytoplankton blooms at the surface may be swept downslope and collect at this depth. The presence in seabed photographs (LOIS Shelf-Edge Study CD-ROM, 1999) of dense populations of the interface-feeding brittle star *Ophiocten gracilis* (see Lamont & Gage, 1998) and cerianthid anthozoans able to intercept particles from near-bed flow, coincides with the slope current (Fig. 10B). Populations of surface deposit feeders, such as the echinoid *Spatangus raschi*, support the view that the area of this current is a zone of enhanced feeding conditions for benthic organisms that take in organic particles, whether these are moving intermittently or continuously over the bed. These high densities in epifaunal populations have largely disappeared by 1000 m (Fig. 10C), but it is not until depths below about 1200–1300 m that hydrodynamic conditions become quiet enough for these conditions to be replaced by a seabed, where at 1500 m current smoothing is absent and the landscape (Fig. 10D) is dominated by biogenic traces, such as mounds, cones and craters, caused by deep-burrowing deposit feeding benthos.

There are few comparative data for the macrobenthos for other sites in the northeast Atlantic. Flach et al. (1999) describe a pattern from box cores taken on the Goban Spur (ca 49° N, 11–14° W) where macrofaunal biomass peaks at about 1000–1500 m depth owing to the high mean individual weight of macrobenthos at this depth. Organic material is resuspended by vigorous flow at this depth and provides good feeding conditions for suspension and interface-feeding organisms. In contrast, macrofaunal abundance seems to decline exponentially with increasing depth on the Goban Spur. That the site-related pattern may very much reflect local conditions is shown by work in the northwest Atlantic where Schaff et al. (1992) record large 100 km scale variability in macrobenthic density at 850 m depth off the Carolinas. This is driven by differences in organic detrital flux from surface productivity, and lateral inputs from currents and mass wasting.

Depth-related pattern in macrofaunal diversity

The curvilinear pattern in species density (Fig. 3) reflects not so much differences in species richness in relation to numbers of fauna, but the declining total density of the macrobenthos (see Fig. 2B). Rarefaction curves generated from the total macrobenthos data from the 10 Group 1 box-core samples, show a rather unclear pattern. The curves for the 400-m stations are much lower than the rest, except for the 1800-m station which lies along the curve for both 402 m and 416 m (Fig. 4). The highest curve is that for the 1400-m station, with the remainder lying at intermediate positions. The expected number of species at a sample size of 180 individuals when plotted against depth (Fig. 3) reinforces this view of depth-related pattern in species richness on the Scottish continental slope, with low values for the stations at around 400 m, and higher, values for the remainder, the highest for 1400 m, and a suggestion of a decline at the two deeper stations. Shannon's H' also shows this pattern (Fig. 5A), but with the high for 1400 m much less obvious against the other deeper stations. However, Simpson's diversity, expressed as $1/D$ (Fig. 5B) shows the low species diversity for the 400-m stations and a pronounced peak for 1400 m against the higher values for the other deeper stations. Simpson's diversity gives the probability of any two individuals being the same species and is, therefore, much weighted towards the abundances of the commonest species. It is, therefore, more an expression of dominance diversity than providing a measure of species richness. Figure 5B indicates much greater dominance among the shallowest stations, with lower dominance (high evenness), peaking at 1400 m, among the deeper stations. This fits in with an expectation of greater dominance among the species abundances (low evenness) for the shallowest stations that are thought to be subject to very high levels of hydrodynamic disturbance. To some extent this pattern is also shown in the plot for Pielou's J (measuring evenness) in Figure 5C, although the peak for 1400 m is less clear. However, the pattern shows up well in values of the Berger-Parker statistic (Fig. 5D) where, in one of the 400 m samples, the most common species comprised almost 40% of the total fauna, and in the others, the most common species was twice as abundant as in most of the deeper ones, although, like Pielou's J, the 1400-m station is barely distinguished from the other deeper stations.

This pattern in downslope macrobenthic diversity is broadly similar to that found by Paterson & Lambshead (1995) based on the polychaete data in the present samples. These authors also included data on polychaetes from the stations worked by SAMS at about 2900 m depth referred to earlier. They concluded the data indicated a parabolic distribution of diversity, based on reduced levels for the stations at 400 and 600 m, higher values for the deeper slope stations, and lower levels for the samples from around 2900 m. For $E(S_{41})$, like that obtained from $E(S_{180})$ for the total fauna minus percarids (Fig. 3) and Simpson's $1/D$ (Fig. 5B), there was a peak at the 1400 m station. We note that this station also shows the highest evenness (see above). We conclude, therefore, that the expression of high diversity in this sample is the result of particularly high evenness in species abundances. This interpretation is supported by a plot of ranked species abundances where the distribution shows the least degree of numerical dominance by common species (Fig. 6).

For the total macrofauna (minus peracarids) considered here, we agree with Paterson & Lambshead's (1995) interpretation for polychaetes as reflecting the effect of hydrodynamic disturbance by 'disrupting feeding or even transporting animals'. However, given the lack of replication to provide an estimate of statistical error, any significance of the peak showing up for the 1400-m sample must remain uncertain.

Although it is reassuring that results for the polychaetes provide a similar pattern to that for the total fauna minus peracarids, a wider significance for the patterns observed is uncertain. A diversity maximum at 800 m, but declining into greater depths, was found with benthic isopods in the Norwegian and Greeland Seas (Svavarsson et al., 1990). A limited data set for gastropods along a benthic transect off southwest Ireland suggests a minimum in diversity at intermediate depths (Rex et al., 1997). However, studies on macrobenthic diversity on the benthic transect referred to earlier on the Goban Spur, shows a slightly different pattern where both species richness and evenness increase with increasing depth down the continental slope and rise, with the greatest rate of increase between the 208 and 670 m depth stations worked (Flach & De Bruin, in press). However, the latter work covers a considerably greater depth range (to 4470 m depth) than in the present study. But the diversity statistics plotted by Flach & De Bruin (1999) suggest a pattern more similar to the present results, with diversity and evenness apparently levelling off over the bathymetric range corresponding to the present study.

Although it might seem possible that a continuation of the present transect into deeper water may show a trend similar to that described by Flach & De Bruin (1999), the opposite result was obtained by Lambshead & Paterson (1995). These authors found polychaete diversity and evenness to decrease to a depth of 2900 m. Perhaps this is related to the differing environmental conditions associated with the 2900 m site being located in mid basin, while even the deepest station of Flach & De Bruin (1999) was still at the base of the continental slope. It is tempting to suggest that such differences are associated with hydrodynamics, with the mid-basin site in the southern Rockall Trough being known to experience more vigorous flow conditions than those on the adjacent lower continental slope and rise. At other sites, diversity of deep-sea macrobenthos has been shown to be at least partially influenced by hydrodynamic disturbance (Gage, 1997).

The parabolic pattern in the northwest Atlantic is different again from both the Hebridean and other sites off northern Europe described above. In the northwest Atlantic, the trend is towards a mid-slope peak in species diversity varying between about 2000 and 3000 m depth depending on taxon (Rex, 1983; Rex et al. 1997). This pattern also has been demonstrated for megafauna and demersal fish. Clearly, there is no global consistency in such patterns suggesting they are more a reflection of local conditions than expressing any fundamental response to depth.

Depth-related pattern in taxonomic composition

The evidence described above on changing environment related mainly to hydrodynamics and food supply seems to provide the key explanation for the changes revealed in the multivariate comparisons. Both Bray–Curtis classification (Fig. 7) and multidimensional scaling (Fig. 8) show a major dichotomy between the 400- and 600-m stations and the deeper ones. Within these sub-groups, levels of similarity, although not particularly high, suggest expression of response to similar environmental conditions. The dichotomy revealed by the multivariate analyses also corresponds to the step-like change in rate of species accumulation (Fig. 9). A similar step-like increase in rate of species accumulation, expressed also as a major boundary in the range of individual species, has been noted for megafaunal echinoderms at this depth (Gage, 1986). For macrobenthos, this change in community composition may result from change from a community dominated by species adapted to feeding on particles transported by currents to one where deposit feeding species, unsuited to the more dynamic conditions at the shallower stations, are common.

We are as yet unable to compare confidently species data sets between sites that have been worked up by different teams. Patterns in species composition, therefore, can only be provided with a local explanation. However, in terms of feeding type, a pattern in dominance by suspension- and interface feeding macrofauna is described by Flach et al. (1999) for the depth zone at 1000–1500 m depth on the transect on the Goban Spur where hydrodynamic conditions are sufficiently dynamic to provide the predominantly sessile fauna with a constant supply of food in the form of organic particles transported by currents. Although this zone is deeper than that influenced by the slope current at the Hebridean shelf edge, the general pattern in depth-related change in predominating feeding is similar.

Acknowledgements

The interpretations of environment in the downslope pattern in the macrobenthic community described here have benefited from advice on bottom mixing by John Huthnance and from discussions with others engaged in the Shelf Edge Study funded by the U.K. Natural Environment Research Council as part of its Land Ocean Interaction Study (LOIS) programme (see

LOIS Shelf Edge Study project CD-ROM, 1999). We thank John Humphery for allowing access to the seabed photographs taken during the Shelf Edge Study. Work on the benthic community requires high-quality taxonomic resolution of often poorly described species, and we are grateful to many taxonomic experts and other helpers, especially David Billett, Ailsa Clark, Edward Cutler, Kristian Fauchald, Peter Gibbs, Robin Harvey, David Heppell, Gordon Paterson, Margaret Pearson, Benjie Rae, Eve Southward, Sheila Smith and Anders Warén, for their valuable contributions to this study. We are grateful to the Atlantic Frontier Environmental Network of oil companies (AFEN), an oil industry/government group operating in the U.K. sector of N.E. Atlantic, for generous financial support in completing the analysis on which this paper is based and for consent to publish the work. We also thank the referees for their helpful comments.

References

Billett, D. S. M., R. S. Lampitt, A. L. Rice, & R. S. M. Mantoura, 1983. Seasonal sedimentaion of phytoplankton to the deep-sea benthos. Nature 302: 520–532.

Booth, D. A. & D. J. Ellett, 1983. The Scottish continental slope current. Contl. Shelf Res. 2: 127–146.

Bray, J. R. & C. T. Curtis, 1957. An ordination of upland forest communities of southern Wisconsin. Ecol. Mon. 27: 325–349

Carney, R. S., R. L. Haedrich & G. T. Rowe, 1983. Zonation of fauna in the deep sea. In Rowe, G. T. (ed.), The Sea. 8. John Wiley & Sons, New York: 371–398.

Flach, E. & W De Bruin, 1999. Diversity patterns in macrobenthos across a continental slope in the NE Atlantic. J. Sea Res. 42: 303–323.

Flach, E., M. Lavaleye, H. Ee Stigter & L. Thomsen, 1998. Feeding types of the bethic community and particle transport across the slope of the N.W. European continental margin (Goban Spur). Prog. Oceanog. 42: 209–231.

Gage, J. D., 1977. Structure of the abyssal macrobenthic community in the Rockall Trough. In Keegan, B.K. P. O. Ceidigh & P. J. S. Boaden (eds), Biology of Benthic Organisms. Pergamon Press, Oxford: 247–260.

Gage, J. D., 1986. The benthic fauna of the Rockall Trough: regional distribution and bathymetric zonation. Proc. r. Soc. Edinb. 88B: 159–174.

Gage, J. D., 1997. High benthic species diversity in deep-sea sediments: the importance of hydrodynamics. In Ormond, R. F. G. J. D. Gage & M. V. Angel (eds), Marine Biodiversity: Patterns and Processes. Cambridge University Press, Cambridge: 148–177.

Gage, J. D. & R. M. May, 1994. A dip into the deep seas. Nature 365: 609–610.

Gage, J. D. & P. A. Tyler, 1991. Deep-sea Biology: a Natural History of Organisms at the Deep-sea Floor. Cambridge University Press, Cambridge: 504 pp.

Gray, J. S., 1994. Is deep-sea species diversity so high? Species diversity of the Norwegian shelf. Mar. Ecol. Prog. Ser. 112: 205–209.

Gunn, G. & B. S. Everitt, 1981. An Introduction to Mathematical Taxonomy. Cambridge University Press, Cambridge: 152 pp.

Hecker, B., 1990. Photographic evidence for the rapid flux of particles to the sea floor and their transport down the continental slope. Deep-Sea Res. 37: 1773–1782.

Hessler, R. R. & P. A. Jumars, 1974. Abyssal community analysis from replicate box cores in the central North Pacific. Deep-Sea Res. 21: 185–209.

Hurlbert, S. H., 1971. The nonconcept of species diversity: a critique and alternative parameters. Ecology 52: 577–586.

Huthnance, J. M., 1986. The Rockall slope current and shelf-edge processes. Proc. r. Soc. Edinb. 88B: 83–101.

Huthnance, J. M., 1995. Circulation, exchange and water mass at the ocean margin: the role of physical processes at the shelf edge. Prog. Oceanog. 35: 353–431.

Lamont, P. A. & J. D. Gage, 1998. Dense brittle star populations on the Scottish continental slope. In Mooi R. & M. Telford (eds), Echinoderms San Francisco (Proceedings of the Ninth International Echinoderms Conference, San Francisco, California, USA, 5–9 August 1996). Balkema, Rotterdam: 377–382.

Lance, G. N. & W. T. Williams, 1967. A general theory of classification sorting strategies 1. Hicarchical systems. Comput. J. 9: 373–380.

Levitus, S., 1982. Climatological atlas of the world ocean. NOAA Prof. Paper 13, U.S. Dept. Commerce.

LOIS Shelf-Edge Study Project CD-ROM, 1999. British Oceanographic Data Centre, Birkenhead, U.K.

Magurran, A., 1988. Ecological Diversity and its Measurement. Chapman & Hall, London: 179 pp.

Murray, J. & J. Hjort, 1912. The Depths of the Ocean. MacMillan, London: 821 pp.

Paterson, G. L. J. & P. J. D. Lambshead, 1995. Bathymetric patterns of polychaete diversity in the Rockall trough, northeast Atlantic. Deep-Sea Res. I 42: 1199–1214.

Rex, M. A., 1983. Geographic patterns of species diversity in the deep-sea benthos. In Rowe, G. T. (ed.), The Sea, 8. John Wiley & Sons, New York: 453–472.

Rex, M. A., R. J. Etter, & C. T. Stuart, 1997. Large-scale patterns of species diversity in the deep-sea benthos. In Ormond, R. F. G. J. D. Gage & M. V. Angel (eds), Marine Biodiversity: Patterns and Processes. Cambridge University Press, Cambridge: 94–121.

Rex, M. A., C. T Stuart, R. R. Hessler, J. A. Allen, Sanders, H. L. & G. D. F. Wilson, 1993. Global-scale latitudinal patterns of species diversity in the deep-sea benthos. Nature 365: 636–639.

Rowe, G. T., 1983. Biomass and production of the deep-sea macrobenthos. In Rowe, G. T. (ed.), The Sea. 8. John Wiley & Sons, New York: 97–121.

Sanders, H. L., 1968. Marine benthic diversity: a comparative study. Am. Nat. 102: 243–282.

Schaff, T., L. Levin, N. Bair, D. DeMaster, R. Pope & S. Boehme, 1992. Spatial herterogeneity of benthos on the Carolina contnental slope: large (100 km)-scale variation. Mar. Ecol. Prog. Ser. 88: 143–160.

Svavarsson, J., T. Brattegard & J.-O. Strömberg, 1990. Distribution and diversity patterns of asellote isopods (Crustacea) in the deep Norwegian and Greenland Seas. Prog. Oceanog. 24: 297–310

Hydrobiologia **440:** 273–279, 2000.
M.B. Jones, J.M.N. Azevedo, A.I. Neto, A.C. Costa & A.M. Frias Martins (eds), Island, Ocean and Deep-Sea Biology.

Lipid profiles of *Nematocarcinus gracilis* a deep-sea shrimp from below the Arabian Sea oxygen minimum zone

Catherine E Allen[1], Paul A Tyler & Mark S Varney
School of Ocean and Earth Sciences, Southampton Oceanography Centre, European Way, Southampton SO14 3ZH, U.K.
[1]*Harbor Branch Oceanographic Institution, 5600 US1 North, Fort Pierce, FL 34946, U.S.A.*
E-mail: callen@hboi.edu

Key words: Arabian Sea, fatty acid, lipid, *Nematocarcinus gracilis*, oxygen minimum zone (OMZ), shrimp

Abstract

Specimens of the deep-sea benthic shrimp *Nematocarcinus gracilis* were collected from 900 m to 1000 m in the Arabian Sea, close to where the permanent oxygen minimum zone meets the sea floor. Lipid profiles, encompassing total lipid, lipid class and fatty acid composition, were compared with previously reported crustacean lipid assays and provided an insight into the life history of the species. The major storage lipid in *N. gracilis* was triglyceride, supporting the supposition that this species exists in benthic regions. Neutral lipid levels were commensurate with *N. gracilis* being an opportunistic feeder. Fatty acid composition was typical of an organism with a diet based on an ultimately photosynthetic source of organic carbon, but also reflected the reduction in the availability of labile organic carbon (in the case of lipid, highly unsaturated fatty acids) in the deep sea.

Introduction

Nematocarcinus gracilis Spence Bate is a bright red deep-sea caridean shrimp that occurs in the Indian Ocean. Nematocarcinids are characterised by three pairs of extremely long pereiopods thought to be used like stilts for walking on soft sediments (Calman, 1911). Relatively little is known about *N. gracilis*, but the congeneric species *Nematocarcinus exilis* (Bate), found on the western Mediterranean slope, has been examined in detail. Stomach content analyses of this latter species (Cartes, 1993) indicated that it is a scavenger and predator, consuming mainly fish remains and slow-moving benthic organisms. *Nematocarcinus gracilis* might be expected to exhibit similar feeding strategies.

The Arabian Sea is one of the most productive areas of the world's oceans and also boasts an extensive oxygen minimum zone (OMZ; Currie et al., 1973). During the south-west monsoon period from June to August, offshore Ekman transport of surface waters results in strong upwelling of nutrient-rich deep waters and, consequently, intense primary production (Currie et al., 1973). The combination of the subsequent high downward flux of organic matter (Nair et al., 1989) and sluggish intermediate water circulation rates (Swallow, 1984; Olsen et al., 1993) is responsible for an intense and permanent oxygen minimum. Dissolved oxygen levels are in the region of 0.2 ml l^{-1}, which corresponds to approximately 2% oxygen saturation at the depths studied (Olsen et al., 1993). The large fluxes of organic carbon are preserved in surface sediments where the oxygen minimum zone intersects the slope (between 150 and 900 m depth). However, it should be pointed out that the high sediment organic carbon content may be a function of the high levels of input and post-depositional reworking by hydrodynamic influences rather than a result of the OMZ (Pedersen et al., 1993).

A high biomass of benthic megafauna is found associated with the high levels of sedimentary organic carbon below the OMZ. The spider crab *Encephaloides armstrongi* Wood-Mason has been observed at densities of >100 m^{-2} at the oxic/anoxic boundary at around 975 m in the Arabian Sea (Gage et al., 1995). This crab is one of a number of organisms that appear adapted to migrate into and out of the oxygen minimum zone (Creasey et al., 1997), taking advantage of the elevated levels of organic matter; polychaetes with bright red gills and tentacles and morid fish with large, bright red gills were caught in trawls in the Arabian Sea oxygen minimum zone (Herring, 1994). The pelagic mysid *Gnathophausia ingens* Dohrn is adapted to survive in an oxygen minimum zone of the South California basin, possessing haemocyanin with elevated affinity and cooperativity for oxygen compared with other cold-water living crustaceans (Sanders & Childress, 1990). *Nematocarcinus gracilis* does not show these adaptations, so is probably less tolerant of hypoxic conditions and avoids the oxygen deficient waters. It appears to inhabit the area below the intersection of the OMZ and the sea floor.

Lipids are a group of biologically important compounds distinguishable by their insolubility in water. They may be broadly divided into classes, according to their solubility and to their function: triglycerides and wax esters are examples of neutral lipid classes used for energy storage. Polar lipid classes comprise mainly structural lipid, although a small proportion is used as transport molecules or hormone precursors. Fatty acids are components of most lipids. Certain fatty acids are produced primarily by photosynthetic organisms, so fatty acid composition can reflect trophic ecology. Functional adaptations to high pressures and low temperatures in the deep sea have resulted in characteristic lipids in deep-sea organisms (Somero, 1992). A lipid profile may, therefore, be used to indicate the physiological relationship between an organism and its environment (Allen, 1998, Pond et al., 2000).

The major questions addressed in this work are: what does the lipid class composition reveal about the life history of *N. gracilis*?, and, how does fatty acid composition reflect the trophic ecology of *N. gracilis*? *Nematocarcinus gracilis* were collected shortly after the southwest monsoon, so we also considered whether total lipid levels were related to inputs of organic matter associated with this event.

Material and methods

Nematocarcinus gracilis were collected during RRS Discovery Cruise 211/94 to the north Arabian Sea in October 1994. Specimens were collected using an Agassiz Trawl just below the area where the oxygen minimum intersects the slope (19° 15′ N, 58° 27′ W, 933–1100 m depth, normoxic conditions) and, on recovery, individuals were frozen immediately to −70 °C. Sex and carapace length were determined before specimens were freeze dried and analysed. Lipid profiles were determined for 28 whole specimens. Males and females were analysed separately, but there were no significant differences in their lipid composition, so all data have been combined in this paper.

Freeze-dried specimens were ground and then rehydrated by adding the mass of water that was lost during lyophilisation to ensure efficient triglyceride extraction (Dunstan et al., 1993). Lipids were extracted after the procedure of Folch et al. (1957) and weighed before re-dissolving in a small volume of chloroform containing 0.01% BHT and storing in a freezer until analysis. Lipid classes were analysed using thin layer chromatography with flame ionisation detection (TLC-FID). Neutral lipids were eluted in 85:15 hexane/chloroform containing 5% iso-propanol and 0.5% formic acid (Shantha & Ackman, 1990), and Chromarods partially scanned using a Mark IV Iatroscan. Polar lipids were then eluted in 70:30:3.5 chloroform/methanol/water (Fraser et al., 1985) before a full Chromarod scan. An aliquot of total lipid was dissolved in toluene and transmethylated in 1.5% sulphuric acid in methanol. Fatty acid methyl esters were extracted and purified by thin layer chromatography before analysis on a very polar (BPX70) column in a Perkin Elmer 8500 gas chromatograph. Fatty acid methyl esters were identified by reference to a cod liver oil standard.

Results

The total lipid percent dry weight determined for *N. gracilis* was 8.2±2.2 S.E. (S.E. = $\sigma/\sqrt{n}$, $n = 28$). This was within the range of values reported previously for deep-sea benthic shrimp (Table 1).

Wax esters were not detected in *N. gracilis*. Triglyceride comprised around 65% total lipid, free fatty acids constituted less than 6% of total lipid and sterols were present at around 5% total lipid. The lipid class composition of *N. gracilis* was compared with that of the deep-sea hydrothermal vent shrimp *Rimicaris exoculata* Williams & Rona and *Alvinocaris markensis* Williams (Fig. 1).

Fatty acid composition for *N. gracilis* is expressed as weight percentage total fatty acids (Table 2). The major fatty acids detected in *N. gracilis* comprised 16:0, 16:1 *n*-7, 18:0, 18:1 *n*-9 and 18:1 *n*-7. The proportions of different types of fatty acids found in *N. gracilis* and other caridean shrimp are presented

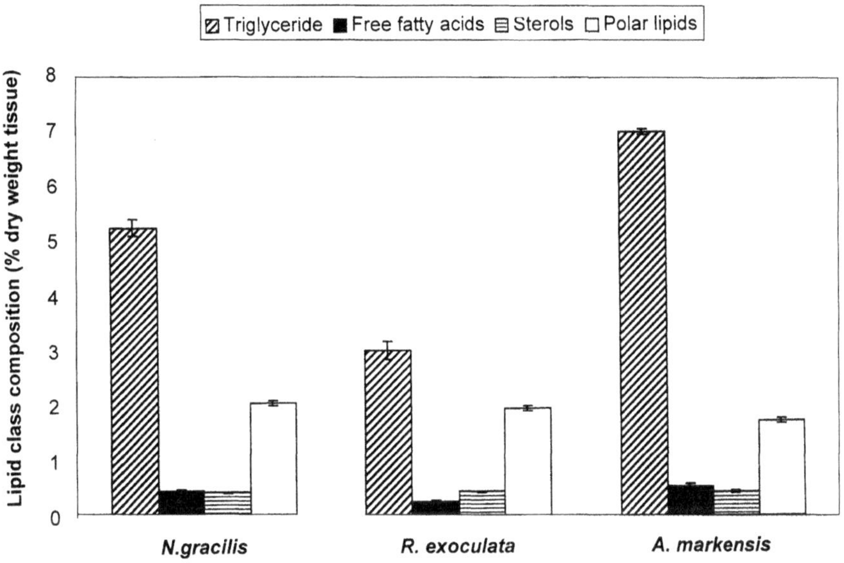

Figure 1. The lipid class composition of *Nematocarcinus gracilis*, *Rimicaris exoculata* and *Alvinocaris markensis* expressed as percentage dry weight tissue. Error bars represent standard error (SE = $\sigma/\sqrt{n}$; *N. gracilis*: $n = 28$; *R. exoculata*: $n = 36$; *A. markensis*: $n = 2$).

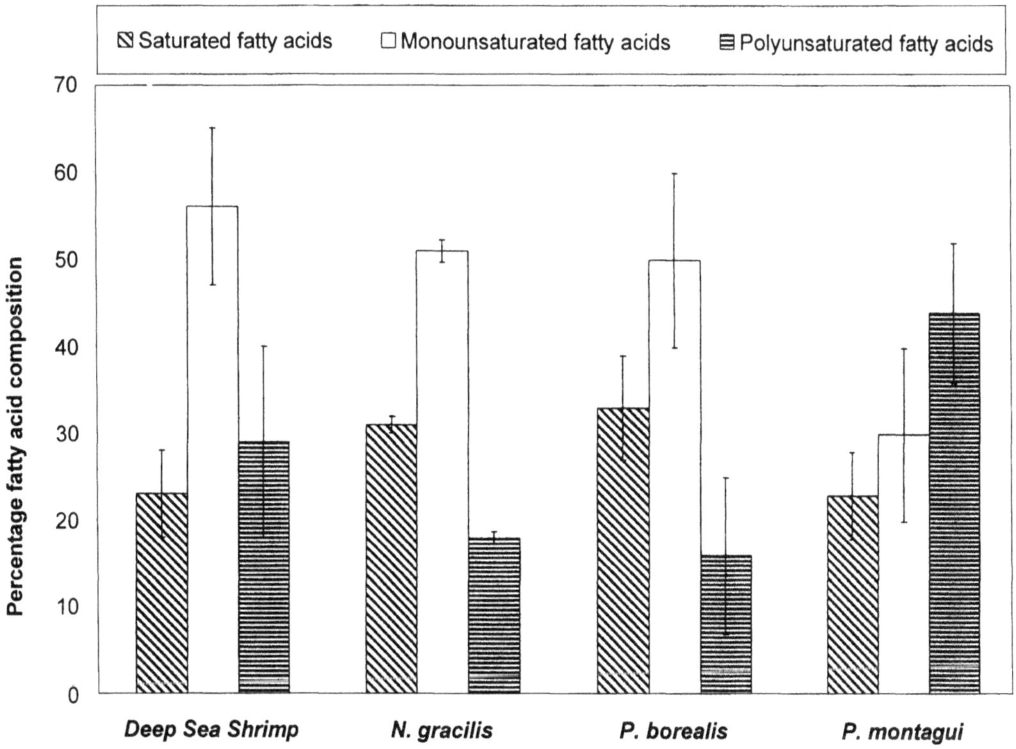

Figure 2. Percentage weights of different types of fatty acid in *Nematocarcinus gracilis* compared with the deep-sea caridean shrimp (Culkin & Morris, 1969), *Pandalas borealis* (Hopkins, 1993) and the shallow-water shrimp, *Pandalas montagui* (Clarke, 1979). Error bars for *N. gracilis* represent standard error (SE: $\sigma/\sqrt{n}$, $n = 28$). For other species error bars represent reported ranges.

Table 1. Summary of total lipid values for whole caridean shrimp. Where percentage wet weight figures were quoted in the literature[†], these have been adjusted to percentage dry weight assuming that dry weight ≈ 1/4 wet weight

Species	Total lipid (percentage dry weight)	Reference	Comments/depths caught
Acanthephyra acanthitelsonis	15–38 24	Culkin & Morris (1969)[†] Herring (1973)[†]	940 m to surface Bathypelagic species
Acanthephyra purpurea	9 13	Culkin & Morris (1969)[†] Herring (1973)[†]	1000 m to surface (shallow species)
Oplophorus spinosus	18	Culkin & Morris (1969)[†]	265 m to surface
Sergestes corniculum	7	Culkin & Morris (1969)[†]	750 m to surface
Systellapsis debilis	22 14	Culkin & Morris (1969)[†] Herring (1973)[†]	750, 265 and 190 m to surface
Parapasiphaea sulcatifrons	60	Herring 1973[†]	Pelagic species
Nematocarcinus cursor	7	Herring (1973)[†]	Benthic or epibenthic species
Acanthephyra aximia	15	Herring (1973)[†]	Benthic or epibenthic species
Plesionika martia	10	Herring (1973)[†]	Benthic or epibenthic species
Chorismus antarcticus	6–20	Clarke (1977)[†]	Varied seasonally and with degree of maturity
Pandalus montagui	4–24	Clarke (1979)[†]	Shallow-water shrimp Varied seasonally
Pandalus borealis	10 – 40	Hopkins et al. (1993)	Varied seasonally Trawled from 185 m
Rimicaris exoculata	6	Allen (1998)	Hydrothermal vent shrimp, 3200 m
Nematocarcinus gracilis	8	This study	933 m – 1100 m

in Figure 2. The most striking feature of this graph is the higher proportion of polyunsaturated fatty acids in shallow-water compared with deep-sea shrimp.

Discussion

Total lipid tissue concentrations of *N. gracilis* were at the low end of the general range of those reported for deep-sea benthic and epibenthic shrimp (Table 1), but similar to those for the congeneric species *N. cursor* Milne Edwards (Herring, 1973). Reasons

Table 2. Mean percentage fatty acid composition of *N. gracilis* expressed as weight percentage. Numbers in parantheses represent standard errors (SE = $\sigma/\sqrt{n}$; n = 28)

Saturates		Monounsaturates		Polyunsaturates	
14:0	3.56 (0.12)	14:1	0.54 (0.07)		
15:0	1.09 (0.04)	15:1	0.64 (0.04)		
16:0	20.4 (0.84)	16:1 *n*-9	1.33 (0.05)	16:2	0.71 (0.04)
		16:1 *n*-7	8.93 (0.31)	16:3	1.05 (0.04)
17:0	0.93 (0.05)				
18:0	4.99 (0.20)	18:1 *n*-9	26.0 (0.96)	18:2 *n*-6	1.23 (0.04)
		18:1 *n*-7	6.67 (0.21)	18:2 *n*-3	0.45 (0.03)
				18:3 *n*-6	0.56 (0.03)
				18:3 *n*-3	0.19 (0.01)
				18:4 *n*-3	0.82 (0.03)
20:0	0.49 (0.04)	20:1 *n*-9	2.44 (0.09)	20:3 *n*-6	0.99 (0.05)
		20:1 *n*-7	1.18 (0.06)	20:3 *n*-3	0.31 (0.02)
				20:4 *n*-6	2.06 (0.15)
				20:4 *n*-3	0.71 (0.09)
				20:5 *n*-3	3.45 (0.53)
		22:1 *n*-11	0.95 (0.08)	22:5 *n*-6	0.52 (0.05)
		22:1 *n*-9	0.97 (0.09)	22:5 *n*-3	0.53 (0.07)
				22:6 *n*-3	4.39 (0.62)
		24:1	0.86 (0.06)		
Total	31.5 (0.84)	Total	50.5 (1.26)	Total	18.0 (1.22)

were not suggested for variations in total lipid contents of deep-sea shrimp reported previously (Culkin & Morris, 1969; Herring, 1973), however, variations in the total lipid content of *Chorismus antarcticus*, (Pfeffer) (Clarke, 1979), *Pandalas montagui* (Leach) (Clarke, 1977) and *P. borealis* (Hopkins et al., 1993) were seasonal and related to reproductive state. The samples studied here were collected about 2 months after the south-west monsoon period, during which extensive upwelling and consequently primary production occurs (Nair et al., 1989). It might, therefore, be expected that the lipid profiles of *N. gracilis* would reflect the resulting input of organic matter in some manner. An increase in the level of total stored lipid might be expected, but seasonal studies of this species are necessary to reveal whether or not this is the case.

The major storage lipid in *N. gracilis* was triglyceride, which is typical of a crustacean with a benthic life style (Sargent, 1976). Triglyceride levels were in the range reported for *P. borealis*, a carnivorous-omnivorous feeder (Hopkins et al., 1993) and also similar to those reported for *A. markensis* (Allen, 1998) that is thought to be a scavenger and opportunistic feeder (Segonzac et al., 1993). Storage lipid levels may, therefore, reflect this kind of life style. Levels of triglyceride were higher in *N. gracilis* than in *R. exoculata* (Allen, 1998). Although wax esters are also present in some (smaller) *R. exoculata*, there are still lower levels of storage lipid in adult *R. exoculata* than in *N. gracilis*. This may be a reflection either of the difference between primary and secondary consumers or the seasonal variability in food supply for *N. gracilis* compared to the relatively constant supply of food at hydrothermal vents.

Free fatty acid levels suggested that the samples were in a relatively good state of preservation (Jeckel et al., 1989). Tissue concentrations of sterols were similar to those found in the deep-sea vent shrimp *R. exoculata* and *A. markensis*, but somewhat lower than those reported for *P. montagui* and *P. borealis*. This is probably a reflection of the depth at which *N. gracilis* exists (*P. montagui* is a shallow-water species and *P. borealis* was collected at depths of ~100 m) and the proportion of its diet that includes unaltered phytoplankton-derived material. Fatty acid composition also reflects a lower dietary input of phytoplankton-derived material and this feature is discussed in more depth later.

Total polar lipid tissue concentrations of *N. gracilis* were similar to those of *R. exoculata* and *A. markensis*. This concurs with polar lipids comprising mainly structural lipid. The five most abundant fatty acids in *N. gracilis* do not include the highly unsaturated fatty acids 20:5 *n*-3 and 22:6 *n*-3 considered characteristic of marine organisms, although these fatty acids are present at around 5% of total fatty acids in *N. gracilis* (Table 2). Levels of different fatty acid types in *N. gracilis* were generally similar to those reported by Culkin & Morris (1969) for deep-sea caridean shrimp (Fig. 2). Saturated fatty acids were present in *N. gracilis* at similar levels to those reported in other caridean shrimp (Fig. 2). Monounsaturated fatty acids comprised around 50% of the fatty acids of *N. gracilis* studied. The ratio of 18:1 *n*-7/*n*-9 fatty acids is low for all *N. gracilis*. This result suggests that anaerobic bacteria, or organisms that feed on anaerobes, do not form a significant part of the diet of *N. gracilis*. This is perhaps surprising considering the proximity of the shrimp to the OMZ and the presumable availability of anaerobic bacteria, but *N. gracilis* were not caught within the OMZ and maybe organic matter derived from anaerobes is not available in the areas where the shrimp were found. Levels of polyunsaturated fatty acids were lower than those detected in the shallow-water species, *P. montagui* (Clarke, 1979), but similar to those reported for *Pandalus borealis* (Hopkins et al., 1993) and deep-sea caridean shrimp (Culkin & Morris, 1969; Fig. 2).

The combination of lower levels of sterols and polyunsaturated fatty acids, neither of which can be synthesised *de novo* by shrimp (Goad, 1981; Teshima & Kanazawa, 1983), in this species of deep-sea shrimp is related probably to the availability of labile organic matter in the deep sea. Studies of fluxes of particulate organic matter in the deep sea have shown that labile organic material, including polyunsaturated fatty acids, is recycled relatively near the surface, in the form of zooplankton, eggs and other buoyant particles (Grimalt et al., 1990). Organic matter reaching the deep sea tends to be more refractory and depleted in highly unsaturated fatty acids (Reemstma et al., 1990). Even in anoxic regions of the ocean, particulate organic matter (POM) from deeper waters has been shown to be depleted in highly unsaturated fatty acids (Wakeham, 1995) compared with surface waters, although to a lesser extent than POM from deep oxic water. In addition to this, the reported lack of enhanced preservation of organic matter in sediments under the oxygen minimum zone in the Arabian Sea (Pedersen et al., 1993), supports the suggestion that labile organic matter is scarce in the region where *N. gracilis* is found.

Conclusions

Seasonal studies are needed to determine whether total lipid levels in *N. gracilis* reflect inputs of organic matter. Lipid class composition supports a benthic life style and conforms to that of an omnivore. Fatty acid composition indicates that *N. gracilis* derives little or no nutrition from anaerobic bacteria and has a similar diet to other deep-sea shrimp.

Acknowledgements

Thanks to David Pond for help with analyses and Pat Hargreaves for assistance with identifying *N. gracilis*. CEA was supported by a University of Southampton PhD studentship.

References

Allen, C. E., 1998. Lipid profiles of deep-sea organisms. University of Southampton, U.K. PhD Thesis.

Allen Copley, C. E., P. A. Tyler & M. S. Varney, 1998. Lipid profiles of hydrothermal vent shrimps. Cah. Biol. Mar. 39: 229–331.

Calman, W. T., 1911. The Life of Crustacea. Methuen & Co., London: 289 pp.

Cartes, J. E., 1993. The diets of two deep-sea decapods – *Nematocarcinus exilis* (Caridea: Nematocarcinidae) and *Munida tenuimana* (Anomura, Galatheidae) on the western Mediterranean Slope. Ophelia 37: 213–29.

Clarke, A., 1977. Seasonal variations in the total lipid content of *Chorismus antarcticus* (Pfeffer) (Crustacea: Decapoda) at South Georgia. J. exp. mar. Biol. Ecol. 27: 93–106.

Clarke, A., 1979. Lipid content and composition of the pink shrimp, *Pandalus montagui* (Leach) (Crustacea: Decapoda). J. exp. mar. Biol. Ecol. 38: 1–17.

Creasey, S., A. D. Rogers, P. A. Tyler, C. M. Young & J. Gage, 1997. The population biology and genetics of the deep-sea spider crab, *Encephaloides armstrongi* Wood-Mason 1891 (Decapoda:majidae). Phil. Trans. r. Soc., Lond. B – Biological Sciences 352: 365–79.

Culkin, F. & R. J. Morris, 1969. The fatty acids of some marine crustaceans. Deep Sea Res. 16: 109–116.

Currie, R. I., A. E. Fisher & P. M. Hargreaves, 1973. Arabian Sea upwelling. In Zeitschel, B. (ed.), The Biology of the Indian Ocean. Chapman & Hall Ltd., London: 37–52.

Dunstan, G. A., J. K. Volkman & S. M. Barrett, 1993. The effect of lyophilisation on the solvent extraction of lipid classes, fatty acids and sterols from the oyster *Crassostrea gigas*. Lipids 28: 937–944.

Folch, J., M. Lees & G. H. Sloane-Stanley, 1957. A simple method for the isolation and purification of total lipid from animal tissue. J. biol. Chem. 226: 497–509.

Fraser, A. J., D. R. Tocher & J. R. Sargent, 1985. Thin-layer chromatography-flame ionisation detection and the quantification of marine neutral lipids and phospholipids. J. exp. mar. Biol. Ecol 88: 91–99.

Gage, J. D., B. J. Bett, A. C. Campbell, J. Crooks, N. J. Debenham, A. Gooday, D. Hoover, L. Levin, A. Meadows, P. S. Meadows, J. Murray, G. Oliver, A. J. Patience, A. Rogers, A. Scheltema, G. Squire, D. Smallman, P. A. Tyler, G. A. Wolff, C. M. Young, 1995. Benthic community and fluxes in relation to the oxygen minimum zone in the Arabian Sea. Discovery 211/94 Cruise Report, 9th October–11th November 1994.

Goad, L. J., 1981. The steroids of marine algae and invertebrate animals. In Malins, D. C. & J. R. Sargent (eds), Biochemical and Biophysical Perspectives in Marine Biology 3: 213–318.

Grimalt, J. O., B. R. T. Simoneit, J. I. Gómez-Belinchón, K. Fischer & J. Dymond, 1990. Ascending and descending fluxes of lipid compounds in North Atlantic and North Pacific abyssal waters. Nature 345: 147–150.

Herring, P. J., 1973. Depth distribution of the carotenoid pigments and lipids of some oceanic animals. 2. Decapod crustaceans. J. mar. biol. Ass. U.K. 53: 539–62.

Herring, P. J., 1994. Cruise Report of the Institute of Oceanographic Science Deacon Laboratory 1994, 244: 55 pp.

Hopkins, C. C. E., J. R. Sargent & E. M. Nilssen, 1993. Total lipid content and lipid and fatty acid composition of the deep-water prawn, *Pandalus borealis* from Balsfjord, northern Norway: growth and feeding relationships. Mar. Ecol. Prog. Ser. 96: 217–28.

Jeckel, W. H., J. E. A. De Moreno & V. J. Moreno, 1989. Biochemical composition, lipid classes and fatty acids in the ovary of the shrimp, *Pleoticus muelleri* Bate. Comp. Biochem. Physiol. B, 92: 271–6.

Nair, R. R., V. Ittekkot, S. J. Manganini, V. Ramaswamy, B. Haake, E. T. Degens, B. N. Desai & S. Honjo, 1989. Increased particle flux to the deep ocean related to monsoons. Nature 338: 749–51.

Olsen, D. B., G. L. Hitchcock, R. S. Fine & B. A. Warren, 1993. Maintenance of the low oxygen layer in the central Arabian Sea. Deep Sea Res. 40: 673–85.

Pedersen, T. F., G. B. Shimmield & N. S. Price, 1993. Lack of enhanced preservation of organic matter in sediments under the oxygen minimum on the Oman margin. Geochim. Cosmochim. Acta. 56: 545–51.

Pond, D. W., J. R. Sargent, A. E. Fallick, C. E. Allen, M. V. Bell & D. R. Dixon, 2000. $\delta^{13}C$ values of lipids from phototrophic zone microplankton and bathypelagic shrimps at the Azores sector of the Mid-Atlantic Ridge. Deep-Sea Res. 47: 121–136

Reemstma, T., B. Haake, I. Ittekkot, R. R. Nair & U. Brockmann, 1990. Downward flux of particulate fatty acids in the central Arabian sea. Mar. Chem. 29: 183–202.

Sanders, N. K.& J. J. Childress, 1990. Adaptations to the oxygen minimum layer: oxygen binding by the haemocyanin of the bathypelagic mysid, *Gnathophausia ingens* Dohrn. Biol. Bull. 178: 286–94.

Sargent, J. R., 1976. The structure, metabolism and function of lipids in marine organisms. In Malins, D. C. & J. R. Sargent (eds), Biochemical and Biophysical Perspectives in Marine Biology. 3: 149–212.

Segonzac, M., M. De Saint Laurent & B. Casanova, 1993. L'enigme du comportement trophique des crevettes Alvinocarididae des site hydothermaux de la dorsale medio-atlantique. Cah. Biol. Mar. 34: 535–71.

Shantha, N. C. & R. G. Ackman, 1990. Advantages of total lipid hydrogenation prior to lipid class determination on Chromarods SIII. Lipids 25: 570–4.

Somero, G. N., 1992. Biochemical ecology of deep-sea animals. Experientia 48: 537–43.

Swallow, J. C., 1984. Some aspects of the physical oceanography of the Indian Ocean. Deep Sea Res. 31: 639–50.

Teshima, S.-I. & A. Kanazawa, 1983. Variation in lipid compositions during the ovarian maturation of the prawn. *Nippon Suisan Gakkaishi* 49: 957–962.

Wakeham, S. G., 1995. Lipid biomarkers for heterotrophic alteration of suspended particulate organic matter in oxygenated and anoxic water columns of the ocean. Deep Sea Res. 42: 1749–1771.

Hydrobiologia **440:** 281–296, 2000.
M.B. Jones, J.M.N. Azevedo, A.I. Neto, A.C. Costa & A.M. Frias Martins (eds), Island, Ocean and Deep-Sea Biology.

Community structure of deep-sea demersal fish in the North Aegean Sea (northeastern Mediterranean)

M. Labropoulou & C. Papaconstantinou
National Centre for Marine Research, Agios Kosmas, Hellinikon, 16 604 Athens, Greece
E-mail: mlabro@ncmr.gr

Key words: demersal fish, depth zonation, species diversity, habitat width, Eastern Mediterranean

Abstract

The spatial structure and seasonal changes of the demersal fish assemblages on the continental shelf (100–200 m) and upper slope (200–500 m) in the North Aegean Sea (Northern Aegean and Thracian Seas, northeastern Mediterranean, Greece) were analysed. Seasonal experimental trawl surveys, carried out from summer 1990 to autumn 1993, provided a total of 151 demersal fish species. Analysis of 259 bottom trawls showed the existence of four groups associated with the continental shelf and the upper slope; each group was dominated by a small number of species. The bathymetric distribution of the species, established using measures of the centre of gravity and habitat width, revealed that most of the species had a wide distributional range within the study area, although a few were restricted to the greatest depths. Density, biomass, species richness and diversity decreased significantly with depth, and were also indicative of distinctive characteristics between these fish assemblages. Mean fish weight exhibited two different trends: a bigger-deeper phenomenon at the continental shelf and a smaller-deeper phenomenon at the upper slope. The variability in assemblage structure was determined mainly by depth and, to a lesser extent, by season and geographical location. For some species, results suggest a pattern of gradual species replacement along the depth gradient coupled with ontogenetic habitat shifts.

Introduction

Information of soft-bottom fish assemblages in the eastern Mediterranean, where demersal fish are heavily exploited as principal targets or by-catch components, is particularly scarce. Demersal fishes of the North Aegean continental shelf and slope are subjected to intensive fishery carried out by trawl, gillnet and longline fleets. The gillnets and longlines catch a small number of species, whereas the trawl fleet exploits a multi-species fishery targeting several demersal and benthic species. Results of experimental trawl fishing in the North Aegean indicated that all the commercially important demersal and inshore stocks suffer from over fishing. As a result, commercial catches consist mainly of young immature individuals and a variety of non-commercial species that are discarded (Stergiou et al., 1997).

Reduction of catch rates and mean size of individuals is a well documented trend in world fisheries (Pitcher, 1996). Consequently, new approaches to the study of exploited populations have been suggested, including the study of the fish assemblage structure in relation to environmental variables, and the characterization of seasonal changes to improve management practices. As pointed out by Caddy & Sharp (1988), this type of study is a necessary step toward understanding multispecies stocks. Such work can then be extended to descriptive community dynamics to find which general patterns of species compositions can be expected under given environmental conditions and fishing effort.

Changes in species composition with depth on continental shelf and slope have been well established for many areas (e.g. Haedrich et al., 1980; Carney et al., 1983; Abelló et al., 1988; Hecker, 1990; Cartes & Sardá, 1993; Koslow, 1993; Smale et al., 1993; Cartes et al., 1994; Sardá et al., 1994; Gordon et al., 1995; Fariña et al., 1997a). Both physical and biological factors have been discussed as causes responsible for faunal zonation with depth. Hydrographic conditions, the steepness of the continental slope and substratum type are amongst the major physical factors considered. Resource availability, predator–prey relationships and interspecific competition are the most important biological factors reported. However, most studies have examined megafaunal assemblages, while little attention has been paid to the structure of the highly exploited demersal fish assemblages, apart from the degree to which different fish species are zoned with depth. Deep-water fish assemblages tend to be more diverse and less dominated by a single species than shelf assemblages, and thus impacts on any one particular species are very likely to be reflected in the responses of others as well as the community as a whole (Merrett & Haedrich, 1997). Only recently, studies of fish community structure have focused on the patterns of spatial and temporal variation on composition, abundance and distribution of demersal fish assemblages of the continental shelf and slope at several latitudes (Markle et al., 1988; Bianchi, 1991, 1992a, 1992b; Fujita et al., 1995; Fariña et al., 1997b; García et al., 1998; Moranta et al., 1998).

In the present study, the trawl catch data obtained seasonally in 3 successive years (1990–1993) from the continental shelf and the upper slope of the north Aegean Sea were analysed to explore the structure of the demersal fish community. The main objectives were to define the faunal composition, the spatial structure and the main faunistic assemblages in the area, and to investigate whether there are any general trends in the distribution of the fish fauna in relation to environmental factors (mainly depth), season and geographical location.

Material and methods

Study area and sampling procedure

The North Aegean Sea is one of the least studied regions of the northern part of the Mediterranean. It is an almost rectangular basin, separated from the South Aegean by the archipelago of the Cyclades Islands. The bottom topography of the North Aegean is characterized by a series of deep trenches and troughs (with depths reaching 1600 m), separated by shallow sills and shelves. The northern part of the sea, the Thracian Sea, lies over an extended shelf and is separated on its west from the Thermaikos Gulf by the peninsula of Chalkidiki. The North Aegean trough, having a SW to NE direction separates the Thracian Sea from the rest of the North Aegean (hereafter: Northern Aegean Sea) and is characterized by a highly irregular coastline, the presence of many islands and a very narrow continental shelf (Fig. 1). The North Aegean basin is the region where the eastern Mediterranean receives the Black Sea outflow. Black Sea water is the major source of fresh/brackish water for the region, as the contribution of all the rivers flowing into the Aegean is smaller than the input of the Black Sea by at least one order of magnitude (Poulos et al., 1997). The North Aegean Sea is also influenced by nutrient inputs from outflowing Black Sea waters as well as by freshwater runoff along its northern rim. Therefore, it is characterized by higher nutrient and plant pigment concentrations, in relation to the rest of eastern Mediterranean, and is one of the most oligotrophic marine regions of the world (Stergiou & Pollard, 1994). Environmental parameters (temperature, salinity) are fairly constant along the slope below 200 m (13.5–14 ° C and 38.0–39.0 ‰, respectively) (D'Onghia et al., 1995), while muddy sediments generally predominate on the shelf and slope from 100 to 1000 m depth (Lykousis & Collins, 1987).

A total of 259 hauls was taken during sixteen experimental bottom trawl survey cruises on a seasonal basis (summer 1990–spring 1992 in the northern Aegean Sea, autumn 1991–winter 1993 in the Thracian Sea) from standard depth stations between 100-500 m (Fig. 1). Sampling stations were selected on a depth-stratified random decision and the otter trawl used (foot-rope length: 65.7 m, headline height: 1.5 m) was equipped with a cod-end bag liner of 16 mm stretched mesh size. Samples were collected during daylight between 0800 and 1700 h. The duration of each trawl (bottom time) was 30–60 min, and the trawling speed fluctuated from 2.5 to 3.0 knots depending on the depth and the nature of the substratum. The catch from each haul was sorted and identified to species; each species was enumerated and weighed individually on board. Part of the catch was frozen or preserved in formalin for later study in the laboratory. Since all hauls were carried out using the commercial trawl vessel 'Ioannis Rossos' and the same fishing gear it was assumed that gear selectivity was constant. Those species regarded as markedly pelagic or semi-pelagic in behaviour were disregarded from the analyses since they had not been sampled quantitatively.

Data analysis

Species abundance and biomass were calculated for each haul after standardization of the data to a 1 h

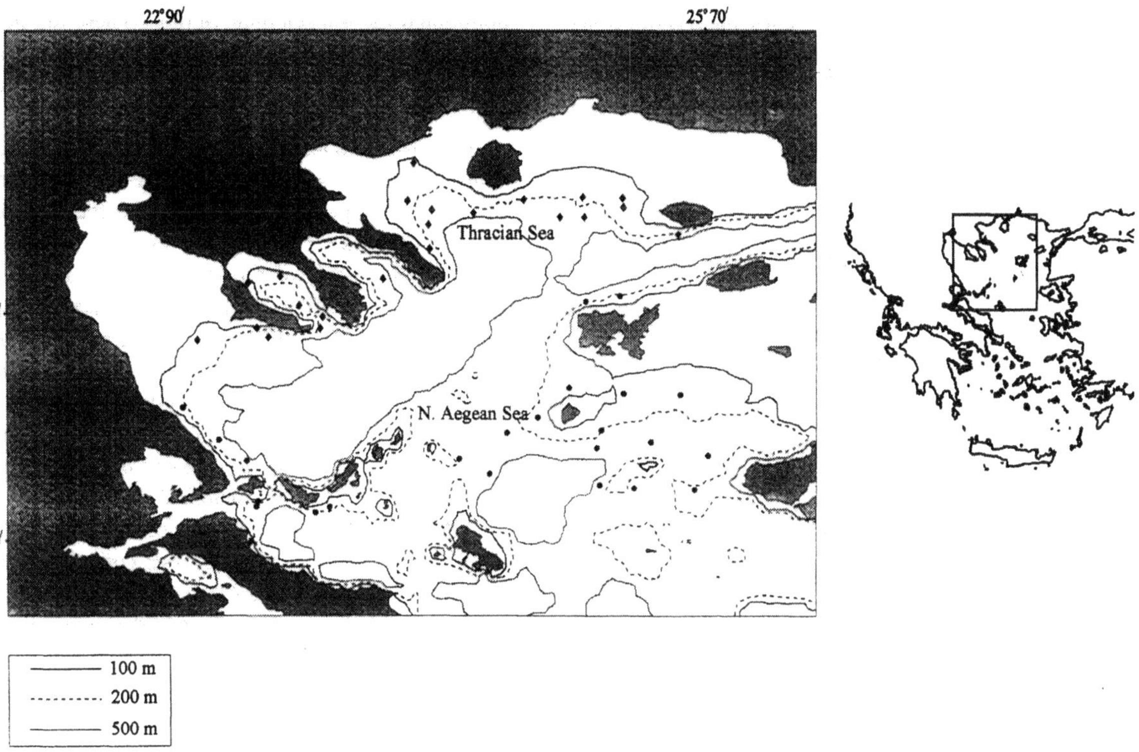

Figure 1. Sampling stations on the continental shelf and the upper slope of the North Aegean Sea. (♦) Thracian Sea, (•) northern Aegean Sea.

tow, making it possible to allow comparisons between sampling stations. To identify zonation patterns, principal components analysis and cluster analysis were applied to the species abundance matrix. Principal component analysis was performed on the 259 sample correlation matrix of 151 species (excluding pelagic and semi-pelagic species). Before the analysis, abundance values were log $(x+1)$ transformed. Cluster analysis (group average) employed the Bray-Curtis similarity index (Field et al., 1982) and was performed on the standardized abundance values of the species using the PRIMER algorithms (Plymouth Marine Laboratory). To normalize the data and avoid skew, a square root transformation was applied to the abundance data prior to cluster analysis (Field et al., 1982). Multidimensional scaling (MDS) ordination analysis was also performed with the same configuration as in cluster analysis with respect to similarity index and transformation. The typifying and discriminating species of the cluster of stations were determined using the SIMPER procedure (Clarke, 1993). This procedure indicates the average contribution of each species to the similarity (typifying species) and dissimilarity (discriminating species) between groups of samples. The bathymetric distribution of the dominant species, according to their abundance rank in each station group identified by cluster analysis, was calculated in a quantitative way using the 'centre of gravity' (COG) (Daget, 1975; Stefanescu et al., 1992a; Moranta et al., 1998) and 'habitat width' (HW) (Pielou, 1969) analyses. The COG model allows the calculation and location with precision of the centre of species distribution by means of a descriptor (such as depth), while the HW model gives a measure of heterogeneity of the species distribution.

The ecological parameters of abundance, biomass, mean fish weight, number of species (S), species diversity, Shannon-Wiener index ($H' = \sum_{i=1}^{n} pi \ln pi$) (Hurlbert, 1978), species richness ($d = \frac{(S-1)}{\log(N)}$ (Margalef, 1968), and evenness ($J' = \frac{H'}{\log(S)}$ (Pielou, 1966) were calculated for each of four 100 m depth intervals, where p_i = proportion of total sample belonging to ith species and N = the number of individuals. Furthermore, regression analyses were used to determine how these attributes, as well as the scores of principal components analysis, changed with water depth. Before using parametric tests, the assumptions of normality and homoscedasticity were tested, and when these assumptions were not met, the data were log $(x+1)$ transformed (Sokal & Rohlf, 1981; Zar, 1984). All statistical inferences were based on the 0.05 significance level.

Results

Overall catch

A total of 151 demersal fish species belonging to 61 families was collected in the study area (see Appendix 1). Of these, 106 species were found at depths between 100 and 200 m, and 94 in the bathymetric range of 200–500 m. Twenty species dominated depths between 100–200 m (78.1% of the overall catches), while 11 dominated the depths below 200 m (82.7%). The remaining species made up a small and inconstant proportion of the catches.

Depth patterns in species composition

Principal components analysis for the pooled data revealed that some gradual changes in species composition occurred along the depth gradient. Many sites were grouped together but, in general, the shallowest (<150 m depth) and the deepest (>400 m depth) stations were set apart in all years by the 1st and the 2nd axis, respectively (Fig. 2). The first 2 principal components, accounting for 48.9% of the total variance, were significantly correlated with depth in linear and polynomial regression, respectively. The score of the 1st axis was greater in deeper water and remarkably well correlated with depth in linear regression. The 1st axis increased with increasing depth because the minimum value was observed for the shallowest depths (Fig. 3a). The score of the 2nd axis was maximum in the intermediate depths around 300 m, indicated by much better fit of the polynomial regression for the pooled data (Fig. 3b). High positive load on the 1st component was found for the deeper-water species (i.e. *Hymenocephalus italicus*, *Nezumia sclerorhynchus*, *Coelorhynchus coelorhynchus* and *Gadiculus argenteus argenteus*) and high negative load for the shallower-water species such as *Serranus hepatus*, *Mullus barbatus*, *Mullus surmuletus* and *Arnoglossus laterna*. High positive load on the 2nd component was observed for the species *Gadiculus argenteus argenteus*, *Argentina sphyraena*, *Lepidorhombus boscii* and *Micromesistius poutassou poutassou* that dominated the intermediate depths. There did not appear to be any significant pattern of seasonal variation in composition of the catches for any of the station groups resulting from this analysis.

Classification and ordination of the 259 trawl catch data in terms of species abundance revealed the existence of four groups associated with the continental shelf and the upper slope (Fig. 4a, b). Groups 1 and 4 consisted entirely of those samples taken from the continental shelf (100–200 m) in the Northern Aegean and Thracian Sea, respectively, while deeper stations for each area were subsequently classified in Groups 2 and 3. However, a relatively small number of species contributed most to the similarity of each group, but their relative abundances varied between adjacent groups (i.e. Groups 1–4, Groups 2–3) (Table 1).

The bathymetric distribution of the species established using the measures of the centre of gravity (COG) and habitat width (HW) revealed that for both areas most of the species presented a wide distribution range within the study area, although a few were restricted to the greatest depths (Fig. 5a, b). From a total of 26 demersal species analysed, only six were restricted to the depth intervals studied. There were three different species groups: (a) species limited to the 100–200 m depth interval (i.e. *Serranus hepatus*, *Mullus barbatus*, *Mullus surmuletus* and *Arnoglossus laterna*), (b) species with a wide bathymetric distribution (i.e. *Merluccius merluccius*, *Micromesistius poutassou poutassou*, *Lophius budegassa*, *Scyliorhinus canicula*), and (c) species restricted to depths greater than 200 m (i.e. *Hymenocephalus italicus*, *Nezumia sclerorhynchus*, *Coelorhynchus coelorhynchus*, *Galeus melastomus*, *Phycis blennoides* and *Molva dipterygia macrophthalma*).

Depth patterns in abundance and species diversity

Significant differences in mean species abundance, biomass and diversity indices existed between the four depth zones (Table 2). The highest values of these parameters were found in samples from the continental shelf (100–200 m) and decreased significantly for those from the upper slope. Significant negative regressions with depth were also found for species abundance, biomass and diversity indices for the pooled data from the 3 years of the study (Figs 6 and 7). However, species abundance appeared to be more or less uniform at depths between 200–400 m, while species biomass decreased sharply with depth. Mean fish weight exhibited two different trends within the

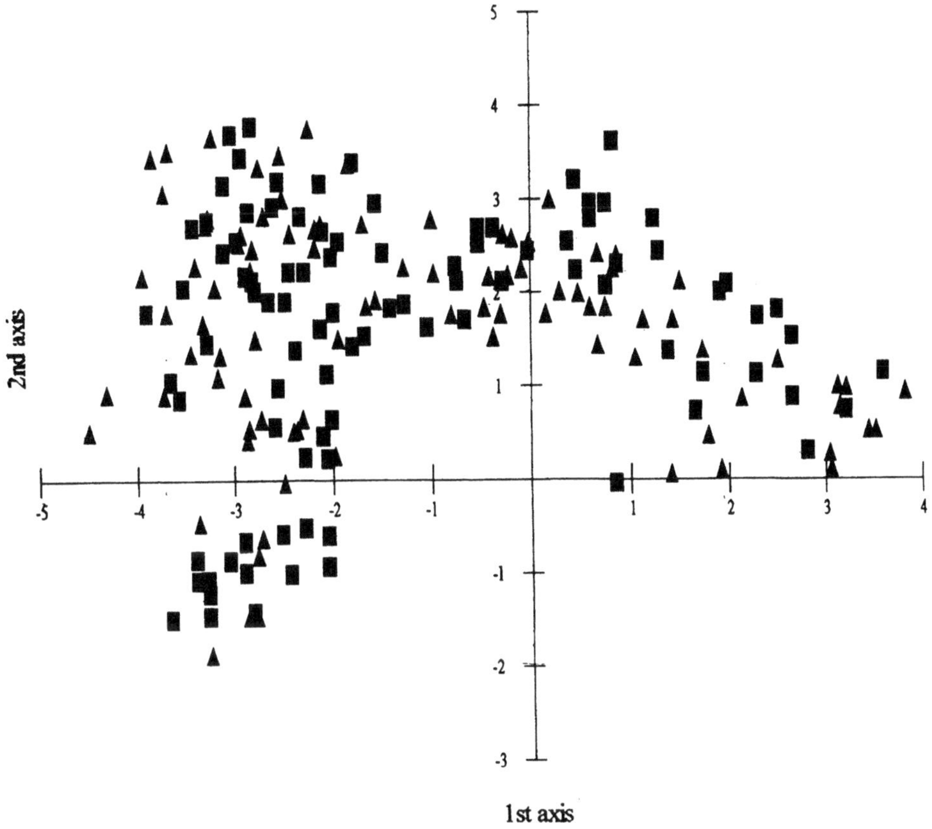

Figure 2. Principal component analysis of 259 samples based on demersal fish species abundance. (▲) Thracian Sea, (■) Northern Aegean Sea.

studied depth-range. A steady increase was observed from 100 to 200 m, while the opposite trend was noted from 200 m down to the maximum depth sampled (Fig. 8).

Discussion

Two distinctive demersal fish assemblages clearly associated with the topography of the study area have emerged from the analysis: a shallow assemblage reaching the 200 m depth limit which represents the continental shelf, and a deep assemblage beyond that depth which represents the upper slope. The first assemblage exhibits greater abundance and biomass, and contains species of larger size and commercial interest (such as *Merluccius merluccius*, *Mullus barbatus*, *Mullus surmuletus* and *Trisopterus minutus capelanus*). The second assemblage is characterized by the presence of species such as *Gadiculus argenteus*, *Hymenocephalus italicus*, *Coelorhynchus coelorhynchus* and *Nezumia sclerorhynchus*, which are small and not commercially important. The main determining feature associated with the structure of the demersal fish assemblages is depth, as it reflects the changes from the continental shelf to the continental slope. However, other bottom and oceanographic characteristics do play a role in structuring the assemblages on the continental shelf between the two areas. The main factors which possibly contribute to such a geographical differentiation are the gradient in eutrophy, fresh/brackish runoff, temperature and salinity of Aegean Sea waters along a NNW to SSE axis, and the differences in the extent of the continental shelf within these areas of the Aegean Sea. Nevertheless, the most important quantitative boundary for both areas was located around 200 m which separated the species of the continental shelf from those of the upper slope. At this depth, a remarkable decline of species richness, abundance and biomass was noted. In contrast, certain factors of the benthic environment were remarkably stable with depth below 200 m. Sediment structure consisted of a dominant silt-clay fraction

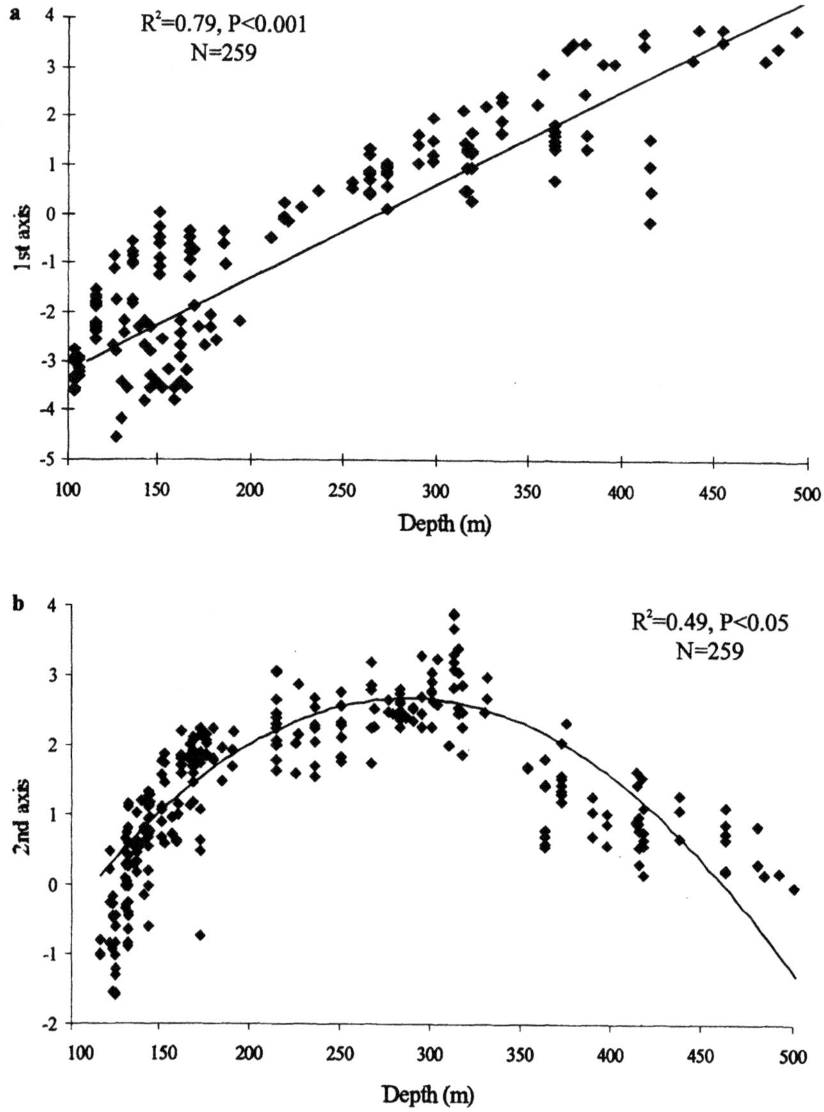

Figure 3. Relationships between the 1st (a) and the 2 nd (b) axes of the principal component analysis and depth.

throughout and temperature above the sediment was fairly constant. Perhaps this is the major reason for the high degree of similarity observed in the faunistic composition of the upper slope for both the Northern Aegean and the Thracian Sea.

Changes in composition of the demersal fish species along the depth gradient were reflected by changes in both abundance and occurrence, even though the species examined were found over relatively broad depth ranges in the study area. The faunistic differences observed were quantitative, at least with regard to the distribution of the characteristic species in the sampled stations. Not only did the abundance of the species examined varied significantly with depth but, when the depth ranges of species' occurrence overlap, they appear to have different depths where they reach their maximum abundance. Hecker (1990) suggested that changes in species composition between different megafaunal assemblages were due to the substitution of the dominant and subdominant species throughout the depth gradient by a continuous faunistic turnover. A somewhat similar distribution pattern, with species showing overlapping depth range but with different depths of maximum density, has been also demonstrated for deep-water crustacean spe-

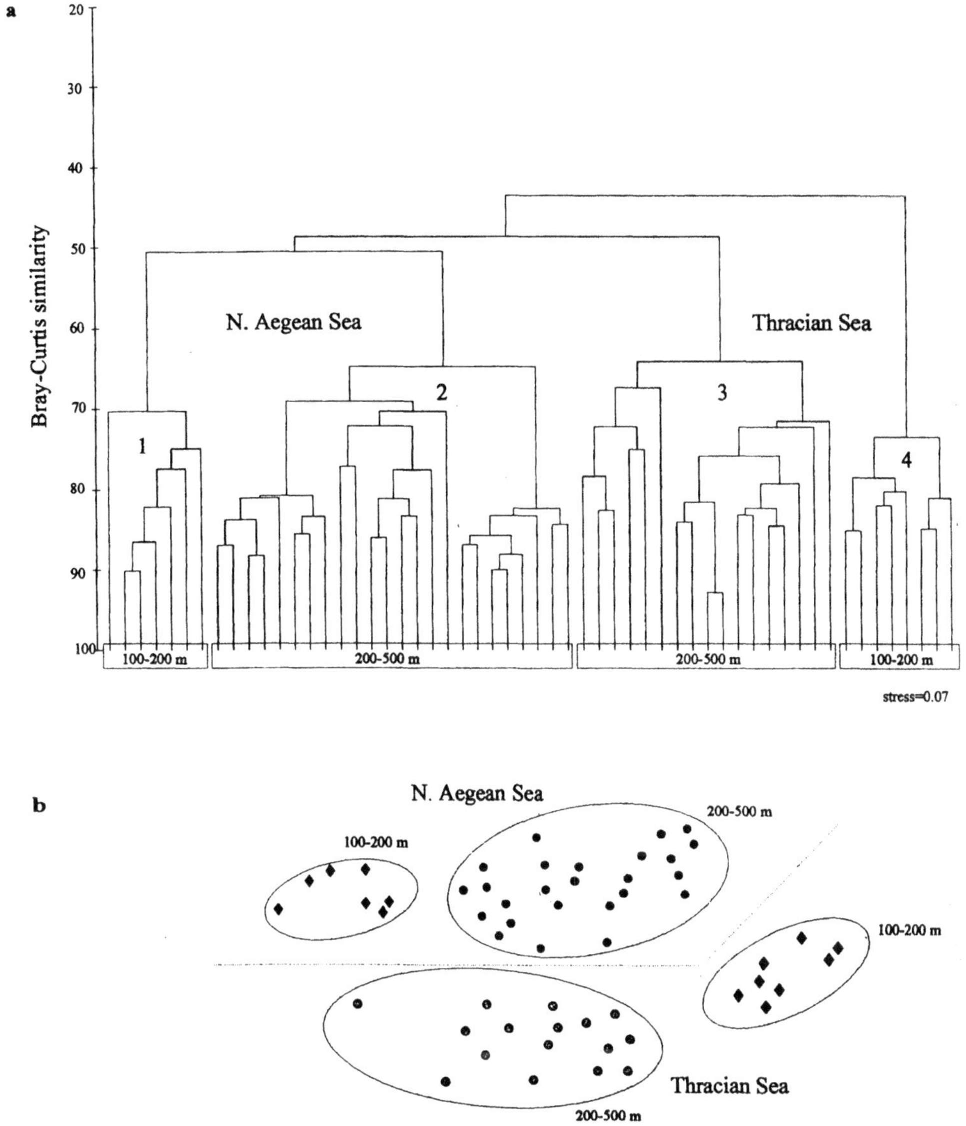

Figure 4. Classification (a) and ordination (b) of the sampling stations based on the species abundance from the North Aegean Sea.

cies in the Atlantic (Wenner & Boesch, 1979) and in the western Mediterranean (Abelló & Cartes, 1992).

Depth zonation and community classification studies should take temporal persistence into account, although until now most studies of the zonation of demersal fish do not consider temporal variation (Markle et al., 1988; Bianchi, 1991, 1992a, 1992b; Moranta et al., 1998). Multivariate analyses did not detect any temporal change in the overall structure of the assemblages between the seasonal cruises. These results suggest that changes in the assemblage composition are weakly associated with seasonal changes at least for our cruise dates. This means that the communities persisted over the three years of the study and that the species ranking also remained constant. Nonetheless, it should be noted that to track temporal persistence of deep-water populations, the length of the time series available must be considered. Comes et al. (1995) identified population shifts during the period 1978 – 1991 in many demersal species on the northeast Newfoundland shelf. Stock size was declining steadily for many years before changes could be detected in the standard survey catches.

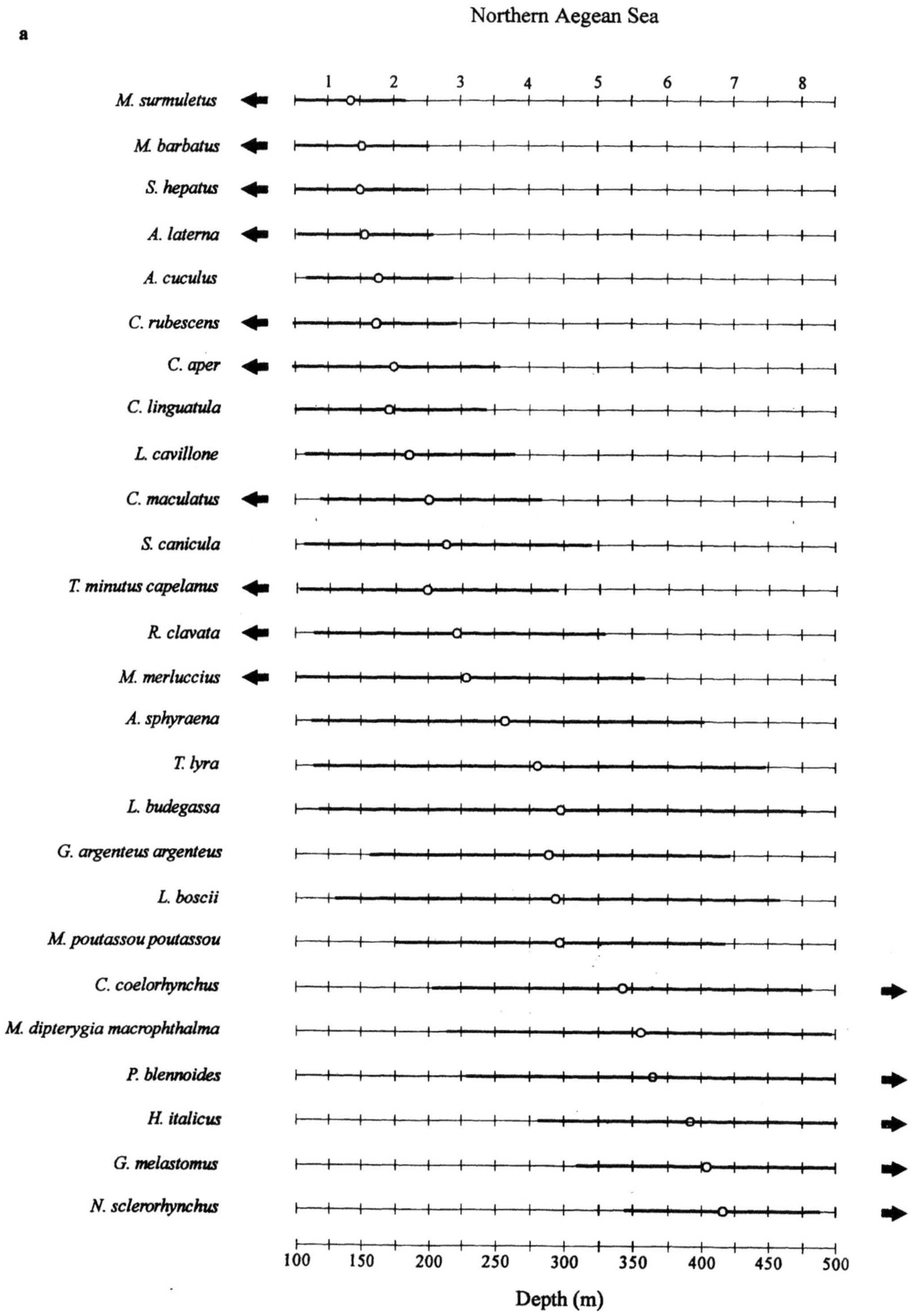

Figure 5a. Bathymetric distribution of the dominant fish species from the Northern Aegean Sea. Black circles represent the centre of gravity (COG), thick black lines correspond to the habitat width (HW). Black arrows indicate a displacement in real terms of COG, but included in the depth range considered. Numbers 1–8 on the top axis correspond to the 8 sectors into which the sampled depth interval was divided.

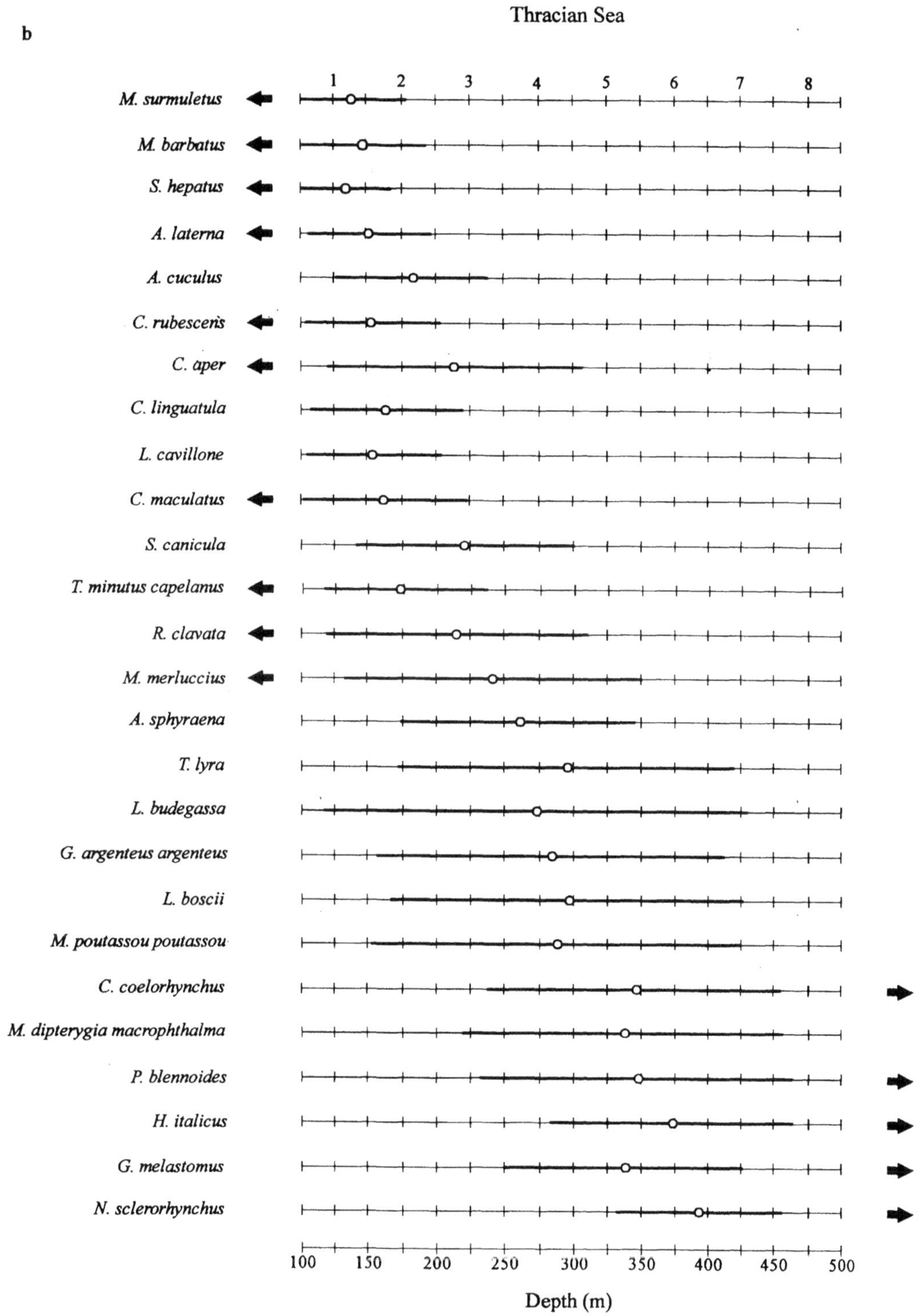

Figure 5b. Bathymetric distribution of the dominant fish species from the Thracian Sea within the depth range studied. Black circles represent the centre of gravity (COG), thick black lines correspond to the habitat width (HW). Black arrows indicate a displacement in real terms of the COG, but included in the depth range considered. Numbers 1–8 on the top axis correspond to the 8 sectors into which the sampled depth interval was divided.

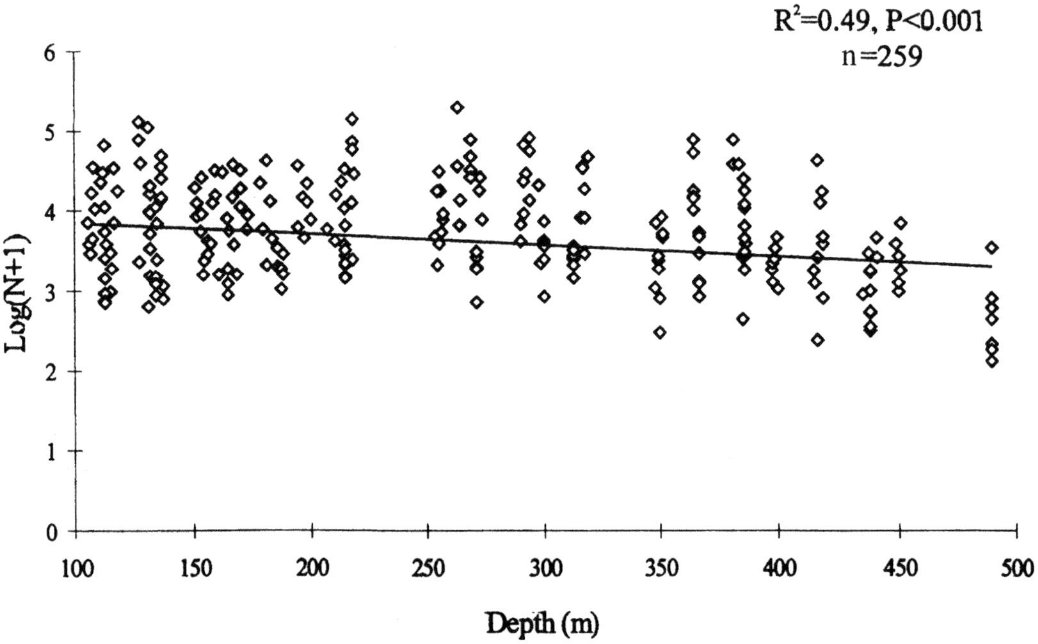

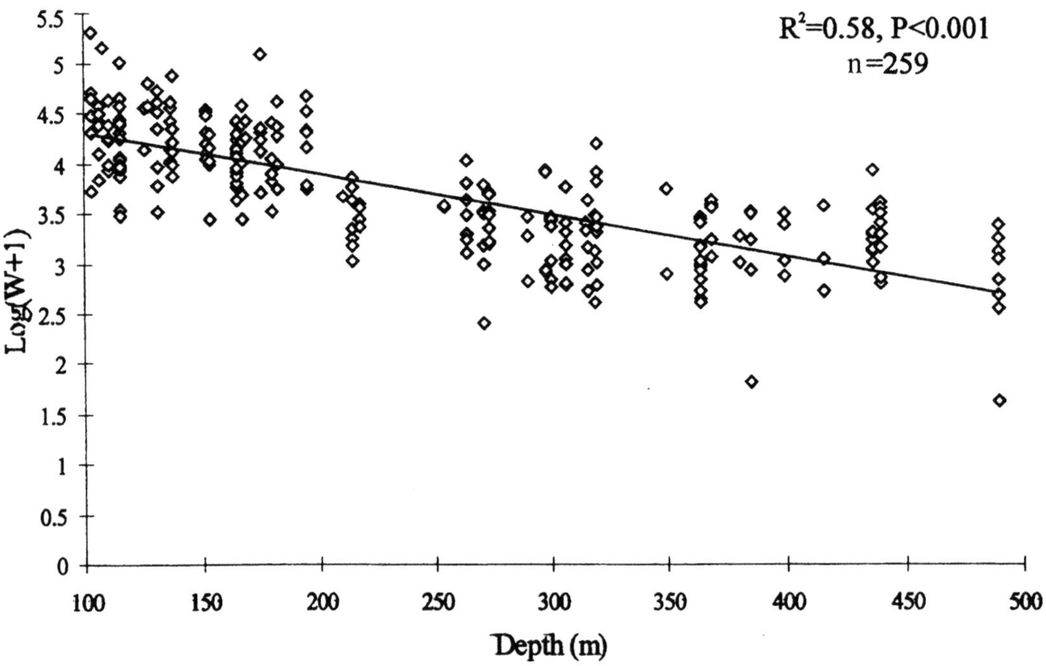

Figure 6. Relationship between species abundance (N) and biomass (W) with depth, calculated by regression analysis for the demersal fish from the North Aegean Sea.

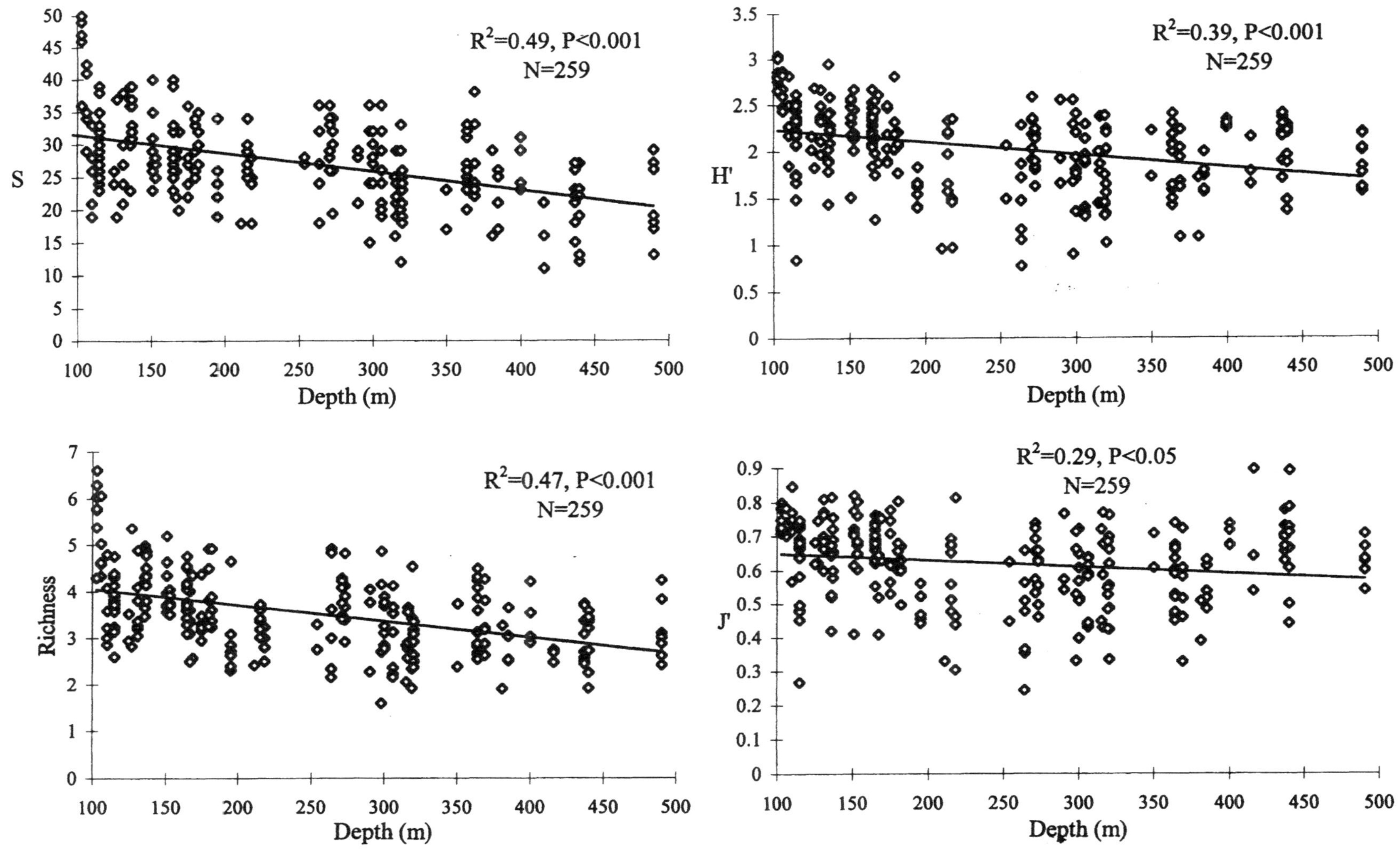

Figure 7. Depth-gradient patterns of community characteristics for the demersal fish species from the North Aegean Sea. S: number of species, species richness, species diversity (H′) and evenness (J′).

Table 1. Dominant fish species based on abundance rank for each station group identified by cluster analysis. Densities (%N) are averaged over all samples in each group

Northern Aegean Sea	%N	Cum.%	Thracian Sea	%N	Cum.%
100–200 m (Group 1)			**100–200 m (Group 4)**		
Argentina sphyraena	16.52	6.73	*Trisopterus minutus capelanus*	32.68	9.62
Merluccius merluccius	10.15	13.02	*Merluccius merluccius*	11.71	17.21
Lepidotrigla cavillone	6.59	18.94	*Argentina sphyraena*	2.49	22.84
Scyliorhinus canicula	6.31	24.70	*Serranus hepatus*	3.77	28.39
Mullus barbatus	8.44	30.32	*Lophius budegassa*	2.01	33.50
Capros aper	10.16	35.77	*Scyliorhinus canicula*	2.36	38.52
Serranus hepatus	3.37	40.35	*Lepidorhombus boscii*	1.75	43.49
Aspitrigla cuculus	2.31	44.87	*Gadiculus argenteus*	16.11	48.41
Trisopterus minutus capelanus	3.86	49.26	*Callionymus maculatus*	1.96	53.30
Trigla lyra	2.28	53.65	*Arnoglossus laterna*	2.11	58.11
Lepidorhombus boscii	3.06	58.01	*Lepidotrigla cavillone*	2.21	62.69
Citharus linguatula	2.18	62.35	*Cepola rubescens*	1.52	67.24
Mullus surmuletus	2.08	65.89	*Mullus barbatus*	2.41	71.36
Lophius budegassa	0.83	69.37	*Capros aper*	1.30	75.42
Raja clavata	0.54	72.46	*Phycis blennoides*	0.38	78.80
200–500 m (Group 2)			**200–500 m (Group 3)**		
Hymenocephalus italicus	20.61	13.64	*Gadiculus argenteus*	24.24	12.03
Coelorhynchus coelorhynchus	17.05	24.03	*Hymenocephalus italicus*	22.40	23.97
Nezumia sclerorhynchus	7.51	32.93	*Lepidorhombus boscii*	5.71	35.11
Lepidorhombus boscii	7.09	40.75	*Micromesistius poutassou*	21.32	46.03
Phycis blennoides	0.95	47.99	*Lophius budegassa*	1.29	53.64
Micromesistius poutassou	16.52	54.67	*Phycis blennoides*	1.65	61.12
Merluccius merluccius	3.50	61.12	*Coelorhynchus coelorhynchus*	2.95	68.53
Lophius budegassa	0.60	67.10	*Merluccius merluccius*	5.19	75.29
Galeus melastomus	0.38	72.91			
Molva dipterygia macrophthalma	0.51	77.72			

Table 2. Ecological parameters by depth zone and summary of statistical tests for the demersal fish community in the North Aegean Sea

Depth	Biomass[a] ($g\ h^{-1}$)	Abundance[a] ($fish\ h^{-1}$)	Mean fish weight (g)	Number of species (S)	Richness (d)	Diversity (H′)	Evenness (J′)
100–200 m	5.48 (± 0.33)	3.99 (± 0.86)	30.99 (± 4.81)	29.98 (± 1.92)	3.94 (±0.68)	2.23 (± 0.34)	0.66 (± 0.97)
200–300 m	5.01 (±0.57)	3.68 (± 0.11)	21.35 (± 3.38)	27.32 (± 1.74)	3.41 (±0.84)	2.07 (± 0.81)	0.62 (± 0.39)
300–400 m	4.73 (±0.46)	3.61 (± 0.96)	13.18 (± 6.20)	24.88 (± 2.09)	3.11 (±0.95)	1.89 (± 0.14)	0.56 (± 0.13)
400–500 m	3.67 (±0.13)	2.66 (± 0.21)[a]	10.69 (± 2.75)	21.47 (± 1.23)	3.01 (±0.16)	1.80 (± 0.48)	0.54 (± 0.27)
$F^b_{(3,255)}$	14.68	12.97	36.26	22.39	22.93	21.12	17.44

[a] Mean and SE were calculated using log $(x + 1)$ -transformed data.
[b] In all cases $P < 0.001$.

Overall in the Greek seas, a total of 447 fish species belonging to 129 families has been reported (Papaconstantinou, 1988). Thus, the number of species recorded in the present study represents 38.9% of the total marine ichthyofauna of Greece. However, species diversity reveals different trends along the entire range of the depths examined. The continental shelf was characterized by high species richness and diversity, but both parameters decreased remarkably with depth. On the other hand, although evenness was negatively

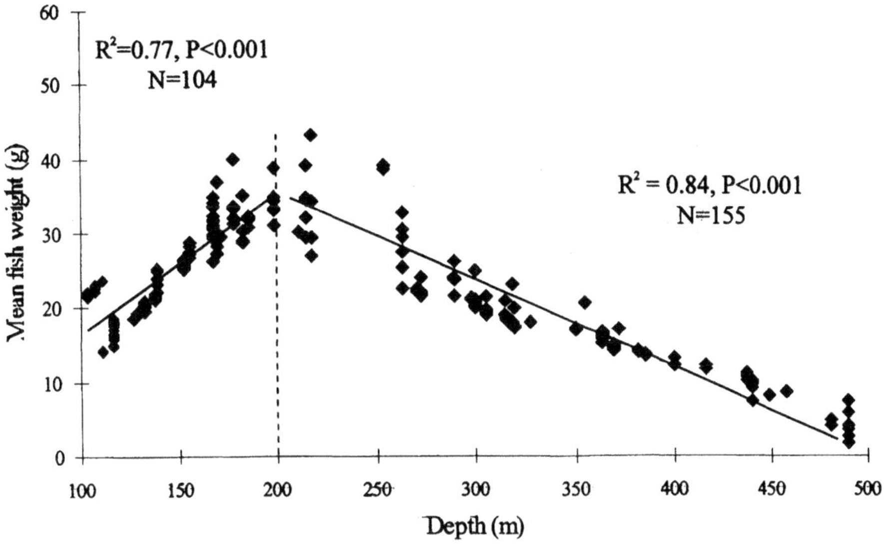

Figure 8. Relationships between mean fish weight and depth calculated by regression analysis for the demersal fish from less than and greater than 200 m depth in the North Aegean Sea.

correlated with depth, it appeared to be more or less uniform along the depth gradient, indicating a small variability in the numerical co-dominance of species over the depth range examined.

The bigger-deeper trend on the continental shelf (100–200 m) resulted primarily from the 'within-species' positive size-depth relationships observed for the dominant species (i.e. *Mullus barbatus*, *Mullus surmuletus, Merluccius merluccius, Trisopterus minutus capelanus*). These species inhabit shallow coastal areas and undertake ontogenetic migrations towards deeper waters as they grow larger (Macpherson & Duarte, 1991). The decrease in biomass below 200 m, coupled with the uniformity in abundance values, resulted in a smaller-deeper trend at these depths. Large-sized species were scarcer and were replaced by smaller ones such as Macrouridae which dominated the upper slope. Depth-size trends in deep-sea fish have been a controversial subject in recent years. A general tendency for larger fish to occur in deeper water has long been known with respect to deep-sea fishes (Haedrich & Rowe, 1977); however, it has also been reported for many demersal fishes of the continental shelf (Macpherson & Duarte, 1991). Contrary to these findings, Stefanescu et al. (1992b) observed a smaller-deeper trend for the demersal fish species in the Catalan Sea between 1000 and 2250 m depth, while Moranta et al. (1988) reported a bigger-deeper phenomenon for fish from the upper and middle slopes of the Balearic Islands and an opposite trend at depths below 1100 m. It appears that bathymetric trends in fish size are not a generalized phenomenon but these relationships are determined largely by factors such as the species under consideration, the study area and the depth ranges examined. Moreover, Merrett et al. (1991) concluded that an accurate indication of ichthyofaunal structure with increased soundings can only be achieved through a detailed knowledge of the size structure of each species over its entire sounding range synthesized from a multi-trawl investigation via analyses of species fidelity.

Despite the enforcement of management regulations for demersal and inshore fisheries in Greek waters (i.e. closed seasons and areas, limited issue of licenses, minimum legal landing size and mesh size regulations), the fishing effort in the North Aegean, as well as in the rest of the Aegean Sea, is very high (Stergiou & Pollard, 1994). Studies in other areas, based on extended time series during which major increases in fishing effort took place, indicated that this factor causes a decrease in the catches of exploited species and increases in other non-commercial species (Overholtz & Tyler, 1985; Rothschild, 1992). In the present study, the fish assemblages under consideration have suffered a long history of fishery exploitation. Therefore, overfishing has affected the population structure and density of the demersal fish communities, at least at depths up to 200 m, where most of the fishing activity is focused. According to Stergiou & Pollard (1994), who analysed the commercial catch weights of

10 fishing sub-areas of the Aegean Sea for the period 1982–1987, the mean catch of the northern group was dominated by grey mullet, hake and soles, representing 10.3%, 8.3% and 5.4% of the mean catch, respectively. However, the results from the present study indicate that most of the dominant species (according to their abundance rank in each station group) are non-commercial species. It is possible that the organization of the demersal fish assemblages analysed are determined to a great extent by an unidirectional trend induced by the fishery, the particular bottom topography and the oceanographic characteristics of the study area.

In summary, depth gradient, with its associated environmental and biological changes, is the main factor responsible for faunal changes in demersal fish community of the North Aegean Sea. Differences noted were more consistent with depth-related changes in community structure than with the geographical location of the study sites. Changes in demersal fish community were reflected by changes in both abundance and occurrence along the depth gradient, but the faunistic differences observed were more quantitative than qualitative, at least with regard to the distribution of the characteristic species in the stations sampled.

Acknowledgements

The authors would like to thank all those involved in demersal trawl surveys in the Northern Aegean and Thracian Seas, carried out by the Fisheries Department of the National Centre for Marine Research, for their valuable help. This work was partially supported by the EU, DG XIV under the Contract No MA-1-90, and the Greek General Secretariat for Research and Technology.

References

Abelló, P. & J. E. Cartes, 1992. Population characteristics of the deep-sea lobsters *Polycheles typhlops* and *Stereomastis sculpta* (Decapoda: Polychelidae) in a bathyal mud community of the Mediterranean Sea. Mar. Biol. 114: 109–117.

Abelló, P., F. J. Valladares & A. Castellón, 1988. Analysis of the structure of decapod crustacean assemblages off the Catalan coast (North-West Mediterranean). Mar. Biol. 98: 39–49.

Bianchi, G., 1991. Demersal assemblages of the continental shelf and slope edge between the gulf of Tehuantepec (Mexico) and the gulf of Papagayo (Costa Rica). Mar. Ecol. Prog. Ser. 73: 121–140.

Bianchi, G., 1992a. Study of the demersal assemblages of the continental shelf and upper slope off Congo and Gabon, based on the trawl surveys of the RV 'Dr Fridtjof Nansen'. Mar. Ecol. Prog. Ser. 85: 9–23.

Bianchi, G., 1992b. Demersal assemblages of the continental shelf and upper slope of Angola. Mar. Ecol. Prog. Ser. 85: 101–120.

Caddy, J. F. & G. D. Sharp, 1988. An ecological framework for marine fishery investigations. FAO Fish. Tech. Pap. 283:155 pp.

Carney, R. S., R. L. Haedrich & G. T. Rowe, 1983. Zonation of fauna in the deep sea. In Rowe, G. T. (ed.), Deep-Sea Biology, The Sea. Vol. 8. Wiley-Interscience, New York: 371–398.

Cartes, J. E. & F. Sardà, 1993. Zonation of deep-sea decapod fauna in the Catalan Sea (Western Mediterranean). Mar. Ecol. Prog. Ser. 94: 27–34.

Cartes, J. E., J. B. Company & F. Maynou, 1994. Deep-water decapod crustacean communities in the Northwestern Mediterranean: influence of submarine canyons and season. Mar. Biol. 120: 221–229.

Clarke, K. L., 1993. Non-parametric multivariate analyses of change in community structure. Aust. J. Ecol. 18: 117–143.

Daget, J., 1975. Ordination des profils ecologiques. Nat. Monspel. Ser. Bot. 26: 109–128.

D'Onghia, G., A. Matarrese, A. Tursi & L. Sion, 1995. Observations on the depth distribution pattern of the small-spotted catshark in the North Aegean Sea. J. Fish Biol. 47: 421–426.

Field J. G., K. R. Clarke & R. M. Warwick, 1982. A practical strategy for analysis of multispecies distribution patterns. Mar. Ecol. Prog. Ser. 8: 37–52.

Farinã, A. C., J. Freire & E. González-Gurriarán, 1997a. Megabenthic decapod crustacean assemblages on the Galician continental shelf and upper slope (north-west Spain). Mar. Biol. 127: 419–434.

Farinã, A. C., J. Freire & E. González-Gurriarán, 1997b. Demersal fish assemblages in the Galician continental shelf and upper slope (NW Spain): spatial structure and long term changes. Estuar. coast. shelf Sci. 44: 435–454.

Fujita, T., T. Inada & Y. Ishito, 1995. Depth-gradient structure of the demersal fish community on the continental shelf and upper slope off Sendai Bay, Japan. Mar. Ecol. Prog. Ser. 118: 13–23.

García, C. B., L. Orlando Duarte & D. Von Schiller, 1998. Demersal fish assemblages of the Gulf of Salamanca, Colombia (southern Caribbean Sea). Mar. Ecol. Prog. Ser. 174: 13–25.

Gomes, M. C., R. L. Haedrich, & M. G. Villagarcia, 1995. Spatial and temporal changes in the groundfish assemblages on the Northeast Newfoundland/Labrador Shelf, Northwest Atlantic, 1978–1991. Fish. Oceanog. 4: 85–101.

Gordon, J. D. M., N. R. Merrett & R. L. Haedrich, 1995. Environmental and biological aspects of slope dwelling fishes. In Hopper, A. G. (ed.), Deep Water Fisheries of the North Atlantic Ocean Slope. Kluwer Academic Publishers, Dordrecht, The Netherlands: 1–30.

Haedrich, R. L. & G. T. Rowe, 1977. Megafaunal biomass in the deep sea. Nature 269: 141–142.

Haedrich, R. L., G. T. Rowe & P. T. Polloni, 1980. The megabenthic fauna in the deep sea south of New England, U.S.A. Mar. Biol. 57: 165–179.

Hecker, B., 1990. Variation in megafaunal assemblages on the continental margin south of New England. Deep-Sea Res. 37: 37–57.

Hurlbert, S. H., 1978. The measurement of niche overlap and some relatives. Ecology 59: 67–77.

Koslow, J. A., 1993. Community structure in North Atlantic deep-sea fishes. Prog. Oceanogr. 31: 321–338.

Lykousis, V. & M. Collins, 1987. Sedimentary environments in the north-western Aegean Sea, identified from sea bed photography. Thalassographica 10: 23–35.

MacPherson, E. & C. M. Duarte, 1991. Bathymetric trends in demersal fish size: is there a general relationship? Mar. Ecol. Prog. Ser. 71: 103–112.
Margalef, R., 1968. Perspectives in Ecological Theory. University of Chicago Press, Chicago: Illinois: 110 pp.
Markle, D. F., M. J. Dadswell & R. G. Hallliday, 1988. Demersal fish and decapod crustacean fauna of the upper continental slope off Nova Scotia from LaHave to St. Pierre Banks. Can. J. Zool. 66: 1952–1960.
Merrett, N. R., J. D. M. Gordon, M. Stehmann & R. L. Haedrich. 1991. Deep demersal fish assemblage structure in the Porcupine Sea Bight (Eastern North Atlantic): slope sampling by three different trawls compared. J. mar. biol. Ass. U.K. 71: 329–358.
Merrett, N. R. & R. L. Haedrich, 1997. Deep-sea Demersal Fish and fisheries. Chapman & Hall, London: 282 pp.
Moranta, J., C. Stefanescu, E. Massuti, B. Morales-Nin & D. Lloris, 1998. Fish community structure and depth-related trends on the continental slope of the Balearic Islands (Algerian basin, western Mediterranean). Mar. Ecol. Prog. Ser. 171: 247–259.
Overholtz, W. J. & A. V. Tyler, 1985. Long-term changes of the demersal fish assemblages of Georges Bank. Fish. Bull. 83: 507–520.
Papaconstantinou, C., 1988. Check-list of marine fishes of Greece. Fauna Graecia. Athens: National Centre for Marine Research and Hellenic Zoological Society: 257 pp.
Pielou, E. C., 1966. The measurement of diversity in differences types of biological collections. J. theor. Biol. 13: 131–144.
Pielou, E. C., 1969. An Introduction to Mathematical Ecology. Wiley, New York: 286 pp.
Pitcher, A., 1996. Reinventing fisheries management. NAGA ICLARM Q 19(3): 15–17.
Poulos, S. E., P. G. Drakopoulos & M. B. Collins, 1997. Seasonal variability in sea surface oceanographic conditions in the Aegean Sea (Eastern Mediterranean): an overview. J. mar. Syst. 13: 225–244.
Rothschild, B. J., 1992. Multispecies interactions on George Bank. ICES Mar. Sci. Symp. 195: 418–423.
Sardá, F., J. E. Cartes & J. B. Company, 1994. Spatio-temporal variations in megabenthos abundance in three different habitats of the Catalan deep-sea (Western Mediterranean) Mar. Biol. 120: 211–219.
Smale, M. J., B. A. Roel, A. Badenhorst & J. G. Field, 1993. Analysis of the demersal community of fish and cephalopods on the Aguilas Bank, South Africa. J. Fish Biol. 43 (Suppl. A): 169–191.
Sokal, R. R. & F. J. Rohlf, 1981. Biometry: the principles and practice of statistics in biological research. San Francisco: Freeman: 859 pp.
Stefanescu C., D. Lloris & J. Rucabado, 1992a. Deep-living demersal fishes in the Catalan Sea (western Mediterranean) below a depth of 1000 m. J. nat. Hist. 26: 197–213.
Stefanescu, C., J. Rucabado & D. Lloris, 1992b. Depth-size trends in western Mediterranean demersal deep-sea fishes. Mar. Ecol. Prog. Ser. 81: 205–213.
Stergiou, K. I. & D. A. Pollard, 1994. A spatial analysis of the commercial fisheries catches from the Greek Aegean Sea. Fish. Res. 20: 109–135.
Stergiou, K. I., E. D. Christou, D. Georgopoulos, A. Zenetos & C. Souvermezoglou, 1997. The Hellenic Seas: physics, chemistry, biology and fisheries. Oceanogr. Mar. Biol. Ann. Rev. 35: 415–538.
Wenner, E. L. & D. F. Boesch, 1979. Distribution patterns of epibenthic decapod Crustacea along the shelf-slope coenocline, middle Atlantic Bight, U.S.A. Bull. Biol. Soc. Wash. 3: 106–133.
Zar, J. H., 1984. Biostatistical analysis. Prentice-Hall, Englewood Cliffs: 718 pp.

Appendix 1. Species list of demersal fish in the North Aegean shelf and upper slope

Hexanchidae	*Heptranchias perlo*	Myctophidae	*Diaphus rafinesqei*	Mullidae	*Mullus surmuletus*	Scorpaenidae	*Scorpaena porcus*
	Hexanchus griseus		*Diaphus* sp.	Sparidae	*Boops boops*		*Scorpaena scrofa*
Scyliorhinidae	*Galeus melastomus*		*Hygophum benoiti*		*Dentex dentex*	Triglidae	*Aspitrigla cuculus*
	Scyliorhinus canicula		*Lampanyctus crocodilus*		*Dentex macrophthalmus*		*Eutrigla gurnardus*
	Scyliorhinus stellaris		*Myctophum punctatus*		*Diplodus annularis*		*Lepidotrigla cavillone*
Triakidae	*Mustelus asterias*		*Notoscopelus elongatus*		*Diplodus vulgaris*		*Lepidotrigla dieuzeidei*
Oxynotidae	*Oxynotus centrina*		*Myctophidae* sp.		*Pagellus acarne*		*Trigla lucerna*
Torpedinidae	*Torpedo marmorata*	Nettastomatidae	*Nettastoma melanurum*		*Pagellus bogaraveo*		*Trigla lyra*
	Torpedo torpedo	Congridae	*Conger conger*		*Pagellus erythrinus*		*Trigloporus lastoviza*
Squalidae	*Dalatias licha*	Ophichithidae	*Echelus myrus*		*Pagrus pagrus*	Peristeiidae	*Peristedion cataphractum*
	Etmopterus spinax		*Ophisurus serpens*		*Sparus aurata*	Dactylopteridae	*Dactylopterus volitans*
	Somniosus rostratus	Notocanthidae	*Notocanthus bonapartei*		*Spondyliosoma cantharus*	Citharidae	*Citharus linguatula*
	Squalus acanthias	Belonidae	*Belone belone gracilis*	Centracanthidae	*Centracanthus cirrus*	Scophthalmidae	*Lepidorhombus boscii*
	Squalus blainvillei	Macroramphosidae	*Macroramphosus scolopax*		*Spicara flexuosa*		*Lepidorhombus whiffiagonis*
Squatinidae	*Squatina oculata*	Syngnathidae	*Syngnathus* sp.		*Spicara maena*		*Phrynorhombus regius*
	Squatina squatina	Macrouridae	*Coelorhynchus coelorhynchus*		*Spicara smaris*		*Scophthalmus rhombus*
Rajidae	*Raja alba*		*Hymenocephalus italicus*	Labridae	*Acantholabrus palloni*	Bothidae	*Arnoglossus laterna*
	Raja asterias		*Nezumia sclerorhynchus*		*Symphodus mediterraneus*		*Arnoglossus rueppelli*
	Raja brachyura		*Trachyrhynchus trachyrhynchus*	Trachinidae	*Trachinus draco*		*Arnoglossus thori*
	Raja clavata	Merlucciidae	*Merluccius merluccius*	Uranoscopidae	*Uranoscopus scaber*	Soleidae	*Buglossidium luteum*
	Raja milaretus	Gadidae	*Gadiculus argenteus argenteus*	Gempylidae	*Ruvettus pretiosus*		*Microchirus ocellatus*
	Raja montagui		*Gaidropsarus* sp.	Trichiuridae	*Lepidopus caudatus*		*Microchirus variegatus*
	Raja naevus		*Merlangius merlangus euxinus*	Gobiidae	*Deltentosteus quadrimaculatus*		*Monochirus hispidus*
	Raja oxyrinchus		*Micromesistius poutassou poutassou*		*Gobius niger*		*Solea kleinii*
	Raja polystigma		*Molva dipterygia macrophthalma*		*Gobius paganellus*		*Solea vulgaris*
	Raja radula		*Phycis blennoides*		*Lesueurigobius friesii*	Cynoglossidae	*Symphurus ligulatus*
Dasyatidae	*Dasyatis pastinaca*		*Phycis phycis*	Callionymidae	*Callionymus lyra*		*Symphurus nigrescens*
Myliobatidae	*Myliobatis aquila*		*Trisopterus minutus capelanus*		*Callionymus maculatus*		*Symphurus* sp.
Chimaeridae	*Chimaera monstrosa*	Moridae	*Gadella maraldi*		*Callionymus risso*	Gobiesocidae	*Lepadogaster lepadogaster*
Sternoptychidae	*Argyropelecus hemigymnus*	Trachichthyidae	*Hoplostethus mediterraneus*		*Synchiropus phaeton*	Lophiidae	*Lophius budegassa*
	Maurolicus muelleri	Zeidae	*Zeus faber*	Blennidae	*Blennius ocellaris*		*Lophius piscatorius*
Chauliodontidae	*Chauliodus sloani*	Caproidae	*Capros aper*	Bythitidae	*Bellottia apoda*		
Stomiidae	*Stomias boa*	Serranidae	*Anthias anthias*	Ophidiidae	*Ophidion rochei*		
Argentinidae	*Argentina sphyraena*		*Callanthias ruber*		*Ophidion vassali*		
	Glossanodon leioglossus		*Serranus cabrilla*	Carapidae	*Carapus acus*		
	Microstoma microstoma		*Serranus hepatus*	Centrolophidae	*Centrolophus niger*		
Synodontidae	*Synodus saurus*	Apogonidae	*Epigonus denticulatus*	Scorpaenidae	*Helicolenus dactylopterus*		
Chlorophthalmidae	*Chlorophthalmus agassizi*		*Epigonus telescopus*		*Scorpaena elongata*		
Myctophidae	*Ceratoscopelus maderen*	Cepolidae	*Cepola rubescens*		*Scorpaena loppei*		
	Diaphus metopoclampus	Mullidae	*Mullus barbatus*		*Scorpaena notata*		

Open Session

Hydrobiologia **440:** 299–305, 2000.
M.B. Jones, J.M.N. Azevedo, A.I. Neto, A.C. Costa & A.M. Frias Martins (eds), Island, Ocean and Deep-Sea Biology.

Influence of light and nitrogen on the phlorotannin content of the brown seaweeds *Ascophyllum nodosum* and *Fucus vesiculosus*

Henrik Pavia & Gunilla B. Toth*
Department of Marine Botany, Göteborg University, Box 461, SE 405 30 Göteborg, Sweden
E-mail: henrik.pavia@marbot.gu.se gunilla.toth@marbot.gu.se

Key words: Ascophyllum, Fucus, CNBM, light, nitrogen, phlorotannins

Abstract

Phlorotannins, C-based defence compounds in brown seaweeds, show a high degree of spatial and temporal variation within seaweed species. One important model explaining this variation is the Carbon Nutrient Balance Model (CNBM), which states that the relative supply of carbon and limiting nutrients will determine the level of defence compounds in plants. Nitrogen is often considered to be the limiting nutrient for marine macroalgal growth and the CNBM thus predicts that when the carbon:nitrogen ratio is high, photosynthetically fixed carbon will be allocated to production of phlorotannins. In the present study, we evaluated the effects of light (i.e. carbon) and nitrogen on the phlorotannin content of two intertidal brown seaweeds, *Ascophyllum nodosum* and *Fucus vesiculosus*. This was done in an observational field study, as well as in a manipulative experiment where plants from habitats with different light regimes were subjected to different nitrogen and light treatments, and their phlorotannin content was measured after 14 days. The results showed that there was a negative relationship between tissue nitrogen and phlorotannin content in natural populations of *F. vesiculosus*, but not in *A. nodosum*. In the short term, the phlorotannin content in both algal species was not affected by changes in nitrogen availability. Exposure to sunlight had a positive effect on the phlorotannin content in natural populations of both algal species but, in the manipulative experiment, only *F. vesiculosus* showed a rapid response to changes in light intensities. Plants subjected to sunlight contained higher phlorotannin content than shaded plants. In conclusion, the results imply that nitrogen availability explains some of the natural variation in the phlorotannin content of *F. vesiculosus*, but the light environment has greater importance than nitrogen availability in predicting the phlorotannin content of each species.

Introduction

Polyphenolic compounds are ubiquitous in plants and have been a major focal point in plant chemical ecology during the last two to three decades. The single group of polyphenolics that has attracted most attention from ecologist is tannins, which are defined as polyphenolic compounds with the ability to precipitate proteins from solutions. The least complex group of tannins is the phlorotannins, which are polymers of phloroglucinol derived entirely from acetate. Phlorotannins are restricted to brown algae (Phaeophyta) where they are found in special vesicles (physodes) within the cells (Ragan & Glombitza, 1986). Phlorotannins have been shown to have antifeeding activity towards several different types of herbivores, although their effectiveness as herbivore deterrents is far from universal (see reviews by Hay & Steinberg, 1992; Targett & Arnold, 1998). It has also been suggested that phlorotannins may be important in the defence against pathogenic organisms (Sieburth & Conover, 1965), as antifouling agents (Sieburth & Conover, 1965; Lau & Qian, 1997; but see Jennings & Steinberg, 1997), as chelators of toxic heavy metal ions (Pedersen, 1984; Karez & Pereira, 1995; but see Toth & Pavia, 2000), in the protection from harmful solar radiation (Berthold, 1882; Pavia et al., 1997), and in the cell-wall construction of brown seaweeds (Crato, 1896; Schoenwaelder & Clayton, 1998). Phlorotannin concentrations vary greatly among different taxa of brown seaweeds, as well as among different geographical areas. Concen-

* Authors appear in alphabetical order

trations are generally higher in fucoid species than in kelps, and higher in species from the Atlantic and the temperate Pacific than in species from the tropical Pacific (see Steinberg, 1992; Targett & Arnold, 1998 for reviews). Besides taxonomic and geographical variation, substantial intraspecific and intraplant variation in phlorotannin content has been documented for several brown algal species (reviewed in Targett & Arnold, 1998; see also Hammerstrom et al., 1998; Van Alstyne et al., 1999a&b).

One influential model proposed to explain intraspecific phenotypic variation in secondary metabolites of plants is the carbon–nutrient balance model (CNBM) (Bryant et al., 1983, 1989). The CNBM states that the production of secondary metabolites is governed by the relative supply of photosynthetically fixed carbon and limiting nutrients. The model assumes that plant growth has higher priority than the production of defence compounds and that excess photosynthate is allocated to carbon-based defence compounds when growth is nutrient limited (Bryant et al., 1983, Karban & Baldwin, 1997, Koricheva et al., 1998). Consequently, the CNBM predicts that environmental conditions that give rise to a high carbon/nutrient ratio (e.g. high light and low nutrient availability) will result in an increased production of carbon-based secondary metabolites. The empirical evidence for the CNBM from studies on vascular plants is mixed, with predictions being either supported or rejected in numerous studies (Herms & Mattson, 1992; Berenbaum, 1995). Most support for the CNBM seems to come from studies on the production of phenolic metabolites in early successional plants (Reichardt et al., 1991; Fajer et al., 1992). For brown algal phlorotannins, there are still only a few studies that have addressed the CNBM, showing that nutrient (nitrogen) availability may (Ilvessalo & Tuomi, 1989; Yates & Peckol, 1993; Arnold et al., 1995; Peckol et al., 1996) or may not (Cronin & Hay, 1996; Pavia et al., 1999; Pavia & Brock, 2000) be related to phlorotannin concentration. Cronin & Hay (1996) found that plants of *Sargassum filipendula* grown in experimentally shaded environments had lower concentrations of phlorotannins, in accordance with the predictions of the CNBM. The experimental data available so far are much too scarce to produce any general conclusions about taxonomic and geographical variation in the relevance of the CNBM to explain phenotypic variation in phlorotannin content of brown macroalgae.

In the present study, we tested the hypotheses, derived from the CNBM, that the availability of light and nutrients (i.e. nitrogen) can affect, separately or interactively, phlorotannin production in the brown seaweeds *Ascophyllum nodosum* and *Fucus vesiculosus*. These two species were chosen since they are dominating and often competing macroalgae in the intertidal zone of temperate rocky shores in the northern Atlantic, but still they have different life spans, with *A. nodosum* being an extremely long-lived species. Nitrogen was chosen as the experimental nutrient since it is commonly the most limiting nutrient for temperate marine macroalgae (Chapman & Craigie, 1977; Hanisak, 1983). The effects of light and N on phlorotannin levels were tested both in an observational field study and in a manipulative aquarium experiment.

Materials and methods

Study organisms and sites

Ascophyllum nodosum and *Fucus vesiculosus* are common perennial seaweeds in the intertidal zone on temperate rocky shores where they can form zones ranging from <1 m in atidal areas to >100 m in tidal areas. Both species have apical growth and are attached to the substratum with a holdfast from which one to several primary shoots arise. In the present study, only the outermost tips of the vegetative shoots were used for chemical analyses. The seaweeds were collected from atidal sunexposed and shaded sheltered sites in the Tjärnö (58° 54′ N, 11° 07′ E) and Strömstad (58° 56′ N, 11° 10′ E) Archipelagos of the Swedish west coast. Low, flat skerries, where the seaweeds are subjected to sunlight most of the day, were chosen as sunexposed sites. Islands with steep, vertical rocks that shade the plants from sunlight were chosen as shaded sites.

Observational field study

Tissue samples were collected in May (1998) from *A. nodosum* and *F. vesiculosus* plants growing in sunexposed and shaded habitats. Ten apical shoots were sampled from each of 16 plants of each species. The seaweed samples were freeze dried, and the shoots originating from the same plant were ground, mixed and analysed for phlorotannin and tissue N content as described below.

Manipulative experiment

The manipulative experiment was carried out in June (1998) at Tjärnö Marine Biological Laboratory. *Ascophyllum nodosum* and *Fucus vesiculosus* plants originating from sunexposed and shaded sites were collected by carefully removing their holdfasts from the substratum with a knife. The plants were transported to the laboratory and placed immediately in Plexiglas aquaria with fresh seawater. The aquaria were placed in a basin with cooling seawater. The basin was built outdoors on a sunexposed bridge where the aquaria were subjected to natural ambient sunlight. Three plants of each species and light origin were subjected to one of the following treatments: (1) shading and additional N, (2) shading and ambient N, (3) ambient sunlight and additional N, and (4) ambient sunlight and ambient N. Shading was achieved by covering the aquaria with black plastic bags, which excluded most of the ambient sunlight. During the day, the plants in the aquaria were supplied continuously with ambient seawater. In the evening, the seawater supply was turned off and N was added in the form of aqueous ammonium nitrate (NH_4NO_3) at a concentration of 20 μm. The seaweeds were allowed to accumulate N during the night and in the morning the seawater supply was turned on again. The same procedure, except for the addition of N, was applied to plants in aquaria subjected to ambient N treatment. Tissue samples, collected after 14 days in the same way as described for the observational field study, were freeze dried and analysed for phlorotannin content as described below.

Chemical analyses

Phlorotannins were extracted from freeze dried, homogenised seaweed tissue in aqueous acetone (60-vol%) in the dark at 4 ° C for 24 h. The acetone was evaporated in vacuo and the remaining water fraction was filtered to remove precipitated lipophilic material. The concentration of phlorotannins was determined using the Folin-Ciocalteus method for quantification of total phenolic compounds (Van Alstyne, 1995). Phloroglucinol (Merck art. 7069) was used as a standard. Samples for tissue N content were ground and analysed in an elemental analyser (FISONS Instruments NA 1500 NC).

Statistical analyses

Analysis of covariance (ANCOVA) was used to test for interactions between the fixed factor Light origin (2 levels) and the covariate N content in the observational field study. No significant interaction was detected for either *A. nodosum* (ANCOVA, $F_{1,12} = 0.67$, $p = 0.43$) or *F. vesiculosus* (ANCOVA, $F_{1,12} = 1.81$, $p = 0.20$), implying that the relationship between tissue N and phlorotannin content was not significantly different amongst sites. Consequently, data of plants from sites with different light conditions were pooled and regression analyses, separate for each species, were used to analyse the relationship between phlorotannin and tissue N content. Data on the phlorotannin content of *A. nodosum* and *F. vesiculosus* plants were also analysed in a one-way analysis of variance (ANOVA) with Light origin (2 levels) as a fixed factor. Data on the phlorotannin content of *A. nodosum* and *F. vesiculosus* plants from the manipulative experiment were analysed in a three-way ANOVA with Light origin (2 levels), Light treatment (2 levels) and N treatment (2 levels) as fixed orthogonal factors. Prior to all statistical analyses, the homogeneity of variances was tested with Cochran's test (Underwood, 1997).

Results

The regression analyses showed that there was a significant negative relationship between phlorotannin and tissue N content in natural populations of *F. vesiculosus* (Fig. 1b), but not in *A. nodosum* (Fig. 1a). The ANOVA revealed that there was no statistically significant difference in the mean phlorotannin content of *A. nodosum* from sunexposed (4.32% dw) and shaded (3.82% dw) sites (ANOVA, $F_{1,14} = 1.35$, $p = 0.26$), but a significant difference was found for *F. vesiculosus* (ANOVA, $F_{1,14} = 7.36$, $p = 0.02$). *Fucus vesiculosus* plants exposed to sunlight in the field had a mean phlorotannin content of 4.88% dw, while shaded plants contained only 3.70% dw.

There were no significant interactions between the factors Light origin, Light treatment and N treatment for any of the species in the manipulative experiment (Table 1, Fig. 2). There was a significant difference in phlorotannin content of *A. nodosum* plants originating from sites with different light conditions (Table 1), with plants from sunexposed and shaded sites having a phlorotannin content of 4.26 and 3.49% dw, respectively (Fig. 2a). The light and N treatments had no apparent effect on the phlorotannin content of *A. nodosum* (Table 1) but, for *F. vesiculosus*, the light treatment had a significant effect (Table 1); plants exposed to ambient sunlight had a higher phlorotan-

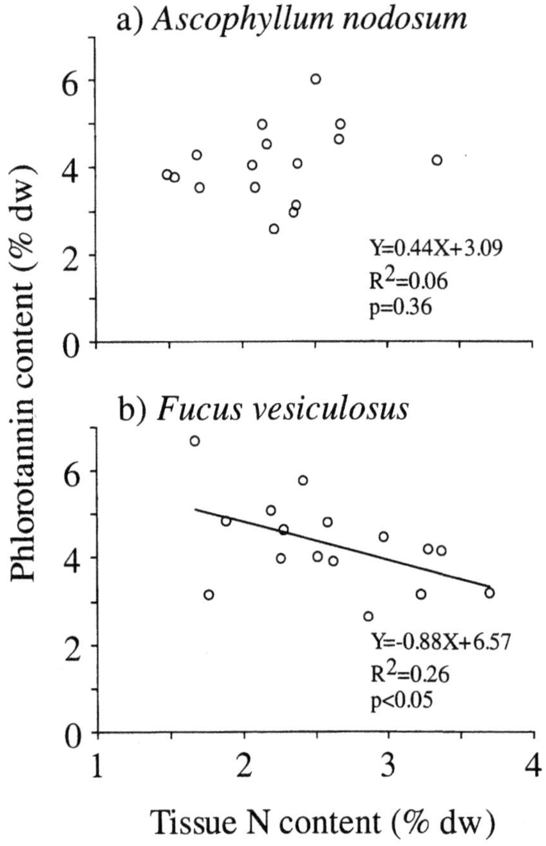

Figure 1. The relationship between phlorotannin and tissue N content (% dw) in natural populations of (a) *Ascophyllum nodosum* and (b) *Fucus vesiculosus* collected from sunexposed and shaded sites in the Tjärnö and Strömstad Archipelagos.

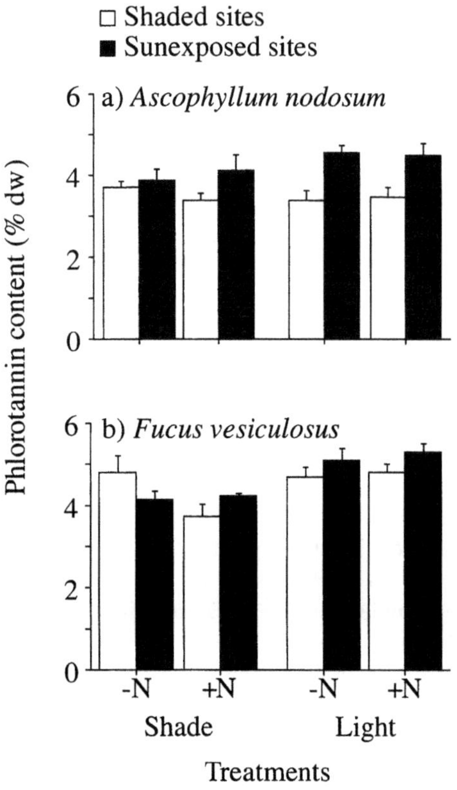

Figure 2. The phlorotannin content (% dw) in (a) *Ascophyllum nodosum* and (b) *Fucus vesiculosus* plants collected from sunexposed and shaded sites in the Tjärnö Archipelago and subjected to light (shade and ambient sunlight) and N treatments (additional and ambient N) in a manipulative aquarium experiment. ANOVA results are presented in Table 1. Error bars show + 1 SE (n = 3).

nin content (4.98% dw) than shaded plants (4.23% dw) (Fig. 2b). In the manipulative experiment, the phlorotannin content in *F. vesiculosus* plants was not affected by N availability (Table 1).

Discussion

The results of the present study indicate that, in general, light has greater influence than N availability on the phlorotannin content of both *A. nodosum* and *F. vesiculosus*. Effects of sunlight intensity on phlorotannin production were found in natural populations of both species. *Ascophyllum nodosum* plants collected from sunexposed sites in the manipulative experiment had higher phlorotannin content than plants from shaded sites, as had *F. vesiculosus* plants collected from sunexposed sites in the observational field study. No statistically significant difference in phlorotannin content was found between *A. nodosum* plants collected from sites with different light conditions in the observational field study. This may be due to the fact that fewer replicates were used in the observational survey than in the manipulative experiment and, thus, the statistical analysis was less powerful. Light intensity had no significant short-term effects on the phlorotannin content of *A. nodosum*, but a significant effect was found for *F. vesiculosus*. In a previous study on *A. nodosum*, increased concentrations of phlorotannins were found in response to additional UV-B radiation after 2 weeks (Pavia et al., 1997). In the present study, however, the light exposure of plants in aquaria covered with black plastic bags was decreased compared to natural levels. Since it is regarded as unlikely that phlorotannins can be re-metabolised (Targett & Arnold, 1998), a decrease in phlorotannin content would implicate that phlorotannins are released to the surrounding water, either by

Table 1. Three-way analysis of variance (ANOVA) of the phlorotannin content (% dw) of *Ascophyllum nodosum* and *Fucus vesiculosus* plants collected from sunexposed and shaded sites, and subjected to light (shade and ambient sunlight) and N treatments (additional and ambient N) in a manipulative aquarium experiment

Source of variation	df	*Ascophyllum nodosum*			*Fucus vesiculosus*		
		MS	*F*	*p*	MS	*F*	*p*
Light origin (LO)	1	3.61	18.57	<0.01	0.20	1.04	0.32
Light treatment (LT)	1	0.23	1.20	0.29	3.40	17.75	<0.01
N treatment (NT)	1	<0.01	0.01	0.91	0.16	0.82	0.38
LO × LT	1	0.67	3.44	0.08	0.44	2.32	0.15
LO × NT	1	0.06	0.28	0.60	0.55	2.87	0.11
LT × NT	1	0.01	0.04	0.84	0.67	3.49	0.08
LO × LT × NT	1	0.91	0.98	0.34	0.43	2.56	0.15
Residual	16	0.19		0.19			

exudation or passive leaking, or transported to other parts of the seaweed. Seaweeds are nonvascular plants and, therefore, the possibility for efficient transport is limited, while several previous studies have found that brown seaweeds can release significant amounts of phlorotannins into the surrounding water (Sieburth, 1969; Carlson & Carlson, 1984; Jennings & Steinberg, 1994). If exudation is the primary way for metabolite turnover of phlorotannins, and if the release rates from healthy seaweed plants are low (see Jennings & Steinberg, 1994), any environmentally induced decreases in phlorotannin content would be relatively slow (cf. Pavia & Brock, 2000). This could explain the lack of a rapid response in *A. nodosum* plants subjected to changing light levels in the manipulative experiment of the present study. In contrast, there was a significant short-term effect of the light treatment on the phlorotannin content of *F. vesiculosus*, where plants subjected to sunlight in the aquaria had higher phlorotannin content than shaded plants. Lack of difference in phlorotannin content of *F. vesiculosus* plants with different light origin in the aquarium experiment could be explained possibly by the fact that the effect of the light treatment was strong enough to mask the initial difference in phlorotannin content of the *F. vesiculosus* plants.

The difference in short-term response in phlorotannin content to changes in light exposure between *A. nodosum* and *F. vesiculosus* may be due to differences in morphology and/or growth rate between the two species. The CNBM predicts that rapidly growing species should show the most pronounced changes in secondary metabolite content in response to changes in nutrient availability and light intensity (Reichardt et al., 1991). However, even if *A. nodosum* is extremely slow growing as a juvenile (Lazo et al., 1994), there does not seem to be any obvious differences in growth rate between adult plants of *F. vesiculosus* and *A. nodosum* during the season (May–June) when the present study was conducted (Carlson, 1991; Stengel & Dring, 1997; H. Pavia & G. Toth, unpublished data). We suggest that the difference in the short-term response of phlorotannin content to changes in light intensity between the two species is more likely to be due to differences in morphology. The apical shoots of *F. vesiculosus* are flat and have a higher surface:volume ratio compared to the apical shoots of *A. nodosum*. Since phlorotannins are stored primarily in the cortical layer of the seaweed thallus (Ragan & Glombitza, 1986; Tugwell & Branch, 1989; Lowell et al., 1991), a light-induced decrease in phlorotannin content, where phlorotannins are lost primarily through exudation, may appear faster in shoots with a high surface:volume ratio. Nitrogen availability had no effects on the phlorotannin content of *A. nodosum* in the present study. There was, however, a significant, negative relationship between tissue N and phlorotannin content of *F. vesiculosus* plants in the field. These results are in accordance with previously reported results on the two species. Nitrogen availability was not correlated to the phlorotannin content of *A. nodosum* in field observations performed in two different geographical areas during two consecutive years (Pavia et al., 1999). Neither did increased N concentrations result in a decreased phlorotannin production in a 4-week manipulative aquarium experiment (Pavia & Brock, 2000). In contrast, significant negative relationships between tissue N and phlorotannin content of *F. vesiculosus* were found both by Yates & Peckol (1993) and by Peckol et al. (1996). The difference in the response to N availability between the two seaweed species is in accordance with results from terrestrial plants with different life histories, where responses in the production of C-based secondary metabolites seem to be less pronounced in long-lived species (Koricheva et al., 1998). *Ascophyllum nodosum* is an extremely long-lived seaweed species (Åberg, 1992), while *F. vesiculosus* individuals usually live for <5 years (Williams, 1990). In conclusion, the results of this study give support to the relevance of the CNBM in explaining long-term effects of N on phlorotannin levels in *F. vesiculosus*, but not in *A. nodosum*, and of light on phlorotannin levels in both species. Furthermore, the manipulative experiment showed that *F. vesiculosus*, but not *A. nodosum*, can respond by a significant short-term decrease in phlorotannin content when light intensity is reduced, in accordance with the predictions of the CNBM. Finally, neither of the two species showed any significant short-term changes in phlorotannin content as a response to increased N availability.

Acknowledgements

We are grateful to staff and students at the Tjärnö Marine Biological Laboratory for their help and hospitality. The Göteborg Marine Research Centre and the foundations of Helge Ax:son Johnson, Kapten Carl Stenholm, Rådman och Fru Ernst Colliander and Paul och Marie Berghaus financially supported this research.

References

Åberg, P., 1992. A demographic study of two populations of the seaweed *Ascophyllum nodosum*. Ecology 73: 1473–1487.

Arnold, T. M., C. E. Tanner & W. I. Hatch, 1995. Phenotypic variation in polyphenolic content of the tropical brown alga *Lobophora variegata* as a function of nitrogen availability. Mar. Ecol. Prog. Ser. 123: 177–183.

Berenbaum, M. R., 1995. The chemistry of defence: theory and practice. Proc. natn. Acad. Sci. U.S.A. 92: 2–8.

Berthold, G., 1882. Beiträge zur Morpholgie und Physiologie der Meeresalgen. Jahrb. Wiss. Bot. 13: 569–717.

Bryant, J. P., F. S. III Chapin & D. R. Klein, 1983. Carbon/nutrient balance of boreal plants in relation to vertebrate herbivory. Oikos 40: 357–368.

Bryant, J. P., J. P. Kuropat, S. M. Cooper, K. Frisby & N. Owen-Smith, 1989. Resource availability hypothesis of plant antiherbivore defence tested in South African savanna ecosystem. Nature 340: 227–228.

Carlson, D. J. & M. L. Carlson, 1984. Reassessment of exudation by fucoid macroalgae. Limnol. Oceanogr. 29: 1077–1087.

Carlson, L., 1991. Seasonal variation in growth, reproduction and nitrogen content of *Fucus vesiculosus* L. in the Öresund, Southern Sweden. Bot. mar. 34: 447–453.

Chapman, A. R. O. & J. S. Craigie, 1977. Seasonal growth in *Laminaria longicruris*: relations with dissolved inorganic nutrients and internal reserves of nitrogen. Mar. Biol. 40: 197–205.

Crato, E., 1896. Beiträge zur anatomie und physiologie das elementarorganismus. Beitr. Biol. Pflanz. 7: 407–535.

Cronin, G. & M. E. Hay, 1996. Effects of light and nutrient availability on the growth, secondary chemistry and resistance to herbivory of two brown seaweeds. Oikos 77: 93–106.

Fajer, E. D., M. D. Bowers & F. A. Bazzaz, 1992. Effects of nutrients and enriched CO_2 environment on production of carbon-based allelochemicals in plantago – a test of the carbon nutrient balance hypothesis. Am. Nat. 140: 707–723.

Hammerstrom, K., M. N. Dethier & D. O. Duggins, 1998. Rapid phlorotannin induction and relaxation in five Washington kelps. Mar. Ecol. Prog. Ser. 165: 293–305.

Hanisak, M. D., 1983. The nitrogen relationships of marine macroalgae. In Carpenter, E. J. & D. G. Capone (eds), Nitrogen in the Marine Environment. Academic Press, New York: 699–730.

Hay, M. & P. D. Steinberg, 1992. The chemical ecology of plant-herbivore interactions in marine *versus* terrestrial communities. In Rosenthal, G. A. & M. R. Berenbaum (eds), Herbivores: Their Interactions with Secondary Plant Metabolites. Academic Press, New York: 371–413.

Herms, D. A. & W. J. Mattson, 1992. The dilemma of plants: to grow or defend. Quart. Rev. Biol. 67: 283–335.

Ilvessalo, H. & J. Tuomi, 1989. Nutrient availability and accumulation of phenolic compounds in the brown alga *Fucus vesiculosus*. Mar. Biol. 101: 115–119.

Jennings, J. & P. D. Steinberg, 1994. In situ exudation of phlorotannins by the sublittoral kelp *Ecklonia radiata*. Mar. Biol. 121: 349–354.

Jennings, J. & P. D. Steinberg, 1997. Phlorotannins versus other factors affecting epiphyte abundance on the kelp *Ecklonia radiata*. Oecologia 109: 461–473.

Karban, R. & I. T. Baldwin, 1997. Induced responses to herbivory. The University of Chicago Press, Chicago: 319 pp.

Karez, C. S. & R. C. Pereira, 1995. Metal contents in polyphenolic fractions extracted from the brown alga *Padina gymnospora*. Bot. mar. 28: 151–155.

Koricheva, J., S. Larsson, E. Haukioja & M. Keinänen, 1998. Regulation of woody plant secondary metabolism by resource availability: hypothesis testing by means of meta-analysis. Oikos 83: 212–226.

Lau, S. C. K. & P.-Y. Qian, 1997. Phlorotannins and related compounds as larval settlement inhibitors of the tube-building polychaete *Hydroides elegans*. Mar. Ecol. Prog. Ser. 159: 219–227.

Lazo, L., J. H. Markham & A. R. O. Chapman, 1994. Herbivory and harvesting: effects on sexual recruitment and vegetative modules of *Ascophyllum nodosum*. Ophelia 40: 95–113.

Lowell, R. B., J. H. Markham & K. Mann, 1991. Herbivore-like damage induces increased strength and toughness in a seaweed. Proc. r. Soc. Lond. B, Biol. Sci. 243: 31–38.

Pavia, H. & E. Brock, 2000. Extrinsic factors influencing phlorotannin production in the brown seaweed *Ascophyllum nodosum*. Mar. Ecol. Prog. Ser. 193: 285–294.

Pavia, H., G. Toth & P. Åberg, 1999. Trade-offs between phlorotannin production and annual growth in natural populations of the brown seaweed *Ascophyllum nodosum*. J. Ecol. 87: 761–771.

Pavia, H., G. Cervin, A. Lindgren & P. Åberg, 1997. Effects of UV-B radiation and simulated herbivory on phlorotannins in the brown alga *Ascophyllum nodosum*. Mar. Ecol. Prog. Ser. 157: 139–146.

Peckol, P., J. M. Krane & J. L. Yates, 1996. Interactive effects of inducible defence and resource availability on phlorotannins in the North Atlantic brown alga *Fucus vesiculosus*. Mar. Ecol. Prog. Ser. 138: 209–217.

Pedersen, A., 1984. Studies on phenol content and heavy metal uptake in fucoids. Hydrobiologia 116/117: 498–504.

Ragan, M. A. & K.-W. Glombitza, 1986. Phlorotannins, brown algal polyphenols. Prog. Phycol. Res. 4: 129–241.

Reichardt, P. B., F. S. III Chapin, J. P. Bryant, B. R. Mattes & T. P. Clausen, 1991. Carbon/nutrient balance as a predictor of plant defence in Alaskan balsam poplar: potential importance of metabolic turnover. Oecologia 88: 401–406.

Schoenwaelder, M. E. A. & M. N. Clayton, 1998. Secretion of phenolic substances into the zygote wall and cell plate in embryos of *Hormosira* and *Acrocarpia* (Fucales, Phaeophyceae). J. Phycol. 34: 969–980.

Sieburth, J. M., 1969. Studies on algal substances in the sea. III. The production of extracellular organic matter by littoral marine algae. J. exp. mar. Biol. Ecol. 3: 290–309.

Sieburth, J. M. & T. J. Conover, 1965. *Sargassum* tannin, an antibiotic which retards fouling. Nature 208: 52–53.

Steinberg, P. D., 1992. Geographical variation in the interaction between marine herbivores and brown algal secondary metabolites. In Paul, V. J. (ed.), Ecological Roles of Marine Natural Products. Cornell University, New York: 51–92.

Stengel, D. B. & M. J. Dring, 1997. Morphology and in situ growth rates of *Ascophyllum nodosum* (Phaeophyta) from different shore levels and responses of plants to vertical transplantation. Eur. J. Phycol. 32: 193–202.

Targett, N. M. & T. M. Arnold, 1998. Predicting the effects of brown algal phlorotannins on marine herbivores in tropical and temperate oceans. J. Phycol. 36: 195–205.

Toth, G. & H. Pavia, 2000. Lack of phlorotannin induction in the brown seaweed *Ascophyllum nodosum* in response to increasing copper concentrations. Mar. Ecol. Prog. Ser. 192: 119–126.

Tugwell, S. & G. M. Branch, 1989. Differential polyphenolic distribution among tissues in the kelps *Ecklonia maxima*, *Laminaria pallida* and *Macrocystis augustifolia* in relation to plant-defence theory. J. exp. mar. Biol. Ecol. 129: 219–230.

Underwood, A. J., 1997. Experiments in ecology: their logical design and interpretation using analysis of variance. Cambridge University Press, Cambridge: 504 pp.

Van Alstyne, K. L., 1995. Comparison of three methods for quantifying brown algal polyphenolic compounds. J. Chem. Ecol. 21: 45–58.

Van Alstyne, K. L., J. J. III McCarthy, C. L. Hustead & D. O. Duggins, 1999a. Geographic variation in polyphenolic levels of Northeastern Pacific kelps and rockweeds. Mar. Biol. 133: 371–379.

Van Alstyne, K. L., J. J. McCarthy, C. L. Hustead & L. J. Kearns, 1999b. Phlorotannin allocation among tissues of northeastern pacific kelps and rockweeds. J. Phycol. 35: 483–492.

Williams, G. A., 1990. The comparative ecology of the flat periwinkles *Littorina obtusata* (L.) and *L. mariae* Sacchi et Rastelli. Field Studies 7: 469–482.

Yates, J. L. & P. Peckol, 1993. Effects of nutrient availability and herbivory on polyphenolics in the seaweed *Fucus vesiculosus*. Ecology 74: 1757–1766.

Hydrobiologia **440:** 307–315, 2000.
M.B. Jones, J.M.N. Azevedo, A.I. Neto, A.C. Costa & A.M. Frias Martins (eds), Island, Ocean and Deep-Sea Biology.

Nutrification impacts on coral reefs from northern Bahia, Brazil

O. S. Costa Jr.[1,2], Z. M. A. N. Leão[2], M. Nimmo[1] & M. J. Attrill[1]
[1]*Plymouth Environmental Research Centre, University of Plymouth, Drake Circus, Plymouth, Devon PL4 8AA, U.K.*
Tel: +44 1752 232951. E-mail: O.Costa@plymouth.ac.uk
[2]*Coastal Studies Laboratory, Geoscience Institute, Federal University of Bahia, Brazil*

Key words: coral reef, eutrophication, nutrients, groundwater contamination, Brazil

Abstract

Coral reefs extend for 20 km along the north coast of the state of Bahia, Brazil. Over the last 15 years, this region has experienced an acceleration of generally unplanned urbanisation, with the irregular and indiscriminate use of septic tanks in urban centres contaminating the groundwater. This infiltration of nutrients and pathogens is facilitated by both the soil permeability and an accented hydraulic head, which eventually leads to the percolation of nutrient-rich groundwater seaward to the reefs. The groundwater nutrient concentrations (nitrate, nitrite, ammonia, phosphate and silicate) from Guarajuba beach (a highly urbanised area) are over 10 times higher than groundwater from Papa Gente beach, an area of low human occupation. The pH values of the groundwater samples also indicate the predominance of reducing conditions in Guarajuba, due to the high availability of organic matter and consequent bacterial activity. Additionally, faecal coliform data indicate domestic wastewater as the source of groundwater contamination. High densities of macroalgae and heterotrophic organisms on the impacted reefs, as well as higher concentrations of nutrients, evoke the effects of eutrophication on this coral reef ecosystem. These data suggest that the high availability of nutrients is affecting the trophic structure in the study area, especially in Guarajuba, with the increased turf and macroalgae growth reducing light penetration to the coral colonies, competing with them for space and inhibiting the settlement of new coral larvae.

Introduction

In marine systems, the eutrophication process is dependent on a complex set of interactions. The accumulation of organic matter (derived from primary production) will occur only if nutrient supply to the system exceeds the losses such as deep-layer sedimentation, denitrification and the export of zooplankton and detritus to adjacent waters. In coral reefs, despite the extremely high gross productivity at low levels of nutrient supply, the concentration of particulate material in the water column is low and biological activity is concentrated in the benthos (D'Elia & Wiebe, 1990). However, with continuous nutrient-enrichment of coastal waters, eutrophication constitutes an increasing threat to the health and biodiversity of coral reefs, either directly (calcification rates – Marubini & Davies, 1996; Marubini & Atkinson, 1999) or by enhancing secondary stresses on corals, such as phytoplankton blooms, sedimentation increase, epiphytic algal growth, coral diseases, light limitation, elevated water temperatures and coral bleaching (Bell, 1992). In coastal coral reefs, eutrophication has primarily anthropogenic causes resulting directly from population growth and development (Hallock et al., 1993). In particular, the last 2 decades has seen an enhanced research effort to assess the significance of nutrient enriched groundwater inputs to coastal coral reefs, such as those reported in Guam (Marsh, 1977), Australia (Johannes, 1980), Hawaii (Bienfang, 1980), Jamaica (D'Elia et al., 1981), Barbados (Lewis, 1985), Bermuda (Jickells et al., 1989), and Florida (Lapointe et al., 1990; Shinn et al., 1994; Szmant & Forrester, 1996).

Brazil has the only coral reefs in the south-western Atlantic Ocean, and the coast of Bahia State is the most extensive and richest area. Although a large number of impacts has been documented on Brazilian reefs

(Belém et al., 1986; Leão, 1996; Leão & Ginsburg, 1997; Leão et al., 1997), there is no information regarding the role of eutrophication and no published literature addressing the problem of nutrient enrichment in these coral reef systems. In the north coast of Bahia (Fig. 1), the coastal coral reefs occur along a small strip of discontinuous carbonate build-ups. In the last 15 years, this region has experienced an accelerated urbanisation process, with the construction of roads and the rapid expansion of small villages, especially in the Guarajuba area, leading to an increase in the number of residents and tourists. This comparatively unplanned human occupation was not accompanied by the implantation of an adequate sewage treatment system, thus stimulating the irregular and indiscriminate use of septic tanks. The construction of a sewage collection system began only in the spring of 1997.

In this study, we examined the contamination of groundwater and its pathway, and attempted to assess the relationship between the contaminated groundwater seepage and the eutrophication of the coastal coral reefs. To produce evidence of such contamination, we analysed the inorganic nutrient (nitrate, nitrite, ammonia, phosphate and silicate) concentration and pathogen content (*E. coli*) of samples taken from coastal lakes, groundwater and coral reefs during both rainy and dry seasons. A quantitative assessment of the benthic macrofauna and macroflora in the reef flat was also performed. To evaluate the anthropogenic influence on the eutrophication process, the study was conducted on two Bahian coastal coral reef systems with different levels of human occupation (urbanised and underdeveloped).

Study area

Situated in the northern portion of the Brazilian State of Bahia (38° 03′ W, 12° 38′ S), the coral reefs extend 20 km along the coastline, between the beaches of Abaí and Papa Gente (Fig. 1). These reef systems are located on the narrowest part of the Eastern Brazilian Shelf (15km average width) in a transition from siliciclastic to carbonatic sediments (Leão et al., 1988). In this area, two transects were established perpendicular to the coastline (Fig. 1): transect 1 was located at Guarajuba beach, an urbanised village; transect 2 was located at Papa Gente beach, an area of low population density.

The coastal belt along the State of Bahia has a tropical humid climate, with an elevated average temperature (25 °C) and a high rainfall (1800 mm/y). The rainy season occurs between April and July, producing 53% of the annual rainfall, whilst the dry season extends from September to December (Rao et al., 1993). Dominant easterly winds occur in January and September, northeasterly winds are frequent from October to December and from February to March, and stronger southeasterly winds can occur during storms (April–August; DHN, 1993). Tides (from the port of Salvador city, 50 km distant from study site) ranged between 2.8 m (maximum high) to −0.2 m (minimum low) (DHN, 1997).

In the study area, the coastal plain is characterised by two generations of sand barriers, formed by quaternary sea level fluctuations (Martin et al., 1979). They run parallel to the coastline, generally in a north east-south west direction. Between the two generations of sand barriers, there is a shallow, sediment-filled lake. Adjacent to the beaches occur emergent reefs, also parallel to the coastline, whose dimensions vary from 50 m to 2 km long and 20 m to 500 m wide. Depth of water around these reefs does not exceed 10 m (Leão, 1996). Typical reef lagoons do not exist and the back reef zone usually slopes downward to the quartz-sand beaches. However, the reef tops include a horizontal flat surface, which becomes exposed at low tide. On these exposed surfaces, large dead coral heads (that have been truncated by erosion) alternate with small pools and meandering channels where small heads of living corals survive along with a variety of soft algae. A few kilometres off the coast, shallow bank reefs occur, with irregular shapes and dimensions. Depth here varies from 5 to 20 m (Nolasco & Leão, 1986).

Materials and methods

The comparative nutrient study between the two sites was based on the determination of nitrogen compounds (nitrate, nitrite and ammonia), phosphate (orthophosphate) and silicate (reactive silica) concentrations. Salinity, temperature, pH and faecal coliform organisms, which were used as tracers of groundwater percolation, were also measured. Water and groundwater sampling was undertaken during May and June (rainy season) and November and December (dry season) 1997. Seven wells were core-drilled (maximum depth 2.34 m) into the external sandy barrier, four at Papa Gente and three at Guarajuba, to determine the groundwater level and to collect water samples. These

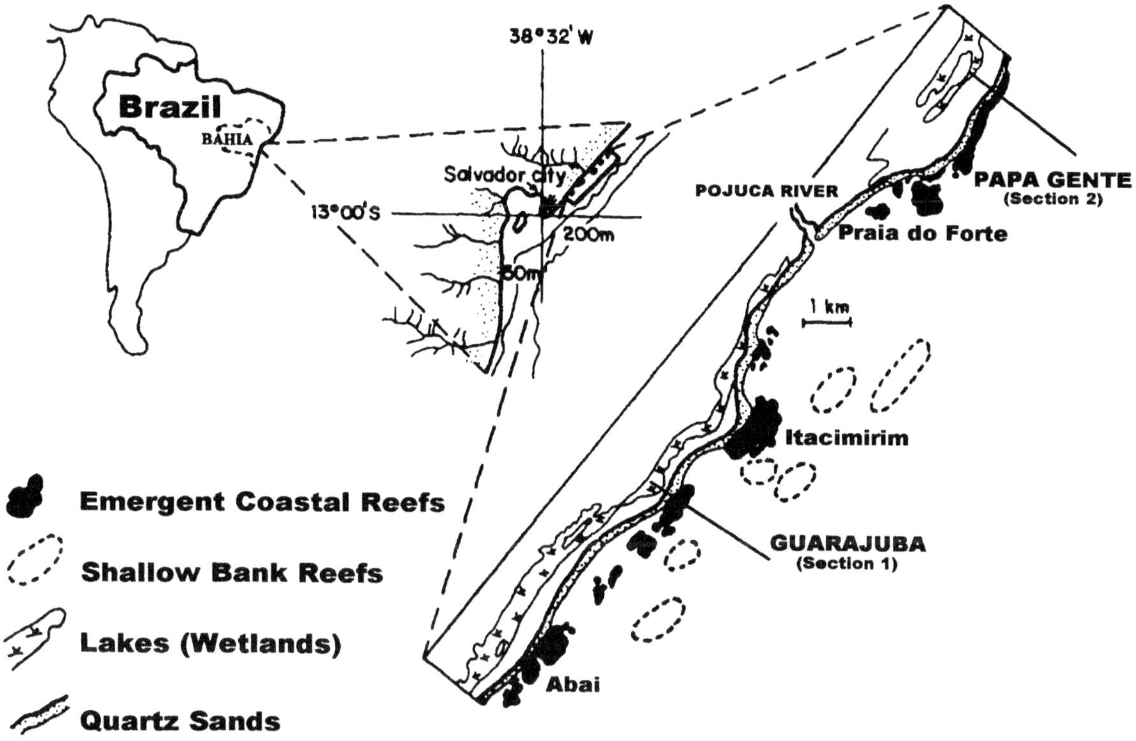

Figure 1. Study area. The Guarajuba reefs are in a dense urbanised region; the Papa Gente reefs are in a underdeveloped region (modified from Leão et al., 1997).

wells were fitted with PVC casings (30 cm diameter) and the samples, collected in glass and polyethylene bottles, were filtered through a GF/C filter and frozen until analysis. For the nutrient analysis, a total of 33 samples was taken from Guarajuba and 37 from Papa Gente, covering the internal lakes, groundwater and coral reefs. For the faecal coliform count, 13 samples were taken from Guarajuba and 12 from Papa Gente.

Inorganic nutrient concentrations were determined according the methods of Grasshoff (1983). The light absorption was measured in a spectrophotometer (Varian, model DMS80). The detection limits of these methods in our experiment were 0.03 μM for PO_4^{3-}, 1 μM for SiO_2, 0.01 μM for NO_3^- and 0.05μM for NH_3. The efficiency of nitrate reduction in the cadmium column was validated regularly to be over 95%. All nutrient values were corrected by using analytical blanks.

Salinity was determined using a Sper Scientific refractometer and temperature with a mercury thermometer. To measure pH the Oakton pH Tester-1 kit (model 35624-00) was used. Faecal coliform count was determined using the membrane filtration technique (APHA, 1992).

The ecological assessment (determination of percent cover of macroalgae and heterotrophs) was performed on the flat for each studied reef utilising a PVC 1 m^2 quadrat with a grid of 10 × 10 cm subdivisions (each corresponding to 1% of the total area) formed using nylon strands. Data were obtained from 66 haphazard quadrats in Guarajuba (30 in the rainy season and 36 in the dry season) and 64 in Papa Gente (28 in the rainy season and 36 in the dry season), distributed throughout the whole reef flat. For each quadrat, the area covered by each studied category was recorded: sponges, Zoanthidea, corals (stony corals as well as hydrocorals and soft corals) and algae (macroalgae and turfs). We also recorded the percentage of bare and/or sand covered areas. According to Foster et al. (1991), these percentage cover estimates may be used as measures of abundance.

The topographic level of the coastal sand barrier in the two transects at Guarajuba and Papa Gente beaches was also measured to estimate the hydraulic head between sea and lake levels, as well the amount of groundwater that reaches the coastal reefs.

Table 1. Mean values of all variables measured at both studied sites: (a) rainy season (n=18) and (b) dry season (n=18)

Papa Gente						Variables	Guarajuba				
Lake	Well 1	Well 2	Well 3	Well 4	Reef		Lake	Well 1	Well 2	Well 3	Reef
(a)											
0.0	0.0	0.0	0.0	0.6	34.0	Salinity (PSU)	0.0	0.0	0.0	0.8	33.3
30.2	27.7	27.3	27.0	27.0	26.0	Temperature (°C)	29.8	26.8	25.8	27.9	28.6
5.7	6.7	7.0	7.5	7.4	8.9	pH	4.7	5.0	5.2	4.4	8.1
0.47	0.25	0.32	0.25	0.17	0.09	Nitrite (μm)	0.22	1.30	1.49	1.42	0.16
6.54	5.14	5.47	5.94	2.16	1.68	Nitrate (μm)	4.38	8.47	16.07	21.20	8.03
13.98	11.24	11.70	10.65	8.73	3.59	Ammonia (μm)	2.73	64.14	76.25	24.75	4.81
3.06	2.02	1.52	1.67	1.98	0.18	Phosphate (μm)	0.25	7.86	7.27	7.59	1.42
35.82	17.48	14.41	12.36	11.67	8.22	Silicate (μm)	48.64	104.08	126.47	91.57	11.21
(b)											
0.0	0.0	0.0	0.1	0.8	35.0	Salinity (PSU)	0.0	0.0	0.0	1.2	33.7
28.5	28.0	26.5	27.0	26.0	27.5	Temperature (°C)	29.5	27.7	26.8	26.2	28.7
6.5	4.6	5.0	6.2	7.0	8.3	pH	4.9	4.3	3.9	4.0	8.1
< 2	< 2	–	–	–	–	Faecal Coliform (colony/100 ml)	< 2	6	8	4	–
0.25	0.18	0.20	0.15	0.13	0.05	Nitrite (μm)	0.24	0.53	0.62	0.59	0.34
2.83	2.65	2.32	1.88	1.27	0.41	Nitrate (μm)	2.57	46.20	65.67	42.14	5.75
3.87	3.30	2.67	2.54	1.86	0.86	Ammonia (μm)	5.62	74.22	88.82	81.23	10.69
2.53	2.07	1.72	0.86	0.44	0.13	Phosphate (μm)	0.59	6.59	4.95	4.78	0.35
122.48	105.04	96.80	64.86	43.40	14.43	Silicate (μm)	203.54	197.77	203.38	201.30	25.86

Results

The topographic profiles for each transect (Guarajuba and Papa Gente) is shown in Figure 2. These profiles refer to the external sandy barrier, located between the lake and the reef, where the groundwater sampling occurred. The Papa Gente transect extended for 840 m, bounded by the lake margin and the reef slope. The vertical distance between the lake water level and mean sea level is 5.02 m (see Fig. 2a) but can reach 5.90 m at low tides, generating a groundwater flux of approximately 45 l m^{-2} day^{-1} towards the coastal reefs. The Guarajuba transect extends for 605 m with a vertical drop of 2.06 m between the lake and sea level (Fig. 2b) which can reach 3.5 m in lower tides. In this case, the flow rate of groundwater to the reefs is approximately 20 l m^{-2} d^{-1}.

Mean values for the physico-chemical variables in both seasons are presented in Table 1. The seasonal and spatial differences were evaluated statistically using a 2-way ANOVA. Water temperatures in lake and groundwater samples were influenced mainly by changes in the insolation and rainfall between the wet and dry seasons, not varying between the two studied sites (Table 2a, 2b). In the reef area, however, the action of waves and currents led to extremely variable temperature behaviour, not only between seasons but also between sites (Table 2c). Salinity of groundwater was consistently less than 2 PSU (Table 1) and appeared to be independent of tidal action. In the lakes, all the samples were equal to 0 PSU, therefore indicating a unidirectional flow of groundwater migrating seaward to the reefs. Conversely, the salinity in the reef area was influenced highly by a combination of tides and groundwater percolation, as the lowest values were found during low tides when the flux of low salinity groundwater was intensified. This pattern was consistent for both study sites and in the two seasons (Table 2c). The pH, however, responded differently and significant differences were recorded, not only between the two sites (ρ=0.000003), but also between seasons (Table 2c). Lake and groundwater samples in Guarajuba presented lower pH values than Papa Gente in all seasons (Table 1).

Faecal coliform counts were performed to provide additional indications of anthropogenic contamination of the groundwater, rather than the occurrence of pathogens *per se*. Sampling was undertaken only in the dry season, when the number of tourists is greater due to the vacation period. These results are also

Table 2. Results of 2-way ANOVA for all investigations. Sampling points (a) lake (wetlands); (b) groundwater (wells); (c) reef pools

Variable	Sites (Guarajuba & Papa Gente)		Seasons (Dry & Rainy)		Interaction (site & season)	
	F (1,32)	p	F (1,32)	p	F (1,32)	p
(a)						
Salinity	All samples = 0 PSU					
Temperature	0.7033	0.4079	8.0572	*0.0078*	4.2616	*0.0472*
pH	21.9298	*<0.0001*	3.4312	0.0732	1.5079	0.2284
NO_3	247.708	*<0.0001*	1292.5130	*<0.0001*	154.3320	*<0.0001*
NO_2	321.1496	*<0.0001*	181.6774	*<0.0001*	273.6422	*<0.0001*
NH_4	8081.11	*<0.0001*	4681.00	*<0.0001*	15145.71	*<0.0001*
PO_4	2293.3890	*<0.0001*	3.8360	*<0.0001*	76.7920	*<0.0001*
SiO_2	4541.71	*<0.0001*	30070.32	*<0.0001*	2399.18	*<0.0001*
(b)						
Salinity	2.4828	0.1249	2.4828	0.1249	0.2759	0.6031
Temperature	3.7448	0.0619	21.8248	*<0.0001*	1.2414	0.2735
pH	314.1617	*<0.0001*	5.4512	0.2599	0.0097	0.9220
NO_3	9954.407	*<0.0001*	1114.513	*<0.0001*	1322.028	*<0.0001*
NO_2	6567.875	*<0.0001*	1726.464	*<0.0001*	1401.201	*<0.0001*
NH_4	570477.3	*<0.0001*	154277.15	*<0.0001*	251611.8	*<0.0001*
PO_4	7258.291	*<0.0001*	1389.548	*<0.0001*	117.760	*<0.0001*
SiO_2	43045.39	*<0.0001*	15231.04	*<0.0001*	4631.10	*<0.0001*
(c)						
Salinity	3.6333	0.0657	1.5973	0.2154	0.4125	0.5253
Temperature	36.0594	*<0.0001*	6.1346	*0.0187*	5.1211	*0.0306*
pH	31.6098	*<0.0001*	8.8049	*0.0057*	7.9024	*0.0084*
NO_3	3782.730	*<0.0001*	346.949	*<0.0001*	28.305	*<0.0001*
NO_2	211.3174	*<0.0001*	107.8150	*<0.0001*	250.1053	*<0.0001*
NH_4	41502.50	*<0.0001*	3363.48	*<0.0001*	25201.72	*<0.0001*
PO_4	4867.098	*<0.0001*	2841.483	*<0.0001*	2365.204	*<0.0001*
SiO_2	430.3752	*<0.0001*	900.9483	*<0.0001*	147.3870	*<0.0001*

shown in Table 1 and suggest that groundwater in Guarajuba has been contaminated by the outflow of septic tanks.

The mean nutrient concentrations between the two sites were distinctly different (Tables 1 and 2), despite their proximity and similar geologic and geomorphologic conditions, with the populated site (Guarajuba) containing nutrients at levels many times higher than the underdeveloped site (Papa Gente). Besides differences from one site to another, the nutrients also behaved distinctly between seasons (Table 2a–c), reflecting the role of rainfall in nutrient dilution and transport.

The silicate concentrations varied between 5 and 17 μm at Papa Gente and between 17 and 104 μm at Guarajuba (Table 1), and the concentrations of ammonia in Guarajuba groundwater were about 10 times higher than Papa Gente, particularly in the dry season. High concentrations of dissolved phosphorus were also found in Guarajuba wells; the mean concentration of phosphorus for all groundwater samples at this site was about three times higher than the Papa Gente site (Table 1). The nitrate concentrations also followed the pattern of the other nutrients, with the highest levels observed in Guarajuba, mainly in the dry season.

Results of the ecological assessment on the reef flats are presented in Figure 3. The algae (macroalgae

Table 3. Comparative nitrate and phosphate values for coastal reefs under land-based nutrient fluxes, including the studied reefs of Guarajuba and Papa Gente

Coral reef sites (references)	Nitrate (μM)	Phosphate (μM)
Tumon Bay, Guam (Marsh, 1977)	4.14–8.04	0.14–0.55
Guarajuba, Brazil (this work)	2.26–12.07	0.12–1.62
Jakarta Bay, Indonesia (Tomascik et al., 1993)	2.71	1.36
Pago Bay, Guam (Marsh, 1977)	0.22–3.31	0.15–0.23
Papa Gente, Brazil (this work)	0.34–2.57	0.09–0.32
Barbados (Tomascik & Sander, 1985, 1987a, 1987b)	0.32–0.71	0.06–0.11
Florida Keys: Inshore (Szmant & Forrester, 1996)	0.46–1.07	0.01–0.17
Florida Keys: Offshore (Szmant & Forrester, 1996)	0.16–0.32	0.01–0.11
Bahamas (Ferrer & Szmant, 1988)	0.29	0.08
Eniwetok, Ilhas Marshall (Pilson & Betzer, 1973; Webb et al., 1975)	0.10–0.30	0.16–0.18

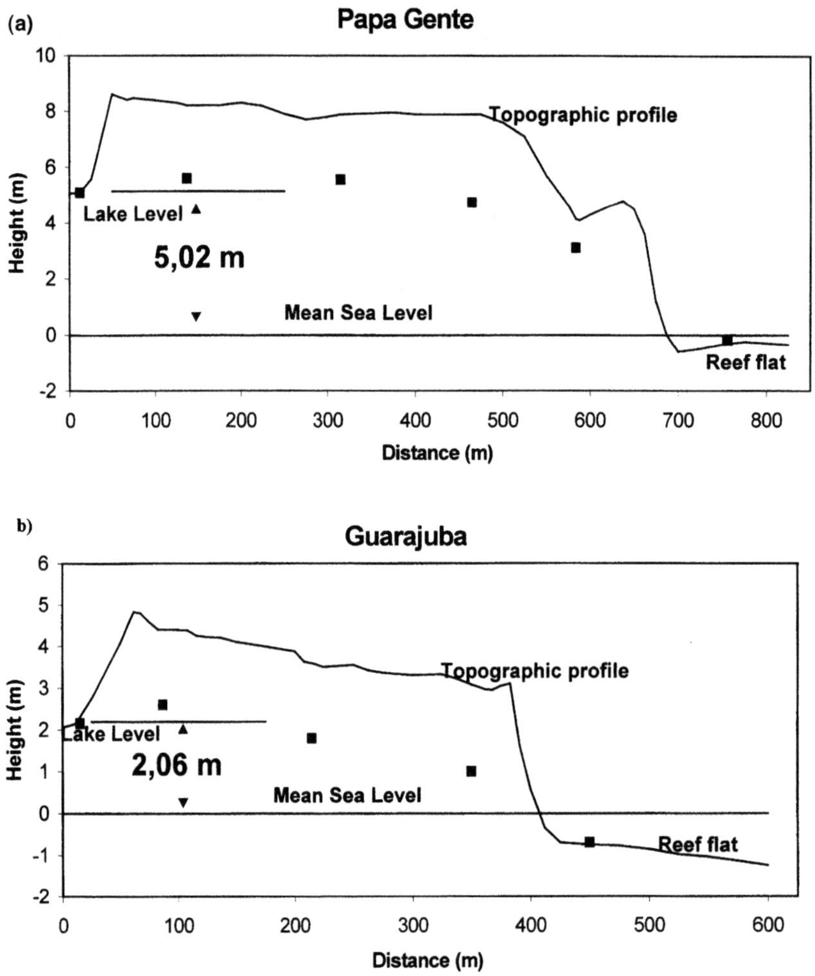

Figure 2. Topographic profiles of the sandy barrier in both study sites showing the fall between lake and sea levels. The squares indicate groundwater sampling points. (a) Papa Gente section; (b) Guarajuba section.

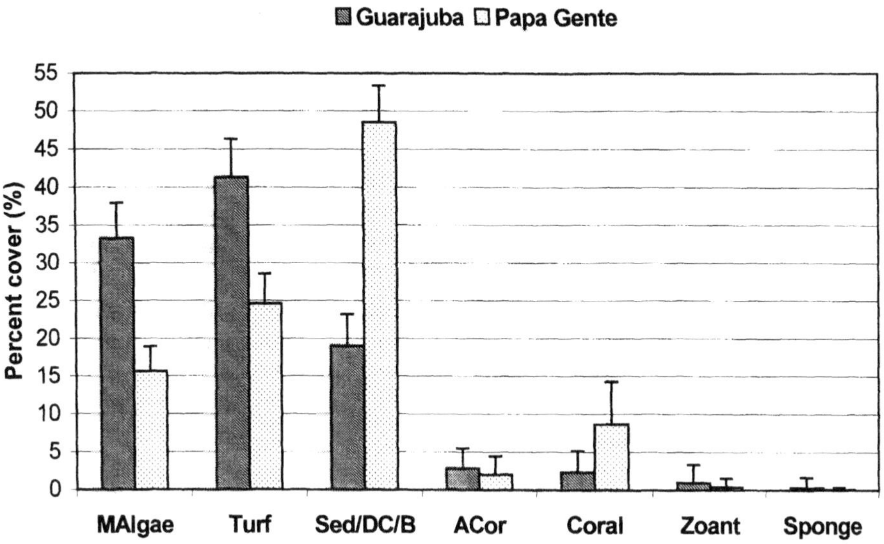

Figure 3. Reef flat percent cover of each major benthic community component in both studied sites. (Malgae = macroalgae; Turf = turf algae; Sed/DC/B = sedimento, dead coral and/or bare areas; Acor = coralline algae; Coral = living coral; Zoant = zoanthids; Sponge = sponges.)

and turf algae) constituted the main component of the benthic community in the studied reef flats with 77% of total cover in Guarajuba and 41% in Papa Gente. Coral cover was extremely low, varying from 2.31% in Guarajuba to 8.62% in Papa Gente. The other heterotrophs (sponges and zoanthids) contributed just over 1.2% of the total cover in Guarajuba but less than 0.5% in Papa Gente.

Discussion

Most of the bedrock underlying the study area is a highly porous sandstone and beach rock. The groundwater level is extremely shallow, as the drilling has shown (average value of 1.7 m depth), and the contaminated wastewater in the septic tanks can flow easily through it. The pH values found in Guarajuba wells highlight the more reducing conditions of groundwater at this site, as a response to higher levels of organic matter. Similarly, faecal coliform counts, although low, do indicate that groundwater in Guarajuba has been contaminated by human waste. In addition, high concentrations of ammonia were found in groundwater, indicating contamination from septic tanks (Pitt et al., 1975). Therefore, reducing conditions dominate this environment. However, differences in data obtained from the two seasons, and the lower levels of ammonia and higher nitrate found in the rainy season, may suggest that a recharge of the permeable aquifer by oxygenated rainfall infiltration can allow an increasing oxidation of ammonia to nitrate. Shinn et al. (1994) found similar patterns while studying percolation of human-waste contaminated groundwater in the Florida Keys.

Elevated concentrations of nitrate can be produced locally by groundwater discharge and surface runoff (Marsh, 1977) and, therefore, can also be used as an indicator of wastewater contamination. Again, the average concentration values in the highly urbanised area were many times that of the undeveloped site, varying from 0.41–1.68 μm in Papa Gente, and from 5.75–8.03 μm in Guarajuba. The relatively low levels of intermediate nitrite, however, could be explained by the fact that ammonia oxidation (to nitrite) and nitrite oxidation (to nitrate) are closely coupled (Webb & Wiebe, 1975).

Silicate concentrations, along with salinity values, can be used as a marker of groundwater discharge (Montaggionni et al., 1993). In the study area, the silicate values indicated that groundwater was likely to be producing a significant input of terrestrial nutrients onto the reef. Marsh (1977) reported that groundwater seepage to the fringing reef at Tumon Bay (Guam) occurred in the form of rills of brackish water (from a subterranean lens) which was observable at low tide. In our study area, a similar seepage occurred at both low and high tide due to the substantial drop between lake and sea water levels (Fig. 2). Therefore, with a flow rate of at least 20 l $m^{-2}d^{-1}$, the groundwater percolation in the study area is likely to constitute one of the main sources of nutrient-rich water to the coastal reefs, especially in Guarajuba.

This model of nutrient enrichment via groundwater seepage is a plausible mechanism that can explain the eutrophication occurring in coral reef systems on the northern coast of Bahia. The nutrient availability seems to be affecting the community structure and, in that condition, the fast-growing turf and macroalgae tend to colonise most of the open substrata, inhibiting the settlement of new coral larvae (Birkeland, 1977; Hallock et al., 1993). As the fishing pressure is the same for both studied sites (there is a fishing community covering the region, so no differential in predator removal) as well as the circulation pattern and geomorphology, the only factor that could explain the difference in the alga/coral cover between the two sites is the differential availability of nutrients. Additionally, the increases in density of bioeroders like sea urchins (e.g. *Echinometer lucunter, Eucidares* sp.) and sponges (e.g. *Anthosigmella varians, Spirastrella* sp., *Cinachirella alboclade* and *Cliona* sp.) reflect the eutrophic conditions. These consequences are in agreement with previous work (Hallock & Schlager, 1986; Hallock, 1988; Montaggionni et al., 1993) in which excess of nutrients appears to be clearly deleterious to reef-building biota and calcification rates. Table 3 presents a comparison between the data produced in this study with those reported in the literature for coastal reefs under the effect of land-based nutrient sources.

A further consequence of eutrophication is the reduction of water transparency as a product of nutrient increase, limiting the depth ranges of zooxanthelate corals, thereby, reducing carbonate production. Thus, in eutrophic conditions, many fast-growing competitors are bioeroders that actively destroy the reef structure and, once the rates of carbonate production and bioerosion are similar, even modest increases in nutrient availability can shift a reef community from net production to net erosion (Ogden et al., 1973; Sammarco et al., 1974). This is probably the case in the northern Bahian reefs, although more studies on bioerosion rates are required to support this hypothesis.

In previous investigations undertaken in the reefs of northern Bahia, the implicit assumption has been that corals declined as a function of historical changes in sea level. However, even being true, this may be a too simplistic explanation, as the nutrification of reefs in this area is obvious and an eventual recovery will be limited by effects such as declining populations of reef building corals, increasing abundance of benthic algae, destabilised herbivore populations, increasing rates of bioerosion and reduction of calcification.

To further understand the process, future work should consider the evaluation of biotic pigments (chlorophyll and ATP) in the water around the reefs, since the biota, especially the benthic biota, play an important role in the uptake and retention of available nutrients (Laws & Redalje, 1979; Szmant-Froelich, 1983; Hallock et al., 1993; Szmant, 1997). In addition, the sediment evaluation (organic matter, carbon content, bacterial activity) should provide complementary information about the nutrification process, as the sediment constitutes the major nutrient source for benthic producers in shallow water systems (Szmant & Forrester, 1996; Szmant, 1997).

Acknowledgements

The authors are grateful to Dr Vera Cancio, Rosane, Erika, Ana Paula and Audrey (Institute of Chemistry/UFBA), for their help with nutrient analysis; to Adriana, Miguel, Kelmo and Leda (Institute of Biology/UFBA), for help with algae, sponges and zoanthids identification; and to staff from Coastal Studies Laboratory/UFBA for incentive and support. This work was supported by grants from CAPES and CNPq (Brazilian Federal Government), and CADCT (Bahian State Government).

References

APHA, 1992. Standard Methods for the Examination of Water and Wastewater. APHA/AWWA/WEF, Washington: 949 pp.

Belem, M. J. C.; C. Rohlfs; D. O. Pires & C. B. Castro, 1986. S.O.S. corais. Ciência Hoje 5(26): 34–42.

Bell, P. R. F., 1992. Eutrophication and coral reefs – some examples in the Great Barrier Reef lagoon. Wat. Res. 26(5):553–568.

Bienfang, P., 1980. Water quality characteristics of Honokohau harbour: a subtropical embayment affected by groundwater intrusion. Pacific Sci. 34: 279–291.

Birkeland, C., 1977. The importance of rate of biomass accumulation in early successional stages of benthic communities to the survival of coral recruits. In Proceedings of International Coral Reef Symposium, 3, Miami, 1977, Florida, 1: 15–21.

D'elia, C. F., K. L. Webb & J. W. Porter, 1981. Nitrate-rich groundwater inputs to Discovery Bay, Jamaica: a significant source of N to local coral reefs? Bull. mar. Sci. 31(4): 903–910.

D'elia, C. F. & W. J. Wiebe, 1990. Biogeochemical nutrient cycles in coral-reef ecosystems. In Dubinsky, Z. (ed), Ecosystems of the World 25: Coral reefs, Elsevier Pub., Amsterdam, 49–74.

DHN, 1993. Atlas de Cartas Piloto. Marinha do Brasil, Diretoria de Hidrografia e Navegação, Rio de Janeiro: 24 pp.

DHN, 1997. Tábuas das Marés. Marinha do Brasil, Diretoria de Hidrografia e Navegação, Rio de Janeiro, 108 pp.

Ferrer, L. M. & A. M. Szmant, 1988. Nutrient regeneration by the endolitic community in coral skeletons. In Proceedings of International Coral Reef Symposium, 6, Townsville, 1988, Australia: 3: 1–4.

Foster, M. S., C. Harrold & D. D. Hardin, 1991. Point vs. Photo quadrat estimates of the cover of sessile marine organisms. J. exp. mar. Biol. Ecol., 146: 193–203.

Grasshoff, K., M. Ehrhardt & K. Kremling, 1983. Methods of seawater analysis. Verlag Chemie, Germany: 419 pp.

Hallock, P., 1988. The role of nutrient availability in bioerosion: consequences to carbonate buildups. Palaeogeography, Palaeoclimatology, Palaeoecology 63: 275–291.

Hallock, P. & W. Schlager, 1986. Nutrient excess and the demise of coral reefs and carbonate platforms. Palaios 1: 389–398.

Hallock, P., F. Muller-Karger & J. C. Halas, 1993. Coral reef decline – anthropogenic nutrients and the degradation of Western Atlantic and Caribbean coral reefs. Research and Exploration 9(3): 358–378.

Jickels, T. D., S. R. Smith, S. Boyd & A. H. Knap, 1989. Phosphorus limitation in the Bermuda inshore waters. Estuar. coast. shelf Sci. 28: 121–132.

Johannes, R. E., 1980. The ecological significance of the submarine discharge of groundwater. Mar. Ecol. Prog. Ser. 3: 365–373.

Lapointe, B. E., J. D. O'Connel & G. S. Garret, 1990. Nutrient couplings between on-site sewage disposal systems, groundwaters and nearshore surface waters of the Florida Keys. Biogeochemistry 10: 289–307.

Laws, E. A. & D. G. Redalje, 1979. Effect of sewage enrichment on the phytoplankton population of a tropical estuary. Pacific Sci. 33: 129–144.

Leão, Z. M. A. N., 1996. The coral reefs of Bahia: Morphology, distribution and the major environmental impacts. Ann. acad. bras. Ci. 68(3): 439–452.

Leão, Z. M. A. N. & R. Ginsburg, 1997. Living reefs surrounded by siliciclastic sediments: the Abrolhos coastal reefs, Bahia, Brazil. In Proceedings of International Coral Reef Symposium, 8, Panama, 1996, Panama, 2: 1767–1772.

Leão, Z. M. A. N., T. M. F. Araújo & M. C. Nolasco, 1988. The coral reefs off the coast of eastern Brazil. In Proceedings of International Coral Reef Symposium, 6, Townsville, 1988, Australia 3: 339–347.

Leão, Z. M. A. N.; R. K. P. Kikuchi; M. P. Maia & R. A. L. Lago, 1997. A catastrophic coral cover decline since 3000 years B.P., Northern Bahia, Brazil. In Proceedings of International Coral Reef Symposium, 8, Panama, 1996, Panama 1: 583–588.

Lewis, J. B., 1985. Groundwater discharge onto coral reefs, Barbados (West Indies). In Proceedings of International Coral Reef Congress, 5, Tahiti, 1985, French Polynesia, 6: 477–481.

Marsh, J. A. Jr., 1977. Terrestrial inputs of nitrogen and phosphorus on fringing reefs of Guam. In Proceedings of International Coral Reef Symposium, 3, Miami, 1977, Florida 1: 331–336.

Martin, L.; J. M. Flexor; G. S. Vilas Boas; A. C. S. P. Bittencourt & M. M. M. Guimarães, 1979. Courbe de variations du niveau relatif de la mer au cours des 7000 dernières années sur un secteur homogène du littoral brésilien (Nord de Salvador - Bahia). In proceedings of international symposium on coastal evolution in the quaternary, São Paulo, 1978, Brazil: 264–274.

Marubini, F. & M. J. Atkinson, 1999. Effects of lowered pH and elevated nitrate on coral calcification. Mar. Ecol. Prog. Ser. 188: 117–121.

Marubini, F. & P. S. Davies, 1996. Nitrate increases zooxanthellae population density and reduces skeletogenesis in corals. Mar. Biol. 127: 319–328.

Montaggioni, L. F., P. Cuet & O. Naim, 1993. Effect of nutrient excess on a modern fringing reef (Reunion Islands, Western Indian Ocean): Geological implications. In Ginsburg, R. N. (compiler), Proceedings of Colloquium on Global Aspects of Coral Reefs: Health, Hazards and History, RSMAS, University of Miami, Florida: 397–403.

Nolasco, M. C. & Z. M. A. N. Leão, 1986. The carbonate buildups along the northern coast of the State of Bahia, Brazil. In: Rabassa, J. (ed.), Quaternary of South America and Antarctic Peninsula. A.A. Balkama Pub, Rotterdam, 4: 159–190.

Ogden, J. C., R. A. Brown & N. Salesky, 1973. Grazing by the echinoid *Diadema antillarum* Philippi: formation of halos around West Indian patch reefs. Science 182: 715–717.

Pilson, M. E. Q. & S. B. Betzer, 1973. Phosphorus flux across a coral reef. Ecology 54: 581–588.

Pitt, W. A.; H. C. Mattraw & H. Klein, 1975. Ground-water quality in selected areas serviced by septic tanks, Dade County, Florida. U.S.Geological Survey Open-File Report 75-607: 82 pp.

Rao, V. B.; M. Lima & S. H. Franchito, 1993. Seasonal and interannual variations of rainfall over Eastern Northeast Brazil. J. Climate 6: 1754–1763.

Sammarco, P. W., J. S. Levinton & J. C. Ogden, 1974. Grazing and control of coral reef community structure by *Diadema antillarum* Philippi (Echinodermata: Echinoidea): A preliminary study. J. mar. Res. 32: 47–53.

Shinn, E. A., R. S. Reese & C. D. Reich, 1994. Fate and pathways of injection-well effluent in the Florida Keys. U.S. Geological Survey Open-File Report 94–276: 116 pp.

Szmant-Froelich, A., 1983. Functional aspects of nutrient cycling on coral reefs. In The Ecology of Deep and Shallow Coral Reefs. NOAA Undersea Research Program 1: 133–139.

Szmant, A. M., 1997. Water quality characterization and the health of coral reefs. Internet WWW page at URL: http://www.epa.gov/OWOW/oceans/coral/s zmant.html (accessed 19/11/1997)

Szmant, A. M. & A. Forrester, 1996. Water column and sediment nitrogen and phosphorus distribution patterns in the Florida Keys, U.S.A. Coral Reefs 15: 21–41.

Tomascik, T. & F. Sander, 1985. Effects of eutrophication on reef-building corals. I – Growth rate of the building coral *Montastrea annularis*. Mar. Biol. 87: 143–155.

Tomascik, T. & F. Sander, 1987a. Effects of eutrophication on reef-building corals. II – Structure of scleractinian coral communities on fringing reefs, Barbados, West Indies. Mar. Biol. 94: 53–75.

Tomascik, T. & F. Sander, 1987b. Effects of eutrophication on reef-building corals. III – Reproduction of the reef-building coral *Porites porites*. Mar. Biol. 94: 77–94.

Tomascik, T.; Suharsono & A. J. Mah, 1993. Case histories: a historical perspective of the natural and anthropogenic impacts in the Indonesian Archipelago with a focus on the Kepulauan Seribu, Java Sea. In Ginsburg, R. N. (ed.), Proceedings of Colloquium on Global Aspects of Coral Reefs: Health, Hazards and History. RSMAS, University of Miami, Florida: 304–310.

Webb, K. L. & W. J. Wiebe, 1975. Nitrification on a coral reef. Can. J. Microbiol. 21: 1427–1431.

Webb, K. L.; K. W. DuPaul, W. J. Wiebe, W. Sottille & R. E. Johannes, 1975. Enewetak (Eniwetok) Attoll: Aspects of the nitrogen cycle on a coral reef. Limnol. Oceanogr. 20: 198–210.

Hydrobiologia **440:** 317–329, 2000.
M.B. Jones, J.M.N. Azevedo, A.I. Neto, A.C. Costa & A.M. Frias Martins (eds), Island, Ocean and Deep-Sea Biology.

Stable isotope evidence of benthic microalgae-based growth and secondary production in the suspension feeder *Cerastoderma edule* (Mollusca, Bivalvia) in the Marennes-Oléron Bay

Pierre-Guy Sauriau[1] & Chang-Keun Kang[2]
[1]*Centre de Recherche en Ecologie Marine et Aquaculture de L'Houmeau (CNRS-IFREMER, UMR-10), BP 5, F-17137 L'Houmeau, France*
E-mail: Pierre.Guy.Sauriau@ifremer.fr
[2]*National Fisheries Research & Development Institute (Pohang Regional laboratory), 616 Duhodong, Buk-gu, Pohang 791-110, South Korea*

Key words: growth, secondary production, *Cerastoderma edule*, stable isotope ratios, microphytobenthos, intertidal, Marennes-Oléron Bay

Abstract

The contribution of natural food sources to the growth and secondary production of the suspension feeding bivalve *Cerastoderma edule* (L.) was estimated under field conditions in the Marennes-Oléron Bay (Atlantic coast, France). Monthly estimates of abundance, biomass and cockle growth were combined with seasonal analyses of $\delta^{13}C$ and $\delta^{15}N$ ratios of juvenile and adult cockles, together with their potential food sources [i.e. suspended particulate organic matter (POM), microphytobenthos, macroalgae and seagrass] sampled at mid-tide level in a muddy sandflat. Adult cockles grew mainly in spring, whereas juveniles grew in summer and autumn, following spat recruitment in early summer. Total annual production and elimination of cockles were estimated to be 32.5 and 34.7 g AFDW m^{-2} yr^{-1}. Relative contributions of each year class to production were ca 40, 41, 11 and 6% for 0-group, 1-, 3- and 4-yr-old cockles in 1995, respectively. Quantitative assessment of proportions of food sources to the annual secondary production of cockles was obtained by using a simple carbon isotope-mixing model with microphytobenthos ($\delta^{13}C$ = -16.0 ± 0.6 ‰) and POM ($\delta^{13}C$ = -22.2 ± 1.1 ‰) as end-members. On average, more than 70% of the total annual cockle production originated from microphytobenthos, with a much higher contribution for the 0-group (88%) than for adult cockles (60%). The between-age difference was induced mainly by changes in the availability of food resources (benthic *versus* planktonic) during the non-synchronous growing seasons of juvenile and adult cockles.

Introduction

The edible cockle *Cerastoderma edule* (L.) is one of the most common benthic suspension feeders on tidal flats along the European Atlantic coast (Tebble, 1966). This species exhibits large year-to-year and between-site fluctuations in larval recruitment success, and of subsequent survival and growth of juveniles (Kristensen, 1957; Dörjes et al., 1986; Ducrotoy et al., 1991). Consequently, relative contributions of each year class to the total annual secondary production of cockle populations are highly variable, as reported previously (Evans, 1977; Möller & Rosenberg, 1983). Amongst the main factors that may induce these variations are abnormal thermal conditions (Orton, 1933; Beukema, 1982; Guillou et al., 1992), predation by shore crabs, shrimps and fish (Reise, 1985; Sanchez-Salazar et al., 1987), intraspecific competition (André & Rosenberg, 1991; Bachelet et al., 1992; Jensen, 1993), and spatfall transport by near-bottom flows (Baggerman, 1953; de Montaudouin & Bachelet, 1996). However, little attention has been paid to the influence of changes in the diet available for suspension feeding cockles during critical periods of their life cycle. Jensen (1992) hypothesised that mismatch in growing seasons of juvenile and adult cockles may induce a differential use of food resources; this may minimise intraspecific competition within the cockle population (Kang et al., 1999). Nevertheless, the hypothesis was not supported by the literature despite several studies on the feeding physiology of *C. edule* describing size-dependent variations (Hawkins et al., 1990; Iglesias & Navarro, 1991; Small et al., 1997) and selective ingestion of organic matter (Prins et al., 1991; Urrutia et al., 1996; Navarro & Widdows, 1997).

Various potential food sources are available to benthic suspension feeders in the seston of coastal and estuarine environments (Newell, 1979; Dame, 1996). It is, however, difficult to assess the relative contribution of each food source to their growth and secondary production because direct observation of feeding behaviour is difficult over long time periods in the field (Bayne et al., 1977; Dame, 1996). Indirect methods, such as gut content analyses (e.g. Verwey, 1952; Ivell, 1981; Kamermans, 1994), are also not satisfactory since they give snapshot information on food items that have been ingested and not what is assimilated over a long period of time. Stable isotope analyses can provide answers to this question (Fry & Sherr, 1984), as the $^{13}C/^{12}C$ and $^{15}N/^{14}N$ ratios of an animal directly reflect the composition of food sources assimilated and incorporated over time (DeNiro & Epstein, 1978, 1981). Stable isotope analyses have been used successfully in many studies of spatial and temporal variations of isotopic composition in invertebrates and their potential diets in various estuarine and saltmarsh food webs (see reviews by Fry & Sherr, 1984; Michener & Schell, 1994). Only Junger & Planas (1994), however, have combined stable carbon isotope mixing models with quantitative inventories of macro-invertebrate biomasses to assess quantitatively the trophic base of lotic ecosystems. Consequently, the contribution of various food sources to the annual growth and secondary production of intertidal marine species has never been estimated *in situ*.

The purpose of the present study was to quantify the relative contribution of natural food sources to the growth and secondary production of the cockle *Cerastoderma edule* in muddy sandflats of the Marennes-Oléron Bay. Our latest results (Kang et al., 1999) indicated that juvenile and adult cockles were able to preferentially ingest and/or assimilate microphytobenthos and that there were only two major sources of organic matter assimilated by cockles i.e. particulate organic matter (POM) advected on these tidal flats through water channels and locally resuspended microphytobenthos (MPB). Therefore, in this study, we used a simple two-source mixing model of stable carbon isotopes (Fry & Sherr, 1984) combined with growth and secondary production estimates to quantify the relative contribution of each food source to the total annual production of the mid-tide level cockle population.

Materials and methods

Study site

The study took place on the muddy intertidal sandflat of Ronce-les-Bains, at the southern end of the Marennes-Oléron Bay, halfway along the French Atlantic coast (Fig. 1a). The bay is characterised by semidiurnal tides with a maximal range of 6.5 m. Marine waters enter mostly via. the north entrance of the bay (Pertuis d'Antioche). They are also introduced through the south-west entrance (Pertuis de Maumusson) but their influence on the southern part of the bay is restricted to flood tides. Interaction between the tidal regime and local bathymetric features induces a southward residual transport of marine water masses, which are drained out of the bay through the Pertuis de Maumusson. The water residence time in the whole bay varies between 5 and 10 d depending on tides (Raillard & Ménesguen, 1994). The River Charente, on the east, is the major source of freshwater input into the bay (Ravail et al., 1988), whereas freshwater input from the River Seudre is poor (Soletchnick et al., 1998). Average salinity within the bay ranges from 28 to 33 (Soletchnick et al., 1998). Seston dynamics are affected strongly by advection of marine water masses, and by wind- and tide-induced resuspension of both muddy sediments (Raillard et al., 1994; Raillard & Ménesguen, 1994) and benthic microalgae (Razet et al., 1990; Zurburg et al., 1994). Additional information on seasonal hydrobiology, sedimentary conditions and shellfish rearing activities in the bay is given by Héral et al. (1983), Sauriau et al. (1989) and Héral et al. (1990), respectively.

At Ronce-les-Bains, cockle beds extend over a 50 ha surface area (Fig. 1b) and the cockle population at mid-tide level appears to be stable over time (Sauriau, 1992). All lower parts of the shore, between MLWN and ELWS, are devoted to farming oysters and cockles are scarce within the oyster parks (Fig. 1b). At mid-tide level, sediments are dominated by fine muddy sands (Guillou et al., 1990) with an organic matter content of ca 1% (loss of weight on ignition at 450 °C). Dominant macrofauna are *Hydrobia ulvae* (Pennant),

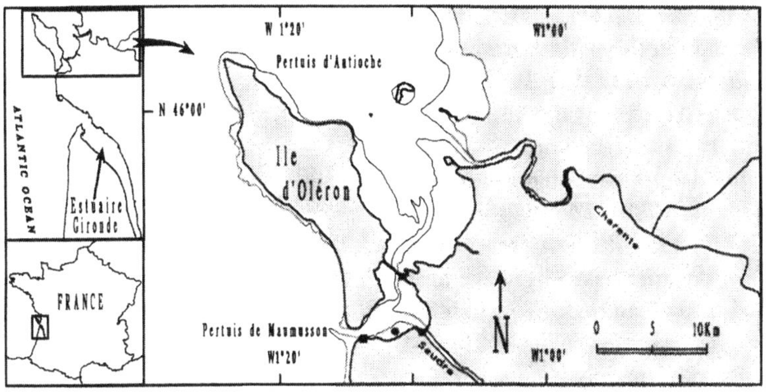

Figure 1a. Map of Marennes-Oléron Bay showing location of the three hydrological sampling stations in channels (■) and the sampling site (•) at Ronce-les-Bains.

Figure 1b. Aerial photograph of Ronce-les-Bains muddy sandflats in August 1993 showing oyster parks on lower parts of the shore and location of the sampling site (•) at mid-tide level. Channels and seashore line are located on the top left-hand corner and the bottom right-hand corner, respectively.

Cerastoderma edule, *Notomastus latericeus* (M. Sars), *Tharyx marioni* (de Saint-Joseph), *Scoloplos armiger* (O.F. Müller) and *Arenicola marina* (L.). In the past, the mid-tide level area was covered by a dense bed of *Zostera noltii* (Hornem), but it vanished in 1985–1986 (Sauriau, 1992), possibly due to local changes in bathymetry (Soletchnick et al., 1998). The seagrass bed has not recovered except for a few *Zostera* spots (<0.1 m^2). Bottom roughness is characterised by wave and current ripple marks of a few centimetres in height. Although several macroalgae species occur in Ronce-les-Bains, they are restricted mainly to the oyster parks. Greening of intertidal sandflats [i.e. summer development of *Ulvae* and *Enteromorpha* algal mats as reported by Thiel & Watling (1998)], is never observed at Ronce-les-Bains. However, some strands of *Chondria dasyphylla* (Woodward), *Enteromorpha* and *Ulva* species have been observed attached to dead cockle shells, mainly on the lower muddy-sand area, throughout the year.

Sample of cockles

Cockles were collected monthly from February 1995 to February 1996. Three replicate plots were sampled randomly using a 0.25 m^2 aluminium corer to determine mean monthly abundance and biomasses at mid-tide level. Sediment was sieved through a 1 mm aperture mesh, and all recovered animals were stored in 4% neutralised formalin and sorted in the laboratory. To determine monthly length weight relationship, cockles (10 for each year class, total >30 individuals) were collected by hand at each sampling date and kept alive overnight in filtered seawater to evacuate gut contents. Shell lengths were measured with Vernier callipers to the nearest 0.05 mm, prior to dissection of individuals. Flesh was separated from shells, stored at −18°C for at least 48 h and freeze-dried to determine flesh dry weight (DW). Subsamples of dry flesh were combusted in a muffle furnace at 450 °C for 6 h to estimate ash contents and ash-free dry weight (AFDW). Flesh organic carbon content was determined using a CHN elemental analyser (Perkin Elmer 2400) after acidification (10% HCl) to remove carbonates. The allometric relationships at each sampling date were expressed as log W = log a + b log L (W = flesh weight and L = shell length). Length data of the samples collected to estimate production were converted into flesh weight and ash-free dry weight using these allometric equations at each sampling date. The age of each cockle was determined by counting the clearly distinguishable annual winter rings, as reported by Sauriau (1992) for this cockle population.

Production and elimination

Production (P) and elimination (E) terms of each year class were calculated according to incremental and removal summation methods reviewed by Crisp (1984) for populations with non continuous recruitment and distinguishable year classes:

$$P = \sum_{t=0}^{t=n} [(N_t + N_{t+1})/2 \cdot (W_{t+1} - W_t)] \text{ and}$$

$$E = \sum_{t=0}^{t=n} [(W_t + W_{t+1})/2 \cdot (N_t - N_{t+1})],$$

where N_t = abundance of a cohort at time t, W_t= mean individual flesh weight in a cohort at time t and n = the number of sampling occasions at different sampling times t.

The mean biomass was calculated by:

$$B = [\sum_{t=1}^{t=n} (N_t \cdot \mathrm{W}_t)]/n,$$

where n = number of sampling dates in a year.

Previous studies have expressed production estimates of *Cerastoderma edule* in ash-free dry weight units (Warwick & Price, 1975; Hibbert, 1976; Evans, 1977; Jensen, 1993) or in organic carbon content (Warwick et al., 1979). To convert flesh dry weight (DW) into ash-free dry weight (AFDW) and organic carbon content (C), we obtained the following two regression equations:

$$\mathrm{AFDW} = 0.915\ \mathrm{DW} - 0.003\ (R^2 = 0.999, n = 729)$$

$$C = 0.321\ \mathrm{DW} + 0.001\ (R^2 = 0.987, n = 578)$$

Sampling and analyses of stable isotopes in cockles and food sources

Cockles for stable isotope analyses were collected seasonally (March, June, September and December) at Ronce-les-Bains. To determine the isotopic value for POM introduced on tidal flats through central channels of the bay, spring and neap tidal samplings were conducted seasonally at three stations: Pertuis de Maumusson, the mouth of the River Seudre and Le Chapus (Fig. 1a), and prepared for analysis as indicated previously by Kang et al. (1999). Macroalgae and seagrass were collected seasonally at Ronce-les-Bains. They were cleaned of epibionts, washed with 10% HCL to remove carbonates, rinsed with Milli-Q water, freeze-dried, ground to powder and kept frozen (−80 °C) until analysis. Microphytobenthos isotope values were given by Riera et al. (1996) and Riera & Richard (1996, 1997) from microphytobenthos samples collected on various occasions on mudflats in the Bay of Marennes-Oléron.

Stable isotope analyses followed the standard protocols used by Riera & Richard (1996, 1997). Samples were combusted at 900 °C using CuO as an oxidant in evacuated quartz tubes. The resulting CO_2 was purified using a cryogenic distillation method similar to that described by Boutton (1991). Before the purification of CO_2, N_2 was trapped on silica gel granules in a stop-cock sample ampoule and analysed immediately after CO_2 collection. The carbon and nitrogen isotope ratios were measured using a Sigma 200 (CJS Sciences) double inlet, triple collector isotope ratio mass spectrometer. Stable isotope ratios are reported in the standard δ unit notation as follows:

$$\delta X = [(R_{\text{sample}}/R_{\text{reference}}) - 1] \times 1000,$$

where X is ^{13}C or ^{15}N, R is $^{13}C/^{12}C$ or $^{15}N/^{14}N$ for carbon and nitrogen, respectively. Values are expressed as units of permil (‰) relative to the Pee Dee Belemnite standard (PDB) for carbon and to atmospheric N_2 for nitrogen. The typical precision of the complete analysis (preparation, combustion and mass spectrometric analysis) was ± 0.1 ‰ for carbon and ± 0.2 ‰ for nitrogen.

Statistical analyses of isotope data were performed using 2-way Analysis of Variance and/or non parametric Kruskal-Wallis test in the case of comparison of samples with a low number of observations (Sokal & Rohlf, 1981). The exact distribution table of the Kruskal–Wallis statistic was used when necessary (Siegel, 1956).

Results

Recruitment, abundance and biomass

At mid-tide level, the mean annual density of *Cerastoderma edule* was 259 ±277 individuals m^{-2} (Fig. 2a). Recruitment started in May/June (Figs 2 and 3), but the main spat fall occurred in July–August with a maximum spat density over 1000 individuals m^{-2} (Fig. 2a). A secondary recruitment occurred during late autumn (November/December), but it was 10 times smaller than during the summer. Densities of cockles decreased quickly from August onwards due to high spat mortality, which appeared to be higher in autumn than in early winter (Fig. 2a). Two cohorts (1994 and 1995 year classes) dominated the abundance during the study period (Fig. 2a). The 1994 year class (1-yr-old cockles) was the more numerous cohort from the beginning of the study (February 1995) to early summer, but was then dominated by the 1995 recruitment from summer to early winter (Fig. 2a).

The 1994 year class attained the highest proportion (over 50%) of total annual biomass during the study period (Fig. 2b). Maximum biomass of the 1994 year class occurred in August (30 g AFDW m^{-2}), but 6 months later, biomass of this year class was less than 2 g AFDW m^{-2}. The contribution of the 1995 recruitment to the total annual biomass peaked in September/October (6 g AFDW m^{-2}), whereas the contribution of the 1992 year class (3-yr-old cockles) was more stable from May to September (Fig. 2b). Age structure of the cockle population in 1995 was characterised by the relative absence of the 1993 year class (2-yr-old cockles), whereas, the 1992 year class (3-yr-old cockles) was well established. Some cockles older than 4 years (1991 year class) were also recorded during this study, but their densities remained lower than 5 – 10 individuals m^{-2}, except for August and September with biomasses ranging from 4 to 5 g AFDW m^{-2}. No cockles older than 5 years were sampled at mid-tide level in 1995 (Fig. 2).

Shell growth

Shell growth of year classes varied with seasons (Fig. 3). The 0-group (1995 year class) grew from May/June (settlement) to November. At the beginning of their first winter, shell length of 0-group cockles averaged 16.2 ± 1.3 mm. Growth rate of the secondary recruitment, that occurred in autumn and early winter, was low and most of these cockles did not survive during their first winter.

The main growing season of 1-yr-old cockles (1994 year class) was from spring to early summer (April until July) and the 1-yr-old cockle growth rate was much lower from July to October (Fig. 3). At the end of their second growing season, shell length of the 1994 year class reached 26.6 ± 1.8 mm. Similar trends were shown by both 1993 year class and cockles older than 3 years. Shell growth of year classes appeared to diminish with increasing age, with an obvious decrease of growth rate in cockles over 2 years (Fig. 3). No cockles more than 34 mm shell length were found during this study.

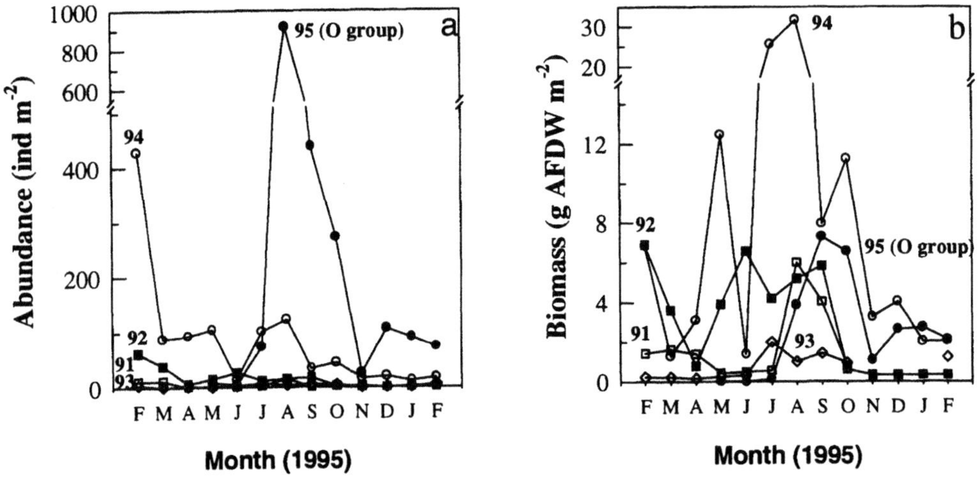

Figure 2. Monthly changes in abundance (a) and biomass (b) of each year classes of *Cerastoderma edule* at mid-tide level. Year class 1995 (•), 1994 (○), 1993 (◇), 1992 (■) and 1991 (□).

Table 1. Production (P), elimination (E), mean biomass (B) and production/mean biomass ratio (P:B) for *Cerastoderma edule* at mide-tide level on the muddy sandflat of Ronce-les-Bains from February 1995 to February 1996. Figures in brackets indicated percentages of total items

	Year class	Production g dry Wt m^{-2} yr^{-1}	Elimination g dry Wt m^{-2} yr^{-1}	Mean biomass g dry Wt m^{-2}	P: B yr^{-1}
	1995 (0-group)	14.12 (40%)	−11.98 (32%)	2.65 (17%)	5.33
	1994	14.60 (41%)	−14.14 (37%)	8.67 (55%)	1.68
	1993	0.65 (2%)	0.01 <(1%)	0.60 (<1%)	1.10
	1992	4.04 (11%)	−8.10 (21%)	2.99 (19%)	1.35
	1991	2.08 (6%)	−3.71 (10%)	1.28 (9%)	1.62
Total	g dry Wt m^{-2} yr^{-1}	35.49	−37.94	15.60	2.28
	g AFDW m^{-2} yr^{-1}	32.47	−34.70	14.27	
	g C m^{-2} yr^{-1}	11.39	−12.18	5.01	

Secondary production

The total annual production estimate was 32.47 g AFDW m^{-2} yr^{-1} (11.39 g C m^{-2} yr^{-1}) at the mid-tide level (Table 1). Mean total biomass was 14.27 g AFDW m^{-2} $yr^{-1}$and thus, the average of production/biomass ratio was 2.28 yr^{-1}. The relative contributions of each year class to the annual secondary production were 40, 41, 2, 11 and 6% for 0-group, 1-, 2-, 3- and 4-yr-old cockles, respectively (Table 1). The 1993 year class (2-yr-old cockles) appeared to be very rare at the mid-tide level, potentially due to poor recruitment in 1993, and did not contribute significantly to either annual production or annual elimination. Production and elimination terms were similar for 0-group and 1-yr-old cockles, whereas elimination terms were two times greater than production terms for 3- and 4-yr-old cockles, with −8.10 versus 4.04 and −3.71 versus 2.08 g AFDW m^{-2} yr^{-1}, respectively (Table 1). Total annual elimination was −34.7 g AFDW m^{-2} yr^{-1} and was similar to total annual production.

Stable isotope ratios for cockles and potential food sources

There were no significant seasonal changes in both $\delta^{13}C$ ($p > 0.6$) and $\delta^{15}N$ values ($p > 0.1$) for POM as indicated by a 2-way ANOVA analysis (unbalanced data with fixed factors 'season' and 'station' and without test of interaction). $\delta^{13}C$ and $\delta^{15}N$ values for POM from the water column averaged −22.2 ±

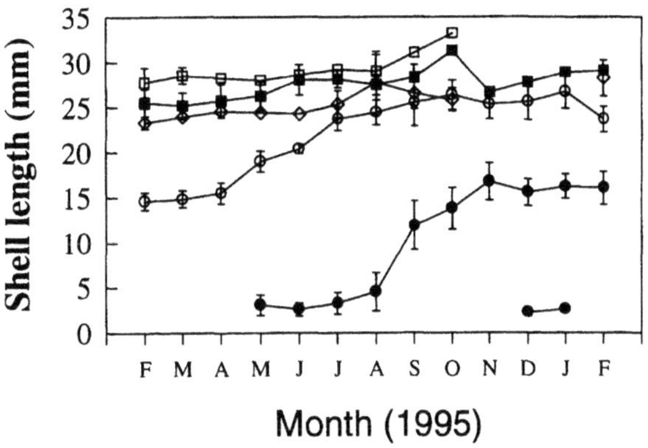

Figure 3. Monthly changes in shell length of each year classes: 1995 (•), 1994 (○), 1993 (◇), 1992 (■) and 1991 (□). Vertical bars are ±1 SD.

1.1 ‰ (n = 19) and 5.0 ± 0.9 ‰ (n = 18) in 1995, respectively.

$\delta^{13}C$ values of macroalgae varied from −20.1 ‰ (*Fucus vesiculosus*, March 1995) to −11.1 ‰ (*Ulva rigida*, December 1995). However, there were no significant differences in $\delta^{13}C$ values between species (Kruskal–Wallis test, $p = 0.19$) due to the great variability of $\delta^{13}C$ values in *Fucus serratus.* No significant differences in $\delta^{15}N$ values between macroalgae were detected (Kruskal–Wallis test, $p > 0.15$) and the mean $\delta^{15}N$ value was 8.2 (± 1.3 ‰).

Zostera noltii $\delta^{13}C$ values ranged from −12.1 ± 1.0 ‰ (March 1995) to −10.3 ‰ (September 1995) and $\delta^{15}N$ values varied from 5.7 ± 0.4 ‰ (March 1995) to 8.7 ‰ (September 1995). There were no significant differences among the sampling dates (Kruskal–Wallis test, p given by exact distribution table of the Kruskal–Wallis statistic for three groups) for both $\delta^{13}C$ ($p > 0.2$) and $\delta^{15}N$ values ($p > 0.1$).

$\delta^{13}C$ values of 0-group cockles varied with seasons between −14.3 ± 1.0 ‰ (spat in June 1995) and −15.7 ± 0.2 ‰ (juvenile in December 1995). As indicated by a Kruskal–Wallis test (exact distribution table for three groups with four, five and four samples for June, September and December, respectively), $\delta^{13}C$ value of spat in June was significantly lower than other $\delta^{13}C$ values of 0-group ($p = 0.037$, Fig. 4). The $\delta^{13}C$ value (− 14.3 ±1.0 ‰) of spat in June 1995 was also 3–3.5 ‰ more positive than that (−18.0 ± 1.2 ‰) of 1–4-yr-old cockles (Fig. 4). Adult cockles showed significant seasonal differences in $\delta^{13}C$ values (Kruskal–Wallis test, $p < 0.025$), $\delta^{13}C$ values from December being about 1.9 ‰ more positive than those of March, June and September (Fig. 4).

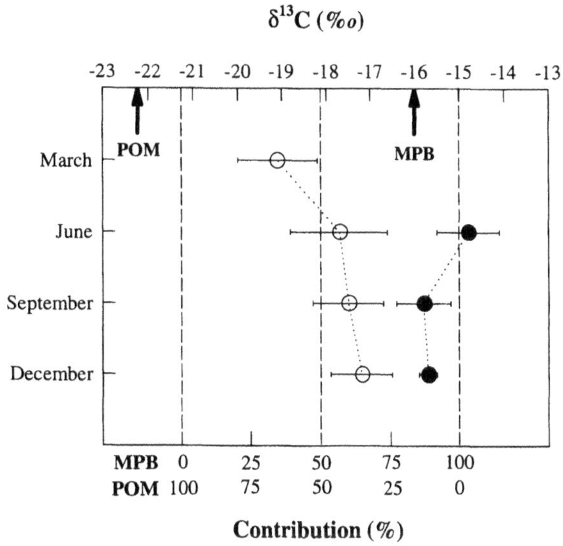

Figure 4. Seasonal changes in the carbon isotopic composition in juvenile (•) and adult (○) cockles (means ±1 SD). Arrows on the upper axis indicate the two food sources used as end-members in the carbon-mixing model viz POM and microphytobenthos (MPB). The lower axis represents the relative contribution of each end-member to carbon incorporated by cockles. Dashed lines indicate the +1.0 ‰ isotopic fractionation per trophic level.

No significant difference was found in the $\delta^{15}N$ values between 0-group and adults (Student's t-test, $p > 0.2$) or among the seasons (Kruskal–Wallis test, $p = 0.07$ and $p = 0.24$ for 0-group and adults, respectively). Thus, mean $\delta^{15}N$ values were 8.0 ± 0.9 ‰ and 8.4 ± 1.1 ‰ for 0-group and adults, respectively.

Use of a carbon isotope two-source mixing model

The isotopic data set from Table 2 was interpreted by Kang et al. (1999). They indicated that, amongst potential available food sources for cockles on Ronce-les-Bains muddy sandflat, only POM advected from water channels to intertidal areas and locally resuspended microphytobenthos constituted the main food sources for cockles (Kang et al., 1999). Prou et al. (1994) indicated that the algal composition of POM advected to intertidal areas is a simple mixture of oceanic and coastal phytoplankton. Therefore, the quantitative proportions of the main food sources incorporated by cockles were estimated by the use of a simple two-source mixing model (Fry & Sherr, 1984) with POM and microphytobenthos as end-members. Since $\delta^{13}C$ values of microphytobenthos were clearly distinct from those of POM (6.2 ‰ of difference) and $\delta^{15}N$ is not as powerful as $\delta^{13}C$ in discriminating food sources due to overlapping of $\delta^{15}N$ values between

Table 2. Mean±1 SD (standard deviation) $\delta^{13}C$ and $\delta^{15}N$ values for POM (from the water column), macroalgae, seagrass and cockles collected at Ronce-les-Bains in 1995 (after Kang et al., 1999). Figures in brackets are the number of analysed samples and, in the case of cockles, the number of analysed pools of 2, 5 and 20 individuals for adults, juveniles and spat, respectively. (-) No data

Sample	March		June		September		December	
	$\delta^{13}C$ (‰)	$\delta^{15}N$ (‰)	$\delta^{13}C$ (‰)	$\delta^{15}N$ (‰)	$\delta^{13}C$ (‰)	$\delta^{15}N$ (‰)	$\delta^{13}C$ (‰)	$\delta^{15}N$ (‰)
POM	-22.3 ± 2.1	4.5 ± 1.4	-22.1 ± 1.2	5.6 ± 0.1	-21.6 ± 0.6	4.9 ± 0.2	-22.4 ± 0.1	5.3 ± 0.4
	(5)	(6)	(5)	(3)	(3)	(3)	(6)	(6)
Phorphyra umbilicalis	-	-	-	-	-19.7 ± 0.57	8.4 ± 0.5	-	-
					(2)	(2)		
Fucus serratus	-20.0 ± 0.2	7.6 ± 1.4	-14.4 ± 0.1	8.1 ± 1.9	-16.9	8.1	-18.8	6.9
	(2)	(4)	(2)	(2)	(1)	(1)	(1)	(1)
Fucus vesiculosus	-20.1	8.6	-	-	-16.9	8.4	-18.5	7.8
	(1)	(1)			(1)	(1)	(1)	(1)
Ulva rigida	-15.0	8.9	-15.8	9.0	-15.9 ± 0.1	8.1 ± 0.1	-11.1	10.6
	(1)	(1)	(1)	(1)	(2)	(2)	(1)	(1)
Enteromorpha compressa	-15.0	8.9	-17.6 ± 2.8	8.2 ± 1.1	-	-	-	-
	(1)	(1)	(2)	(2)				
Zostera noltii	-12.1 ± 1.0	5.7 ± 0.4	-10.6 ± 0.4	6.7 ± 0.6	-10.3	8.7	-	-
	(2)	(3)	(2)	(2)	(1)	(1)		
Cerastoderma edule								
Spat	-	-	-14.3 ± 1.0	7.6 ± 0.8	-15.8 ± 0.4	7.8 ± 0.6	-	-
(size ≤ 5 mm)			(4)	(4)	(2)	(2)		
Juveniles	-	-	-	-	15.7 ± 0.7	8.7 ± 0.8	-15.7 ± 0.2	8.8 ± 1.2
(5 < size ≤ 12 mm)					(3)	(3)	(4)	(4)
Adults	-19.1 ± 0.9	8.0 ± 0.5	-17.7 ± 1.1	7.6 ± 0.8	-17.5 ± 0.8	7.7 ± 1.0	-17.2 ± 0.7	9.3 ± 1.9
(size > 12 mm)	(7)	(7)	(11)	(11)	(5)	(5)	(4)	(4)

cockle tissues and food sources, the mixing model was based on $\delta^{13}C$ values.

Assuming that $\delta^{13}C$ values in cockles should be corrected from fractionation due to metabolic processes, the proportion of carbon assimilated by cockles from microphytobenthos (pMPB) at a sampling time (t) can be calculated using the following equation:

$$\text{pMPB}\ (t) = \frac{\delta^{13}C_{\text{cockle}} - f - \delta^{13}C_{\text{POM}}}{\delta^{13}C\ \text{MPB} - \delta^{13}C\ \text{POM}} \times 100,$$

where pMPB is the proportion of microphytobenthos that contributes to cockle diet and $f = +1$ ‰ is the average enrichment of animal carbon relative to their diet (DeNiro & Epstein, 1978; Fry & Sherr, 1984). The proportion of particulate organic matter (pPOM) that contributes to cockle diet is: pPOM = 100 − pMPB.

In Figure 4, both pMPB and pPOM were reported, together with seasonal changes in $\delta^{13}C$ values for spat, juvenile and adult cockles. March $\delta^{13}C$ values of adults (-19.1 ± 0.9 ‰) indicated that only 34% of their organic carbon, incorporated during the preceding winter months, was derived from microphytobenthos and 66% from POM. Relatively high $\delta^{13}C$ values for adult cockles between June and December 1995 indicated that they assimilated between 57% (in June) and 65% (in December) of microphytobenthic carbon during their spring growing season (Figs 3 and 4). Consequently, the proportion of carbon incorporated from POM by adult cockles varied between 43 (in June) and 35% (in December). On the other hand, the $\delta^{13}C$ values for spat (-14.3 ± 1.0 ‰ in June) and juvenile cockles (-15.7 ± 0.2 ‰ from September to December 1995) must reflect a large contribution of microphytobenthic carbon to 0-group growth; the contribution varied during their growing season from 100% in the summer to approximately 85% in the autumn (Figs 3 and 4).

Contribution of microphytobenthos to annual secondary production

The contribution of microphytobenthic carbon C_{MPB} to each year class production was calculated using the following equation [where $C_{MPB}\ (y)$ is expressed as a

Table 3. Relative contribution of carbon derived from microphytobenthos to the annual secondary production and elimination of *Cerastoderma edule*. 0-group (1995 cohort) and adult cockles (1991 to 1994 cohorts)

Year class (age group)	Production $g\,C\,m^{-2}\,yr^{-1}$	Elimination $g\,C\,m^{-2}\,yr^{-1}$	Contribution of microphytobenthos
1995 (0-group)	4.53	−3.85	88%
1994–1991 (adults)	6.86	−8.33	60%
Total	11.39	−12.18	71%

percentage of the production P(y) of a given year class (y)]:

$$C_{\text{MPB}}(y) = 100 \cdot \text{SUM/P}(y)$$

with

$$\text{SUM} = \sum_{t=0}^{t=n}[(N_t + N_{t+1})/2 \cdot (W_{t+1} - W_t)] \cdot \text{pMPB}(t+1).$$

N_t, W_t and n are the terms used in the computation of the production P(y) of a year class (y) and pMPB (t+1) is the proportion of microphytobenthos that contributed to the year class diet between the two sampling dates t and t+1.

Calculation of the terms C_{MPB} (y) for the 0-group and adult cockles are given in Table 3 and indicate that the relative contribution of microphytobenthos carbon to the annual secondary production of cockle carbon on average is 88% and 60% for the 0-group and adult cockles, respectively. The weighted contribution of microphytobenthos to the annual production of the whole population was more than 70% (Table 3).

Discussion

Seasonal recruitment patterns

The first spawning period of cockles at Ronce-les-Bains occurs in late spring, following winter and spring gonadal development in adult cockles, but a second spawning period can also occur in late summer (Guillou et al., 1990; Sauriau, 1992). As a consequence, the major spatfall in early summer contributes to the maintenance of the cockle stock, whereas later recruitments are insignificant. These seasonal patterns are in agreement with results of most studies performed along the North European coasts (Boyden, 1971; Seed & Brown, 1975; Newell & Bayne, 1980; Möller & Rosenberg, 1983; Navarro et al., 1989). However, Guillou et al. (1992) and Guillou & Tartu (1994) found opposite results in an intertidal population in northern Brittany. They reported, from a 5 year study, that summer spawnings were followed by early autumn to winter recruitment and that, in spite of high post-larval mortality during winter, juveniles born during the foregoing autumn largely contributed to the population biomass during the following spring. It is also interesting to note that Seed & Brown (1978) in Northern Ireland found both opposite recruitment patterns in the same cockle population, but during different years: autumnal to early winter recruitment in 1972, 1973 and 1975 and spring recruitment in 1974. Year-to-year changes in local thermal and food conditions, competition and predation pressures could explain such variabilities.

Production and elimination estimates

Total annual production and elimination of cockles in Ronce-les-Bains fell within the range 0–200 g AFDW $m^{-2}\,yr^{-1}$, as reported by Warwick & Price (1975), Hibbert (1976), Evans (1977), Ivell (1981) and Jensen (1993) for unexploited stocks. In the sandy mudflat of Ronce-les-Bains, the cockle population suffers human predation on cockles larger than 30 mm in shell length, mainly during the summer tourist season (Sauriau, 1992). Such a selective mortality rate leads to a decrease in density and elimination exceeding production in all oldest cohorts (Table 1). This also induces underestimation of growth rate estimates and longevity. Hibbert (1976) reported 6-year-old cockles in Southampton Water and the species is known to live for up to 10 years in Northern Europe (Hancock & Urquhart, 1965; Jones, 1979). Consequently, total production estimates, calculated here, are underestimates because methods used for estimating annual production (Crisp, 1984) assume that mortality is not size dependent. It should, however, be pointed out that selective removal of the oldest adult cohorts may favour growth and survival of remaining juveniles and adults (de Montaudouin & Bachelet, 1996) and recruits of the following year (Bachelet et al., 1992) through intraspecific density-dependent relationships. Since P:B ratios of the youngest cohorts are generally much higher than those of oldest cohorts (e.g. Hibbert, 1976; see also Brey, 1990), selective removal of oldest cohorts may also result, by feedback, in a relative over-contribution of production of youngest cohorts in total annual production estimates.

Age-related differences in growing season and links with microphytobenthos availability

Due to recruitment patterns of the cockle population, a shift in growth seasons between juveniles and adults (> 1 yr old) was found at Ronce-les-Bains (Fig. 3). A similar pattern has been reported for other cockle populations by various studies (Farrow, 1971; Barnes, 1973; Jensen, 1992). Jensen (1992) hypothesised that juvenile and adult cockles exploit different food resources during their different growth periods in relation to seasonal food availability. A similar mismatch-hypothesis was made by Ankar (1980) in the case of *Macoma balthica*. Hydrological surveys performed within the Marennes-Oléron Bay (Héral et al., 1983; Razet et al., 1990; Soletchnik et al., 1998; Kang et al., 1999) and our stable carbon and nitrogen isotope results (Kang et al., 1999) support this hypothesis. From long-term monitoring in the Marennes-Oléron Bay, Soletchnik et al. (1998) indicated that maximum values of chlorophyll *a* in the water column appear in May and June. Results of the 1995 hydrological survey (performed at three locations surrounding the Ronce-les-Bains sandy mudflat, Figure 1), were also in agreement with this seasonal pattern: values up to $3.7 \pm 1.7\ \mu g\ l^{-1}$ of chlorophyll *a* were found in POM during June 1995, whereas POM chlorophyll *a* values ranged from 0.5 to 0.9 $\mu g\ l^{-1}$ during the other seasons (Kang et al., 1999). However, surveys of sedimentary chlorophyll *a* on the muddy sandflat at Ronce-les-Bains in 1995 (Kang et al., 1999) indicated that sedimentary chlorophyll *a* values were not significantly different from $10.9 \pm 3.5\ \mu g\ g^{-1}$ dry sediment, all year round. Since seston dynamics within the Marennes-Oléron Bay are largely governed by the resuspension of intertidal sediments (Razet et al., 1990; Zurburg et al., 1994) under the influence of tidal currents and wind-driven waves (Raillard et al., 1994), it may be expected that resuspended benthic diatoms are available for suspension feeding species most of the time. Therefore, Zurburg et al. (1994) showed that, in the central water channel, benthic diatom species formed about 50% of chlorophyll *a* in the water column at spring tide in May, and most of the chlorophyll *a* in winter during both spring and neap tides. These observations are also in agreement with various reports indicating that resuspended microphytobenthos from large intertidal areas exposed to current and wave actions becomes a major algal component in the water column (Baillie & Welsh, 1980; Delgado et al., 1991; De Jonge & Van Beusekom, 1995). These can be ingested subsequently by suspension feeders (De Jonge & Van Beusekom, 1992; Kamermans, 1994).

The age-related differences in the incorporation of microphytobenthos carbon by cockles in the Marennes-Oléron Bay is partly explained by the mismatch hypothesis. After settlement, 0-group cockles grew from summer to early autumn, a period with low phytoplankton but high microphytobenthos abundance, whereas growth of adults mostly occurred in spring during phytoplankton blooms. Furthermore, juveniles have incorporated a higher proportion of microphytobenthos in their tissues than adults. However, significant differences in the $\delta^{13}C$ signature between spat in June (size $\leq$5 mm) and juveniles (size >5 mm) from the other seasons are not satisfactorily explained. Similarly, the spring to autumn $\delta^{13}C$ signature of adults still appeared to be intermediate between POM and microphytobenthos, and also significantly different from the $\delta^{13}C$ signature of juvenile even after a 6 month exposure to similar microphytobenthic food source. Therefore, several possible explanations were proposed by Kang et al. (1999): (1) size-related changes in particle capture mechanism within the benthic boundary layer, (2) size-related selective feeding capabilities, (3) size-related differential assimilation of dissolved organic matter *versus* particulate organic matter and (4) higher lipid contents of adults during reproductive winter to spring periods. Due to the lack of evidence, to our knowledge, in the literature on *Cerastoderma edule*, for testing the validity of the three latter hypotheses, only the first explanation will be dealt with here. A closer connection of juvenile cockles to near-bottom layers compared with adult cockles, which have potentially longer inhalant siphon and higher strength of inhalant/exhalant jets (André et al., 1993), may explain that spat and juveniles are fueled mainly by microphytobenthos, while adult growth is sustained by a higher proportion of planktonic items.

Role of microphytobenthos in cockle production

Estimates of the microphytobenthos proportion in the annual production of both 0-group and adult cockles indicated that over 70% of the secondary production of the whole population was fueled by microphytobenthos. The fact that *Cerastoderma edule* feed on epibenthic diatoms has been established by gut content analyses (e.g. Verwey, 1952; Ivell, 1981; Kamermans, 1994). However, our results suggest a much stronger trophic link between microphytobenthos and the infaunal suspension feeder *Cerastoderma edule* than would be expected from these studies. On the one hand, examination of cockle gut contents generally gave qualitative results and, for instance, Ivell (1981) indicated that "in addition to their normal diet of phytoplankton, in the Limfjord the cockles were also taking in benthic diatoms and other material including sand grains, bivalve larvae and Foraminifera". On the other hand, quantitative results of Kamermans (1994) indicated that cockle stomach contents appeared to be slightly enriched in benthic algae compared to algal composition of water samples taken at 1 cm above the bottom (70% pelagic algae, 30% benthic algae). The over-proportion of benthic algae in cockle stomachs remained small and highly variable (between −30 and +50%) amongst cockles (Kamermans, 1994).

Similarly, our results appear to contradict the generally accepted paradigm that marine phytoplankton is a major component of the diet of suspension feeding cockles on sandy habitats (Newell, 1979; Loo & Rosenberg, 1989). It is not known whether the strength of the trophic links between benthic diatoms and cockles in the Marennes-Oléron Bay depends on ecophysiological capabilities of the species and/or trophic/hydrodynamic characteristics of the local muddy sandflats. *Cerastoderma edule* is reported to be able to selectively ingest microflora from a wide variety of food items (Iglesias et al., 1992; Navarro et al., 1992). The importance of microphytobenthos to the epifaunal suspension feeder *Crassostrea gigas* (Thunberg), living in the vicinity of the River Charente plume in the north-east of the Marennes-Oléron Bay, has also been proven by isotope techniques (Riera & Richard, 1996, 1997). Microphytobenthos is, therefore, expected to be the major component of the total annual primary production in the Marennes-Oléron Bay (Guarini et al., 1998), where primary production of phytoplankton is limited by water turbidity (Ravail et al., 1988; Raillard & Ménesguen, 1994). Other primary producers, such as macroalgae, seagrass and marsh vascular plants, are restricted to relative small areas (Guillaumont, 1991) and most organic matter is provided by resuspension of intertidal muddy sediments (Razet et al., 1990; Zurburg et al., 1994).

In conclusion, this study highlights new hypotheses on age-related processes that may influence the suspension feeding *Cerastoderma edule* trophic base primarily driven by temporal changes in the availability of food resources in coastal ecosystems.

Acknowledgements

This study was supported by CNRS and IFREMER. C.-K. K. was supported by a grant from the French Embassy in the framework of a French-Korean cooperation programme between IFREMER and NFRDI (National Fisheries Research & Development Institute). Thanks to M. Héral, G. Blanchard and P. Richard for their support and to A. Knutsen for her comments on the English. We gratefully acknowledge two anonymous referees for their helpful comments on the manuscript.

References

André, C. & R. Rosenberg, 1991. Adult-larval interactions in the suspension-feeding bivalves *Cerastoderma edule* (L.) and *Mya arenaria* L. Mar. Ecol. Prog. Ser. 71: 227–234.

André, C., P. R. Jonsson & M. Lindegarth, 1993. Predation on settling bivalve larvae by benthic suspension feeders: the role of hydrodynamics and larval behaviour. Mar. Ecol. Prog. Ser. 97: 183–192.

Ankar, S., 1980. Growth and production of *Macoma balthica* (L.) in a Northern Baltic soft bottom. Ophelia, Suppl. 1: 31–48.

Bachelet, G., J. Guillou & P. J. Labourg, 1992. Adult-larval and juvenile interactions in the suspension-feeding bivalve, *Cerastoderma edule* (L.): field observations and experiments. In Colombo, G., I. Ferrari, V. U. Ceccherelli & R. Rossi (eds), Marine Eutrophication and Population Dynamics. Olsen & Olsen, Fredensborg: 175–182.

Baillie, P. W. & B. L. Welsh 1980. The effect of tidal resuspension on the distribution of intertidal epipelic algae in an estuary. Estuar. coast. shelf Sci. 10: 165–180.

Baggerman, B., 1953. Spatfall and transport of *Cardium edule* L. Arch. Néerl. Zool. 10: 315–342.

Barnes, R. S. K., 1973. The intertidal lamellibranchs of Southampton Water, with particular reference to *Cerastoderma edule* and *C. glaucum*. Proc. malac. Soc. Lond. 40: 413–433.

Bayne, B. L., J. Widdows & R. I. E. Newell, 1977. Physiological measurements on estuarine bivalve molluscs in the field. In Keegan, B. F., P. O. Ceidigh & P. J. S Boaden (eds), Biology of Benthic Organisms. Proc. 11th Europ. Mar. Biol. Symp., Pergamon Press, Oxford: 57–68.

Beukema, J. J., 1982. Annual variation in reproductive success and biomass of the major macrozoobenthic species living in a tidal flat area of the Wadden Sea. Neth. J. Sea Res. 16: 37–45.

Boyden, C. R., 1971. A comparative study of the reproductive cycles of the cockles *Cerastoderma edule* and *C. glaucum*. J. mar. biol. Ass. U.K. 51: 605–622.

Boutton, T. W., 1991. Stable carbon isotope ratios of natural materials: I. Sample preparation and mass spectrometric analysis. In Coleman, D. C. & B. Fry (eds), Carbon Isotope Techniques. Academic Press, San Diego: 155–171.

Brey, T., 1990. Estimating productivity of macrobenthic invertebrates from biomass and mean individual weight. Meeresforsch 32: 329–343.

Crisp, D. J., 1984. Energy flow measurements. In Holme, N. A. & A. D. McIntyre (eds), Methods for the Study of Marine Benthos, IBP Handbook 16. Blackwell Sci. Publ., Oxford: 284–372.

Dame, R. F., 1996. Ecology of marine bivalves. An ecosystem approach. CRC Marine Science Series, CRC Press, Boca Raton, Florida: 254 pp.

De Jonge, V. N. & J. E. E. Van Beusekom, 1992. Contribution of resuspended microphytobenthos to total phytoplankton in the Ems estuary and its possible role for grazers. Neth. J. Sea Res. 30: 91–105.

De Jonge, V. N. & J. E. E. Van Beusekom, 1995. Wind- and tide-induced resuspension of sediment and microphytobenthos from tidal flats in the Ems estuary. Limnol. Oceanogr. 40: 766–778.

Delgado, M., V. N. De Jonge & H. Peletier, 1991. Experiments on resuspension of natural microphytobenthos populations. Mar. Biol. 108: 321–328.

De Montaudouin, X. & G. Bachelet, 1996. Experimental evidence of complex interactions between biotic and abiotic factors in the dynamics of an intertidal population of the bivalve *Cerastoderma edule*. Oceanol. Acta. 19: 449–463.

DeNiro, M. J. & S. Epstein, 1978. Influence of diet on the distribution of carbon isotopes in animals. Geochim. Cosmochim. Acta. 42: 495–506.

DeNiro, M. J. & S. Epstein, 1981. Influence of diet on the distribution of nitrogen isotopes in animals. Geochim. Cosmochim. Acta. 45: 341–351.

Dörjes, J., H. Michaelis & B. Rhode, 1986. Long-term studies of macrozoobenthos in intertidal and shallow subtidal habitats near the island of Norderney (East Frisian coast, Germany). Hydrobiologia 142: 217–232.

Ducrotoy, J. P., H. Rybarczyk, J. Souprayen, G. Bachelet, J. J. Beukema, M. Desprez, J. Dörjes, K. Essink, J. Guillou, H. Michaelis, B. Sylvand, J. G. Wilson, B. Elkaïm & F. Ibanez, 1991. A comparison of the population dynamics of the cockle (*Cerastoderma edule,* L.) in North-Western Europe. In Elliott, M. & J. P. Ducrotoy (eds), Estuaries and Coasts: Spatial and Temporal Intercomparisons. ECSA 19 Symposium, Olsen & Olsen, Fredensborg: 173–184.

Evans, S., 1977. Growth, production and biomass release of a non-stable population of *Cardium edule* L. (Bivalvia). Zoon 5: 133–141.

Farrow, G. E., 1971. Periodicity structures in the bivalve shell: experiments to establish growth controls in *Cerastoderma edule* from the Thames Estuary. Ibid 14: 571–588.

Fry, B. & E. B. Sherr, 1984. δ^{13}C measurements as indicators of carbon flow in marine and freshwater ecosystems. Contrib. mar. Sci. 27: 13–47.

Guarini, J.-M., G. F. Blanchard, C. Bacher, P. Gros, P. Riera, P. Richard, D. Gouleau, R. Galois, J. Prou & P.-G. Sauriau, 1998. Dynamics of spatial patterns of microphytobenthic biomass: inferences from a geostatistical analysis of two comprehensive surveys in Marennes-Oléron Bay (France). Mar. Ecol. Prog. Ser. 166: 131–141.

Guillaumont, B., 1991. Utilisation de l'imagerie satellitaire pour les comparaisons spatiales et temporelles en zone intertidale. In Elliott, M. & J. P. Ducrotoy (eds), Estuaries and Coasts: Spatial and Temporal Intercomparisons. ECSA 19 Symposium, Olsen & Olsen, Fredensborg: 63–68.

Guillou, J., G. Bachelet, M. Desprez, J.-P. Ducrotoy, I. Madani, H. Rybarczyk, P.-G. Sauriau, B. Sylvand, B. Elkaim & M. Glémarec, 1990. Les modalités de la reproduction de la coque (*Cerastoderma edule*) sur le littoral français de la Manche et de l'Atlantique. Aquat. Living Resour. 3: 29–41.

Guillou, J., G. Bachelet & M. Glémarec, 1992. Influence des fluctuations de température sur la reproduction et le recrutement de la coque *Cerastoderma edule* (L.). Ann. Inst. Océanogr. Paris 68: 65–74.

Guillou, J. & C. Tartu, 1994. Post-larval and juvenile mortality in a population of the edible cockle *Cerastoderma edule* (L.) from northern Brittany. Neth. J. Sea Res. 33: 103–111.

Hancock, D. A. & A. E. Urquhart, 1965. The determination of natural mortality and its causes in an exploited population of cockles (*Cardium edule* L.). M. A. F. F., Fish. Invest. 24: 1–40.

Hawkins, A. J. S., E. Navarro & J. I. P. Iglesias, 1990. Comparative allometries of gut-passage time, gut content and metabolic faecal loss in *Mytilus edulis* and *Cerastoderma edule*. Mar. Biol. 105: 197–204.

Héral, M., D. Razet, J.-M. Deslous-Paoli, J.-P. Berthomé & J. Garnier, 1983. Caractéristiques saisonnières de l'hydrobiologie du complexe estuarien de Marennes-Oléron (France). Rev. Trav. Inst. Pêches. Marit. 46: 97–119.

Héral, M., C. Bacher & J.-M. Deslous-Paoli, 1990. La capacité biotique des bassins ostréicoles. In Troadec, J.-P. (ed.), L'homme et les Ressources Halieutiques. IFREMER, Brest: 225–259.

Hibbert, C. J., 1976. Biomass and production of a bivalve community on an intertidal mud-flat. J. exp. mar. Biol. Ecol. 25: 249–261.

Iglesias, J. I. P. & E. Navarro, 1991. Energetics of growth and reproduction in cockles (*Cerastoderma edule*): seasonal and age-dependent variations. Mar. Biol. 111: 359–368.

Iglesias, J. I. P., E. Navarro, P. Alvarez Jorna & I. Armentia, 1992. Feeding, particle selection and absorption in cockles *Cerastoderma edule* (L.) exposed to variable conditions of food concentration and quality. J. exp. mar. Biol. Ecol. 162: 177–198.

Ivell, R., 1981. A quantitative study of a *Cerastoderma – Nephthys* community in the Limfjord, Denmark, with special reference to production of *Cerastoderma edule.* J. Moll. Stud. 47: 147–170.

Jensen, K. T., 1992. Dynamics and growth of the cockle, *Cerastoderma edule*, on an intertidal mud-flat in the Danish Wadden Sea: effects of submersion time and density. Neth. J. Sea Res. 28: 335–345.

Jensen, K. T., 1993. Density-dependent growth in cockles (*Cerastoderma edule*): evidence from interannual comparisons. J. mar. biol. Ass. U.K. 73: 333–342.

Jones, A. M., 1979. Structure and growth of a high-level population of *Cerastoderma edule* (Lamellibranchiata). J. mar. biol. Ass. U.K., 59: 277–287.

Junger, M. & D. Planas, 1994. Quantitative use of stable carbon isotope analysis to determine the trophic base of invertebrate communities in a boreal forest lotic system. Can. J. Fish. aquat. Sci. 51: 52–61.

Kamermans, P., 1994. Similarity in food source and timing of feeding in deposit- and suspension-feeding bivalves. Mar. Ecol. Prog. Ser. 104: 63–75.

Kang, C. K., P.-G. Sauriau, P. Richard & G. F. Blanchard, 1999. Food sources of the infaunal suspension-feeding bivalve *Cerastoderma edule* in a muddy sandflat of Marennes-Oléron Bay, as determined by analyses of carbon and nitrogen stable isotopes. Mar. Ecol. Prog. Ser. 187: 147–158.

Kristensen, I., 1957. Differences in density and growth in a cockle population in the Dutch Wadden sea. Arch. Néerl. Zool. 12: 350–453.

Loo, L. O. & R. Rosenberg, 1989. Bivalve suspension-feeding dynamics and benthic-pelagic coupling in an eutrophicated marine bay. J. exp. mar. Biol. Ecol. 130: 253–276.

Michener, R. H. & D. M. Schell, 1994. Stable isotope ratios as tracers in marine aquatic food webs. In Lajtha, K. & R. H. Michener (eds), Stable Isotopes in Ecology and Environmental Science. Blackwell Scientific Publications, Oxford: 138–157.

Möller, P. & R. Rosenberg, 1983. Recruitment, abundance and production of *Mya arenaria* and *Cardium edule* in marine shallow waters, western Sweden. Ophelia 22: 33–35.

Navarro, E., J. I. P. Iglesias & A. Larranaga, 1989. Interannual variation in the reproductive cycle and biochemical composition of the cockle *Cerastoderma edule* from Mundaca Estuary (Biscay, North Spain). Mar. Biol. 101: 503–511.

Navarro, E., J. I. P. Iglesias & M. M. Ortega, 1992. Natural sediment as a food source for the cockle *Cerastoderma edule* (L.): effect of variable particle concentration on feeding, digestion and the scope for growth. J. exp. mar. Biol. Ecol. 156: 69–87.

Navarro, J. M. & J. Widdows, 1997. Feeding physiology of *Cerastoderma edule* in response to a wide range of seston concentrations. Mar. Ecol. Prog. Ser. 152: 175–186.

Newell, R. C., 1979. Biology of Intertidal Animals. 3rd edn. Marine Ecological Surveys Ltd., Faversham, Kent: 781 pp.

Newell, R. I. E. & B. J. Bayne, 1980. Seasonal changes in the physiology, reproductive condition and carbohydrate content of the cockle *Cardium* (= *Cerastoderma*) *edule* (Bivalvia: Cardiidae). Mar. Biol. 56: 11–19.

Orton, J. H., 1933. Summer mortality of cockles on some Lancashire and Cheshire Dee beds in 1933. Nature, London 132: 314–315.

Prins, T. C., A. C. Smaal & A. J. Pouwer, 1991. Selective ingestion of phytoplankton by the bivalves *Mytilus edulis* L. and *Cerastoderma edule* (L.). Hydrobiol. Bull. 25: 93–100.

Raillard, O. & A. Ménesguen, 1994. An ecosystem box model for estimating the carrying capacity of a macrotidal shellfish system. Mar. Ecol. Prog. Ser. 115: 117–130.

Raillard, O., P. Le Hir & P. Lazure, 1994. Transport de sédiments fins dans le bassin de Marennes-Oléron: mise en place d'un modèle mathématique. Houille Blanche 4: 63–71.

Ravail, B., M. Héral, S. Maestrini & J. M. Robert, 1988. Incidence du débit de la Charente sur la capacité biotique du bassin ostréicole de Marennes-Oléron. J. Rech. Océanogr. 13: 48–52.

Razet, D., M. Héral, J. Prou, J. Legrand & J.-M. Sornin, 1990. Variations des productions de biodépôts (feces et pseudofeces) de l'huître *Crassostrea gigas* dans un estuaire macrotidal : baie de Marennes-Oléron. Haliotis 10: 143–161.

Reise, K., 1985. Tidal flat ecology. An experimental approach to species interactions. Springer-Verlag, Berlin: 191 pp.

Riera, P. & P. Richard, 1996. Isotopic determination of food sources of *Crassostrea gigas* along a trophic gradient in the estuarine bay of Marennes-Oléron. Estuar. coast. shelf Sci. 42: 347–360.

Riera, P. & P. Richard, 1997. Temporal variation of $\delta^{13}C$ in particulate organic matter and oyster *Crassostrea gigas* in Marennes-Oléron Bay (France): effect of freshwater inflow. Mar. Ecol. Prog. Ser. 147: 105–115.

Riera, P., P. Richard, A. Grémare & G. Blanchard, 1996. Food source of intertidal nematodes in the Bay of Marennes-Oléron (France), as determined by dual stable isotope analysis. Mar. Ecol. Prog. Ser. 142: 303–309.

Sanchez-Salazar, M. E., C. L. Griffiths & R. Seed, 1987. The effect of size and temperature on the predation of cockles *Cerastoderma edule* (L.) by the shore crab *Carcinus maenas* (L.). J. exp. mar. Biol. Ecol. 111: 181–193.

Sauriau, P.-G., 1992. Les mollusques benthiques du bassin de Marennes-Oléron : estimation et cartographie des stocks non cultivés, compétition spatiale et trophique, dynamique de population de *Cerastoderma edule* (L.). Unpub. PhD. Thesis, Université de Bretagne Occidentale, Brest: 309 pp.

Sauriau, P.-G., V. Mouret & J.-P. Rincé, 1989. Organisation trophique de la malacofaune benthique non cultivée du bassin ostréicole de Marennes-Oléron. Oceanol. Acta. 12: 193–204.

Seed, R. & R. A. Brown, 1975. The influence of reproductive cycle, growth and mortality on population structure in *Modiolus modiolus* (L.), *Cerastoderma edule* (L.) and *Mytilus edulis* L., (Mollusca: Bivalvia). In Barnes, H. (ed.), The Biochemistry, Physiology and Behaviour of Marine Organisms in Relation to their Ecology. Proc. 9th Europ. Mar. Biol. Symp., Aberdeen University Press, Aberdeen: 257–274.

Seed, R. & R. A. Brown, 1978. Growth as a strategy for survival in two marine bivalves, *Cerastoderma edule* and *Modiolus modiolus*. J. anim. Ecol. 47: 283–292.

Siegel, S., 1956. Nonparametric statistics for the behavioral sciences. McGraw-Hill, New York: 312 pp.

Smaal, A. C., A. P. M. A. Vonck & M. Bakker, 1997. Seasonnal variation in physiological energetics of *Mytilus edulis* and *Cerastoderma edule* of different size classes. J. mar. biol. Ass. U. K. 77: 817–838.

Sokal, R. & F. J. Rohlf, 1981. Biometry. 2nd edn. W.H. Freeman & Co, New York: 859 pp.

Soletchnik, P., N. Faury, D. Razet & Ph. Goulletquer, 1998. Hydrobiology of the Marennes-Oléron bay. Seasonal indices and analysis of trends from 1978 to 1995. Hydrobiologia 386: 131–146.

Tebble, N., 1966. British bivalve seashells. A handbook for identification. Trustees of the British Museum (Natural History), London: 212 pp.

Thiel, M. & L. Watling, 1998. Effects of green algal mats on infaunal colonization of a New England mud flat – long-lasting but highly localized effects. Hydrobiologia 375/376: 177–189.

Urrutia, M. B., J. I. P. Iglesias, E. Navarro & J. Prou, 1996. Feeding and absorption in *Cerastoderma edule* under environmental conditions in the Bay of Marennes-Oléron (western France). J. mar. biol. Ass. U. K. 76: 431–450.

Verwey, J., 1952. On the ecology of distribution of cockle and mussel in the Dutch Wadden Sea, their role in sedimentation and the source of their food supply. Arch. Néerl. Zool. 10: 171–239.

Warwick, R. M., I. R. Joint & P. J. Radford, 1979. Secondary production of the benthos in an estuarine environment. In Jefferies, R. L. & A. J. Davy (eds), Ecological Processes in Coastal Environments. Blackwell Scientific Publications, Oxford: 429–450.

Warwick, R. M. & R. Price, 1975. Macrofauna production in an estuarine mud-flat. J. mar. biol. Ass. U. K. 55: 1–18.

Zurburg, W., A. A. D. Smaal, M. Héral & N. Danker, 1994. Seston dynamics and bivalve feeding in the bay of Marennes-Oléron (France). Neth. J. aquat. Ecol. 26: 459–466.

Hydrobiologia **440:** 331–338, 2000.
M.B. Jones, J.M.N. Azevedo, A.I. Neto, A.C. Costa & A.M. Frias Martins (eds), Island, Ocean and Deep-Sea Biology.

Annual growth of the cockle *Clinocardium ciliatum* in the Norwegian Arctic (Svalbard area)

Minna Elisabeth Tallqvist[1] & Jan Henry Sundet[2]
[1]*Department of Biology & Husö Biological Station, Åbo Akademi University, FIN-20500 Åbo, Finland*
[2]*Norwegian Institute of Fisheries and Aquaculture, N-9002 Tromsø, Norway*

Key words: Bivalvia, growth, Arctic, *Clinocardium ciliatum*, Svalbard

Abstract

The Svalbard Islands are influenced by warm Atlantic water in the south and west, and cold Arctic water in the east. Ice cover, and hence the location of the highly productive marginal ice zone, varies both intra and interannually. Part of the primary production accumulates on the bottom and is utilized by the benthos. In this study, the annual growth of the cockle *Clinocardium ciliatum* (Fabricius, 1780) from three sites in Svalbard waters is reported. Moffen, the site in the north (80° 01′ N, 13° 48′ E) is located in the northernmost areas influenced by Atlantic water. The Storfjorden site (77° 10′ N, 20° 09′ E) is situated in cold Arctic water masses, and the Bear Island site (74° 50′ N, 18° 54′ E) is in the Polar front area where Atlantic and Arctic water masses meet. Annual growth of cockles was analysed retrospectively by measuring external growth increments, which gave annual growth records from the 1970s to 1996. Shell height for age for different year classes was highest at the Storfjorden site, and lowest at Bear Island. Periods of high growth occurred at Storfjorden and Bear Island during the 1980s while the beginning of 1990s was characterized by low growth. At Moffen, growth was more variable between single years. Several factors are influencing the growth of *C. ciliatum* in the Svalbard area and growth cannot be coupled to only one environmental factor like ice cover.

Introduction

Generally, bivalve growth is dependent upon age and size, but is affected also by environmental factors such as temperature, current velocity and food availability (Richardson et al., 1980; Appeldoorn, 1981; MacDonald & Bourne, 1987; Jones et al., 1989; Navarro et al., 1992; Arsenault et al. 1997; Witbaard et al., 1997), and biotic factors such as competition (Peterson & Andre, 1980). In Arctic areas, food supply is probably one of the most important factors affecting growth of benthic animals and species are forced to adjust to a short growth season (Picken, 1979; Sagar, 1980; Clarke, 1988). Polar bivalves grow slower than similar species in more temperate areas (Ralph & Maxwell, 1977; Clark, 1983) and the growth in winter can be close to zero (Richardson, 1979). In a study of the bivalve *Macoma balthica*, growth was more dependent on food supply than on temperature (Beukema et al., 1985) and the same has been shown for other species living in environments with marked seasonality (Clarke, 1988).

In the Arctic, suspension feeders are subjected to large seasonal and annual variations in food supply, and the marginal ice zone is considered to have a great impact on the primary production in these areas (Wassman et al., 1991). In our study, we sampled cockles from three sites with different characteristics of ice coverage and water masses. Moffen, in the north of Spitsbergen Island, is located in the marginal ice zone where ice conditions vary even over short periods of time (days/weeks); Storfjorden, on the east of Spitsbergen Island, has the longest period of ice cover and is situated in Arctic water masses; Bear Island is situated in the Polar front area where Atlantic and Arctic water masses meet (Fig. 1). We hypothesized that long periods of annual ice may decrease growth in benthic organisms due to depressed primary production compared with areas with less ice. Consequently, sites influenced by Atlantic water will show higher growth rates than sites in Arctic water masses. It has been suggested that there is a strong coupling between the pelagic and the benthic environment in the Arctic

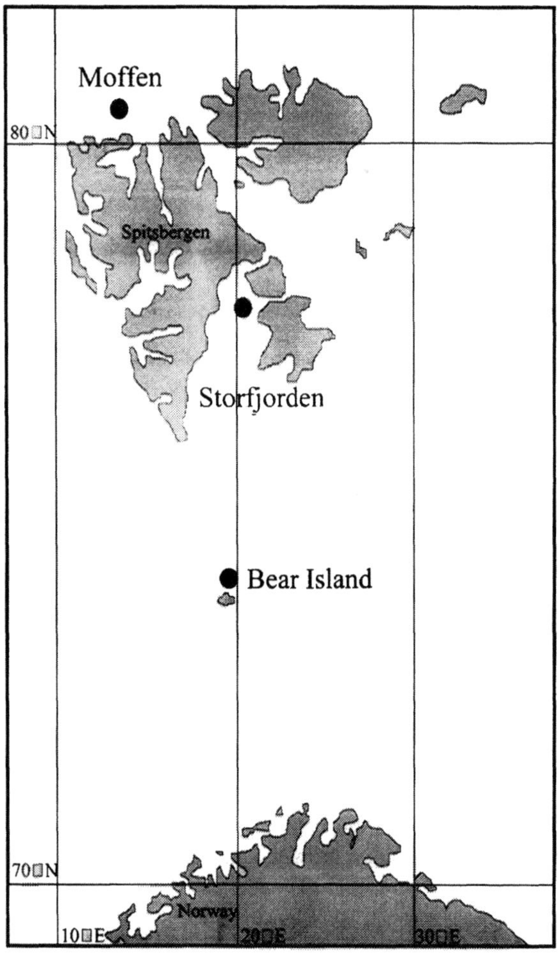

Figure 1. A map of the study area. All sites (indicated with black dots) are situated in Svalbard Island waters with Moffen in the north, Storfjorden in the east of the island Spitsbergen, and the southernmost site north of Bear Island.

due to a slow response of zooplankton to increased primary production in the spring (Grebmeier et al., 1988; Peinert et al., 1989).

In this paper, the annual growth of the common Arctic cockle, *Clinocardium ciliatum* (Fabricius, 1780), is compared between three Arctic sites in the Svalbard area. At each sampling site, bivalve growth patterns are linked to ice coverage. *Clinocardium ciliatum* is one of the common larger bivalve species in the Svalbard area (Rozycki, 1987). It prefers clay sediment (Ockelmann, 1958) and is found in a wide depth range from the intertidal down to about 500 m (Bernard, 1979). Maximum size may be up to 70 mm (Bernard, 1979), but cockles of up to 40 mm are most common in our study area. It is a circumpolar species of Pacific origin and is found at water temperatures below 5 ° C. Together with another common cockle species in this area, *Serripes groenlandicus*, it is important as food for walrus and seals (Fay & Burnes, 1988).

Description of sampling sites

Generally, Arctic waters are characterized by high seasonality (due to the light regime) with a short growth season and annual variation in ice coverage. The polar night (i.e. when the sun is constantly under the horizon) is from 22nd October to 20th February at the northernmost site and from 6th November to 5th February at the southernmost site. The midnight sun (i.e. when the sun is constantly above the horizon) is from 15th April to 27th August at the northernmost site and from 30th April to 12th August at the southernmost site. In the Svalbard area, the maximum ice extent is in March and the minimum in September. Moffen (80° 01′ N, 13° 48′ E) was sampled on 7th September 1997 at 112–130 m water depth (Fig. 1). This site is dominated by gravel and scallop beds, and *C. ciliatum* were found only in hauls where clay was the dominant sediment. Of the 64 individuals sampled, 63 were intact and used in the analyses. Probability for ice in March is about 70–80% (Vinje & Kvambekk, 1991).

The Storfjorden site (77° 10′ N, 20° 09′ E) was sampled on 14th September 1997 at a sampling depth of 80–118 m. This site was dominated by clay and 162 cockles were sampled, of which 154 were intact and used in the analyses. The probability of ice cover in March at Storfjorden is 100% (Vinje & Kvambekk, 1991).

Bear Island site (74° 50′ N, 18° 54′ E) was sampled on 11th November 1997 at a depth range of 100–130 m. The substratum was clay with a fraction of empty cockles; 146 cockles were sampled, of which 143 were included in the analyses. The probability of ice cover in March at Bear Island is about 30% (Vinje & Kvambekk, 1991).

Materials and methods

Bivalves were sampled with a triangular dredge (opening 1 m∗1 m∗1 m, inner net size 1 cm) from the research vessel 'Jan Mayen' (Tromsø University) during two cruises in 1997. Sediment was sieved through a 5 mm mesh and *C. ciliatum* were picked out. Cockles were washed and shell height was measured with vernier callipers to the nearest 0.01 mm. The body was

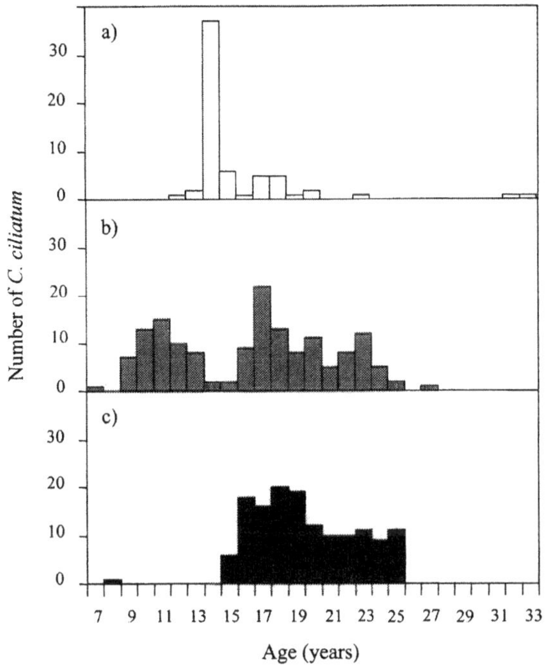

Figure 2. Age distribution of *Clinocardium ciliatum* from (a) Moffen, (b) Storfjorden and (c) Bear Island.

separated from the shell and the shells were stored for further analyses. Age determination was done by counting external growth rings on the shells. This species has very marked external growth rings, and these were assumed to be formed annually due to the pronounced seasonality in the area and on the basis of results from other polar mollusc species (Ralph & Maxwell, 1977; Richardson, 1979; Picken, 1980; Bray & Hain, 1992; Brey & Mackensen, 1997). Incremental widths were measured with callipers and the height from umbo to each increment was also measured. Incremental width data were used in Figures 3 and 4, and height to each increment data was used in results presented in Figure 5.

To remove the ontogenetic growth trend, all incremental widths were standardized according to Witbaard (1996a), with the modification that we used 5 year averages instead of 7 due to the shorter life span of *C. ciliatum* compared with *Arctica islandica* in Witbaard's (1996a) study. The standardized increments were, therefore, calculated as

$$I_t = (\omega_t - \varpi_{t-2,t+2})/\text{s.d.}_{t-2,t+2},$$

where I = the standardized increment, ω= increment width, ϖ= average of increment width over the adjacent 5-year period and s.d. = standard deviation of the mean. In this way, different year classes of bivalves can be analysed together and growth at different calendar years can be compared. Standardization of growth rings is a method widely used in dendrochronology (Fritts, 1976; Schweingruber, 1989) and has also been applied in growth analysis of marine bivalves (Witbaard, 1996a,b). Standardization is done by comparing expected values with actual values. Expected values of growth (increment size) can be calculated by different methods, most of them based on curve fitting. The time series of indices describe higher or lower growth than expected for different years. The standardization was done only from the 6th increment (i.e. the largest increment). Therefore, the standardization was done only for ages with decreasing increment size and the increase in growth in young ages (0–6) was excluded from the analyses. Studies in dendrochronology show that the period of years with increasing growth at younger ages is usually not very sensitive to environmental factors and these years are often excluded to facilitate the standardization calculations (Fritts, 1976). We present indices only for the time period 1980–1994, excluding the 1970s because of the low number of cockles representing this time period.

The ice data are presented as the number of days with more than 70% ice coverage at the sites between 15th March to 1st October (1980–1994).

Results

The age distribution of the bivalves sampled varied between sites (Fig. 2). The sample from Moffen was dominated by the 1983 year class (the 14-year-olds), with other year classes sparsely represented. *Clinocardium ciliatum* from Storfjorden were distributed over several year classes, and, at Bear Island, the bivalves were evenly distributed over the year classes 1972–1982 (15–25-year-old individuals). The youngest year classes were absent at all sites, which could be due to the sampling method.

Incremental size at different ages is shown in Figure 3 for all sites. In this figure, all bivalves are included independent of year class. Annual incremental width increased with age up to about 6 years, after which there is a decrease with age. To compare the growth of the bivalves at different sites, we plotted height against age for different year classes (Fig. 4). This was done for year classes where we had appropriate cockles from at least two sites. For the year classes 1979 and 1980, all three sites were

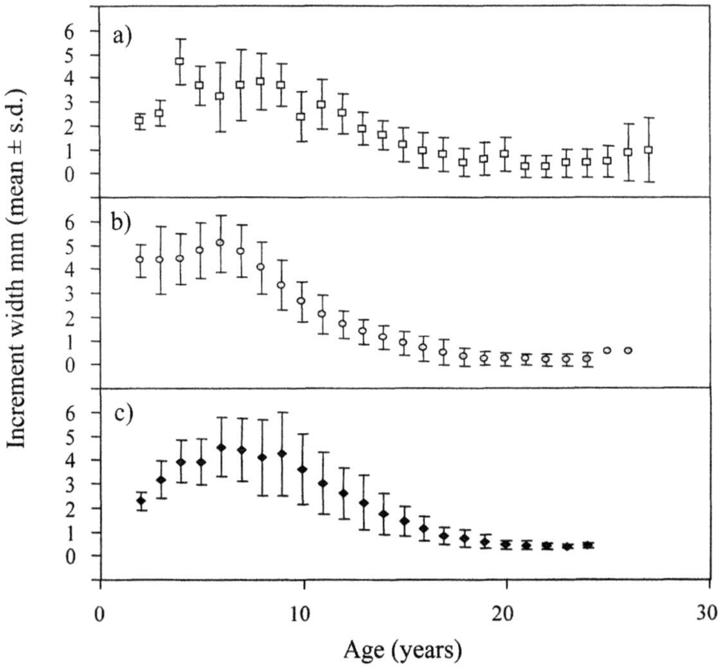

Figure 3. Width of the annual growth increment (mean ± s. d.) at different ages of *Clinocardium ciliatum* from (a) Moffen, (b) Storfjorden and, (c) Bear Island.

compared. The results show that the Storfjorden site has the highest height at age, Bear Island the lowest for several year classes, and Moffen has a height at age between these two. To analyse growth in different calendar years, independent of the age of the bivalves, we standardized the material to get time series of mean standardized indices (Fig. 5). High or low mean index values indicates that most of the cockles showed the same growth trend (i.e. higher or lower growth than expected). For Storfjorden and Bear Island, the values do not vary much from 0, due to low similarity in the growth pattern of the cockles. At Moffen, most of the cockles showed either positive or negative index values at certain years. These results are, however, based on one dominant year class of *C. ciliatum*. From the Storfjorden and Bear Island results, we can distinguish some periods of years with higher or lower growth than expected. Storfjorden has a period of high growth in the beginning of and in the late 1980s, and a period of low growth in the beginning of the 1990s. Bear Island *C. ciliatum* show a period of high growth from 1982 to 1985 and a period of low growth from 1989 to 1992.

The ice data (Fig. 6) show that there was a high number of days with ice coverage at the Moffen site in the beginning of 1980s. In the late 1980s and during the beginning of the 1990s, no trend in ice coverage can be distinguished and the variation is large between years. The Storfjorden ice data indicate that the 1980s was a period with a high number of days with ice and the 1990s had almost half the number of days with ice coverage compared to the 1980s. Bear Island showed peaks with a high number of days with ice (compared to other years at the same site) at 1982 and 1989. The beginning of 1980s is characterized by a low number of days with ice, and the end of 1980s and the beginning of 1990s have a high number of days with ice.

Discussion

The age distribution of the bivalve samples shows that there is a lack of younger specimens at all sites. The triangular dredge is quite robust and with a minimum inner mesh size of 1 cm, small specimens can be lost. When the dredge is hauled up, the uppermost sediment layer is lost with the small bivalves inhabiting this habitat. Another possibility is that the younger specimens were not present in the hauls because of the small area sampled. There is also the possibility that there is a

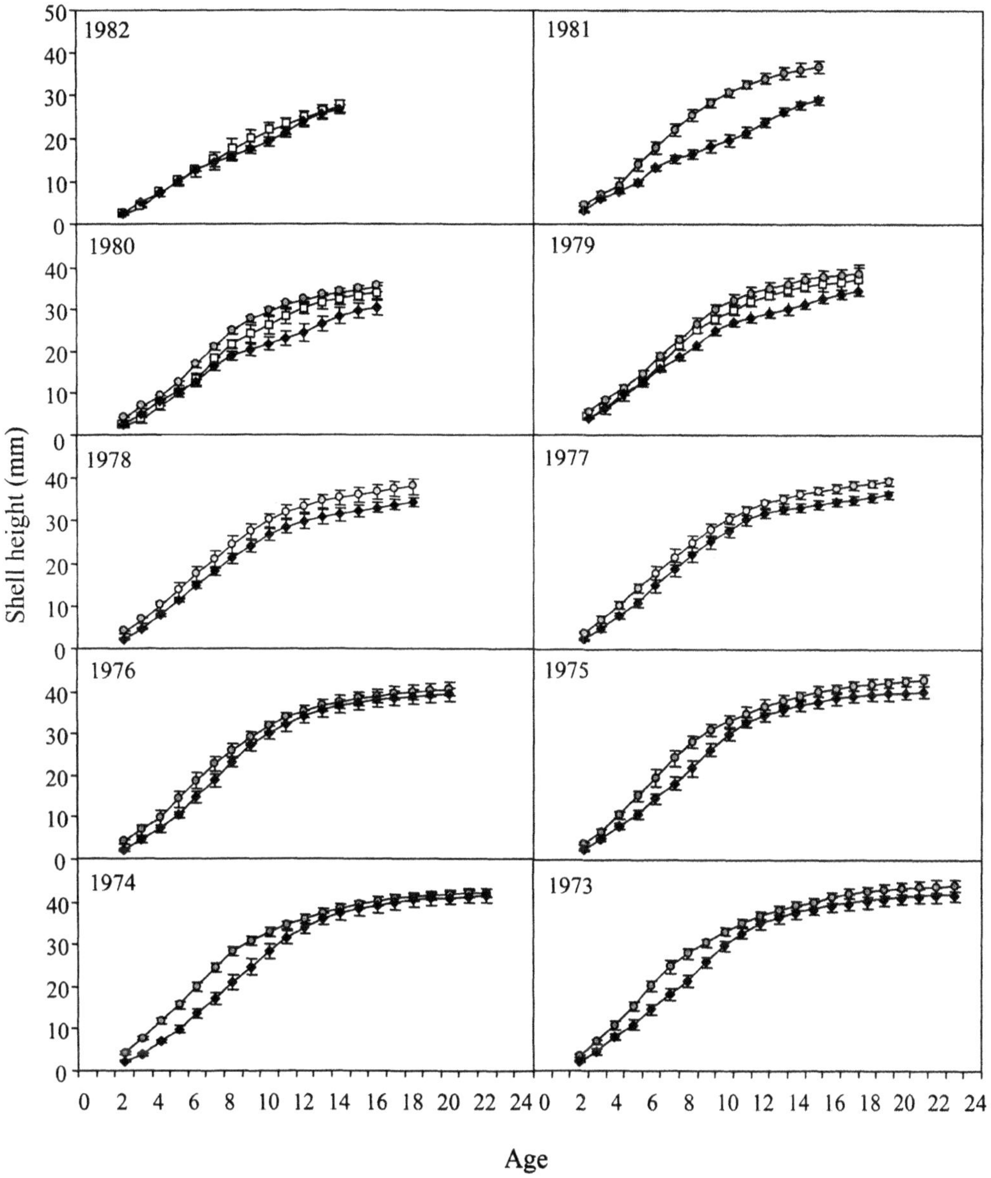

Figure 4. Shell height at age for *Clinocardium ciliatum* of different year classes from Moffen (—□—), Storfjorden (—○—) and Bear Island (—◆—).

lack of younger specimens at all three sites but more sampling is needed to support this hypothesis.

From incremental width results, it is clear that *C. ciliatum* has a nonlinear growth pattern with increase in growth until an age of 6 years after which annual growth decreases with age. This break point in growth may be linked to other morphological changes and be coupled to different factors, such as maturity. No information of the age of maturation of this species in the Svalbard area is available at present.

Storfjorden has the smallest variability in incremental width at different ages, indicating that the environmental conditions influencing growth are more stable here than at the two other sites. Ice data support the growth variation data, as the variation in number of days with ice cover between years is low at Storfjorden compared to the two other sites.

To compare the annual growth at the three sites, the height at different ages for separate year classes was analysed (Fig. 4). Interestingly, Storfjorden has

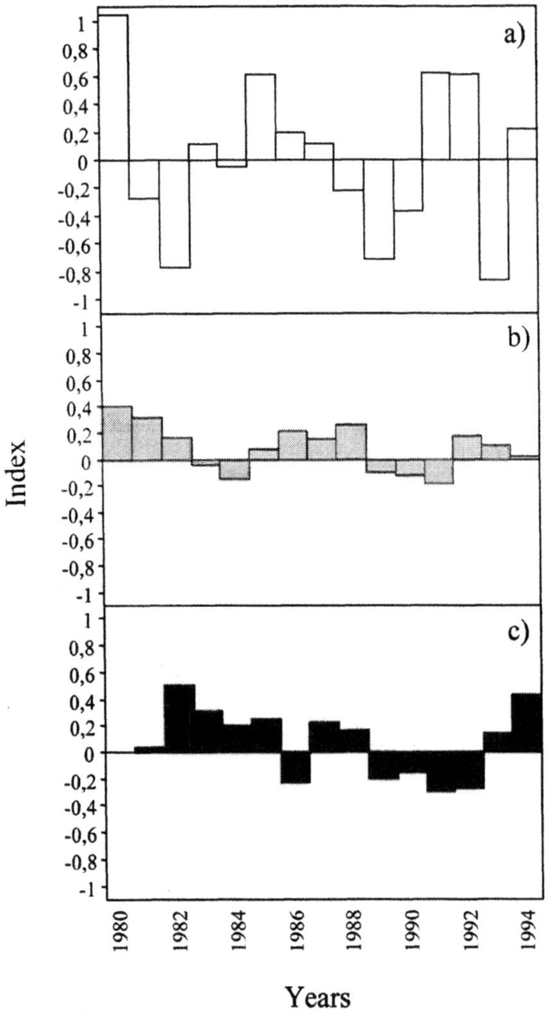

Figure 5. Mean standardized indices for *Clinocardium ciliatum* from (a) Moffen, (b) Storfjorden and (c) Bear Island for the period 1980 to 1994.

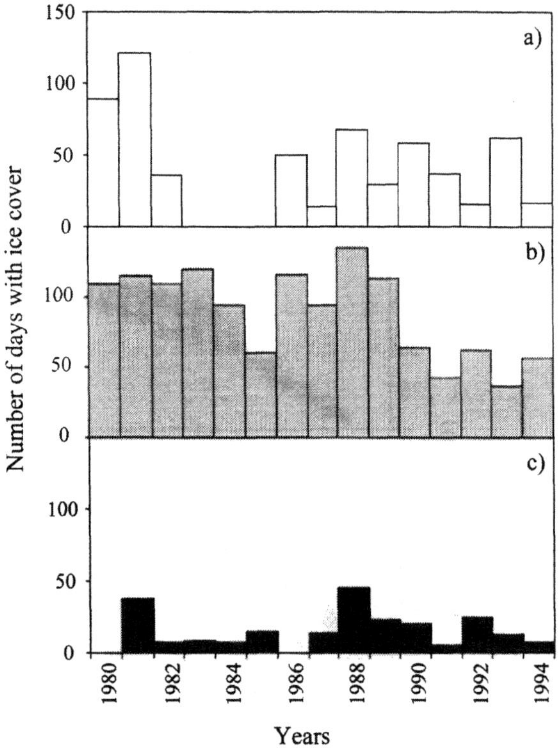

Figure 6. Number of days with ice cover at (a) Moffen, (b) Storfjorden and (c) Bear Island from the 15th of March to the 1st of October for the period 1980 to 1994.

the highest annual growth rate (i.e. highest height at age) and Bear Island the lowest. The longer period of annual ice cover at Storfjorden, compared with Moffen and Bear Island, does not appear to affect the total annual growth of the species. Studies from the Antarctic show that seasonality and ice cover have an effect on the feeding and growth of suspension feeders, with lower growth during the dark, ice-covered winter period (Picken, 1979, 1980; Clarke, 1988; Barnes & Clarke, 1995). This does not mean that the growth is zero during the dark periods and it has been suggested that suspension feeders with low feeding thresholds can use the low number of algal cells present in the water column for feeding (Barnes & Clarke, 1995). Resuspension of benthic microflora can also be an important source of food during this period of time with low cholorophyll content in the water column (Ahn, 1993). Storfjorden is also characterized by a high frequency of polynyas covering a significant part of the area (Vinje & Kvambekk, 1991), which can be an important factor affecting the production patterns found here.

Water depth can affect the growth of *C. ciliatum* and the growth is higher at shallower depths (Høpner-Petersen, 1978). Høpner-Petersen (1978) also showed that there may be large variations in growth between different sites close to each other (all samples taken in Disco Bay, West Greenland). Variation in height at the age of 5 varied from 10 mm in deep waters (100 m) to about 25 mm height at the same age at shallower depths (25–40 m). The height at age in our populations were almost the same values as found by Høpner-Petersen (1978) at the deeper stations. Subarctic populations of *C. ciliatum* grow slower than Arctic populations and the growth has been shown to be related to both temperature and salinity (Andrews,

1972). Andrews (1972) analysed both recent and fossil growth rate of *C. ciliatum* in the high Arctic and height at age for the recent populations were higher than our records, but consistent with the results of Høpner-Petersen (1978) from the shallower sites [the sampling depth is not mentioned by Andrews (1972)].

To be able to compare growth at different years, independent of bivalve year classes, we standardized the annual growth data. The standardized indices are presented as mean index values at different years for the different sites but we did not do any further correlations or other statistical analyses on the data. The comparison with ice data is subjective and not based on any analysis. At the Moffen site, most of the cockles showed the same growth variation at different years (i.e. either negative or positive index values) demonstrated by the higher mean index values than at the two other sites. This is probably due to the fact that Moffen is dominated by just one year class of bivalves, minimizing the variability between year classes. At Storfjorden and Bear Island, the synchrony between index values was lower, demonstrated by the index values not varying much around zero. Witbaard (1996a) found high synchrony between index values for the bivalve *Arctica islandica*, with several years with 100% of the bivalves showing the same growth variation (i.e. either positive or negative values). Our results indicate that there are several factors influencing the growth of *C. ciliatum* in the Svalbard area and that different year classes, and different specimens within a year class, can respond differently to these factors. Variability in growth within and between sites, and within and between year classes, can be an interesting factor that must be studied in more detail to be able to explain the growth patterns found. Witbaard (1996a) correlated standardized growth increments with phytoplankton data, and data on influx of Atlantic water with the Fair Isle current and the East Shetland Atlantic Inflow, but did not get any significant correlations. The feeding activity of Antarctic bryozoan species could not be related to environmental cues like ice cover, temperature, chlorophyll or sedimentation, underlining further the difficulty of linking growth to specific environmental parameters in the field, (Barnes & Clarke, 1994). The problem can be scale-dependent since laboratory experiments often show relationships between growth and environmental factors that can be difficult to obtain in the field, and natural conditions are also far more complex than laboratory environments. For future field studies in pelagic benthic coupling, more specific data on the food supply are needed to be able to study the relationship between production in the water column and growth of benthic animals. In one study on the growth of an Antarctic bryozoan species, *Cellarinella watersi*, positive correlation could be detected between annual growth and summer plankton blooms (Barnes, 1995). It has also been shown that the period of growth decreases at higher latitudes resulting in lower growth rates of the mollusc *Protothaca* (Harrington, 1986, in Clarke, 1988).

Ice coverage is known to affect the production in the water and can be considered as one of several factors determining the growing environment for benthic animals in polar areas. Storfjorden and Bear Island seem to have a better growth in the 1980s than in the beginning of the 1990s. Ice data from the Storfjorde site show more days of ice coverage in the 1980s than in the 1990s, whilst ice data from Bear Island do not match *C. ciliatum* growth. We cannot see any match between growth and ice coverage on the Moffen site either. When comparing the three sites, Storfjorden and Bear Island show similar annual growth trends in mean standardized indices, with Moffen differing from these two.

We have no quantitative data on the potential competitors of *C. ciliatum* at the three sites, but species that were caught together with *C. ciliatum* were the cockle *Serripes groenlandicus* and the scallop *Chlamys islandica*. These two species were low in numbers but are common at all three sites. Water depth and substratum seemed to determine the distribution of these three species at the sites, and *C. ciliatum* was dominating in soft clay substrata.

Our results indicate that higher Arctic latitudes (Moffen) and Arctic water masses (Storfjorden) do not seem to reduce growth of the bivalve *C. ciliatum* compared to the more southern site (Bear Island). The most Arctic site (Storfjorden) had the highest annual growth, indicating that factors other than duration of ice cover and water temperature are important, and that generally high Arctic areas should not be considered less productive than other areas.

References

Ahn, I.-Y., 1993. Enhanced particle flux through the biodeposition by the Antarctic suspension-feeding bivalve *Laternula elliptica* in Marian Cove, King George Island. J. exp. mar. Biol. Ecol. 171: 75–90.

Andrews, J. T., 1972. Recent and fossil growth rates of marine bivalves, Canadian Arctic, and Late-Quaternary Arctic marine environments. Palaeogeogr., Palaeoclimatol., Palaeoecol. 11: 157–176.

Appeldoorn, R. S., 1981. Variation in the growth rate of *Mya arenaria* and its relationship to the environment as analyzed through principal component analysis and the omega parameter of the von Bertanlanffy equation. Fish. Bull. U. S. 81: 75–84.

Arsenault, D. J., P. Girard & J. H. Himmelman, 1997. Field evaluation of the effects of refuge use and current velocity on the growth of juvenile Iceland scallops, *Chlamys islandica* (O. F. Müller, 1776). J. exp. mar. Biol. Ecol. 217: 31–45.

Barnes, D. K. A., 1995. Seasonal and annual growth in erect species of Antarctic bryozoans. J. exp. mar. Biol. Ecol. 188: 181–198.

Barnes, D. K. A. & A. Clarke, 1994. Seasonal variation in the feeding activity of four species of Antarctic bryozoan in relation to environmental factors. J. exp. mar. Biol. Ecol. 181: 113–129.

Barnes, D. K. A. & A. Clarke, 1995. Seasonality of feeding activity in Antarctic suspension feeders. Polar Biol. 15: 335–340.

Bernard, F. R., 1979. Bivalve mollusks of the western Beaufort Sea. Contrib. Sci. Natur. Hist. Mus. Los Angeles County 313: 1–80.

Beukema, J. J., E. Knol & G. C. Cadée, 1985. Effects of temperature on the length of the annual growing season in the tellinid bivalve *Macoma balthica* (L.) living on tidal flats in the Dutch Wadden Sea. J. exp. mar. Biol. Ecol. 90: 129–144.

Brey, T. & S. Hain, 1992. Growth, reproduction and production of *Lissarca notorcadensis* (Bivalvia: Philobryidae) in the Weddell Sea, Antarctica. Mar. Ecol. Prog. Ser. 82: 219–226.

Brey, T. & A. Mackensen, 1997. Stable isotopes prove shell growth bands in the Antarctic bivalve *Laternula elliptica* to be formed annually. Polar Biol. 17: 465–468.

Clarke, A., 1983. Life in cold water: the physiological ecology of polar marine ectotherms. Oceanogr. Mar. Biol. Ann. Rev. 21: 341–453.

Clarke, A., 1988. Seasonality in the Antarctic marine environment. Comp. Biochem. Physiol. 90B: 461–473.

Fay, F. H. & J. J. Burns, 1988. Maximal feeding depth of walruses. Arctic 41: 239–240.

Grebmeier, J. M., C. P. McRoy & H. M. Feder 1988. Pelagic.benthic coupling on the shelf of the northern Bering and Chukchi Sea I. Food supply and benthic biomass. Mar. Ecol. Prog. Ser. 48: 57–67.

Høpner Petersen, G., 1978. Life cycles and population dynamics of marine benthic bivalves from the Disko Bugt area of west Greenland. Ophelia 17: 95–120.

Jones, D. S., M. A. Arthur & D. J. Allard, 1989. Sclerochronological records of temperature and growth form shells of *Mercenaria mercenaria* from Narragansett Bay, Rhode Island. Mar. Biol. 102: 225–234.

MacDonald, B. A. & N. F. Bourne, 1987. Growth, reproductive output, and energy partitioning in weathervane scallops, *Patinopecten caurinus*, from British Columbia. Can. J. Fish. aquat. Sci. 44: 152–160.

Navarro, E., J. I. P. Iglesias & M. M. Ortega, 1992. Natural sediment as a food source for the cockle *Cerastoderma edule* (L.): effect of variable particle concentration on feeding, digestion and the scope for growth. J. exp. mar. Biol. Ecol. 156: 69–87.

Ockelmann, K. W., 1958. Marine lamellibranchiata. Meddr Grønland 122: 256 pp.

Peinert, R., B. V. Bodungen & V. Smetacek, 1989. Food web structure and loss rates. In Berger, W., V. Smetacek & G. Wefer (eds), Productivity of the Ocean: Present and Past. John Wiley and Sons, New York: 34–48.

Peterson, C. H. & S. V. Andre, 1980. An experimental analysis of interspecific competition among marine filter feeders in a soft-sediment environment. Ecology 6: 129–139.

Picken, G. B., 1979. Growth, production and biomass of the antarctic gastropod *Laevilacunaria antarctica* Martens 1885. J. exp. mar. Biol. Ecol. 40: 71–79.

Picken, G. B., 1980. The distribution, growth, and reproduction of the Antarctic limpet *Nacella (Patinigera) concinna* (Strebel, 1908). J. exp. mar. Biol. Ecol. 42: 71–85.

Ralph, R. & J. G. H. Maxwell, 1977. Growth of two Antarctic lamellibranchs: *Adamussium colbecki* and *Laternula elliptica*. Mar. Biol. 42: 171–175.

Richardson, C. A., D. J. Crisp, N. W. Runham & L. D. Gruffydd, 1980. The use of tidal growth bands in the shell of *Cerastoderma edule* to measure seasonal growth rates under cool temperate and sub-arctic conditions. J. mar. biol. Ass. U. K. 60: 977–989.

Richardson, M. G., 1979. The ecology and reproduction of the brooding Antarctic bivalve *Lissarca miliaris*. Br. Antarctic Survey Bull. 49: 91–115.

Ròzycki, O., 1987. Marine bivalve molluscs of Svalbard. Polar Res. 5: 257–260.

Sagar, P. M., 1980. Life cycle and growth of the Antarctic gammarid amphipod *Paramoera walkeri* (Stebbing, 1906). J. R. Soc. N. Z. 10: 259–270.

Schweingruber, F. H., 1989. Tree Rings, Basics and Applications of Dendrochronology. Kluwer Academic Publishers, Dordrecht: 273 pp.

Vinje, T. & Å. S. Kvambekk, 1991. Barents Sea drift ice characteristics. In Sakshaug, E., C. C. E. Hopkins & N. A. Øritsland (eds), Proceedings of the Pro Mare Symposium on Polar Marine Ecology, Trondheim 1990. Polar Res. 10: 59–68.

Wassman, P., R. Peinert & V. Smetacek, 1991. Patterns of production and sedimentation in the boreal and polar Northeast Atlantic. In Sakshaug, E., C. C. E. Hopkins & N. A. Øritsland (eds), Proceedings of the Pro Mare Symposium on Polar Marine Ecology, Trondheim 1990. Polar Res. 10: 209–228.

Witbaard, R., 1996a. Growth variations in *Arctica islandica* L. (Mollusca): a reflection of hydrography-related food supply. ICES J. mar. Sci. 53: 981–987.

Witbaard, R., 1996b. Tree of the Sea. The use of the internal growth lines in the shell of *Arctica islandica* (Bivalvia, Mollusca) for the retrospective assessment of marine environmental change. PhD thesis, Groningen University: 149 pp.

Witbaard, R., R. Franken & B. Visser, 1997. Growth of juvenile *Arctica islandica* under experimental conditions. Helgoländer wiss. Meeresunters. 51: 417–431.

Hydrobiologia **440:** 339–345, 2000.
M.B. Jones, J.M.N. Azevedo, A.I. Neto, A.C. Costa & A.M. Frias Martins (eds), Island, Ocean and Deep-Sea Biology.

Distribution and abundance of co-occurring chthamalid barnacles *Chthamalus montagui* and *Chthamalus stellatus* (Crustacea, Cirripedia) on the southwest coast of Portugal

E. B. Sousa, T. Cruz & J. J. Castro
Laboratório de Ciências do Mar, Universidade de Évora, Apartado 190, 7520-903 Sines, Portugal

Key words: barnacles, *Chthamalus*, distribution, abundance, intertidal

Abstract

Two co-occurring chthamalid barnacle species occur on the rocky shores of continental Portugal: *Chthamalus stellatus* (Poli) and *Chthamalus montagui* Southward. In the present study, patterns of distribution and abundance (density and percentage cover), both vertically (tidal level) and horizontally (wave action), were studied for each species on headlands (more wave action) and not on headlands (less wave action) and at two spatial scales (shore and site) and three tidal levels (low, mid and upper). *Chthamalus montagui* was significantly more abundant than *C. stellatus* in both degrees of wave action and at all tidal levels. Despite this great difference in abundance and considering each species individually, *C. montagui* was more abundant at mid tide-level, while *C. stellatus* was more abundant at the low tide-level, and *C. montagui* was less abundant on headlands, while *C. stellatus* was more abundant on headlands. Small-scale variability (between sites, tens of meters) of abundance of each species was detected, while differences on a larger scale (between shores, tens of kilometres) were not significant.

Introduction

Two species of Chthamalid barnacles co-occur on rocky shores of continental Portugal: *Chthamalus montagui* Southward and *C. stellatus* (Poli) (Crustacea: Cirripedia). Each species is temperate and their geographical ranges overlap extensively (from the Atlantic coast of Britain to northwest Africa and in the Mediterranean Sea) (Crisp et al., 1981). *C. montagui* was distinguished from *C. stellatus* rather recently (Southward, 1976).

Vertical (in relation to tidal level) and horizontal (in relation to degree of wave action) patterns of distribution and abundance of intertidal barnacles have been studied for several species in different parts of the world (e.g. Connell, 1961a; Foster, 1971a; Southward, 1976; Crisp et al., 1981; Miyamoto et al., 1999). In Britain, *C. montagui* is more common in sheltered places, while *C. stellatus* is more abundant on the more exposed coasts of the southwest (Southward, 1976). On several shores in Britain, Spain and France, and the Mediterranean, where both species exist together, *C. montagui* tends to occupy slightly higher tidal levels and *C. stellatus* lower tidal levels (Southward, 1976; Crisp et al., 1981). In Portugal, *C. montagui* is very abundant in the intertidal, but studies on the ecology of *Chthamalus* species are rare.

In the present study, patterns of distribution and abundance of both *Chthamalus* species on the Southwest coast of Portugal were studied at different tidal levels, different degrees of wave action and different spatial scales. Based on vertical and horizontal patterns described for each species in Britain (Southward, 1976; Crisp et al., 1981), the following hypotheses were tested:

1. *C. stellatus* is more abundant than *C. montagui* on shores with more wave action, while *C. montagui* is more abundant on shores with less wave action;
2. *C. montagui* is more abundant at mid-tide level, while *C. stellatus* is more abundant at low-tide level.

Material and methods

Patterns of density and percentage cover of *C. montagui* and *C. stellatus* were studied during winter

1997/1998. Three tidal levels (low, mid and upper) were selected within the vertical zone where *C. montagui* is more abundant. Two degrees of wave action were considered: more (shores on headlands) and less (shores not on headlands). Both types of shores face directly the prevailing winds with no kind of protection. The assumption that shores on headlands are subject to more wave action than shores not on headlands is based on the influence on waves and tides that prominent headlands on cliffed coasts can have (Carter, 1989) and that headlands capture wave energy (Robinson, 1983 *in* Carter, 1989). Also, the increased energy of the waves breaking on headlands is reflected in an increased wave height (Thurman, 1993). The two shores located on headlands were Cabo de Sines (37° 58′ N) and Cabo Sardão (37° 38′ N) and those not on headlands were Oliveirinha (37° 53′ N) and Nascedios (37° 41′ N). On each shore, two different sites (10–20 m wide; 50–100 m apart) were considered.

Density of each species was estimated by counting the total number of individuals in a 10 × 10 cm quadrat [sub sampling of ten random 1 cm^2 quadrats was performed when high densities were observed ($>3\ cm^{-2}$)]. Percentage cover of each species was estimated using a 10 × 10 cm quadrat with 25 regular intersection points. For each species, six random replicates were sampled (independent quadrats for each species).

Patterns of density and percentage cover were tested by ANOVA (Underwood, 1997) with GMAV5 software (Institute of Marine Ecology, University of Sydney). Both analyses included 5 factors (with treatments in brackets): species (Sp) (*C. montagui* and *C. stellatus*), a fixed factor; vertical tidal level (V) (low, mid and upper), a fixed factor; degree of wave action (W) (headlands and not headlands), a fixed factor; shore (Sh) (two shores per degree of wave action), a random factor nested in W; and site (Si) (two sites per shore), a random factor nested in Sh and W. Cochran's test (Winer et al., 1991) was used to test the assumption of homogeneity of variances. When appropriate, SNK tests were used for multiple comparisons of the means (at $\alpha = 0.05$).

Results

Density

Chthamalus montagui was much denser than *C. stellatus* on all shores (see Figs 1 and 2). Mean number of individuals per 10 × 10 cm of *C. montagui* varied between 99 and 3228, while mean density of *C. stellatus* varied between 0 and 20 individuals per 10 × 10 cm (Figs 1 and 2). The significant three-way interaction of Species × Tide-level × Site [Shore (Degree of wave action)] indicated that vertical distribution of density changed with species and also from site to site (Table 1). Despite this small-scale variability, Species × Tidal level × Degree of wave action interaction was also significant, indicating that vertical distribution changed with species but also with degree of wave action (Table 1). SNK tests performed on this interaction (Table 1) revealed the following patterns: (1) *C. montagui* was significantly denser than *C. stellatus* on all shores and at all tidal levels; (2) *C. montagui* was denser at mid-tide level (on headlands) or at mid- and upper-tidal levels (not on headlands) than on the other tidal levels; while *C. stellatus* density was always higher at low-tide level compared with the other tidal levels; (3) at low-tide level, *C. montagui* and *C. stellatus* densities were similar between shores on headlands and shores not on headlands; (4) at mid- and upper-tidal levels, density of *C. montagui* was lowest on headlands; while the opposite was observed with *C. stellatus*. There were no significant differences between shores (Table 1).

Percentage cover

Mean percentage cover of *C. montagui* varied between 8 and 64.3 and mean percentage cover of *C. stellatus* varied between 0 and 7.3 (Figs 1 and 2). The three-way interaction of Species × Tide-level × Degree of wave action was significant, indicating that the vertical distribution of percentage cover changed with species but also with degree of wave action (Table 2). SNK tests performed on this interaction (Table 2) revealed the following patterns: (1) percentage cover of *C. montagui* was significantly higher than percentage cover of *C. stellatus* on all shores and at all tidal levels; (2) percentage cover of *C. montagui* was always higher at mid-tide level than on the other tidal levels; (3) at shores not on headlands, percentage cover of *C. stellatus* was higher at low-tide level than at other tidal levels, while no significant differences were detected among tidal levels at shores on headlands; (4) horizontal distribution (in relation to degree of wave action) of percentage cover of *C. montagui* was not consistent across different tidal levels; (5) at low- and mid-tidal levels, percentage cover of *C. stellatus* was highest at shores on headlands, while at upper-tidal

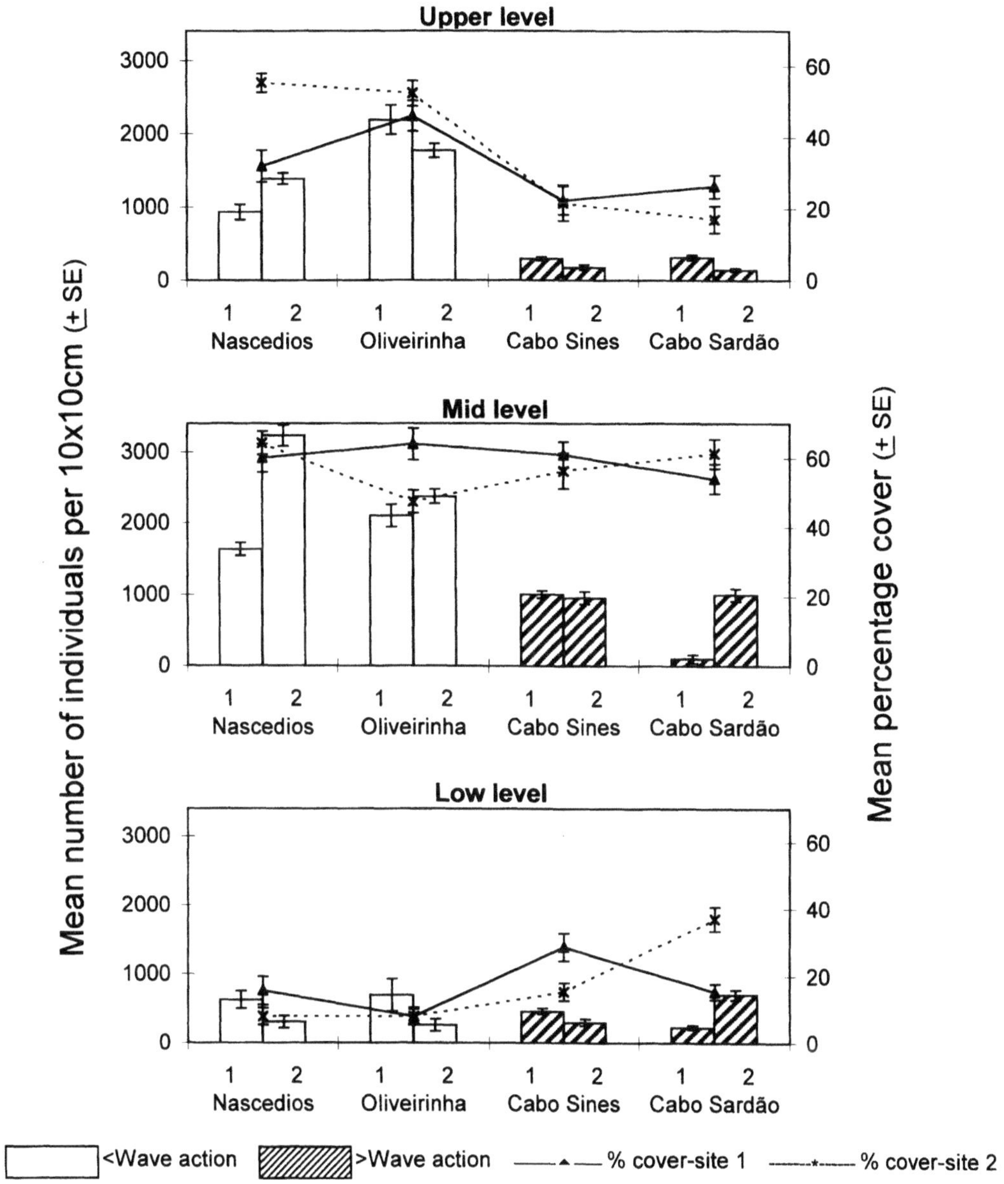

Figure 1. Density and percentage cover of *C. montagui* per 10×10 cm (mean ± SE) on shores with less (Nascedios and Oliveirinha) and more (Cabo Sines and Cabo Sardão) wave action, at upper-, mid- and low-tidal levels.

levels, there were no significant differences between different types of wave action. There were no significant differences in percentage cover between shores but there were significant differences between sites (Table 2).

Discussion

On the southwest coast of Portugal, *C. montagui* was significantly more abundant than *C. stellatus* on shores with different degrees of wave action and at all tidal levels. The reduced number, and percentage cover, of *C. stellatus* adults may be due to lower egg production and limited supply of settling larvae compared with *C. montagui*. Recent observations of recruitment to one month of both species on this coast revealed that all recruits were *C. montagui* (unpublished data).

Maximal density of both species on the southwest of England (Southward, 1976; Burrows, 1988; Pannacciulli, 1995) differ considerably when compar-

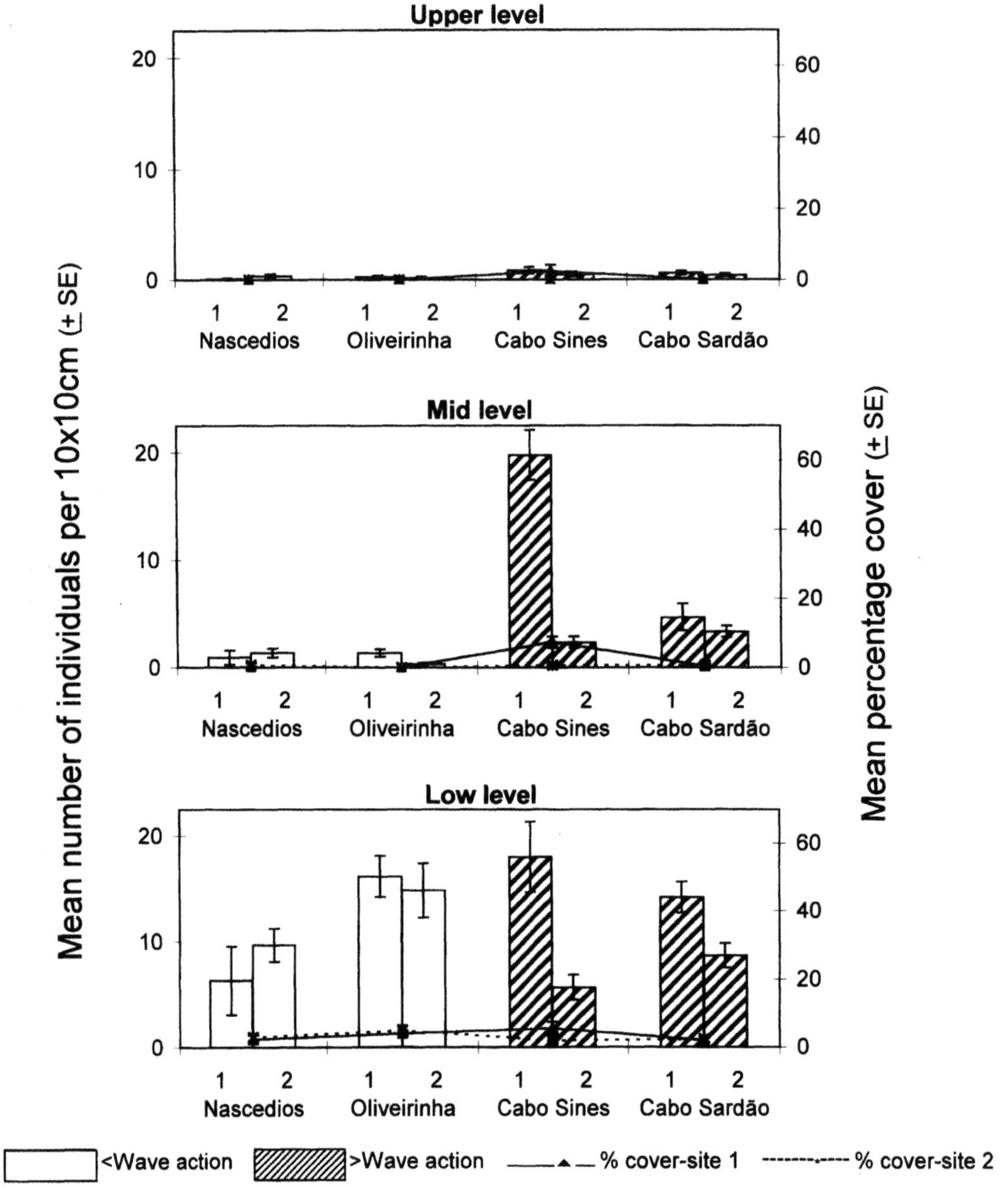

Figure 2. Density and percentage cover of *C. stellatus* per 10×10 cm (mean ± SE) on shores with less (Nascedios and Oliveirinha) and more (Cabo Sines and Cabo Sardão) wave action, at upper-, mid- and low-tidal levels.

ing with Southwest of continental Portugal (present study): *C. montagui* values are lower in Southwest of England (up to 6 animals per 1 cm^2) than on southwest of Portugal (mean number up to 32 animals per 1 cm^2); while maximal density of *C. stellatus* is considerably higher in the southwest of England (up to 3 animals per 1 cm^2) than on the southwest of Portugal (mean number up to 0.2 animals per 1 cm^2). In Britain, *C. stellatus* can be more abundant than *C. montagui* on exposed shores and at lower tidal levels (Southward, 1976; Crisp et al., 1981). The patterns of vertical distribution of *C. montagui* and *C. stellatus* on the Southwest coast of Portugal were specifically different. Despite the low density on all shores, *C. stellatus* was denser at the low-tide level than at mid- and upper-tidal levels. In Britain, Southward (1976) and Burrows (1988) have observed a similar pattern for *C. stellatus*. Burrows (1988) linked this vertical pattern with the recruitment of this species which was more abundant at the low shore. Another possible explanation is that

Table 1. ANOVA (a) and SNK tests (b) results of number of individuals per 10 × 10 cm in relation to species (Sp), tide-level (T), wave exposure (W), shore (Sh), and site (Si). *N*= 6. Tests of sources of variation involved in significant higher-order interactions are not presented because there are no hypotheses related to them

(a) ANOVA. Data transformed [log (x + 0.1)]; variances homogenous by Cochran's test, P>0.05

Source of variation	d.f.	MS	*F*	*P*	*F versus*
Sp	1	2737.79	307.07		
T	2	55.17	11.11		
W	1	1.08	0.13		
Sh(W)	2	8.21	1.11	0.4136 NS	Si(Sh(W))
Si(Sh(W))	4	7.40	9.37		
Sp × T	2	100.14	326.50		
Sp × W	1	84.26	71.01		
Sp × Sh(W)	2	1.18	0.31	0.7504 NS	Sp × Si(Sh(W))
Sp × Si(Sh(W))	4	3.84	4.86		
T × W	2	5.84	1.18		
T × Sh(W)	4	4.96	1.66	0.2511 NS	T × Si(Sh(W))
T × Si(Sh(W))	8	2.99	3.79		
Sp × T × W	2	20.30	66.20	0.0009 **	Sp × T × Sh(W)
Sp × T × Sh(W)	4	0.30	0.09	0.9822 NS	Sp × T × Si(Sh(W))
Sp × T × Si(Sh(W))	8	3.32	4.20	0.0001 **	Residual
Residual	240	0.79			
Total	287				

(b) SNK tests to Sp × T × W [= denotes not significantly different (P>0.05); > or < (P<0.05)]
CM=*C.montagui*, CS=*C.stellatus*,(-)W=less wave action, (+)W=more wave action

Sp (T×W)	T (Sp × W)	W (Sp × T)
Low, (-)W: CM>CS	CM, (-)W: Upper=Mid>Low	CM, Low: (+)W=(-)W
Low, (-)W: CM>CS	CM, (+)W:Mid>Low>Upper	CM, Mid: (+)W<(-)W
Mid, (-)W: CM>CS	CS, (-)W: Low>Mid>Upper	CM, Upper: (+)W<(-)W
Mid, (-)W: CM>CS	CS, (+)W: Low>Mid>Upper	CS, Low: (+)W=(-)W
Upper, (-)W: CM>CS		CS, Mid: (+)W>(-)W
Upper, (-)W: CM>CS		CS, Upper: (+)W>(-)W

NS=Not significant (P>0.05); * P<0.05; ** P<0.001

the upper limit of vertical distribution of *C. stellatus* can be determined by post-settlement mortality due to physical stress or biological disturbance (e.g.Connell, 1961a, b; Foster, 1971b; Denley & Underwood, 1979).

On headlands, *C. montagui* was denser at mid-tide level than at other tidal levels, while on shores not on headlands, this species was denser at mid- and upper-tide levels. In contrast to *C. stellatus*, Burrows (1988) observed that the vertical distribution of recruits of *C. montagui* did not match with the vertical distribution of adults. Burrows (1988) reported that, as in *C. stellatus*, recruitment of *C. montagui* was also higher on the low shore, but density of *C. montagui* adults was higher at mid-tide levels. This author also observed increasing mortality of *C. stellatus* recruits higher on the shore, while the highest mortality of *C. montagui* occurred at low-tide level, and suggested that the difference of post-recruitment mortality between both species could be linked with differences in the morphology of metamorphs of both species. In fact, the tighter fit of the opercular plates to the parietes in *C. montagui* compared with *C. stellatus* may confer a greater resistance to desiccation stress (Burrows, 1988).

Horizontal (in relation to degree of wave action) patterns of distribution of both species on the south-west coast of Portugal were also different. At the low-tide level, density of each species was similar on shores with different degrees of wave action, but percentage cover of *C. montagui* and of *C. stellatus* was

Table 2. ANOVA (a) and SNK tests (b) results of percentage cover using a 10 × 10 cm quadrat in relation to species (Sp), tide-level (T), wave exposure (W), shore (Sh), and site (Si). *N*= 6. Tests of sources of variation involved in significant higher-order interactions are not presented because there are no hypotheses related to them

(a) ANOVA. Data transformed; [ArcSin (%)]; variances homogenous by Cochran's test, *P*>0.05

Source of variation	d.f.	MS	*F*	*P*	*F versus*
Sp	1	73474.23	795.48		
T	2	4464.58	23.34		
W	1	0.96	0.02		
Sh(W)	2	50.33	0.19	0.8367 NS	Si(Sh(W))
Si(Sh(W))	4	269.99	4.58	0.0014 *	RES
Sp × T	2	6271.86	213.76		
Sp × W	1	55.78	6.68		
Sp × Sh(W)	2	8.35	0.09	0.9193 NS	Sp x Si(Sh(W))
Sp × Si(Sh(W))	4	97.20	1.65	0.1625 NS	RES
T × W	2	716.66	3.75		
T × Sh(W)	4	191.25	1.91	0.2024 NS	T × Si(Sh(W))
T × Si(Sh(W))	8	100.18	1.70	0.0990 NS	RES
Sp × T × W	2	1637.93	55.82	0.0012 *	Sp × T × Sh(W)
Sp × T × Sh(W)	4	29.34	0.25	0.8999 NS	Sp × T × Si(Sh(W))
Sp × T × Si(Sh(W))	8	115.97	1.97	0.0512 NS	RES
Residual	240	58.92			
Total	287				

(b) SNK tests to Sp × T × W [= denotes not significantly different (*P*>0.05); > or < (*P*<0.05)]
CM=*C.montagui*, CS=*C.stellatus*,(-)W=less wave action, (+)W=more wave action

Sp (T×W)	T (Sp × W)	W (Sp × T)
Low, (-)W: cm>CS	CM, (-)W: Mid>Upper>Low	CM, Low: (+)W>(-)W
Low, (-)W: CM>CS	CM, (+)W:Mid>Upper=Low	CM, Mid: (+)W=(-)W
Mid, (-)W: CM>CS	CS, (-)W: Low>Mid=Upper	CM, Upper: (+)W<(-)W
Mid, (-)W: CM>CS	CS, (+)W: Low=Mid=Upper	CS, Low: (+)W>(-)W
Upper, (-)W: CM>CS		CS, Mid: (+)W>(-)W
Upper, (-)W: CM>CS		CS, Upper: (+)W=(-)W

NS=Not significant (*P*>0.05); * *P*<0.05; ** *P*<0.001.

higher on headlands suggesting that barnacles were bigger in this habitat. At mid-tide level, density of *C. montagui* was lower on headlands but percentage cover was similar on all shores, suggesting again that individuals of this species were larger on headlands. At upper-tide level, density and percentage cover of *C. montagui* were lower on headlands. At mid- and upper-tide levels, density of *C. stellatus* was higher on headlands. At mid-tide level, percentage cover of this species was also higher on headlands, while at upper-tide levels it was similar and very low on all shores. Possible differences in size of *C. montagui* and *C. stellatus* in relation to wave action may be due to different growth patterns in these habitats. On more exposed shores, water circulation is more intense and the amount of food can be higher (Sanford et al., 1994), resulting in higher rates of barnacle growth.

While at mid- and upper-tide levels density patterns of both species differed in relation to degree of wave action, at low-tide level, densities species were similar irrespective of degree of wave action. One possible explanation is that the physical stress due to wave action could be more variable at mid- and upper-tide levels than at low tide, due to the longer immersion period of the latter. Also, different horizontal patterns at mid- and upper-tide levels of both species may be due to differences in settlement and post-settlement processes between shores with different degrees of wave action. Crisp et al. (1981) suggested that either cyprids of *C. stellatus* are better adapted to settle or the adults are able to survive on wave-exposed areas.

Differences in the abundance of both species between shores were not detected but variability at the smaller scale (between sites) was significant. This small-scale variability can be due to small-scale variation in settlement and in post-settlement processes (e.g. Caffey, 1982; Hunt & Scheibling, 1997). Settlement rate depends on the rate of arrival of larvae on the shore (which depends on the larval pool and physical transport processes), on suitable settlement substrata, and on micro-hydrodynamic and behavioral processes (Pineda, 1994). In several studies, small-scale variability of settlement (amongst sites, within sites) have been observed (e.g. Connell, 1961b; Denley & Underwood, 1979; Gaines & Roughgarden, 1985; Pineda, 1994).

Acknowledgements

This work was partially funded by a MAST project MAS3-CT95-0012 (EUROROCK). We wish to thank two anonymous referees for critical comments on the manuscript.

References

Burrows, M., 1988. The comparative biology of *Chthamalus stallatus* (Poli) and *Chthamalus montagui* Southward. Ph.D. thesis, University of Manchester: 319 pp.

Caffey, H., 1982. No effect of naturally occurring rock types on settlement or survival in the intertidal barnacle, *Tesseropora rosea* (Krauss). J. exp. mar. Biol. Ecol. 63: 119–132.

Carter, R. W. G., 1989. Coastal environments. Academic Press, London: 617 pp.

Connell, J. H., 1961a. The influence of interspecific competition and other factors on the distribution of the barnacle *Chthamalus stellatus*. Ecology 42: 710–723.

Connell, J. H., 1961b. Effects of competition, predation by *Thais lapillus*, and other factors on natural populations of the barnacle *Balanus balanoides*. Ecol. Monogr. 31: 61–104.

Crisp, D., A. J. Southward & E. Southward, 1981. On the distribution of the intertidal barnacles *Chthamalus stellatus*, *Chthamalus montagui* and *Euraphia depressa*. J. mar. biol. Ass. U. K. 61: 359–380.

Denley, E. & A. Underwood, 1979. Experiments on factors influencing settlement, survival and growth of two species of barnacles in New South Wales. J. exp. mar. Bio. Ecol. 36: 269–293.

Foster, B. A., 1971a. On the determinants of the upper limit of intertidal distribution of barnacles (Crustacea: Cirripedia). J. anim. Ecol. 40: 33–48.

Foster, B. A., 1971b. Desiccation as a factor in the intertidal zonation of barnacles. Mar. Biol. 8: 12–29.

Gaines, S. & J. Roughgarden, 1985. Larval settlement rate: a leading determinant of structure in an ecological community of the marine intertidal zone. Proc. natl. Acad. Sci. U.S.A. 82: 3707–3711.

Hunt, H. & R. Scheibling, 1997. Role of post-settlement mortality in recruitment of benthic marine invertebrates. Mar. Ecol. Prog. Ser. 155: 269–301.

Miyamoto, Y., T. Noda & S. Nakao, 1999. Zonation of two barnacle species not determined by competion. J. mar. biol. Ass. U. K. 79: 621–628.

Pannacciulli, F., 1995. Population ecology and genetics of European species of intertidal barnacles. Ph.D. thesis, University of Liverpool: 143 pp.

Pineda, J., 1994. Spatial and temporal patterns in barnacle settlement rate along a southern California rocky shore. Mar. Ecol. Prog. Ser. 107: 125–138.

Sanford, E., D. Bermudez, M. D. Bertness & S. D. Gaines, 1994. Flow, food supply and acorn barnacle population dynamics. Mar. Ecol. Prog. Ser. 104: 49–62.

Southward, A. J., 1976. On the taxonomic status and distribution of *Chthamalus stellatus* (Cirripedia) in the northeast Atlantic region: with a key to the common intertidal barnacles of Britain. J. mar. biol. Ass. U. K. 56: 1007–1028.

Underwood, A. J., 1997. Experiments in Ecology: their Logical Design and Interpretation using Analysis of Variance. Cambridge University Press, Cambridge, England: 504 pp.

Thurman, H. V., 1993. Essentials of Oceanography. Macmillan Publishing Company, New York: 393 pp.

Winer, B. J., D. R. Brown & K. M. Michels, 1991. Statistical Principles in Experimental Design. Mc-Graw-Hill, New York, New York: 1057 pp.

Hydrobiologia **440**: 347–355, 2000.
M.B. Jones, J.M.N. Azevedo, A.I. Neto, A.C. Costa & A.M. Frias Martins (eds), Island, Ocean and Deep-Sea Biology.

Predation by juvenile *Platichthys flesus* (L.) on shelled prey species in a bare sand and a drift algae habitat

Katri Aarnio & Johanna Mattila
Husö Biological Station & Department of Biology, Åbo Akademi University, FIN-22220 Emkarby, Åland Islands, Finland

Key words: predation, juvenile fish, *Platichthys flesus*, drifting algae, shelled zoobenthic prey, experiment

Abstract

Due to increasing eutrophication of the coastal Baltic waters, drifting algae are a common phenomenon. Drifting algal mats accumulate on shallow sandy bottoms in late summer and autumn, and affect the ambient fauna. Juvenile flounder, *Platichthys flesus*, utilize these habitats during their first few years. They feed on benthic meio- and macrofauna; part of their diet consists of shelled species, such as Ostracods, and juvenile *Hydrobia* spp. and *Macoma balthica*. Earlier studies have shown that up to 75% of ostracods and 92% of hydrobiids survive the gut passage of juvenile flounder, while all *M. balthica* are digested by the fish. We conducted laboratory experiments to study how the shelled prey responded to a drift algal mat, and the predation efficiency of juvenile *P. flesus* on these prey species on bare sand and with drifting algae (50% coverage). *Hydrobia* spp. utilized the drift algae as a habitat and, after 1 h, 50% had moved into the algae; ostracods and *M. balthica* were more stationary and, after 96 h, only 23 and 12%, respectively, were found in the algae. For the predation efficiency of *P. flesus*, a two-way ANOVA with habitat (algae, bare sand) and predation (fish, no fish) as factors revealed that both algae and predation affected negatively the survival of all three prey species. The algae, thus, affected the predation efficiency of juvenile *P. flesus* and the consumption of prey was much reduced in the algal treatments compared to the bare sand. This was due probably to increased habitat complexity and the ability of prey, especially hydrobiids, to use the algal mat as a refuge. Altered habitat structure due to drift algae, together with the resultant changes in habitat (refuge) value for different prey species, may profoundly change the structure of benthic communities.

Introduction

Eutrophication is a major factor affecting near-shore areas in the Baltic Sea (Larsson et al., 1985; Cederwall & Elmgren, 1990; Bonsdorff et al., 1991; Bonsdorff et al., 1997a,b). Increased concentrations of nutrients in the water have lead to higher primary production and increased production of filamentous annual algae (Bonsdorff, 1992; Bonsdorff et al., 1997b). When these algae die, they detach and form dense mats which drift in the water column and, eventually, end up on shallow bottoms in late summer and autumn (Ólafsson, 1988; Bonsdorff, 1992). Normally, shallow sandy bottoms are homogenous habitats with low habitat complexity, and are important feeding and nursery areas for juvenile and small fish. Occurrence of moderate amounts of algae in these habitats increases habitat complexity and gives an opportunity for species to coexist (Heck & Thoman, 1981 and references therein). On the other hand, the presence of large, uniform algal mats decreases habitat complexity, and the algal mats may have serious effects on the benthic fauna, depending on how long the algae persist in the area (Norkko, 1997). Severe oxygen deficiency occurs rapidly under these mats and moderate hypoxic conditions may occur in few days (Norkko & Bonsdorff, 1996a). In the northern Baltic Sea, juvenile flounders, *Platichthys flesus* (L.), utilize shallow sandy areas during their first few years. They feed mainly on meiofauna until they reach a lenght of about 45 mm, thereafter their diet consists of macrofaunal species (Aarnio et al., 1996). Some species in their diet are equipped with a shell, for example Ostracoda, and juvenile gastropods (*Hydrobia* spp.) and juvenile bivalves [*Macoma balthica* (L.)] (Aarnio et al., 1996). Some shelled species may survive the gut passage of

Table 1. Species composition of the algae in the drifting algal mats used in the experiments

Division	Species
Rhodophyta	*Ceramium tenuicorne* (Kütz.)Waern
Phaeophyta	*Stictyosiphon tortilis* (Rupr.) Reinke sensu Rosenv.
	Pilayella littoralis (L.) Kjellm.
	Ectocarpus siliculosus (Dillwyn) Lyngb.
	Sphacelaria arctica Harv.
Chlorophyta	*Cladophora rupestris* (L.) Kütz.
	Cladophora glomerata (L.) Kütz.
	Acrosiphonia centralis (Lyngb.) Kjellm.
	Enteromorpha intestinalis (L.) Nees
	Percursaria percursa (C. Agardh) Bury
	Zygnema sp.
	Ulothrix sp.
	Spirogyra sp.
	Mougeotia sp.
Cyanophyta	*Planktolyngbya* sp.

predators (Vinyard, 1979; Cadee, 1994). In a study by Aarnio & Bonsdorff (1997), more than 90% of *Hydrobia* spp. and more than 50% of Ostracoda survived the gut passage of juvenile flounder, while all *M. balthica* died. Algal mats are preferred habitats for hydrobiids and ostracods which utilize the algae for feeding (Levinton, 1985; Uiblein et al., 1996; Norkko et al., 2000). Consequently, these species may increase in number in shallow areas due to the presence of algae, whereas most other benthic fauna decrease (Hull 1987; Norkko, 1997; Raffaelli et al., 1998). As a result, the share of hydrobiids and ostracods in the diet of juvenile flounders may increase, affecting their nutritional status and growth. In this study, we investigated the movement and distribution of ostracods, *Hydrobia* spp. and *Macoma balthica* in aquaria with an algal mat covering 50% of the bottom. In a separate experiment, we studied the predation by juvenile *Platichthys flesus* on ostracods, *Hydrobia* spp. and *M. balthica* in aquaria with bare sand and drift algal mats.

Materials and methods

Habitat choice of invertebrates

Experimental animals were sampled with an Ockelmann sledge from a nearby sea area. The experimental densities of each species were comparable to natural densities found in shallow areas in the northern Baltic and the proportional densities were similar to natural densities found in the collection area (Bonsdorff et al., 1991; Aarnio et al., 1998). The experiment was conducted in plastic aquaria measuring 2.5 l (surface area: 256 cm^2). Each aquarium had a layer of azoic, sieved (0.5 mm) sand (1 cm depth) covered with seawater (salinity 5.5). A plastic wall separated each aquarium into two halves; 100 ostracods, 50 *Hydrobia* and 40 *M. balthica* were added to each aquarium, so that half of them were added on one side and the other half on the other side of the aquarium. When all animals had sunk to the bottom, the wall was removed and an algal mat covering half of the bottom surface was placed in all aquaria. Each aquarium was equipped with an aerating device which was placed just beneath the water surface above the algal mat. The experiment was examined after 1 h, 12 h, 24 h, 48 h and 96 h; five replicate aquaria were terminated at each time (a total of 25 aquaria was used). The algae used were drifting, filamentous algae consisting mainly of annual brown and green algal species (Table 1). They were collected 1 week prior to the experiment and kept in seawater with an aerating device to prevent degradation. The algae were rinsed with fresh water and defaunated by picking out all animals using a magnifying glass and stereomicroscope. One mat consisted of 13 g algae (corresponding to appr. 1000 g wwt m^{-2}) which was put in a loose net bag (12 mm mesh) to prevent the algae floating around in the aquaria. Physico-chemical conditions were measured in all aquaria both above and under the algae at the start and at the end of the experiment. At the termination of the experiment, the algal mats were removed from the aquaria and all animals in the algae counted using a magnifying glass and stereomicroscope. Sand was removed from the aquaria from one half at a time by means of the plastic wall. Both halves were treated separately, sieved through 0.5 mm mesh sieve, and all animals counted using stereomicroscope.

Predation experiment

This experiment used glass aquaria measuring 11 l (surface area: 609 cm^2) and the same protocol as described above. Four treatments, each with five replicates were used: sand control (hereafter referred to as C; only prey), algal control (A; prey and algae), predator treatment (P; prey and fish), and predator + algae treatment (PA; prey, algae and fish). Of each prey species, 150 individuals were added to all aquaria, the total number of prey adding up to 450 individuals per aquarium. The size of prey ranged from 0.8 to 1.2 mm and they were added to the aquaria as described above. Algal mats (each of 30.5 g algae, corresponding to 1000 g wwt m^{-2}) were put in one half of 10 aquaria (A and PA) and all aquaria were equipped with an aerating device. After 14 h, two juvenile *P. flesus* (4.5–5.5 cm TL) were added to 10 aquaria; five with algae (PA) and five without algae (P). The fish were allowed to forage for 8 h after which they were removed and the experiment was terminated. Physico-chemical conditions were measured in all aquaria (in aquaria with algae both above and under algae) when prey were added to the aquaria, when fish were added to the aquaria, and when fish were removed from the aquaria. At the end of the experiment, the fish were killed and the gut contents analysed for prey using a stereomicroscope. As the fish were allowed to forage for only 8 h, all prey consumed were still in the gastrointestinal tract (De Groot, 1971). Algae and sand were removed from the aquaria, and the prey in each half were counted as described above.

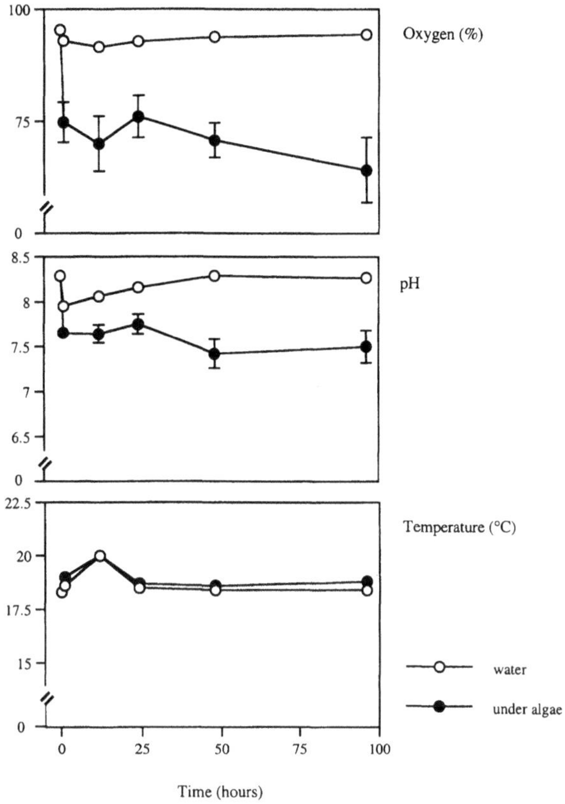

Figure 1. Physico-chemical conditions (oxygen, pH and temperature) in the water column and under algae in the habitat choice experiment. Error bars = ±1 standard error of the mean.

Numerical and statistical analyses

In the habitat choice experiment, the proportion of prey on the sand and on the algal side was compared for each time interval using paired Student's *t*-test. An arcsin-transformation of the percentage values was made for the analyses. The number of prey found in the algae at the end was counted to determine the rate of movement of different kinds of prey into the algal mat. These results are presented only descriptively, i.e. no statistical analyses were done on them. In the predation experiment, all treatments (C, A, P, PA) were compared using a two-way ANOVA with habitat (algae, sand) and predation (fish, no fish) as main factors. If no interaction was found between the main factors, the Student–Neuman–Keuls *post hoc* test was used to make pairwise comparisons between the treatments. For *Hydrobia*, the data revealed a non-normal distribution and heterogeneity of variances, and thus a non-parametric two-way ANOVA for ranked data (the Scheirer–Ray–Hare extension of the Kruskall–Wallis test) was used to compare the treatments (Sokal & Rohlf, 1995). To analyse if respective prey species were distributed equally over each half of the aquaria (sand vs algae) in the treatments with algae (A and PA), the proportions of each prey on both halves of the aquaria were compared using paired Student's *t*-test. Furthermore, the distribution of prey over the aquarium surface in the same treatments (A, PA) were compared with each other using *G*-tests of independence (Sokal & Rohlf, 1995) to see if the distribution of prey in the algal treatments was similar regardless of the presence of a predator. To study if prey choice by the fish was affected by the algae, the number of each prey species found in the stomachs of flounders in P and PA treatments was compared using a one-way ANOVA and the proportional consumption of prey with R × C test of independence using a *G*-test (Sokal & Rohlf, 1995).

Results

Habitat choice experiment

Physico-chemical conditions in the habitat choice experiment are illustrated in Figure 1. Oxygen saturation under the algae decreased rapidly during the first hour (from 95±0.3% to 75±4%). After the initial decline,

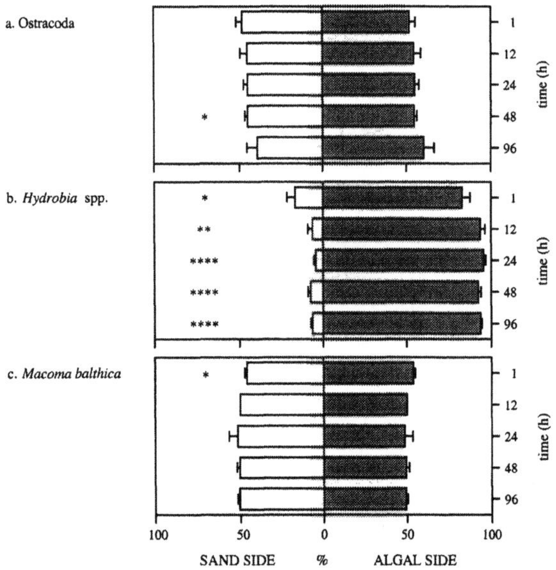

Figure 2. Proportion of Ostracoda (a) *Hydrobia* spp. (b) and *Macoma balthica* (c) on the sand and algal sides of the aquaria after 1, 12, 24, 48 and 96 h in the habitat choice experiment. Levels of significance: *= $p<0.05$, **= $p<0.01$, ****= $p<0.0001$. Error bars = 1 standard error of the mean.

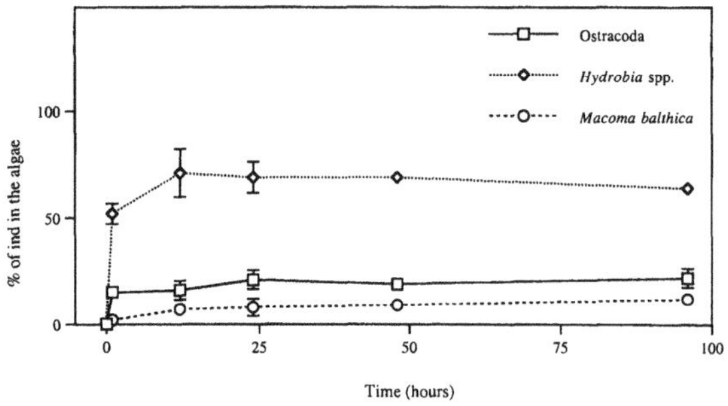

Figure 3. Rate of movement of Ostracoda, *Hydrobia* spp. and *Macoma balthica* into algal mat in the habitat choice experiment. Error bars = ±1 standard error of the mean.

Table 2. Physico-chemical conditions in the aquaria in the predation experiment. Error bars = ±1 Standard error of the mean

	Start	14 h	22 h
Temperature, °C	15.1±0.3		
Above algae		16.7±0.1	17.6±0.2
Under algae		16.8±0.1	17.8±0.1
pH	8.1±0.01		
Above algae		8.0±0.01	7.8±0.01
Under algae		7.5±0.08	7.5±0.05
Oxygen, %	85±0.5		
Above algae		93±0.6	91±0.4
Under algae		61±7.9	61±6.2

oxygen saturation decreased slowly and 64±7% oxygen was measured under the algae after 96 h. In the water column, oxygen saturation was 92–95% throughout the experiment. A rapid initial decline occured also for pH under the algae (8.3±0.01 at start and 7.6±0.04 after 1 h). After 1 h, the pH level was quite stable and was 7.5±0.2 after 96 h. In the water column, pH was between 8 and 8.3 during the entire experiment. Temperature varied between 18.3 and 20 °C both above and under the algae.

There was no significant difference in the distribution of ostracods between sand and algal sides during the first 24 h (Fig. 2). After 48 h significantly more ostracods (55%; $p<0.05$) were found on the algal sides. After 96 h, more ostracods were found on the algal side (61%), but the difference between sides was not significant due to a relatively large variation between the replicate aquaria (Fig. 2a). *Hydrobia* showed a clear preference for algae and, after 1 h, 83% were found on this side (significant difference; $p<0.05$). Significantly more hydrobiids were found on algal side after 12 h ($p<0.01$) and 24, 48 and 96 h ($p<0.0001$). From 12 h on, more than 90% were found on the alga side (Fig. 2b). A significant difference was found for the distribution of *Macoma* between the aquarium halves after 1 h ($p<0.05$), when 46% of the individuals were on sand side. No differences were found in the distribution of *Macoma* at later times (Fig. 2c).

Figure 3 illustrates the rate of movement of ostracods, *Hydrobia* and *Macoma* into the algal mat. After 1 h, 15% of ostracods had moved into the algae. This movement into the algae increased steadily during the experiment and, after 96 h, 23% were in the algae. *Hydrobia* showed a very rapid movement into the algal mat and, after 1 h, 50% were found in the algae. Between 12 and 96 h, 64–71% of the hydrobiids were found in the algal mat. *Macoma* showed a slow movement into the algae; after 1 h, 2% were found and, after 96 h, 12% were found in the algae.

Predation experiment

The physico-chemical conditions in the predation experiment are illustrated in Table 2. Temperature and pH showed a similar pattern above and under algae throughout the experiment. Oxygen declined under the

Table 3. Summary of two-way Analysis of Variance (ANOVA) results for Ostracoda, *Hydrobia* spp. and *Macoma balthica*

Source of variation	df	MS	F	*p*	SNK	Sign
Ostracoda						
Habitat (algae, bare sand)	1	5511.2	24.792	0.0001		
Predation (fish, no fish)	1	14364.8	64.619	<0.0001		
Interaction	1	5780.0	26.001	0.0001		
Residual	16	222.3				
Total	19	1537.52				
Hydrobia spp.						
Habitat (algae, bare sand)	1	204.8	5.874	<0.05	A-P	<0.05
Predation (fish, no fish)	1	231.2	6.631	<0.05	A-PA	ns
Interaction	1	0.2	0.006	ns	A-C	ns
Residual	16	14.14			C-P	<0.05
Total	19	34.87			C-PA	ns
					PA-P	<0.05
Macoma balthica						
Habitat (algae, bare sand)	1	54.450	0.51283	0.4842		
Predation (fish, no fish)	1	2486.45	23.41841	0.0002		
Interaction	1	1140.05	10.73746	0.0047		
Residual	16	106.175				
Total	19	283.145				

algae to about 60%, but no severe hypoxia was detected in any aquaria. When the experiment was ended, all flounders, except one, were found on the sand side in the treatments with algae (PA).

Two-way ANOVA analysis ot the survival of prey in all treatments (Table 3, Fig. 4) showed that, for ostracods, both the habitat (p=0.0001) and predation ($p<0.0001$) had significant effects on prey survival. An interaction (p=0.0001) between these two factors, indicating that the algae affected the predation efficiency of juvenile *P. flesus*, allowed no further comparisons (Table 3, Fig. 4a). For *Hydrobia*, habitat ($p<0.05$) and predation ($p<0.05$) affected prey survival, and pairwise tests between the treatments revealed significant differences ($p<0.05$; SNK-test) between predation (P) and sand control (C), predation (P) and algal control (A), and between predation (P) and predation+algae (PA) (Table 3, Fig. 4b). For *Macoma*, habitat had no effect on prey survival, while predation had a significant (p=0.0002) effect on survival. An interaction (p=0.0047) was found between these two factors, indicating that the algae affected the predation efficiency of juvenile *P. flesus*, and thus no further comparisons were performed (Table 3, Fig. 4c).

When the distribution of prey over sand and algal halves of the aquaria was compared in the treatments with algae (A and PA; Fig. 5), no difference was found in the distribution of Ostracoda in the A treatment, whereas in the PA treatment, significantly ($p<0.05$) more ostracods were found on the algal side. Significantly more *Hydrobia* were found on the algal side in both treatments (A: $p<0.01$, PA: $p<0.05$), while the distribution of *Macoma* was not significantly different in either treatment. When the distribution of prey in treatment A was compared with the distribution in treatment PA (Fig. 5), the only difference was found for Ostracoda ($p<0.01$; G-test), indicating that their distribution over the aquarium halves (sand, algae) was dependent of the presence of *P. flesus*.

Comparing the consumption of prey by *P. flesus* in the sand habitat (treatment P), showed that ostracods were consumed significantly more ($p<0.05$) than *Macoma*. In the algal habitat (treatment PA), no significant difference was found in consumption of the three prey species (Fig. 6a). The proportional consumption of prey differed significantly ($p<0.01$;

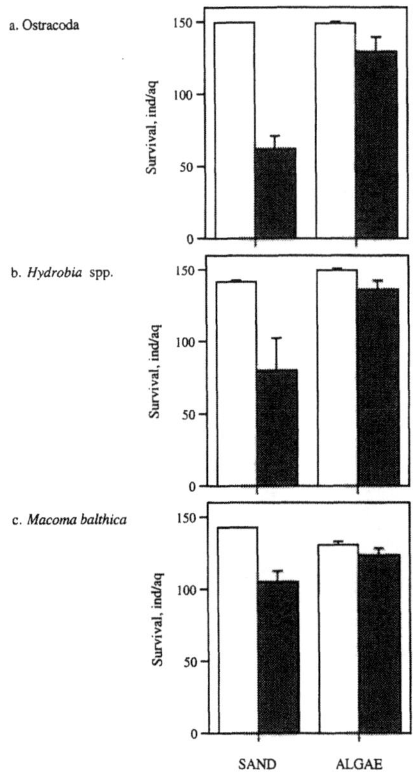

Figure 4. Survival (mean number ± S.E.) of Ostracoda (a), *Hydrobia* spp. (b), and *Macoma balthica* (c), in bare sand and drifting algae in the predation experiment. White bars: no predator, shaded bars: with *Platichthys flesus*.

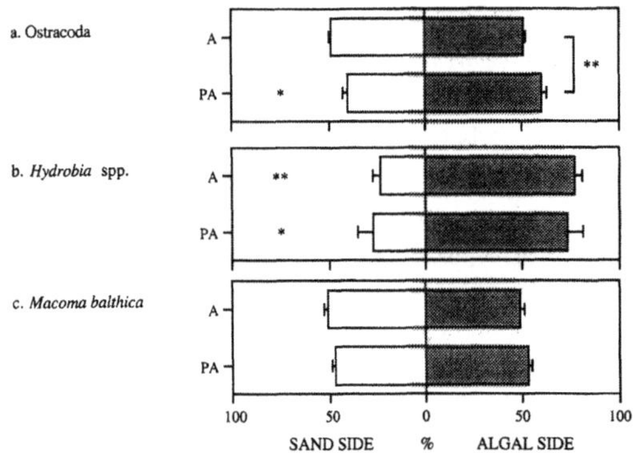

Figure 5. Distribution of Ostracoda (a), *Hydrobia* spp. (b), and *Macoma balthica* (c) over sand and algal sides of the aquaria in treatments A (algal control) and PA (algae + *Platichthys flesus*) in the predation experiment. Significant differences (paired Student's *t*-test) in the distribution between aquarium halves within one treatment are indicated with asterisks. Significant differences (G-test) between control (A) and predator (PA) treatments are indicated with asterisks connected with brackets. Levels of significance: *= $p<0.05$, **= $p<0.01$. Error bars = 1 standard error of the mean.

R × C test) between P and PA treatments (Fig. 6b). The proportion of *Macoma* in the diets was higher in treatment PA (22%) compared to treatment P (13%), while the proportion of *Hydrobia* was lower in treatment PA (29%) than in treatment P (38%). Fourty nine percent of the consumed prey were ostracods in each treatment.

Discussion

Results from these experiments showed that the presence of algae significantly reduced the predation efficiency of juvenile flounders for all three prey species. This is probably due to the increased habitat complexity provided by the algal mat. Increased habitat complexity has been shown to reduce predation by visual predators on zoobenthic prey and fish (Nelson, 1979; Heck & Thoman, 1981; Coull & Wells, 1983; Gotceitas & Colgan, 1989; Nelson & Bonsdorff, 1990; Mattila, 1992; Isaksson et al., 1994). Increased complexity may moderate predation through reduced foraging success by the predators, as the movement of the predator is inhibited or the visibility of the prey is reduced (Savino & Stein, 1982; Orth et al., 1984). For the prey to find a refuge in algae, they have to be able, through their behaviour, to utilize the algae as a shelter (Pierce, 1988; Savino & Stein, 1989). In this study, *Hydrobia* used the algae as a habitat, but the reason for this behaviour was not only to hide from predators, since the same behaviour was observed when no predators were present. More likely, the algae, and epiphytes growing on them, served as food for the herbivorous hydrobiids. In the first experiment, the hydrobiids moved rapidly into the algae, illustrating that the algae is their preferred habitat. Ostracods and *Macoma* were more stationary and did not use the algae to the same extent as *Hydrobia.* The presence of flounders did not affect the distribution of *Hydrobia* over the aquarium halves in the treatment with algae and, obviously, they were consumed to the same extent on both halves. The distribution of ostracods changed as a repsonse to flounders, such that more ostracods were found on the algal side of the aquaria. Fish probably consumed ostracods on the sand side of the aquaria, leading to a changed distribution over the aquaria. Ostracods may also have responded to the presence of predators by moving to the algal side of the aquaria. They have been shown to have a well developed capability for exploring their surroundings and responding to external factors in a

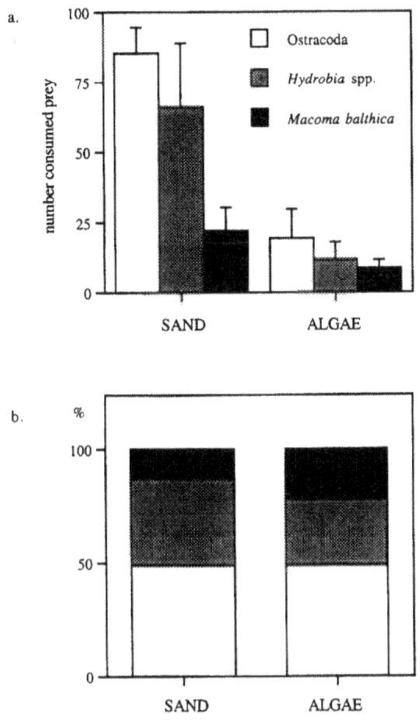

Figure 6. Consumption of prey by juvenile *Platichthys flesus* in bare sand and with algae; presented as: mean number consumed prey ± S.E. (a), and as proportional consumption (b).

flexible and behaviourally dynamic manner (Uiblein et al., 1996). Mbahinzireki et al. (1991) showed that, when given a choice, ostracods chose conditions of dark and shadow, and were especially vulnerable to predation by fish when exposed to light. Thus, the shadow given by the algal mat could have attracted some of the ostracods to that side of the aquaria. The algae may also provide the ostracods with food, explaining why they were attracted to the algal side. The distribution of *Macoma* did not change due to the presence of the predator, exemplifying their slow mobility as shown by Norkko (1998). Juvenile *Macoma* are filter feeders, and it is unlikely that they seek food in the algae. More probably, poor oxygen conditions may drive them into the algal mat, as *Macoma* ar e sensitive to stress induced by algae (Norkko & Bonsdorff, 1996b). Although, the oxygen conditions under the algae were relatively good in each experiment, it cannot be excluded that the oxygen conditions were patchily lower and that this was the reason why *Macoma* was found in the algae.

As the fish did not consume prey more effectively over sand, as would have been expected (Bonsdorff et al., 1995), some other reason along with the presence of algal mat may have affected the foraging of the fish. Light is an important factor affecting predation success in visual predators (James & Heck, 1994), and the algal mats may have shaded the aquaria and inhibited foraging by the flounders. However, this seems unlikely, as the mats were lying on the sand surface, they were not very thick and the light came from straight above the aquaria. Another reason could be that algal exudates may have affected the fish in a way that decreased their consumption rate. Aneer (1987) suggested that toxic exudates released from filamentous brown algae cause mortality of herring eggs. In our study, filamentous brown algae of the same species as in Aneer's (1987) study (*Pilayella littoralis* and *Ectocarpus siliculosus*), comprised a large part of the algal mats, and may have affected the flounders negatively. Also, as flounders usually live in bare sand habitats, they were possibly not used to this new habitat type offered to them in the aquaria and may have needed a longer time to forage normally. A longer experimental time could, however, not be allowed because the experiment had to be ended before the prey had passed through the gastrointestinal tract of the fish, as ostracods and *Hydrobia* may survive the gut passage. Accordingly, no estimates on prey consumption could have been made. However, juvenile flounders have also been observed to occur near algal mats in the field (pers. obs.).

According to the optimal foraging theory, predators should maximize their net energy gain (Pyke et al., 1977). A large proportion of both *Hydrobia* and Ostracoda are able to survive the gut passage of fish (Vinyard, 1979; Aarnio & Bonsdorff, 1997), and thus the consumption of them cannot benefit the predators energetically. As predators should eat only the highest ranking prey if the encounter rate with it is sufficiently high, the flounders should select *Macoma* over the two other prey types, as *Macoma* is an easily-digestible prey (Aarnio & Bonsdorff, 1997). This was not the case even in the bare sand habitat (treatment P), where all prey were equally available to the fish. *Macoma* is an infaunal species, but juveniles less than 2 mm in size have short siphons and are found at the sediment surface. On the other hand, ostracods and hydrobiids are more mobile than *Macoma* and are probably encountered more easily by the flounders as prey mobility is an important factor affecting prey selection by visual predators. *Hydrobia* are also darker making them more visible to the flounders. Also, the feeding behaviour of flounders may have affected the visibility of *Macoma* to them, since flounders move over the bottom biting at prey when encountered, and this behaviour causes disturbance which may stimulate adjacent prey to retract their appendages (Ansell & Gibson, 1990). Thus, *Macoma* may retract its siphons which may reduce its visibility for the fish. However, Mattila & Bonsdorff (1998) showed that juvenile *P. flesus* selected juvenile *Macoma* over the more mobile crustacean *Bathyporeia pilosa*. This was due to the capability of *B. pilosa* to escape from the predators. In our experiment, the ostracods and hydrobiids had probably no fast way of escaping predation. The proportional consumption of prey was different when algae was present in the aquaria than in the bare sand. *Macoma* was consumed proportionally more when algae was present, while the opposite was found for the consumption of hydrobiids. This further illustrates the protective role of the algae for hydrobiids.

In conclusion, all prey species were protected from predation by the algae. Hydrobiids moved actively into the algae and may benefit from the algal mats as long as the oxygen conditions remain high. However, if the algal mats are thicker, hypoxic conditions may occur under and in the algal mats (Norkko & Bonsdorff, 1996a), leading to more negative effects on all prey species, resulting in a lower protective effect of the mats. Juvenile *P. flesus* seem to be less successful foragers in the presence of drifting algae than on bare sand. In the northern Baltic Sea, drifting algae in the shallow sandy areas are a common phenomenon, and a drift algal coverage of more than 50% is normal in late summer and autumn (Berglund, 1998; Norkko, 1998). The uncovered shallow sandy bottoms are thus changing to more heterogeneous habitats leading to alterations in species composition which may have important consequences for trophic interactions in the area. This may have severe effects on juvenile *P. flesus* which normally utilize and forage in uncovered areas, and eventually entire flounder populations in these areas may be affected negatively.

Acknowledgements

We thank Husö Biological Station for providing excellent working conditions. This study was financed by Maj & Tor Nesslings Foundation.

References

Aarnio, K. & E. Bonsdorff, 1997. Passing the gut of juvenile flounder, *Platichthys flesus*: differential survival of zoobenthic prey species. Mar. Biol. 129: 11–14.

Aarnio, K., E. Bonsdorff & A. Norkko, 1998. Role of *Halicryptus spinulosus* (Priapulida) in structuring meiofauna and settling macrofauna. Mar. Ecol. Prog. Ser. 163: 145–153.

Aarnio, K., E. Bonsdorff & N. Rosenback, 1996. Food and feeding habits of juvenile flounder, *Platichthys flesus* (L.), and turbot, *Scophthalmus maximus* L., in the Åland archipelago, northern Baltic Sea. J. Sea Res. 36: 311–320.

Aneer, G., 1987. High natural mortality of Baltic herring (*Clupea harengus*) eggs caused by algal exudates? Mar. Biol. 94: 163–169.

Ansell, A. & R. N. Gibson, 1990. Patterns of feeding and movement of juvenile flatfishes on an open sandy beach. In Barnes M. & R. N. Gibson (eds), Trophic Relationships in the Marine Environment. Proc. 24th Europ. Mar. Biol. Symp. Aberdeen University Press: 191–207.

Berglund, J., 1998. Survey of macrophytes and drifting algae on shallow soft bottoms in the Åland archipelago. Forskningsrapporter från Husö biologiska station 97: 1-29 (in Swedish with English summary).

Bonsdorff, E., 1992. Drifting algae and zoobenthos – effects of settling and community structure. Neth. J. Sea Res. 30: 57–62.

Bonsdorff, E., K. Aarnio & E. Sandberg, 1991. Temporal and spatial variability of zoobenthic communities in archipelago waters of the northern Baltic Sea – consequences of eutrophication. Int. Rev. ges. Hydrobiol. 76: 433–450.

Bonsdorff, E., A. Norkko & C. Boström, 1995. Recruitment and population maintenance of the bivalve *Macoma balthica* (L.) – factors affecting settling success and early survival on shallow sandy bottoms. In Eleftheriou, A., A. D. Ansell & C. J. Smith (eds), Biology and Ecology of Shallow Coastal Waters. Proc. 28th Europ. Mar. Biol. Symp. Olsen & Olsen, Fredensborg: 253–260.

Bonsdorff, E., E. M. Blomqvist, J. Mattila & A. Norkko, 1997a. Coastal eutrophication: causes, consequences and perspectives in the archipelago areas of the northern Baltic Sea. Estuar coast. shelf Sci. 44 (Suppl. A): 63–72.

Bonsdorff, E., E. M. Blomqvist, J. Mattila & A. Norkko, 1997b. Long-term changes and coastal eutrophication. Examples from the Åland Islands and the Archipelago Sea, northern Baltic Sea. Oceanol. Acta 20: 319–329.

Cadee, G. C., 1994. Eider, shelduck and other predators, the main producers of shell fragments in the Wadden Sea: Palaeoecological implications. Palaeontology 37: 181–202.

Cederwall, H. & R. Elmgren, 1990. Biological effects of eutrophication in the Baltic Sea, particularly the coastal zone. Ambio 19: 109–112.

Coull, B. C. & J. B. J. Wells, 1983. Refuges from fish predation: experiments with phytal meiofauna from the New Zealand rocky intertidal. Ecology 64: 1599–1609.

De Groot, S. J., 1971. On the interrelationships between morphology of the alimentary tract, food and feeding behaviour in flatfishes (Pisces: Pleuronectiformes). Neth. J. Sea Res. 5: 121–196.

Gotceitas, V. & P. Colgan, 1989. Predator foraging success and habitat complexity: quantitative test of the threshold hypothesis. Oecologia 80: 158–166.

Heck, K. L., Jr. & T. A. Thoman, 1981. Experiments on predator prey interactions in vegetated aquatic habitats. J. exp. mar. Biol. Ecol. 53: 125–134.

Hull, S. C., 1987. Macroalgal mats and species abundance: a field experiment. Estuar. coast. shelf Sci. 25: 519–532.

Isaksson, I., L. Pihl & J. Van Montfrans, 1994. Eutrophication-related changes in macrovegetation and foraging of young cod (*Gadus morhua* L.): a mesocosm experiment. J. exp. mar. Biol. Ecol. 177: 203–217.

James, P. L. & K. L. Heck Jr., 1994. The effects of habitat complexity and light intensity on ambush predation within a simulated seagrass habitat. J. exp. mar. Biol. Ecol. 176: 187–200.

Larsson, U., R. Elmgren & F. Wulff, 1985. Eutrophication and the Baltic Sea: causes and consequences. Ambio 14: 9–14.

Levinton, J. S., 1985. Complex interactions of a deposit feeder with its resources: roles of density, a competitor and detrital addition in the growth and survival of the mudsnail *Hydrobia totteni*. Mar. Ecol. Prog. Ser. 22: 31–40.

Mattila, J., 1992. The effect of habitat complexity on predation efficiency of perch (*Perca fluviatilis* L.) and ruffe (*Gymnocephalus cernuus* (L.)). J. exp. mar. Biol. Ecol. 157: 55–67.

Mattila, J. & E. Bonsdorff, 1998. Predation by juvenile flounder (*Platichthys flesus* L.): a test of prey vulnerability, predator preference, switching behaviour and functional response. J. exp. mar. Biol. Ecol. 227: 221–236.

Mbahinzireki, G., F. Uiblein & H. Winkler, 1991. Microhabitat selection of ostracods in relation to predation and food. Hydrobiologia 222: 115–119.

Nelson, W. G., 1979. Experimental studies of selective predation on amphipods: consequences for amphipod distribution and abundance. J. exp. mar. Biol. Ecol. 38: 225–245.

Nelson, W. G. & E. Bonsdorff, 1990. Fish predation and habitat complexity: are complexity thresholds real? J. exp. mar. Biol. Ecol. 141: 183–194.

Norkko, A., 1997. The role of drifting macroalgal mats in structuring coastal zoobenthos. PhD Thesis, Åbo Akademi University, 41 pp.

Norkko, A., 1998. The impact of loose-lying algal mats and predation by the brown shrimp *Crangon crangon* (L.) on infaunal prey dispersal and survival. J. exp. mar. Biol. Ecol. 221: 99–116.

Norkko, A. & E. Bonsdorff, 1996a. Population responses of coastal zoobenthos to stress induced by drifting algal mats. Mar. Ecol. Prog. Ser. 140: 141–151.

Norkko, A. & E. Bonsdorff, 1996b. Altered benthic prey-availability due to episodic oxygen deficiency caused by drifting algal mats. P.S.Z.N.I Mar. Ecol. 17: 355–372.

Norkko, J., E. Bonsdorff & A. Norkko, 2000. Drifting algal mats as an alternative habitat for benthic invertebrates: species specific responses to a transient resource. J. exp. mar. Biol. Ecol. 248: 79–104.

Olafsson, E. B., 1988. Inhibition of larval settlement to a soft bottom benthic community by drifting algal mats: an experimental test. Mar. Biol. 97: 571–574.

Orth, R. J., K. L. Heck Jr. & J. Van Montfrans, 1984. Faunal communities in seagrass beds: a review of the influence of plant structure and prey characteristics on predator prey relationships. Estuaries 7: 339–350.

Pierce, C. L., 1988. Predator avoidance, microhabitat shift, and risk-sensitive foraging in larval dragonflies. Oecologia 77: 81–90.

Pyke, G. H., H. R. Pulliam & E. L. Charnov, 1977. Optimal foraging: a selective review of theory and tests. Q. Rev. Biol. 52: 137–154.

Raffaelli, D. G., J. A. Raven & L. J. Poole, 1998. Ecological impact of green macroalgal blooms. Oceanogr. mar. biol. Annu. Rev. 36: 97–125.

Savino, J. F. & R. A. Stein, 1982. Predator-prey interaction between largemouth bass and bluegills as influenced by simulated submersed vegetation. Trans. am. Fish. Soc. 111: 255–266.

Savino, J. F. & R. A. Stein, 1989. Behavior of fish predators and their prey: habitat choice between open water and dense vegetation. Envir. Biol. Fishes 24: 287–293.

Sokal, R. R. & F. J. Rohlf, 1995. Biometry. W.H. Freeman and Company, New York: 887 pp.

Uiblein, F., J. R. Roca, A. Baltanas & D. L. Danielopol, 1996. Tradeoff between foraging and antipredator behaviour in a macrophyte dwelling ostracod. Arch. Hydrobiol. 137: 119–133.

Vinyard, G., 1979. An ostracod (*Cypriodopsis vidua*) can reduce predation from fish by resisting digestion. Am. midl. Nat. 102: 188–190.

Hydrobiologia **440**: 357–367, 2000.
M.B. Jones, J.M.N. Azevedo, A.I. Neto, A.C. Costa & A.M. Frias Martins (eds), Island, Ocean and Deep-Sea Biology.

Feast or famine for intertidal grazing molluscs: a mis-match between seasonal variations in grazing intensity and the abundance of microbial resources

R.C. Thompson[1], M.F. Roberts[1], T.A. Norton[2] & S.J. Hawkins[1,3]
[1]*Division of Biodiversity and Ecology, School of Biological Sciences, University of Southampton SO16 7PX, U.K.*
[2]*Port Erin Marine Laboratory, University of Liverpool, Isle of Man IM9 6JA, U.K.*
[3]*Marine Biological Association of the United Kingdom, Citadel Hill, Plymouth PL1 2PB, U.K.*

Key words: top-down, bottom-up, biofilm, microphytobenthos, density dependence, intra-specific competition

Abstract

Distinct seasonal variations in the abundance of photosynthetic microbiota and limpet grazing intensity were recorded at Port St Mary, Isle of Man between January 1994 and June 1996. Microbial abundance was negatively correlated with insolation stress, while grazing intensity was positively correlated with sea and air temperature. These patterns result in a mis-match between the supply of and the demand for microbial resources with maximal grazing intensity during the summer and autumn, but maximal microbial standing stock during the winter and early spring. The importance of top-down control of microbial assemblages by grazing was demonstrated by experimental exclusion of limpets during autumn 1993. This resulted in a four-fold increase in the abundance of cyanobacteria within 6 days, followed by a more gradual proliferation of ephemeral algae during the next 4 weeks. The abundance of diatoms remained relatively constant and was not influenced by the removal of grazers at this time of year. The influence of microbial resource availability on the growth and mortality of limpets was examined using experimental enclosures of differing densities of either *Patella vulgata* or *P. depressa*. After 6 months, there were significant relationships between grazer density and both mortality and growth with increased mortality and reduced growth for *P. vulgata* at increased densities, and reduced growth for *P. depressa* at increased densities. Hence, the availability of microbial resources may also influence the biomass of grazers on rocky shores from the bottom upwards. A conceptual model is presented which describes seasonal and annual variations in microbial resources and grazing intensity and their potential consequences for other shore dwellers.

Introduction

Biofilms consisting of mucus, bacteria and microalgae form on any hard surface immersed in water, and are an important component of the benthos in marine (e.g. Castenholz, 1963; Underwood, 1984c; Wahl, 1989; Hill & Hawkins, 1991) and freshwater ecosystems (e.g. Lock, 1993). These microbial films form an interface between the underlying substratum and the surrounding water, and can be important for habitat selection by settling larvae (Crisp, 1974; Thompson et al., 1998) and the survival of macroalgal propagules (Huang & Boney, 1984). They also provide a food resource for a variety of grazers (molluscs, Steneck & Watling, 1982; Underwood, 1984a; Hawkins et al., 1989; Mak & Williams, 1999; echinoderms, Ogden et al., 1989). In marine sediments, photosynthetic micro-organisms are usually described as microphytobenthos and can contribute over 20% of local primary production (Fielding et al., 1988; Kristensen, 1993). Although less extensively studied on rocky shores, photosynthetic micro-organisms may make an equally important contribution to primary production (Workman, 1983; Dye & White, 1991; Fuji et al., 1991).

On rocky shores, molluscs such as limpets, which graze on biofilms, also remove macroalgal propagules and invertebrate larvae, thus influencing recruitment to macrobiotic communities (Southward, 1964; Hawkins & Hartnoll, 1983; Jernakoff, 1985). Numerous researchers have manipulated the abundance of grazers on rocky shores and, without exception, this leads to an increase in macroalgal growth (Southward, 1964;

Hawkins & Hartnoll, 1983; Farrell, 1988; Mak & Williams, 1999) and may also have indirect consequences for communities over the shore as a whole. For example, on rocky shores in the northeast Atlantic, an increase in algal cover leads to a reduction in barnacle cover and an increase in the abundance of anemones and dog whelks (Southward & Southward, 1978; Hartnoll & Hawkins, 1985). Macroalgal propagules are, however, a small component of microbial standing stock and regulation of macroalgal abundance appears to be a consequence of molluscs grazing on the more permanent constituents of the film such as diatoms and cyanobacteria. Regulation of these microbial communities by molluscan grazers has been less extensively studied (but see Castenholz, 1963; Nicotri, 1977) and will be assessed here.

In addition to the possibility that microbial abundance may be regulated from the top-down by grazers, an alternative view is that availability of microbial resources regulates grazer biomass and the structure of intertidal communities from the bottom upwards (e.g. Bustamante et al., 1995; Menge et al., 1997, 1999). Hence, the availability of microalgal food could set the density or total biomass of grazers at a given location. The relative importance of top-down and/or bottom-up control may be constant through time or, if the supply of and the demand for resources are not synchronised, could vary seasonally with periods of either feast or famine for the grazers. The abundance of epilithic micro-organisms shows considerable seasonal variation (Underwood, 1984a; Hill & Hawkins, 1991), yet most studies of interactions between grazers and microbiota on rocky shores have focused on top-down regulation of microbial communities by grazers (but see Bustamante et al., 1995). Here we will also consider the extent to which intertidal communities are regulated from the bottom-up, by microbial resource availability.

In this paper, we describe seasonal variations in the abundance of epilithic microflora on a northern European shore. We tested the hypotheses that observed patterns in microbial abundance were correlated with physical forcing factors such as insolation, sea and air temperatures, and seasonal patterns in grazing intensity recorded using a new technique (Thompson et al., 1997). The hypotheses that grazing intensity was correlated with putative determining factors (air and sea temperatures, and the abundance of molluscs) were also examined. We experimentally tested the hypotheses that microbial biomass would increase in the absence of grazing, and that for molluscs growth and survival have positive relationships with the availability of microbial food resources. The results are integrated with other published literature to provide a conceptual model. This speculates on how bottom-up forcing factors and top-down grazing control interact to set both microbial and grazer biomass. Implications for other intertidal organisms are also discussed.

Materials and methods

Study sites

Descriptive surveys and manipulative experiments were carried out at Port St Mary, Isle of Man, U.K. (4° 44′ W, 54° 4′ N) and Swanage in Dorset, U.K. (1° 58′ W, 50° 37′ N). At both sites, the shores were moderately exposed to wave action, and were colonised by fucoid algae, barnacles and limpets (mainly *Patella vulgata* L., on the Isle of Man, and *P. vulgata* together with *P. depressa* Pennant, at Swanage) interspersed with areas of open rock. The macrobiota and, to a lesser extent, the microbiota on these shores have been described previously (for site descriptions see: Isle of Man: Southward, 1953; Swanage: Grubb, 1936; Aleem, 1950).

Seasonal variations in microbial standing stock and grazing intensity

Between January 1994 and June 1996, the abundance of photosynthetic microbiota was monitored at approximately monthly intervals at Port St Mary by extracting chlorophyll from intertidal rock chips as an index. On each sampling occasion, 18 rock chips (each 2 cm × 2 cm) were chiselled from randomly selected areas of rock which were bare of macrobiota and encrusting algae. Pigment was extracted from these using methanol (for methods see Thompson et al., 1999).

Limpet foraging intensity was estimated using discs of dental wax recessed into pre-drilled holes in the shore. These discs became scratched by limpet radulae during foraging excursions and provided an index of grazing intensity (Thompson et al., 1997). Discs were deployed for 2-week periods every month, in each of three replicate arrays of 16 discs. The percentage area scraped was scored for each disc and average grazing intensity calculated for each array and for the mid shore as a whole (average of the three arrays) on each occasion. The abundance of limpets around each array was recorded at the same time. Microbial abundance, grazing intensity, air temperature (monthly average of maximum daily values), sunshine hours (monthly average of sunshine hours per day from the Meteorological Office at Ronaldsway Airport, nearby) and sea temperature data (from the Cypris Monitoring Station, maintained by Port Erin Marine Laboratory) were compared using Pearson's Correlation Coefficient after Bonferroni correction for multiple tests (Sokal & Rohlf, 1995).

Experimental manipulation of grazing intensity

The influence of grazing activity on microbial abundance was examined experimentally by excluding grazers from areas on the mid-lower shore (just above MLWN) at Port St Mary during September and October 1993. An area of shore was selected (approximately 20 × 5 m) with patches of open rock and limpets. Fifteen 50 cm diameter circular plots were chosen within this area which were predominately freely draining open rock not colonised by macrobiota nor swept by algae. Five replicate plots were assigned randomly to each of three treatments. The treatments were 'complete fences' to exclude limpets, 'partial fence controls' which had the same physical structure as complete fences, but with openings so that limpets could enter, and 'unfenced controls' marked only by stainless steel screws. 'Fences' and 'partial fences' were approximately 5 cm high and made of plastic-coated garden mesh fastened to the substratum with rawlplugs and stainless steel screws.

Rock chips, complete with epilithic microbiota, were collected from each plot for chlorophyll extraction (n = 3, see above) as an index of microbial standing stock, and for direct counts (n = 2) of diatoms, cyanobacteria and ephemeral algae. For direct counts, the rock chips were fixed in gluteraldehyde (2.5% in filtered seawater), air dried, coated in gold palladium and viewed at ×480 magnification using a Philips XL 30 scanning electron microscope (SEM, see Hill & Hawkins, 1990). Six randomly located black and white photographs were taken on the surface of each rock sample; photographs were processed into negatives and examined with a binocular microscope. Data from these photographs were averaged to provide an estimate of abundance on each chip. Samples were collected 7 days before experimental treatments were applied, immediately before treatments were applied, and 6, 24 and 37 days thereafter.

Experimental manipulation of microbial resource availability

The availability of microbial resources per grazer was manipulated by creating enclosures of differing densities of either *Patella vulgata* or *P. depressa* on the mid shore at Peveril Point near Swanage. These two species have virtually identical radulae (feeding apparatus) and are believed to exploit similar microbial food resources (Hawkins et al., 1989). The objective of this experiment was to establish the effects, in terms of growth and mortality, of intra-specific competition for microbial resources. Three replicates of each of four densities were created for each species, representing either 1/4, 1/2, 1 or 2 times natural 'background' density (9.14±6.75 per 25 × 25 cm quadrat, average ± SD, n = 58,). These treatments were created by fixing 25 × 25 cm square enclosures constructed of garden mesh around natural groups of limpets. To ensure that the biomass of grazers was reasonably constant among treatments, limpets with a shell length of approximately 30 mm were selected. Limpets were removed, where necessary, to achieve the desired size and density treatments. This method avoided the need to transplant limpets as this may have damaged the muscular foot of the animals and biased the outcomes of experimental treatments. Experiments began during April 1998 and each individual limpet within the plots was uniquely tagged using micro-markers fixed to the apex of the shell with super glue. The size of each limpet and any mortalities (missing limpets) were recorded at monthly intervals between May and October 1998. Data from grazer exclusions were compared among treatments using Analysis of Variance followed by S.N.K. multiple comparisons using GMAV for Windows (Underwood et al., 1997). Growth and mortality data from grazer inclusions were compared using linear regression. These analyses were one-tailed and tested the hypotheses that mortality and growth would have positive relationships with the availability of microbial resources. For mortality, the data were weighted (weighted least squares) to compensate for differences in the number of individuals used at each density (weighting for each treatment = number of individuals used in treatment/number of individuals used highest density treatment).

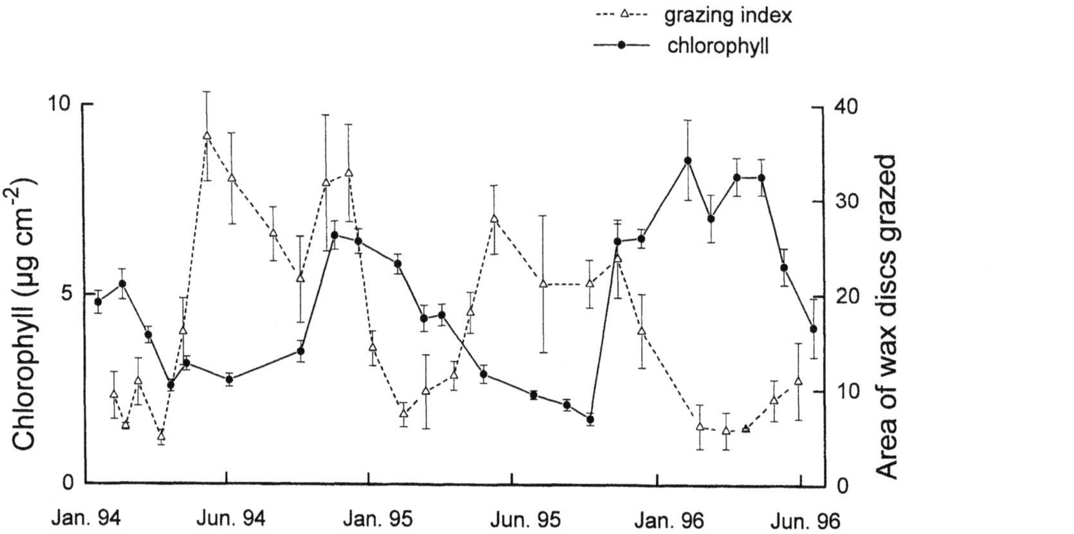

Figure 1. Seasonal changes in epilithic microbial standing stock (using extracted chlorophyll as an index, n = 18) and grazing intensity (using feeding marks on arrays of wax discs as an index, n = 3) on the mid shore at Port St Mary, Isle of Man (average ± 1 SE) see text for details.

Results

Seasonal variations in the abundance of microbiota and grazing intensity

Clear seasonal trends in the abundance of microbiota were evident, and were characterised by reduced abundance in the spring and summer (May–September) during 1994 and 1995. A similar pattern was also evident from May to June 1996 when sampling ended (Fig. 1). There was some variation in absolute values among years, but the general pattern was of a gradual decline in extracted chlorophyll from around 5–7 μg cm^{-2} in the late winter (January–February) to 2–4 μg cm^{-2} early in the summer (May–June). These values remained relatively stable during the summer, then increased again during the autumn (October and November).

Seasonal patterns in grazing intensity were also apparent with around five times more grazing in the summer and autumn than during the winter and early spring. A depression in grazing intensity was also evident during the late summer and autumn of 1994 and to a lesser extent during 1995. These patterns did not follow the abundance of microbial resources with grazing intensity and microbial biomass showing both similar and opposing trends at various times during the monitoring period. There was a consistent mis-match between the availability of and the apparent demand for microbial resources – food was most abundant during the winter and early spring, but grazing intensity was maximal during the summer and autumn.

Comparisons between putative physical and biological regulators of the seasonal patterns in chlorophyll and grazing intensity revealed a significant negative correlation between microbial standing stock and insolation (sunshine hours), and significant positive correlations between grazing intensity and both sea and air temperatures (Table 1). However, microbial standing stock was not correlated with grazing intensity, air temperature, sea temperature or limpet abundance, and grazing intensity was not correlated with sunshine hours, limpet abundance or chlorophyll.

Experimental manipulation of grazing intensity

Microbial standing stock was similar among plots prior to, and for 1 week after, the experimental treatments were initiated. However, 24 days after initiation, microbial abundance was around 50% greater in areas where limpets were excluded than in open areas or areas with control fences (Fig. 2a). These differences were significant (Table 2). After a further 2 weeks (37 days after initiation), areas where limpets had been excluded were predominantly covered by *Enteromorpha* sp., which was clearly visible without the aid of a microscope, while the two control treatments remained unaltered. Chlorophyll analysis was discontinued in the exclusion treatments after 24 days, as these areas were no longer dominated by microbiota.

Table 1. Pearson's correlation of chlorophyll concentration (as an index of microbial standing stock) and grazing intensity with monthly averages of various meteorological and biological variables. All data were recorded on the Isle of Man (U.K.) between January 1994 and June 1996. Significant values after Bonferroni correction for multiple tests are shown ($P < 0.05$ = *), n=22, except for limpet abundance where n=18

	Chlorophyll	Grazing	Sunshine hours	Air temperature	Sea temperature	Limpet abundance
			r values			
Chlorophyll	-	−0.29	**−0.68** *	−0.52	−0.29	−0.50
Grazing	−0.31	-	0.29	**0.73** *	**0.73** *	0.08

Table 2. Analysis of variance of extracted chlorophyll (microbial standing stock) from limpet exclusion treatments and controls 24 days after initiation of the experiment on the mid shore at Port St Mary, Isle of Man (U.K.). Significant differences are shown ($P < 0.05$ = *), n=3. Data were untransformed, Cochran's Test for homogeneity C = 0.72 (not significant). The outcomes of S.N.K. multiple comparisons tests are shown together with treatment means (± 1 SE)

Source	DF	MS	*F*	*P*
Treatment	2	19.20	5.96	**0.02** *
Residual	12	3.22		
Total	14			

Treatment means and SNK multiple comparisons	6.51	=	7.16	<	10.18
				*	
	open unfenced areas	=	**partial fences**	<	**fenced exclusion areas**

SEM analyses were made only for the grazer exclusion treatment. These data provide an illustration of changes in the microbial community as a result of removing grazers. Here, the abundance of cyanobacteria increased approximately four-fold (13.0% ± 4.3 to 50.0% ± 6.1, mean ± 1 SE) during the first 6 days after limpets had been excluded, they then declined in abundance and were steadily replaced by *Enteromorpha* (Fig. 2b). After 24 days, ephemeral algae covered around 33% of the rock surface when viewed with the SEM, but were barely visible to the naked eye. *Enteromorpha* continued to grow rapidly and after 37 days was clearly visible in plots on the shore and covered 78±17.9% (mean ± 1 SE) of the rock surface. Excluding limpets had little effect on the abundance of diatoms, which remained relatively constant throughout (Fig. 2c).

Manipulation of microalgal resources

Enclosing groups of limpets provided an efficient method of manipulating grazer density. There was no evidence of greater mortality at the start of the experiment due to disturbance caused by setting up the experimental treatments (average mortality 3.4% in the first month). After 6 months, there was a clear trend of increasing mortality with density of grazers. This relationship was significant for *P. vulgata* but not for *P. depressa* (linear regression, $P < 0.01$ and $P < 0.06$, respectively; Fig. 3a). Elevated mortality in increased density plots reduced the absolute level of differences in the number of limpets among the various density treatments. Nevertheless, treatments retained the same relative order of increasing density throughout the course of the experiment.

A clear relationship between density and growth was also apparent with limpets in reduced density treatments growing approximately 50–60% faster than those in treatments with greater densities. The relationship between limpet density and percentage change in length (May–October) was significant for both *P. vulgata* and *P. depressa* (linear regression, $P <$ 0.01; Fig. 3b). Comparisons between species were not made as the *P. depressa* used in the enclosures were

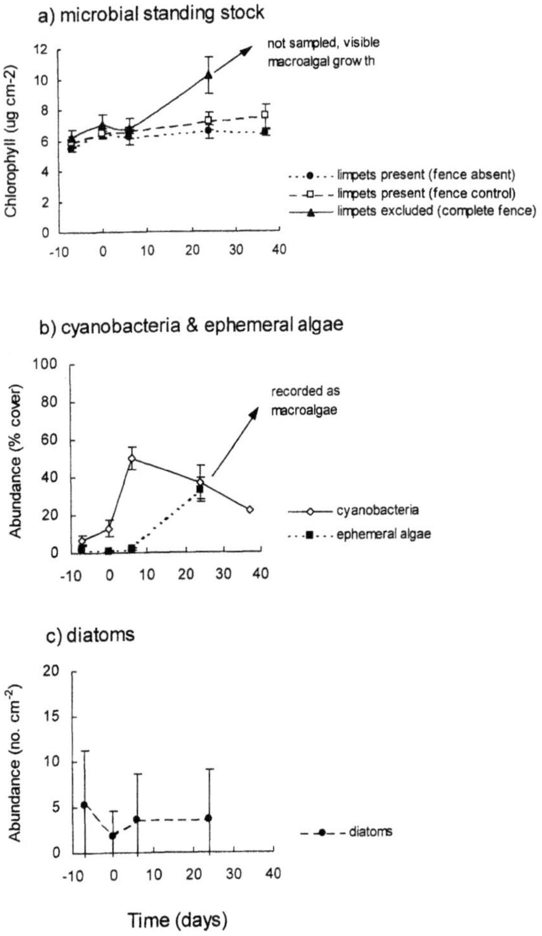

Figure 2. Changes in epilithic microbiota (average ± 1 SE, n = 5) as a result of excluding limpets at time = 0, assessed as: (a) photosynthetic microbial standing stock (extracted chlorophyll as an index) in areas where limpets had been excluded compared to control treatments, (b) percentage cover of cyanobacteria and ephemeral algae, and (c) abundance of diatoms (for b and c samples were viewed using a scanning electron microscope, data are presented for exclusion areas only).

nearer to their maximal size (on this shore 41 mm) than were *P. vulgata* (53 mm).

Discussion

Seasonal variations in microalgal standing stock and grazing intensity

There were distinct seasonal trends in the standing stock of epilithic microflora. These were characterised by reduced abundance during late spring and summer. Similar seasonal variations have been recorded in other locations, but the timing of microfloral biomass maxima and minima appear to vary with latitude. In temperate regions, maximal abundance was recorded in the winter and early spring (Aleem, 1950; Castenholz, 1963; Underwood, 1984a; MacLulich, 1987; Hill & Hawkins, 1991), while in the tropics, many species were most abundant during the winter (Nagarkar, 1998). Hence, it would appear that environmental conditions favouring microfloral growth vary seasonally and that the period when conditions are optimal may also vary according to latitude. The cause(s) of these patterns are not clear. Micro-organisms lack many of the structural and physiological adaptations which enable macrobiota to tolerate extremes of temperature, desiccation and insolation experienced during low tide. They are photo-adapted to low light levels and can become photo-inhibited during emersion (Stahl et al., 1985; Lamontagne et al., 1989). Consequently, in temperate and tropical regions, increased levels of emersion stresses may be responsible for the decline in microalgal abundance during the summertime (e.g. Castenholz, 1963; Jenkins et al., in press). Further research is required to confirm this relationship but experimental work on the Isle of Man and tropical shores has shown that shading can enhance microbial productivity during the summertime (Williams, 1994; Thompson, 1996).

Seasonal patterns in grazing intensity were also evident. These were characterised by considerably reduced intensity during the wintertime and were not correlated with seasonal trends in the abundance of microbial food resources (see above). The abundance of limpets remained reasonably constant throughout the study, increasing slightly during the spring as a result of juveniles moving into the monitoring areas, but these patterns were not correlated with microbial standing stock. Similar findings have been reported at other locations (Underwood, 1984c; Dye & White, 1991). The reduction in limpet grazing intensity during winter could be a response to increased availability of food per unit area at that time. Similar responses have been demonstrated experimentally for *Littorina littorea* (L.) grazing on areas with differing microalgal biomass (Sommer, 1999). Hence, to some extent, the mis-match between resource availability and grazing intensity may reflect compensation, in terms of foraging effort by the limpets, for changes in food availability. Grazing intensity, however, appeared to be most strongly influenced by seasonal changes in temperature. The metabolic rate and feeding activity of other marine invertebrates is generally recognised to increase with temperature (for review see Newell

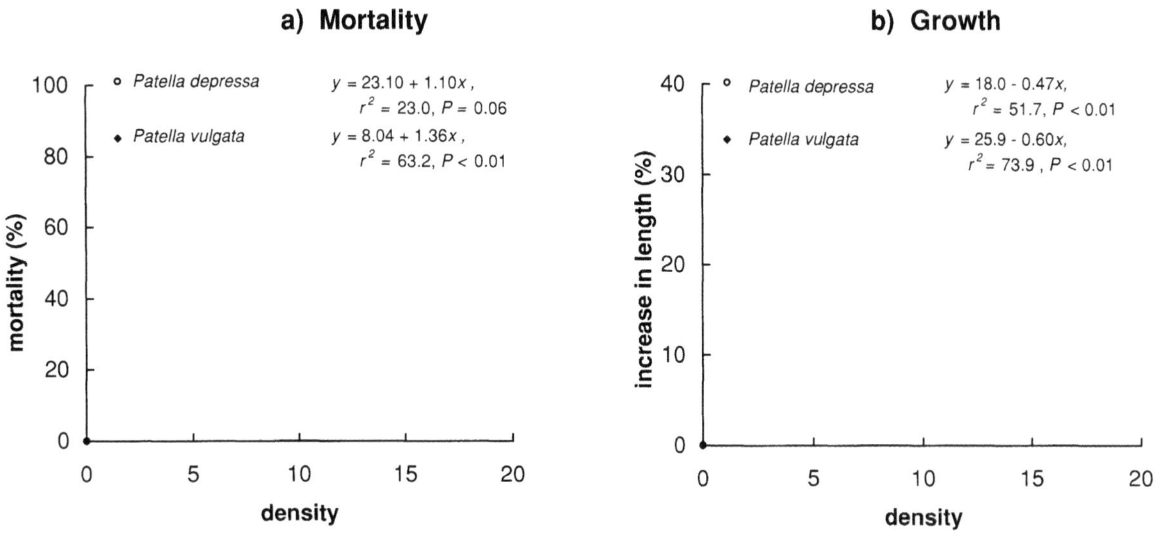

Figure 3. (a) Percentage mortality and (b) percentage increases in length (after 6 months) of *Patella vulgata* and *P. depressa* in 25 cm × 25 cm enclosures of differing densities (n =3) at Swanage (U.K.). Fitted lines obtained by linear regression and the significance of these relationships are shown (N.B. the relationship between mortality and density was, just, non-significant at $P = 0.05$ for *P. depressa*, and a line has been fitted to illustrate the trend in the data only). The average background density of limpets on the mid shore as a whole is also shown (arrowed).

& Branch, 1980). The heart rate of the siphonarian limpet *Siphonaria oculus* (Kr.) (Marshall & McQuaid, 1994) and the feeding rate of the patellid limpet *Cellana ornata* (Dillwyn) also increase with temperature (Boyden & Zeldis, 1979). Hence, variations in feeding intensity in the present study were probably the result of increased metabolic activity during the warmer months of the year. A slight decline in grazing intensity was observed during mid-summer, the cause of which is not clear. However, data recently collected on the Isle of Man indicate that this suppression in feeding activity may be related to maturation of the gonads towards the end of the summer (Jenkins et al., in press).

Experimental manipulation of grazing intensity

The majority of studies on interactions between grazers and algae on rocky shores have focused on the regulation of macroalgal abundance by grazers. The present study has shown that exclusion of grazers initially leads to microalgal proliferation with, in this case, rapid growth of cyanobacteria within 6 days. The species composition of microbial films varies seasonally and geographically (Underwood, 1984b; Hill & Hawkins, 1991; Nagarkar, 1998), and a reduction in grazer density in other locations or seasons would probably lead to different pattern of early succession. Seasonal variations in successional sequences of this kind have been recorded for macroalgal species, with grazer exclusions during July leading to algal assemblages dominated by *Fucus* germlings, while exclusions during April led to assemblages dominated by ephemeral algae (Hawkins, 1981).

Food resources per individual

Increasing the density of limpets resulted in reduced growth for *Patella vulgata* and *P. depressa* and increased mortality for *P. vulgata.* A similar trend of increasing mortality at elevated densities was also apparent for *P. depressa* but was, just, not significant ($P = 0.06$). Variability in lower density treatments was considerable, however, because of the small numbers of individuals used (Fig. 3) and this reduced our ability to detect differences among treatments. For example, the death of one individual in a lower density enclosure changed the percentage mortality for that enclosure by up to 50%. This difficulty was partly overcome by using a weighted analysis for mortality data (see 'Methods'). A better approach, however, would have been to adjust the number of experimental units used for each treatment density such that the total number of individuals monitored in each treatment was approximately the same. Differences in mortality amongst treatments are likely to have resulted from food shortages at higher densities. *Patella vulgata* has sufficient food reserves to survive for short periods of

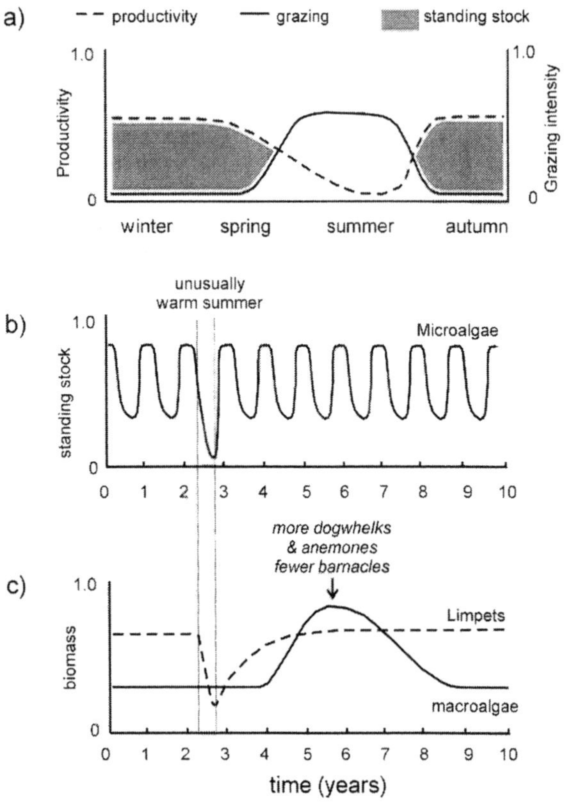

Figure 4. Schematic model of possible bottom-up regulation of macrobiota by epilithic microflora on rocky shores: (a) mis-match in supply of and demand for epilithic microbial resources during an annual cycle, (b) effect of an unusually warm summer on microalgal standing stock, and (c) consequences for intertidal macrobiota (see text for details).

time (Santini et al., 1995), but beyond this they will starve and their abundance will decline. Similar patterns have been recorded for other intertidal molluscs, with a shortage of food reducing growth and leading to increased mortality (for review see Underwood, 1979; Branch, 1981). Selective predation could also have caused increased mortality in the higher density treatments. The percentage mortality did, however, increase progressively during the experiment (3.4, 4.5, 5.8, 8.5, 7.4, 6.0, monthly percentage mortality May–October); a pattern which is more consistent with starvation than predation. In addition, avian predators have been shown to attack solitary limpets more frequently and with a greater success rate than they attacked aggregated limpets (Coleman et al., 1999).

Synthesis and conceptual model

These experiments demonstrated the direct effect of grazing on the composition of species and the standing stock of epilithic biofilms, and the effect of availability of resources per individual on the growth and survival of limpets. Seasonal patterns in the intensity of grazing and abundance of microbiota indicate that these factors were not coupled directly and suggest a more complex pattern of grazing intensity regulated by temperature (positive correlation) and microbial resources driven by insolation stress (negative correlation). These differing responses to environmental conditions could lead to a mis-match in the supply of and the demand for microbial resources. Hence, 'bottom-up' forcing of the availability of resources per individual, demonstrated in the grazer inclusion experiments, and 'top-down' control of micro and macroalgal succession, demonstrated by the grazer exclusion experiment, are both important in regulating the structure of intertidal communities, and the relative importance of each may vary between seasons and between years. Seasonal variations in microbial resources and grazing intensity, and possible consequences of the mis-match between these variables, are described in a conceptual model (Fig. 4) for limpets, algae and other rocky shore organisms.

The uncoupling of food requirements and food resources is represented schematically (Fig. 4a). In winter and early spring, there is a surplus of food – 'feast', while in summer there is a shortage – 'famine'. The pattern of microalgal abundance is analogous to seasonal cycles of phytoplankton in the N. Atlantic, which typically show a dramatic increase in abundance during the early spring. Epilithic microflora and phytoplankton are comprised mostly of single-celled organisms, which are capable of rapid rates of cell division when conditions are favourable. For phytoplankton, the principal grazers are zooplankton, which also have relatively short generation times. An increase in the abundance of phytoplankton typically leads to a increase in the abundance of zooplankton on a time scale of weeks to months (see Parsons et al., 1984). On rocky shores, however, the principal grazers are limpets which are long lived (individuals of over 15 years have been recorded), reproduce annually (Bowman & Lewis, 1977) and do not reach sexual maturity until three or more years of age (Orton et al., 1956). Clearly, short-term increases in food supply will not be reflected by changes in the density of these grazers.

Short-term reductions in resources may, however, influence the density and biomass of limpets. We suggest that the shortfall in resources during the summertime may set the absolute density of limpets on a shore. During typical years, the density and reproductive output of a population may be set at the level that can be supported by standing stock during the summertime. Assuming the seasonal pattern of resource availability recorded here could be extrapolated to unusually warm or cool summers, then bottom-up control would lead to fluctuations in mortality between years according to conditions during the summertime. For example, microbial productivity will be especially poor during unusually warm summers (Fig. 4b) and food shortages may lead to considerably elevated mortality of limpets (Fig. 4c). Increased limpet mortality has been observed during hot summers, but it remains uncertain whether this resulted from heat stress, food shortage or a combination of these factors (Bowman & Lewis, 1977). Intermittent or sporadic bottom-up regulation of grazer abundance, as a result of changes in food availability, has also been observed, over various time scales, for fishes in temperate streams and birds in forests (Karr et al., 1992).

Changes in the population density of limpets will have consequences for the structure of intertidal communities as a whole. For instance, an increase in the mortality of limpets, as the result of a hot summer, will lead to an associated increase in the abundance of macroalgae during the following years (Fig. 4c). Because limpets and fucoid algae are slow growing and recruit annually, this shift in the composition of the community will persist for several years, long after patterns of microalgal abundance have returned to more typical levels (Fig. 4b). The relative abundance of other shore dwellers is also likely to change with increases in macroalgal cover leading to an associated increase in the abundance of anemones and dogwhelks, but a reduction in the abundance of barnacles (Hartnoll & Hawkins, 1985). Annual variations in climatic conditions may also influence macroalgal recruitment and species composition, with cool damp periods in either the spring or the autumn providing good conditions for microbial and macroalgal growth with minimal losses to grazing. Hence, ephemeral green algae may be more prolific in years with cool conditions during the spring while growth of *Fucus* germlings would be favoured by a cool summer (see Hawkins, 1981).

The abundance of grazers and macroalgae on rocky shores are spatially and temporally very variable (Hartnoll & Hawkins, 1985), and cycles in abundance similar to those described in Figure 4c have been recorded over similar time scales. A variety of explanations have been offered for these patterns and some of these have been demonstrated experimentally (Hartnoll & Hawkins, 1985; Hawkins et al., 1992). These theories focus on top-down regulation of intertidal communities in the northeast Atlantic by grazers; the model presented here is not suggested as an alternative, but rather as an additional regulatory influence, operating from the bottom-upwards.

Acknowledgements

RCT was funded by a NERC studentship (GT4/92/201/A, supervised by SJH and TAN) and Leverhulme Grant (F/180/AO), SJH was supported by EC contract (MAS3 – CT95 – 0012: EUROROCK) and MFR was supported by the Grundy Educational Trust. The authors also thank Ian Allen, Michelle Tobin, Candace Browne and Kees Veltkamp for assistance with analysis, and Dave Brown at Ronaldsway Meteorological Office for weather data.

References

Aleem, A. A., 1950. Distribution and ecology of British marine littoral diatoms. J. Ecol. 38: 75–106.

Bowman, R. S. & J. R. Lewis, 1977. Annual fluctuations in the recruitment of *Patella vulgata* L. J. mar. biol. Ass. U.K. 57: 793–815.

Boyden, C. R. & J. R. Zeldis, 1979. Preliminary observations using an attached microphone sensor to study feeding behaviour of an intertidal limpet. Estuar. Coast. mar. Sci. 9: 759–769.

Branch, G. M., 1981. The biology of limpets: physical factors, energy flow and ecological interactions. Oceanogr. mar. biol. Ann. Rev. 19: 235–379.

Bustamante, R. H., G. M. Branch, S. Eekhout, B. Robertson, P. Zoutendyk, M. Schleyer, A. Dye, N. Hanekom, D. Keats, M. Jurd & C. McQuaid, 1995. Gradients of intertidal primary productivity around the coast of South Africa and their relationships with consumer biomass. Oecologia 102: 189–201.

Castenholz, R. W., 1963. An experimental study of the vertical distribution of littoral marine diatoms. Limnol. Oceanogr. 8: 450–462.

Coleman, R. A., J. D. Goss-Custard, S. E. A. LeV. Dit Durell & S. J. Hawkins, 1999. Limpet *Petella* spp. consumption by oystercatchers *Haematopus ostralegus*: a preference for solitary prey items. Mar. Ecol. Prog. Ser. 183: 253–261.

Crisp, D. J., 1974. Factors influencing the settlement of marine invertebrate larvae. In Grant, P. T. & A. M. Mackie (eds), Chemoreception in Marine Organisms. Academic Press, London: 177–265.

Dye, A. H. & R. A. White, 1991. Intertidal microalgal production and molluscan herbivory in relation to season and elevation on two rocky shores on the east coast of Southern Africa. S. Afr. J. mar. Sci. 11: 483–489.

Farrell, T. M., 1988. Community stability: effects of limpet removal and reintroduction in a rocky intertidal community. Oecologia 75: 190–197.

Fielding, P. J., K. S. J. Damstra & G. M. Branch, 1988. Benthic diatom biomass, production and sediment chlorophyll in Langebaan lagoon, South Africa. Estuar. coast. shelf Sci. 27: 413–426.

Fuji, A., H. Watanabe, K. Ogura, T. Noda & S. Goshima, 1991. Abundance and productivity of microphytobenthos on a rocky shore in southern Hokkaido. Bull. Fac. Fish. Hokkaido Univ. 42: 136–146.

Grubb, V. M., 1936. Marine ecology and the exposure factor at Peveril Point, Dorset. J. Ecol. 24: 392.

Hartnoll, R. G. & S. J. Hawkins, 1985. Patchiness and Fluctuations on Moderately Exposed Rocky Shores. Ophelia 24: 53–64.

Hawkins, S. J., 1981. The Influence of Season and Barnacles on the Algal Colonization of Patella-Vulgata Exclusion Areas. J. mar. biol. Ass. U.K. 61: 1–15.

Hawkins, S. J. & R. G. Hartnoll, 1983. Grazing of intertidal algae by marine invertebrates. Oceanogr. mar. Biol. Ann. Rev. 21: 195–285.

Hawkins, S. J., R. G. Hartnoll, J. M. Kain, & T. A. Norton, 1992. Plant-animal interactions on hard substrata in the North-east Atlantic. In John, D. M., Hawkins, S. J. & Price, J. H. (eds), Plant-Animal Interactions in the Marine Benthos. Oxford University Press, Oxford, U.K.: 1–32.

Hawkins, S. J., D. C. Watson, A. S. Hill, S. P. Harding, M. A. Kyriakides, S. Hutchinson & T. A. Norton, 1989. A comparison of feeding mechanisms in microphagous, herbivorous, intertidal, prosobranchs in relation to resource partitioning. J. Mollusc. Stud. 55: 151–165.

Hill, A. S. & S. J. Hawkins, 1990. An investigation of methods for sampling microbial films on rocky shores. J. mar. biol. Ass. U.K. 70: 77–88.

Hill, A. S. & S. J. Hawkins, 1991. Seasonal and spatial variation of epilithic microalgae distribution and abundance and its ingestion by *Patella vulgata* on a moderately exposed rocky shore. J. mar. biol. Ass. U.K. 71: 403–423.

Huang, R. & A. D. Boney, 1984. Growth interactions between littoral diatoms and juvenile marine algae. J. exp. mar. Biol. Ecol. 81: 21–45.

Jenkins, S. R., F. Arenas, J. Arrontes, J. Bussell1, J. Castro, R. A. Coleman, S. J. Hawkins, S. Kay, B. Martínez, J. Oliveros, M. F. Roberts, R. C. Thompson & R. G. Hartnoll, In Press. Seasonal variation in limpet grazing activity and microalgal abundance over a European scale. Mar. Ecol. Prog. Ser.

Jernakoff, P., 1985. Interactions between the limpet *Patelloida latistrigata* and algae on an intertidal rock platform. Mar. Ecol. Prog. Ser. 23: 71–78.

Karr, J. R., M. Dionne & I. J. Schlosser, 1992. Bottom-up versus top-down regulation of vertebrate populations: lessons from birds and fish. In Hunter, M. D., T. Ohgushi & P. W. Price (eds), Effects of resource distribution on animal plant interactions. Academic Press, London: 243–286.

Kristensen, E., 1993. An *In situ* method for measuring the primary productivity and standing crop of the epilithic periphyton community in lentic systems. Estuar. coast. shelf Sci. 36: 565–586.

Lamontagne, I., A. Cardinal & L. Fortier, 1989. Environmental forcing versus endogenous control of photosynthesis in intertidal epilithic microalgae. Mar. Ecol. Prog. Ser. 51: 1–2.

Lock, M. A., 1993. Attached microbial communities in rivers. In Ford, T. A. (ed.), Aquatic Microbiology. Blackwells, Oxford: 113–138.

MacLulich, J. H., 1987. Variations in the density and variety of intertidal epilithic microflora. Mar. Ecol. Prog. Ser. 40: 285–293.

Mak, Y. M. & G. A. Williams, 1999. Littorinids control high intertidal biofilm abundance on tropical, Hong Kong rocky shores. J. exp. mar. Biol. Ecol. 233: 81–94.

Marshall, D. J. & C. D. McQuaid, 1994. Seasonal and Diel Variations of *in-Situ* Heart-Rate of the Intertidal Limpet Siphonaria-Oculus Kr (Pulmonata). J. exp. mar. Biol. Ecol. 179: 1–9.

Menge, B. A., B. A. Daley, J. Lubchenco, E. Sanford, E. Dahlhoff, P. M. Halpin, G. Hudson & J. L. Burnaford, 1999. Top-down and bottom-up regulation of New Zealand rocky intertidal communities. Ecol. Monogr. 69: 297–330.

Menge, B. A., B. A. Daley, P. A. Wheeler, E. Dahlhoff, E. Sanford & P. T. Strub, 1997. Benthic-pelagic links and rocky intertidal communities: bottom-up effects on top-down control? Proc. natn. Acad. Sci. U.S.A. 94: 14530–14535.

Nagarkar, S., 1998. New records of marine cyanobacteria from rocky shores of Hong Kong. Bot. mar. 41: 527–542.

Newell, R. C. & G. M. Branch, 1980. The influence of temperature on the maintenance of metabolic energy balance in marine invertebrates. Adv. mar. Biol. 17: 329–396.

Nicotri, M. E., 1977. Grazing effects of four marine intertidal herbivores on the microflora. Ecology 58: 1020–1032.

Ogden, N. B., J. C. Ogden & I. A. Abbott, 1989. Distribution, abundance and food of sea urchins on a leeward Hawaiian reef. Bull. mar. Sci. 45: 539–549.

Orton, J. H., A. J. Southward & J. M. Dodd, 1956. Studies on the biology of limpets II. The breeding of *Patella vulgata* L. in Britain. J. mar. biol. Ass. U.K. 35: 149–176.

Parsons, T. R., M. Takahashi & B. Hargrave, 1984. Biological oceanographic processes. Butterworth-Heinemann. Oxford, U.K.: 332 pp.

Santini, G., G. Chelazzi & P. Della Santina, 1995. Size-related functional and energetic constraints in the foraging of the limpet Patella vulgata (Mollusca, Gastropoda). Funct. Ecol. 9: 551–558.

Sokal, R. R. & F. J. Rohlf, 1995. Biometry. Freeman & Co. New York: 850 pp.

Sommer, U., 1999. The susceptibility of benthic microalgae to periwinkle (Littorina littorea, Gastropoda) grazing in laboratory experiments. Aquat. Bot. 63: 11–21.

Southward, A. J., 1953. The ecology of some rocky shores in the south of the Isle of Man. Proc. Trans. Liverpool Biol. Soc. 59: 1–50.

Southward, A. J., 1964. Limpet grazing and the control of vegetation on rocky shores. In Crisp, D. J. (ed.), Grazing in Terrestrial and Marine Environments. Blackwell Scientific Publications, Oxford: 265–273.

Southward, A. J. & E. C. Southward, 1978. Recolonisation of rocky shores in Cornwall after use of toxic dispersants to clean up the Torrey Canyon spill. J. Fish. Res. Bd Can. 35: 682–706.

Stahl, L. J., H. Van Gemerded & W. E. Krumbein, 1985. Structure and development of benthic marine microbial mat. Fems Microbiol. Ecol. 31: 111–125.

Steneck, R. S. & L. Watling, 1982. Feeding capabilities and limitation of herbivorous molluscs: a functional group approach. Mar. Biol. 68: 299–319.

Thompson, R. C., 1996. The Ecology of Epilithic Microalgae on Manx Shores. School of Biological Sciences. Liverpool University, Liverpool: 244 pp.

Thompson, R. C., L. E. Johnson & S. J. Hawkins, 1997. A method for spatial and temporal assessment of gastropod grazing intensity in the field: the use of radula scrapes on wax surfaces. J. exp. mar. Biol. Ecol. 218: 63–76.

Thompson, R. C., S. J. Hawkins & T. A. Norton, 1998. The influence of epilithic microbial films on the settlement of *Semibalanus balanoides* cyprids – a comparison between laboratory and field experiments. Hydrobiologia 375/376 (Dev. Hydrobiol. 132): 203–216.

Thompson, R. C., M. L. Tobin, S. J. Hawkins & T. A. Norton, 1999. Problems in extraction and spectrophotometric determination of chlorophyll from epilithic microbial biofilms: towards a standard method. J. mar. biol. Ass. U.K. 79: 551–558.

Underwood, A. J., 1979. The ecology of intertidal gastropods. Adv. mar. Biol. 16: 111–210.

Underwood, A. J., 1984a. Microalgal food and the growth of the intertidal gastropods *Nerita atramentosa* Reeve and *Bembicium nanum* (Lamarck) at four heights on a shore. J. exp. mar. Biol. Ecol. 79: 277–291.

Underwood, A. J., 1984b. Vertical and seasonal patterns in competition for microalgae between intertidal gastropods. Oecologia 64: 211–222.

Underwood, A. J., 1984c. The vertical distribution and seasonal abundance of intertidal microalgae on a rocky shore in New South Wales. J. exp. mar. Biol. Ecol. 78: 199–220.

Underwood, A. J., M. G. Chapman, S. A. Richards & M. B. Sage, 1997. GMAV for Windows. Institute of Marine ecology, Sydney.

Wahl, M., 1989. Marine epibiosis. I. Fouling and antifouling: Some basic aspects. Mar. Ecol. Prog. Ser. 58: 1–2.

Williams, G. A., 1994. The relationship between shade and molluscan grazing in structuring communities on a moderately-exposed tropical rocky shore. J. exp. mar. Biol. Ecol. 178: 79–95.

Workman, C., 1983. Comparisons of energy partitioning in contrasting age-structured populations of the limpet *Patella vulgata* (L). J. exp. mar. Biol. Ecol. 68: 81–103.

Hydrobiologia **440:** 369–377, 2000.
M.B. Jones, J.M.N. Azevedo, A.I. Neto, A.C. Costa & A.M. Frias Martins (eds), Island, Ocean and Deep-Sea Biology.

Bioturbation as a mechanism for setting and maintaining levels of diversity in subtidal macrobenthic communities

S. Widdicombe[1,2], M.C. Austen[1], M.A. Kendall[1], R.M. Warwick[1] & M.B. Jones[2]
[1] *Plymouth Marine Laboratory, Prospect Place, West Hoe, Plymouth PL1 3DH, U.K.*
Tel: + 01752 633100. Fax: 01752 633101. E-Mail: s.widdicombe@pml.ac.uk
[2] *Plymouth Environmental Research Centre (Department of Biological Sciences), University of Plymouth, Drake Circus, Plymouth PL4 8AA, U.K.*

Key words: bioturbation, diversity, macrofauna, meiofauna, disturbance

Abstract

Over 2 years, experiments were conducted to compare the effects of sediment disturbance by different bioturbating, macrofaunal organisms on the diversity and structure of the associated infaunal community. The four species investigated were the bivalves *Nuculoma tenuis* (Montagu, 1808) and *Abra alba* (Wood, 1802), the heart urchin *Brissopsis lyrifera* (Forbes, 1841), and the burrowing decapod *Calocaris macandreae* (Bell, 1846). These organisms were chosen to allow assessment of the effects of contrasting feeding activities and body sizes of the bioturbating species on the diversity of the macrobenthic communities. Bioturbation by the sub-surface deposit feeders *N. tenuis* and *B. lyrifera* promoted higher levels of α and β diversity in treatments exposed to intermediate levels of disturbance. Whilst no such 'intermediate response' was demonstrated for *A. alba* or *C. macandreae*, it was evident that changes in the associated fauna were influenced by the feeding type of the bioturbating organism responsible. It was also shown that different elements of the associated community responded differently to biotic disturbance. The results indicate that the variability in density and distribution of such bioturbators are important factors in structuring infaunal communities, and in setting and maintaining levels of diversity in apparently homogeneous areas.

Introduction

The structure and diversity of communities stems from interactions between biological and physical processes acting at a range of spatial and temporal scales nested within each other (Huston, 1994 and references therein). To understand how communities are assembled and maintained, it is imperative that these interactions should be identified and their effects examined. In sediment community dynamics, disturbance is an important driving force and can vary in scale from ray feeding pits (Thrush et al., 1991) to storms (Posey et al., 1996). The definition of the term 'disturbance', however, has received much attention. Menge & Sutherland (1987) differentiated between the effects of stress and disturbance, defining 'stress' as a factor which prevented an organism from operating at its optimal level due to either mechanical forces (e.g. being buffeted by strong currents) or through inducing physiological responses in the organism (e.g. temperature); 'disturbance' was reserved for processes that exerted lethal levels of stress on the organism or population (e.g. currents that remove organisms from a habitat or temperatures outside an organism's tolerance range). In the experiments reported in this paper, it is difficult to state categorically whether the effects of the bioturbating organisms constitute environmental stress or environmental disturbance. Therefore, we use the widely accepted definition of Pickett & White (1985) who described disturbance as "a discrete event in time that disrupts ecosystem, community or population structure and changes resources, substratum availability, or the physical environment". This definition encompasses processes which act directly on the organisms themselves (e.g. predation) as well as those which act indirectly, such as sediment conditioning (i.e. changes in sediment granulometry, organic content, oxygen penetration).

Once an assemblage has been subjected to disturbance, the processes that determine the identity and abundance of the organisms that ultimately recolonise defaunated areas shape the community structure and diversity of an area. Subsequent community development is determined by the identity and biology of the colonists plus interactions with subsequent immigrants. Competition was central to the 'Intermediate Disturbance Hypothesis' (IDH) of Connell (1978), which emphasised the importance of the frequency and severity of disturbance. The IDH proposed that at intermediate levels of disturbance, competitive exclusion was impeded, promoting co-existence in potentially competing species. Support for the IDH was provided by Huston (1979) who proposed that diversity was set by processes which acted on rates of competitive displacement. His 'Dynamic Equilibrium Theory' stated that "diversity represented a balance between growth rates and disturbance" (Huston, 1979). The implication of both the IDH and the 'Dynamic Equilibrium Theory' is that diversity will be highest in areas subjected to some intermediate level of disturbance. This prediction has not been accepted fully for marine, sub-littoral, sediment communities, even though evidence of the competitive pressure required to comply with this hypothesis has been provided (Branch, 1984). Additionally, growing evidence suggests that a response similar to that predicted by the IDH is elicited by low levels of biotic disturbance (bioturbation) in soft sediment environments (e.g. Kukert, 1991; Widdicombe & Austen, 1998, 1999).

To date, the study of bioturbation has failed to examine adequately the role played by different types of bioturbator, concentrating instead on bioturbation disturbance as a single, quantifiable unit of damage. It was assumed that any response to physical disturbance is solely a result of the size and frequency of disturbance (Petraites et al., 1989). Such an assumption cannot be justified until the effects of different bioturbating species are compared directly. Intuitively, it seems unlikely that the activities of bioturbating organisms, which differ in size, feeding strategy, burrowing activity and mobility, will have identical effects on the chemical, physical and biological processes within the sediment. The effects of such differences of feeding strategy and mobility on sediment disturbance were addressed by Thayer (1983), who concluded that 'biological bulldozing' is the most effective form of bioturbation and that the effects of sediment ingestion or egestion are probably less important. Indeed, a review by Hall (1994) concluded that, with respect to sediment disturbance, species must be judged on their individual characteristics.

This paper reports the effects of sediment disturbance by different bioturbating, macrofaunal organisms on the diversity and structure of the associated macroinfaunal community. The experiments used different densities of bioturbators to produce different intensities of disturbance and addressed the following issues:

1. Does the response of a benthic community to bioturbation support the 'Intermediate Disturbance Hypothesis'?
2. Is the effect of bioturbation the same regardless of the identity of the bioturbating organism, with any differences being purely a function of disturbance intensity?
3. From an examination of the first two issues, we explore the effects of bioturbators on scales of heterogeneity and discuss how such pattern may influence monitoring strategies.

Materials and methods

Bioturbating species

The burrowing heart urchin *Brissopsis lyrifera* is a non-selective, infaunal deposit feeder (Nichols, 1959). By 'bulldozing' through the sediment, it has a major impact on sediment structure and stability. In addition, by creating a respiratory funnel in the upper 3 cm of sediment (De Ridder & Lawrence, 1982), *B. lyrifera* increases the depth of oxygenation. Experimental work (Austen & Widdicombe, 1998; Widdicombe & Austen, 1998) suggests that the presence of this species enhances both local heterogeneity and β diversity. The bivalve *Nuculoma tenuis* is also a sub-surface deposit feeder and its feeding activities disturb the top 2–3 cm of the sediment. It may, therefore, be expected that bioturbation caused by *N. tenuis* and *B. lyrifera* will have the same effect on the community structure of their associated fauna. However, *B. lyrifera* (mean length 3 cm, mean volume 10.61 cm^3) is much larger than *N. tenuis* (mean length 0.9 cm, mean volume 0.21 cm^3) and, as its movements displace more sediment than the bivalve, any community response to its bioturbation may differ in magnitude from *N. tenuis*. In contrast, the bivalve *Abra alba* (mean length 1.5 cm, mean volume 0.75 cm^3) is a surface deposit/suspension feeder that reworks the surface 0.5 cm of the sediment and has limited lateral movement. In an aquarium, one *A. alba* reworked an area 7 cm in diameter and raised the sediment surface by 2 cm during a 4-week period (Eagle, 1975). A third type of sediment disturbance was provided by the thalassinidean shrimp *Calocaris macandreae* (Decapoda) which constructs deep, complex burrows with multiple surface openings (Nash et al., 1984). These shrimps rarely emerge onto the sediment surface, obtaining the majority of their food from within the burrow; sub-surface sediment from as deep as 215 mm may be expelled during normal feeding activities and burrow maintenance (Nash et al., 1984).

Experimental design

Experiments were carried out in the mesocosm facility of the Norwegian Institute for Water Research (NIVA) marine research station Solbergstrand, Norway (Berge et al., 1986). Details of collection and experimental protocol for the experiments reported here have been described previously (Widdicombe & Austen, 1998, 1999). In summary, muddy sand, containing a natural assemblage of macrofaunal organisms, was collected from 30 m depth in Bjørhodenbukta, a sheltered bay in the inner part of Oslofjord, and was used to fill square boxes (1 m × 1 m) to a depth of 30 cm. Boxes were placed in a 5 m × 7 m indoor, epoxy resin-coated concrete basin. A continuous flow of seawater, pumped from 60 m depth in the fjord, maintained a constant water depth of 100 cm. Circular areas of sediment were partitioned off and seeded with different densities and types of bioturbators (Table 1). Treatment densities were replicated both within and between boxes (four replicates per treatment per box for Experiments 1 and 3; two replicates per treatment per box for Experiment 2; all experiments used four boxes). Due to constraints of space, the larger bioturbating species (*Brissopsis lyrifera* and *Calocaris macandreae*) were studied at fewer density treatments than the two bivalves (Table 1). Another consequence of space constraints resulted in the overall study being addressed by three separate experiments (Table 1). Experiments 1 and 2 were run in parallel, each was set up on 12th April 1995 and sampled on 29th August 1995. Experiment 3 was set up on 2nd September 1995 and sampled on 8th May 1996. Experiment 1 examined the effect of presence and density of *B. lyrifera* on the macrofaunal community of sediment collected from Bjørhodenbukta. Experiment 2 compared the effects on the same macrofaunal community of different species (*Nuculoma tenuis*, *Abra alba* and *B. lyrifera*) and densities of bioturbating organisms. Experiment 3 used the same macrofaunal community as Experiments 1 and 2, and examined the effect of the presence of *C. macandreae* on this community. All three experiments were sampled and processed in the same way. Plastic cores (10 cm diameter) were used to sample each treatment and control area. These samples were sieved over a 250 μm mesh to retain the juvenile macrofauna, since these young individuals may be more susceptible to the effects of disturbance than larger, more established animals. The residue was fixed in a 10% formaldehyde solution and sorted under a binocular microscope. All animals were extracted and identified to the lowest possible taxonomic level. The remainder of the sediment in each treatment and control area was coarsely sieved to quantify the bioturbators remaining in these areas.

Data analysis

Multivariate data analyses followed the methods described by Clarke (1993) using the PRIMER software package. Using the Bray-Curtis similarity measure, analysis was carried out on untransformed and double square root ($\surd\surd$) transformed data. The former identified changes in dominant species whilst $\surd\surd$ transformation is a balanced approach which increases the influence of low abundance species. Although using $\surd\surd$ transformed data is not a direct observation of the rare fauna, comparison with the results obtained from the two transformations allows examination of the influence of the rare species on the overall community pattern. Two-way crossed ANOSIM (Clarke, 1993) was carried out to test for box effects. Where none was detected, one-way pairwise ANOSIM was used to test for significant differences (p<0.05) between macrofaunal assemblages in different treatments. The PRIMER software package was used also to generate the diversity indices (number of species, species richness, eveness and Shannon–Wiener). A Kruskal–Wallace test was used to test the means for significant differences between treatments for the means of these indices. Species area curves were constructed using the BioDiversity Pro software package (Natural History Museum and the Scottish Association of Marine Biology). The slope of these curves indicate differences in species turnover, a measure of β diversity, with a steeper curve being representative of higher diversity. The mean Bray–Curtis dissimilarities within treatments were calculated as an indication of the extent of within treatment variability. This can be used

Table 1. Identity, type and treatment densities of the bioturbating species used in each of the three experiments

Duration	Species	Bioturbating type	Treatment densities (individuals m^{-2})
Experiment 1 20 weeks	*Brissopsis lyrifera*	Large 'Bulldozer'	0, 28, 71
Experiment 2 20 weeks	*Nuculoma tenuis*	Small 'Bulldozer'	0, 612, 1836, 3304
	Abra alba	Surface resuspension, limited mobility	0, 612, 1469, 3304
	Brissopsis lyrifera	Large 'Bulldozer'	0, 57
Experiment 3 32 weeks	*Calocaris macandreae*	Deep burrowing with surface deposition	0, 14

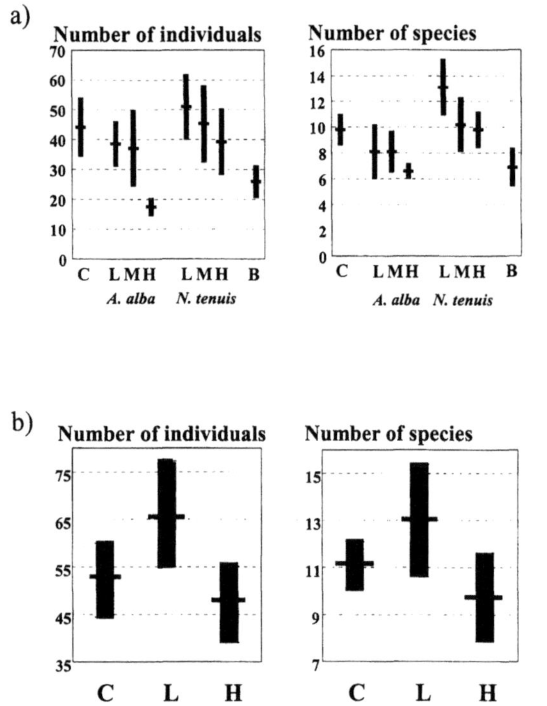

Figure 1. Mean numbers (±95% confidence intervals) of individuals and species among controls and treatments for (a) multi-species experiment (Experiment 2), (b) *Brissopsis* experiment (Experiment 1) (B = *Brissopsis*, C = Control, L = Low, M = Mid, H = High).

as a measure of β diversity with greater heterogeneity being indicative of higher β diversity. An index of multivariate dispersion was calculated to quantify the variability within, and between, treatments.

Results

Univariate measures and diversity indices

Figure 1 compares the results obtained from the *Nuculoma tenuis* and *Abra alba* treatments (Experiment 2) with those from *Brissopsis lyrifera* treatments (Experiment 1). Experiment 2 demonstrated that for *N. tenuis*, the α diversity (number of species) of the associated macrofauna in the low density treatments was significantly greater than the diversity in the high density (p = 0.026) and control (p = 0.014) treatments. This was not the case in the *A. alba* treatments, where the only significant difference was seen between the high density and control treatments (p = 0.001). Whilst it appears that the response of α diversity to different densities of *B. lyrifera* is similar to the response of α-diversity to different densities of *N. tenuis*, no significant differences were demonstrated between either the low *B. lyrifera* density and the high *B. lyrifera* density treatments, or between the low *B. lyrifera* density and control treatments. In Experiment 3, the number of species in treatments that contained *Calocaris macandreae* was 9.5 (±0.9) and was not significantly different from the number of species in the control treatments (8.7±0.7). For number of individuals (Fig. 1), the only significant results observed were that the abundance in high density *A. alba* treatments was lower than in all other treatments (Experiment 2) and the abundance in low *B. lyrifera* treatments was higher than high *B. lyrifera* treatments (Experiment 1).

The β diversity (species turnover as demonstrated by species area curves) of the associated macrofauna in the low *Nuculoma tenuis* density treatments was greater than the β diversity in the control, mid and

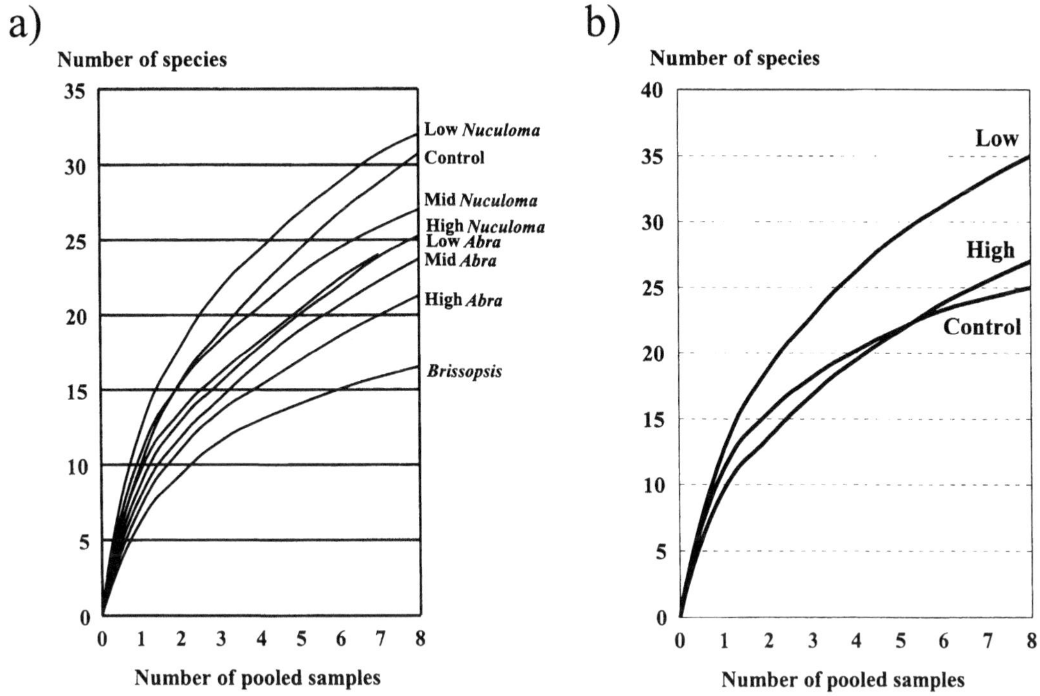

Figure 2. Species area curves from combined replicates for treatments and controls, based on 100 random sorts, for (a) multi-species experiment, and (b) *Brissopsis* experiment.

Table 2. Multivariate dispersion results and mean Bray-Curtis dissimilarities indicating variability within treatments using untransformed data

Bioturbating species	Treatment level	Multivariate dispersion sequence	Mean Bray–Curtis dissimilarities
Brissopsis lyrifera	Control	0.93	32.47
	Low	1.21	37.49
	High	0.86	31.42
Abra alba	Control	0.69	36.33
	Low	1.19	48.08
	Mid	0.86	39.51
	High	1.27	49.24
Nuculoma tenuis	Control	0.91	36.33
	Low	1.21	38.57
	Mid	1.02	41.25
	High	0.76	38.96

Table 3. R-values from pairwise comparisons of treatments containing bioturbating organisms and control treatments using 1-way ANOSIM. Bold values indicate significant differences, $p < 0.05$ (p-values in parentheses)

	Untransformed	$\sqrt{}\sqrt{}$ transformed
Control v low *A. alba*	**0.159 (0.035)**	**0.241 (0.013)**
Control v mid *A. alba*	0.108 (0.091)	**0.349 (0.001)**
Control v high *A. alba*	**0.569 (0.000)**	**0.493 (0.001)**
Control v low *N. tenuis*	**0.251 (0.004)**	**0.273 (0.008)**
Control v mid *N. tenuis*	**0.185 (0.040)**	**0.485 (0.000)**
Control v high *N. tenuis*	0.175 (0.094)	**0.511 (0.002)**
Control v low *B. lyrifera*	**0.246 (0.018)**	**0.182 (0.027)**
Control v high *B. lyrifera*	**0.205 (0.021)**	**0.325 (0.001)**
Control v *C. macandreae*	**0.167 (0.015)**	0.046 (0.219)

high *N. tenuis* density treatments (Fig. 2a). A similar pattern was observed in Experiment 1, where β diversity was highest in the low *Brissopsis lyrifera* density treatments (Fig. 2b). Other measures of β diversity (multivariate dispersion and mean Bray-Curtis) displayed the same pattern, with diversity highest in the low *N. tenuis* density and low *B. lyrifera* density treatments (Table 2).

Community structure

Multivariate analysis indicated that the presence of bioturbating organisms resulted in significant differences in community structure between bioturbated and non-bioturbated treatments (Table 3). Only three

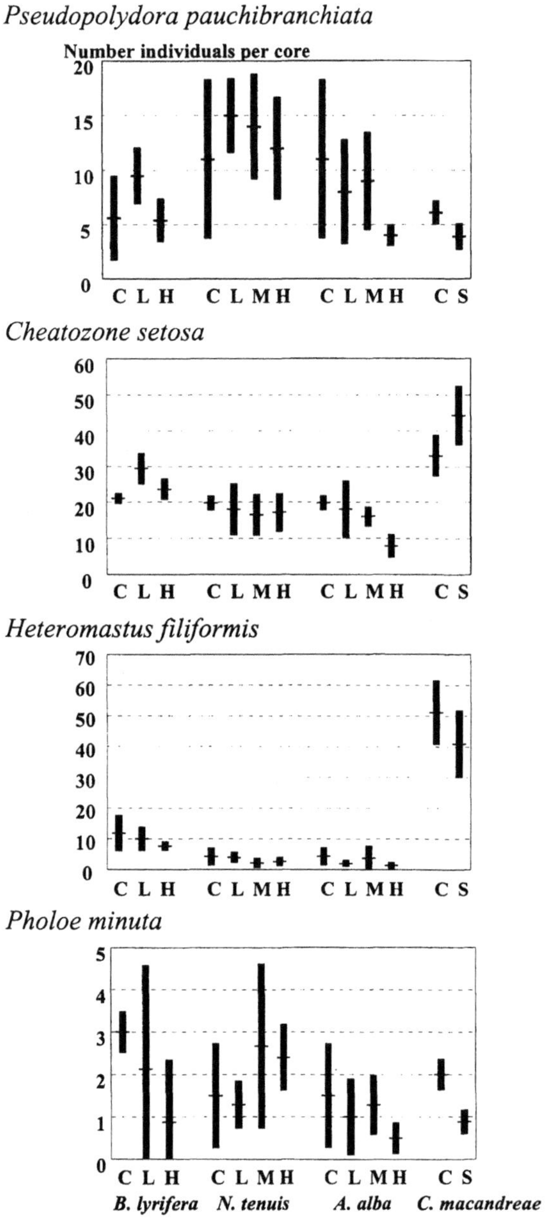

Figure 3. Mean abundances (number individuals per core ±95% C.I.) in each treatment for four important species of macrofauna, as identified by SIMPER analysis (C = Control, L = Low, M = Mid, H = High, S = Shrimp).

pairwise comparison produced results indicating treatments that were not significantly different. These were control v mid *Abra alba* and control v high *Nuculoma tenuis* using untransformed data, and control v *Calocaris macandreae* using $\sqrt{}\sqrt{}$ transformed data.

Species responses

The abundance of many associated macrofaunal species changed in response to different densities of bioturbator. Additionally, the manner and extent to which these abundances changed was affected by the identity of the bioturbating organism present. For example, the polychaete *Pseudopolydora pauchibranchiata* showed an increase in abundance in the presence of the bulldozing species *Nuculoma tenuis* and *Brissopsis lyrifera* compared with the control treatments; however, in the presence of *Abra alba* and *Calocaris macandreae*, *P. pauchibranchiata* abundances were lower than in the control treatments (Fig. 3). Other polychaetes showing different responses between treatments subjected to different bioturbators included *Chaetozone setosa*, *Heteromastus filiformis* and *Pholoe minuta* (Fig. 3). In general, differences in abundances of these species in response to changes in bioturbator density were most strongly observed when comparing *N. tenuis* and *B. lyrifera* treatments to either of the other two bioturbating species. The smallest differences were observed when comparing abundance changes between treatments containing either of the two bulldozing species, *N. tenuis* and *B. lyrifera*.

Discussion

For the bulldozing bioturbators (*Brissopsis lyrifera* and *Nuculoma tenuis*), either (or both) the α diversity or β diversity of the associated community was highest in the intermediate treatments. While this agrees well with the predictions of the IDH, it would be unwise to dismiss other potential causes without consideration. The most plausible alternative is that the bioturbators inhibit some activity or function of the dominant infauna. This is unlikely, however, as such inhibition would lead to a reduction of the dominant species in the intermediate experimental treatments and this did not happen. In some of these latter treatments, not only did some of the high abundance-ranking species increase in abundance, but the total number of individuals in the sample was also greater than in controls. The most likely explanation for the elevation of diversity in the intermediate treatments is that a moderate increase in the activity of bulldozing bioturbators increased some potentially limiting resource, permitting the number of organisms to increase. There appears to be a trade-off between the increased availability of such resources and the severity of physical disturbance/damage. This trade-off limits the extent to which disturbance can be seen as favorable. In muddy sediments, the oxic zone is shallow and it may be that, following disturbance, the limited volume is quickly recolonised to the point where species interactions begin to influence the composition of the assemblage. Such a disturbance could be caused by *B. lyrifera* ploughing though this surface layer and increasing the depth to which oxygen can penetrate (Widdicombe & Austen, 1998). The effective volume of the sediment is, therefore, greater and, consequently, the time between a disturbance event occurring and the point at which species interactions begin, is increased. The exclusion of species by competition would be delayed accordingly. This would fit well with the IDH. Nevertheless, not all present data support the hypothesis. For example, for *Abra alba,* the greater the number of bivalves used in a treatment, the lower was the infaunal diversity. For a given volume, the physical sediment disturbance caused by this type of organism is far less than that caused by an equivalent volume of bulldozing species (Thayer, 1983). Consequently, high densities of *A. alba* may result in overwhelming competition for food before any beneficial effects on diversity, through physical disturbance, can be realised.

Paradigms such as the IDH are based on diversity measures that take no account of species identity, but the response of a community to disturbance is compounded by the response of individual species. Whilst it is important to understand the processes which set the number of different species able to inhabit an environment, it is also vital to understand that bioturbation may influence the identity of the species which make up the community. It has been suggested that, in some circumstances, the species colonising an area can be predicted on the basis of their mobility, feeding or reproductive biology. Studies investigating the effects of burrowing organisms on benthic community structure (Brenchley, 1981; Posey, 1986, 1991) demonstrated a reduction in the number of tube-dwelling polychaetes and other sedentary organisms coupled with an increase in the abundance of other more mobile organisms in bioturbated areas. This led Brenchley (1981) to develop the 'Mobility Mode Hypothesis' which predicted the consequences of interaction between animals with contrasting modes of life. In the experiments of Widdicombe & Austen (1998, 1999), multivariate analysis of community structure demonstrated that changes in the intensity of bioturbation altered the relative abundance of the numerically-dominant species. It cannot be claimed, however, that these results totally support the 'Mobility Mode Hypothesis' as different bioturbators elicited different community responses. Much of the evidence for the 'Mobility Mode Hypothesis' is based on the effects of burrowing shrimps such as *Callianassa californiensis* (Brenchley, 1981; Posey, 1986) and *Upogebia pugettensis* (Posey et al., 1991). The thalassinidean shrimp used in the current study, *Calocaris macandreae*, conformed to the typical response predicted by this hypothesis, with the abundance of the tube-building, spionid polychaete *Pseudopolydora pauchibranchiata* being lower in shrimp treatments than in the controls. A reduction in the number of tube-building species in response to increased disturbance was also observed in the experiments with *Abra alba*. However, no such reduction was seen in response to increased disturbance by either of the two bulldozing species, *B. lyrifera* and *N. tenuis*. Indeed, the abundance of this tube-building polychaete species was higher in some of the *N. tenuis* treatments than in the controls. It is clear that the 'quality' of bioturbatory activity, not simply its intensity, is the determining factor in a community response to disturbance. This bioturbation 'quality' is a complex association of many factors which include the spatial extent to which sediment is moved, the periodicity and intensity of disturbance, and changes in the physical and chemical properties of the sediment. Benthic assemblages seldom contain a single dominant bioturbator, usually a number of such species with a range of body sizes and mobility will occur. If different bioturbation 'qualities' have distinctive effects on community structure, then it is probable that there will be a relationship between the diversity of bioturbators and the diversity of the sediment infauna.

Changes in community structure and diversity can manifest themselves in at least two ways. Firstly, patterns of eveness alter, due to changes in the relative abundance of the numerically-dominant organisms. Secondly, species richness measures are altered, as a result of changes in total species number and identities. The importance of distinguishing between bioturbation quality and intensity is highlighted further when examining these different community elements. Widdicombe & Austen (1999) showed that different elements of diversity and community structure are influenced by different aspects of bioturbation. While the intensity of disturbance can determine the relative abundance of the dominant species, the type of bioturbator (quality) can influence the identity of the rare species. The idea that different species of large macrofauna will impact on the associated fauna in different ways complies with the 'Grain-matching model of community organisation' (Jumars, 1976). The basis of this model is that individual benthic organisms create habitat heterogeneity that is exploited by other species. Additional evidence for this model was provided by Thistle et al. (1993) who combined polychaete species into functional groups based on feeding mode and mobility. These authors found that different functional polychaete types displayed positive correlations with specific groups of harpacticoid copepods (Thistle et al., 1993).

In the 'Spatio-temporal Mosaic Model' (Grassle & Morse-Porteous, 1987), the dynamics within which disturbances are operating can be understood by visualising a spatially structured framework of patches creating a mosaic of different assemblages. The way in which these patches fit together will dictate the faunal structure observed over larger scales and bioturbation may be considered as an important patch forming process within such a model. However, before applying disturbance/diversity theories to community data, it is vital to consider the nature of the disturbance used to construct a particular theory. In the 'Spatio-temporal Mosaic Model', disturbance is a discrete event that removes all the organisms in the affected area. Bioturbating animals do not have this effect on macrofaunal animals and the disturbance that they cause is more of a continuous, sub-lethal process that leads to the re-organisation of community structure. Within a community, however, the spatial distribution of large bioturbating organisms is often aggregated. Many bioturbating species also display either periodicity or seasonality in their activities (e.g. Rowden & Jones, 1993). Hence, natural biological disturbance is usually spatially and temporally patchy. Furthermore, the patches are not discrete defaunations of the type more generally considered in the literature (e.g. epifauna feeding pits) (Hall et al., 1991; Thrush et al., 1991). Hence, areas of high bioturbation grade into areas of low or no bioturbation. In any area where they co-occur, bioturbators will have different spatial and temporal distributions. Widdicombe & Austen (1999) concluded that each bioturbating species may have a distinct effect on the fauna around it and, therefore, the existence of a diverse bioturbating assemblage will increase both the total number of species and β diversity. This process appears to be analogous to the Spatio-temporal Mosaic model with areas of different community structure being maintained by the patchy distributions and variable activities of a number of bioturbating organisms.

In a separate study (Austen et al., 1999), meiofaunal samples were taken with the macrofaunal samples. Direct comparisons indicated differences between the meiofaunal response and that of the macrofauna. The response of meiofaunal diversity to disturbance by both *Abra alba* and *Nuculoma tenuis* fitted the IDH (Austen et al., 1999), however, the highest diversities were observed in the low *A. alba* and mid *N. tenuis* treatments. This indicates that the intensity of disturbance considered to be of maximum benefit in promoting diversity is higher for meiofauna than for macrofauna, which is in concurrence with the findings of Warwick et al. (1990). It may be argued that the disturbance intensity in the lowest *A. alba* treatments may have been greater than the intensity required to elicit maximum diversity. This assumes that the processes acting on the meiofauna are the same as those acting on the macrofauna. It is perhaps more plausible that bioturbation acts on the two elements of the community in different ways. We suggest that the major interactions between the bioturbators and the macrofauna derive from physical disruption (e.g. displacement of tubes, burial) and direct competition for resources, whilst the meiofauna will suffer more from direct non-selective predation as well as from secondary effects such as the inhibition of macrofaunal predators and competitors. This is a consequence of the size and ambit of the bioturbating organisms. We propose that whilst both the macro- and meiofaunal response to disturbance is as predicted by the IDH, important differences in the exact mechanism by which competitive pressure is reduced will mean that a single species or type of bioturbating organism may affect the two elements of the fauna differently.

In conclusion, aspects of a number of biodiversity hypotheses may be combined to explain patterns within sediment assemblages influenced by bioturbation. The effects of different bioturbators interact within patch mosaics to create both faunistic and sedimentary heterogeneity. Consequently, when attempting to understand the dynamics of soft sediment communities, it is essential to take into account the identity, abundance and spatial distribution of bioturbating species. Unfortunately, traditional methods of grab sampling greatly underestimate the larger-bodied species (Kendall & Widdicombe, 1999). To obtain full information on bioturbators within an area, large samples must be taken with good spatial referencing.

It is already becoming common practice to adopt a stratified sampling design to include both meio- and macro-benthos in faunal surveys. It seems logical to extend this procedure to the mega-infauna to sample the entire benthic community.

Acknowledgements

This work was enabled by a grant from the EC Large Scales Facility Programme, was funded in part by the UK Ministry of Agriculture, Fisheries and Food (project No AE1113), and is a contribution to the PML Coastal Biodiversity research project. We thank Torgeir Bakke, John Arthur Berge, Liv Berge and Joanna Maloney of NIVA Oslo for helping to facilitate this work, to Håkon Oen, Einar Johannesen and Oddbjørn Pettersen for their technical support at the Solbergstrand Marine Station and to the crew of R.V. Trygve Braarud. We gratefully acknowledge the constructive input of Erik Bonsdorff and an anonymous referee.

References

Austen, M. C. & S. Widdicombe, 1997. Experimental evidence of effects of the heart urchin *Brissopsis lyrifera* on associated subtidal meiobenthic nematode communities. J. exp. mar. Biol. Ecol. 222: 219–238.

Austen, M. C., S. Widdicombe & N. Villano, 1998. Effect of biological disturbance on diversity and structure of a nematode meiobenthic community. Mar. Ecol. Prog. Ser. 174: 233–246.

Berge, J. A., M. Schaanning, T. Bakke, K. Sandøy, G. M. Skeie & W. G. Jr. Ambrose, 1986. A soft-bottom sublittoral mesocosm by the Oslofjord: description, performance and examples of application. Ophelia 26: 37–54.

Branch, G. M., 1984. Competition between marine organisms: Ecological and evolutionary implications. Oceanogr. mar. Biol. ann. Rev. 22: 429–593.

Brenchley, G. A., 1981. Disturbance and community structure: an experimental study of bioturbation in marine soft-bottom environments. J. mar. Res. 39: 767–790.

Clarke, K. R., 1993. Non-parametric multivariate analyses of changes in community structure. Aust. J. Ecol. 18: 117–143.

Connell, J. H., 1978. Diversity in tropical rain forests and coral reefs. Science 199: 1302–1309.

De Ridder, C. & J. M. Lawrence, 1982. Food and feeding mechanisms: Echinoidea. In: Jangoux, M. & J. M. Lawrence (eds), Echinoderm Nutrition. A.A. Balkema, Rotterdam: 57–115.

Eagle, R. A., 1975. Natural fluctuations in a soft bottom benthic community. J. mar. biol. Ass. U.K. 55: 865–878.

Grassle, J. F. & H. L. Sanders, 1973. Life histories and the role of disturbance. Deep Sea Res. 20: 643–659.

Grassle, J. F. & L. S. Morse-Porteous, 1987. Macrofaunal colonization of disturbed deep-sea environments and the structure of deep-sea benthic communities. Deep Sea Res. 34: 1911–1950.

Hall, S. J., 1994. Physical disturbance and marine benthic communities: life in unconsolidated sediments. Oceanogr. mar. Biol. ann. Rev. 32: 179–239.

Hall, S. J., D. J. Basford, M. R. Robertson, D. G. Raffaelli & I. Tuck, 1991. Patterns of recolonisation and the importance of pit-digging by the crab *Cancer pagurus* in a subtidal sand habitat. Mar. Ecol. Prog. Ser. 72: 93–102.

Huston, M. A., 1979. A general hypothesis of species diversity. Am. Nat. 113: 81–101.

Huston, M. A., 1994. Biological Diversity. Cambridge University. 681 pp.

Jumars, P. A., 1976. Deep-sea species diversity: does it have a characteristic scale? J. mar. Res. 34(2): 217–246.

Kendall, M. A. & S. Widdicombe, 1999. Small scale patterns in the structure of macrofaunal assemblages of shallow soft sediments. J. exp. mar. Biol. Ecol. 237: 127–140.

Kukert, H., 1991. *In situ* experiments on response of deep-sea macrofauna to burial disturbance. PAC-SCI 45(1): 95 pp.

Menge, B. A. & J. P. Sutherland, 1987. Community regulation: variation in disturbance, competition and predation in relation to environmental stress and recruitment. Am. Nat. 130: 730–757.

Nash, R. D. M., C. J. Chapman, R. J. A. Atkinson & P. J. Morgan, 1984. Observations on the burrows and burrowing behaviour of *Caocaris macandreae* (Crustacea: Decapoda: Thalassinoidea). J. Zool. Lond. 202: 425–439.

Nichols, D., 1959. Changes in the chalk heart-urchin *Micraster* interpreted in relation to living forms. Phil. Trans. r. Soc. B., Vol. 242: 347–437.

Petraitis, P. S., R. E. Latham & R. E. Niesenbaum, 1989. The maintenance of species diversity by disturbance. Quart. Rev. Biol. 64: 393–418.

Pickett, S. T. A. & P. White (eds), 1985. The Ecology of Natural Disturbance and Patch Dynamics. Orlando, Florida: Academic Press.

Posey, M. H., 1986. Changes in a benthic community associated with dense beds of a burrowing deposit feeder, *Callianassa californiensis*. Mar. Ecol. Prog. Ser. 31: 15–22.

Posey, M. H., B. R. Dumbauld & D. A. Armstrong, 1991. Effects of a burrowing mud shrimp, *Upogebia pugettensis* (Dana), on abundances of macro-infauna. J. exp. mar. Biol. Ecol. 148: 283–294.

Posey, M., W. Lindberg, T. Alphin & F. Vose, 1996. Influence of storm disturbance on an offshore benthic community. Bull. mar. Sci. 59: 523–529.

Rowden, A. A. & M. B. Jones, 1993. Critical evaluation of sediment turnover estimates for Callianassidae (Decapoda: Thalassinidea). J. exp. mar. Biol. Ecol. 173: 265–272.

Thayer, C. W., 1983. Sediment-mediated biological disturbance and the evolution of marine benthos. In Tevesz, M. J. S. & P. L. McCall (eds), Biotic Interactions in Recent and Fossil Benthic Communities. New York: Plenum Press: 479–625.

Thistle, D., B. Hilbig & J. E. Eckman, 1993. Are polychaetes sources of habitat heterogeneity for harpacticoid copepods in the deep sea? Deep Sea Res. I. 40: 151–157.

Thrush, S. F., R. D. Pridmore, J. E. Hewitt & V. J. Cummings, 1991. Impact of ray feeding disturbances on sandflat macrobenthos: do communities dominated by polychaetes or shellfish respond differently? Mar. Ecol. Prog. Ser. 69: 245–252.

Warwick, R. M., H. M. Platt, K. R. Clarke, J. Agard & J. Gobin, 1990. Analysis of macrobenthic and meiobenthic community structure in relation to pollution and disturbance in Hamilton Harbour, Bermuda. J. exp. mar. Biol. Ecol. 138: 119–142.

Widdicombe S. & M. C. Austen, 1998. Experimental evidence for the role of *Brissopsis lyrifera* (Forbes, 1841) as a critical species in the maintenance of benthic diversity and the modification of sediment chemistry. J. exp. mar. Biol. Ecol. 228: 241–255.

Widdicombe, S. & M. C. Austen, 1999. Mesocosm investigation into the effects of bioturbation on the diversity and structure of a subtidal macrobenthic community. Mar. Ecol. Prog. Ser. 189: 181–193.

Hydrobiologia **440:** 379–391, 2000.
M.B. Jones, J.M.N. Azevedo, A.I. Neto, A.C. Costa & A.M. Frias Martins (eds), Island, Ocean and Deep-Sea Biology.

Detecting anthropogenic disturbance in an environment with multiple gradients of physical disturbance, Manukau Harbour, New Zealand

J.I. Ellis[1], D.C. Schneider[2] & S.F. Thrush[1]
[1]*National Institute of Water & Atmospheric Research Ltd., P.O. Box 11-115, Hamilton, New Zealand*
Fax: +64-7-8560151. Tel: +64-7-8567026 (ext 849). E-mail: j.ellis@niwa.cri.nz
[2]*Ocean Sciences Centre, 4 Clark Place, Memorial University of Newfoundland, St. John's, Newfoundland, Canada A1C 5S7*

Key words: sampling design, macrobenthic community, gradient analysis, organic enrichment, physical disturbance

Abstract

Demonstrating spatial or temporal gradients of effects on macrobenthic communities can be a useful way of providing strong empirical evidence of natural or anthropogenic disturbance. Gradient designs for environmental assessment are sensitive to change for point source data, enabling the scale of the effects of a disturbance to be readily identified. If the spatial scale that is sampled from the point source is adequate, problems of selecting control sites can be avoided. However, sources of spatial variation in macrobenthic communities, which are not related to the impact, can confound the use of gradient designs. This can occur if the natural spatial structure overlaps that of the gradient and cannot be identified either as a location or environmental covariable. The ability to detect point source impacts using a gradient design against natural spatial variability was tested using benthic macrofaunal data collected from Manukau Harbour, New Zealand. Treated sewage wastewater is discharged into the north-west area of the Manukau Harbour. Sandflats in the vicinity of the outfall are also subject to physical disturbance from wind-waves and strong tides. Ordination techniques and the testing of *a priori* predictions were used to try and separate the relative effects of organic and physical disturbance on the benthic communities. While the occurrence of other environmental disturbances along a gradient of anthropogenic disturbance makes interpretation of community pattern more difficult, the use of a gradient sampling layout, ordination analysis and the testing of *a priori* predictions enabled impacts of the anthropogenic and natural environmental disturbances to be interpreted. Gradient designs, therefore, provide a method of assessing complex impacts that operate over broad spatial and temporal scales.

Introduction

The development of sampling designs for impact assessment has been based on a stratified design where random samples are collected within strata (treatment and controls). These Before After Control Impact (BACI) designs are based on the principle that if two locations (control and impact) are monitored before an anthropogenic disturbance, the impact location will show a different pattern after the disturbance than will the control location. The evolution of stratified designs has been considered in full previously (Underwood, 1992, 1993).

Ellis & Schneider (1997) highlight problems when stratified BACI designs are applied to detect point source disturbances where the scale of the disturbance is unknown. If a stratified BACI design is applied to point source disturbances, placement of control sites becomes problematic as the control areas must be far enough away that they will not be affected by the potential disturbance, and yet close enough that the areas are comparable. Furthermore, the ability to assess the spatial scale of a disturbance with time is reduced by sampling an impact and a control stratum rather than quantifying the gradient of contamination (Ellis & Schneider, 1997). When a contaminant disperses with distance from a point source, gradient designs can be more sensitive to change than randomised placement of samples. Gradient designs also enable the scale of a disturbance to be readily identified and, given sufficient spatial extent, a gradient layout avoids the problem of selecting adequate control sites. Another advantage of gradient designs is that they lend themselves to constructing mechanistic models that generate testable predictions of attenuating effects with increasing distance.

Gradient designs, however, assume that natural spatial patterns that occur will not confound the ability to detect an anthropogenic disturbance. The presence of large salinity gradients, sedimentary particle size gradients or other environmental gradients associated with an anthropogenic disturbance can distort the pattern of ecological change observed. Pearson & Rosenberg (1978) also noted that in estuaries, or other areas where there are salinity gradients, organisms may be exposed to the combined stress of organic enrichment and reduced and/or fluctuating salinities. Separation of the relative effects of these two factors is often difficult, even more so when other environmental factors (e.g. variation in tidal flow, sedimentary structure, temperature, etc.) also influence the species distribution. Furthermore, as the scale of a study or environmental impact increases, complex interactions or encountering spatial variation not related to the impact are more likely to occur. This is an issue for most sampling designs. For example, broad-scale environmental impacts mean block control impact designs are likely to have increased difficulties in finding adequate control sites and are more likely to encompass a broader range of changing environmental variables. Gradient designs are more likely to encompass spatial variation not related to the impact.

Many environmental impacts are occurring over increasingly large scales. For example, runoff due to changes in land use, dredging and modification of coastlines have resulted in alterations in rates of sedimentation and the extent of coastal areas that are depositional environments. In recent years, there has been growing recognition of the potential for changes in marine ecosystems as a consequence of commercial fishing resulting in broad-scale changes to continental shelf regions (Dayton et al., 1995; Jennings & Kaiser, 1998; Watling & Norse, 1998; Auster & Langton, 1999). BACI designs have been proposed to detect changes in single response variables, which are particularly useful for documenting localised impacts. However, in order to study broad scale disturbance, it is imperative that we supplement these techniques with time-series data as well as by developing new concepts and statistical tools (Thrush et al., 1999; Ellis et al., 2000). Demonstrating spatial or temporal gradients of effects on macrobenthic communities can be a useful way of providing strong empirical evidence of effects where the scale of disturbance is unknown (Ellis & Schneider, 1997). Determining broad-scale effects may be more difficult, however appropriate gradient designs can enable such disturbances to be assessed. Hence, we have tested the utility of these designs to detect an anthropogenic impact against natural spatial variability when multiple gradients of disturbance occur.

Data of benthic macrofaunal abundance from the vicinity of the Manukau Harbour sewage outlet, New Zealand, were collected to test this assumption. Manukau Harbour is a dynamic and rigorous environment (Turner et al., 1995). This harbour is subject to physical wind-wave and tidal disturbance (Commito et al., 1995; Thrush et al., 1996), as well as the anthropogenic disturbance created by the sewage outlet. The organic enrichment gradient decreases with distance from the outfall while, in contrast, the wind-wave disturbance increases with distance (Bell et al., 1998). Adequately assessing spatial and temporal variability across environmental gradients requires consideration of the sampling grain, interval and extent. A trade off between replication to describe site variability *versus* sampling more distances requires careful consideration preferably from what is known about the scale of the pattern or process, e.g. from a pilot study (Thrush et al., 1994; Legendre & Legendre, 1998).

Canonical correspondence analysis can be used to separate effects of multiple gradients. Another method is to make *a priori* predictions (Hewitt et al., 1996) of the expected community pattern as a response to the various gradients. The use of ordination techniques and the testing of *a priori* predictions were used to try and separate the relative effects of organic and physical disturbance on the benthic communities. The aim of this paper is, therefore, to determine how effective gradient designs are for detecting an anthropogenic disturbance in an environment with natural physical gradients operating over similar spatial scales to those of potential environmental effects.

Predictions

Organic enrichment

Pearson & Rosenberg (1978) reviewed 47 publications of effects of organic enrichment and pollution on marine benthic communities. A consistent pattern of faunal changes observed along a gradient of increasing organic input to marine sediments was observed.

Close to a disturbed area, high densities of small, fast growing and rapidly colonising opportunistic species reach high abundance. Moving further from the disturbed area, a transition zone occurs, which is still dominated by opportunistic species although not at such high densities. At the end of the successional trajectory, a diverse assemblage dominated by suspension feeders and large and slow growing burrowing organisms is found. Further studies of effects of organic enrichment on marine communities support the community change described by Pearson & Rosenberg (1978) (Mirza & Gray, 1981; Essink, 1984; Whitlatch & Zajac, 1985; Pearson et al., 1986; Weston, 1990); however, Pearson & Rosenberg (*op. cit.*) also note that in areas where there are other natural gradients this species distribution may be altered. The Manukau is a physically dynamic environment. Physical disturbance is known to affect benthic community composition (see Prediction 4). Hence, we did not expect to see a diverse assemblage comprised of large slow growing species at the far sites that were affected by wind-wave disturbance. We, therefore, predicted the following: (1) The effects of organic enrichment on community structure will be most pronounced in the vicinity of the outfall and decrease progressively with distance from the discharge source due to mixing and dilution. (2) Consistent patterns to the faunal changes associated with an organic enrichment gradient occur. High numbers of a few small opportunistic (tolerant) species will occur adjacent to the discharge source. A maximum in the number of species will be reached with distance from the outfall, however, due to physical disturbance, we did not expect the far sites to be dominated by large and slow growing organisms. (3) A decline in suspension feeding organisms and an increase in deposit feeding organisms will occur as organic input increases.

Physical wind-wave disturbance

The hydrodynamic regime (tidal currents and waves) largely determines the sedimentary characteristics of an area. Hydrodynamics determine the nature of the bottom substratum, influence the stability of the sediment, and affect the nature of the food supply for benthic organisms (Sanders, 1958; Warwick & Uncles, 1980). In shallow-water coastal systems, environmental factors such as bottom topography, sediment characteristics and hydrodynamic processes play a definite role in structuring benthic communities and producing patchiness through a variety of mechanisms. Biotic interactions have traditionally been considered in the context of 'static' physical factors such as bottom topography and roughness, sediment characteristics and tidal inundation and exposure which in turn are largely determined by hydrodynamic processes (Bell et al., 1997). The importance of hydrodynamic variables such as current velocities, bed shear stress and wind-wave activity have also been recognised as influencing larval settlement and post-settlement transport (Grant, 1983; Commito et al., 1995), availability of particulate food resources (Warwick & Uncles, 1980) and sediment stability of the substratum (Warwick et al., 1991). Quantifying the linkages and interactions between physical factors and benthic community structure is difficult and continues to be debated (Hall, 1994; Raffaelli et al., 1994). However, significant differences in density, diversity and species composition of benthic communities have been correlated with various physical factors most of which can be related to exposure (Eleftheriou & Nicholson, 1975; Dexter, 1983, 1984). The relationship between the number of species observed and the degree of exposure to wave action is evident, with increases in species richness occurring with decreasing exposure to wave action (Dexter, 1992). Density, diversity and dominance of polychaetes relative to crustaceans also increase along this gradient. Similar correlations between physical parameters and faunal attributes have been noted in other comparisons of sandy beach communities (McLachlan et al., 1981; Dexter, 1983; Brown & McLachlan, 1990). Therefore, we predicted: (4) Species diversity, total abundance and species richness is reduced with increasing exposure to wave action.

Materials and methods

Manukau Harbour is located adjacent to Auckland city on the west coast of the North Island of New Zealand. Treated sewage wastewater is discharged from a system of oxidation ponds into the north-west area of the Harbour. The Wastewater Treatment Plant discharges c. 3.9 $m^3 s^{-1}$ (Vant & Williams, 1992). Sampling sites were arranged along two transects taken with distance from the outfall (Fig. 1). Five sites were sampled on transect one, and six sites were taken on transect two. At each site, five core samples (10 cm diameter by 15 cm depth) were taken at 10 m intervals. Samples were sieved (500 μm mesh), preserved in 70% isopropyl alcohol and stained with 0.1% Rose Bengal. In the laboratory, macrofauna were sorted, identified to

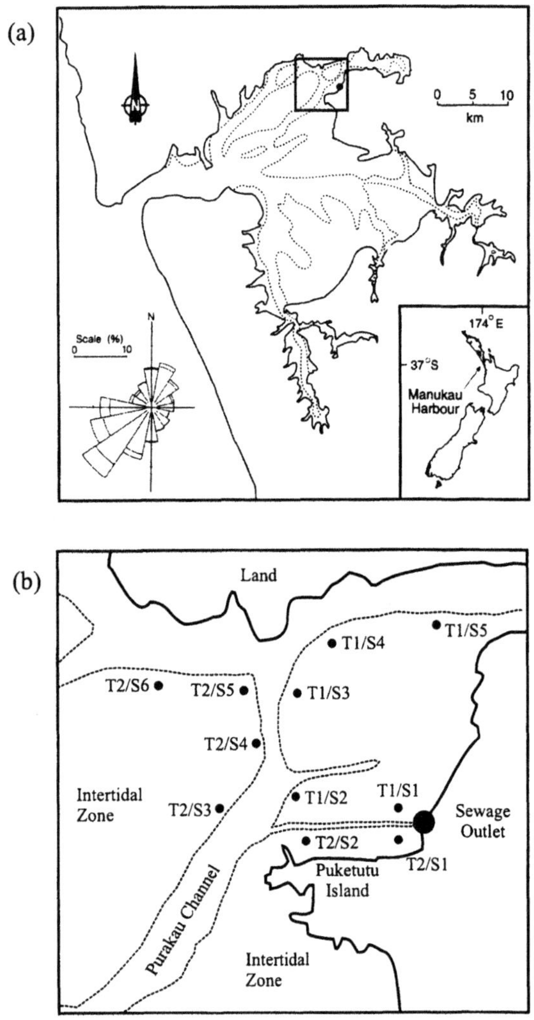

Figure 1. (a) Manukau Harbour, North Island, New Zealand, showing the location of the Manukau sewage outfall. Dotted line indicates area of sand flat exposed at spring low tide. Wind rose data are for Auckland International Airport (Manukau Harbour); (b) position of the sampling sites with distance from the Manukau sewage outfall (T = Transect, S = Site).

the lowest possible taxonomic level, counted and preserved in 70% IPA. Surficial sediment (0–2 cm) was collected at the same location as each core sample. Replicate samples from a particular sampling location were pooled to assess sediment particle size and organic matter content. Sediment samples for particle size analysis were digested in 6% hydrogen peroxide for 24 h. Samples were then wet sieved (2 mm, 500 μm, 125 μm and 63 μm sieves) and each fraction was dried at 60 °C in an oven and weighed. Organic matter was determined from ash-free dry weight. Samples were combusted for 6 h at 400 °C.

To assess wind-wave disturbance, tidal current values, as well as wave orbital speeds at the bed, were calculated. This is because peak tidal currents in isolation are incapable of eroding bottom sediments, but in combination with near-bed orbital currents generated by only small wind waves, sediment transport can be initiated (Bell et al., 1997). Depth-averaged water velocity (cm/s) values for mean ebb and flood tides were generated from an oceanographic model of the harbour's tidal regime (Bell et al., 1998). A numerical model WGEN (Black, 1997) was used to predict wave growth subject to wind stress. The model generates time series of wave height and wave period calculated from an input record of wind measurements or wind rose. The model can account for spatially variable bathymetry, shoaling, wave dissipation by bed friction and depth limited wave breaking. Linear wave theory was used to compute the orbital speed at the bed ($U_b{}^3$, cm/s).

Data analysis

Changes in macrofaunal abundance as a function of distance from the outfall, transect, replicate (within site variation) and interaction terms were analyzed using general linear models. Distance was treated as a continuous variable. The statistical package SAS was used to calculate p values under the assumption of independent residuals with equal variance. Residuals were plotted against expected values. If no association between the residuals and expected values were evident, the model was assumed to be an acceptable description of the data (Draper & Smith, 1981).

The variability in macrofaunal community structure in relation to the changing wind-wave disturbance and organic loading gradients were examined using multivariate analyses. Data were analyzed using non-metric multidimensional scaling (MDS), correspondence analysis (CA), canonical correspondence analysis (CCA) and partial canonical correspondence analysis (PCCA). Both CA and non-metric MDS gave similar results, therefore, only MDS is presented. Non-metric MDS was performed on the species abundance matrices using PRIMER (Clarke, 1993). Canonical correspondence analysis and PCCA were used to identify relationships between the community structure and the environmental variables using CANOCO (Ter Braak, 1986, 1987). Environmental variables included the organic content, sediment grain size (% <63 μm, 63 μm, 125 μm, 500 μm, 2 mm), wave orbital speed at the bed, and the

Table 1. ANOVA between environmental variables and macrofauna for Manukau Harbour

Source	DF	*F* Value	Pr > *F*
Distance	1	43.46	0.0001
Transect	1	1.32	0.25
Replicate	1	0.24	0.6218
Dis*Trans	1	0.16	0.6911
Trans*Rep	1	1.21	0.2723
Dis*Rep	1	0.24	0.6222
Dis*Trans*Rep	1	0.81	0.3693

n = 3,795.
SSTot = 400 070.6.

mean depth averaged flow speed of ebb and flood tides (cm/s).

Another method to identify species responsible for changes in community structure is to calculate the magnitude and sign of spatial gradients. Analysis of Covariance (ANCOVA) for species abundance as a function of distance was used to test whether gradients were heterogeneous among species. A significant interaction term indicates whether species differ in pattern of changes in abundance with distance from the pollution source. Regression coefficients were calculated for all species and then ranked. A negative coefficient indicated tolerant species, a positive coefficient indicated sensitive species. Mid-ranking species showed little or no gradient relative to the contamination source. The statistical model to calculate species coefficient was

$$Y = \beta_o + \beta_{Dis} + \text{Dis} + \epsilon,$$

where Y = benthic densities, Dis = distance and ϵ = residuals.

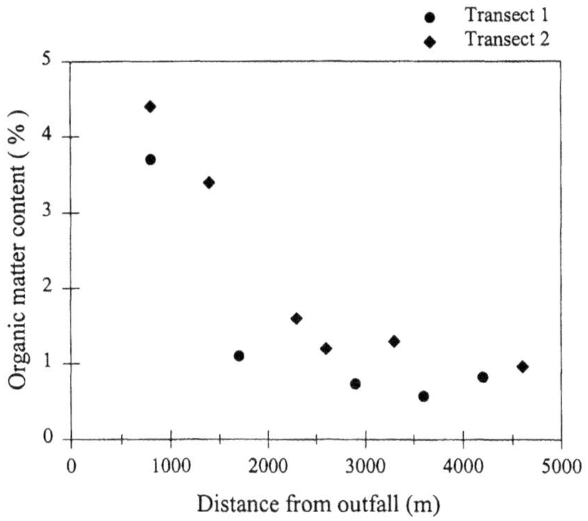

Figure 2. Change in sediment organic content as a function of distance from the Manukau Harbour sewage outfall.

Table 2. Depth-averaged mean tidal currents and orbital speed at the bed for Manukau Harbour survey sites

Transect	Distance (m)	Peak ebb (cm/s) mean tide	Peak flood (cm/s) mean tide	U_b^3 (cm/s)
1	800	25	16	9.1
	1700	30	36	14.0
	2900	19	30	7.6
	3600	25	27	16.0
	4200	26	23	5.0
2	800	16	9	2.4
	1400	25	35	1.5
	2300	44	38	34.0
	2600	43	19	33.0
	3300	33	9	12.0
	4600	31	19	14.0

Results

Analysis of variance indicated that distance from the sewage outfall was the only explanatory variable tested that was related to density of benthic macrofaunal organisms (Table 1). Changes in macrofaunal abundance for transect, replicate and interaction terms were not significant.

Organic content of the sediment was elevated within 1400 m (T1/S1, T2/S1, T2/S2) of the outfall (Fig. 2). Sediment grain size indicated an increase in wind-wave disturbance with distance from the outfall (Fig. 3). Mud content (silt/clay fraction) was elevated within 1400 m of the outfall. These sites are sheltered from the predominant wind-wave disturbance by Puketutu Island (Fig. 1b). The percentage of sand (125 μm – 2 mm) and tidal velocity (Table 2) increased for sites beyond Puketutu Island.

Total abundance of benthic organisms decreased as a function of distance from the outfall (Fig. 4). The relationship between number of species and distance was not a simple linear function (Fig. 4). Species richness was reduced at sites 1 and 2 within 1400 m of the outfall. High numbers of a few tolerant species characterized these sites. There appeared to be a peak of high diversity at site 3, transect 2 (distance=2300 m). Diversity dropped to levels similar to sites adjacent to the outfall at the far sites (4200 m and 4600 m).

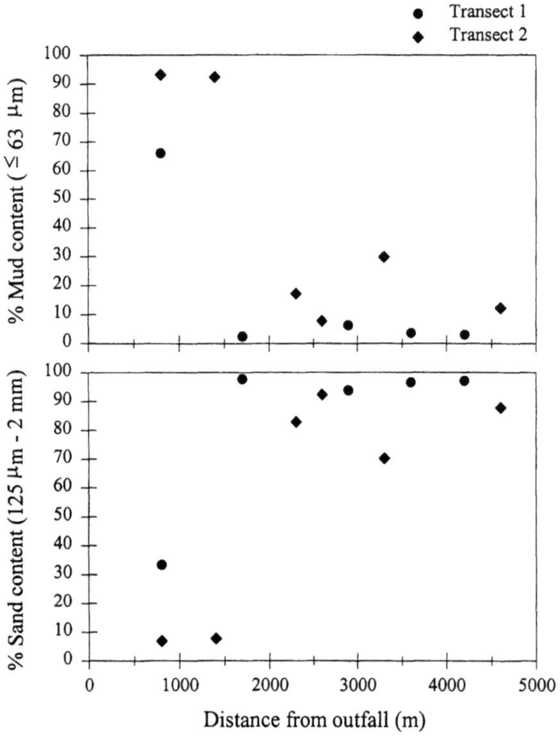

Figure 3. Change in sediment grain size with distance from the Manukau Harbour sewage outfall.

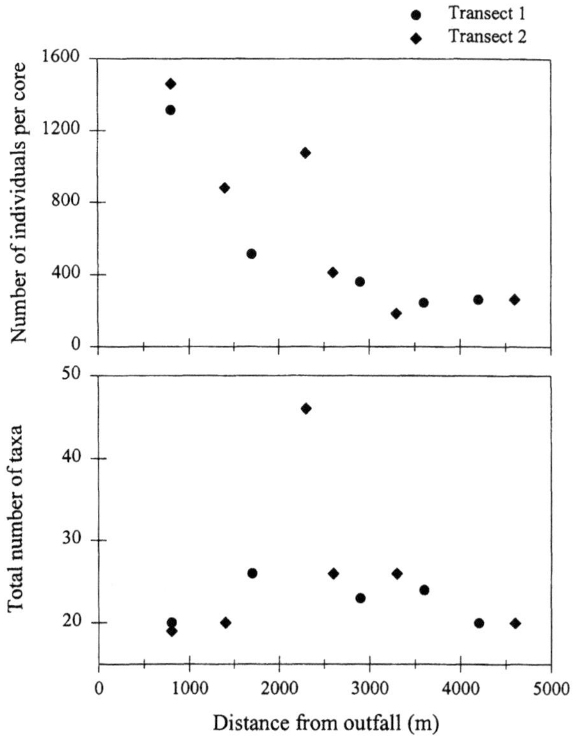

Figure 4. Change in macrofaunal abundance and species richness with distance from the Manukau Harbour sewage outfall. Note that standard error bars cannot be seen as they were less than the symbol size.

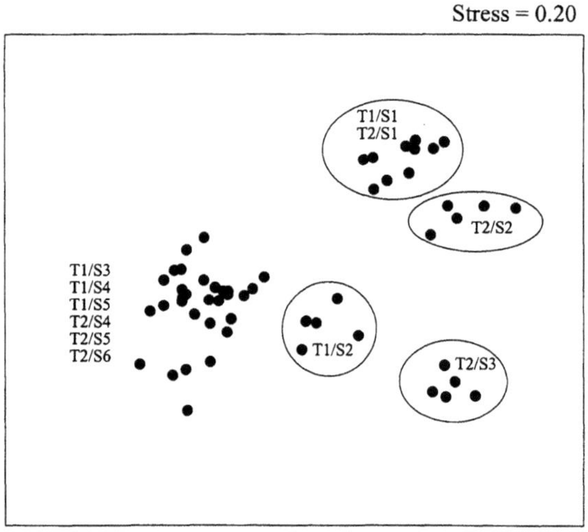

Figure 5. Non-metric MDS ordination by station of the macrobenthos species abundance data for Manukau Harbour sewage outfall (T = Transect, S = Site).

MDS grouped sites as a function of the organic and physical disturbance (Fig. 5). Sites with high organic content, low wind-wave disturbance (high mud content) and low tidal velocity were grouped together (T1/S1, T2/S1, T2/S2). Unpolluted sites with high physical disturbance (low organic content, high% sand) were grouped together.

The first four axes of the initial unconstrained correspondence analysis accounted for 42% of the variation in the species site data. Figure 6 indicates the position of the sites on the first two axes. Figure 7 shows the position of the species on the first two axes. Partial canonical correspondence analysis suggests that variability was related significantly to organic loading and physical disturbance or sediment characteristic covariables (correlation = 0.9325, F-ratio = 14.01, $P \approx 0.00$). Organic loading accounted for 22% of the variability in community structure. Species identified as occurring in mud-enriched habitats were: *Heteromastus filiformis*, *Aquilaspio aucklandica*, *Arthritica bifurca*, *Boccardia syrtis*, *Torridaharpinia hurleyi*, Nereidae and *Halicarcinus whitei* and so on (Fig. 7). Although these animals occur throughout the Manukau Harbour, they were found at higher densities close to the outfall. These groupings are consistent with the species groupings identified using covariance analysis or plots of sensitive and tolerant species as discussed below.

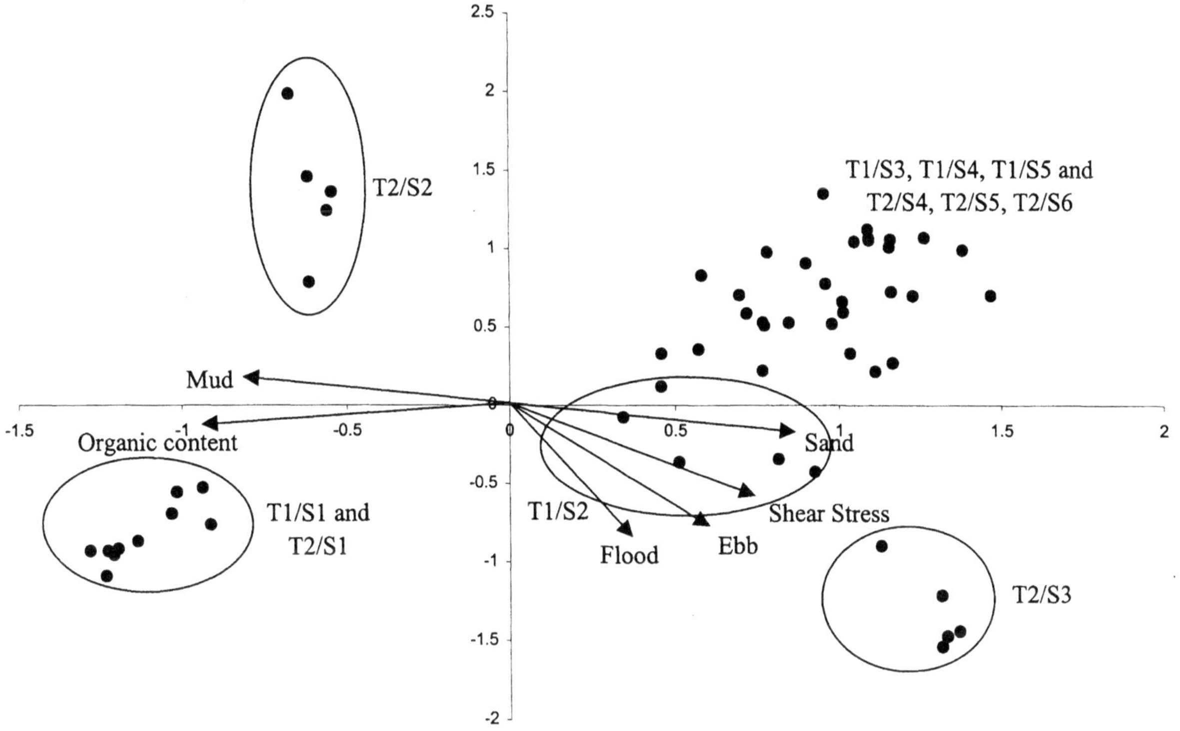

Figure 6. Canonical correspondence analysis of environmental variables (lines) and sampling sites (•) using macrofaunal abundance data from Manukau Harbour (T=Transect, S=Site).

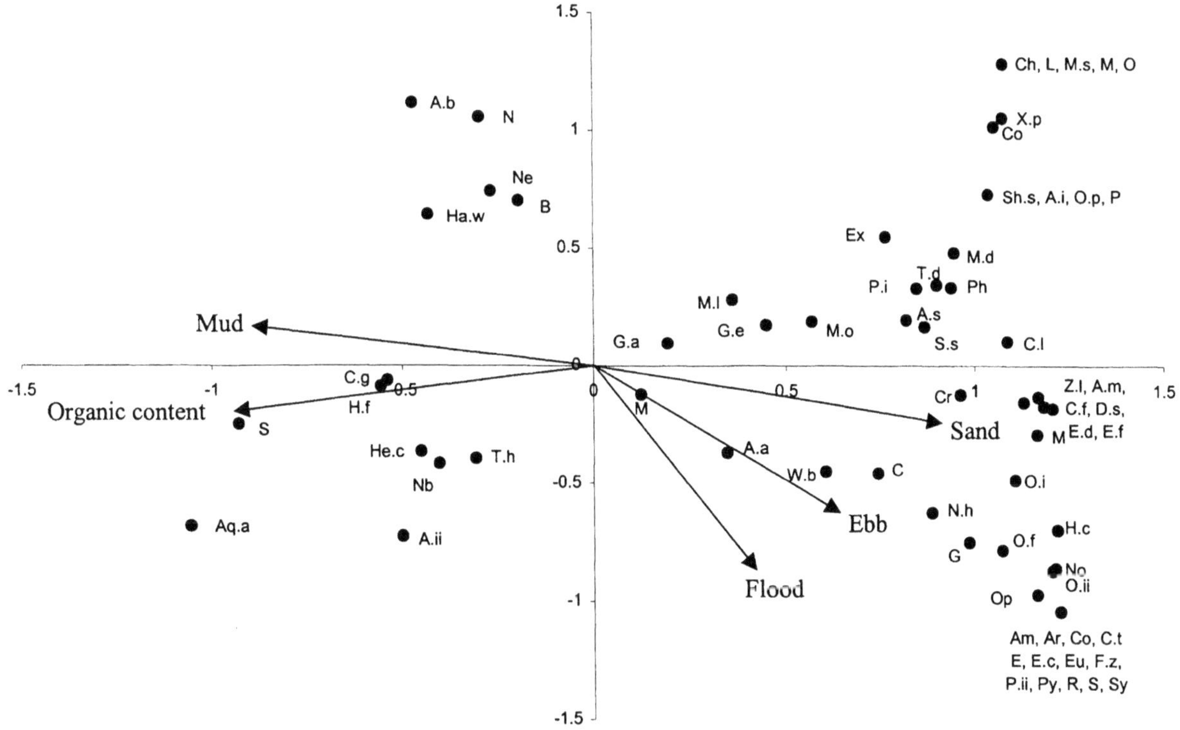

Figure 7. Canonical correspondence analysis of environmental variables (lines) indicating species positions for Manukau Harbour. Species codes are provided in Table 3.

Table 3. Species information and strength of density gradient, as estimated by regression for Manukau Harbour (n = 3795)

Species		Average	Sum	Coefficient	Feeding Guild
Heteromastus filiformis	H.f	30.38	1671	−0.02230	Surface deposit feeder, motile, non-jawed
Aquilaspio aucklandica	Aq.a	16.89	929	−0.01637	Surface deposit feeder, discretely motile, tentaculate
Arthritica bifurca	A.b	9.76	537	−0.00977	Suspension feeder
Boccardia syrtis	B	10.55	580	−0.00934	Surface deposit feeder, discretely motile, tentaculate
Torridaharpina hurleyi	T.h	3.55	195	−0.00252	
Owenia fusiformis	O.f	8.89	489	−0.00215	Filter &deposit feeder, discretely motile, tentaculate
Nucula hartivigiana	N.h	6.40	352	−0.00123	Deposit feeder
Nereididae	Ne	1.51	83	−0.00106	Carnivore, motile, jawed
Halicarcinus whitei	Ha.w	1.11	61	−0.00100	
Nemertean	N	0.96	53	−0.00076	
Waitangi brevirostris	W.b	2.25	124	−0.00059	
Amphipod ii	A.ii	0.33	18	−0.00022	
Helice crassa	He.c	0.36	20	−0.00020	Deposit feeder
Cominella glandiformis	C.g	0.25	14	−0.00017	Carnivore
Anthopleura aureoradiata	A.a	0.58	32	−0.00016	
Ostracod ii	O.ii	1.65	91	−0.00014	
Austrovenus stutchburyi	A.s	2.51	138	−0.00014	Suspension feeder
Notoacmea spp.	No	1.53	84	−0.00011	
Glycera americana	G.a	0.25	14	−0.00007	Carnivore
Soletellina siliqua	S.s	0.85	47	−0.00003	Deposit feeder
Nebalacea	Nb	0.20	11	−0.00002	
Scolecolepides sp.	S	0.05	3	0.00000	Surface deposit feeder
Methalimendon sp.	M	0.44	24	0.00001	
Ophelliidae	Op	0.15	8	0.00002	Non-selective deposit feeder
Mactra ovata	M.o	0.16	9	0.00002	
Gastropod i	G.a	0.07	4	0.00002	
Exosphaeroma chilensis	E.c	0.11	6	0.00003	
Ostracod i	O.i	0.16	9	0.00003	
Cyclaspis thomsoni	C.t	0.09	5	0.00003	
Cirolana sp.	C	0.02	1	0.00004	
Phoronid	Ph	0.11	6	0.00004	Surface deposit feeder, motile, non-jawed
Halicarcinus cookii	H.c	0.09	5	0.00004	
Scaleworm	S	0.04	2	0.00004	Carnivore, motile, jawed
Ruditapes sp.	R	0.04	2	0.00004	
Corophid	Co	0.04	2	0.00004	
Edwardsii	E	0.04	2	0.00004	
Phyllodocid	Py	0.02	1	0.00005	Carnivore
Syllid	Sy	0.02	1	0.00005	Carnivore
Euchone	Eu	0.02	1	0.00005	Filter feeder, sessile, tentaculate
Aricidea	Ar	0.02	1	0.00005	Non-selective sub-surface deposit feeder
Aminotrypan	Am	0.02	1	0.00005	
Felaniella zelandica	F.z	0.02	1	0.00005	
Paraonid ii	P.ii	0.02	1	0.00005	Non-selective sub-surface deposit feeder
Exosphaeroma falcatum	E.f	0.02	1	0.00005	
Zeacumantus lutulentus	Z.l	0.02	1	0.00005	Surface deposit feeder
Hermit crab	Cr	0.04	2	0.00005	
Goniada emerita	G.e	0.73	40	0.00005	Carnivore, dicretely motile, jawed
Macroclymenella stewartensis	M.s	0.02	1	0.00006	Sub-surface deposit feeder
Lepidodontidae	L	0.02	1	0.00006	
Eteone durantiaca	E.d	0.05	3	0.00006	Carnivore

Continued on p. 387

Table 3. contd.

Chitin	Ch	0.04	2	0.00007	
Maldonidae	M	0.04	2	0.00007	
Exogonidae	Ex	0.02	1	0.00007	Carnivore
Aglaphamus macroura	A.m	0.04	2	0.00007	Carnivore, motile, non-jawed
Cominella fusiformis	C.f	0.04	2	0.00007	
Xymene plebeius	X.p	0.04	2	0.00007	Carnivore
Amphipod i	A.i	0.02	1	0.00007	
Polydora sp.	P	0.02	1	0.00007	Filter & surface feeder
Sphaerosyllis semiverrocosa	Sh.s	0.02	1	0.00007	Carnivore, motile, non-jawed
Oligochaete	O	0.05	3	0.00007	
Microspio sp.	M	0.11	6	0.00008	Surface deposit feeder
Diloma subrostrata	D.s	0.49	27	0.00008	
Paraonid i	P.i	0.05	3	0.00009	
Colurostylis lemurum	C.l	0.18	10	0.00010	
Orbinia papillosa	O.p	0.04	2	0.00010	Non-selective sub-surface deposit feeder
Trochodota dendyi	T.d	0.11	6	0.00011	
Cossura sp.	Co	0.49	27	0.00030	Sub-surface deposit feeder, motile, non-jawed
Macomona liliana	M.l	5.93	326	0.00053	Surface deposit feeder
Magelona dakini	M.d	15.73	865	0.00770	Surface deposit feeder, discretely motile, tentaculate

Another method of determining which species accounted for the change in community structure, is to rank density gradients from negative to positive (Table 3). The negative coefficients indicate the species most tolerant to the contamination gradient. Species identified as tolerant, with high abundance adjacent to the outfall in enriched areas, were *Heteromastus filiformis*, *Aquilaspio aucklandica*, *Arthritica bifurca*, *Boccardia syrtis*, *Torridaharpinia hurleyi*, *Owenia fusiformis*, Nereidae, Nemertean etc. Species identified as sensitive were *Magelona dakini*, *Macomona liliana*, *Cossura* sp., *Trochodota dendyi*, *Orbinia papillosa*, *Colurostylis lemurum* etc. Benthic organisms identified as sensitive or tolerant from the regression coefficients were plotted to investigate biological change associated with the enrichment and physical disturbance gradient. Figure 8 provides examples of three tolerant species identified using regression analysis. *Heteromastus filiformis*, *Boccardia syrtis* and *Aquilaspio aucklandica* occur at high densities within 1700 m of the outfall. These sites are those identified as being enriched with low physical disturbance. Figure 9 provides examples of two sensitive species, *Magelona dakini* and *Cossura* sp. These species occur at sites identified as having low organic content and high physical disturbance. The species identified as sensitive or tolerant to organic enrichment using density gradients are the same as those identified using canonical correspondence analysis. Where possible, taxa were allocated to feeding guilds principally following Fauchald & Jumars (1979) with further information provided by Pridmore et al. (1990). A change from suspension feeders to deposit feeders along a gradient of increasing organic enrichment was not observed (Table 3).

Discussion

Prediction 1

As expected, elevated levels of organic content were most pronounced in the vicinity of the outfall. However, a progressive decrease in the level of organic enrichment with distance did not occur. Wind-wave disturbance created a sharp boundary of enriched *versus* non-enriched areas, rather than the predicted gradient of attenuation to background levels. The plots of organic content and sediment grain size for Manukau Harbour indicate that elevated levels of organic enrichment occur within 1400 m of the outfall. These sites were sheltered from the predominant south-west wind-wave disturbance by Puketutu Island. Beyond Puketutu Island, the wind-wave disturbance and tidal currents increase. The hydrodynamic regime largely determines the sedimentary characteristics of an area (Sanders, 1958; Warwick & Uncles, 1980). Water

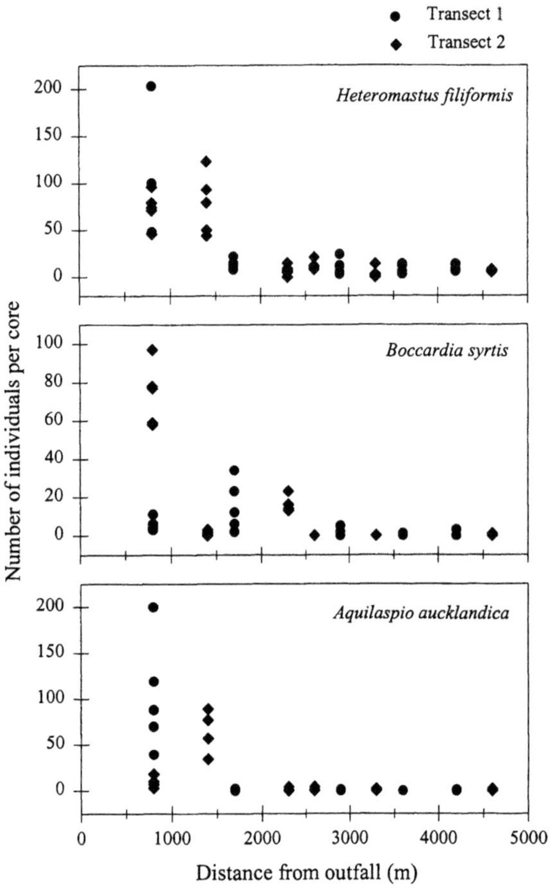

Figure 8. Change in abundance of three tolerant species with distance from the Manukau Harbour sewage outfall.

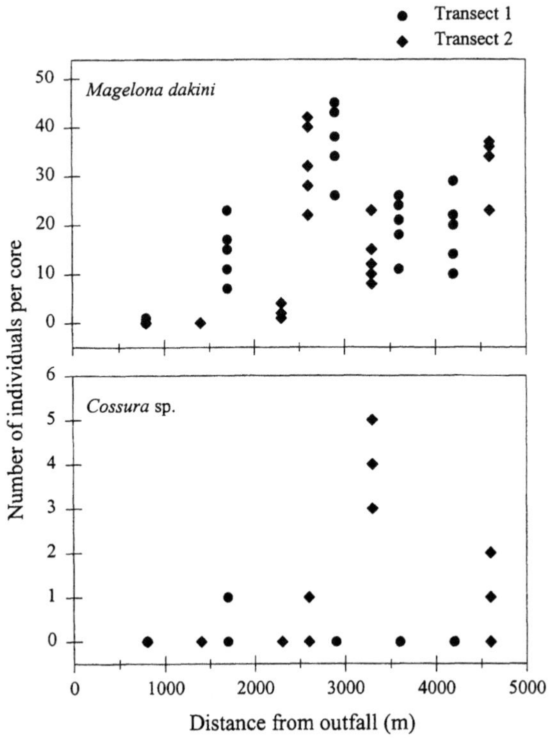

Figure 9. Change in abundance of two sensitive species with distance from the Manukau Habour sewage outfall.

movement is one of the critical factors controlling the deposition of organic material at the sediment-water interface. In areas of strong currents, little deposition will take place. The predicted gradient of impact was, therefore, modified by sediment transport due to increased hydrodynamic factors for sites beyond Puketutu Island.

Predictions 2 and 4

A gradient of community change was observed for Manukau Harbour. High numbers of a few small tolerant species such as *Heteromastus filiformis*, *Aquilaspio aucklandica* and *Boccardia syrtis* were recorded within 1400 m of the outfall. The Pearson & Rosenberg model (1978) suggests that a maximum in species richness occurs at a transitory distance from the discharge source. At Manukau Harbour, the number of species rose to 46 at 2300 m away from the outfall, compared with 20 species in the enriched zone. The species diversity at the far wind-wave exposed sites was between 20 and 26 species per core, similar to the diversity recorded adjacent to the outfall. Rigorous hydrodynamic conditions capable of mobilising sediment can, ultimately, preclude the development of complex benthic communities with high densities of large, long-lived species. The most obvious impacts of hydrodynamic conditions on the macrobenthic communities are likely to arise from wind-generated wave activity that increases physical disturbance of the surface sediment (Turner et al., 1995; Dolphin & Green, 1997). Wind will also contribute to desiccation stress at times of low tide, with potential effects on the survival of some species (Turner et al., 1995). Less obvious impacts are the effects of sedimentary and hydrodynamic conditions on the relative success of larval, juvenile and adult transport and settlement (Turner et al., 1995). The lower species diversity recorded at the wind-wave exposed sites is consistent with previous studies of the effects of exposure on macrofaunal communities supporting Prediction 4 (McLachlan et al., 1981; Dexter, 1984; Dexter, 1992).

Prediction 3

The prediction that a decline in suspension feeding organisms and an increase in deposit feeding organisms occurs as organic enrichment increases was not supported. The ratio of filter feeders: deposit feeders did not change noticeably with distance from the outfall. Decreases in suspension feeders in enriched areas are predicted due to physical clogging of ciliary and siphonal mechanisms, an inability to withstand lower oxygen tensions, and the increasing sedimentary instability brought about by deposit feeders (Pearson & Rosenberg, 1978). In Manukau, the enrichment-tolerant species identified using spatial gradients were mainly deposit feeders. However, there was no noticeable increase in suspension feeders with distance from the outfall. Pridmore et al. (1990) highlight problems with inferring relationships between trophic structure and grain size in estuarine systems. Firstly, allocating animals to trophic groups can be difficult. Studies of some taxa have indicated that feeding behaviour is too variable to allow allocation to distinct groups. Some organisms are capable of switching their mode of feeding; examples of plasticity in feeding behaviour as a function of flow and sediment transport are provided in Fauchald & Jumars (1979), Pridmore et al. (1990) and by Snelgrove & Butman (1994). For example, many species of surface deposit feeders are now known to be facultative suspension feeders (Hughes, 1969; Buhr & Winter, 1977; Dauer et al., 1981), evidently in response to flow and elevated fluxes of suspended particles (Levinton, 1991; Taghon & Greene, 1992). Finally, in shallow and turbid environments, where surface sediments and associated diatom mats are frequently resuspended, differentiation between deposit feeders, suspension feeders and grazers etc. is further confounded.

Conclusions

Canonical correspondence analysis and the estimation of spatial gradients were both effective at identifying species-environment relationships. Multivariate analysis of community composition demonstrated that 22% of the variability in community composition was attributable to organic enrichment, and that sites beyond Puketutu Island were influenced by physical wind-wave disturbance. Both methods identified similar sensitive and tolerant species driving community composition at the enriched *versus* far sites. The use of *a priori* predictions was also effective at inferring relevant processes. *A priori* predictions about effects of organic enrichment and physical wind-wave disturbance on benthic community composition were supported. However, the *a priori* predictions of the scale of disturbance from enrichment, and the predictions of a change to suspension feeding communities at the far non-enriched sites, were not supported. Physical wind-wave disturbance modified the scale of enrichment with elevated organic content within 1400 m of the outfall.

Sampling down spatial gradients enables indications of the scale of processes to be postulated and enables the role of particular processes in accounting for site-to-site variation to be tested. By ordering studies down gradients, an understanding of the changing role of small-scale processes within the increasing spatial extent can be gained (Thrush et al., 1999). This understanding could not be achieved by simple random sampling or by block designs.

Demonstrating gradient trends through time can also address problems of a lack of temporal information. A lack of adequate baseline data or controls can affect the ability to provide strong empirical evidence of effects (Dayton et al., 1998). For example, appropriate benchmark data describing natural standards are generally not available or, in the case of environmental accidents, 'before' data sets cannot be collected. Wiens & Parker (1995) consider appropriate sampling designs and assumptions for analysing the effects of accidental environmental impacts. Because they have the potential to document both initial and subsequent recovery, the impact level by time and impact trend by time designs seem particularly well suited to the analysis of unplanned impacts. Designs that treat contamination as continuous variables have the potential to document contamination effects with greater precision and to detect non-linearity's in responses of the community (Wiens & Parker, 1995). Suter (1996) also recommends monitoring gradients or levels of exposure and matching in space and time the measures of effects to these gradients or levels. Hence, ordering studies down gradients enables both temporal changes and changes in processes with increasing spatial extent to be identified.

While the occurrence of other environmental disturbances along a gradient of anthropogenic disturbance makes interpretation of community pattern more difficult, knowledge of the differences in the expected and observed patterns increases greatly ecological understanding of how to link patterns and processes. The use of a gradient sampling layout together with testing *a priori* predictions enabled impacts of the anthropogenic and natural environmental disturbances to be interpreted. However, this is an area requiring further research. Efforts to include analysis of spatial patterns (density variations and habitat gradients) in both field experiments and survey programmes, and the collection of time series data and spatial pattern analysis of larger scale phenomena, may enable general mechanistic hypothesis to be generated and tested.

Acknowledgements

We thank Rob Bell for generating mean ebb and flood values, and John Oldman for generating orbital speed values. Thanks to Bob Whitlatch who provided invaluable comments on this work. We also wish to thank Vonda Cummings and Judi Hewitt for reviewing earlier drafts of this manuscript.

References

Auster, P. J. & R. W. Langton, 1999. The effects of fishing on fish habitat. Am. Fisheries Soc. Symp. 22.

Bell, R. G., T. M. Hume, T. J. Dolphin, M. O. Green & R. A. Walters, 1997. Characterisation of physical environmental factors on an intertidal sandflat, Manukau Harbour, New Zealand. J. exp. mar. Biol. Ecol. 216: 11–31.

Bell, R. G., S. V. Dumnov, B. L. Williams & M. J. N. Greig, 1998. Hydrodynamics of Manukau Harbour, New Zealand. New Zealand J. mar. Freshwat. Res. 32(1): 81–100.

Black, K. P., 1997. WGEN3DD, Wave generation model for enclosed water bodies and support software. University of Waikato.

Brown, A. C. & A. McLachlan, 1990. Ecology of sandy shores. Elsevier Amsterdam.

Buhr., K. J. & J. E. Winter, 1977. Distribution and maintenance of a *Lanice conchilega* association in the Weser Estuary (FRG), with special reference to the suspension-feeding behaviour of *Lanice Conchilega*. In Keegan, B. F. (ed.), Biology of Benthic Organisms. Pergamon Press, Oxford: 101–113.

Clarke, K. R., 1993. Non-parametric multivariate analyses of changes in community structure. Aust. J. Ecol. 18: 117–143.

Commito, J. A., S. F. Thrush, R. D. Pridmore, J. E. Hewitt & V. J. Cummings, 1995. Dispersal dynamics in wind-driven benthic system. Limnol. Oceanogr. 40(8): 1513–1518.

Dauer, D. M., C. A. Maybury & R. M. Ewing, 1981. Feeding behaviour and general ecology of several spionid polychaetes from the Chesapeake Bay. J. exp. mar. Biol. Ecol. 54: 21–38.

Dayton, P. K., S. F. Thrush, T. M. Agardy & R. J. Horman, 1995. Environmental effects of fishing. Aquatic Conservation: Mar. Freshwat. Ecos. 5: 205–232.

Dayton, P. K., M. J. Tegner, P. B. Edwards & K. L. Riser, 1998. Sliding baselines, ghosts and reduced expectations in kelp forest communities. Ecol. Apps. 8: 309–322.

Dexter, D. M., 1983. Community structure of intertidal sandy beaches in New South Wales, Australia. In McLachlan, A. & Erasmus (eds), Sandy Beaches as Ecosystems. Dr W. Junk Publishers, The Hague: 461–472.

Dexter, D. M., 1984. Temporal and spatial variability in the community structure of the fauna of four sandy beaches in south-eastern New South Wales. Aust. J. mar. Freshwat. Res. 35: 663–672.

Dexter, D. M., 1992. Sandy beach community structure: the role of exposure and latitude. J. Biogeog. 19: 59–66.

Dolphin, T. J. & M. O. Green, 1997. Sediment dynamics of an estuarine 'turbid fringe'. Pacific Coasts and Ports, 1997. Proceedings Vol 1. Centre for advanced engineering, University of Canterbury: 113–118.

Draper, N. R. & H. Smith, 1981. Applied Regression Analysis. 2nd edn. Wiley, New York.

Eleftheriou, A. & M. D. Nicholson, 1975. The effects of exposure on beach fauna. Cah. Biol. mar. 16: 695–710.

Ellis, J. I. & D. C. Schneider, 1997. Evaluation of a gradient sampling design for environmental impact assessment. Envir. Mon. and Assess. 48(2): 157–172.

Ellis, J. I., A. Norkko & S. F. Thrush, 2000. Broad scale disturbance of intertidal and shallow sublittoral soft sediment habitats: effects on benthic macrofauna. J. Aquat. Ecos. Health 7(1): 57–74.

Essink, K., 1984. The discharge of organic waste into the Wadden sea - Local effects. Netherlands Inst. Sea Res. Pub. Ser. 10: 165–177.

Fauchald, K. & P. A. Jumars, 1979. The diet of worms: A study of polychaete feeding guilds. Oceangr. mar. Biol. ann. Rev. 17: 193–284.

Grant, J., 1983. The relative magnitude of biological and physical sediment reworking in an intertidal community. J. mar. Res. 41: 673–689.

Hall, S. J., 1994. Physical disturbance and marine benthic communities: life in unconsolidated sediments. Oceanogr. mar. Biol. ann. Rev. 32: 179–239.

Hewitt. J. E., S. F. Thrush, V. J. Cummings & R. D. Pridmore, 1996. Matching patterns with processes: predicting the effect of size and mobility on the spatial distributions of the bivalves *Macomona liliana* and *Austrovenus stutchburyi*. Mar. Ecol. Prog. Ser. 135: 57–67.

Hughes, R. N., 1969. A study of feeding in *Scrobicularia plana*. J. mar. biol. Ass. U.K. 49: 805–823.

Legendre, P. & L. Legendre, 1998. Numerical Ecology. 2nd edn. Elsevier, The Netherlands.

Levinton, J. S., 1991. Variable feeding behavior in three species of *Macoma* (Bivalvia: Tellinacea) as a response to water flow and sediment transport. Mar. Biol. 110: 375–383.

Jennings, S. & M. J. Kaiser, 1998. The effects of fishing on marine ecosystems. Adv. mar. Biol. 34: 203–314.

McLachlan, A., T. Wooldridge & A. H. Dye, 1981. The ecology of sandy beaches in southern Africa. Sth. african J. Zool. 16: 219–231.

Mirza, F. B. & J. S. Gray, 1981. The fauna of benthic sediments from the organically enriched Oslofjord, Norway. J. exp. mar. Biol. Ecol. 54: 181–207.

Pearson, T. H. & R. Rosenberg, 1978. Macrobenthic succession in relation to organic enrichment and pollution of the marine environment. Oceanogr. mar. Biol. ann. Rev. 16: 229–311.

Pearson, T. H., A. D. Ansell & L. Robb, 1986. The benthos of the deeper sediments of the Firth of Clyde, with particular reference to organic enrichment. Proc. r. Soc. Edinburgh 90B: 329–350.

Pridmore, R. D., S. F. Thrush, J. E. Hewitt & D. S. Roper, 1990. Macrobenthic community composition of six intertidal sandflats in Manukau Harbour, New Zealand. New Zealand J. mar. Freshwat. Res. 24: 81–96.

Raffaelli, D. G., A. G. Hildrew & P. S. Giller, 1994. Scale, pattern and process in aquatic systems: concluding remarks. In Giller, P. S., A. G. Hildrew & D. G. Raffaelli (eds), Aquatic Ecology: Scale, Pattern and Process. Blackwell Scientific Publications, London: 601–606.

Sanders, H. L., 1958. Benthic studies in Buzzards Bay. I. Animal-Sediment relationships. Limnol. Oceanogr. 3: 245–258.

Snelgrove, P. V. R. & C. A. Butman, 1994. Animal–Sediment relationships revisited: cause *versus* effect. Oceangr. mar. Biol. ann. Rev. 32: 111–177.

Suter, G. W., 1996. Abuse of hypothesis testing statistics in ecological risk assessment. Human Ecol. Risk Assess. 2(2): 331–347.

Taghon, G. L. & R. B. Greene, 1992. Utilization of deposited and suspended particulate matter by benthic 'interface' feeders. Limnol. Oceanogr. 37: 1370–1391.

Ter Braak, C. J. F., 1986. Canonical correspondence analysis: A new eigenvector technique for multivariate direct gradient analysis. Ecology 67(5): 1167–1179.

Ter Braak, C. J. F., 1987. The analysis of vegetation–environment relationship by canonical correspondence analysis. Vegetatio 69: 69–77.

Thrush, S. F, R. D. Pridmore & J. E. Hewitt, 1994. Impacts on soft-sediment macrofauna: the effects of spatial variation on temporal trends. Ecol. Apps. 4(1): 31–41.

Thrush, S. F., R. B. Whitlatch, R. D. Pridmore & J. E. Hewitt, 1996. Scale-dependent recolonization: the role of sediment stability in a dynamic sandflat habitat. Ecology 77(8): 2472–2487.

Thrush, S. F., S. M. Lawrie, J. E. Hewitt & V. J. Cummings, 1999. The problem of scale: uncertainties and implications for soft-bottom marine communities and the assessment of human impacts. In Gray, J. S. (ed.), Biogeochemical Cycling and Sediment Ecology. Kluwer Academic Publishers, Dordrecht, The Netherlands: 195–210.

Turner, S. J., S. F. Thrush, R. D. Pridmore, J. E. Hewitt, V. J. Cummings & M. Maskery, 1995. Are soft-sediment communities stable? An example from a windy harbour. Mar. Ecol. Prog. Ser. 120: 219–230.

Underwood, A. J., 1992. Beyond BACI. The detection of environmental impacts on populations in the real, but variable, world. J. exp. mar. Biol. Ecol. 161: 145–178.

Underwood, A. J., 1993. The mechanics of spatially replicated sampling programmes to detect environmental impacts in a variable world. Aust. J. Ecol. 18: 99–116.

Vant, W. N. & B. L. Williams, 1992. Residence times of Manukau Harbour, New Zealand. New Zealand J. mar. Freshwat. Res. 26: 393–404.

Warwick, R. M. & R. J. Uncles, 1980. Distribution of benthic macrofauna associations in the Bristol channel in relation to tidal stress. Mar. Ecol. Prog. Ser. 3: 97–103.

Warwick, R. M., J. D. Goss-Custard, R. Kirby, C. L. George, N. D. Pope & A. A. Rowden, 1991. Static and dynamic environmental factors determining the community structure of estuarine macrobenthos in SW Britain: Why is the Severn Estuary different. J. appl. Ecol. 28: 329–345.

Watling, L. & E. A. Norse, 1998. Disturbance of the seabed by mobile fishing gear: a comparison to forest clearcutting. Conserv. Biol. 12: 1180–1197.

Weston, D. P., 1990. Quantitative examination of macrobenthic community changes along an organic enrichment gradient. Mar. Ecol. Prog. Ser. 61: 233–244.

Wiens, J. A. & K. R. Parker, 1995. Analyzing the effects of accidental environmental impacts: approaches and assumptions. Ecol. App. 5(4): 1069–1083.

Whitlatch, R. B. & R. N. Zajac, 1985. Biotic interactions among estuarine infaunal opportunistic species. Mar. Ecol. Prog. Ser. 21: 299–311.

GPSR Compliance

The European Union's (EU) General Product Safety Regulation (GPSR) is a set of rules that requires consumer products to be safe and our obligations to ensure this.

If you have any concerns about our products, you can contact us on ProductSafety@springernature.com

In case Publisher is established outside the EU, the EU authorized representative is:

Springer Nature Customer Service Center GmbH
Europaplatz 3
69115 Heidelberg, Germany

Batch number: 10800211

Printed by Printforce, the Netherlands